NITROGEN FIXATION

# Developments in Plant and Soil Sciences

VOLUME 48

# Nitrogen Fixation

Proceedings of the Fifth International Symposium on
Nitrogen Fixation with Non-Legumes,
Florence, Italy, 10–14 September 1990

*Edited by*

M. POLSINELLI, R. MATERASSI and M. VINCENZINI
*University of Florence,
Florence, Italy*

SPRINGER-SCIENCE+BUSINESS MEDIA, B.V.

Library of Congress Cataloging-in-Publication Data

International Symposium on Nitrogen Fixation with Non-legumes (5th :
    1990 : Florence, Italy)
    Nitrogen fixation : proceedings of the Fifth International
Symposium on Nitrogen Fixation with Non-legumes, Florence, Italy,
10-14 September, 1990 / edited by M. Polsinelli, R. Materassi, and
M. Vincenzini.
        p.    cm. -- (Developments in plant and soil sciences ; v. 48)
    ISBN 978-94-010-5541-3      ISBN 978-94-011-3486-6 (eBook)
    DOI 10.1007/978-94-011-3486-6
    1. Nitrogen--Fixation--Congresses.   2. Nitrogen-fixing
microorganisms--Congresses.   I. Polsinelli, M. (Mario)
II. Materassi, R.   III. Vincenzini, M.   IV. Title.   V. Series.
    QR89.7.I59   1990
    589.9'0133--dc20                                        91-28411

ISBN 978-94-010-5541-3


*Printed on acid-free paper*

# Contents


**Opening lecture**

* A. Quispel, A critical evaluation of the prospects for nitrogen fixation with non-legumes    1

**Free living diazotrophs**

C. Kennedy et al., Regulation of expression of genes for three nitrogenases in *Azotobacter vinelandii*    13

R. Kreutzer et al., Genetic characterization of nitrogen fixation in *Enterobacter* strains from the rhizosphere of cereals    25

N. Abe et al., Analysis of nitrogenase reaction using monoclonal antibodies against -subunit of component I of *A. vinelandii*    37

D. Govezensky et al., Involvement of GroEL in *Klebsiella pneumoniae nif* gene expression and nitrogenase assembly    43

* M. Simek et al., The effect of some terrestrial oligochaeta on nitrogenase activity in the soil    49

**Posters**

M.J. Merrick and J.R. Coppard, Cassette mutagenesis of a proposed DNA-binding motif in *klebsiella pneumoniae* rpon (sigma 54)    55

N. Miclaus et al., Characterization of a mutant of *Azotobacter chroococcum* resistant to some fungicides    57

G. Blanco et al., A chromosomal linkage map of *Azotobacter vinelandii*    59

L.A. Syrtsova et al., Energy coupling and electron transfer in nitrogenase    61

F. Bermudez de Castro et al., Fluctuation of the ara under *elaeagnus angustifolia* canopy    63

A. Sawicka, The influence of root exudates on dinitrogen fixation of chosen free-living bacteria    65

A. Pidello and L. Menendez, Soil redox potential in presence of azospirillum    67

N. Miclaus et al., Biofertilization of maize with free-living nitrogen fixing bacteria: results of field trials on grain production    69

D.M.S. Mano et al., Kinetics of kelthane uptake and distribution in azospirillum lipoferum    71

T. Langenbach et al., Bioaccumulation of the acaricide kelthane (dicofol) by *azospirillum lipoferum*    73

Y.Z. Ishac et al., Effect of biofertilizers on controlling *fusarium solani* f. Sp. *phaseoli*    75

P. Montaini et al., Phyllospheric nitrogen fixing bacteria in tillandsia leaves    77

**Root-associated diazotrophs**

C. Elmerich et al., *Nif* and *nod* genes in *Azospirillum*    79

* R.H. Burris et al., Control of nitrogenase in *Azospirillum* sp.    89

* J. Boonjawat et al., Biology of nitrogen-fixing bacteria    97


* Reprinted from *Plant and Soil*, Volume 137, No. 1, 1991.

* R.M. Boddey et al., Biological nitrogen fixation associated with sugar cane — 105

Y. Okon et al., Physiological properties of *Azospirillum brasilense* involved in root growth promotion — 113

C. Schipani et al. Regulation of the *nifHDK* genes transcription in *Azospirillum brasilense* — 127

M.A. Knopik et al., Cloning of the *nifA* and *nifB* genes of *Azospirillum brasilense* strain FP2 — 133

A.K. Tripathi et al. Regulation of nitrogen fixation genes in *Azospirillum*: characterization of A *nif* regulatory region — 139

A. Haegi et al., *Azospirillum*-plant interaction: a biochemical approach — 147

* A. Hartman et al., Osmotollerance of diazotrophic rhizosphere bacteria — 155

P. Barbieri et al., *Azospirillum*-gramineae interaction: effect of indole-3-acetic acid — 161

S. Ventura et al., DNA restriction fingerprint for the identification of Azospirilla — 169

* Y. Bashan et al., Alteration in membrane potential and in proton efflux in plant roots induced by *Azospirillum brasilense* — 175

* M. Del Gallo et al., Effect of soil organic matter on chickpea inoculated with *Azospirillum brasilense* and *Rhizobium leguminosarum* bv. *ciceri* — 181

* H. Levanony et al., Active attachment of *Azospirillum brasilense* to root surface of non-cereal plants and to sand particles — 187

* C.B. You et al., Association of *Alcaligenes faecalis* with wetland rice — 195

* J. Fages et al., Sunflower inoculation with *Azospirillum* and other P.G.P.R. — 201

* K.M. Khammas et al., Characterization of a pectinolytic activity in *Azospirillum irakense* — 205

* K.A. Malik et al., Associative $N_2$-fixation in plants growing in saline sodic soils and its relative quantification based on $^{15}N$ natural abundance — 211

K. Michiels et al., Genetic analysis of the *Azospirillum* plant-root interaction — 219

* J.P. Pimentel et al., Dinitrogen fixation and infection of grass leaves by *Pseudomonas rubrisubalbicans* and *Herbaspirillum seropedicae* — 225

C. Fritzche et al., Growth parameters of microaerobic diazotrophic rhizobacteria determined in continuous culture — 231

T. Hurek et al., Infection of intact roots of kallar grass and rice seedlings by *Azoarcus* — 235

* R. Rai, Strain-specific salt tolerance and chemotaxis of *Azospirillum brasilense* and their associative N-fixation with finger millet in saline calcareous soil — 243

J. Ueckert et al., Nitrate reductase activity of *Azospirillum brasilense* sp7 and sp245 V – and C – forms in continuous culture — 249

H. Vandenhove et al., Microcalorimetric study of the growth of *Azospirillum brasilense*: relation between heat production and other growth parameters — 255

* Y.T. Tchan et al., Nitrogen fixation in *P*- nodules of wheat roots by introduced free-living diazotrophs — 269

A.J. Palomares et al., Firefly luciferase expression on nitrogen fixation with non-legumes — 275


**Posters**


L. Civardi et al., Nucleotide sequence of regulatory regions of *Azospirillum brasilense* — 283

M. Delledonne et al., Nucleotide sequence and expression of an *Azospirillum nod* homologous region — 285

I. Giannellini et al., Isolation and characterization of TN5 induced *Azospirillum brasilense* mutants alterated in calcofluor straining    287

A. Milcamps, In vitro construction of Lacz gene fusions with the nifhdk operon of azospirillum brasilense    289

M. Bazzicalupo et al., Isolation and characterization of *Azospirillum Brasilense* mutants altered in the production of siderophores    291

M. Gillis et al., Taxonomic relationships between [*Pseudomonas*] *Rubrisubalbicans*, some clinical isolates (EF group 1), *Herbaspirillum Seropedicae* and [*Aquaspirillum*] *Autotrophicum*    293

I. Giannellini et al., Study on the binding of azospirillum strains to plant tissues    295

M. Fulchieri et al., Effect of *Azospirillum lipoferum* and $GA_3$ on root growth in corn    297

M.A. Paula et al., Interactions of VA-mycorrhizae with *Acetobacter diazotrophicus* inoculated into sweet potatoes (*Ipomoea batatas* L. Lam.)    299

E. Ruckdäschel et al., Genetic and biochemical studies on the production of indole-3-acetic acid (IAA) in azospirillum lipoferum    301

M. R. Alonso et al., Autoregulation of pH by Azospirillum spp.    303

L. Halda et al., Nitrogen fixing bacteria isolated from maize root and antagonistic activity against *Fusarium* sp    305

Nguyen Ngoc Dung et al., Some characters of *Azospirillum spp.* isolated from the rice fields near Hanoi, Vietnam    307

A. Givaudan et al., Melanin production by *Azospirillum Lipoferum* strains    311

E. Paul et al., Effect of $pO_2$, $pCO_2$ and agitation on growth kinetics of *Azospirillum lipoferum* under fermentor conditions    313

S. Urquiaga et al., Selection of wetland rice genotypes for associated biological nitrogen fixation    315

M. Monib et al., Enrichment of tomato sand cultures with composite inocula of associative dinitrogen fixers, p-dissolving bacilli and vam.    317

B.E. Baca et al., Partial purification and properties of two aromatic amino acid aminotransferases from the *A. brasilense* UAP 14 strain indole-3-acetic acid producer    321

B.E. Baca et al, Characterization and regulation of two enzymatic activities involved in biosynthesis of indoleacetic acid in *Azospirillum Brasilense* UAP 14 strain    323

P. de Troch et al., Analysis of extracellular polysaccharides of *Azospirillum Brasilense*    325

C. Baggio et al., $NO_3^-$ and $Fe^{++}$ uptake in durum wheat is affected by *Azospirillum brasilense* inoculation    327

B. Krstić et al., Efficiency of Azotobacter strains depending of nitrogen level and sugar beet genotypes    329

M. Sarić et al., Specific responses of azotobacter strains and sugar beet genotypes    333

L.F. Vasyuk et al., Associative diazothrops of different systematic groups and their effect on the productivity of agricultural crops    337

S. Waschütza et al., Construction of a growth chamber for long-termed axenic plant cultivation    339

C. Scotti et al., N2 fixing bacteria in the Rhizosphere of natural populations of Italian Ryegrass (*Lolium multiforum* L.)    341

C. Cassinelli et al., Studies on few strains of azospirilla colonizing maize rhizosphere: their establishment on roots and effects on plant physiology    343 viii

V. Catska et al., Root-associated diazotrophs used as plant growth stimulators   345

N. Andriollo et al., Screening program for the isolation of N2-fixing bacteria of the genus azospirillum   347

D. Lippi et al., Interactions between *azospirillum* and *sorghum rhizosphere* isolates under different
cultural condiitons   349

P. Quatrini et al., Competition in sorghum rhizosphere among *Azospirillum* and indigenous bacteria   351

P. Montaini et al., Nitrogenase activity in the tobacco rhizosphere   353

E. Gallori et al., Effect of different pesticides on *Azospirillum Brasilense*   355

P. Iacopini et al., Experiments of *Azospirillum* bacterization of maize seeds (zea mais L,) in open
field   357

**Nitrogen-fixing photosynthetic microorganisms**

*1 – Free living and symbiotic cyanobacteria*

R. Haselkorn et al., Nitrogen fixation in filamentous cyanobacteria.   359

H. Bothe et al., Recent aspects on the hydrogenase-nitrogenase relationship in cyanobacteria   367

* G.A. Peters, *Azolla* and other plant-cyanobacteria symbioses: aspects of form and function   377

E. Tel-Or et al., Structure, metabolism and nitrogenase regulation in the *Azolla-Anabaena* association   389

A. Herrero et al. Regulation of nitrogen assimilation in the cyanobacterium *Synechococcus*   399

L.A. Onek et al., Calcium, dinitrogen fixation and calmodulin in a *Nostoc*   405

* G.A. Peschek et al., Respiratory protection of the nitrogenase in dinitrogen-fixing cyanobacteria   411

K.P. Bader et al., Photosynthetic oxygen evolution in the filamentous cyanobacterium *Oscillatoria
chalybea:* interrelationship between water splitting, hydrogen peroxide decomposition and nitrate
metabolism   419

E. Fernandez Valiente et al., Photosynthesis and respiration in mixotrophic cultures of the cyanobacterium *Anabaena variabilis*   425

M.C. Làzzaro et al., A possible role of flavodoxin in nitrogen fixation in heterocysts from *Anabaena*   431

L.J. Stal et al., Photosynthesis and nitrogen fixation in the unicellular cyanobacterium *Gloeothece*
PCC 6909   437

* M. Villbrandt et al., Diurnal and seasonal variation of nitrogen fixation and photosynthesis in
cyanobacterial mats   443

A. Canini et al., Superoxide dismutase detection in *Anabaena azollae* strasb. isolated from *Azolla
filiculoides* LAM   447

* F. Carrapiço, Are bacteria the third partner of the *Azolla-Anabaena* symbiosis?   453

* C. Forni et al., Effects of antibiotic treatments on *Azolla-Anabaena* and *Arthrobacter*   457

* W.J. Zimmerman et al., The *Anabaena-Azolla* symbiosis: diversity and relatedness of neotrophic
host taxa   463

* T. de Waha Baillonville et al., Assessment and attempt to explain the high performances of *Azolla*
in subdesertic trophics versus humid tropics   467

S.A. Kulasooriya, Constraints for the widespread use of *Azolla* in rice production   473

* M.C. Margheri et al., Heterotrophic metabolism and regulation of uptake hydrogenase activity in

symbiotic cyanobacteria    481

* S. Boussiba, Nitrogen fixing cyanobacteria potential uses    487

*2 – Photosynthetic anoxygenic bacteria*

S. Nordlund, Metabolic regulation of nitrogen fixation in anoxygenic phototrophic bacteria    491

A. Colbeau et al., Organization of the genes encoding uptake hydrogenase in the photosynthetic bacterium *Rhodobacter capsulatus*    503

* J. Oelze, Diazotrophic mixed culture of *Azotobacter vinelandii* and *Rhodobacter capsulatus*    509

Y. Jouanneau et al., Structural and genetic studies on four ferredoxins from the photosynthetic bacterium *Rhodobacter capsulatus:* probable involvement in nitrogen fixation    513

**Posters**

J.S. Yunes et al., Growth and amino acid liberation by a mutant strain of *Anabaena variabilis* resistant to the amino acid analogue azetidine 2-carboxylic acid    519

L. Bolaños et al., Interaction B-Ca on the nitrogenase activity in nitrogen-fixing cyanobacteria *(Anabaena* PCC 7119)    521

J.R. Milam et al., The effect of ammonia-excreting cyanobacteria on the growth of rice    523

J.R. Gallon et al., Photoautotrophic and photoheterotrophic $N_2$ fixation in *Oscillatoria* spp exposed to alternating light and darkness    525

P. Lindblad et al., Occurrence and localization of an uptake hydrogenase in the filamentous heterocystous cyanobacterium *nostoc* PCC 73102    527

B. Marsalek et al., The effect of phytohormones on nitrogenase activity and growth of *nostoc muscorum* agardh    529

M. Jha et al., Effect of ammonium on the nitrogenase activity of *Anabaena variabilis*    531

I. Böhm et al., Modification and in-vitro activation of dinitrogenase reductase from anabaena variabilis    533

A. Quesada et al., Nitrogen fixation in Spanish ricefields    535

'L. Tomaselli et al., Production of dinitrogen-fixing cyanobacteria for soil inoculation    537

B. Pushparaj et al., Productivity of the marine nitrogen fixing cyanobacterium *nodularia* sp. in outdoor culture    539

J. R. Dias et al., Studies on rice production using *Azolla* as biofertilizer in Guinea-Bissau    541

Cang Lin et al., The roles of *Anabaena Azollae* in *Anabaena-Azolla* association    543

N. Esiobu et al., Field studies using $^{15}N$ dilution techniques on *Azolla* N contribution to rice in Nigeria    545

P. Albertano et al., Localization of nitrogen in the *Azolla-Anabaena* symbiosis by eftem    551

S.O. Sanni, The pattern of release of azolla-nitrogen in flooded and unflooded soils at three temperatures    555

R. Caudales et al., Factors controlling morphological changes and the life cycle of nitrogen-fixing cyanobionts in different *Azolla* species    561

R. Caudales et al., Fatty acid composition of free-living anabaena and *nostoc* cyanobacteria and of cyanobionts from different *azolla* species    563

M.A. Monsalve et al., Dinitrogen fixation (ara) in *Lobaria Pulmonaria*    565

A. Endal et al., Biological nitrogen fixation in arctic soil and vegetation determined by the acetylene reduction methode — 567

E. Soderback et al., The *Nostoc-Gunnera Magellanica* symbiosis: developmental patterns related to nitrogen fixation — 569

L. Vagnoli et al., Morphological and physiological characterization of $N_2$-fixing symbiotic cyanobacteria — 571

S. Shestakov et al., Genetic organization of *Rhodobacter sphaeroides* chromosome region containing gene *GLNA* — 573

J. Pierrard et al., Factors involved in selection of Nif-mutants in N-limited cultures of the photosynthetic bacterium *Rhodobacter capsulatus* — 575

C. Duport et al., Genetic characterization of the gene encoding ferredoxin II from the nitrogen fixing phototrophic bacterium *rhodobacter capsulatus* — 577

C. Grabau et al., DNA sequence of a new ferredoxin gene mapping in a *nif* Region of *Rhodobacter capsulatus*, and expression in *Escherichia coli* — 579

A. Gori et al., Analysis of DNA restriction pattern in the genus *ectothiorhodospira* — 581

A.F. Yakunin et al., Effect of Mo, V and W on the growth and nitrogenase synthesis in phototrophic bacteria — 583

R. Borghese et al., Nitrogen metabolism in *Rhodobacter capsulatus* — 585

E. Brostedt et al., Purification of a pyruvate: oxidoreductase from nitrogen fixing *Rhodospirillum rubrum* — 587

H.Y. Song et al., Derepression of nitrogenase by 2-ketoglutarate in a glutamate auxotroph of *Rhodobacter sphaeroides* — 589

G. Tassinato et al., Outdoor Hydrogen Photoevolution by immobilized cells of *Rhodopseudomonas palustris* — 591

F. Milicia et al., Azolla Biomass as green manure for rice in calabria — 593


## Diazotrophic actinomycetes


* A.D.L. Akkermans et al., Molecular ecology of *Frankia*: advantages and disadvantages of the use of DNA probes — 595

A. Séguin et al., Expression of actinorhizins in the development of the *Frankia-Alnus* symbiosis — 601

* J. Schwencke, Rapid exponential growth and increased biomass yield for some *Frankia* strains in buffered and stirred mineral medium (BAP) added of phosphatidyl choline — 615

M.G. Girgis et al., Enhancing nodulation and growth of *Casuarina glauca* planted in substrate already infected with *Frankia* — 621

M.P. Fernandez et al., Structure of the genus *Frankia* — 629


## Posters


V.N. Akimov et al., Grouping of *Frankia* strains by DNA-DNA homology: how many genospecies are in the genus *Frankia*? — 635

M. Bosco et al., Genetic diversity of *elaeagnus*-isolated *Frankia* strains infective on both *alnus* and *elaeagnus* — 637

E. Cervantes et al., The infectivity and symbiotic performance of a Frankia strain as affected by its growth medium towards *Hippophae rhamnoides L.* — 639

A. Brioua et al., Utilisation and metabolism of C1 compounds by *Frankia* — 641

P.-O. Lundquist et al., Nitrogenase activity and nitrogenase protein levels in a *Frankia-Alnus incana* symbiosis subjected to darkness — 643

A. Müller et al., Secretion of aminopeptidases and proteinases by *Frankia* sp BR. — 645

D.K. Jah et al., Interaction between VAM and actimorhizal symbionts in *Alnus nepalensis* — 647

E. Buresti et al., Comparison between *Quercus Robur / Alnus Cordata* mixed plantation and *Quercus Robur* in monoculture — 651


## Closing lectures


G. Del Bino, EEC regulatory requirements for the use and release of genetically modified organisms — 653

M.J. Dowson-Day et al., Studies of the potential for expression of nitrogenase Fe-protein in cells of higher plants — 659

Index — 671

FOREWORD


The Fifth International Symposium on Nitrogen Fixation with Non-legumes was held in Florence (Italy) on 10-14 September, 1990. Earlier Symposia of this series were held in Piracicaba (Brazil), Banf Alberta (Canada), Helsinki (Finland) and Rio De Janeiro (Brazil).

The Symposium's main objectives were to bring together scientists working in many different fields of nitrogen fixation, to stimulate discussion on this important process and to have an appraisal of the most recent studies concerning nitrogen fixation with non-legumes. The Symposium was attended by 230 scientists from 32 different countries.

This volume collects the contributions of 65 lectures and 87 posters, which are an up-to-date account of the state of knowledge on biological nitrogen fixation with non-legumes. The book provides a valuable reference source not only for specialists in nitrogen fixation, but also for researchers working on related aspects of agronomy, biochemistry, genetics, microbiology, molecular biology and plant physiology.

It is with great pleasure that we aknowledge the contributions of the authors in assuring the prompt pubblication of this volume. We would also like to express our thanks to Kluwer Academic Publishers B.V. for the publication of these Proceedings.

M. Polsinelli
R. Materassi
M. Vincenzini

# ORGANIZING COMMITTEE

| | |
|---|---|
| *M. Polsinelli* | President |
| *M. Vincenzini* | Secretary |
| *F. Favilli* | Treasurer |
| *E. Galli* | |
| *E. Gallori* | |
| *L. Giovannetti* | |
| *R. Materassi* | |
| *M.P. Nuti* | |
| *M.R. Tredici* | |

# SCIENTIFIC COMMITTEE

| | |
|---|---|
| *M. Bazzicalupo* | Florence, Italy |
| *H. Bothe* | Cologne, West Germany |
| *R.H. Burris* | Madison, U.S.A. |
| *J. Döbereiner* | Rio de Janeiro, Brazil |
| *W. Klingmüller* | Bayreuth, West Germany |
| *R. Materassi* | Florence, Italy |
| *A. Quispel* | Leiden, The Netherlands |
| *C. Rodriguez-Barrueco* | Salamanca, Spain |
| *S. Shestakov* | Moscow, U.S.S.R. |
| *B.E. Smith* | Brighton, U.K. |
| *N. Tandeau de Marsac* | Paris, France |

# ACKNOWLEDGEMENTS

The Organizing Committee gratefully acknowledges the following for their valuable support:

— Agrifutur S.r.l. - Biotecnologie per l'agricoltura - Brescia
— Agrimont - Enichem Agricoltura. Gruppo Enimont
— Applikon Italia S.r.l. - Genova
— Beckman Analytical S.p.A. - Milano
— Biorad Laboratories S.r.l. - Milano
— Carlo Erba Strumentazione S.p.A. - Milano
— Cassa di Risparmio di Firenze
— Chemisint S.p.A. - Servizi per l'ambiente e l'agricoltura - Milano
— Commissione delle Comunità Europee DG XI A2 Bruxelles
— Comune di Firenze
— Consiglio Nazionale delle Ricerche - Roma
   Comitato Nazionale per le Biotecnologie e la Biologia Molecolare
   Comitato Nazionale per le Scienze Agrarie
   Comitato Nazionale per le Scienze Biologiche e Mediche
— Consorzio Biotecnologie Bioprogram - Ravenna
— A. De Mori S.p.A. - Milano
— ENEA. Comitato Nazionale per la Ricerca e per lo Sviluppo dell'Energia Nucleare e delle Energie Alternative - Roma
— ENEL - Ente Nazionale Energia Elettrica - Roma
— Kontron Instruments S.p.A. - Milano
— Ministero dell'Agricoltura e delle Foreste - Roma
— Ministero dell'Università e della Ricerca Scientifica e Tecnologica - Roma
— Perkin Elmer Italiana S.p.A. - Monza
— Regione Emilia Romagna - Assessorato Agricoltura ed Alimentazione - Bologna
— Università degli Studi di Firenze

# A critical evaluation of the prospects for nitrogen fixation with non-legumes

A. QUISPEL
*Department of Plant Molecular Biology, Leiden University, Botanical Laboratory, Nonnensteeg 3, 2311 VJ Leiden, The Netherlands*

*Key words:* Azospirillum, cyanobacteria, endosymbionts, *fix* genes. *Frankia*, klebsiella, *nif* genes, *nod* genes, Rhizobium (Bradorhizobium and Azorhizobium), rhizosphere, root nodules

## Abstract

Some twenty years ago many speculations were made about possibilities to improve the use of the well-known nitrogen-fixing plant symbioses and to extend the use of atmospheric nitrogen to the most important agricultural plants, like cereals. Since then our understanding of the molecular biology of nitrogenase and the symbiotic interactions during root nodule formation and activity enable a reconsideration of these original speculations. Besides the possibilities for better practical use of the existing systems some more far-reaching speculations will be discussed: the introduction of *nif* genes into cells of higher plants, the extension of the host range of *Rhizobium* or *Frankia* and the possibilities to transform rhizosphere bacteria, like *Azospirillum*, into more efficient endosymbionts. In all cases it will be evident that the more we know, the more we realize what we still have to understand.

## Introduction

Students of biological nitrogen fixation have invented the most simple taxonomic system for the plant kingdom: a mere division between legumes and non-legumes. Nobody wonders when special conferences are organized to discuss progress with legumes and their nodule-forming bacteria. The plants are definitely taxonomically related, the problems of nodulation and nitrogen fixation comparable, research on leguminous root nodules has given the most exciting new results and their practical importance has been appreciated from antiquity. But why should we have a special symposium on non-legumes? Besides the activity of a nitrogenase system there is not much in common. The plants belong to quite different groups. The micro-symbionts may be eubacteria, Actinomycetes or Cyanobacteria. The symbiotic relations vary from loose associations to intercellular or intracellular endosymbioses. A comparison between so many highly divergent systems may be interesting from an evolutionary point of view, but why then should we exclude the most successful of all the legumes? Do we want to come together without our big brothers watching us?

The main and convincing argument to organize symposia on non-legumes is based on the desire to overcome the limitations to legumes in the exploitation of the agronomical benefits of biological nitrogen fixation. However important legumes may be, the most important plants for the nutrition of this worlds population still are the cereals like wheat, corn or rice. The use of atmospheric nitrogen for the nitrogen nutrition of these plants might reduce the economic and ecological problems of the use of organic and inorganic nitrogen fertilizers.

In his introduction to a symposium on world food supply during the international botanical congress in Seattle, Van Overbeek (1969) suggested: 'I would urge botanists to tackle the problem of nitrogen deficiencies with boldness. We need to devise more efficient ecosystems, put root nodules on cereal crops, and put nitrogen fixing chloroplasts in their leaves'. Such ideas, inspired by the rapid progress of molecular genetics, were further developed and discussed at several meetings. At the Pullman conference in 1974, Hardy (1976) gave a realistic state of the problem. He doubted 'that any alternative technology for improved provision of fixed nitrogen will be implemented in crop production in the next ten years. . . while some solutions lie in the realm of an undefined future, possibly 25 years'. 'Scientific progress will occur in several areas in a shorter time, but practical success is the only criterion relevant to food production'.

Now, sixteen years after, we must admit that Hardy was right though his prognosis, considered rather pessimistic at that time, now must be considered as still too optimistic. The scientific knowledge increased in a fascinating way. Yet the practical applications did not yet correlate with the fundamental progress.

I want to reconsider the most far-reaching claims of transfer of *nif* genes to higher plants in the light of our present knowledge of these genes and their regulation. We then will discuss the progress of research of the existing symbiotic systems and ask the question whether this research will enable an improved practical use. This may enable us to reformulate some more far-reaching goals: the extension of the host range of *Rhizobium* or *Frankia* to other plants than their regular hosts and finally the existing possibilities to improve the reliability and efficiency of nitrogen fixation by the still less successful associations.


**The transfer of nif genes to other organisms**

When the first suggestions for transfer of *nif* genes to cells of higher plants were made nothing was known about the complexity of the *nif* gene region. Today extensive studies, first in *Klebsiella*, have shown that about 21 genes are involved, many of which have now been identified in other diazotrophic organisms like *Rhizobium*, *Bradyrhizobium*, *Frankia*, *Anabaena* and *Azospirillum* (Long, 1989). The structural genes K, D and H for the apoproteins I and II respectively, are highly conserved among all diazotrophic organisms studied. Other genes are involved in the uptake and transfer of Mo, and in some not yet identified functions, while the genes *nif* L and *nif* A regulate the transcription of the other *nif* genes. In *Klebsiella nif* L has a repressing, *nif* A an activating function. Especially interesting are the differences between the regulation in *Klebsiella* and in *Rhizobium*. In *Rhizobium* the *nif* L has not been identified. While in *Klebsiella* the expression of *nif* A, is regulated by the internal nitrogen level through a two-component system Ntr B/Ntr C, in *Rhizobium* bacteroids the expression of *nif* A is uncoupled from this Ntr B/Ntr C system but is regulated by the external signal oxygen through another two-component system Fix L/Fix J (De Philip et al., 1990). This different regulation may be an adaptation to the free-living or symbiotic situation in *Klebsiella* and *Rhizobium* bacteroids respectively (Roelvink and Van den Bos, 1989).

Notwithstanding the complexity of the *nif* regions the whole cluster of genes may be transferred from *Klebsiella pneumonia* to *Escherichia coli* in such a way that a new diazotrophic transgenic bacterium is formed (Pühler et al., 1979). This success was achieved between two related bacteria and it may be doubted whether similar success is possible with a less related prokaryotic host. The different regulations of *nif* A in *Klebsiella* and *Rhizobium* bacteroids warn us that the introduced *nif* A must be compatible with the prevailing situation in the new host cell. Still greater problems have to be expected when we want to introduce this whole *nif* region into the eukaryotic cells of higher plants. Methods for the introduction of foreign genes like the use of the Ti and Ri vectors of *Agrobacterium* species or the bombardment with DNA-coated microprojectiles are available. But will it be possible to transfer such a vast region of DNA in a functional way? Experiments on the transfer of *nif* genes to higher plants will be discussed at this symposium by Merrick.

Successful transcription of the *nif* genes is not yet sufficient for an active nitrogenase. Many auxiliary reactions are essential, part of them depending on bacterial enzymes encoded by specific *fix* genes. This means that these *fix* genes too have to be transferred to the host cell. On the other side the plant has to provide reducing conditions and sufficient ATP, so that the nitrogenase is integrated in the metabolic system of the cell. Moreover the plant has to provide a situation in which the nitrogenase is protected against too high concentrations of oxygen, while the fixation products must be removed by the transport system and by assimilation in the protein synthesis of the plant.

Since nitrogen fixation is limited to prokaryotic organisms it might be more promising to insert the *nif*-genes not in the eukaryotic plant nucleus but in the prokaryotic-like chloroplasts (Van Overbeek, 1969; Postgate, 1987). The micro-projectiles method indeed allows the, albeit transient, introduction of foreign genes into plant chloroplasts (Daniell et al., 1990) but we are still far removed from the successful preparation of transgenic diazotrophic chloroplasts. And how can we protect nitrogenase in an oxygen producing chloroplast?

In the light of our present knowledge the prospect of active nitrogenase in plant cells seems farther removed than it appeared in 1969.

## The improvement of existing diazotrophic symbiotic systems

Since nitrogen fixation by plant cells is an aim which will only be fulfilled, if ever, in a far-away future it is realistic to focus our attention on the existing cooperation of higher plants with diazotrophic prokaryotes. Already during the discussions of the early seventies it was repeatedly stated that short-time practical results could only be expected from studies with the well-known symbiotic systems, the search after new or less know plant-bacteria associations and the improvement of their efficiency. This formed the main impetus for the stimulation of research on biological nitrogen fixation. The fascinating results of this research have been summarized in recent reviews and conference proceedings (Djordjevic et al. 1987; Quispel, 1988; Bothe et al., 1988; Greshoff et al., 1990).

*Symbioses with* Rhizobium (*including* Bradyrhizobium *and* Azorhizobium)

Though up till now the only known non-legume hosts for these bacteria are the species of the Ulmaceae *Parasponia* it would be unwise to exclude the legumes in this discussion. The outstanding scientific progress in our understanding of symbiotic nitrogen fixation is based on the leguminous root nodules and no real progress with non-legumes is feasible without the inspiration from the advanced frontier of the legume-*Rhizobium* symbioses. Fascinating progress was made in our understanding of the infection and nodulation process, the genetics of bacteria and hosts, the biochemistry and molecular biology of nitrogen fixation and nitrogen assimilation, synthesis and function of leghemoglobins and other nodulins (nodule specific proteins) and the integration of nitrogen fixation in the metabolic and transport physiology of the whole nodulated plants.

Our increased, but certainly far from sufficient, knowledge concerning the factors which determine the efficiency of nodulation with *Rhizobium* is not yet correlated with its impact on practical applications. We are able to select and modify bacterial strains, host plants and bacteria-plant combinations, which are highly efficient under laboratory conditions. Successful field inoculation methods have been developed. Yet the results in the field are mostly disappointing since the introduced selected bacteria have to compete with the indigenous *Rhizobium* population in the soil. Studies with marked strains have shown that the factors which determine the success in this competition are different from those involved in infection, nodulation and effectivity (Dowling and Broughton, 1986).

*Symbioses with* Frankia

Since the first successful isolations and cultivations of *Frankia* strains great progress in the study of actinorhizae has been made (Schwintzer and Tjepkema, 1990). Though there are obvious differences between *Frankia* and *Rhizobium* root nodules, these differences must be related to the differences within both groups. It then appears that both belong to one group of symbioses: the diazotrophic root nodules. It therefore is highly undesirable to consider root nodule symbioses with *Rhizobium* and *Frankia* as separate fields of study. In studies on e.g. the mechanism of infection (Quispel and Burggraaf, 1986) or molecular biology progress iş inspired by data and theories about the *Rhizobium*-legume symbioses but unfortunately still lags far behind. On the other hand the special aspects of *Frankia* like the multicellular hyphae, the formation and form of the vesicles and the formation of sporangia and spores inside the nodules indicate interesting host-microbe interactions which might be easily overlooked in symbioses with a morphologically more simple microsymbiont.

For further studies on fundamental as well as practical aspects it is very important that *Frankia* strains can be compared which are isolated from nodules of different hosts and different environmental conditions and which are representative for the main endophytic population in these nodules. Though many strains of *Frankia* have been isolated there still are actinorhizae from which no isolations were successful. This problem must not be underestimated since its solution may elucidate certain as yet unknown aspects of symbiotic interactions. Many strains were isolated from root nodule fragments from which only a few hyphae developed after several months of incubation. This was the case with spore-negative root nodules from *Alnus glutinosa* in the Netherlands on the usual nutrient media. However, if a root lipid fraction was added abundant outgrowth of hyphae was observed within 12 days. It is evident that only these hyphae can be considered to be representative for the main endophytic population (Burggraaf, 1984). The active factor could be identified as the triterpenoid dipterocarpol (Quispel et al., 1989). After primary isolation it can be replaced by less specific lipids like Tween 80 or lecithin. Doubts about the identity of isolated strains with the main endophytic population are possible when the isolated strains no longer infect the hosts from which they were isolated (Nazaret et al., 1989) or changed their Spore-positive or Spore-negative character (Burggraaf, 1984; Torrey, 1987). Moreover we have to realize that most strains have been isolated from field material and appear to be genetically highly diverse. This diversity is reduced by selection of monoclonal strains (Burggraaf and Valstar, 1984; Burggraaf, 1984).

These aspects have to be critically considered before studies on molecular genetics can be done and before selection of superior *Frankia* strains will lead to success. Application of superior strains under field conditions will meet the same problems of competition with indigenous strains as are encountered with *Rhizobium*. The availability of differently marked strains will enable a further study of these competitions (Akkermans, 1991). The potential for practical applications is already illustrated by the use of *Casuarinaceae* in tropics and subtropics and of *Alnus* and other plants from temperate regions in forestry and soil management (extensively discussed in the book of Schwintzer and Tjepkema, 1990).

*Symbiosis with cyanobacteria*

The symbioses in which diazotrophic Cyanobacteria are involved form a most heterogenous group with hosts ranging from fungi, mosses and liverworts to waterferns (*Azolla*), some Gymnosperms (e.g. *Cycas* and *Macrozamia*) and angiosperms (*Gunnera*). They cannot be compared with the root nodules of *Rhizobium* and *Frankia* since there are no indications that the microsymbionts affect the morphology of their hosts. In all plant-hosts the endophytes are intercellular, inhabiting preformed structures in the host. On the other hand the microsymbionts are markedly affected by the hosts. The formation of nitrogen-fixing heterocysts is stimulated while the assimilation of the fixation product ammonium is more or less reduced. They are fully dependent on the photosynthesis of their hosts while their own photosynthesis is inhibited.

Isolation and re-infection of the endophytic Cyanobacteria has been successful in most types of host plants and the re-infection experiments indicated a great though not absolute promiscuity of the isolated strains (Bonnet and Silvester, 1981; Enderlin and Meeks, 1983).

Quite different were the results with the endosymbiont of *Azolla*. It is now increasingly evident that the available isolates of *Anabaena* from *Azolla* species are not representative for the main endophytic population and at best represent a minor population or were considerably modified after isolation and cultivation (Zimmerman et al., 1989). Even in nature it may be doubted whether *Azolla* is ever infected from the environment. During growth the *Anabaena* becomes associated 'with the apical meristem and from there is partitioned into the pre-formed cavities at the basis of the aerial dorsal lobes.

It thus appears to be impossible to select strains of the microsymbionts of *Azolla* species but promising results have been obtained by inoculation of *Anabaena*-free plantlets with isolated induciums (Plazinski et al., 1989). This may become of great practical importance for the selection of the most efficient *Azolla-Anabaena* combinations with regard to the appreciated use of *Azolla* as green fertilizer in flooded rice fields.

*Associations with diazotrophic bacteria*

Studies on nitrogen fixation in the rhizosphere were considerably stimulated since Döbereiner and Day (1976) obtained indications for substantial fixation of nitrogen in the association between *Paspalum notatum* with *Azobacter paspalii*. Further studies led to the rediscovery of Beyerinck's *Spirillum lipoferum*, soon renamed as *Azospirillum*, while the list of diazotrophic rhizosphere bacteria is still increasing. At present these bacteria and especially the species of *Azospirillum* are among the most studied microorganisms (Boddey and Döbereiner, 1988; Skinner et al., 1989).

Today, overviewing the immense amount of data on nitrogen fixation in the rhizosphere of different plants, with different bacteria under different environment conditions using different methods for the determination of nitrogen fixation, we can only state that progress has been disappointing and confusing. There can be no doubt that, especially in tropical conditions, considerable amounts of nitrogen may be fixed e.g. in ricefields and kallar grass vegetations. Studies with the isotope dilution method showed that e.g. in maize atmospheric nitrogen is taken up by the plant and similar results were obtained by the direct $^{15}N_2$ method. In many studies, however, results appeared to be controversial and highly variable. Moreover, growth stimulation after the addition of rhizosphere bacteria, was not always correlated with nitrogen fixation, but could be attributed to production of growth stimulating substances and effects on root development, increased ammonium or nitrate up-

take and N-assimilation or effects on iron uptake through the production of siderophores.

In studies where real provision of atmospheric nitrogen to the plant could be correlated with the amount of bacteria in the rhizosphere, such correlations were only established for the endorhizosphere (Döbereiner, 1988). Bacteria like *Azospirillum* not only adhere to the surface of root tips and root hairs but, as was already observed by Döbereiner and Day (1976), can enter the roots intercellularly (e.g. Levanony et al., 1989). If it turns out that efficient use of fixed nitrogen depends on a more intense contact between bacteria and root cells research can be directed to the stimulation of these contacts. Far more fundamental research is needed before the unmistakable use of atmospheric nitrogen fixation by rhizosphere bacteria for plant growth can be made more reproducable and more efficient. As long as this is not yet achieved efforts to inoculate rhizosphere bacteria in field experiments are premature. Even when sometimes positive results were obtained they mostly did not meet the postulates of Hardy (1976). Farmers demand clear-cut, reproducable results and are not interested in incidental increases of plant growth which can only be evaluated after application of estimations for statistical significance.

*General remarks*

In conclusion we may state that in all symbioses and associations with diazotrophic bacteria considerable progress has been made, but that the practical applications are not yet related to the increase of our fundamental insight.

When comparing the different plant-microbe combinations it is evident that successful and fully efficient use of atmospheric nitrogen for plant growth can only be expected in endosymbiotic systems. Only here the prerequisites for symbiotic nitrogen fixation can be fulfilled: reliable supply of metabolic substrates by the host photosynthesis providing sufficient energy and reducing conditions, protection against too high oxygen concentrations, transport of the $N_2$-fixation products to the host, development of membrane systems for bidirectional transport between host and endosymbiont and protection

6

against competitive or antagonistic bacteria in the environment.

When we re-open the question: could the capacity for symbiotic nitrogen fixation be extended to other plants, we may consider two possibilities:
1. The use of well-established successful symbiotic nitrogen fixing bacteria like *Rhizobium* or *Frankia* and trying to extend their host-range to other plants like cereals.
2. The use of bacteria like *Azospirillum*, which already establish some relation with grasses and cereals and trying to modify their associations with roots into an endosymbiosis.

**Extension of the host range of Rhizobium strains**

The formation of root nodules is a complicated process consisting of different phenotypic steps as discussed extensively by Sprent (1989). Every step depends on the activity of genes, both in the bacteria as in the host plants. Every step may be interrupted when there is no compatibility between the host plant and the bacterial strain.

Studies of the molecular genetics of *Rhizobium* (and *Bradyrhizobium*) have led to the discovery of a number of *nod* genes, whose transcription is essential for at least the earlier steps like root hair deformation, infection thread initiation, initiation of cell divisions in the cortex and the induction of some early nodulins by the plant cells. In *Rhizobium* they are clustered on big plasmids, together with the *nif* and some *fix* genes. In *Bradyrhizobium* they have been identified as clusters on the bacterial chromosome (reviews by Rolfe and Greshoff, 1988; Long, 1989). The genes *nod* A, B and C can be exchanged between *Rhizobium* strains without affecting host specificity and thus are called the common *nod* genes. Other genes change host specificity after transfer to other *Rhizobium* strains and are called the host specific nodulation (*hsn*) genes. All *nod* genes are grouped in a series of operons each of which with a highly conserved sequence, the *nod* box, in the operator regions. The activation of all operators depends on the product of the only constitutive genes *nod* D of which there are one to three

different types (e.g. one in *R. leguminosarum*, three in *R. meliloti*). The protein *Nod* D can only activate the different *nod* operons after a specific reaction with flavones or flavonones in the exudate of the plant roots. This specific cooperation with plant-specific flavonoids is another aspect of *Rhizobium* host specificity. Spaink et al. (1989a) succeeded in constructing hybrid *nod* D genes from one of the *nod* D genes (*nod* D1) of *R. meliloti* and the *nod* D from *R. leguminosarum* bv *trifolii*. In strains with such *nod* D hybrids all *nod* operons were activated without the cooperation of flavonoids. Moreover such strains showed an extended host range. Similar results were obtained after point mutations in the *nod* D of a strain of *R. leguminosarum* bv *trifolii* in which the host range was extended even to the non-legume *Parasponia* (Melver et al., 1989). Bender et al. (1988) transferred a *nod* D from a *Rhizobium* strain with an extended host range, including *Parasponia*, to a strain from clover after which this *Rhizobium* too could nodulate *Parasponia*. In this transgenic strain the transcription of *nod* A was induced by many more flavonoids than the original strains and by plant extracts, not only from *Parasponia*, but as well from *Trema*, *Casuarina*, wheat, corn and rice. Further understanding of the effects of the different *nod* genes depends on the progress in the elucidation of the nature of the products. Particularly interesting are the observations that some *nod* genes are responsible for the production of an excreted extra-cellular signal. In *R. meliloti* Faucher et al. (1989) observed that the *hsn*-genes H and Q determined changes in the production of plant-specific extra-cellular signals. According to their hypothesis the proteins *Nod* A, B, C are responsible for the synthesis of a factor of low MW which is specifically modified by the proteins *Nod* H and *Nod* Q. Lerouge et al. (1990) recently succeeded in isolating an extra-cellular factor from *R. meliloti* produced by the cooperative activity of the *nod* genes A, B, C with H and Q, induced after activation of their *nod* D genes with luteoline. The extracts which contained these signals were able to elicit root hair deformation, cell dedifferentiation in the cortex and nodule organogenesis in roots of alfalfa, even in the absence of *Rhizobium* cells. The main factor in these extracts (NodRm-1) was purified and identified as sulphated $\beta$-1, 4-tetrasaccharide of D-glucosamine in which three amino acids were acetylated and one was acylated with a $C_{16}$ bis-unsaturated fatty acid.

The *hsn* gene *nod* E appears to be especially involved in the highly specific step of infection thread formation. Experiments with *nod* E gene hybrids showed that a central region in the protein *Nod* E of 185 amino acids is responsible for differences in host specificity (Spaink et al. 1989). For further steps still other genes are important. In nodulation of *Vicia sativa* by *R. leguminosarum* the *nod* genes A, B, C and D were sufficient for root hair curling and the side-effect of thick, short roots, for infection thread formation the minimum requirements were formed by the activity of the *nod* genes A, B, C and D together with E and F while for nodule induction all *nod* genes were required (Van Brussel et al., 1988).

Besides the production of soluble products, like NodRM-1, the bacterial cell surface is important. Mutants in genes responsible for the synthesis of exo-polysaccharides (EPS) may be deficient in normal infection thread formation (Keller et al., 1988). Mutants in genes for the production of Lipo Polysaccharides (LPS) too may be deficient in normal nodulation e.g. by the absence of bacterial release from the infection thread (De Maagd et al., 1989). The genes for EPS and LPS production are not situated on the *sym*-plasmids and thus cannot be identical with some of the *nod* genes. This does not exclude the possibility that certain *Nod* products affect some cell wall substances.

All bacterial signals, whether soluble, like NodRm-1, or cell surface components, must be recognized by the host plant. Adhesion of bacteria to the surface of the root epidermis or root hairs consists of different but non-specific steps (Smit et al., 1987). *Rhizobium* may even bind to wheat or rice (Shimschick and Heber, 1979). The much discussed role of plant lectins in specific binding of bacterial polysaccharides has obtained a new impetus by the work of Diaz et al. (1989). The gene for pea lectin was introduced into white clover roots using *Agrobacterium rhizogenes* as a vehicle for transformation. The hairy roots of these transgenic plants

now could be occasionally infected and nodulated by *R. leguminosarum* bv *viciae*. These results put an end to the doubts about a role of plant lectins in the determination of host specificity in the legume-*Rhizobium* symbiosis though the nature of their role still has to be elucidated. Still more important for our problem is the conclusion that here for the first time a host-specificity barrier has been broken by transfer of plant genes.

We still must realize that many more steps are involved before a functional effective nodule is obtained. Some steps in the normal nodulation pathway of legumes may not be essential, like the bacterial release (see Sprent (1989)) but other steps will be. Will other plants be capable of reacting in such a way that a functional endosymbiont is formed? We now know that genes for the synthesis of leghemoglobin are present in non-nodulated plants like *Trema* (Bogusz et al., 1988). Since *Trema* is closely related with the nodulating *Parasponia* it is interesting to know whether leghemoglobin genes are found in quite unrelated plants as well. Other essential plant nodulins might be products of genes which still play a less obvious role in other plants, but will they be activated after infection in the same way as in legumes? Will they have the possibilities for morphogenetic reactions leading to root nodules?

There are some promising new approaches. Root nodules have been obtained in wheat or rice after infection with *Rhizobium* strains with simultaneous application of either cell-wall degrading enzymes (Al-Mallah et al., 1987) or the synthetic auxin 2-4 D (Tchan and Kennedy, 1989), (most recent ref. Greshoff et al., 1990). These nodules contain bacteria but nitrogen fixation is absent or doubtful. It may be interesting to continue such experiments with genetically modified bacterial strains.

We certainly have a long way to go before the extension of the host range to plants like cereals can – if ever – be realized, but we do possess the tools to proceed.

## Stimulation of an endophytic way of life by Azospirillum

When we restrict ourselves to the best studied

and most promising rhizosphere bacterium *Azospirillum* the first question is: how could we stimulate *Azospirillum* to infect the roots and develop into an endosymbiont? *Azospirillum* has some encouraging characteristics: they are absorbed to roots and root hairs and may develop intercellularly, surface polysaccharides bind to wheat germ agglutinin (Del Gallo and Neyra, 1988), they contain all necessary *nif* and *fix* genes and hybridization experiments showed homology with some *nod* genes. If essential *nod* genes or other genes, e.g. for synthesis of specific cell-surface polysaccharides, are missing it must be possible to introduce them. Do cereals form the necessary flavonoids for their induction in their rhizosphere? If this might be a problem the *Azospirillum* strain might be provided with a broad host-range flavonoid-independent *nod* D. Since the result of Diaz et al. (1989) we may appreciate the possiblity to prepare transgenic plants with the right type of lectins. Similar possibilities may exist for the induction of the synthesis of essential nodulins if cereal hosts might fail to produce them in sufficient amounts. It is encouraging that there appear to be common genetic plant systems controlling certain steps in the infection process both of *Rhizobium* and the very different VA fungi (Duc et al., 1989). Finally we might try to introduce *Azospirillum* after simultaneous addition of cell-wall degrading enzymes, 2-4-D or the addition of *Rhizobium* signals like NodRM-1. There are indeed ample possibilities to dream about a golden future.

Certainly all such possibilities need to be explored. But let us remain realistic. What we finally have to obtain is the formation of a real mutualistic endosymbiont, fully adapted to an endophytic way of life. The development of bacteria in necrotic root cells has nothing in common with an endosymbiosis.

There are several prerequisites for the establishment of an endosymbiotic situation. Hostile reactions from the host, like hypersensitive reactions and production of phytoalexins have to be prevented. This even may fail in some mutants of Rhizobium (Djordjevic et al., 1988; Parniske et al., 1990). The bacteria must find an inter- or intracellular niche, where sufficiently reducing conditions are provided and the $pO_2$ is

regulated by diffusion barriers and/or leghemoglobins. Finally, and most essential for all endosymbioses (Smith and Smith, 1990), membrane systems around the endosymbiont of microbial and host origin must be adapted in such a way that a bidirectional transport of substances is possible. This whole structure must be integrated in the long-distance transport system of the plant. It may not be necessary to obtain the highly evolved system as found in the *Rhizobium* bacteroid system with their peribacteroid membranes. In some Leguminosae, in *Parasponia* and in the *Frankia* actinorhizae efficient nitrogen fixation is possible in an apparently less far evolved system. But in all cases the endosymbiosis must be the result of a fine network of mutual adaptations and interactions.

Specific molecular signals may affect gene regulation in the partner. We already discussed such signals in the *Rhizobium* symbiosis with regard to the earlier steps in the infection and nodulation process. Far less is known about the signals which play a role during the later steps leading to the endophytic situation. It has been claimed that bacteroids contain high amounts of the *Nod* C protein (John et al., 1988), but other results indicated no transcription of the early nod genes (Sharma and Signer, 1990). *exo-Genes* are actively transcribed (Keller et al., 1988). Other genes are involved in steps like bacterial release (Ward et al., 1989) and bacteroid formation (Rossbach et al., 1989). The low oxygen concentration inside the nodules may affect the regulation of essential genes like this has been demonstrated for the regulation by *nif* A.

The more or less hostile environment around an endosymbiont may have direct effects. This environment has been compared with a lytic compartment (Mellor, 1989). The peribacteroid membrane forms a barrier against proteases (Pladys and Rigaud, 1988). A disfunctioning of such barriers will lead to senescence and may be one of the main causes for non-effectivity.

We have to realize that during formation of bacteroids and their surrounding membranes the endosymbionts are surrounded by interfering substances in the intracellular (or intercellular) host environment. The effect of dipterocarpol on the viability of the endosymbiontic state in some *Frankia* root nodules (Quispel et al., 1989) might be explained by a direct effect on membranes. Dipterocarpol belongs to a group of substances with marked effects on cell membranes (Nes and Heftmann, 1981). Since dipterocarpol is present in the host cells the developing endosymbiont must be affected by its presence and can only survive after adaptation of its membranes.

All such effects, and certainly many others, must be considered before we will understand the mutual adaptations which lead to a functional endosymbiont. Without this understanding nobody can predict whether a functional endosymbiotic state of e.g. *Azospirillum* in cereals will ever be obtained.

## Conclusions

Studies on the existing symbiotic diazotrophic systems still are the most promising for better use of biological nitrogen fixation in agriculture. Yet we must realize that, even in the legumes, progress in the practical areas still lags far behind the progress with fundamental studies. The possibilities for extension of nitrogen fixation to other plants are still speculative. In view of recent progress the possibilities for the extension of the host range of *Rhizobium* may be the most realistic. The development of new endosymbionts will be faced with still more aspects which need further clarification. Introduction of *nif* genes directly into higher plants still appears to be a very far-away goal. When such speculations were first made around 1970 the studies of molecular aspects of symbiotic nitrogen fixation were still in their very beginnings. Since then progress was considerable and fascinating but the more we know, the more we realize how much we still do not understand.

The prospect of the extension of nitrogen fixation to other plants was originally formulated to stimulate the possibilities for the biological use of atmospheric nitrogen in order to overcome the ecological and economical problems of nitrogenous fertilizers. Will this extension, if it is ever realized, really have ecological and economical advantages or will it give new problems? As to the ecological aspects: the fear that new diazotrophic plants might develop as dangerous weeds is not realistic. Biological nitrogen fixation is characterized by many built-in regulations and the biochemical reduction of di-nitrogen is so energy-consuming that no epidemics of new nitrogen-fixing weeds can be expected. Caution is necessary when we really might succeed in extending the host range of *Rhizobium* of *Frankia* since it might be possible that highly promiscuous strains are symbionts on some plants and parasites on many others. As to the economic side we must hope that the costs for the development of new forms of nitrogen fixation in agronomical plants will not have been so outstanding that such plants can not be used by the poor farmers in developing countries for which they were originally intended. The realisation of new diazotrophic systems will be so far in the future that there must be ample time and, let us hope, wisdom to foresee and to prevent such problems. Meanwhile the prospects of extension of nitrogen fixation, whether it will be fulfilled or not, stimulate our research programmes. If around 1970 the question was asked whether such ideas belonged to science or science fiction, we know with certainty: they belong to fascinating science.

## References

Akkermans A D L, Hahn D and Mirza M S 1991 Molecular ecology of *Frankia*: Advantages and disadvantages of the use of DNA probes. Plant and Soil 137, 49–54.

Al-Mallah M K, Davey M R and Cocking E C 1987 Formation of nodular structures on rice seedlings by Rhizobia. J. Exp. Bot. 140, 473–478.

Bender G L, Nayudu M, Le Strange K K and Rolfe B G 1988 The *nod* D1 gene from Rhizobium strain NGR 234 is a key determinant in the extension of host range to the non-legume *Parasponia*. Mol. Plant. Microbe Interact. 1, 259–266.

Boddey R M and Döbereiner J 1988 Nitrogen fixation associated with grasses and cereals recent result and perspectives for future research. Plant and Soil 108, 53–65.

Bogusz D, Appleby C A, Landsmann J, Dennis E S, Trinick M J and Peacock W J 1988 Functioning haemoglobin genes in non-nodulating plants. Nature (London) 331, 178–180.

Bonnet H T and Silvester W B 1981 Specificity in the *Gunnera – Nostoc* endosymbiosis. New Phytol. 89, 121–128.

Bothe H, De Bruijn F J and Newton W E 1988 Nitrogen Fixation: Hundred Years After. G. Fischer, Heidelberg.

Burggraaf A J P 1984 Isolation, cultivation and characterization of *Frankia* strains from actinorhizal root nodules. University Thesis Leiden.

Burggraaf A J P and Valstar J 1984 Phenotypic heterogeneity in *Frankia* LDAgp1 studied on clones and reisolates. Plant and Soil 78, 29–44.

Daniell H, Vivekananda J, Nielson B L, Ye G N, Tewari K K and Sanford J C 1990 Transient foreign gen expression in chloroplasts of cultured tobacco cells after biolistic delivery of chloroplast vectors. Proc. Natl. Acad. Sci. USA 87, 88–92.

Del Gallo M M and Neyra C A 1988 Polysaccharides production and cell surface interactions in *Azospirillum* sp. *In* Nitrogen Fixation: Hundred Years After. Eds. H Bothe, F J de Bruijn and W E Newton. G. Fischer, Heidelberg.

De Maagd R A, Rao A S, Mulders I H M, Goosen-de Roo L, Van Loosdrecht M C M, Wijffelman C A and Lugtenberg B J. 1989 Isolation and characterization of mutants of Rhizobium leguminosarum by viciae 248 with altered lipopolysaccharides: Possible role of surface charge or hydrophobicity in bacterial release from the infection thread. J. Bacteriol. 171, 1143–1150.

De Philip P, Batur J and Boistard P 1990 *Rhizobium meliloti Fix* L is an oxygen sensor and regulates *R. meliloti nif* A and *fix* K genes differently in *Escherichia coli*. J. Bacteriol. 172, 4255–4262.

Diaz C L, Melchers L S Hooykaas P J J, Lugtenberg E J J and Kijne J W 1989 Root lectin as a determinant of host plant specificity in the *Rhizobium*-legume symbiosis. Nature (London) 338, 579–581.

Djordjevic M A, Gabriel M A and Rolfe B G 1987 *Rhizobium* – the refined parasite of legumes. Annu. Rev. Phytopathol. 25, 145–168.

Djordjevic S P, Ridge R W, Chen H, Redmond J W, Batley M and Rolfe B G (1988) Induction of pathogenic-like responses in the legume *Macroptileum atropurpureum* by a transposon-induced mutant of the fast-growing, broad host range *Rhizobium* strain MGR 234. J. Bacteriol. 170, 1848–1857.

Döbereiner J 1988 Isolation and identification of root-associated diazotrophs. Plant and Soil 110, 207–212.

Döbereiner J and Day J M 1976 Associative symbioses in tropical grasses: Characterization of micro-organisms and dinitrogen-fixing sites. *In* Proc. 1st Intern. Symp. on Nitrogen Fixation. Eds. W E Newton and C J Nyman. pp. 518–538. Washington State University Press.

Dowling D N and Broughton W J (1986) Competition for nodulation of legumes. Ann. Rev. Microbiol. 40, 131–157.

Duc G, Trouvelot A, Gianinazzi-Pearson V, and Gianinazzi S 1989 First report of non-mycorrhizal plant mutants (Myc) obtained in pea (*Pisum sativum* L.) and fababean (*Vicia faba* L.). Plant Science 60, 215–222.

Enderlin C S and Meeks J C 1983 Pure culture and reconstitution of the *Anthoceros-Nostoc* symbiotic association. Planta 158, 157–165.

Faucher C, Camut S, Dénarié J and Truchet G 1989 The *nod* H and *nod* Q host range genes of *Rhizobium meliloti* behave as avirulence genes in *R. leguminosarum* bv. *viciae* and determine changes in the production of plant-specific extracellular signals. Mol. Plant Microbe Interact. 2, 291–300.

Greshoff P M, Newton W E, Roth E C and Stacey G (Eds.) 1990 Nitrogen Fixation: Achievement and Objectives. Chapman-Hall Publ., New York.

Hardy R W F 1976 Potential impact of current abiological and biological research on the problem of providing nitrogen. *In* Proc. 1st Intern. Symp. on Nitrogen Fixation. Eds. W E Newton and C J Nyman. pp 693–717. Washington State University Press.

John M, Schmidt J, Wieneke U, Krüssman H D and Schell J 1988 Transmembrane orientation and receptor-like structure of the *Rhizobium meliloti* common nodulation protein *Nod* C. EMBO J. 7, 583–588.

Keller M, Müller P, Simon R and Pühler A 1988 *Rhizobium meliloti* genes for exopolysaccha ride synthesis and nodule infection located on megaplasmid 2 actively transcribed during symbiosis. Mol. Plant Microbe Interact. 1, 267–274.

Lerouge P, Roche Ph, Faucher C, Maillet F, Truchet G, Promé J C and Dénarié J 1990 Symbiotic host-specificity of *Rhizobium meliloti* is determined by a sulphonated and acylated glucosamine oligosaccharide signal. Nature (London) 344, 781–784.

Levanony H, Bashan Y, Romano B and Klein E 1989 Ultrastructural localization and identification of *Azospirillum brasilense* Cd on and within wheat roots by immunogold labelling. Plant and Soil 117, 201–218.

Long S 1989 *Rhizobium* genetics. Ann. Rev. Genet. 23, 483–506.

Mellor R B 1989 Bacteroids in the *Rhizobium*-legume symbiosis inhabit a plant internal lytic compartment: Implication for other microbial endosymbioses. J. Exp. Bot. 40, 831–846.

Melver J, Djordjevic M A, Weinman J J, Bender G L and Rolfe B G 1989 Extension of host range of *Rhizobium leguminosarum* bv. *trifolii* caused by point mutations in *nod* D that result in alterations in regulatory function and recognition of inducer molecules. Mol. Plant Microbe Interact. 2, 97–106.

Nazaret S, Simonet P., Normand Ph and Bardin R 1989 Genetic diversity among *Frankia* isolated from *Casuarina* nodules. Plant and Soil 118, 241–247.

Nes, W D and Heftmann E 1981 A comparison of triterpenoids with steroids as membrane components. J. Nat. Prod. 44, 377–400.

Parniske M, Zimmermann Chr, Cregan P B and Werner D 1990 Hypersensitive reaction of nodule cells in the *Glycine* sp. *Bradyrhizobium japonicum* symbiosis occurs at the genotype-specific level. Bot. Acta 103, 143–148.

Pladys D and Rigaud J 1988 Lysis of bacteroids in vitro and during the senescence in *Phaseolus vulgaris* nodules. Plant Physiol. Biochem. 26, 179–186.

Plazinsky J, Zheng Q, Taylor R, Rolfe B G and Gunnins B E S 1989 Use of DNA/DNA hybridization technique to authenticate the production of new *Azolla-Anabaena* symbiotic associations. FEMS Microbiol. Lett. 65, 199–209.

Postgate J, 1987 Prospects for the improvement of biological nitrogen fixation. J. Appl. Bacteriol. Symp. Suppl. 85S–91S.

Pühler A, Burkardt H J and Klipp W 1979 Cloning of the entire region for nitrogen fixation from Klebsiella pneumoniae on a multicopy plasmid vehicle in *Escherichia coli*. Mol. Gen. Genet. 176, 17–24.

Quispel A 1988 Bacteria-plant interactions in symbiotic nitrogen fixation. Physiol. Plant. 74, 783–790.

Quispel A and Burggraaf A J P 1988 Infection, initiation and structure of actinorhizal root nodules *In* Biological Nitrogen Fixation: Recent Development. Ed. N S Subba Rao. pp 255–282. Oxford IBH Publ. Co., New Delhi.

Quispel A, Baerheim Svendsen A, Schripsema J, Baas W J, Erkelens C and Lugtenberg J 1989 Identification of dipterocarpol as isolation factor for the induction of primary isolation of *Frankia* from root nodules of *Alnus glutinosa* (L.), Gaertn. Mol. Plant Microbe Interact. 2, 107–112.

Roelvink P W and Van den Bos R C 1989 Regulation of nitrogen fixation in diazotrophs: The regulatory *nif* A gene and its characteristics. Acta Bot. Neerl. 38, 233–252.

Rolfe B G and Greshoff P M 1988 Genetic analysis of legume nodule initiation. Annu. Rev. Plant Physiol. and Mol. Biol. 39, 297–319.

Rossbach S, Gloudemans T, Bisseling T, Stader B, Kaluza N, Ebeling S and Hennecke H. 1989 Genetic and physiologic characterization of a *Bradyrhizobium japonicum* mutant defective in early bacteroid development. Mol. Plant Microbe Interact. 2, 233–240.

Schwintzer Chr R and Tjepkema J D 1990 The biology of *Frankia* and actinorhizal plants. Acad. Press, San Diego.

Sharma S B and Signer E R 1990 Temporal and spatial regulation of the symbiotic genes of *Rhizobium meliloti* in planta revealed by transposon Tn5-gusA. Genes Development 4, 344–356.

Shimschick E J and Heber R R 1979 Binding characteristics of $N_2$-fixing bacteria to cereal roots. Appl. Environ. Microbiol. 38, 447–453.

Skinner F A, Boddey R M and Fendrick I (Eds.) 1989 Nitrogen Fixation with Non-Legumes. Kluwer Academic Publishers, Dordrecht, The Netherlands.

Smit G, Kijne J W and Lugtenberg E J J 1987 Involvement of both cellulose fibrils and a $Ca^{2+}$ dependent adhesin in the attachment of *Rhizobium leguminosarum* to pea root hairs. J. Bacteriol. 169, 4294–4301.

Smith S E and Smith F A 1990 Structure and function of the interface in biotrophic symbiosis and their relation to nutrient transport. New Phytol. 114, 1–38.

Spaink H P, Okker R J H, Wijffelman C A, Tak T, Goosen-de Roo L, Pees E, Van Brussel A A N and Lugtenberg E J J 1989a Symbiotic properties of *Rhizobium* containing a flavonoid-independent hybrid *nod* D product. J. Bacteriol. 171, 4045–4053.

Spaink H P, Weinman J, Djordjevic M A, Wijffelman C A, Okker R J H and Lugtenberg B J J 1989b Genetic analysis and cellular localization of the *Rhizobium* host-specificity-determining Nod E protein. EMBO J. 8, 2811–2818.

Sprent J I 1989 Which steps are essential for the formation of functional legume nodules. New Phytol. 111, 129–153.

Tchan Y and Kennedy I R 1989 Possible $N_2$-fixing root nodules induced in non-legumes. Agric. Science N. S. 2, 57–59.

Torrey J G 1987 Endophyte sporulation in root nodules of actinorhizal plants. Physiol. Plant. 70, 279–288.

Van Brussel A A N, Pees E, Spaink H P, Tak T, Wijffelman C A, Okker R J H, Truchet G and Lugtenberg B J J 1988 Correlation between *Rhizobium leguminosarum nod* genes and nodulation phenotypes on *Vicia*. *In* Nitrogen Fixation: Hundred Years After. Eds. H Bothe, F J de Bruijn and W E Newton. 483 p.

Van Overbeek J 1969 What botanists can do, Introduction to the 'All congress symposium World Food Supply' XIth Int. Bot. Congress, Seattle, WA.

Ward L J H, Rockman E S, Ball Ph, Jarvis B D W and Scott D B 1989 Isolation and characterization of a *Rhizobium meliloti* gene required for effective nodulation of *Lotus pedunculatus*. Mol. Plant Microbe Interact. 2, 224–232.

Zimmerman W J, Rosen B H and Lumpkin Th A 1989 Enzymatic, lectin and morphological characterization and classification of presumptive Cyanobionts from *Azolla* Lam. New Phytol. 113, 497–503.

# REGULATION OF EXPRESSION OF GENES FOR THREE NITROGENASES IN AZOTOBACTER VINELANDII


Christina Kennedy, Anil Bali, Gonzalo Blanco,
Asunscion Contreras, Martin Drummond, Mike Merrick,
Jean Walmsley, and Paul Woodley
AFRC IPSR Nitrogen Fixation Laboratory
University of Sussex
Brighton BN1 9RQ, UK



ABSTRACT. Three genetically distinct nitrogenases can be synthesized in
A. vinelandii: conventional molybdenum nitrogenase encoded by nifHDK, a
vanadium nitrogenase encoded by vnfHDGK, and a third enzyme which
contains neither Mo nor V encoded by anfHDGK. A complex array of
regulatory proteins controls expression of the three sets of nitrogenase
structural genes. These include the three specific activators NIFA, VNFA,
and ANFA; NIFL, which as in K. pneumoniae is necessary for ammonium
repression of Mo nitrogenase genes; and NFRX which is required for
expression from the nifH and anfH promoters. The nfrX gene has been found
to be similar to an E. coli gene, glnD, which encodes uridylyl
transferase. A model is presented showing the way these interacting
regulatory products are thought to regulate synthesis of the three
nitrogenases. Two other observations are of more practical significance.
Firstly, the A. vinelandii nifL mutants excreted more than 5mM ammonium
when fixing nitrogen. Secondly, Mo fully repressed the alternative
nitrogenase genes at 30°, repressed much less at 20°, and not at all at
14°. This result suggests that A. vinelandii has the capacity to
synthesize non-molybdenum nitrogenases regardless of environmental Mo
content at temperate soil temperatures.


## 1. Introduction

Azotobacters have featured long and large in the history and development
of non-symbiotic nitrogen fixation physiology, biochemistry and genetics.
Early interest in Azotobacter physiology was due to its uniquely high
tolerance to oxygen while fixing nitrogen. Azotobacters can fix
dinitrogen not only in air, but in laboratory conditions where oxygen is
present at greater than 20%. This continues to be a field of active
research, an example of which is the recent discovery that the cytochrome
bd branch of terminal respiration in A. vinelandii is necessary for
nitrogen fixation in air but not under microaerobic (1-2% $O_2$) conditions
(Kelly et al., 1990). The cloning and site-directed mutagenesis of cyd
genes followed by physiological characterization of mutants has
established a role for cytochrome d oxidase in respiratory protection of
nitrogenase.

14

Nitrogenase biochemistry and genetics took an unexpected turn in 1980 when Bishop et al (1980) in North Carolina reported that A. vinelandii could, in the absence of molybdenum, synthesize a version of nitrogenase that did not apparently require that metal. Shortly after this, Robson et al. (1986) showed that the related A. chroococcum produced a vanadium-containing enzyme when grown in Mo-free medium supplemented with vanadium. The presence of nitrogenases without molybdenum had been earlier inferred by Bortels (1936) who observed that vanadium could replace molybdenum as a requirement in medium for Azotobacter nitrogen fixation; later, McKenna et al. (1970) reported that a crude preparation of nitrogenase from A. vinelandii contained vanadium but not molybdenum. In the ensuing years, however, the dogma that molybdenum was always an essential metal in nitrogenase pushed aside the earlier ideas about vanadium. Ironically, the growth in Mo-free (and V-free) medium that Bishop and co-workers observed was probably due to a third type of nitrogenase which contains iron, as do the Mo and V nitrogenases, but no other metal. Thus, in growth medium with Mo, molybdenum nitrogenase is synthesized (phenotype Nif$^+$); with V, vanadium nitrogenase is made (phenotype Vnf$^+$); with neither metal, the third (possibly an iron nitrogenase) is present (phenotype Anf$^+$). Mo represses the Vnf and Anf systems while V represses synthesis of the Anf enzyme under normal laboratory conditions.


## 2. Genes for Three Nitrogenases: nif, vnf, and anf

The three nitrogenases are coded for by three distinct but similar sets of structural genes: nifHDK,(Brigle et al., 1989) vnfHDGK (Joerger et al., 1990) and anfHDGK (Joerger et al., 1989a) (Fig. 1). The H genes encode the subunit of the dimeric component II (or Fe) proteins of all three enzymes. The DK genes, respectively, determine the $\alpha$ and $\beta$ subunits of the tetrameric component I proteins, also called MoFe protein (nifDK), VFe protein (vnfDK), and FeFe protein (anfDK). The vnfG and anfG genes encode approx. 12,000 Mr proteins whose function in the non-molybdenum nitrogenases is not yet established. Five of the other nif gene products, those of nifM, nifU, nifS (necessary for maturation or stability of nitrogenase components) and nifV nifB (involved in cofactor synthesis) are required for activity of all three nitrogenase enzymes (Kennedy et al., 1986; Joerger and Bishop 1988; Kennedy and Dean, unpublished results). While the nifEN genes are only required for molybdenum nitrogenase activity (FeMocofactor synthesis), the vnfEN gene products are necessary for activity of both non-molybdenum nitrogenases (Bishop and Joerger, 1990) although no direct role in cofactor synthesis has been demonstrated.

The aim of the work presented here is to understand how expression of genes involved in structure or activity of one, two, or all of three systems is controlled by regulatory gene products and by environmental factors such as fixed nitrogen, metals and temperature.

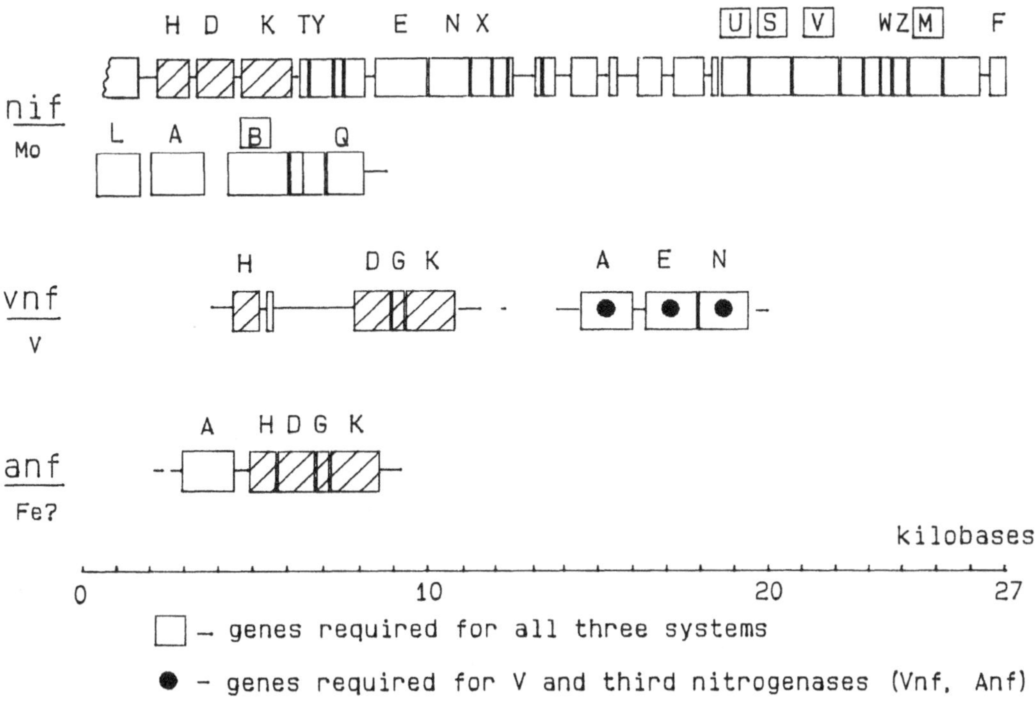



Figure 1. Genes for three nitrogenases in <u>A</u>. <u>vinelandii</u>. See text for the mapping and sequencing of these regions. Unlettered boxes correspond to open reading frames of unknown function.

## 3. Regulation: Three NIFA–like Activators for Three Nitrogenases

Expression of <u>nif</u> genes has been best characterized in <u>Klebsiella pneumoniae</u>. Two major regulatory components are required: firstly, a minor RNA polymerase sigma factor different from the abundant $\sigma^{70}$, for recognition of sequences located at $-12$ (GC) and $-24$ (GG) upstream of the transcription start site; and secondly, an activator protein that is essential for conversion of the closed to the transcriptionally active open form of the complex between RNA polymerase and <u>nif</u> promoter DNA. The novel sigma factor, $\sigma^{54}$, is encoded by a single gene variously designated as <u>rpoN</u>, <u>ntrA</u>, or <u>glnF</u>. The activator protein can be either the <u>ntrC</u> gene product (NTRC), required for expression from the promoter adjacent to the <u>nifLA</u> operon, or the <u>nifA</u> gene product (NIFA), required for expression from all the other <u>nif</u> promoters (adjacent to <u>nifB</u>, <u>nifF</u>, <u>nifM</u>, <u>nifU</u>, <u>nifE</u>, <u>nifH</u> and <u>nifJ</u>). NTRC binds to two sequences similar to the consensus $-GCAC-N7-GTGC$ located at $-142$ and $-163$ in the <u>nifL</u> promoter while NIFA binds to the DNA sequence $-TGT-N10-ACA-$ located approximately 100 bases upstream of the RNA polymerase-$\sigma^{54}$ recognition site in the other <u>nif</u> promoters. The upstream DNA bound to an activator protein (either NIFA or NTRC) probably loops around to meet the downstream region bound to RNA polymerase $-\sigma^{54}$ where protein-protein interactions lead to open complex formation necessary for transcription initiation. Thus, the major regulatory genes required for nitrogen fixation in K. pneumoniae are rpoN, ntrC and nifA, (for review see Merrick, 1990). In addition, the nifL gene product (NIFL) inactivates NIFA in the presence of excessive fixed nitrogen or $O_2$. The requirement for $\sigma^{54}$ and NIFA for expression of nif genes in several other genera (Azotobacter, Rhodobacter, Thiobacillus, Azospirillum, Herbaspirillum, Rhizobium, Bradyrhizobium, Azorhizobium) has been indicated by characterizing the phenotype of nifA and ntrA mutants and/or by the presence of the appropriate consensus sequences in the promoter regions of nif genes in these organisms.

We have examined the requirement in A. vinelandii of NIFA, NTRC and $\sigma^{54}$ for expression of several of the nif, vnf, and anf operons by constructing lacZ fusions in the appropriate genes. The fusions were transformed into A. vinelandii strains carrying mutations in nifA, ntrC or rpoN which had been previously identified (Toukdarian and Kennedy, 1987). After selection and purification, transformants were grown in N-deficient medium and β–galactosidase activity was measured. In addition, two other nifA-like genes, vnfA and anfA had been identified and sequenced by Joerger et al. (1989). These genes show extensive sequence homology with nifA but the encoded proteins have significant differences from NIFA in the DNA binding domain thought to be important for recognition of upstream activator sequences. Mutants defective in vnfA and anfA were also transformed with the lacZ fusions.

In the wild-type strain UW136, full expression of the nifH-lacZ fusion required molybdenum while the vnfH, vnfD, and anfH fusions were repressed by molybdenum. The anfH-lacZ fusion was repressed by vanadium. Expression of the structural gene operons showed specificity for activator: NIFA for nifH-lacZ; VNFA for vnfH-lacZ, vnfD-lacZ, vnfE-lacZ; and ANFA for anfH-lacZ. In the rpoN mutant background, none of the fusions produced significant b-galactosidase; this was expected since all of the promoters examined contain consensus $\sigma^{54}$ binding sites (Fig.1). None of the fusions showed a requirement for NTRC for expression, consistent with previous work showing that A. vinelandii ntrC mutants are Nif$^+$, Vnf$^+$ and Anf$^+$ (Toukdarian and Kennedy, 1987).

The direct involvement of VNFA in expression of the vnf genes was demonstrated by constructing strains of Escherichia coli containing plasmids with vnfE-lacZ, vnfH-lacZ, or vnfD-lacZ and introducing vnfA (and nifA as control) expressed from an inducible lac promoter. Expression of all three fusions was significant only when the vnfA-containing plasmid was also present. Experiments are currently under way to examine common sequences present in the vnfE, vnfH and vnfD promoter regions that will determine their importance for binding and activation by the VNFA protein.

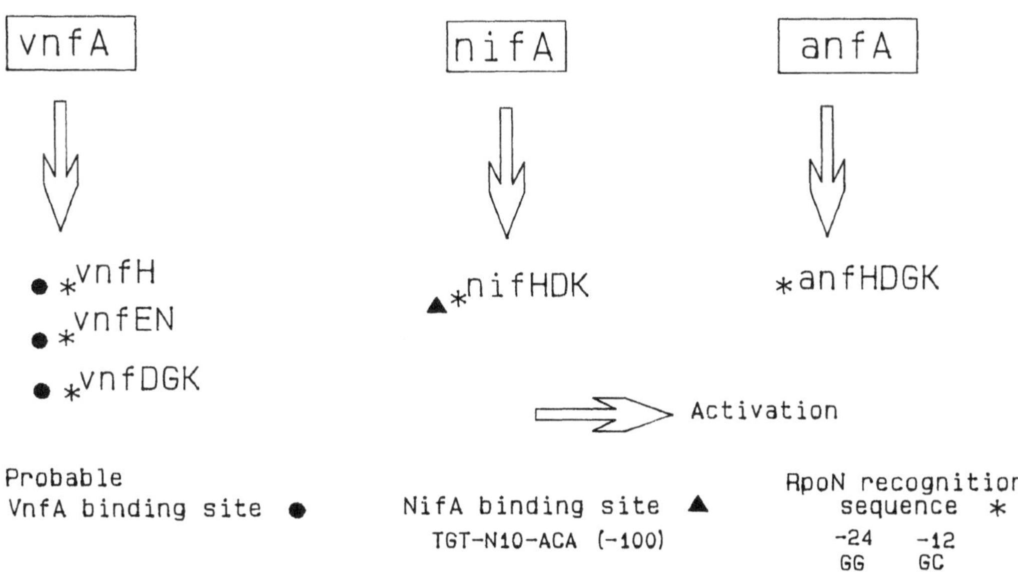



Figure 2. Activators for expression of nitrogenase genes in <u>A.</u> <u>vinelandii</u>.

## 4. <u>nifL</u> in <u>A. vinelandii</u>:  Mutants Excrete Ammonia

Sequencing of the region upstream of <u>nifA</u> in <u>A. vinelandii</u> indicates the presence of a gene sharing some homology to the <u>nifL</u> gene of <u>K. pneumoniae</u> (Bennett <u>et al</u>., 1988). Up to now, a <u>nifL</u> equivalent has not been found in other nitrogen fixing organism. The KIXX cartridge, encoding kanamycin (Km) resistance, was inserted in this gene by cloning into the SalI site (Fig. 3). The mutated <u>nifL</u>-KIXX gene was transformed into two <u>A. vinelandii</u> strains: wild-type UW136 and the <u>nifH</u>-<u>lacZ</u> strain MV101. The resulting mutants had high nitrogenase or β-galactosidase activities, respectively, in the presence of 15mM ammonium acetate in contrast to the NifL$^+$ parent strains which were repressed. Thus, as in <u>K. pneumoniae</u>, the <u>nifL</u> gene product of <u>A. vinelandii</u> mediates $NH_4^+$ repression of nitrogenase genes.

The <u>A. vinelandii</u> <u>nifL</u> mutants were also found to excrete ammonia in significant amounts when growing on $N_2$. Concentrations of ammonia in excess of 5mM were found in the supernatants of well-grown nifL mutant cultures. This is possible because, unlike in <u>K. pneumoniae</u>, $NH_4^+$ does not repress expression of the <u>nifA</u> gene in <u>A. vinelandii</u>; this was demonstrated with a <u>nifA</u>-<u>lacZ</u> fusion (A. Bali, unpublished results). Therefore, in <u>K. pneumoniae</u> <u>nifL</u> mutants, while $NH_4^+$ does not immediately repress expression of the <u>nifHDK</u> genes as it does in NifL$^+$ strains, expression is ultimately prevented because <u>nifA</u> is not transcribed in cells grown with high $NH_4^+$. This is due to the conversion of active phosphorylated NTRC to the  dephosphorylated form, which is inactive in the presence of high NH$_4^+$. Unlike in K.pneumoniae, A. vinelandii does not require NTRC (nor any other NH$_4^+$-controlled activator) for nifA gene expression (A. Bali, unpublished results). Whether the ability of nifL mutants of A. vinelandii to excrete ammonium while fixing nitrogen has any potential practical application must be determined.

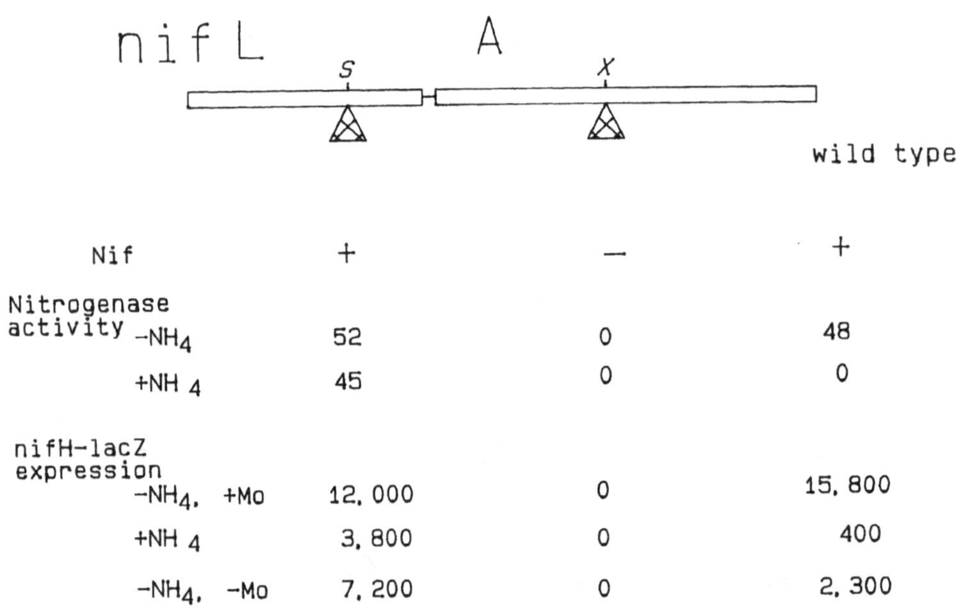


Figure 3. Features of nifL- and nifA-Kixx mutants of A. vinelandii.

## 5. nfrX, a Gene Required for nif and anf Expression, is Similar to glnD in E. coli

A Tn5-induced regulatory mutant described by Santero et al. (1988) defined the nfrX gene in A. vinelandii. Although this mutant, MV3, is corrected to Nif$^+$ by introducing constitutively-expressed NIFA from either A. vinelandii or A. chroococcum, the mutated gene is not homologous to nifA. Also, neither nif genes nor anf genes are expressed in nfrX mutants while nifA mutants express anf but not nif genes (J. Walmsley, unpublished results). The nfrX sequence shows an open reading frame of 2.7Kb encoding a large protein of approx. 102,000 Mr. A similarity was noticed in molecular weight and amino acid content of the nfrX gene product with those published for the enzyme uridylyl transferase (UTase) from E. coli, encoded by glnD (Garcia and Rhee, 1983). A comparison of the partial sequence of E. coli glnD (E. Garcia, personal communication) with that of nfrX showed considerable homology at the N-terminal end of the protein products. Finally, the cloned nfrX gene could correct the Ntr$^-$ phenotype (growth on arginine or nitrate as N source) of both E.coli and K.pneumoniae glnD mutants while the E.coli glnD gene cloned on a wide host range plasmid restored a Nif+ phenotype to an nfrX mutant of A. vinelandii (Table 1). We conclude therefore that NFRX is structurally and functionally similar to GLND (UTase).

Table 1. Cross-complementation of glnD and nfrX in E. coli and A. vinelandii

| Strain | plasmid | growth on N source | | |
|---|---|---|---|---|
| | | Arginine | $N_2$ | $NH_4$ |
| E. coli | | | | |
| YMC26 (glnD⁻) | pDK6 (vector) | ± | nt | +++ |
| " | pCC62 (Av nfrX cloned in pDK6) | ++ | nt | +++ |
| A. vinelandii | | | | |
| MV16 (nfrX⁻) | pJRD215 (vector) | nt | − | ++ |
| " | pAB41 (Ec glnD cloned in pJRD215) | nt | + | ++ |

(nt: not tested)

UTase has a role in the regulatory cascade that controls the activities of glutamine synthetase (GS) enzyme and NTRC activator (Rhee et al., 1985). In the presence of limiting $NH_4^+$, UTase uridylylates PII protein. Uridylylated PII protein promotes adenylyl removal from the inactive adenylylated form of GS; thus glnD mutants of E. coli require glutamine for good growth. Deuridylylated PII promotes dephosphorylation of NTRC, the transcriptional activator of nifLA and glnA (Keener & Kustu, 1988). Other operons of nitrogen assimilation such as nitrate reductase (nar) and proline utilization (put) in Klebsiella species appear to be under ntr gene control, and it is significant that a glnD mutant of K. aerogenes is unable to grow on nitrate (Foor et al., 1978). In contrast to the phenotype of E. coli or K. aerogenes glnD mutants which require glutamine for growth and fail to grow on nitrate as an N source, respectively, the A. vinelandii nfrX mutants neither required glutamine for growth nor were unable to grow on nitrate. Therefore control of adenylylation of GS and presumptive phosphorylation of NTRC in A. vinelandii may occur by a somewhat different regulatory cascade than occurs in enteric bacteria or there may be another glnD-like gene which fulfills this function.

A role for NFRX in nitrogen fixation may be to modify the NIFL protein from active to inactive forms in conditions of nitrogen deficiency. Evidence for this function is that when the A. vinelandii nifL-KIXX mutation described above was transformed into the nfrX mutant, it became Nif$^+$. Therefore in the absence of NIFL, the nfrX mutation had no effect on expression of nif genes. Whether the NFRX-mediated inactivation of NIFL is by uridylylation of NIFL itself or another protein in a regulatory cascade is not yet known. Also not known is whether glnD mutants of K. pneumoniae would be Nif$^-$ and suppressible by mutations in nifL.


## 6. Molybdenum Does not Repress vnf and anf Genes at Low Temperatures

The observation of Miller & Eady (1988) that vanadium nitrogenase is more active than molybdenum enzyme at 5° to 10°C prompted us to examine whether Mo represses expression of the vnf genes at lower growth temperatures. Again the lacZ fusion strains were used to examine expression from the nifH, vnfH, and anfH promoters. Molybdenum repressed the vnfH and anfH operons to a lesser extent at 20°C than at 30°C, and at 14°C molybdenum did not repress these genes at  all. In metal uptake experiments, much less Mo was transported into cells grown at 14°C than into those grown at 30°C, which may explain why this metal does not repress expression of the two non-molybdenum nitrogenases at lower temperatures. We have been unable to demonstrate better growth of A. vinelandii on vanadium than on molybdenum at lower growth temperatures (minimum growth temperature for this mesophilic organism is 10°C). Since vanadium nitrogenase is more $O_2$ sensitive than is molybdenum nitrogenase, better $O_2$ control in growth experiments may be required to show any potential growth difference. Nevertheless, A. vinelandii has the potential to synthesize any or all of the three nitrogenases at low growth temperatures regardless of the presence of Mo. Although it initially seemed that the importance of non-Mo nitrogenases might be to provide a capacity for nitrogen fixation in Mo deficient soils, it is possible that one of them, vanadium nitrogenase, allows more efficient nitrogen fixation in low temperature soils.


## 7. Summary

A model showing the relationship of regulatory gene products and expression of the three nitrogenases in A. vinelandii is shown below.

The major points presented in this paper are
1) three different activators, nifA, vnfA, and anfA are required for expression of nif, vnf, and anf genes. VNFA directly activates expression from the vnfH, vnfD and vnfE promoters in an E. coli background.
2) a nifL gene is present upstream of nifA, as in K. pneumoniae. nifL mutants of A. vinelandii are able to express nitrogenase constitutively in the presence of $NH_4^+$. The nifL mutants excrete $NH_4^+$ while fixing nitrogen.
3) The nfrX gene, necessary for expression of nif and anf operons, is

homologous to the <u>glnD</u> gene of enteric bacteria. The Nif⁻ phenotype of
<u>nfrX</u> mutants is suppressed by mutation in <u>nifL</u>.
4) Mo represses <u>vnfH</u> and <u>anfH</u> expression at 30°C but not at 14° or 20°C.
Vanadium nitrogenase may be significant in organisms growing at low
temperature.

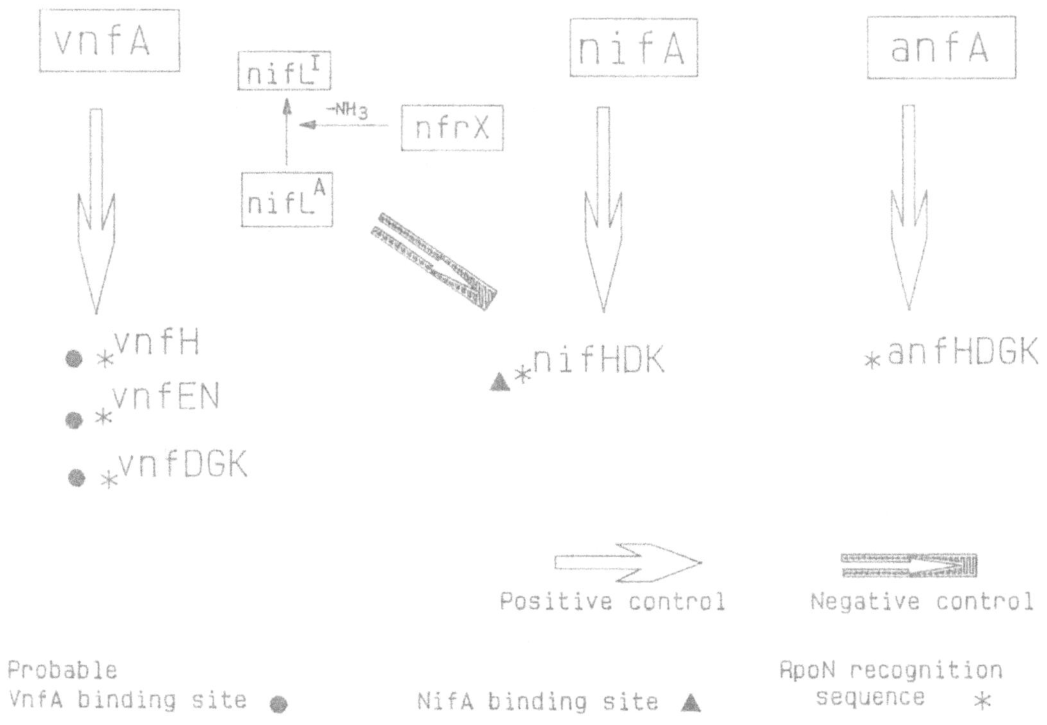


Figure 4. A model for regulation of three nitrogenases in <u>A</u>. <u>vinelandii</u>.

## 8. Acknowledgements


We thank R. Pau, P. Bishop, D. Dean and their co-workers for useful
discussions and sharing unpublished results that were helpful in the
development of this work. E. Garcia kindly provided the sequence of the
N-terminal region of the <u>glnD</u> gene. R. Foote is thanked for help in
preparing the manuscript.


## 9. References


Bennett, L.T., Cannon, F.C., and Dean, D. (1988) Nucleotide sequence and
    mutagenesis of the <u>nifA</u> gene from <u>Azotobacter</u> <u>vinelandii</u>. <u>Mol Microbiol</u>

2: 315-321.

Bishop, P.E., Jarlenski, D.M.L., and Hetherington, D.R. (1980) Evidence for an alternative nitrogen fixation system in Azotobacter vinelandii. Proc Natl Acad Sci USA 77: 7342-7346.

Bishop, P.E., and Joerger, R.D. (1990) Genetics and molecular biology of alternative nitrogen fixation systems. Ann Rev Pl Physiol Pl Mol Biol 41: 109-125.

Bortels, H. (1936) Weitere Untersuchungen uber die Bedeutung von Molybdan, Vanadium, Wolfram und anderen Erdascenstoffen fur stickstoffbindende und andere Mikroorganismen. Zentr Bakteriol Parasitenk Abt II 95: 193-218.

Foor, F., Cedergren, R.J., Streicher, S.L., Rhee, S.G., and Magasanik, B. (1978) Glutamine synthetase of Klebsiella aerogenes: properties of glnD mutants lacking uridylyltransferase. J Bacteriol 134: 562-568.

Garcia, E., and Rhee, S.G. (1983) Cascade control of Escherichia coli glutamine synthetase. J Biol Chem 258: 2246-2253.

Jacobson, M.R., Brigle, K.E., Bennett, L., Setterquist, R.A., Wilson, R.A., Cash, V.L., Beynon, J., Newton, W.E., and Dean, D.R. (1989) Physical and genetic map of the major nif gene cluster from Azotobacter vinelandii. J Bacteriol 171: 1017-1027.

Joerger, R.D., and Bishop, P.E. (1988) Nucleotide sequence and genetic analysis of the nifB-nifQ region from Azotobacter vinelandii. J Bacteriol 170: 1475-1487.

Joerger, R.D., Jacobson, M.R., Premakumar, R., Wolfinger, E.D., and Bishop, P.E. (1989a) Nucleotide sequence and mutational analysis of the structural genes (anfHDK) for the second alternative nitrogenase from Azotobacter vinelandii. J Bacteriol 171: 1075-1086.

Joerger, R.D., Jacobson, M.R., and Bishop, P.E. (1989b) Two nifA-like genes required for the expression of alternative nitrogenases in Azotobacter vinelandii. J Bacteriol 171: 3258-3267.

Joerger, R.D., Loveless, T.M., Pau, R.N., Mitchenall, L.A., Simon, B.H., and Bishop, P.E. (1990) Nucleotide sequence and mutational analysis of the structural genes for nitrogenase 2 of Azotobacter vinelandii. J Bacteriol 172: 3400-3408.

Keener, J., and Kustu, S. (1988) Protein kinase and phosphoprotein phosphatase activities of nitrogen regulatory proteins NTRB and NTRC of enteric bacteria: Roles of the conserved amino-terminal domain of NTRC. Proc Natl Acad Sci USA 85: 4976-4980.

Kelly, M.J.S., Poole, R.K., Yates, M.G., and Kennedy, C. (1990) Cloning and mutagenesis of genes encoding the cytochrome bd terminal oxidase complex in Azotobacter vinelandii: mutants deficient in the cytochrome d complex are unable to fix nitrogen in air. J Bacteriol 172: (in press).

Kennedy, C., Gamal, R., Humphrey, R., Ramos, J., Brigle, K., and Dean, D. (1986) The nifH, nifM and nifN genes of Azotobacter vinelandii: Characterisation by Tn5 mutagenesis and isolation from pLAFR1 gene banks. Mol Gen Genet 205: 318-325.

McKenna, C.E., Benemann, J.R., and Traylor, T.G. (1970) A vanadium containing nitrogenase preparation: implications for the role of molybdenum in nitrogen fixation. Biochem Biophys Res Commun 41: 1501-1508.

Merrick, M.J. (1990) Regulation of nitrogen fixation genes in free-living and symbiotic bacteria. In <u>Fifty years of nitrogen fixation research</u>. Stacey, G., Evans, H.W., and Burris, R.H. (eds). Chapman and Hall, New York.

Rhee, S.G., Chock, P.B., and Stadtman, E.R. (1985) Nucleotidylations involved in the regulation of glutamine synthetase in <u>Escherichia coli</u>. In <u>The enzymology of post-translational modifications of proteins</u>. Volume 2. Freedman, R.B., and Hawkins, H.C. (eds). Academic Press Inc., New York, pp. 273-297.

Robson, R.L., Eady, R.R., Richardson, T.H., Miller, R.W., Hawkins, M., and Postgate, J.R. (1986) The alternative nitrogenase of <u>Azotobacter chroococcum</u> is a vanadium enzyme. <u>Nature</u> **322**: 388-390.

Santero, E., Toukdarian, A., Humphrey, R., and Kennedy, C. (1988) Identification and characterisation of two nitrogen fixation regulatory regions <u>nifA</u> and <u>nfrX</u> in <u>Azotobacter vinelandii</u> and <u>Azotobacter chroococcum</u>. <u>Mol Microbiol</u> **2**: 303-314.

Toukdarian, A., and Kennedy, C. (1986) Regulation of nitrogen metabolism in <u>Azotobacter vinelandii</u>: isolation of <u>ntr</u> and <u>glnA</u> genes and construction of <u>ntr</u> mutants. <u>EMBO J</u> **5**: 399-407.

# GENETIC CHARACTERIZATION OF NITROGEN FIXATION IN *ENTEROBACTER* STRAINS FROM THE RHIZOSPHERE OF CEREALS


R. KREUTZER, H.-D. STEIBL, S. DAYANANDA, R. DIPPE, L. HALDA,
M. BUCK, W. KLINGMÜLLER
*Department of Genetics*
*University of Bayreuth*
*8580 Bayreuth, FRG*



ABSTRACT. Nitrogen-fixing *Enterobacter agglomerans* strains isolated from the rhizosphere of wheat harbour well arranged *nif*-gene clusters on large indigenous plasmids. The *nif*-gene organization is similar to that found on the chromosome of *Klebsiella pneumoniae*. However, the two electron transport protein genes *nifJ* and *nifF* form one operon in *E. agglomerans*. The plasmids of two *E. agglomerans* strains have been examined. They differ strikingly in size and restriction pattern, but their *nif*-gene clusters are identical in gene arrangement and almost identical in nucleotide sequence. No homogy was detected with non-*nif*-regions of the two plasmids. The amino acid sequences of NifT, NifW and NifZ show homologies with the corresponding proteins of other nitrogen fixing bacteria, implying their important roles in nitrogen fixation. The regulation of the *E. agglomerans nif*-gene cluster involves the proteins NifL and NifA. *E. agglomerans* and *Klebsiellae* are thus the only examples of species to date with *nifL*-genes. As in several other bacteria, *nif*-promoters of *E. agglomerans* have -24,-12 promoter elements, upstream activator sequences and integration host factor binding sites. When the sequence of the *E. agglomerans nifH* -24,-12-promoter was changed, a higher promoter activity was observed under conditions where NifA cannot bind to the upstream activator sequence, probably due to the preformation of a closed promoter complex. No difference in the mechanisms of *nif*-gene regulation was found between *K. pneumoniae* and *E. agglomerans*. A *K. pneumoniae nifHDK* probe hybridized with the 30kb indigenous plasmid of a nitrogen fixing *Enterobacter cloacae* strain at low stringency conditions. Apart from this, however, no other *nif*-gene probes hybridized to that plasmid.


## 1. Introduction

*Enterobacter* species are widespread in the rhizosphere of grasses grown in tropical and temperate climates (1,2,3,4,5,6). The association of nitrogen-fixing *Enterobacter* strains with cereals such as wheat, maize and barley is of special interest. The taxonomic analysis of nitrogen-fixing bacteria in the rhizosphere of wheat grown around Bayreuth, West Germany, revealed a large proportion of *Enterobacter agglomerans* (7). Among the several strains which were isolated there in 1980, five grew very well under culture conditions free of combined nitrogen and reduced acetylene to a high extent. Surprisingly, all five strains were found to harbour the nitrogenase structural genes *nifHDK* on large plasmids of 100 to 200 kb in size (8).

Plasmid-borne *nif*-genes had at that time only been reported for symbiotic organisms such as rhizobia. As *E. agglomerans*, in contrast, belongs to the free-living group of nitrogen-fixing bacteria, it was of special interest to examine the organization and regulation of these plasmid-borne *nif*-genes. A considerable amount of genetic data on the *nif*-genes of *E. agglomerans* is available. This proves that the genus *Enterobacter* not only contributes several more species to the long list of nitrogen-fixing bacteria but also exhibits properties which may be of importance to understanding more about the function and evolution of nitrogen fixation genes.

## 2. Materials and Methods

### 2.1. BACTERIAL STRAINS AND PLASMIDS

*E. coli* strains CB454 (9), JM103 (10), BMH71-18 (11), ET8894 (12), *E. agglomerans* strains 243, 333, 334, 335, 339 (7) and *E. cloacae* strain ZP101 (13) were used. Cloning vectors were pUC18, M13mp18, M13mp19 (14), pRK415 (15) and the promoter probe plasmids pCB182 (9) and pMC1403 (16). Plasmids pCK3 (17), pMB163 (18) and pMJ220 (19) were used to provide NifA or its deletion derivative. The *K. pneumoniae nifH-lacZ* fusion plasmid pMB1 (20) was used as a control in the $\beta$-galactosidase test. The $\Omega$-fragment was obtained from plasmid pHP45$\Omega$ (21).

### 2.2. GROWTH CONDITIONS AND RECOMBINANT DNA TECHNIQUES

Growth conditions were as reported (22). For the preparation of RNA, *E. agglomerans* was grown in nitrogen-free medium (NFDM) either in the absence or in the presence of 2mg ammonium sulfate per ml. All recombinant DNA-techniques were applied as described (22). Southern hybridizations with labeled pEA3 and pEA9 probes were performed using the non-radioactive digoxygenin labeling and detection kit from Boehringer Mannheim, FRG, as specified by the manufacturer. For other Southern hybridizations radioactively labeled probes were used. Standard stringency conditions were 68°C without formamide in 2xSSC. It is noted in the text, where other stringency conditions were used.

### 2.3. $\beta$-GALACTOSIDASE ASSAYS

Deviating from the earlier described procedure (22), the cells were lysed using a lysis mixture consisting of 1 part (v/v) 20mM $MnSO_4$, 1 part 10% sodiumdodecylsulfate, 1 part toluene and 5 parts $\beta$-mercaptoethanol.

### 2.4. DNA SEQUENCING

Defined restriction fragments were subcloned either in pUC18 or in M13mp18 and M13mp19. Plasmid sequencing was done as described earlier (22), M13 sequencing was performed as described in the Amersham *M13 Cloning and Sequencing Handbook*. The sequence data were processed either in the GENMON program package by the Gesellschaft für Biotechnologische Forschung, Braunschweig, FRG, or in the University of Wisconsin Genetics Computer Group software package.

### 2.5. NUCLEASE S1 PROTECTION ASSAYS

A *Sal*I-*Bam*HI-fragment covering most of the *nifJ* and *nifF* genes (see Fig. 1) was eluted from an agarose-gel and dephosphorylated as described (22). 10pmoles of the purified fragment were endlabeled with 6pmoles [$\gamma$-$^{32}$P] ATP (3000Ci/mmole) adding 2U of T4-polynucleotide-kinase (23).

RNA from *E. agglomerans* was prepared using Qiagen-pack 500 anion exchange columns according to the manufacturer's protocol (Qiagen Inc. USA). 250mg of RNA were hybridized with 1pmole of the endlabeled restriction fragment, treated with nuclease S1, run on an 1% agarose gel and subjected to autoradiography according to the protocol of (23).

## 3. Results and Discussion

### 3.1. *Nif*-GENE ORGANIZATION OF THE PLASMID pEA3

In our first approach to study the plasmid-borne *nif*-genes we performed Southern hybridization experiments with lysates of five of the isolated *E. agglomerans* strains using several fragments of the *Klebsiella pneumoniae nif*-gene cluster as labeled probes. It appeared that in each of the five strains one plasmid hybridized not only with *nifHDK* probes but also with all other *nif*-gene probes (24). One of these plasmids, which was designated pEA3, had been cloned in cosmids (25). It has a size of 111kb. Physical mapping revealed that four recombinant cosmids comprise the entire plasmid. In order to determine the organization of the *nif*-genes, all of which were localized on the recombinant cosmid peaMS2-2, we again performed Southern hybridizations using various restriction fragments of the *K. pneumoniae nif*-gene cluster as probes. The result of these detailed studies was the finding that all *nif*-genes are clustered on the plasmid pEA3 in a region of approximately 23kb (24, Fig. 2). Moreover the arrangement of the *nif*-genes is similar to that of *K. pneumoniae* (26). In the latter, however, the *nif*-genes are located on the chromosome (27).

Thus, *E. agglomerans* is the first example of a naturally occuring, plasmid-borne *nif*-gene cluster from any bacterium showing such extensive homology to the *nif*-gene cluster of *K. pneumoniae*. But nevertheless, there are remarkable differences in the arrangements of the two *nif*-gene clusters which relate to the *nifJ* and *nifF* genes. *NifJ* is located at the opposite end of the *E. agglomerans nif*-gene cluster when compared to its location in *K. pneumoniae* (29). *NifF* is situated immediately downstream of *nifJ* in *E. agglomerans*, while it is located between *nifM* and *nifL* in *K. pneumoniae*. It should be emphasized that the two electron transport protein genes *nifJ* and *nifF* are oriented opposite to all other *nif*-genes both in *E. agglomerans* and *K. pneumoniae* (28, 26).

There existed two possibilities to explain the transcription of the *nifJ* and *nifF* genes in *E. agglomerans*. (i) Both genes are cotranscribed from a promoter upstream of *nifJ*, which has been sequenced and shown to be active (22). (ii) The genes are seperately transcribed involving a sequence found upstream of *nifF* (CTGG-N$_8$-GTGCG), which displays some similarities with the recognition sequence for RNA-polymerase-$\sigma^{54}$. To determine which of the possibilities is correct we cloned different sized fragments of the *nifJF*-region as transcriptional fusions with the promoterless *Escherichia coli lacZ*-gene. A strong transcriptional terminator, an $\Omega$-fragment, was also inserted upstream of the possible promoter, in order to terminate transcription from the *nifJ*-promoter (29, Fig. 1).

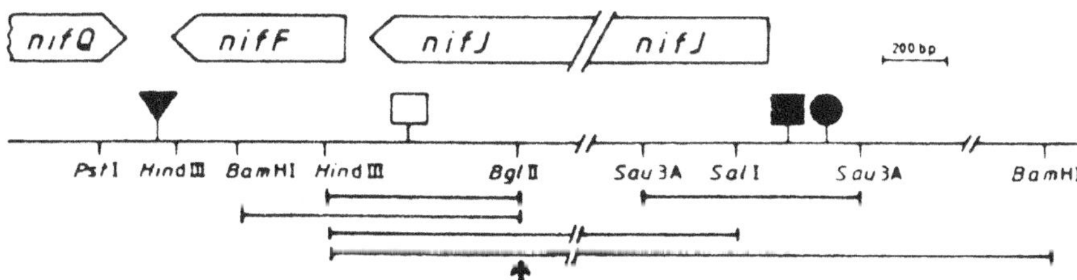



Fig. 1: Restriction map of the right end of the *nif*-gene group of pEA3 and the positions of the genes *nifQ*, *nifF* and *nifJ*. The postions of the *nifJ*-promoter and a promoter-like motif are indicated by filled and open boxes, respectively. The *nifJ*-upstream activator sequence is represented by a circle and the putative transcriptional terminator by a triangle. The restriction fragments cloned as transcriptional fusions with *lacZ* are marked by horizontal bars. The position of the inserted $\Omega$-fragment is indicated by an arrow.

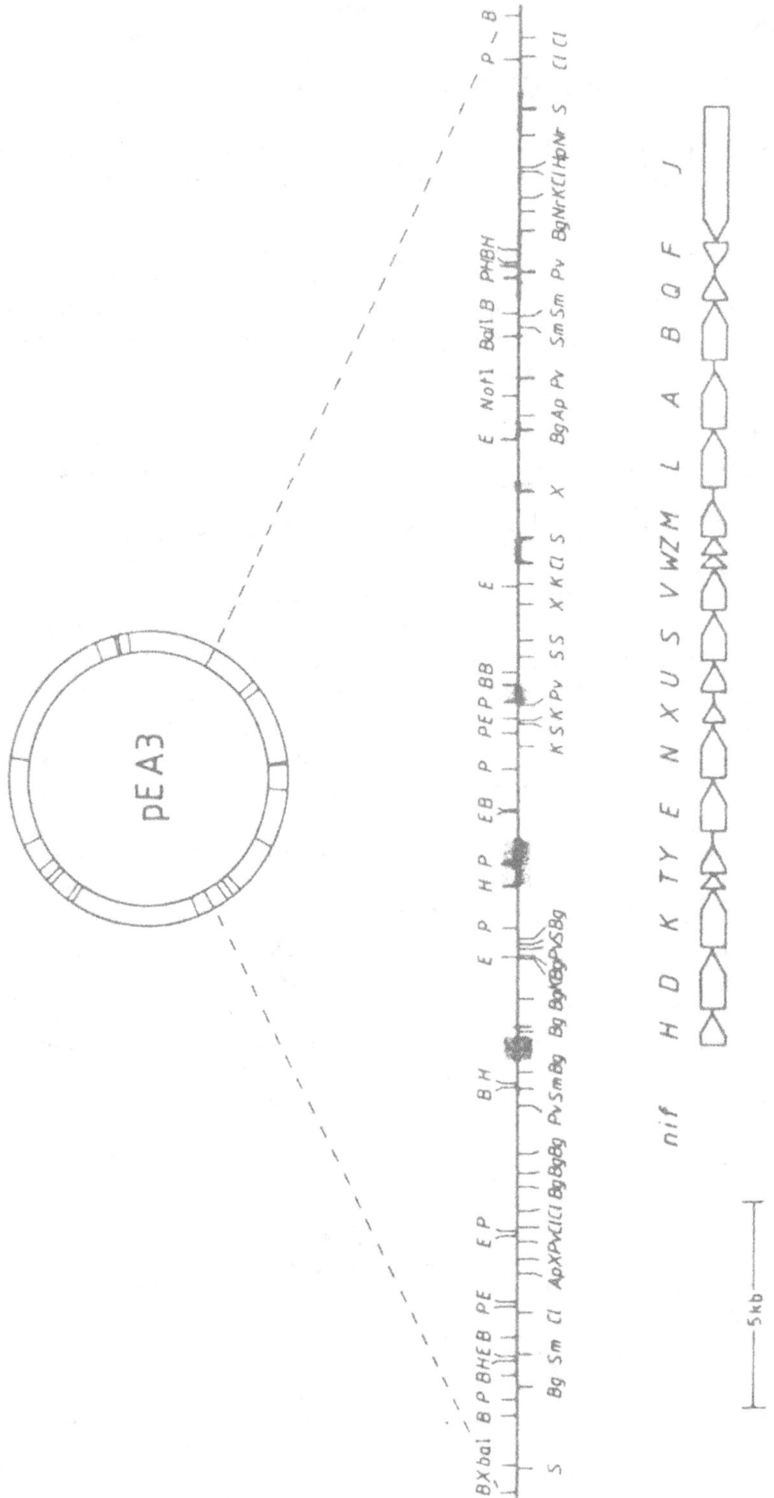

Fig. 2: *Bam*HI restriction map of the entire plasmid pEA3 and detailed restriction map of the *nif*-gene region (24) (*Ap, Apa*I; *B, Bam*HI; *Bg, Bgl*II; *Cl, Cla*I; *E, Eco*RI; *H, Hin*dIII; *Hp, Hpa*I; *K, Kpn*I; *Nr, Nru*I; *P, Pst*I; *Pv, Pvu*II; *S, Sal*I; *Sm, Sma*I; *X, Xho*I). Shaded bars represent parts which have been sequenced. The positions of the *nif*-genes are indicated below .

However, when the *K. pneumoniae* NifA-protein was provided from the constitutively expressed *nifA*-gene on the plasmid pCK3, β-galactosidase activity was only detected when the *nifJ*-promoter was on the cloned fragment. No activity was found, when transcription from the *nifJ*-promoter was blocked by the inserted Ω-fragment. This shows that no transcription is initiated downstream of the Ω-fragment. The promoter-like motif therefore does not represent an active promoter. Also, in a nuclease S1 transcript mapping assay a DNA-RNA-hybrid was found corresponding in size to the *nifJF*-cotranscript. No hybrid was detected when RNA was prepared from bacteria which were grown under repressing culture conditions. In conclusion the two electron transport protein genes *nifJ* and *nifF* form one transcriptional unit in *E. agglomerans*, whose transcription is initiated from the *nifJ*-promoter and terminated by a putative transcriptional terminator downstream of *nifF* (29). Thus, the *E. agglomerans nif*-gene cluster represents the tightest arrangement of all *nif*-gene clusters found so far. It is still controversial whether clustering of the *nif*-genes represents an evolutionary archaetype or the temporary result of an evolutionary process in which originally dispersed *nif*-genes were arranged together (30, 31, 32).

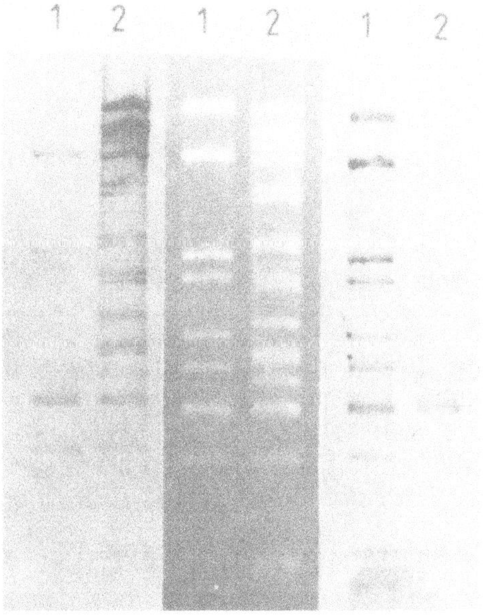


Fig. 3: Southern hybridization of *Eco*RI digested plasmid DNA from pEA3 (lane 1) and pEA9 (lane 2). The gel (center) was blotted on a filter which was used for probing with either labeled pEA9 (left) or pEA3 (right).

## 3.2. *NIF*-GENE ORGANIZATION OF THE PLASMID pEA9

It is noteworthy that exactly the same *nif*-gene arrangement as on pEA3 was found on another *E. agglomerans* plasmid (33). This plasmid, pEA9, has a size of 200kb and therefore is about 90kb larger than pEA3. The restriction patterns of both plasmids differ remarkably (Fig. 3). Southern hybridizations, using one of each of the two plasmids as labeled probes, showed that only restriction fragments of both plasmids carrying *nif*-genes hybridized to each other. This means that the homology of both plasmids is low, with the exception of their *nif*-genes, for which the nucleotide sequence homology is over 98% (determined for *nifF*). We therefore assume that a horizontal *nif*-gene transfer between these strains may have occurred in evolution. Evidently, the plasmid pEA9 is selftransmissible among related *E. agglomerans* strains (34). A clustered *nif*-gene group may be advantagous for transfer via selftransmissible plasmids.

## 3.3. *Nif*-PLASMID IN *E. CLOACAE*

The existence of *nif*-plasmids in *Enterobacter* may not be restricted to the species *E. agglomerans*. We also identified a *nif*-plasmid (pLT1) of about 30kb in size in a nitrogen-fixing *E. cloacae* strain (ZP101) isolated from the rhizosphere of maize in Yugoslavia. The uncut plasmid-DNA as well as certain restriction fragments of this remarkably small *nif*-plasmid hybridized with the *K. pneumoniae nifHDK*-probe (Fig. 4).

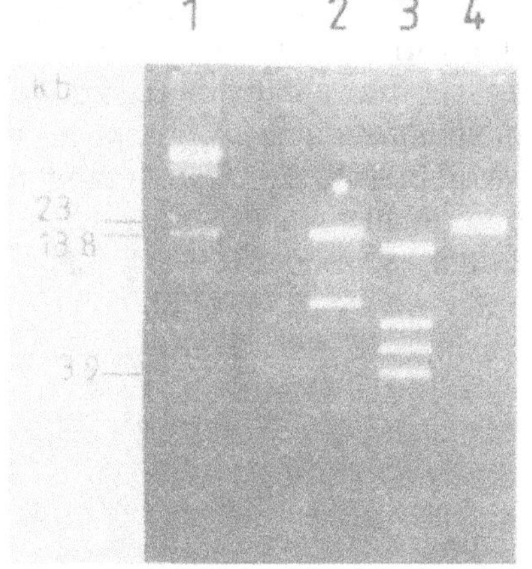

Fig. 4: Southern hybridization of pLT1 with the *K. pneumoniae nifHDK* probe. pLT1 uncut (lane 1), pLT1 hydrolysed with *Eco*RI, *Nru*I and *Pst*I (lanes 2 to 4), respectively.

However, the signal was only obtained under low stringency conditions (50°C without formamide, 2xSSC). We still do not know whether other *nif*-genes are located either on this plasmid or on the chromosome since no hybridization signals with other *nif*-genes were obtained even under very low stringency conditions. This could mean that the nitrogen fixation system is different from those of *E. agglomerans* and *K. pneumoniae*.

## 3.4. SEQUENCE COMPARISONS OF SELECTED NIF-PROTEINS

Several parts of the pEA3 *nif*-gene group have been sequenced. These are marked as shaded bars on the restriction map in Fig. 2. The average homology with corresponding regions of *K. pneumoniae* is about 70% at DNA level (22, 29, 35). As mentioned above, the *nif*-gene arrangement of both clusters is similar. Also, as at least the regulatory genes can substitute for each other (36), both *nif*-gene clusters may work in a similar manner. On the other hand, the sequence homology of the *E. agglomerans* and *K. pneumoniae nif*-genes is significantly lower than that found, for instance, among different *Klebsiella* species (37). *E. agglomerans* therefore fills a space in the phylogeny of nitrogen-fixing enterobacteria and may be a well-suited organism for sequence comparisons. Conserved regions in the Nif-proteins of both organisms could indicate those parts of the proteins which are relevant for their function. Comparisons among proteins who's functions are not yet well understood are of special interest. We sequenced the small genes *nifT*, *nifW* and *nifZ* of *E. agglomerans*. *NifWZ*-mutants of *K. pneumoniae* were reported which show almost normal growth on medium free of combined nitrogen but need a longer time for derepression (38). It has been suggested that the respective proteins are involved in maturation or accumulation of an active dinitrogenase. For *A. vinelandii* evidence has been reported that NifZ is required for the biosynthesis of the iron-molybdenum-cofactor (39). Comparison of the NifW-protein of *E. agglomerans* with those of *K. pneumoniae*, *Azotobacter vinelandii* (40), *A. chroococcum* (41) and *Rhodobacter capsulatus* (42) shows a few amino acids which are conserved among the proteins but no obvious features (Fig. 5a). In the case of NifZ the comparison shows that all the proteins are relatively well conserved (Fig. 5b). There is a notable number of conserved arginine residues especially in the N-terminal parts. Also, a cysteine is conserved within a conserved stretch of several amino acids in the center of the proteins. Altogether the conservation of many charged residues is obvious, which could indicate that NifZ is an ion binding protein. The gene *nifT* was identified downstream of the *nifDHK*-genes in *K. pneumoniae* and *A. vinelandii*. We also found this gene at the corresponding position in *E. agglomerans*. Its function is completely unknown thus far. The alignment of the amino acid sequences, deduced from the genes, shows a significant degree of homology (Fig. 5c). Four conserved proline residues in the C-terminal portions of the proteins are particularly remarkable.

**(a)** NifW

```
Ea                 MDWFTRIEGVDELESAQSFFDFFELEVDPVLLRSRHLHIMAQF
Kp                 MMEWFYQIPGVDELRSAESFFQFFAVPYQPELLGRCSLPVLATF
Av       MTVQPFSPDSDLTLDEAMDELVSAEDFLEFFGVPFDQDVVHVNRLHIMQRY
Ac       MTVQPFSPDSELTLDEAMDELVSAEDFLEFFGVPFDQTVVHVNRLHIMQRY
Rc       MTPES....PTLA...ALTKLSSAEEIFAFLGVEPIREVLNSSRLHIMKRF
Cons.    MT :        :    : ::L SA: :::F: :        ::     L ::  :


Ea       NQRLTAAVPVHFVDEEESDRADWRLARRLLAESYQHTVAGPLNTQSGLAVY
Kp       HRKLRAEVPLQ.NRLEDNDRAPWLLARRLLAESYQQQFQESGT*
Av       HDYLSKAGDL....DEHDDQARYAVFQKLLARAYLDFVESDALTEKVFKVF
Ac       HDYLTKAGDL....DEHDDQARYAVVPAA.ARAYLDFVESDALTEKVFKVF
Rc       GAYL.RETDM....TELTEDGIFERARDALLRAQADFVASTPLKEKVFKVF
Cons.       L    :      E : : :                :        :  : V:


Ea       QRNNG..SFIGWNDLLEVRP*
Av       RMHEPQKTFVSIDQLLS*
Ac       RLHEPQKTFVSIDQLLS*
Rc       ETEAAKRKARFVGLETLKVIKS*
Cons.        :
```

**(b)** NifZ

```
Ea       MKPVFEVDQQVRVTRIVRDDGTFAGKTRGDLLLRRGSLGYVREWGVFLQDT
Kp       MRPKFTFSEEVRVVRAIRNDGTVAGFAPGALLVRRGSTGFVRDWGVFLQDQ
Av       MLPQFEYGDEVRLIRNVRNDGIYPGMNIGALLMRRGAVGCVYDVGIYLQDQ
Ac       MIPQFFYGDEVRIIRNVRNDGTYPGMDTGALLIRRGAVGCVYDVGTYLQDQ
Cons.    M P F   ::VR: R :R:DGT :G    G LL:RRG  G V : G :LQD


Ea       IVYQVHFLDDDLIIGCREQELIAGDAPWLAGAFQYGDRVSSRRSLVIRGEV
Kp       IIYQIHFPETDRIIGCREQELIPITQPWLAGNLQYRDSVTCQMALAVNGDV
Av       LIYRVHFLDEGRTIGCREEELILASAPWIPNLFEFRDDVIATRSLAVRGOV
Ac       LIYRVHFLNEGRTVGCREEELILASAPWIPNLFEFRDNVIATRSLAVRGQV
Cons.    ::Y :HF : :  :CCRE:ELI    PW::  ::: D V     L : C:V
```

```
Ea      VVPQA.YGRVWAV..NRWTA/
Kp      VVSAGQRGRVEAT..DRGELGDSYTVDFSGRW.FRVPVQAIALIEEREE*
Av      LVKRGQLGSIMKVLRDEPELGIQYHVHFGDGLVLQVPEQSLAMADSTAAIE
Ac      LVTRGQLGSIMKVLRDESELGIQYHVHFGDGLVLQVPEQSLVMAETEAAME
Cons.   :V : G :     :   G  Y V F:: : : VP Q : : :   A:E


Av      EVLDGI*
Ac      .VLDEL*
Cons.   VLD :



(c) NifT


Ea      MPMVIFRLRNNS.LYCYIAKQDMEAKVVQLEHNHPQQWGGRVCLEGGKAYY
Kp      MPIVIFRERGAD.LYAYIAKQDLEARVIQIEHNDAERWGGAISLEGGRRYY
Av      MPSVMIRRNDEGQLTFYIAKKDQEEIVVSLEHDSPELWGGEVTLGDGSTYF
Cons.   MP V::R      L  YIAK D E  V: :EH: :: WGG : L::G  Y:


Ea      VEPQPGIPSFPVSLRATRSEL*
Kp      VHPQPGRPVFPISLRATRNTLI*
Av      IEPIPQ.PKLPITVRAKRAGEA*
Cons.   : P P  P :P: :RA R
```

Fig. 5: Comparison of the NifW proteins of *E. agglomerans*, *K. pneumoniae*, *A. vinelandii*, *A. chroococcum* and *Rhodobacter capsulatus* (a), the NifZ proteins of *E. agglomerans*, *K. pneumoniae*, *A. vinelandii* and *A. chroococcum* (b), and the NifT proteins of *E. agglomerans*, *K. pneumoniae* and *A. vinelandii* (c). : indicates related residues. The *E. agglomerans* sequence of NifZ is not complete. Conserved arginine residues are shaded in (b), charged amino acids are underlined and a conserved cystein is in bold letters. Conserved proline residues are shaded in (c).

---

Although the comparison of amino acid sequences of the proteins NifW, NifZ and NifT has not yet revealed direct indications on their functions, the observation that the proteins are conserved to a significant extent implies their important roles in nitrogen fixation.

## 3.5. *Nif*-GENE REGULATION IN *E. AGGLOMERANS*

As expected by the similar organization, even the regulation of the *E. agglomerans nif*-gene group is very similar to that of *K. pneumoniae*. The ability to reduce acetylene is only observed in the absence of both a combined nitrogen source and oxygen in the medium. The two regulatory genes *nifL* and *nifA* were identified (24). Thus *E. agglomerans* represents the only other known nitrogen-fixing organism besides *Klebsiella* species where *nifL* was found.

In order to study the influence of the two regulatory genes on the expression of other *nif*-genes, we used a system of two compatible plasmids in *E. coli*. One of these plasmids carries an *nifH-lacZ* transcriptional fusion. The other harbours different sized fragments of the *nifLA*-operon under control of a *lac*-promoter (Fig. 6).

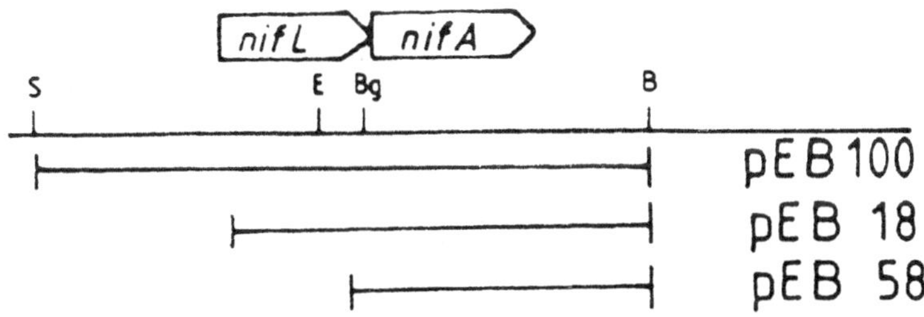


Fig. 6: Restriction map of the *nifLA* -region. The bars below indicate the extensions of the fragments cloned under control of the *lac*-promoter in pRK415.

While pEB100 comprises the entire *nifLA*-operon, in pEB18 and pEB58 N-terminal portions of *nifL* were deleted. When the *nifH*-promoter was induced by NifA $\beta$-galactosidase was produced. This was recognized by blue colour due to the hydrolysis of added X-Gal in a qualitative test where cultures were grown with each of the two plasmids in semisolid medium in capped serum tubes. In strains with pEB100, which contain the complete *nifLA*-operon, the *nifH*-promoter was only induced when the ammonium concentration was below 4mM before inoculation. In strains with pEB18 or pEB58, however, where parts of *nifL* were deleted, the induction of the *nifH*-promoter was detected both in the absence and in the presence of normally repressing ammonium concentrations. This behaviour fully resembles that which we know from *K. pneumoniae* (for reviews see 43, 44) and shows that NifA is an activating and NifL a repressing protein for nitrogen fixation in *E. agglomerans* as well. A *nifA*-like gene was also reported for an *Enterobacter cloacae* strain isolated from the rhizosphere of rice (45).

TABLE 1. Assumed *nif*-promoter consensus sequences and the relevant nucleotide sequences of three *E. agglomerans nif*-promoters.

| Promoter | UAS | | IHF-binding site | | -24,-12-promoter |
|---|---|---|---|---|---|
| Cons. | TGT$^{AA}_{GG}$GNNNNC$^{CC}_{TT}$ACA | $N_X$ | $^{A}_{T}$ATCAANNNNTT$^{G}_{A}$[a] | $N_X$ | TGGCACNNNNTTTGC$^{T}_{A}$ |
| nifH | TGTTTGTTTTATAACA | $N_{65}$ | TTTCAATGGGTTT | $N_{16}$ | TGGTACAAACACTGCA |
| nifJ | TGTCAGTGTAACGACA | $N_{77}$ | ATTCAATATATTA | $N_{16}$ | TGGCATGTCCCATGCT |
| nifU | TGTCATCTGTATGACA | $N_{53}$ | TTTCAACCACATA[b] | $N_{18}$ | TGGCATCACAATTGCT |

Conserved residues are underlined. [a] Sequence taken from (46). [b] Sequence complementary and reverse to the others.

We cloned and sequenced three of six expected *E. agglomerans nif*-promoters. In their nucleotide sequences, features were identified that are typical for *nif*-promoters. Table 1 shows the assumed *nif*-promoter consensus sequence (Cons.) and the relevant sequences of three *E. agglomerans nif*-promoters. One of the features is the -24,-12-promoter, which was originally characterized in *K. pneumoniae* as a conserved CTGG-N$_8$-TTGCA motif by Beynon *et al.* (47). This site has been reported to be recognized by the alternative $\sigma$-factor $\sigma^{54}$ (48) in complex with RNA-polymerase, thus forming a closed promoter complex (for reviews see 49, 50). The second feature is an upstream activator sequence (UAS), originally characterized as a highly conserved TGT-N$_{10}$-ACA motif by Buck *et al.* (51). This site has been shown to bind the transcriptional activator NifA (18, 19, 52). The upstream bound NifA is believed to make contacts with the closed RNA-polymerase-$\sigma^{54}$-complex at the downstream promoter through the formation of a DNA loop (53). Subsequently this closed promoter complex is isomerized to an open complex to initiate transcription. However, the closed complex is apparently weak at *K. pneumoniae nif*-promoters and may not readily form in the absence of the activator protein NifA (54, 55). Site-directed promoter mutations in *K. pneumoniae* and *E. agglomerans* indicated that in addition to the highly conserved bases, specific residues in the -24,-12-promoter are necessary for the formation of a stable closed initial recognition complex before NifA binds to the UAS (54). In promoters which are not fitted with those residues, the closed complex formation is probably dependent on NifA binding to the UAS. T to C mutations at position -23 in the *E. agglomerans* and *K. pneumoniae nifH*-promoters resulted in 7 to 11 times higher promoter activities under conditions where NifA could not bind to the upstream sites. This was achieved using a truncated NifA (provided from plasmid pMB163) which lacks the DNA-binding domain (Table 2).

TABLE 2. Effect of T to C mutations in the *nifH*-promoters on promoter induction.

| | | $\beta$-Galactosidase activity (units) | |
| --- | --- | --- | --- |
| Plasmid | Characteristcs | ET8894 | ET8894(pMB163) |
| | | | |
| pMK310 | *E. agglomerans nifH* wild-type | 2 | 27 |
| pMK88078 | *E. agglomerans nifH* -23 T to C | 2 | 300 |
| pMB1 | *K. pneumoniae nifH* wild-type | n.d. | 38 |
| pMK88075 | *K. pneumoniae nifH* -23 T to C | 7 | 288 |

Obviously, in the mutant promoters a closed complex was preformed more frequently. This result and other results from mutagenesis and *in vivo* footprinting experiments with *K. pneumoniae* and *Rhizobium meliloti nif*-promoters (55) clearly show that a cytosine at position -23 and thymidine residues at positions -17 to -15 facilitate the formation of a closed promoter complex under conditions where NifA cannot bind to the upstream activator sequence. Most of the wild-type *nif*-genes make apparent use of promoter sequences that are non-optimal for the ready formation of a closed complex. This phenomenon is also observed with other promoters which are induced by activating proteins, i.e. the catabolite activator protein, and may be explained by the advantage of strictly preventing initiation of transcription in the absence of the activating protein and also by increasing the specificity for one particular avtivator protein.

Recently the involvement of the integration host factor (IHF) in *nif*-promoter induction was reported for the *K. pneumoniae nifU*-promoter and was also suggested for the *E. agglomerans nifU*-promoter (56). The IHF was originally found to be essential for the integration of phage lambda into the *E. coli* chromosome (57). Later it was found to be involved in several other cellular processes (58, 59). Its function is to bend the DNA at its binding sites, thus facilitating the interaction with sites or other proteins bound to sites which would be well separated on a linear DNA molecule (60). Very recent data indicate that the IHF is not only involved in the control of the *nifU*-promoters and several *K. pneumoniae nif*-promoters other than *nifF*, *nifL* and *nifM* (46). It seems to be a general mechanism in purple bacteria that NifA-dependent promoters are stimulated by the binding of the IHF. The IHF binding sites are located between the UASs and the -24,-12-promoters. Also, in the *E. agglomerans* promoters, putative binding sites for the IHF were identified (Table 1). Since in *E. agglomerans nifF* transcription originates from the *nifJ*-promoter (see above), its expression is likely (in contrast to *K. pneumoniae*) to be IHF-dependent. The function of the IHF in *nif*-promoter induction may be to facilitate DNA loop formation so that upstream bound NifA can establish productive interactions with the downstream RNA-polymerase-$\sigma^{54}$-complex.

## 4. Acknowledgements

S. Dayananda and R. Kreutzer were financially supported by DAAD. We thank N. Schönbrunner for reading the manuscript.

## 5. References

(1) Pedersen, W.L., Chakrabarty, K., Klucas, R.V. and Vivader, A.K. (1978) 'Nitrogen fixation (acetylene reduction) asssociated with roots of winter wheat and sorgum in Nebraska', Appl. Environ. Microbiol. 35, 129-135.
(2) Haahtela, K., Wartiovara, Sundman, V. and Skujin, J. (1981) 'Root associated N₂ fixation (acetylene reduction) by Enterobacteriaceae and *Azospirillum* strains in cold-climate spodsols, Appl. Environ. Microbiol. 41, 203-206.
(3) Belly, R., Thomas-Bauzon, D., Heulin, T., Balandreau, J. and Richard, C. (1983) 'Determination of the most frequent N₂-fixing bacteria in a rice rhizosphere', Canad. J. Microbiol. 29, 881-887.
(4) Subba Rao, N.S. (1983) 'Nitrogen fixing bacteria associated with plantation of orchard plants', Canad. J. Microbiol. 29, 863-866.
(5) Jagnow, G. (1988) 'Enterobacteriaceae in the rhizosphere of wheat, barley and ryegrass: fractions of *nif*-positive strains in field experiments with different doses of N-fertilizers. In: Bothe, H., de Bruijn, F.J. and Newton, W.E. (eds.) Nitrogen Fixation: Hundred Years After. Fischer Verlag, Stuttgart, p. 795.
(6) Väisänen, O., Haahtela, K., Bask, L., Kari, K. Salkinoja-Salonen, M. and Sundman, V. (1985) 'Diversity of *nif* gene location and nitrogen fixation among root-associated *Enterobacter* and *Klebsiella* strains, Arch. Microbiol. 141, 123-127.
(7) Kleeberger, A., Castorph, H. and Klingmüller, W. (1983) 'The rhizosphere microflora of wheat and barley with special reference to Gram-negative bacteria', Arch. Microbiol. 136, 306-311.
(8) Singh, M., Kleeberger, A. and Klingmüller, W. (1983) 'Location of nitrogen fixation (*nif*) genes on indigenous plasmids of *Enterobacter agglomerans*', Mol. Gen. Genet. 190, 373-378.
(9) Schneider, K. and Beck, C.F (1986) 'Promoter-probe vectors for the analysis of divergently arranged promoters', Gene 42, 37-48.
(10) Messing, J., Crea, R. and Seeburg, P.H. (1981) 'A system for shotgun DNA sequencing', Nucl. Acids Res. 9, 309-321.
(11) Gronenborn, B. (1976) 'Overproduction of phage repressor under control of the *lac* promoter of *E. coli*', Mol. Gen. Genet. 148, 243-250.
(12) MacNeil, T., Roberts, G.P., MacNeil, D. and Tyler, B. (1982) 'The products of *gln*L and *gln*G are bifunctional regulatory proteins', Mol. Gen. Genet. 188, 325-333.
(13) Halda, L., Levic, J., Denic, M., Pencic, V. and Neyra, C.A. (1990) 'Nitrogen-fixing bacteria isolated from maize roots and antagonistic activity against *Fusarium* species'. Poster contribution, this meeting.
(14) Yanisch-Perron, C., Vieira, J. and Messing, J. (1985) 'Improved M13 phage cloning vectors and host strains: nucleotide sequence of the M13mp18 and pUC19 vectors', Gene 33, 103-119.
(15) Keen, N.T., Tamaki, S., Kobayashi, D. and Trollinger, D. (1988) 'Improved broad-host-range plasmids for DNA cloning in Gram-negative bacteria', Gene 70, 191-197.
(16) Casadaban, M.J., Chou, J. and Cohen, S.N. (1980) 'In vitro gene fusions that join an enzymatically active ß-galactosidase segment to amino-terminal fragments of exogenous proteins: *Escherichia coli* plasmid vectors for the detection and cloning of translation initiation signals', J. Bacteriol. 143, 971-980.
(17) Kennedy, C. and Drummond, M. (1985) 'The use of cloned *nif* regulatory elements from *Klebsiella pneumoniae* to examine *nif* regulation in *Azotobacter vinelandii*', J. Gen. Microbiol. 131, 1787-1795.
(18) Morett, E., Cannon, W. and Buck, M. (1988) 'The DNA-binding domain of the transcriptional activator protein NifA resides in its carboxy terminus, recognizes the upstream activator sequences of *nif* promoters and can be separated from the positive control function of NifA', Nucl. Acids Res. 16, 11469-11488.
(19) Morett, E. and Buck, M. (1988) 'NifA-dependent *in vivo* protection demonstrates that the upstream activator sequence of *nif* promoters is a protein binding site', Proc. Natl. Acad. Sci. USA 85, 9401-9405.
(20) Buck, M., Khan, H. and Dixon, R. (1985) 'Site directed mutagenesis of the *Klebsiella pneumoniae nifL* and *nifH* promoters and *in vivo* analysis of promoter activity', Nucl. Acids Res. 13, 7621-7638.
(21) Prentki, P. and Krisch, H.M. (1984) 'In vitro insertional mutagenesis with a selectable DNA fragment.' Gene 29, 303-313.
(22) Kreutzer, R., Singh, M. and Klingmüller, W. (1989) 'Identification and characterization of the *nifH* and *nifJ* promoter regions located on the *nif*-plasmid pEA3 of *Enterobacter agglomerans* 333', Gene 78, 101-109.
(23) Sambrook, J., Fritsch, E.F. and Maniatis, T. (1989) 'Molecular Cloning. A Laboratory Manual. Cold Spring Harbor Laboratory. Cold Spring Harbor, NY.
(24) Singh, M., Kreutzer, R., Acker, G. and Klingmüller, W. (1988) 'Localization and physical mapping of a plasmid-borne 23kb *nif*-gene cluster from *Enterobacter agglomerans* showing homology to the entire *nif*-gene cluster of *Klebsiella pneumoniae* M5a1', Plasmid 19, 1-12.
(25) Singh, M. and Klingmüller, W. (1986) 'Cloning of pEA3, a large plasmid of *Enterobacter agglomerans* containing nitrogenase structural genes', Plant and Soil 90, 235-242.
(26) Arnold, W., Rump, A., Klipp, W., Priefer, U. and Puhler, A. (1988) 'Nucleotide sequence of a 24,206bp DNA fragment carrying the entire nitrogen fixation cluster of *Klebsiella pneumoniae*', J. Molec. Biol. 203, 715-738.

(27) Dixon, R.A. and Postgate, J.R. (1972) 'Genetic transfer of nitrogen fixation from *Klebsiella pneumoniae* to *Escherichia coli*', Nature 237, 102-103.

(28) Cannon, M., Deistung, J., Hill, S. and Cannon, F. (1984) '*Klebsiella nifF* and *nifJ*'. In: Advances in Nitrogen Fixation Research. Martinus Nijhoff/Junk., The Hague, NL.

(29) Kreutzer, R., Dayananda, S. and Klingmüller, W.(1990) 'Cotranscription of the electron transport genes *nifJ* and *nifF* in *Enterobacter agglomerans* 333.' J. Bacteriol., submitted.

(30) Postgate, J.R. (1982) 'The Fundamentals of Nitrogen Fixation', Cambridge University Press, UK.

(31) Postgate, J.R. and Eady, R.R. (1988) 'The evolution of nitrogen fixation'. In: Bothe, H., de Bruijn, F.J. and Newton, W.E. (eds.) Nitrogen Fixation: Hundred Years After. Fischer Verlag, Stuttgart, pp.31-40.

(32) Mancinelli, R.L. and McKay, C.P. (1988) 'The evolution of nitrogen cycling', Origins of Live and Evolution of the biosphere 18, 311-325.

(33) Steibl, H.-D., Dayananda, S. and Klingmüller, W. (in preparation) 'Comparative genetic analysis of the two *nif*-plasmids pEA3 and pEA9 from *Enterobacter agglomerans*'.

(34) Klingmüller, W., Herterich, S. and Min, B.-W. (1989) 'Self-transmissible *nif*-plasmids in *Enterobacter*'. In: Nitrogen Fixation with Non-legumes. Kluwer Academic Publishers, Dordrecht, NL, pp.173-178.

(35) Dippe, R. and Klingmüller, W. (in preparation) 'Nucleotide sequences of the *Enterobacter agglomerans* 333 genes *nifT*, *nifW* and *nifZ*'.

(36) Kreutzer, R., Brandmüller, E. and Klingmüller, W. (1988) 'NifA-mediated positive regulation of *nif*-genes in *Enterobacter*'. In: Bothe, H., de Bruijn, F.J. and Newton, W.E. (eds.) Nitrogen Fixation: Hundred Years After. Fischer Verlag, Stuttgart, p. 311.

(37) Wang, P.-L., Hoh, S.K., Chung, K.-S., Uozumi, T. and Beppu, T. (1985) 'Cloning and Expression in *E. coli* of the whole *nif* genes of *Klebsiella oxytoca*, a nitrogen fixer in the rhizosphere of rice', Agric. Biol. Chem. 49, 1469-1477.

(38) Paul, W. and Merrick, M. (1989) 'The roles of the *nifW*, *nifZ* and *nifM* genes of *Klebsiella pneumoniae* in nitrogenase biosynthesis', Eur. J. Biochem. 178, 675-682.

(39) Jacobson, M.R., Cash, V.L., Weiss, M.C., Laird, N.F., Newton, W.E., Dean, D.R. (1989). 'Biochemical and genetic analysis of the *nifUSVWZM* cluster from *Azotobacter vinelandii*', Mol. Gen. Genet. 219, 49-57.

(40) Jacobson, M.R., Brigle, K.E., Bennett, L.T., Setterquist, R.A., Wilson, M.S., Cash, V.L., Beynon, J., Newton, W.E. and Dean, D.R. (1989) 'Physical and genetic map of the major *nif* gene cluster from *Azotobacter vinelandii*', J. Bacteriol. 171, 1017-1027.

(41) Evans, Robson, unpublished.

(42) Moreno-Vivian, C., Masepohl, B., Schmehl, M., Klipp, W. and Pühler, A. (1988) 'Nucleotide sequence of *Rhodobacter capsulatus nif* gene regions carrying homologous genes tio *Klebsiella pneumoniae nifE*, *nifN*, *nifX*, *nifQ*, *nifS* and *nifV*.' In: Bothe, H., de Bruijn, F.J. and Newton, W.E. (eds.) Nitrogen Fixation: Hundred Years After. Fischer Verlag, Stuttgart, p. 177.

(43) Gussin, G.N., Ronson, C.W. and Ausubel, F.M. (1986) 'Regulation of nitrogen fixation genes', Ann. Rev. Genet. 20, 567-591.

(44) Dixon, R. (1988) 'Genetic regulation of nitrogen fixation'. In: Cole, J.A. and Ferguson, S.J. (eds.) The Nitrogen and Sulphur Cycles. Cambridge University Press, Cambridge, pp. 417-438.

(45) Zhu, J.-B., Li, Z.-G., Wang, L.-W., Shen, S.-S. and Shen, S.-C. (1986) 'Temperature sensitivity of a *nifA*-like gene in *Enterobacter cloacae*', J. Bacteriol. 166, 357-359.

(46) Hoover, T.R., Santero, E., Porter, S. and Kustu, S. (1990) 'The integration host factor (IHF) stimulates interaction of RNA polymerase with NifA, the transcriptional activator for nitrogen fixation operons'. In press.

(47) Beynon, J., Cannon, M., Buchanon-Wollaston, V. and Cannon, F. (1983) 'The *nif* promoters of *Klebsiella pneumoniae* have a characteristic primary structure', Cell 34, 665-671.

(48) Hirschman, J., Wong, P.-K., Sei, K., Keener, J. and Kustu, S. (1985) 'Products of the nitrogen regulatory genes *ntrA* and *ntrC* of enteric bacteria activate *glnA* transcription in vitro: evidence that the *ntrA* product is a r factor', Proc. Natl. Acad. Sci USA 82, 7525-7529.

(49) Kustu, S., Santero, E., Keener, J., Popham, D. and Weiss, D. (1989) 'Expression of $\sigma^{54}$ (*ntrA*)-dependent genes is probably united by a common mechanism', Microbiol. Rev. 53, 367-376.

(50) Thöny, B. and Hennecke, H. (1989) 'The -24/-12 promoter comes of age', FEMS Microbiol. Rev. 63, 341-358.

(51) Buck, M., Miller, S., Drummond, M. and Dixon, R. (1986) 'Upstream activator sequences are present in the promoters of nitrogen fixation genes', Nature 320, 374-378.

(52) Popham, D., Szeto, D., Keener, J. and Kustu, S. (1989) 'Function of bacterial activator protein that binds to transcriptional enhancers', Science 243, 629-634.

(53) Buck, M., Cannon, W. and Woodcock, J. (1987) 'Transcriptional activation of the *Klebsiella pneumoniae* nitrogenase promoter may involve DNA loop formation', Mol. Microbiol. 1, 243-249.

(54) Buck, M. and Cannon, W. (1989) 'Mutations in the RNA polymerase recognition sequence of the *Klebsiella pneumoniae nifH* promoter permitting transcriptional activation in the absence of NifA binding to upstream activator sequences', Nucl. Acids Res. 17, 2597-2612.

(55) Morett, E. and Buck, M. (1989) 'In vivo studies on the interaction of RNA polymerase-σ54 with the *Klebsiella pneumoniae* and *Rhizobium meliloti nifH* promoters: the role of NifA in the formation of an open promoter complex', J. Mol. Biol. 210, 65-77.

(56) Cannon, W.V., Kreutzer, R., Kent, H.M., Morett, E. and Buck, M. (1990) 'Activation of the *Klebsiella pneumoniae nifU* promoter: identification of multiple and overlapping upstream NifA binding sites', Nucl. Acids Res. 18, 1693-1701.

(57) Nash, H.A. and Robertson, C.A. (1981) 'Purification and properties of the *Escherichia coli* protein factor required for λ integrative recombination', J. Biol. Chem. 256:9246-9253.

(58) Friden, P., Voelkel, K., Sternglanz, R. and Freundlich, M. (1984) 'Reduced expression of the isoleucin and valin enzymes in integration host factor mutants of *Escherichia coli*', J. Mol. Biol. 172:573-579.

(59) Friedmann, D.J., Olson, E.J., Carver, D. and Gellert, M. (1984) 'Synergistic effect of *himA* and *gyrB* mutations: evidence that Him functions control expression of *ilv* and *xyl* genes', J. Bacteriol. 157:484-489.

(60) Robertson, C.A. and Nash, H.A. (1988) 'Bending of the bacteriophage λ attachment site by *Escherichia coli* integration host factor', J. Biol. Chem. 263:3554-3557.

# ANALYSIS OF NITROGENASE REACTION USING MONOCLONAL ANTIBODIES AGAINST α-SUBUNIT OF COMPONENT I OF *A.vinelandii*


Naoto Abe, the late Yoshiharu Maruyama and Kazukiyo Onodera
Department of Agricultural Chemistry, The University of Tokyo
1-1-1 Yayoi, Bunkyo-ku, Tokyo 113
Japan



ABSTRACT. Conversion of atmospheric nitrogen to ammonia by nitrogen-fixing microorganisms is catalysed by the enzyme nitrogenase. Nitrogenase consists of two components, component I and component II. Component I is a Fe-S protein containing Mo and is a tetramer($\alpha_2\beta_2$). Component I catalysed the reduction of nitrogen with component II. Reduced component II gives electrons to component I with the consumption of ATP. The reaction mechanism fo nitrogenase is complex and has been studied by kinetic analysis. We have generated monoclonal antibodies against the nitrogenase of *A.vinelandii* which show inhibitory effects agaist either acetylene reduction or hydrigen evolution. We have identified the peptide fragments to which each monoclonal antibody binds.


## Isolation of monoclonal antibodies against nitrogenase

Mice were immunized with purified component I. When the titer increaced the mice were sacrified. The spreen cells of the immunized mice and myeloma cells(NS-1) were mixed. After fusion, selected hybridomas were cloned by the limiting dilution method. Cloned cells were injected intraperitoneally into BALB/C mice for the production of ascitic fluid. Eight monoclonal antibodies were obtained. The subclasses of five antibodies were $IgG_1$ and the other were IgM. Western blot analysis revealed that all monoclones reacted with the α-subunit of component I.

## Effect of monoclonal antibodies on nitrogenase activity

We used a heat treated supernatant of *A.vinelandii*(heat-sup) grown in medium without a nitrogen source as an enzyme. Reaction was carried out in 0.1 ml of heat-sup under anaerobic conditions for 1hr. The reaction mixture was supplemented with each monoclonal antibody. Inhibition fo acetylene reduction and hydrogen evolution was measured. As a control, bovine serum albumin was added in the place of monoclonal antibody.

Result was summerized in Table 1. It was shown that each monoclonal antibody exhibited different effects on nitrogenase activity. Some inhibited acetylene reduction only and the others inhibited only hydrogen evolution. One monoclonal antibody inhibited both reactions.

It is reported that the active centers of acetylene reduction and hydrogen evolution are different.

Table 1. Effect of monoclonal antibodies against nitrogenase activity

| N a m e | MA-1 | MA-2 | MA-3 | MA-4 | MA-5 | MA-6 | MA-7 | MA-8 |
|---|---|---|---|---|---|---|---|---|
| $C_2H_2$ reduction (% remained) | 29 | 90 | 80 | 79 | 94 | 104 | 72 | 69 |
| $H_2$ evolution (% remained) | 92 | 61 | 63 | 89 | 111 | 109 | 97 | 65 |
| $C_2H_2/H_2$ ratio | 0.31 | 1.48 | 1.27 | 0.89 | 0.85 | 0.95 | 0.74 | 1.06 |

Reduction of nitrogen is inhibited by carbon monooxide but hydrogen evolution is not inhibited.

Our data seemed to confirm these previous results. Monoclone MA-8 inhibited 30% of acetylene reduction and inhibited 40% of hydrogen evolution. We have chosen monoclones MA-1, MA-2 and MA-8 for further study.

## Identification of peptide fragments of nitrogenase recognized by monoclonal antibodies

To determine the peptide fragments recognized by monoclonal antibodies, we have employed genetic engineering techniques. It is known that component II is coded for by nifH and the $\alpha$ and $\beta$ subunits of component I are coded for by the nifD and nifK genes respectively. Both of these genes have been sequenced.

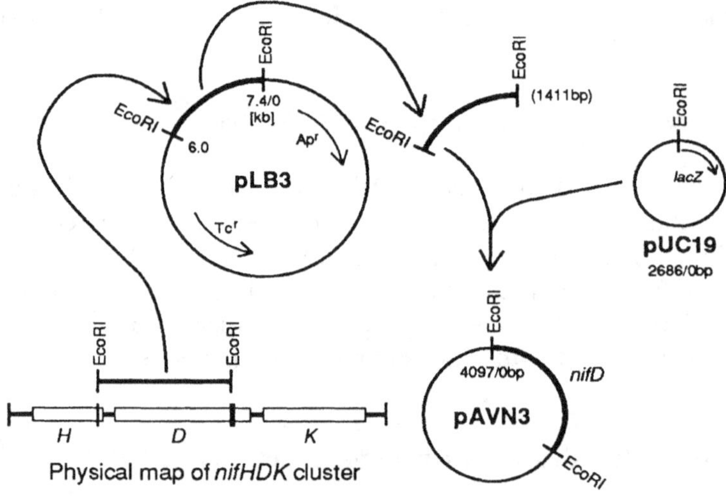


Fig.1 Construction of pAVN3.

For our purpose, we have constructed a clone, pAVN3, as shown in Fig.1. It was necessary to prepare various DNA clones with different size DNA fragments containing the *nif* gene and to determine whether each DNA fragment contained the peptide fragments recognized by each monoclonal antibody. We constructed deletion mutants by limit digestion of restriction fragments with exonuclease III, for example, the upper portion of *nifD* was deleted from PstI-SalI fragment of pAVN3.

We have selected mutants with the correct reading frame by screening for the expressed peptide using polyclonal antibody against component I subunit. The deletion mutants were classified into four groups by the reactivity with each monoclonal antibody as shown in Fig.2.

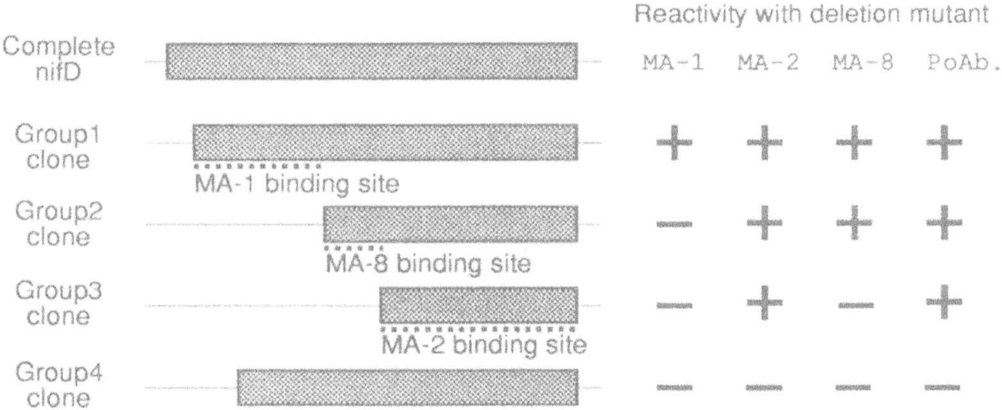



Fig. 2 Screening of pAVN3 deletion mutants by anti-nitrogenase Component I monoclonal antibodies.

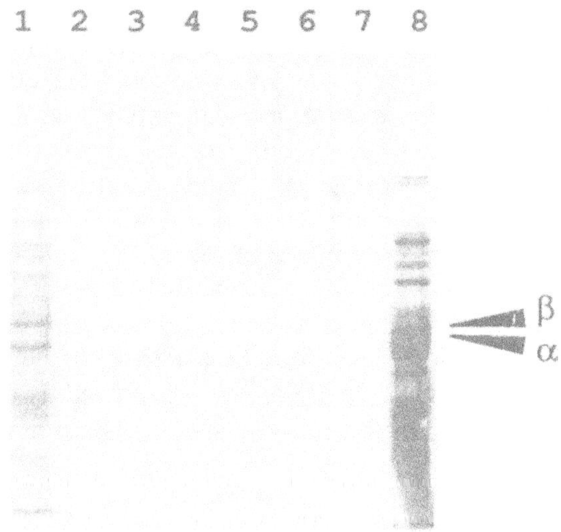



Fig. 3 Immunoblotting of pAVN3 deletion mutant by anti-nitrogenase Component I polyclonal antibody.   Lane 1:gold colloidal staining of pAVN3; 2:pUC19; 3:pAVN3; 4:group1 clone; 5:group2 clone; 6:group3 clone; 7:group4 clone; 8:gold colloidal staining of *A.vinelandii* heat-sup.; α:Component I α subunit; β:Component I β subunit.

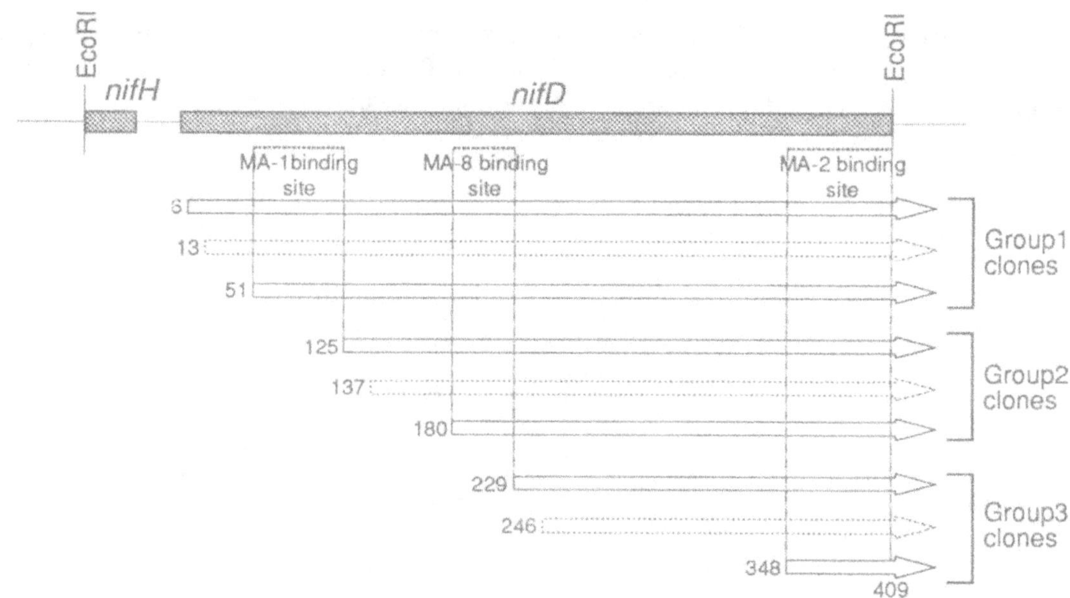



Fig. 4 Recoginition sites of anti-nitrogenase Component I monoclonal antibodies on *nifD* coding peptide.

The western blot analysis fo cell lysates from each type of deletion mutant is shown in Fig.3. The peptide fragments recognized by each monoclonal antibody are shown in Fig.4. It is concluded that the peptide fragment recognized by MA-1, the Mab which inhibited acetylene reduction strongly, was mapped to the region located in the N terminal region. The fragment recognized by MA-2, the Mab which inhibited hydrogen evolution was mapped to the C terminal region. The fragment recognized by MA-8, the Mab which inhibited both acetylene reduction and hydrogen evolution was mapped to the region between the two regions. The total amino acid sequence of the *nif* genes and the binding sites of each monoclonal antibody is shown in Fig.5.

It is cosidered that metal clusters are important for the function of nitrogenase and therefore the cystein residues are essential. It is interesting to note that two cystein residues were found in the region rcognized by MA-1 and one cystein was found in the region recognized by MA-8. The size of peptide thus identified was still large. Therefore the reactivity of synthetic peptides from each region will clarify the precise and minimum size of peptide recognized by the inhibitory Mab's. The results using these synthetic peptide will contribute to the understanding of nitrogenase reaction.

Fig. 5 DNA and amino acid sequence of *nifD*.(1)   Underlines indicate recognition sites of monoclonal antibodies.

```
          850       860       870       880       890       900
TACATCTCCCGTCACATGGAAGAGAAGTACGGTATCCCATGGATGGAGTACAACTTCTTC
TyrIleSerArgHisMetGluGluLysTyrGlyIleProTrpMetGluTyrAsnPhePhe

          910       920       930       940       950       960
GGCCCGACCAAGACCATCGAGTCGCTGCGTGCCATCGCCGCCAAGTTCGACGAGAGCATC
GlyProThrLysThrIleGluSerLeuArgAlaIleAlaAlaLysPheAspGluSerIle

          970       980       990      1000      1010      1020
CAGAAGAAGTGCGAAGAGGTCATCGCCAAGTACAAGCCCGAGTGGGAAGCGGTGGTCGCC
GlnLysLysCysGluGluValIleAlaLysTyrLysProGluTrpGluAlaValValAla

         1030      1040      1050      1060      1070      1080
AAGTACCGTCCGCGCCTGGAAGGCAAGCGCGTCATGCTCTACATCGGTGGCCTGCGTCCG
LysTyrArgProArgLeuGluGlyLysArgValMetLeuTyrIleGlyGlyLeuArgPro
                        MA-2 Binding Site
         1090      1100      1110      1120      1130      1140
CGCCACGTGATCGGCGCCTACGAAGACCTGGGCATGGAAGTGGTGGGTACCGGCTACGAG
ArgHisValIleGlyAlaTyrGluAspLeuGlyMetGluValValGlyThrGlyTyrGlu

         1150      1160      1170      1180      1190      1200
TTCGCCCACAACGACGACTATGACCGCACCATGAAAGAAATGGGTGACTCCACCCTGCTG
PheAlaHisAsnAspAspTyrAspArgThrMetLysGluMetGlyAspSerThrLeuLeu

         1210      1220      1230      1240      1250      1260
TACGATGACGTGACCGGCTACGAATTCACTGGCCGTCGTTTTACAACGTCGTGACTGGGA
TyrAspAspValThrGlyTyrGluPheThrGlyArgArgPheThrThrSer***Leu
```

Fig. 5 DNA and amino acid sequence of *nifD*.(2)

# Involvement of GroEL in *Klebsiella pneumoniae nif* gene expression and nitrogenase assembly

David Govezensky, Tsvika Greener and Ada Zamir,
Biochemistry Department, Weizmann Institute of Science, Rehovot 76100, Israel.

## Abstract

An *E. coli groEL* mutant transformed with the *Klebsiella pneumoniae nif* gene cluster accumulated very low to non-detectable levels of Kp1 and Kp2 as compared to the isogenic wild type strain. In *K. pneumoniae*, overexpression of the *E. coli groE* operon markedly accelerated the rate of appearance of Kp1 and its constituent polypeptides after the start of derepression. NifA-dependent expression of β-galactosidase from the *nifH* promoter followed by an in-frame *nifH'-'lacZ* fusion was threefold lower in the *groEL* mutant compared to the wild type strain. A direct interaction between NifA and GroEL was indicated by co-immunoprecipitation of NifA with anti-GroEL antibodies. Pulse-chase analysis of derepressed *K. pneumoniae* indicated assembly was rapid for Kp2 and slow for Kp1. Immunoprecipitation of the pulse-chased cell extracts with anti-GroEL antibodies revealed a transient association of newly synthesized NifH and NifDK with GroEL, constituting the first step of nitrogenase assembly. Additional rate-limiting steps are involved in the assembly of Kp1.

## Introduction

Genetic and biochemical studies have led to the identification and characterization of the *nif* genes involved in the biogenesis of nitrogenase components (Merrick, 1988; Smith *et al.*, 1988). However, the potential contribution of non-*nif* genes to the process has been largely overlooked so far.

We have previously addressed the problem of the mechanism of nitrogenase assembly by following the interaction between the Nif structural polypeptides in foreign hosts expressing selected *nif* genes (Berman *et al*, 1985; Holland *et al.* 1987). When *E. coli* and *Saccharomyces cerevisiae* were compared for the interaction between NifD and NifK, the results indicated the two polypeptides formed an assembly product in *E. coli* but remained in their monomeric form in yeast (Holland *et al.*, 1987). These results have led us to consider the possibility that a host function(s) played a role in this interaction.

A clue for the potential nature of such a host function(s) has been provided with the emerging role of molecular chaperones in assisting correct folding and supramolecular assembly of other proteins (reviewed in Ellis, 1990). The best characterized group of molecular chaperones, classified as chaperonins (Ellis, 1990; Georgopoulos and Ang, 1990), includes evolutionary conserved proteins encoded by heat shock induced genes (Hemmingsen *et al.*, 1988; Ellis, 1990). The prototype bacterial chaperonin is the product of the *E. coli groEL*, a gene essential for cell viability (Fayet *et al.*, 1989). Together with the co-transcribed *groES, groEL* has been implicated in diverse processes such as phage morphogenesis, stress protection, protein folding, oligomer assembly, mRNA turnover and general protease activity (reviewed in Georgopoulos and Ang, 1990). According to the current model, GroEL, a decatetramer in its native form, binds unfolded or misfolded polypeptides. After ATP-dependent release, mediated by GroES or functional homologues thereof, the polypeptides assume a correct conformation with respect to both intra and intermolecular interactions (Ellis, 1990; Georgopoulos and Ang, 1990).

The present study examines by a variety of experimental approaches the role played by GroEL in nitrogenase biogenesis.


## Materials and Methods

The experimental procedures employed in this study have been detailed elsewhere (Berman *et al*, 1985; Holland *et al.* 1987; Govezensky and Zamir, 1989; Govezensky *et al.*, 1990).


## Results and Discussion

EFFECT OF GroEL ON THE ACCUMULATION OF NITROGENASE COMPONENTS AND THEIR SUBUNITS

An *E. coli groEL* mutant and isogenic wild type strain were transformed with the entire *nif* cluster from *K. pneumoniae* and derepressed transformants were compared, by gel electrophoresis and immunoblot analysis, for the level of native or denatured Kp1 and Kp2 (Berman *et al.*, 1985; Holland, 1987; Govezensky *et al.*, 1990). The *nif*-transformed *groEL* mutant was also compared to co-transformants with a plasmid containing the wild type *groE* operon. Representative results (Fig. 1) show the mutant cells contain very low to non-detectable levels of native Kp1 and Kp2 as compared to the wild type cells or to the mutant co-transformed with the *groE* operon. A similar reduction was also observed for the denatured NifDK and NifH polypeptides.

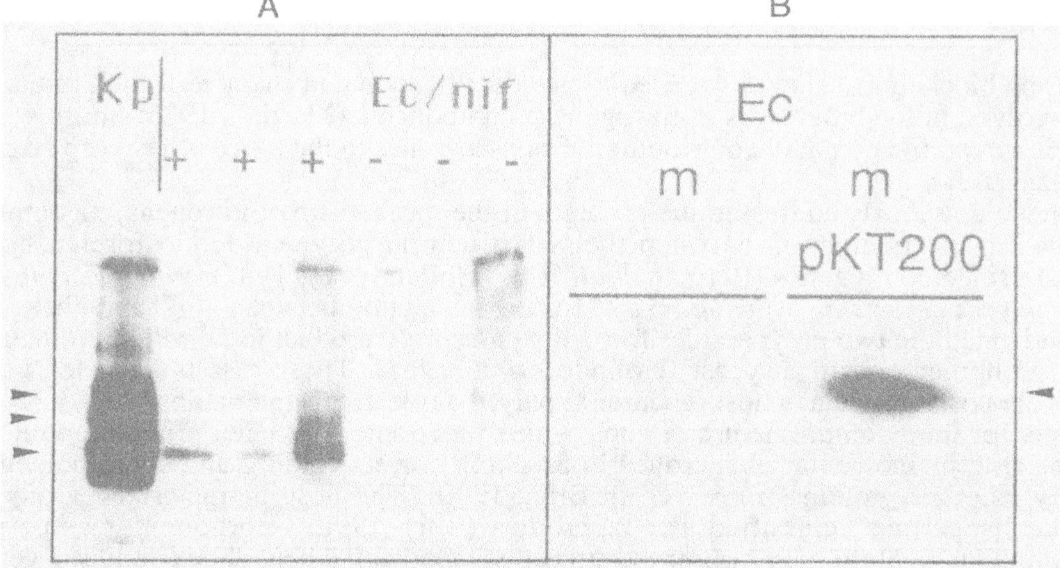


Figure 1. Effect of GroEL on the accumulation of Kp1 and Kp2. Immunoblot analysis with anti-Kp1 (A) and anti-Kp2 (B) antibodies of crude cell extracts resolved by electrophoresis under non-denaturing conditions. Kp, derepressed *K. pneumoniae*; Ec or Ec/nif, *nif*-transformed *E.coli* strain T850 (*groEL⁻*, Takano and Kakefuda, 1972) (designated m, or -) or strain KL355 (designated +); pKT200, plasmid containing the *E. coli groE* operon. Extracts of independent *E. coli* transformant clones were analyzed on each lane. Arrowheads, position of the three electrophoretic variants of native Kp1 (A), or Kp2 (B).

In view of its evolutionary conservation (Hemmingsen *et al.*, 1988, Ellis, 1990) a homologue of groEL could be expected to be present in *K. pneumoniae* . Nonetheless, the level of the protein could be less than necessary to saturate the requirements of the *nif* system. Consequently, overexpression of *groEL* might possibly promote nitrogenase accumulation even in wild type *K. pneumoniae* cells. To test this possibility, transformants with the *groE* operon from wild type *E. coli* were compared to non-transformed *K. pneumoniae* cells at different times after the start of derepression. The results showed that Kp1 and NifDK accumulation started earlier and progressed more rapidly in the transformed as compared to the non-transformed cells. An immunoblot analysis with anti-GroEL antibodies confirmed the existence of endogenous GroEL in *K. pneumoniae,* and demonstrated the overexpression of the protein in the transformed cells.

GroEL FUNCTIONS IN *nif* REGULATION

The results described so far indicate a function for GroEL in the synthesis, stability and/or assembly of the Nif structural polypeptides. These three aspects are mechanistically interrelated since the assembly into functional nitrogenase components is thought to enhance the intracellular stability of Nif polypeptides as well as to promote *nif* derepression (Roberts *et al.*, 1978). To directly examine the potential regulatory function of GroEL it was therefore necessary to use an assay system which was independent of the synthesis and assembly of Nif polypeptides. The mutant and wild type *E. coli* strains were transformed with a plasmid containing the *nifH* promoter followed by an in-frame *nifH'-'lacZ* fusion and with another plasmid containing a constitutively expressed *nifA*. The results indicated that the *groEL* mutation caused a threefold decrease in the $\beta$-galactosidase expressed from the *nifHDK* promoter in the presence of NifA. This difference did not reflect a strain difference in plasmid copy number nor a general decrease in transcriptional activity in the mutant cells, but rather pointed to a direct involvement of GroEL in *nif* gene expression.

The positive regulatory function of GroEL could arise from its activity in the correct folding and maintenance in a functional state of NifA, as based on the following observations: (i) temperatures above 37°C inhibit *in vivo* transcription from *nif* promoters due to NifA inactivation (Buchanan-Wollaston, *et al.*, 1981); (ii) *in vivo* heat inactivation of NifA is reversible (Brooks *et al.*, 1984); (iii) when overproduced in *E. coli,* NifA is very insoluble (Tuli and Merrick, 1988) and (iv) NifA is very rapidly inactivated in *in vitro nif* transcription assays (Santero *et al.*, 1989). The thermal lability and poor solubility of NifA indicated by these observations make it highly likely that the *nif* transcriptional activator needs to interact with the *groE* system in order to initially assume an active conformation and/or to correct structural damage during its course of action.

An outcome of this proposal is that at any given time a part of NifA should be physically associated with GroEL. Immunoprecipitates with anti-GroEL antibodies indeed indicated that some NifA was co-immunoprecipitated with the chaperonin. The specificity of immunoprecipitation was indicated by the absence, or drastic reduction in the amount of precipitated NifA when instead of the antiserum against GroEL, the extracts were mixed with the Protein A-Sepharose beads without pretreatment with antiserum, or when preimmune serum was used. These results provided direct evidence for the interaction of NifA with GroEL and thereby support the possibility that the chaperonin promotes *nif* expression via its effect on NifA.

ROLE OF GroEL IN NITROGENASE ASSEMBLY

In view of its functional diversity, we considered it possible that GroEL in addition to its regulatory role might also participate in the assembly of nitrogenase components. To explore this possibility, the process of nitrogenase assembly was followed by pulse-chase and co-immunoprecipitation analyses.

Pulse-labeling of *K. pneumoniae* cells with [$^{35}$S] methionine followed by different periods of chase indicated a continuous increase in labeling of native Kp1 for at least 1 h of chase, but no corresponding decline in radioactivity of any other component visible on the gel. These results indicated that MoFe protein assembly occurred by a slow process that entailed the formation of intermediates undetected in the gel-electrophoretic analysis. For Kp2, the results showed maximal labeling already by the end of the pulse period indicating a rapid assembly process.

Extracts of *K. pneumoniae* cells pulse-chased as above were also immunoprecipitated with anti-GroEL antibodies and the precipitates were analyzed by denaturing gel electrophoresis and autoradiography. The results showed that newly synthesized NifH, NifD, and NifK, and a number of other proteins co-immunoprecipitated with the anti-GroEL antibodies. The association of GroEL with the Nif structural polypeptides, as well as with most other co-immunoprecipitated proteins, was short-lived and decayed within 2 min of chase, in agreement with *in vivo* turnover values of other GroEL-assisted processes such as protein export (e.g., Kusukawa *et al.*, 1989). The results of this analysis are compatible with the suggestion that GroEL interacts with many newly synthesized polypeptides, but still exhibits a preference for a specific set of substrates (Georgopoulos and Ang, 1990). In the present case, the set is likely to include not only the Nif structural polypeptides, but also other *nif* gene products which are preferrentially synthesized in the derepressed *K. pneumoniae* cells. More specifically, the results indicate that newly synthesized NifH, NifD and NifK transiently associate with GroEL. Probably due to their rapid turnover these complexes remain undetected in the analysis on native gels.

For the Fe protein these results suggest that the interaction with GroEL is the only intermediate step in NifH intersubunit interaction. However, the duration of the association of NifD and NifK with GroEL is far too short to account for the overall rate of Kp1 assembly. Therefore, the process must include additional, rate-limiting steps during which the NifD and NifK polypeptides are likely to be bound to large complexes of as yet unknown composition.


## Conclusions

This study indicates that GroEL fulfills at least two functions in the *K. pneumoniae nif* system (Fig. 6). The positive role played by GroEL in *nif* expression, most likely by assisting the correct folding of NifA, provides the first example for a specific effect of a ubiquitous chaperonin on gene regulation. The interaction of GroEL with the newly synthesized Nif structural polypeptides sheds new light on the mechanism and genetic requirements of the assembly of nitrogenase components.

A possible regulatory link between *nif* regulation and nitrogenase assembly has been suggested for *K. pneumoniae* based on the observation that many mutations in structural and auxiliary *nif* genes involved in nitrogenase biogenesis caused a general decrease in the level of Nif products (Roberts *et al.*, 1978). The present results provide a possible basis for this effect. According to the proposed model (Fig. 6), mutations affecting the structure or function of NifHDK or other Nif products might slow down the release of the newly synthesized proteins from the GroEL complex. As a consequence, the amount of chaperonin available for NifA binding might become limiting resulting in a decrease in the level of functional NifA and thereby reducing the efficiency of *nif* transcription.

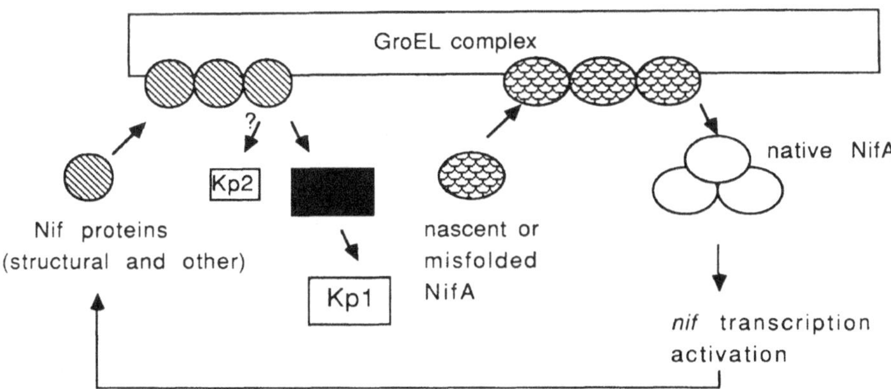


Fig. 2. An integrated model for *nif* regulation and nitrogenase assembly. The black box represents the rate-limiting steps in Kp1 assembly following the transient interaction of the newly-synthesized Nif polypetides with GroEL.

## Acknowledgements

We are grateful to Angela Mehlert for purified GroEL and anti-GroEL antibodies, Irun Cohen for anti-GroEL antibodies, Eduardo Santero for purified NifA and anti-NifA antibodies. The study was supported by research grants from the Anthony Weldon Research fund in Nutritional Sciences, and the Leo and Julia Forchheimer Center for Molecular Genetics, The Weizmann Institute of Science.

## References

Berman,J., Gershoni,J.M. and Zamir,A. (1985) 'Expression of nitrogen fixation genes in foreign hosts: assembly of nitrogenase Fe protein in *Escherichia coli* and in yeast' J. Biol. Chem. 260, 5240-5243.

Brooks,S.J., Collins,J.J. and Brill,W.J. (1984) 'Repression of nitrogen fixation in *Klebsiella pneumoniae* at high temperatures' J. Bacteriol. 157, 460-464.

Buchanan-Wollaston,V., Cannon,M.C., Beynon,J.L. and Cannon,F.C. (1981) 'Role of the *nifA* gene product in the regulation of *nif* expression in *K. pneumoniae*' Nature 294, 776-778.

Ellis,R.J. (1990) 'The molecular chaperone concept' Seminars in Cell Biology 1, 1-9.

Fayet,O., Ziegelhoffer,T. and Georgopoulos,C. (1989) 'The *groES* and *groEL* heat shock gene products of *Escherichia coli* are essential for bacterial growth at all temperatures' J. Bacteriol. 171, 1379-1385.

Georgopoulos,C., and Ang,D. (1990) 'The *Escherichia coli groE* chaperonins' Seminars in Cell Biology 1, 19-25.

Govezensky,D. and Zamir,A.(1989) 'Structure function relationships in the *alpha* subunit of *K. pneumoniae* nitrogenase MoFe protein from analysis of *nifD* mutants' J. Bacteriol., 171, 5729-5735.

Govezensky,D., Greener,T., Segal,G. and Zamir,A. (1990) submitted.

Hemmingsen,S.M., Woolford,C., van der Vies,S., Tilly,K., Dennis,D.T., Georgopoulos,C.P., Hendrix,R.W. and Ellis,R.J. (1988) 'Homologous plant and bacterial proteins chaperone oligomeric protein assembly' Nature 333, 330-334.

Holland,D., Zilberstein,A., Govezensky,D., Salomon,D. and Zamir,A. (1987) 'Nitrogenase MoFe protein subunits from *Klebsiella pneumoniae* expressed in foreign hosts: characteristics and interactions' J. Biol. Chem. 262, 8814-8820.

Kusukawa,N., Yura,T., Ueguchi,C., Akiyama,Y. and Ito,K. (1989) 'Effects of mutations in heat-shock genes *groES* and *groEL* on protein export in *Escherichia coli*' EMBO J. 8, 3517-3521.

Merrick,M.J. (1988) 'Organisation and regulation of nitrogen fixation genes in *Klebsiella* and *Azotobacter*' n Bothe,H., de Bruijn,F.J. and Newton,W.E. (eds) Nitrogen fixation: hundred years after, Gustav Fischer Verlag, Stuttgart, Federal Republic of Germany,   pp.293-302.

Roberts,G.P., MacNeil,T., MacNeil,D. and Brill,W.J. (1978) 'Regulation and characterization of protein products coded by the *nif* (nitrogen fixation) genes of *Klebsiella pneumoniae*' J. Bacteriol. 136, 267-279.

Santero,E., Hoover,T., Keener,J. and Kustu,S. (1989) '*In vitro* activity of the nitrogen fixation regulatory protein NIFA' Proc. Natl. Acad. Sci. USA 86, 7346-7350.

Smith,B.E, Buck M., Eady,R.R., Lowe,D.J., Thorneley,R.N.F., Ashby,G., Deitsung,J., Eldridge,M.,  Fisher,K., Gormal,C., Ioannidis I., Kent,H., Arber,J., Flood,A., Garner,C.D., Hasnain,S. and Miller,R. (1988) 'Recent studies on the structure and function of molybdenum nitrogenase' in Bothe,H., de Bruijn,F.J. and Newton,W.E. (eds) Nitrogen fixation: hundred years after, Gustav Fischer Verlag, Stuttgart, Federal Republic of Germany, pp 91-100.

Takano,T. and Kakefuda,T. (1972) 'Involvement of a bacterial factor in morphegenesis of bacteriophage capsid' Nature New Biol. 239, 34-37.

Tuli,R. and Merrick,M.J. (1988) 'Over-production and characterization of the *nifA* gene product of *Klebsiella pneumoniae*-the transcriptional activator of *nif* gene expression' J. Gen. Microbiol., 134, 425-432.

# The effect of some terrestrial oligochaeta on nitrogenase activity in the soil

M. ŠIMEK, V. PIŽL and J. CHALUPSKÝ
*Institute of Soil Biology, Czechoslovak Academy of Sciences, Ná sadkách 7, CS-370 05 České Budějovice, Czechoslovakia*

*Key words:*   earthworms, enchytraeids, nitrogenase activity, nitrogen fixation, soil casts

## Abstract

Nitrogenase activity, assayed by the acetylene reduction technique, was estimated on control soils and soils which contained several species of earthworms. The worms and casts were separately assayed for nitrogenase activity. The results indicated that nitrogenase activity was associated with gut content or body surface of worms and also their casts. Nitrogen fixation was also enhanced in soils enriched with enchytraeids in the presence of leaf litter.

## Introduction

In the soil, nitrogen fixation often depends on the presence of easily available energy sources. Hence, diazotrophs are clumped in sites possessing high amount of these sources, e.g. the rhizosphere. Similarly, soil animals producing mucus and other exudates rich in carbon compounds can enhance $N_2$-fixation and/or form more suitable microhabitats for diazotrophs' activity. Many investigations have shown that nitrogen fixing microorganisms are common in the gut contents of various soil animals (Barois et al., 1987; Breznak et al., 1973; Raw, 1967; Satchell, 1983), where they probably find favourable conditions for $N_2$-fixation: high moisture and anaerobiosis. There are also some indications, that the numbers of $N_2$-fixers in soil were increased by the activity of soil animals, as well as by the addition of organic wastes worked by earthworms. However, it seems that the numbers of diazotrophs are not always directly correlated with nitrogenase activity. So, while Kaplan and Hartenstein (1977) testing a range of soil invertebrates found no evidence of nitrogen fixation, Citernesi et al. (1977) showed detectable nitrogen fixation in a variety of soil animals,

including the earthworm *Eisenia fetida*. More recently, Barois et al. (1987) found nitrogenase activity in microorganisms isolated from the gut and casts of the tropical earthworm *Pontoscolex corethrurus*, and Striganova et al. (1988) measured several times higher nitrogenase activity in the gut content of *Aporrectodea caliginosa* compared with that in surrounding soil.

The aim of the present study was to estimate the nitrogenase activity associated with terrestrial oligochaetes and their activity in the soil, as influenced by worm species, soil type and/or presence of organic matter.

## Materials and methods

*Measurement of nitrogenase activity*

Nitrogenase activity (NA) was determined by acetylene reduction assay, in details described previously (Šimek and Pižl, 1989). NA was taken to be the difference between the total production of ethylene (TPE = amount of ethylene in the atmosphere of samples incubated in the presence of acetylene) and the endogenous production of ethylene (EPE = amount of ethylene in the atmosphere of samples incubated without acetylene). For determination of NA, samples were incubated at 22°C for 24 hours.

Resulting data were analysed by t-test or by analysis of variance ANOVA, the means being separated by Student-Newman-Keuls test.

### Experiment 1

This experiment was designed with earthworms *Lumbricus rubellus* Hoffmeister, 1843 and soil collected from the woodlot (pH/KCl 4.7, total N 0.102%) to measure: a) the NA directly associated with earthworms, b) the NA of the casts in comparison with that of the surrounding soil (both rewetted to 30%), c) the NA of the soil modified by earthworms (3 worms per 50 g of the soil for 8 days) in comparison with that of undisturbed soil (both with the moisture of 27.9%).

### Experiment 2

In this experiment NA was measured in the soil and in the casts of 8 earthworm species: *Allolobophora chlorotica* (Savigny, 1826); *Aporrectodea caliginosa* (Savigny, 1826); *Aporrectorea rosea* (Savigny, 1826); *Eisenia fetida* (Savigny, 1826); *Lumbricus castaneus* (Savigny, 1826); *Lumbricus rubellus* (Hoffmeister, 1843); *Lumbricus terrestris* (Linnaes, 1758); *Octolasion lacteum* (Orley, 1881). Soil and worms were taken from the same woodlot as in Exp. 1. 5 worms were placed into each of 8 pots with 150 g of the soil with moisture 28.5%, one pot contained only soil. All pots were kept for 14 days in a dark chamber at 15°C. Casts were then removed, and subsamples were assayed.

### Experiment 3

This experiment was designed to measure the NA in the casts of 3 species of earthworms kept for 14 days in 3 different soils: a) from a woodlot (see Exp. 1), b) from a field (pH/KCl 6.8, total N 0.203%), c) from a fallow (pH/KCl 4.3, total N 0.118%). Earthworms *Aporrectodea caliginosa* (Savigny, 1826); *Lumbricus rebellus* (Hoffmeister, 1843) and *Octolasion lacteum* (Orley, 1881) were assayed.

### Experiment 4

In this experiment the influence of terrestrial enchytraeids on the NA in the soil was studied. Enchytraeids and soil were taken from the meadow (pH/KCl 6.6, total N 0.368%). Parallel study showed the enchytraeid community to consist of 14 *Fridericia* species, 3 *Henlea* species, *Enchytraeus buchholzi* species complex. 2 *Enchytronia* species, and 1 *Achaeta* species. Four variants were prepared: a) control soil (435 g per pot, moisture 21.0%), b) soil and litter (the same soil received 3 g dw of a leaf litter-mixture of *Tilia platyphyllos* Scop. and *Sambucus nigra* L.), c) soil and enchytraeids (the same soil received 260 individuals of enchytraeids), d) soil and litter and enchytraeids (the same soil received litter and echytraeids as in b) and c), respectively). Pots were kept for 4.5–7 weeks (see Table 3) in a dark chamber at 18°C. then subsamples were assayed.

## Results

*Nitrogenase activity associated with* L. rubellus, *its casts and with the soil modified by this worm*

a) High rate of TPE was detected in adult worms, while EPE was much lower. The difference between TPE and EPE ($0.508$ pmol $C_2H_4$ $g^{-1}$ dw day$^{-1}$) was statistically significant (t-test, $p = 0.01$). It represented the nitrogenase activity of the diazotrophs on the body surface and/or in the gut of worms.

b) No EPE was evident from the casts or the control soil. Therefore, the TPE may be taken as NA. The difference between the NA of the casts ($44.8$ pmol $C_2H_4$ $g^{-1}$ dw day$^{-1}$) and that of the soil ($36.6$ pmol $C_2H_4$) was not great, but due to the low variability of the results, it was statistically significant (t-test, $p = 0.05$).

c) Earthworms significantly influenced both EPE and TPE in the soil. The significant (t-test, $p = 0.05$) difference between TPE and EPE represents NA. It amounted $485$ pmol $C_2H_4$ $g^{-1}$ dw day$^{-1}$ in the soil modified by worms, but only $93$ pmol $C_2H_4$ in control soil.

*Nitrogenase activity in the casts of 8 species of earthworms and in surrounding soil. compared with that in control soil*

Positive effect of all earthworm species on the NA both in casts and surrounding soil was found (Table 1). Soils inhabited by worms showed in all cases a higher NA (39.6–94.1 pmol $C_2H_4$ $g^{-1}$ dw $day^{-1}$, i.e. 159–378%) than the control soil (24.9 pmol $C_2H_4$, i.e. 100%), in some cases significantly. However, there were no significant differences between soils inhabited by different species. Casts were also higher in NA (53.4–153.2 pmol $C_2H_4$ $g^{-1}$ dw $day^{-1}$, i.e. 214–615%) compared with the control soil. Despite great differences in NA among the casts (and between casts and control soil), no significant difference

was found due to great variability of the results within the casts. NA in the casts was in most cases higher than that in the surrounding soil, differences being though insignificant.

*Nitrogenase activity in the casts of 3 species of earthworms kept in 3 different soils*

Higher (1.2–4.8 times) NA was generally detected in the casts in comparison with the soil (Table 2), differences being mostly significant (ANOVA, SNK-test, $p = 0.5$). However, differences among the casts of different species kept in the same soil were not so obvious. The soils studied were found to differ slightly in NA. Some insignificant differences (not indicated in Table 2) were also among the casts of individual worm species kept in various soils.

*Nitrogenase activity in the soil enriched with leaf litter or enchytraeids. and/or with both, compared with that in control soil*

The results from three similar experiments are shown in Table 3. The enrichment of the soil by enchytraeids did not result in any significant change of NA, while the addition of leaf litter enhanced NA slightly and in September 1988 significantly. On the other hand, when enchytraeids were added together with the litter, NA was very high and differed significantly from all other variants in all the experiments. It reached up to 711% (Exp. 4/1), 334% (Exp. 4/2) and 930% (Exp. 4/3) in comparison with the control soil.

*Table 1.* Nitrogenase activity in the casts of 8 species of earthworms and in surrounding soil inhabited by earthworms. and in control soil without earthworms (pmol $C_2H_4$ $g^{-1}$ dw $day^{-1}$)

| Species | Casts | | Soil | |
|---|---|---|---|---|
| *A. caliginosa* | 53.4 | (16.7)a | 39.6 | (45.8)ab |
| *L. terrestris* | 55.1 | (9.9)a | 64.1 | (44.3)ab |
| *A. rosea* | 78.4 | (99.6)a | 70.8 | (8.4)ab |
| *L. castaneus* | 82.9 | (72.5)a | 94.1 | (9.5)b |
| *A. chlorotica* | 95.3 | (112.7)a | 75.9 | (28.6)ab |
| *L. rubellus* | 109.5 | (45.5)a | 80.2 | (15.7)ab |
| *O. lacteum* | 123.2 | (88.8)a | 88.9 | (16.3)b |
| *E. fetida* | 132.2 | (80.3)a | 90.0 | (12.2)b |
| Control soil | 25.4 | (16.8)a | 25.4 | (16.8)a |

Note: 1. Means of 4 replicates, in parentheses standard errors.
2. Different letters indicate significant difference ($p = 0.05$) according to ANOVA and SNK-test (within the frame of the casts and/or the soil).

*Table 2.* Nitrogenase activity in the casts of three species of earthworms and in surrounding soils (pmol $C_2H_4$ $g^{-1}$ dw $day^{-1}$)

| Sample | Soil | | |
|---|---|---|---|
| | a | b | c |
| Casts of | | | |
|   *A. caliginosa* | 136.8 (54.3)b | 61.6 (45.6)a | 216.5 (122.6)b |
|   *L. rubellus* | 135.2 (43.2)b | 176.0 (46.4)b | 206.3 (103.9)b |
|   *O. lacteum* | 168.9 (99.8)b | 80.5 (20.6)a | 116.7 (25.2)a |
| Soil | 34.94(3.7)a | 49.7 (9.2)a | 57.7 (20.8)a |

Note: 1. Means of 4 replicates. in parentheses standard errors.
2. Different letters indicate significant difference ($p - 0.05$) according to ANOVA and SNK-test (within the frame of particular soil).

*Table 3.* Nitrogenase activity in the soil, and in the soil enriched with leaf litter or enchytraeids, or both (pmol $C_2H_4$ $g^{-1}$ dw day$^{-1}$)

| Sample | Experiment | | |
|---|---|---|---|
| | 4/1-June 1988 4.5 weeks pre-incubation | 4/2-Sept. 1988 7 weeks pre-incubation | 4/3-May 1989 4.5 weeks pre-incubation |
| Soil | 32.2 (3.7)a | 44.7 (7.1)a | 4.1 (2.7)a |
| Soil and litter | 56.4 (5.5)a | 67.2 (5.3)b | 15.9 (8.6)a |
| Soil and enchytr. | 26.4 (3.6)a | 41.9 (15.6)a | 7.1 (9.7)a |
| Soil and litter and enchytr. | 229.0 (220)b | 149.4 (10.6)c | 37.9 (7.6)b |

Note: 1. Means of 5 replicates, in parentheses standard errors.
2. Different letters indicate significant difference ($p = 0.05$) according to ANOVA and SNK-test (within the frame of particular experiment).

## Discussion

Nitrogenase activity associated with gut content of earthworms or that of the microorganisms isolated from guts have been reported by few authors (Barois et al., 1987; Citernesi et al., 1977; Striganova et al., 1988). Our results indicate that NA is associated not only with gut content (or the body surface) of earthworms, but also with their casts and with the soil modified by earthworms. In addition we observed an enhancement of NA of the soil enriched with enchytraeids in the presence of leaf litter.

Nitrogenase activity associated with soil animals can be explained in terms of colonization of the gut, the body surface and the casts by the diazotrophs. Here they can find more favourable conditions than in the soil: both the casts (Svensson et al., 1986) and the guts (Barois et al., 1987) sustain higher level of anaerobiosis and probably also moisture. Moreover complex organic compounds ingested by worms are partly decomposed during their passage through the gut, thus becoming more available for microorganisms (Syers et al., 1979). Another favourable site for bacteria including diazotrophs could be the body surface due to the production of a considerable amount of the cutaneous mucus. On the other hand, mucus contains also various N-substances which could decrease $N_2$-fixing activity of the diazotrophs.

The possible cause of the enhanced NA in the soil modified by earthworms is the higher content of the available carbon substances. Worms also influence the occurence of diazotrophs disseminating them in the soil.

Enchytraeids, much smaller but more numerous soil animals could influence NA by similar mechanisms which involve: decomposition of organic compounds, enhancing the anaerobiosis in the guts and casts, disseminating the diazotrophs. However, there is lack of information about such relationships between enchytraeids and soil microflora.

We found evident nitrogenase activity associated with *L. rubellus* body and casts; soil modified by worms showed also a much higher NA. The positive effect of worms on the NA was confirmed in experiment with other eight species of earthworms (Table 1) and three different soils (Table 2). In the last case there were no evident relations between soil chemical properties and soil NA (and NA of the casts of several worms). An interesting matter for discussion is the fact, that also enchytraeids increased NA, even though only after the addition of leaf litter (Table 3).

However, our data were obtained using acetylene reduction assay (ARA). Regardless of an indirect character of this method, to assess the $N_2$-fixation, there are further problems including the following: 1. Accuracy of ARA. It may overestimate NA owing to endogenous production of ethylene. 2. Long-term incubations. In contrast with pure cultures of diazotrophs or legumes, up to 24-h incubation times are common with soils or similar samples. This may cause damage to diazotrophs or this may increase their numbers (Rennie and Rennie, 1983).

Endogenous production of ethylene (EPE) seems to be a source of major error in assaying soils and similar samples. Ethylene is a normal gaseous component of soil air produced by microorganisms, especially by bacteria and fungi, and released by plant roots although other abiotic sources cannot be excluded. To eliminate EPE, the subtraction of ethylene formed without the addition of acetylene from the amounts formed in the presence of acetylene is usually recommended. However, ethylene metabolism in the soil may depend on the the presence or absence of acetylene (Knowles, 1980). Because bacterial ethylene oxidation may proceed in air, but not in the presence of acetylene, EPE in the presence of acetylene leads to an overestimate of NA (Hendrickson, 1989).

All these features of ethylene metabolism in soil contribute to the difficulty of interpreting the results mentioned in this paper, and make the reliability of the computation of NA in some cases questionable. Therefore, more accurate methodology based on $^{14}C/C_2H_2$ or on the inhibition of NA by carbon monoxide (see Nohrstedt, 1983; Hendrickson, 1989) should be evaluated in similar investigations. Despite of this fact, the large number of differences between samples related to earthworms and enchytraeids and the control samples can not be explained as a mistake. Thus, soil animals can influence the nitrogenase activity of diazotrophs and in some environments may play an interesting role in nitrogen cycling due to nitrogen fixation.

## References


Barois I, Verdier B, Kaiser P, Mariotti A, Rangel P and Lavelle P 1987 Influence of tropical earthworm *Pontoscolex corethrurus* (Glossoscolecidae) on the fixation and mineralization of nitrogen. *In* On earthworms. Selected Symposia and Monographs Eds. A M Bonvicini Pagliani and P. Omodeo pp 151–158. U.Z.I., 2, Mucchi, Modena, Italy.

Breznak J A, Brill W J, Mertins J W and Coppel H C 1973 Nitrogen fixation in termites. Nature 244, 577–580.

Citernesi U, Neglia R, Serriti A, Lepidi A A, Filippi C, Bagnoli G, Nuti M P and Galluzii R 1977 Nitrogen fixation in the gasteroenteric cavity of soil animals. Soil Biol. Biochem. 9, 71–72.

Hendrickson O Q 1989 Implications of natural ethylene cycling processes for forest soil acetylene reduction assay. Can. J. Microbiol. 35, 713–718.

Kaplan D L and Hartenstein R 1977 Absence of nitrogenase and nitrate reductase in soil microinvertebrates. Soil Sci. 124, 328–331.

Knowles R 1980 Nitrogen fixation in natural plant communities and soils. *In* Methods for Evaluating Biological Nitrogen Fixation. Ed. F J Bergersen. pp 557–581. Wiley, New York.

Nohrstedt H O 1983 Natural formation of ethylene in forest soils and methods to correct results given by the acetylene-reduction assay. Soil Biol. Biochem. 15, 281–286.

Raw F 1967 Arthropoda (except Acari and Collembola). *In* Soil Biology. Eds. A Burges and F Raw. pp. 323–362. Academic Press, London.

Rennie R J and Rennie D A 1983 Techniques for quantifying $N_2$ fixation with nonlegumes under field and greenhouse conditions. Can. J. Microbiol. 29, 1022–1035.

Satchell J E 1967 Lumbricidae. *In* Soil Biology. Eds. A Burges and F Raw pp 259–322. Academic Press, London.

Šimek M and Pižl V 1989 The effect of earthworms (Lumbricidae) on nitrogenase activity in soil. Biol. Fertil. Soils 7, 370–373.

Striganova B R, Pantosh-Derimova T D, Mazantseva G P and Tiunov A V 1988 Effect of earthworms on biological nitrogen fixation in soil. Izv. AN SSSR. ser. Biol. 6, 878–884 (*In Russian*).

Svensson B H, Bostrom U and Klemendson L 1986 Potential for higher rates of denitrification in earthworm casts than in the surrounding soil. Biol. Fertil. Soils 2, 147–149.

Syers J K, Sharpley A N and Keeney D R 1979 Cycling of nitrogen by surface casting earthworms in a pasture ecosystem. Soil Biol. Biochem. 11, 181–185.

# CASSETTE MUTAGENESIS OF A PROPOSED DNA-BINDING MOTIF IN KLEBSIELLA PNEUMONIAE RPON (SIGMA 54).


M.J. Merrick and J.R. Coppard
AFRC Nitrogen Fixation Laboratory
University of Sussex
Brighton BN1 9RQ, UK


## 1. Introduction

The majority of _nif_ genes sequenced to date in a variety of diazotrophs are apparently dependent for their transcription upon the _nif_-specific activator, NifA, and a specific form of RNA polymerase containing a sigma factor ($\sigma^{54}$) encoded by _rpoN_.

Promoters which are dependent on RNA polymerase containing $\sigma^{54}$ ($E\sigma^{54}$) for their expression are characterised by a highly conserved sequence (TGGCAC-N5-TTGC) between -26 and -12 with respect to the start of transcription. In promoters which readily form a closed complex with $E\sigma^{54}$, e.g. _glnAp2_, this region is protected by the polymerase and hence one or more residues in this sequence are likely to make specific contacts with $E\sigma^{54}$ and probably with $\sigma^{54}$ itself.

We have identified a potential helix-turn-helix (HTH) motif near the C-terminus of $\sigma^{54}$, and in order to determine whether this motif in RpoN is involved in recognition of the -26/-12 promoter sequence we have carried out an extensive cassette mutagenesis of the second (recognition) helix in this motif and analysed the phenotypes of the resultant mutants.

## 2. Methods

By introducing a novel SpeI site in the K.pneumoniae rpoN gene we flanked the region encoding the proposed DNA-recognition helix by two unique restriction sites (Fig.1)

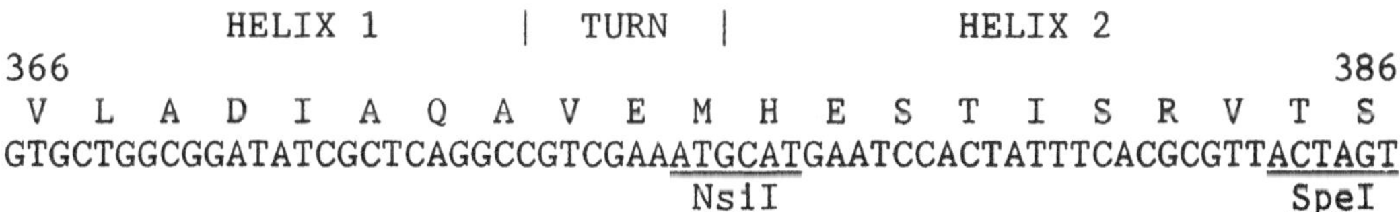



Fig.1 The proposed HTH motif in K.pneumoniae $\sigma^{54}$

We then used cassette mutagenesis to modify each of the six codons encoding Glu378 to Arg383. For each mutagenesis the codon to be changed was replaced by NNG/C so that all potential amino acid substitutions could be generated. A total of 50 different mutations were recovered, representing 7 to 9 different substitutions at each of the six positions. All twenty amino acids were represented amongst the 50 mutants and 30 out of a possible 32 codons were recovered.


## 3. Results and Discussion

The activity of each $\sigma^{54}$ mutant was determined by introducing the mutant gene, on a low copy-number plasmid, into an <u>rpoN</u> deletion strain and measuring $\sigma^{54}$-dependent expression from a <u>pnifL-lacZ</u> translational fusion.

The results obtained are quite consistent with a model in which the region between Val366 and Thr385 in <u>K.pneumoniae</u> $\sigma^{54}$ adopts a helix-turn-helix structure which is involved in recognition of either the conserved GG at -25,-24 or the conserved GC at -13,-12 of the promoter sequence.

In the second helix (Glu378 to Thr385) Glu378, Ser379, Ser382 and Arg383 are predicted to be solvent exposed and by analogy with well-characterised HTH motifs any of these four residues could be involved in base-specific contacts with the DNA.

Glu378 was substituted with nine other residues including proline and lysine and in all cases the mutant $\sigma^{54}$ was active.

Ser379 was only efficiently substituted by Thr or Ala, all other substitutions being inactive. This could indicate that Ser379 makes a base-specific contact through the hydroxyl group (also present on threonine) but in this case the contact would have to be dispensable as alanine is also active. Alternatively the requirement at residue 379 may simply be for a small residue.

Ser382 was efficiently replaced by a number of amino acids and hence cannot make a base-specific contact. However certain substitutions at this position resulted in promoter-specific activity suggesting that residues at this position may modulate the activity of another residue e.g. the adjacent Arg383.

All substitutions of Arg383 were inactive, suggesting that this residue might well make a base-specific contact.

Thr380 and Ile381 should face into the molecule and Ile381 is predicted to make a van der Wall's contact with Ala 371 in the first helix. As predicted Ile381 can only be substituted by another hydrophobic residue such as valine or, somewhat less well, by a small residue such as alanine or glycine. Likewise Thr380 is only efficiently substituted by serine. Hence the data for Thr380 and Ile381 are consistent with both these residues playing a structural role in the recognition helix.

Preliminary <u>in vivo</u> methylation protection experiments have shown that, as predicted, inactive $\sigma^{54}$ mutants modified in helix 2 fail to protect the -24/-12 region of the <u>glnAp2</u> promoter.

# CHARACTERIZATION OF A MUTANT OF <u>Azotobacter</u> <u>chroococcum</u> RESISTANT TO SOME FUNGICIDES


N. MICLAUS, C. VANNINI, E. GREGORI, G. CHINI-ZITTELLI, P. COIANIZ
Experimental Institute for Soil Conservation
P.za D'Azeglio 30
50121 Firenze (Italy)


The use of fungicides for seed and shoot preservation resulted in a strong selection of micro-organisms in the spermosphere. Many free-living nitrogen fixing bacteria have been proved to be affected by these agrochemicals (i.e. Captan, Carboxin at concentration of 50 - 100 $\mu$g/ml and up), thus being discriminated in the early stage of root colonization.
The inoculation of bacteria with the aim to reduce the input of chemical nitrogen fertilizers may be incompatible with the presence of the pesticides applied to prevent phytopathies during seed germination.
The selection of diazotrophic bacteria resistant to fungicides may open new perspectives in this area.
In this context we have isolated a spontaneous mutant strain of <u>Azotobacter</u> <u>chroococcum</u> (Azcap1), resistant up to 300 and 1500 $\mu$g of Captan / ml in liquid and solid medium respectively. This character turned out also against Carboxin and persisted after "curing" and subculturing in absence of selective pressure.
No particular effects of Captan concentration have been found on the growth (protein and ATP production) and ARA of Azcap1 in liquid medium, whereas the parent strain was proved to be strongly affected starting from the concentration of 25 $\mu$g/ml.
When growing in absence of fungicide, Azcap1 produced an amount of glutathione (involved in cell detoxification processes) 3-4 times higher than the wild type.
The mobilization of "cap-markers" has been done by means of the plasmid RP4, that promotes the conjugation and chromosome transfer in many Gram-negative bacteria. The presence of pRP4 alone allowed the growth of the recipient cells up to 100 $\mu$g of Captan / ml; this limit has been trespassed by mating Azcap1(RP4) with other strains of <u>Azotobacter</u> and <u>E. coli</u>.
Consequently, a fragment of Azcap1 chromosome was probably

transferred to the transconjugants. The frequency of "cap-markers" transfer in intra and interspecific crosses was $10^{-5}$ and $10^{-4}$ respectively.

Compared with the parent strain, the mutant electrophoretical pattern of the soluble cytoplasmatic proteins showed a new band of about 35 K and the disappearance of the 23 K band. When growing in presence of Captan, Azcap1 lost three bands at 64.5, 38 and 25.5 K.

Up to now the screening of some 1000 clones of the genomic library has not succeeded in the isolation of the gene(s) responsible for the Captan resistance.

The genomic origin of the resistance is validated by the plasmid analysis and by the easy transfer capacity of "cap-markers".

The resistance mechanism is far to be satisfactorily understood; nevertheless it seems that the level of glutathione production as well as the biosynthesis of some target(?) proteins are involved.

However the possibility to transfer this character to other diazotrophic bacteria, suitable for agro-biotechnologies, appears to be promising.

# A CHROMOSOMAL LINKAGE MAP OF *AZOTOBACTER VINELANDII*

G. Blanco[1], F. Ramos[1], J.R. Medina[2], J.C. Gutierrez[1] and Mª. Tortolero[1].
Department of Microbiology (1) and Department of Genetics. University of
Seville. P.O. Box 1095. 41080 Seville. Spain.

**ABSTRACT.** We have attempted to obtain a genetic map of *A. vinelandii* UW strain, constructing
multimarked strains and using them for mapping experiments. Three mapping techniques have been
employed. 1. Cotransfer of markers mediated by plasmids R68.45 and pJB3JI. 2. Isolation of R-
prime derivatives of plasmid pJB3JI and analysis of markers included in the chromosomal fragment.
3. Polar transfer of chromosomal markers mediated by Tn5-Mob.

**INTRODUCTION.** The main problem in applying classical genetics in
*Azotobacter* is the shortage of mutants and multi-marked strains. The *A.
vinelandii* chromosome differs little from that of others prokaryotes in size, but
the number of chromosomal copies per cell may be 40 (Sadoff *et al.* 1979) or 80
(Nagpel *et al.* 1989). Thus recessive mutations are expressed only after a long
segregation period at the price of losing mutants trough dilution and
recombinational repair. Moreover, some auxotrophic mutations may be lethal
because the required external metabolite cannot be transported ( Kennedy *et al.*
1987, F. Luque *et al.* 1987).

**MATERIALS AND METHODS.** Strains, media and growth conditions are
described in Blanco *et al.* (1990). Tn5 mutagenesis was carried out by
transferring the suicide vector pSUP5011 to *A. vinelandii* strains. Strains bearing
two or more Tn5-insertions were constructed by replacing, in sucessive steps,
the wild type Tn5 transposon by a defective, non-transposing element as
described previously (Blanco *et al.* 1990). Matings were performed on solid
media at 30ºC for 15-24 hours. The alkalyne lisis method was used both for
minipreparations and large scale isolation of plasmids.

**RESULTS AND DISCUSSION.**

**Crosses between *A. vinelandii* strains mediated by R68.45 and pJB3JI.**
Derivatives of the UW strain carrying the plasmids R68.45 or pJB3JI were used
as donors in crosses with multimarked strains. Transfer of individual wild type
markers occurred at frequencies ranging trom $10^{-5}$ to $10^{-7}$. The relative order of
eleven markers could be established from the coinheritance values of a set of pair
of markers (Figure 1). In *A. vinelandii* as in others systems plasmid R.68.45

promotes gene transfer from several chromosomal sites and may be useful in determining the relative order of genes. Nevertherless, this method has two serious inconvenients when applying in a polyploid bacterium. First, the high number of chromosomes per cell is a handicap because coinheritance of unselected recessive alleles cannot be directly observed. Thus, we had accumulate in the recipient strains all the mutations that were to be mapped and to analyse only the transfer of dominant alleles. Second, the plasmid R68.45 promotes the "retrotransfer" of chromosomal markers, i.e. the transfer of chromosomal markers from the initially recipient bacterium to the donor bacterium giving rise to false linkage values (Mergeay *et al* . 1987). In our case the difficulties derived from the "retrotransfer" could be overcame not allowing the conjugation to proceed beyond 24 hours and non selecting the simultaneous transfer of two alleles.

**Construction of R-prime derivatives of plasmid pJB3JI and their further use for linkage mapping.** R-prime derivatives were constructed by conjugal mobilization, mediated by plasmid pJB3JI, of Tn5 induced mutations from *A. vinelandii* to *E. coli*. Several Ap$^r$ Km$^r$ isolated clones were analysed, using agarose gel electrophoresis,to determine the presence of R- primes. The R- primes were retransferred to the appropriate *A. vinelandii* strains to test whether the chromosomal fragments included wild type genes linked to the Tn5 insertions. Results confirmed the relative order of markers *str-met-ura-ntrC-mtl-ace* and allowed to map the markers *srb* and *ntrA* (Figure 1).

**Transfer of chromosomal markers mediated by Tn5-Mob.** In the presence of plasmid RP4, *A. vinelandii* strains with Tn5-Mob inserted into the chromosome behaved like Hfr donors. Frequencies of mobilization for markers near the insertion are very high ($10^{-4}$-$10^{-5}$ per recipient). The high transfer frequencies depended on Tn5-Mob. These Hfr-like strains have been very useful in confirming the relative position of some genes previously established from R68.45 and pJB3JI conjugal experiments and in locating two new markers, *inl* and *suc,* on the chromosome (Figure 1)

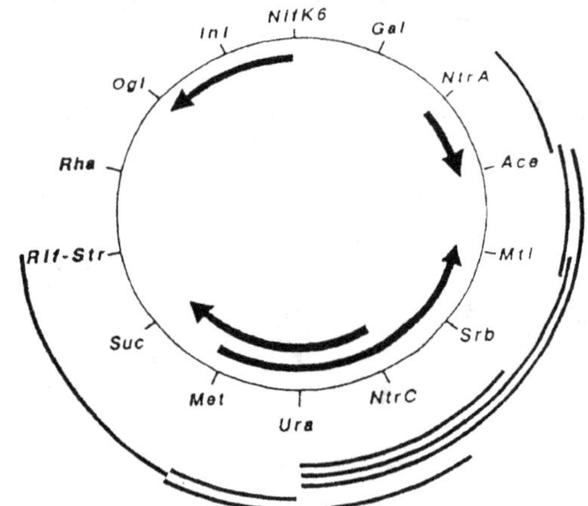


Figure 1. Circular linkage map of *A. vinelandii*. Arrows indicate markers mapped by polar transfer mediated by Tn5-Mob. Lines outside the circle represent chromosomal fragments mapped by R-primes.


**REFERENCES**
- Blanco G, Ramos F, Medina JR and Tortolero M (1990). Mol. Gen. Genet (in press).
- Kennedy C and Toukdarian A (1987). Ann. Rev of Microbiol. 41: 227-258.
- Luque F, Santero E, Medina JR and Tortolero M (1987). Microbiologia SEM 3:45-48.
- Mergeay M, Lejeune P, Sadouk A, Gerits J, Fabry J (1987). Mol. Gen. Genet: 209:61-70.
- Nagpal P, Jafri S, Reddy MA, Das HK (1989) J. Bacteriol. 171: 3133-38.
- Sadoff HL, Shimel B, Ellis S (1979). J. Bacteriol. 138: 871-877.

# ENERGY COUPLING AND ELECTRON TRANSFER IN NITROGENASE


L.A. SYRTSOVA, S.Yu. DRUJININ, A.M. USENSKAYA,
G.I. LIKHTENSTEIN
The Institute of Chemical Physics of the Academy
of Sciences of the USSR
Chernogolovka, Moscow Region, 142432 USSR



ABSTRACT. The consecution of nitrogenase turnover intermediate stages and the mechanism of energy coupling in nitrogenase are suggested.


RESULTS

The consecution of nitrogenase turnover intermediate stages and the mechanism of energy coupling in nitrogenase are suggested (the scheme).

$$E_{1,r}E_{2,r} + ATP \xrightleftharpoons{(1)} (E_{1,r}E_{2,r})^*ATP \xrightleftharpoons{(2)} (E_{1,r}E_{2,r})^{**} : \frac{P_i,H^+}{ADP} \xrightleftharpoons{(3)}$$

$$(E_{1,sr}E_{2,ox})^{**} : \frac{P_i,H^+}{ADP} \xrightleftharpoons{(4)} (E_{1,sr}E_{2,ox})^{**} ADP \xrightleftharpoons{(5)} (E_{1,sr}E_{2,ox}) ADP$$

$$\xrightleftharpoons{(6)} E_{1,sr}E_{2,ox} + ADP$$

(r - reduced; sr - superreduced; ox - oxidized).
Stage (1) follows from the NMR-results obtained [1], that ATP-binding center is formed by nitrogenase complex, yielding the ternary complex with $Mg^{2+}$. These data shows also the different conformation states of nitrogenase being in complex with MgATP or MgADP, and the change (increase) of water access to ATPase center in the course of ATP to ADP hydrolysis. Stage (2) - formation of phospho-intermediate complex follows from data on catalysis of O-exchange reactions $P_i \rightleftharpoons H_2O$ by nitrogenase [2,3]. Stage (3) - really coupling stage - follows from the possibility of artificial uncoupling between ATP hydrolysis and electron transfer reactions by photolysis [4]. Stage (4) follows from the stopped-flow experiments on proton release kinetics in the pro-

cess of nitrogenase reaction. It was established that proton release rate ($k=20-23$ $s^{-1}$) coincides with that of electron transfer and $P_i$-release rate ($k=24+2$ $s^{-1}$) in coupling system (pH 7.3; t=14°), but is lower ($k=3.5$ $s^{-1}$) in uncoupling system [5]. Proton delay indicates the nitrogenase protonation. Rate-limiting stage is probably the biphase reaction of MgADP release (stages 5,6) as in the majority of biological energy coupling syestems. Full reaction is limited by the relaxation of protein conformation, changed in complex with ADP, to previous state [6]. The possibility of $N_2$ and $C_2H_2$ photoreduction without dithionite, acceleration of $C_2H_2$ photoreduction in full system at illumination was useful for understanding of rate-limiting stage on nitrogenase turnover [7].

The energy coupling in nitrogenase can be represented as follows. ATP forms the phospho-intermediate complex and ATPase centre becomes more hydrophobic than in initial state. Intermediate formally loses the macroergic properties due to hydrolysis products being less solvated than in more polar medium. The proton - product of ATP hydrolysis before to be released into the medium protonates the Fe-cluster or its nearest ligand (with $pK < 7$) of MoFe-protein inside. The protonation provides an increase in acceptor capacity of the cluster (adapting of electron level) and provokes the transfer of an extra electron from $E_2$.

REFERENCES
1. Syrtsova, L.A., Nazarova, I.I., Pysarskaya, T.N., Nazarov, V.B. (1972) 'The study of nitrogenase ATPase active center by NMR-method', Dokl.Akad.Nauk USSR 206, 367-369.
2. Skvortsevich, E.G., Syrtsova, L.A. et al.(1979) 'Isotopic exchange $^{18}O-P_i \rightleftharpoons H_2O$ in Azotobacter vinelandii nitrogenase system', ibid. 244, 241-244.
3. Tertyshnaya,N.I., Skvortsevich,E.G., Syrtsova,L.A. et al.(1982) 'Study of localization of the ATPase site of nitrogenase by isotopic exygen exchanhe $^{18}O-P_i \rightleftharpoons H_2O$', Biokhimiya 47, 1741-1746.
4. Khramov,A.V., Syrtsova,L.A. et al.(1988) 'Kinetics of proton release in the process of nitrogenase electron transfer coupled with ATP hydrolysis', Biofyzyka 33, 391-395.
5. Syrtsova,L.A.(1988) 'Rate-limiting step of nitrogenase reaction', Izv.Akad.Nauk SSSR.Ser.Biol.N4, 625-627.
6. Likhtenstein,G.I., Panteleeva,N.S., Skvortsevich,E.G. et al.(1980) 'Upon the role of ATP on nitrogenase function', Molekul.Biol. 14, 147-156.
7. Drujinin,S.Yu., Syrtsova,L.A., Likhtenstein,G.I. et al. (1989) 'Eosin-photosensitized catalytic reduction of acetylene by nitrogenase', Biokhimiya 54, 1638-1645.

# FLUCTUATION OF THE ARA UNDER <u>ELAEAGNUS ANGUSTIFOLIA</u> CANOPY


F. BERMUDEZ DE CASTRO[1] , F. LLINARES[2] , J.M. POZUELO[2] AND
F.J. GUTIERREZ MAÑERO[2]
[1]*Univ.Complutense, Fac.Biología, Dpt.Ecología, 28040 Madrid*
[2]*Col.Univ. San Pablo (CEU), Lab. Ciencias.28660 Boadilla del Monte, Madrid (Spain)*


## Introduction

The free fixing microorganisms acting as intermediaries between the atmospherical and edaphic nitrogen carry out regulating action on the total nitrogen available at soil level. On the other hand, plants can influence the physiological activity of these microorganisms through the formation of microclimates by canopy effect and by means of organic material input. Thus, in the rhizosphere there are from 5-10 times more microorganisms than in the soil because the organic compounds liberated by the roots are used directly by the microbial population as substratum for their growth. Dinitrogen fixation in the rhizosphere of some plants also increases because this area is a most favourable place for diazotrophic microflora development.

In previous works it has been studied the influence of <u>Alnus glutinosa</u> and <u>Myrica gale</u> on edaphic microflora (Bermúdez de Castro and Schmitz, 1990). Here it is shown density and ARA variations of the free dinitrogen fixing microorganisms under <u>E. angustifolia</u> canopy and their relation to N, C and pH of the soil, in autumn, in a Russian-olive grove near Madrid (coordinates UTM, 30TUK448492), on an alkaline gley gypsaceous solonchack.

## Materials and Methods

Soil samples were taken under the canopy of <u>E. angustifolia</u> at 0-10, 10-30 and 30-50 cm and at 5 and 10 m far from the trees. The ARA was analysed in soil sieved through a 0.2 mm net. Replicas of 250g soil were placed in a 1 l bottle with a 10% acetylene atmosphere. They were incubated for 24 hours at ambient temperature. They ethylene was measured in gas chromatograph KONIC 3000 HRGC with column of Porapak R. Microbial density was evaluated by the method of Arima and Yoshida (1982). Physicochemical variables were evaluated by usual methos in soil studies.

## Results

The obtained results oscillate between the following values:

pH, 7.5–7.8; total–N, 0.4–1.5%; $NH_4^+$, 109.2–339.0 µg/g; $NO_3^-$, 80.0–228.6 µg/g; organic–C, 2.9–9.7%; log MPN microorganisms, 3.90–4.38; ARA, 0.5–14.2 pM $C_2H_2$/g.h.

Soils are slightly basic and pH increases in terms of the distance to the tree. As for depth, the maximum pH corresponds with the medium distance and the minimum with the surface. N and C also decrease from the foot of the trees but total–N maximum is at 5 m. All of them reduce from the surface. The smallest microbial density is obtained at a distance of 10 m and the biggest one under canopy. In all the cases there are more microorganisms at a depth of 10–30 cm and less in the surface sample. Although the ARA acts in the same way with regard to the distance to the trees, it has the maximum value in the surface and it decreases according to depth.

The anova, for $p \leq 0.01$, shows significative differences in the distances to the trees for pH, $NO_3^-$ and organic–C and in depth for total–N, $NH_4^+$, organic–C and MPN, and, at the level $p \leq 0.05$, for $NH_4^+$ and ARA in the distances. The ARA is correlated negatively with $NH_4^+$, $NO_3^-$, total–N and organic–C and positively with pH. All correlations are significant ($p \leq 0.01$ or $p \leq 0.05$).

## Discussion

The lowest pH near _E. angustifolia_ confirms that actinorhizal plants cause soil pH fall. Similarly, nitrogen and carbon increase under the canopy is related to organic material contributions, mainly through litter fall, wich, in this place, provides 34.5 kg N/ha.year (Bermúdez de Castro et al., 1990). Therefore, under the canopy the microbial density of the nitrogen fixing microorganisms and of other functional groups increases (unpublished dates). Diazotrophic microorganisms density is bigger at a depth of 10–30 cm, which agrees with the data about the distribution of _Rhizobium leguminosarum_ bv. _trifolii_, attributable to the filtration of bacterial cells towards less porous horizons. Like in other cases microorganisms activity does not go together with MPN. The negative correlation of the ARA with total–N and organic–C suggests the predominance of heterotrophic microbial populations.

## References

Arima, Y. and Yoshida, T. (1982) 'Nitrogen fixation and denitrification in the roots of flooded crops' Soil Sci. Plant Nutr. 24, 483–489.

Bermúdez de Castro, F. and Schmitz, M.F. (1990) '_Alnus glutinosa_ y _Myrica gale_ ¿Plantas pioneras en suelos ribereños?' in M. Mejías (ed.) Aspectos Generales de la Fijación de Nitrógeno Atmosférico, Univ. Sevilla (In press).

Bermúdez de Castro, F., Aranda, Y. and Schmitz, M.F. (1990) 'Acetylene-reducing activity and nitrogen inputs in a bluff of _Elaeagnus angustifolia_' Orsis (In press).

# THE INFLUENCE OF ROOT EXUDATES ON DINITROGEN FIXATION OF CHOSEN FREE-LIVING BACTERIA


A.SAWICKA
Department of Soil Microbiology
Academy of Agriculture
ul.Wołyńska 35
60-637 Poznań
Poland


Different species of plants growing under the same climatic and soil conditions can have strongly differentiated rhizosphere nitrogen fixation rates. The cause of these differentiations in nitrogen fixation activity among plant species is probably the spcific composition of root exudates. Rovira (1956) showed that there are considerable quantitative and qualitative differences in the components excreted by the roots of plants growing under identical conditions.

The purpose of our experiment was to study the extent to which root exudates can influence non-symbiotic nitrogen fixation in two different soils inoculated with nitrogen fixing bacteria.

Two kinds of soils were used in our experiments: peat-earth soil from a meadow and sandy-loam soil from a field. Synthetic root exudates that have been proposed by Kunc and Macura (1966) into each soil (sterilized and non-sterilized) were added in complete composition where $C:N=11.4$ and root exudates contained only carbon compounds. Soils were inoculated with effective nitrogen fixing bacteria: Azotobacter chroococcum, Azospirillum brasilense, Clostridium sp. and mixture of bacteria. Soils were incubated in glass jars. Soils without the root exudates and non-inoculated with bacteria were used as a control.

Nitrogenase activity was measured directly in jars with soil by the acetylene reduction technique after 24, 48, 72 hours and after 1, 2, 3, 4, 6, 8 weeks. Nitrogen fixation was estimated using theoretical conversion factor 3.

Nitrogenase activity was not found in the control soils at all the dates of measurements.

In the soils with added root exudates , nitrogenase activity was noted only after 24, 48, and 72 hours.

The non-nitrogen root exudates compounds added to the non-sterilized soils increased nitrogenase activity free-living nitrogen fixing bacteria especially in the peat-earth. However, **non-sterilized** sandy-loam soil enriched in synthetic root exudates containing carbon and nitrogen compounds was more effective in nitrogenase activity as compared to the peat-earth (Table 1) .

TABLE 1. The highest nitrogen fixation in soils with root exudates (kg N·ha$^{-1}$·season$^{-1}$)

| Experimental combinatios | Soil with added | | | |
| --- | --- | --- | --- | --- |
| | $C^x$ | | $C^x$ and $N^{xx}$ | |
| | a | b | a | b |
| **Peat-earth soil:** | | | | |
| non-inoculated with bacteria | 0 | 0.59 | 0 | 0 |
| inoculated with Azotobacter | 0.09 | 33.70 | 0 | 0.32 |
| inoculated with Clostridium | 0 | 2.22 | 0 | traces |
| inoculated with Azospirillum | 0.12 | 8.02 | 0 | traces |
| inoculated with bacteria mix. | 0.16 | 1.96 | 0 | traces |
| **Sandy-loam soil:** | | | | |
| non-inoculated with bacteria | 0 | 0 | 0 | 0.09 |
| inoculated with Azotobacter | 0 | traces | 0 | 1.91 |
| inoculated with Clostridium | 0 | 0 | 0 | 3.12 |
| inoculated with Azospirillum | 0 | traces | 0 | 0.56 |
| inoculated with bacteria mix. | 0 | traces | 0.29 | 0.29 |

Explanations: $C^x$-root exudates containing only carbon compounds, $C^x$ and $N^{xx}$-root exudates containing carbon and nitrogen compounds; a-sterilized soil, b-non-sterilized soil; bacteria mix. = Azotobacter+Clostridium+Azospirillum

The sterilized soils almost foiled to fix dinitrogen irrespective of the kind of added root exudates and bacterial inoculation. The root exudates added to the soil influenced most significantly Azotobacter chroococcum nitroge fixation activity (up to 33kg N·ha$^{-1}$·season$^{-1}$) next, Azospiririllum brasilense (about 8kg N) . But these numbers of nitrogen were fixed only in the presence of other non-fixing nitrogen soil microflora. Some researches point out that free-living nitrogen fixing bacteria can fix this element more effectively when they are living in associations with other soil microorganisms.

References
Kunc F.,Macura J. (1966). Decomposition of root exudates in soil. Folia Microbiol. 11, 239-247
Rovira A.D. (1956). Plant root excretions in relation to the rhizosphere effect. Plant and Soil 7, 178-194

# SOIL REDOX POTENTIAL IN PRESENCE OF AZOSPIRILLUM


A.PIDELLO  and L.MENENDEZ
Laboratorio de Química Biológica I.Facultad Cs.Vet.-UNR
Bv.O.Lagos y Ruta 33
2170 Casilda (Santa Fe)
Argentine



ABSTRACT.  The  following  results confirm the Azospirillum sp7 capacity
of  increasing the redox potential (Eh) of soil in anaerobic conditions.
The  effect  appeared  related with a lower quantity of reducing equivalents in the inoculated soil with respect to control soil. The $N_2O$ poten
tial  production  (DEC,  Denitrifier  Enzyme  Concentration) was 6 times
higher  in  control  soil and the $CO_2$ 3 times higher in inoculated soil.


## INTRODUCTION

Eh  affects  the  degree of oxidation of many substrates and metabolites
related with the biological activity /1/ /2/ and may condition the energetic  yield  of  biochemical  reactions  /3/. This report describes the
results  of  experiments  carried out to determine whether the effect of
Azospirillum  sp7  on Eh /4/ appeared under more severe anaerobic conditions and if the presence of bacteria affected any soil biological activities related with Eh.

## MATERIALS AND METHODS

The  experiments  were  done  in a hermetic glass columm through which a
$N_2$  current  passed.  Twelve  Pt electrodes wers introduced radially in
the  soil  (Vertic  molisol)  which helped to determine the Eh evolution
(fig.1-A)$CO_2$production was determined by titration of the alkaline solution in a $CO_2$trap. In the inoculated columm $10^8$cells/g soil of Azospirillum sp7 were added.After 140 hs.of anaerobic conditions soil oxidable-C
/6/was determined.Potential $N_2O$ and $CO_2$production /5/ were also determined after  incubating  the  soil with glucose and nitrate (1 µg/g soil).
The quantity of gas was determined with a gas chromatograph.

## RESULTS AND DISCUSSION

After  an abrupt inicial fall, bacterial inoculum seems to stabilize the
Eh  values, situation not present in control soil where, during the studied period, a slope of 70 mV/day was observed. The evolution of $CO_2$

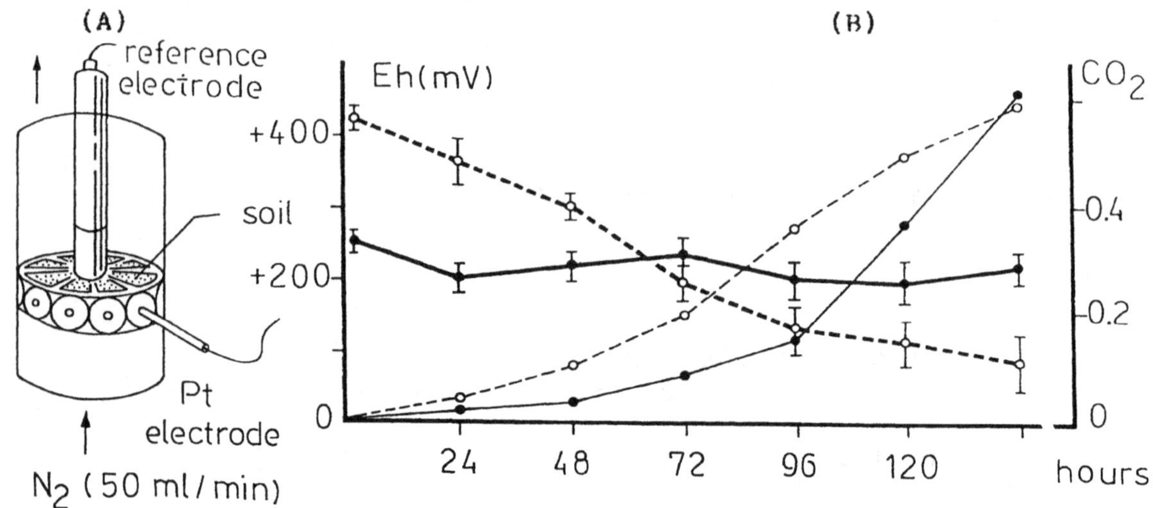



FIGURE 1: (A) scheme of utilized columm; (B) Eh evolution (thick lines) and $CO_2$ accumulation ( µg/g soil)(thin lines) in inoculated soil ( •——• ) and control soil ( o---o ).

shows that, though there were different kinetics, gas production was similar in both soils (Fig.1-B). Table 1 shows that in coincidence with Eh behaviour, in control soil oxidable-C and $N_2O$ production were significantly higher than in inoculated soil where a higher $CO_2$ was observed.

TABLE 1: Parameters set at the end of the experiments

| Treatment | oxidable-C (%) | $N_2O$ ( µmol/g soil) | $CO_2$ ( µmol/g soil) |
|---|---|---|---|
| Inoculated | 1.22** | 5** | 36** |
| Control | 1.52 | 35 | 9 |

** significant difference from control soil at $P < 0,01$ level

References

1-Bohn, H.L. (1971) Soil Sci, 112:39-45
2-DeLaune, R.D;Reddy,C.N.;Patrick,W.H.(1981) J.Environ.Qual.,10:276-279
3-Fenchel,T;Blackburn,T.H.(1979)Bacteria and mineral cycling.Academic Press
4-Pidello,A.;Lensi,R.;Chapo,G.;Lescure,C.;Gamard,P.(1988)in:Coll.S.F.M. "Dynamique des populations microbiennes dans le sol".Lyon,Sept.1988. 64-68
5-Tiedje,J.M.;Sextone,A.J.;Myrold,D.D.;Robinson,J.A.(1982)Antonie Van Loeuwenhoek, 48:569-583.
6-Walkley,A.(1947)Soil Sci.63:251:264

Part of this work was supported by grants from CEE, SECYT and CIUNR.

# BIOFERTILIZATION OF MAIZE WITH FREE-LIVING NITROGEN FIXING BACTERIA: RESULTS OF FIELD TRIALS ON GRAIN PRODUCTION


N. MICLAUS, E. GREGORI, C. PIOVANELLI, G. CHINI-
ZITTELLI, C. VANNINI, P. COIANIZ, E. GALLORI
Experimental Institute for Soil Conservation
P.za D'Azeglio 30
50121 Firenze (Italy)


A useful method to reduce the input of chemical fertilizers in agriculture and to control the soil and water pollution may be represented by the use of asymbiotic nitrogen fixing micro-organisms as biological fertilizer. Extensive experimentation so far carried out under different environmental conditions has generally put in evidence some positive effects of inoculation on the production of various cereals and other crops.
The nature of these effects is not yet completely defined and might involve both an hormonal enhancement of the root growth and an overall increase of nitrogen balance.
The effects of the inoculation with <u>Azospirillum brasilense</u> (Cd) and <u>Azotobacter chroococcum</u> (Azcap1) on maize production were evaluated in plot trials on Levisone sandy loam (typic Udifluvent, fine loamy, mixed calcareous, mesic), under temperate climate with dry summer.
The <u>Azotobacter</u> strain was resistant to some fungicides (Captan and Carboxin), which are currently used in seed dressing.
A randomised not completed block design, by which the effects of three nitrogen fertilizer levels were confused with the soil variability (supposed to be poor), was adopted. Each block was divided into six plots of 60 m$^2$ with two treatments (Control and Inoculated).
Pure liquid cultures of the selected strain were diluted in water plus 1.5 g/l of K phosphates to a final concentration of about 4-5 x 10$^6$ cells / ml. The inoculum was sprayed along the rows. Controls were treated with the same amount of diluted medium without bacteria.
Grain yield data collected during five years of measurement are summarized in table 1 for <u>Azospirillum</u> and <u>Azotobacter</u>. The production is expressed as q/ha of grain at 15.5 % water content. The % increase of yield due to the inoculation is indicated in brackets.

## TABLE 1. Grain yield of maize and Anova table.

| Treatment | N(kg/ha) | Azospirillum | | | | Azotobacter | | |
|---|---|---|---|---|---|---|---|---|
| | | 1985 | 1986 | 1987 | Mean | 1988 | 1989 | Mean |
| Control | 150 | 96 | 74 | 110 | 93 | 92 | 89 | 90 |
| | 225 | 104 | 87 | 123 | 104 | 109 | 109 | 109 |
| | 300 | 106 | 96 | 138 | 113 | 130 | 110 | 120 |
| Inoculated | 150 | 100(+4) | 81(+9) | 119(+8) | 100(+7) | 100(+9) | 90(+1) | 95(+6) |
| | 225 | 112(+8) | 89(+2) | 129(+5) | 110(+6) | 116(+6) | 111(+2) | 114(+5) |
| | 300 | 114(+7) | 98(+2) | 140(+1) | 117(+4) | 129(-1) | 113(+3) | 121(+1) |

| Source of var. | D.F. | Azospirillum | | | Azotobacter | | |
|---|---|---|---|---|---|---|---|
| | | $\Sigma$ squares | $\bar{X}$ square | F | $\Sigma$ squares | $\bar{X}$ square | F |
| Among blocks | 2 | 2838.7 | 1419.4 | 31.4[**] | 4799.4 | 2399.7 | 40.5[**] |
| Within blocks | 33 | 15366.8 | | | 2606.2 | | |
| 1986 vs 1987 | 1 | 13767.1 | 13767.1 | 304.2[**] | | | |
| 1988 vs 1989 | 1 | | | | 711.1 | 711.1 | 12.0[**] |
| Inoc. vs Cont. | 1 | 235.1 | 235.1 | 5.2[*] | 100.0 | 100.0 | 1.7 |
| Years x Treat. | 1 | 7.1 | 7.1 | <1 | 16.0 | 16.0 | <1 |
| Error | 30 | 1357.5 | 45.2 | | 1779.1 | 59.3 | |
| Total | 35 | 18205.6 | | | 7405.6 | | |

Therefore the inoculation resulted in a general increase of grain production: +6 and +4 % respectively for <u>Azospirillum</u> and <u>Azotobacter</u>, particularly at the lowest rate of N fertilization.

The inoculation effect was not statistically significant as a whole; nevertheless the comparison Inoculated vs Control for the years 1986-87 (<u>Azospirillum</u>) resulted significant for P<0.05. Due to the fact that the nitrogen fixation activity (measured between plants as "in situ" ARA) was unaffected by the inoculum, the yield increase is probably related to the production of  root growth factors by the bacteria.

After all, the biofertilization appears as a suitable practice for more sustainable agricultural systems, also in temperate environment. However, better results could be expected by using bacterial strains with some host specificity and/or derepressed for combined nitrogen.

# KINETICS OF KELTHANE UPTAKE AND DISTRIBUTION IN AZOSPIRILLUM LIPOFERUM


D.M.S.MANO*, K.BUFF** & T.LANGENBACH*
*I.Microbiologia, UFRJ, CCS,I.Fundäo,RJ,Brazil
**GSF,Ingolstäder Landstr.1,D-8042,München,FRG



ABSTRACT. The organochlorine Kelthane EC (a comercial formulation of Dicofol) inhibits *A.lipoferum* growth and nitrogenase activity. We investigated the distribution and uptake kinetics of Dicofol on *Azospirillum* cells. The photo-induced binding method was used in order to prevent loss of Dicofol weakly bound to *Azospirillum* during the washing and fractionation procedures. The results showed that Dicofol bound to *Azospirillum* cells immediately after addition. Binding was higher when the cells were exposed to Dicofol in "Premix" (the commercial formulation solvent) instead of ethanol. Cell fractionation showed that Dicofol remained mainly in the envelope rather than in the cytosol, independent of the solvent vehicle.


## Introduction

The insecticide-acaricide Kelthane showed inhibition effects on *Azospirillum* growth, nitrogenase activity and induced appearance of giant cysts with 50ppm and higher doses (1). The objective of this work was to study the uptake kinetics of Dicofol binding to *A.lipoferum* ATCC 29709 cells and its distribution in cell fractions. For this purpose we used the photo-induced binding method which is able to strengthen reversible pesticide-cell interactions by generating covalent bonds.

## Materials and Methods

Kelthane EC (a commercial Dicofol formulation), 14C-Dicofol [1,1-bis(p-chlorophenyl)2,2,2-trichloroethanol] and "Premix"(the commercial formulation solvent) were obtained from Rohm & Hass. For kinetics studies, log phase cells were incubated in a shaker, at 32 C, in Nfb medium enriched of ammonium nitrate(2), with 14C-Dicofol. Aliquots of differents periods of incubation were irradiated with a Phillips HPK 125-W mercury lamp for 15 sec (3). Cells were washed with aceton to estimate protein content(4) and radioactivity by scintillation counting. To verify Dicofol binding to cell fractions, cells were disrupted by sonification after the photo-induced binding and washing procedures. Envelope and cytosol were separated by ultracentrifugation (105000 g/1h), treated with TCA and washed with aceton to estimate protein (4) and radioactivity content by liquid scintillation counting.

## Results and Discussion

Photo-induced binding showed to be an efficient method to prevent Dicofol loss during drastic cell treatments (fig.1). Dicofol bound to *Azospirillum* cells immediately after addition (fig.1 and 2). When Dicofol was dissolved in ethanol the number of molecules bound remained constant but it increased in time, when cells were incubated with Dicofol dissolved in "Premix"(fig 2). The distribution data showed that 80% of Dicofol was found in the cell envelope and only 20% in the cytosol, independent of the solvent vehicle. Probably the higher pesticide uptake observed from "Premix" was due to a inclusion into the cell membrane.

Binding of Dicofol to Azospirillum lipoferum

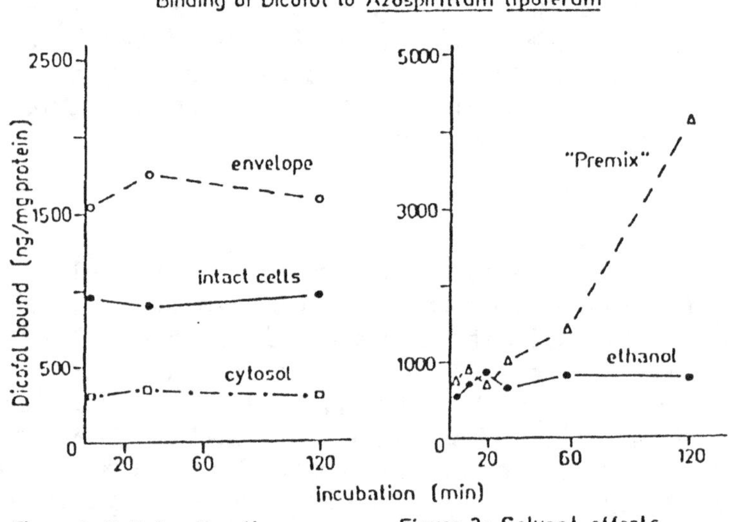


Figure 1: Cell fractionation

Figure 2: Solvent effects

The authors thank financial support of CNPq, DAAD and EC.

## References

1-Mano,D.M.S.,Matos,A.C.M.& Langenbach,T.1988. Azospirillum IV;Gen.,Phys.,Ecol.,159-165.Ed.Klingmüller,Springer Verlag.
2-Döbereiner,J., Mariel,I,E. & Nery,M. 1976. Can. J. Microbiol., **22**: 1464-1473.
3-Bründl, A. & Buff, K. 1988. Bioch.Pharm.,**37**(8):1601-1608.
4-Lowry, O.H., Rosenbrough, N.T., Farr, A.L. & Randall, R.J.1951. Biol. Chem. **193**:265-275.

# BIOACCUMULATION OF THE ACARICIDE KELTHANE (DICOFOL) BY *AZOSPIRILLUM LIPOFERUM*


T. Langenbach*, E. Clausen** & D.M.S. Mano*
* I. Microbiologia, UFRJ, CCS, I. Fundão R.J. Brazil
** GSF, Ingolstädter Landstr.1, D-8042, München, FRG.



ABSTRACT. *A. lipoferum* tolerates Dicofol up to 50ppm. Growth and nitrogenase activity are inhibited by higher doses. Our aim was to verify whether Dicofol degradation occurs as a detoxification route in *Azospirillum*. Analysis with GC-EC and TLC showed that Dicofol residues were accumulated in *Azospirillum* with low amounts remaining in the medium. Since degradation occured more intensively at oxidation conditions in the absence of cells, we suggest that uptake by *Azospirillum* protected Dicofol against degradation.


## Introduction

In *A. lipoferum* growth and nitrogenase activity tolerate up to 50ppm of the organochlorine Kelthane EC, the technical formulation of Dicofol, above which both are inhibited (1). Our objective was to verify if A. lipoferum has a detoxification capacity as observed in others microorganism (2) through Dicofol degradation. To avoid the degradation conditions encountered in gas chromatography (GC-EC), Dicofol residues were analyzed by thin layer chromatography (TLC) despite the lower resolution.

## Materials and Methods

A.*lipoferum* were inoculated in Nfb medium supplemented with ammonium nitrate plus 50ppm of Dicofol and incubated in a shaker at 32 °C for 5 days. Control assays were performed in Nfb without cells but with 50ppm Dicofol, incubated under the above growth conditions (presence of oxygen) and in tighly closed screw cap tubes wich were filled completely with medium without shaking (absence of oxyxen). Dicofol was extracted from lyophilized samples with hexan/ethyl ether 96:4 (3) and the extracts were analysed by EC-GC and by TLC. In the latter case, 14C-Dicofol was added before inoculation. Fractions of the silica plates were scraped into vials with scintillation cocktail to measure radioactivity.

## Results and Discussion

Recovery of Dicofol was lower in cell free medium in the presence of oxygen than in the absence of oxygen (table I, sample 3, 4). Oxydation conditions allow hydrolysis of Dicofol (3). From the 50ppm Dicofol, 35% was accumulated in the cells and 1% remained in the supernatant (sample 1, 2, column A). TLC analysis confirmed that Dicofol degradation was higher in the medium than in the cells. The 97% recovery of Dicofol from the supernatant suggest that it remained inbedded in non precipitated cellular debris, because in medium alone higher degradation would occur (sample 3, column A, B). Dicofol binding to the cell was increased during the first 2 days and decreased after 5 days (5). The results suggest that cell accumulation protect this molecule against hydrolytic degradation. This excludes biodegradation as an effective detoxification mechanism in *A.lipoferum*.

Table I - Percentage of Dicofol recovery. Column A: amount of Dicofol measured by  GC-EC, % related to 50ppm. Column B: % of Dicofol related to the total radioactivity on TLC.

| | | Column A<br>% Dicofol<br>in samples | Column B<br>%Dicofol on<br>TLC plate |
|---|---|---|---|
| inoculated<br>medium | cell(1)<br>supernatant(2) | 35<br>1 | 72<br>97 |
| non inoculated<br>medium | (+) oxygen(3)<br>(-) oxygen(4) | 7<br>22 | 29<br>89 |

Acknowlegments: The authors are greatful to Dr. I.Gebefugi and for financial support of CNPq, DAAD, EC and FAPERJ.

## References:

1- Mano, D.M.S., Matos,A.C.M. & Langenbach,T. 1988. Azospirillum IV; Genetics, Physiology, Ecology, 159-165. Ed. Klingmüller, Springer Verlag.
2-Golovleva,L.A.,Pertsova,R.N.,Boronin,A.M., Traukin,U.M. & Kozlovsky,S.A.1988. Appl.Environm. Microbiol.,54:1587-1590.
3-Ives, N.F. 1973. J. Ass. Off.Anal.Chem., 56(6):1335-1338.
4-Döbereiner,J., Mariel,I.E. & Nery, M. 1976. Can.J. Microbiol., 22:1464-1473.
5-Mano, D.M.S., Buff, K. & Langenbach,T.1990. Proceedings of the "Fifth Int. Symp. Nitrogen.Fix. in Non-legumes".

EFFECT OF BIOFERTILIZERS ON CONTROLLING <u>FUSARIUM SOLANI</u> f. Sp. <u>PHASEOLI</u>


Ishac* Y.Z., Ahmed** M.A., El-Deeb** S.H.
*Unit of Biofertilizers & Agric. Botany and Plant Path. Dept.
Faculty of Agric. Ain-Shams University, Egypt.



ABSTRACT.

A pot experiment was conducted to study the effect of seed and soil inoculation with a mixture of VAM-spores and/or mixture of <u>Azotobacter chrococcum</u> and <u>Azospirillum lipoferum</u> on controlling <u>Fusarium Solani</u> f. sp. <u>phaseoli</u>. <u>Golmus</u> was the most VAM-genus while <u>Sclerocysts</u> and <u>Gigaspora</u> were obtained in low frequency.
Number of VAM-spores increased with time and reached maxima at fruiting stage. Disease severity of root-rot was decreased when seed and soil were treated with tested biofertilizers. Root infection with VAM-fungi increased by ageing. Rotted roots of soybean plants (unnodulated) showed less infection with native VAM-fungi.

Resistance of soybean roots treated with biofertilizers against <u>F. solani</u> f. sp. <u>phaseoli</u> was histologically studied. Many lysed dark cells appeared in the epidermal and outer cortical cells due to root-rot disease. This produced an irregular dark epidermal layer. Less changes were showed when seeds and soil were treated with biofertilizers. Thickening of the cell walls through lignification and higher number of xylem vessels decreased the penetration and growth of the pathogenic fungus.

PHYLLOSPHERIC NITROGEN FIXING BACTERIA IN TILLANDSIA LEAVES


P.MONTAINI, L. BRIGHIGNA*, S. FEDI and F. FAVILLI
Dipart.mo Scienze e Tecnologie Alimentari e Microbiologiche.
*Dipart.mo di Biologia Vegetale.
Università degli Studi di Firenze, Italia.



ABSTRACT. Nitrogen fixation of leaf surface of 12 species of Tillandsia was determined by ARA assay. The data showed that positive nitrogenase activity occurred on leaves put on enrichment cultures only. Bacillus and Pseudomonas were observed to be the predominant bacteria on the leaves, but other genera as Erwinia sp., Agrobacterium sp., Vibrio sp., Xanthomonas sp. and Rhanella sp. were found. Among all the isolates only Bacillus megaterium fixed nitrogen in pure culture.


## 1. INTRODUCTION

The surface of the leaves constitues a distinct ecological environment, known as the phyllosphere, which provides niches for a variety of microorganisms influencing the nutrition of the plants. In Tillandsia plants, which have developped one of the most extreme level of epiphytic life (Benzing, 1980; Brighigna et al., 1989, 1990), the influence of the phyllospheric microflora on the plant nitrogen uptake represents a considerable field of research. The aim of this work was to investigate on nitrogen fixation in Tillandsia leaves and to isolate and characterize the leaf surface nitrogen fixing mioroorganiomo in ordor to confirm and cxtcnd our prcvioua invcotigationo (Favilli ct al., 1075).

## 2. MATERIALS AND METHODS

Plants of 12 species of Tillandsia were collected in different habitat of Mexico.
The leaf surface morphology of the Tillandsia was evalued by S.E.M. (Philips 515).
Small pieces of leaves were prepared according to the usual method of

fixation, dehydration, critical point drying for S.E.M. observations. Loopfuls from a <u>Bacillus megaterium</u> culture were prepared for T.E.M. (Philips 300) observations.

Nitrogen fixing activity was estimated, by ARA method (Burris, 1974), on small pieces of leaves directly or in enrichment cultures.

Nitrogen fixing bacteria were isolated from enrichment cultures by successive planting in N free solid Rennie medium (Rennie, 1981). The isolates were identified according the criteria recommended in Bergey's Manual of Systematic Bacteriology (1984) and using the API tests 20 E, 20 NE and 50 CHB.

## 3. RESULTS

8 out of 12 Tillandsia species showed phyllospheric nitrogen fixation in leaf samples put in enrichment cultures only. The nitrogenase activity, ranging from $65,61 \pm 7.2$ to $8311,21 \pm 850.5$ nmoles $C_2H_4$ $h^{-1}$ $g^{-1}$ (d.w.), occurred in all the Tillandsia plants collected in semiarid environments and in the half of the plant samples collected in humid and arid environments. Among all the microbial species occurring on the leaves, the genus Bacillus was present in more of the 75% of the plant collected. The species <u>Bacillus megaterium</u>, the sole species fixing nitrogen in pure culture, was isolated from 5 of the 8 Tillandsia showing nitrogenase activity. The other bacterial isolates do not show ARA in pure culture.

## 4. CONCLUSION

The results obtained in these investigations suggest that nitrogen fixation in Tillandsia phyllosphere occurs only under suitable climatic conditions and most of the nitrogen fixing bacteria living on the leaves are not well adapted to fix nitrogen in the leaf environment. The investigations also suggest that a connection exist between the composition of the microbial population grown on the leaf and the different morphological development of the absorbing trichomes in Tillandsia.

## 5. REFERENCES


Benzing, D. (1980) The biology of Bromeliads. Mad River Press, Eureka, California, U.S.A.

Brighigna, L. et al. (1989) Caryologia 41, 111-129.

Brighigna, L. et al. (1990) Caryologia 43, 27-42.

Favilli, F. et al. (1975). Atti XVI Congr.Naz.Soc.I.Microbiol. 2, 793-799, Padova.

Burris, R.H. (1974) 'Methodology', in A. Quispel (eds.), The biology of nitrogen fixation, North Holland Publ.Co., Amsterdam, pp.9-33.

Rennie R. (1981) Can.J. Microbiol. 27, 8-14.

Bergey's Manual of Systematic Bacteriology (1984), Williams and Wilkins, Baltimora, U.S.A., vol.1.

# *NIF* **AND** *NOD* **GENES IN** *AZOSPIRILLUM*

C. ELMERICH, M. de ZAMAROCZY, C. VIEILLE, F.
DELORME, I. ONYEOCHA, Y.Y. LIANG and W. ZIMMER
*Département des Biotechnologies*
*CNRS URA 1300*
*Institut Pasteur*
*25 rue du Dr. Roux*
*75724 Paris cedex 15*
*France.*

ABSTRACT. *Azospirillum* can fix nitrogen under free-living conditions or in association with grasses. Genetics of nitrogen fixation was initiated in *A. brasilense* Sp7. A 30 kb DNA region containing *nifHDK*, the structural genes for nitrogenase, *nifE*, *nifUS*, and *fixABC* has been analyzed. DNA regions containing *nifA-nifB* and *ntrC* have been cloned. These regions are not adjacent to the main nitrogen fixation cluster nor to the structural gene for glutamine synthetase (*glnA*). The regulation of *nifH*, *glnB* and *glnA* transcription was studied. Three types of nitrogen-regulated promoter regions were characterized. *Azospirillum* strains contain large plasmids. A 90 MDa plasmid was purified from strain Sp7 and its physical map was established. This plasmid carries two ORFs highly homologous to the *Rhizobium meliloti* nodulation genes *nodPQ* as well as loci that complement EPS mutants of *R. meliloti*. Another region carrying an ORF homologous to *R. meliloti nodG* was characterized.

## 1. Cloning and Characterization of *nif* and *fix* Genes

Nitrogen fixation genetics was developed in many bacteria. However, current knowledge on *nif* gene organization and function is limited to a few species. The 20 nitrogen fixation (*nif*) genes of *Klebsiella pneumoniae*, clustered on a 22 kb DNA region, are organized in 8 transcription units, *nifJ*, *nifHDKTY*, *nifENX*, *nifUSVWZ*, *nifM*, *nifF*, *nifLA* and *nifBQ*. In most cases, genes equivalent to those found in *K. pneumoniae* are present in other diazotrophs, but additional genes were discovered both in free-living and symbiotic nitrogen fixers.

1.1. THE MAIN *NIF* CLUSTER OF *A. BRASILENSE* Sp7

Genetics of nitrogen fixation in *Azospirillum* was initiated with *A. brasilense* Sp7. Hybridization studies with *K. pneumoniae* nitrogenase structural genes (*nifHDK* probe) led to the cloning of the corresponding genes from strain Sp7 (Quiviger et al., 1982; Schrank et al., 1987). Further analysis of a 30 kb DNA region containing *nifHDK* led to the characterization of other *nif* genes, including *nifE* and *nifUS* homologues and an unidentified region (Fig. 1.1). The genes have been identified on the basis of hybridization with heterologous probes and by inactivation with Tn5 transposon (Perroud et al., 1985; Galimand *et al.*, 1989). In addition, we characterized a DNA region involved in nitrogen fixation which shared homology with the *fixABC* region present in *Rhizobiaceae* (Fogher *et al.* 1985; Galimand *et al.*, 1989). This region is located downstream of *nifUS* (Fig. 1.1). It contains functional genes in *Azospirillum*, since inactivation by Tn5 led to a Nif⁻ phenotype (Galimand *et al.*, 1989).

## 1.2. IDENTIFICATION OF *NIFA* AND *NIFB*

Pedrosa and Yates (1984) isolated, after chemical mutagenesis, *A. brasilense* Sp7 Nif⁻ mutants with NtrC-like or NifA-like phenotypes. The isolation of another NifA-like mutant, obtained after random Tn5 mutagenesis of *A. brasilense* strain ATCC 29710, has been reported (Singh *et al.*, 1989). A DNA region containing *nifA-nifB* was cloned from *A. brasilense* Sp7 by hybridization with *Bradyrhizobium japonicum* probes (this laboratory, unpublished). As shown in Figure 1.2, the *nifA* gene overlaps two *Eco*RI fragments of 5.6 and 3.6 kb. Identity of *nifA* was confirmed by the nucleotide sequence. Inactivation of *nifA* and *nifB* by site-directed Tn5 mutagenesis led to mutants with a Nif⁻ phenotype. The physical map of the *nifBA* clone from Sp7 was different from that of the NifA-like clone from strain ATCC 29710. Moreover, Sp7 *nifA* hybridized with DNA fragments of identical size in the two strains. This strongly suggests that in addition to *nifA, Azospirillum* contains another regulatory gene confering a NifA-like phenotype, a situation also encountered in *Azotobacter* (Santero et al., 1986).

## 1.3. CLONING OF THE *GLNB-GLNA* STRUCTURAL GENES

The structural gene for glutamine synthetase (GS) in enteric bacteria, *glnA*, belongs to a complex regulon, which contains the *ntrBC* genes. *NtrBC* gene products, together with sigma-54, are responsible for the transcriptional activation of a number of operons including nitrogen fixation genes (reviewed by Merrick, 1988). *A. brasilense* Sp7 glutamine auxotrophs, strains 7029 and 7028, impaired both in GS activity and in nitrogen fixation have been isolated (Gauthier and Elmerich 1977). A plasmid that restores a wild-type phenotype to one of these mutants has been isolated from a gene bank of strain Sp7. This plasmid contained the GS structural gene, *glnA* (Bozouklian *et al.* 1986). The nucleotide sequence of *glnA* was established, and is very similar to that of other *glnA* genes (Bozouklian and Elmerich, 1986). It was found that *glnA* alone was sufficient to restore a wild phenotype to 7029 and 7028 mutants. Upstream of *glnA* we found another ORF, ORF*glnB* (Fig. 1.3), whose translation product is very homologous to the $P_{II}$ protein, which in enteric bacteria is involved both in adenylylation and in the regulation of the GS synthesis (de Zamaroczy *et al.*, 1990). The physical map of the *glnBA* region is shown in Figure 1.3. As this region does not contain *ntrBC* genes, it appears that the organization of *gln* and *ntr* genes in *Azospirillum* is similar to that observed in *Rhizobiaceae*, where *glnB* and *glnA* are contiguous (Colonna-Romano *et al.*, 1987; Martin *et al.*, 1989). Genes of A. *brasilense* Sp7, sharing similarity with *B. japonicum ntrBC*, were recently cloned (this laboratory, unpublished), but their involvement in $N_2$ fixation has not yet been established.

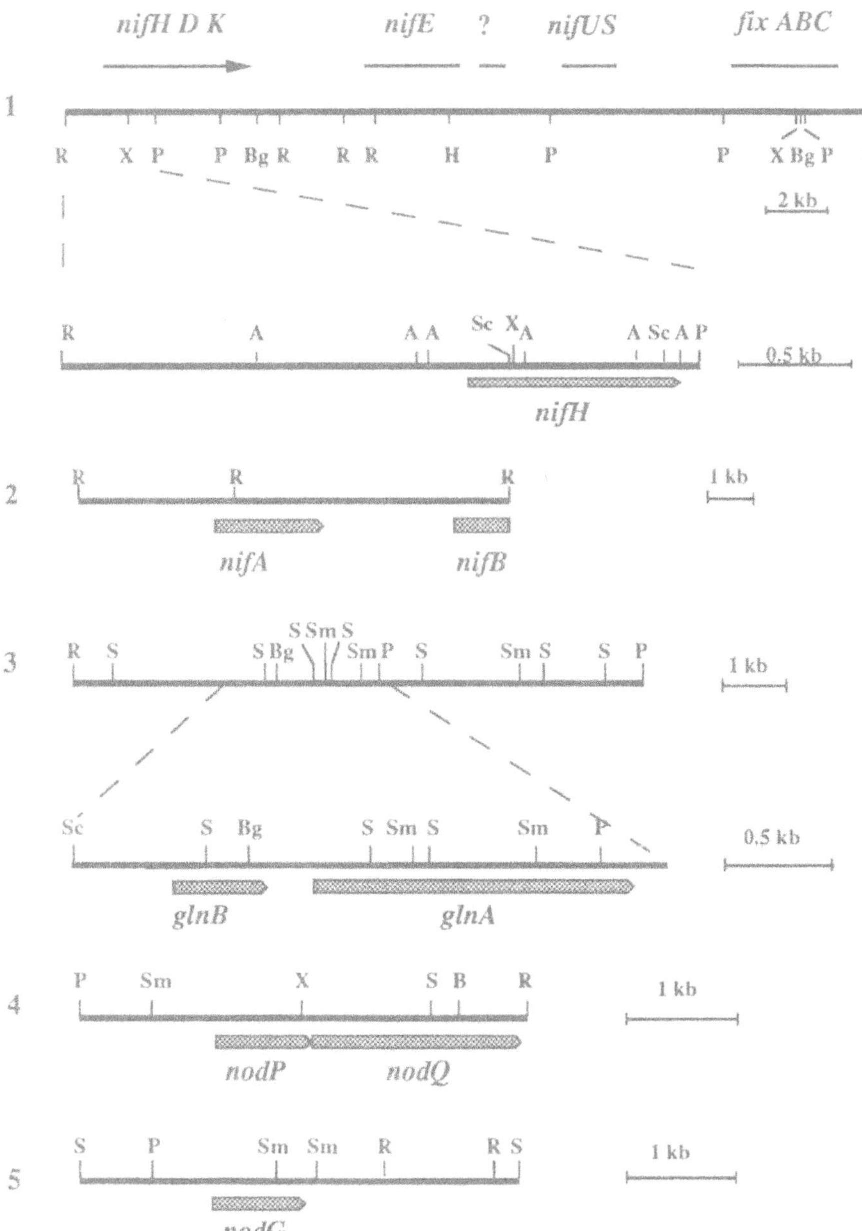

Figure 1. Physical map of the *gln*, *nif*, *fix*, *nod* gene regions of *A. brasilense* Sp7. Restriction sites: A: *Ava*II, B: *Bam*HI, Bg: *Bgl*II, P: *Pst*I, R: *Eco*RI, S: *Sal*I, Sc: *Sac*I, Sm: *Sma*I, X: *Xho*I. 1/ Major *nif* cluster; 2/ *nifA-nifB* region; 3/ *glnB-glnA* region  4/ *nodPQ*; 5/ *nodG*.

1.4. TRANSCRIPTIONAL ORGANIZATION OF *GLNBA*

Genetic evidence suggested that *glnA* was transcribed from its own promoter. Northern blot analysis was performed with total RNA extracted from bacteria grown under different physiological conditions, using *glnB* and *glnA* probes. It was found that *glnA* can be transcribed from its own promoter, as a 1.4 kb RNA, and cotranscribed with *glnB*, as a 2.4 kb RNA, depending on the physiological conditions (de Zamaroczy *et al.*, 1990). In particular, under conditions of nitrogen fixation with molecular nitrogen as sole nitrogen source, transcription from the *glnA* promoter is almost abolished whereas in ammonia or glutamate containing media transcription proceeds mainly from this promoter (de Zamaroczy *et al.*, 1990).

## 2. Characterization of Promoter Regions

2.1. PROMOTER REGION OF *NIFH*

The *nifH* gene of strain Sp7 was sequenced and it displayed the classical features of other *nifH* genes (de Zamaroczy *et al.*, 1989; Fani *et al.*, 1989). Transcription of *nifHDK* was analyzed using Northern blots of total RNA. Two transcripts of 1.1 and 5.6 kb corresponding to *nifH* RNA and *nifHDK* RNA were characterized (de Zamaroczy *et al.*,1989). This confirmed that the *nifHDK* genes are organized in a single transcription unit and suggested a regulation of *nifH* expression at the level of termination. The *nifH* promoter region was studied by S1 nuclease mapping. Two RNA starts were found at 10 bp and 40 bp respectively from a sequence reminescent of NtrA-like dependent promoter (Fig. 2) (de Zamaroczy *et al.*, 1989). This promoter is preceded by two upstream elements (TGT-N10-ACA), typical of genes controlled by *nifA* (de Zamaroczy *et al.*,1989; Fani *et al.*, 1989).

2.2. PROMOTER REGIONS OF *GLNB*

The transcription starts of *glnB* and *glnA* were determined by S1 nuclease mapping with RNA purified from cells grown under various physiological conditions. *GlnB* transcription is regulated by tandem promoters, *glnB*p1 and *glnB*p2 very similar to promoters found upstream of *glnA* in enteric bacteria (Fig. 2). *GlnB*p1 sequence is reminescent of the *E. coli* canonical promoter, whereas *glnB*p2 is reminiscent of NtrA-like dependent promoters. Transcription at p2 occurs mainly under limiting nitrogen conditions and the conserved sequence is very close to that found for the *nifH* promoter. In contrast, transcription at p1 is weak and the promoter overlaps a NtrC binding site as in enteric bacteria (de Zamaroczy *et al.*, 1990).

2.3. PROMOTER REGIONS OF *GLNA*

No promoter sequence reminescent of the *E. coli* sigma[70] promoter was detected upstream of *glnA* (Bozouklian and Elmerich, 1986). RNA mapping enabled to identify a putative promoter region, located 20 bp upstream of the mRNA start site (de Zamaroczy *et al.*, 1990). As shown in Figure 2, a sequence reminescent of the NtrA-dependent promoter consensus was found, except for a T which replaces a G at position -33. This sequence overlaps a putative weak NtrC-binding site, similar to that identified upstream of *glnA* in enteric bacteria (de Zamaroczy *et al.*, 1990). Thus, several features may account for the finding that *glnA* transcription functions poorly in the absence of fixed nitrogen:

i) the replacement of a G by a T at a fundamental position, ii) the unusual location of the promoter sequence, 20 bp upstream of the RNA transcription start site, iii) the binding of the promoter region by NtrC-P which could prevent transcription.

| | |
|---|---|
| *nifH* promoter | CT<u>GG</u>CACG $N_4$ AT<u>GC</u>A |
| *glnBp1* promoter | TT<u>GG</u>CACG $N_4$ CT<u>GC</u>TT |
| *glnA* promoter | CT<u>G</u>TCACG $N_4$ CT<u>GC</u>TT |
| *glnBp2* promoter | TGCGC $N_{24}$ ATATT |

Figure 2. Nucleotide sequence of the putative promoter regions idenfied 10 bp upstream of the mRNA start sites of *nifH*, *glnB* and 20 bp upstream of the mRNA start site of *glnA*.

2.4. REGULATIION OF THE NITROGEN METABOLISM

Regulation of nitrogen fixation and assimilation in *Azospirillum* shares common pathways to that established in *K. pneumoniae*. *Azospirillum* contains genes homologous to *nifA,* and *ntrBC*. *Nif* genes are probably transcribed from NtrA-dependent promoters, as it has been established for *nifH*, which suggests the presence of a gene encoding an NtrA-like product in *Azospirillum*. The role of *glnBA* gene products in regulation of nitrogen fixation remains to be elucidated. GlnBA transcription is under the control of an NtrA promoter and glnA is controlled by a novel type of nitrogen regulated promoter.

## 3. Characterization of *nod* gene homologues

The genes necessary for nodule organogenesis and host specifity are carried by megaplasmids in several fast growing *Rhizobia*. Twelve genes clustered not far from the nitrogenase structural genes have been identified in *R. meliloti*. The *nodDABCIJ* genes are referred to as the common nodulation genes and the host specific nodulation genes *nodQPGEFH* are also designated *hsn* region (reviewed by Long, 1989). As *Azospirillum* does not form nodules, it is difficult to screen for mutants impaired in the bacteria-plant interaction. We made the assumption that some of the genes involved in the early steps of the bacteria-plant interaction might be common with other soil bacteria such as *Rhizobium* and *Agrobacterium*. It was previously found that the *hsn* region hybridized with total DNA of several *Azospirillum* strains (Fogher *et al.*, 1985). Further characterization of the hybridizing fragments was performed. From *A. brasilense* Sp7, a 10 kb *Eco*RI and a 4 kb *Sal*I fragments sharing similarity with the "*hsn*" region of *R. meliloti* were cloned in pUC18, to yield (Elmerich *et al.*, 1987).

### 3.1. IDENTIFICATION OF *NODPQ*

The physical map of the *nodPQ* region is shown in Figure 1.4. The nucleotide sequence of a 3.5 kb *Eco*RI-*Sma*I fragment was established (Vieille *et al.*, 1990). It revealed 67 and 61 % similarity with *R. meliloti nodP* and *nodQ* respectively (Cervantes *et al.*, 1989; Schwedock and Long, 1989). The two genes overlapped by one bp, suggesting that they belong to the same operon. *NodP* encodes a 34 kDa protein. *NodQ* encodes a 67 kDa protein, which presents a significant level of homology with a family of initiation and elongation factors. In particular, the consensus for the GTP binding site is well conserved. A *nodP-lacZ* translational fusion was constructed in the broad host range vector pGD926. The gene was functional in *Azospirillum* and constitutively expressed. To elucidate the role of *nodPQ*, mutants of pAB502 containing Tn*5-mob* insertions, kanamycin cartridges or deletions have been constructed in *Escherichia coli* and recombined in *Azospirillum* genome. The resulting strains had the same growth rate as the wild type, they were not impaired in nitrogenase activity, indole acetic acide production or interaction with plant seedlings (Vieille and Elmerich, 1990).

### 3.2. IDENTIFICATION OF *NODG*

The physical map of the *nodG*-containing fragment is shown in Figure 1.5. The nucleotide sequence of a 1.4 kb *Pst*I-*Sma*I fragment was established (Vieille and Elmerich, in preparation). It revealed 50 % similarity with *R. meliloti nodG* (Debellé and Sharma, 1986). A *nodG-lacZ* fusion is expressed in *Azospirillum* suggesting that *nodG* is also functional. Mutants of *nodG* have been constructed. No associated phenotype has yet been characterized.

### 3.3. LOCALIZATION OF *NODPQ* AND *NODG* IN *AZOSPIRILLUM* GENOME

*Azospirillum* strains contain plasmids (reviewed by Elmerich, 1986). In general, the strains examined contain from 1 to 6 plasmids with $M_r$ ranging from a few MDa to more than 300 MDa. The presence of a 90 MDa plasmid was noted frequently in *A. brasilense* strains, whereas *A. lipoferum* carries a 150 MDa plasmid (Vieille *et al.*, 1989; Onyeocha *et al.*, 1990) To determine the chromosomal or plasmid location of the *nodPQ* and *nodG* homologous regions, hybridization was performed with plasmid preparations from several *A. brasilense* and *A. lipoferum* strains. Homology with *nodPQ* was detected with a plasmid of 90 MDa in most strains (Vieille *et al.*, 1989), whereas *nodG* was chromosomal.

## 4. Physical and Genetic Analysis of the 90 MDa Plasmid

The medium size plasmids and megaplasmids in *Azospirillum* were assumed to carry important information for the interaction with the host plant, by analogy with other soil bacteria harboring plasmids, in particular *Agrobacterium* and *Rhizobium*. The 90 MDa plasmid (p90) of strain Sp7 was purified, its physical map was established and *nodPQ* were localized on the map (Onyeocha *et al.*, 1989). Other functions have also been mapped on p90, two loci that correct *exoB* and *exoC* mutations in *R. meliloti* (Michiels *et al.*, 1989) and two ampicillin resistance genes. Hybridization experiments with fragments of p90 as probes showed that each fragment hybridized with a plasmid of similar size in other *Azospirillum* strains, suggesting that all 90 MDa the plasmids of in this genus contain conserved regions, which presumably code for common functions. In addition, attempts to obtain strains cured of this plasmid were unsuccessful. It is tempting to speculate that p90 carries essential functions, which remain to be determined.

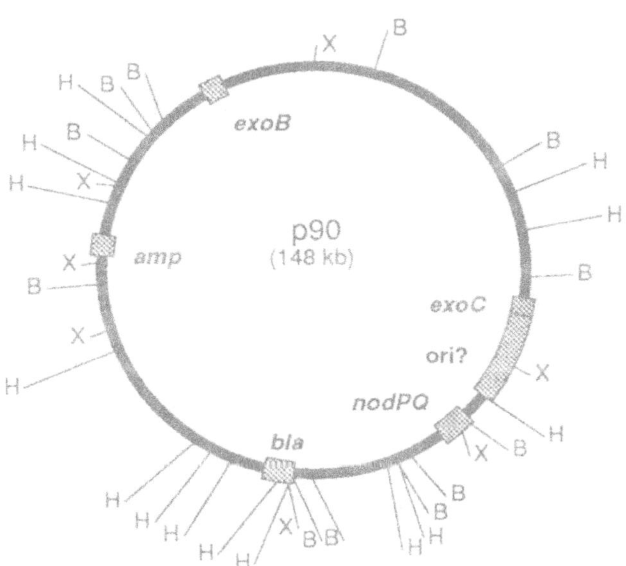


Figure 3. Physical map of the 90 MDa plasmid (p90) of *A. brasilense* strain Sp7 and localization of a few markers according to Onyeocha *et al.*, (1990). Restriction sites: B: *Bam*HI, H: *Hind*III, X: *Xho*I. *nodPQ* is the locus which hybridizes with the *nodPQ* genes of *R. meliloti*, *bla* and *amp* are non-homologous loci involved in ampicillin resistance, *exoB* and *exoC* are loci which functionally complement mutations in the *exoB* and *exoC* genes of *R. meliloti*, ori indicates a DNA region necessary for the maintenance of the plasmid as an independent replicon.

## 5. Acknowledgements

The authors wish to thank Prof. J.-P. Aubert for helpful discussions and Ms Aparicio and Paquelin for skillful technical assistance. CV was appointed by the Ministry of Agriculture, France; F.D was a recipient of a fellowship from MRES, France, I.O. was a recipient of a fellowship from the French Government, Y.Y.L. was the recipient of a fellowship from the P. R. China Government and W.Z was the recipient of a postdoctoral fellowship from the D. F. G., Federal Republic of Germany. This work was supported by research funds from the Paris 7 University, France.

## 6. References


Bozouklian, H. and Elmerich, C. (1986) 'Nucleotide sequence of the *Azospirillum brasilense* Sp7 glutamine synthetase structural gene', Biochimie 68, 1181-1187.

Bozouklian, H., Fogher, C. and Elmerich, C. (1986) 'Cloning and characterization of the *glnA* gene of *Azospirillum brasilense* Sp7', Ann. Inst. Pasteur/Microbiol. 137, 3-18.

Cervantes, E., Sharma, S.B., Maillet, F., Vasse, J., Truchet, G. and Rosenberg, C. (1989) 'The *Rhizobium meliloti* host-range *nodQ* gene encodes a protein which shares homology with translational elongation and initiation factors', Molecular Microbiology 3, 745-755.

Colonna-Romano, S., Riccio, A., Guida, M., Defez, R., Lamberti, A., Iaccarino, M., Arnold, W., Priefer, U. and Pühler, A. (1987) 'Tight linkage of *glnA* and a putative regulatory gene in *Rhizobium leguminosarum*', Nucl. Acids Res. 15, 1951-1964.

Debellé, F. and Sharma, S.B. (1986) 'Nucleotide sequence of *Rhizobium meliloti* RCR2011 genes involved in host specificity of nodulation', Nucl. Acids Res. 14, 7453-7472.

de Zamaroczy, M., Delorme, F. and Elmerich, C. (1989) 'Regulation of transcription and promoter mapping of the structural genes for nitrogenase (*nifHDK*) of *Azospirillum brasilense* Sp7', Mol. Gen. Genet. 220, 88-94.

de Zamaroczy, M., Delorme F. and Elmerich, C. (1990) 'Characterization of three different nitrogen regulated promoter regions for the expression of *glnB* and *glnA* in *Azospirillum brasilense*', Molec. Gen. Genet. in press.

Elmerich, C. (1986) '*Azospirillum*', in W.J. Broughton and A. Pühler (eds.), Nitrogen fixation-Molecular biology, Clarendon Press, Oxford, pp.106-126.

Elmerich, C., Bozouklian, H., Vieille, C., Fogher, C., Perroud, B., Perrin, A. and Vanderleyden, J. (1987) '*Azospirillum*: genetics of nitrogen fixation and interaction with plants', Phil. Trans. R. Soc. 317, 183-192.

Fani, R., Allotta, G., Bazzicalupo, M., Ricci, F., Schipani, C. and Polsinelli, M. (1989) 'Nucleotide sequence of the gene encoding the nitrogenase iron protein (*nifH*) of *Azospirillum brasilense* and identification of a region controlling *nifH* transcription', Mol. Gen. Genet. 220, 81-87.

Fogher, C., Dusha, I., Barbot, P. and Elmerich, C. (1985) 'Heterologous hybridization of *Azospirillum* DNA to *Rhizobium nod* and *fix* genes', FEMS Microbiol. Lett., 30, 245-249.

Galimand, M., Perroud, B., Delorme, F. Paquelin, A., Vieille, C., Bozouklian, H. and Elmerich, C. (1989) 'Identification of DNA regions homologous to nitrogen fixation genes *nifE*, *nifUS* and *fixABC* in *Azospirillum brasilense* Sp7', J. Gen. Microbiol. 135, 1047-1059.

Gauthier, D. and Elmerich, C. (1977) 'Relationship between glutamine synthetase and nitrogenase in *Spirillum lipoferum*', FEMS Microb. Lett. 2, 101-104.

Long, S.R. (1989) '*Rhizobium*-Legume nodulation: Life together in the underground', Cell 56, 203-214.

Martin, G.B., Thomashow, M.F. and Chelm, B.K. (1989) *'Bradyrhizobium japonicum glnB*, a putative nitrogen-regulatory gene, is regulated by NtrC at tandem promoters', J. Bacteriol. 171, 5638-5645.
Merrick, M.J. (1988) 'Regulation of nitrogen assimilation by bacteria', in J.A. Cole and S.J. Ferguson (eds.), The Nitrogen and Sulphur Cycle, Cambridge University Press, Cambridge, pp. 331-361.
Onyeocha, I., Vieille, C., Zimmer, W., Baca, B.E., Flores, M., Palacios, P. and Elmerich, C. (1990) 'Physical map and properties of a 90 MDa plasmid of *Azospirillum brasilense* Sp7', Plasmid, 23, in press.
Pedrosa, F.O. and Yates, M.G. (1984) 'Regulation of nitrogen fixation (*nif*) genes of *Azospirillum brasilense* by *nifA* and *ntr* (*gln*) type gene products', FEMS Microbiol. Lett. 23, 95-101.
Perroud, B., Bandhari, S.K. and Elmerich, C. (1985) 'The *nifHDK* operon of *Azospirillum brasilense* Sp7', In W. Klingmüller (ed.) Azospirillum III: Genetics, Physiology, Ecology, Springer Verlag, Berlin, pp. 10-19.
Quiviger, B., Franche, C., Lutfalla, G., Rice, D., Haselkorn, R. and Elmerich, C. (1982) 'Cloning of a nitrogen fixation (*nif*) gene cluster of *Azospirillum brasilense*', Biochimie 64, 495-502.
Santero, E., Toukdarian, A., Humphrey, R., Kennedy, C. (1988) 'Identification and characterization of two nitrogen-fixation regulatory regions, *nifA* and *nfrX*, in *Azotobacter vinelandii* and *Azotobacter chroococcum*', Mol. Microbiol. 2, 303-314.
Schrank, I., Zaha, A., De Arujo, E.F. and Santos, D.S. (1987) 'Construction of a gene library from *Azospirillum brasilense* and characterization of a recombinant clone containing the *nif* structural genes', Brazil. J. Med. Biol. Res. 20, 321-330.
Schwedock, J. and Long, S.R. (1989) 'Nucleotide sequence and protein products of two new nodulation genes of *Rhizobium meliloti*, *nodP* and *nodQ*', Molec. Plant-Microbe Interact. 2, 181-194.
Singh, M., Tripathi, A.K. and Klingmüller, W. (1989) 'Identification of a regulatory *nifA* type gene and physical mapping of cloned new *nif* regions of *Azospirillum brasilense*', Mol. Gen. Genet. 219, 235-240.
Vieille, C. and Elmerich, C. (1990) 'Characterization in *Azospirillum brasilense* Sp7 of two plasmid genes homologous to *Rhizobium meliloti nodPQ*', Molec. Plant-Microbe Interact. in press.
Vieille, C., Onyeocha, I., Galimand, M. and Elmerich, C. (1989) 'Homology between plasmids of *Azospirillum brasilense* and *Azospirillum lipoferum*', in F.A. Skinner, R.M., Boddey and I. and Fendrik, I. (eds.) Nitrogen Fixation with Non-Legumes, Kluwer Academic Publishers, The Netherlands, vol. 35, pp.165-172.

# Control of nitrogenase in *Azospirillum* sp.


ROBERT H. BURRIS, ANTON HARTMANN, YAOPING ZHANG and HAIAN FU
*Department of Biochemistry, University of Wisconsin, Madison, WI 53706, USA*



*Key words:*    ammonium, *Azospirillum* sp., dinitrogenase reductase, 'switch off'



**Abstract**

Many $N_2$-fixing organisms can turn off nitrogenase activity in the presence of $NH_4^+$ and turn it on again when the $NH_4^+$ is exhausted. One of the most interesting systems for accomplishing this is by covalent modification of one subunit of dinitrogenase reductase by dinitrogenase reductase ADP-ribosyltransferase (DRAT). The system can be reactivated when $NH_4^+$ is exhausted, by dinitrogenase reductase activating glycohydrolase (DRAG) which removes the inactivating group. It is fascinating that some species of the genus *Azospirillum* possess the DRAT and DRAG systems (*A. lipoferum* and *A. brasilense*), whereas *A. amazonense* in the same genus lacks DRAT and DRAG. *A. amazonense* responds to $NH_4^+$ but does not exhibit modification of dinitrogenase reductase characteristic of the action of DRAT. However, it has been possible to clone DRAT and DRAG and to introduce them into *A. amazonense*, whereupon they become functional in this organism. The DRAT and DRAG system does not appear to function in *Acetobacter diazotrophicus*, an organism isolated from sugar cane, that fixes $N_2$ at a pH as low as 3.0. *A. diazotrophicus* does show a rather sluggish response to $NH_4^+$. A level of about $10\ \mu M\ NH_4^+$ is required to 'switch off' the system. The response to $NH_4^+$ is influenced by the dissolved oxygen concentration (DOC) as has been reported for *Azospirillum* sp. A DOC in equilibrium with 0.1 to 0.2 kPa $O_2$ seems optimal for the response in *A. diazotrophicus*.


## Introduction

Nitrogen fixation is very energy demanding. Chemical fixation via the Haber-Bosch process achieves breakage of the N–N bond and reduction of $N_2$ to $2\ NH_3$ by employing high temperature and high pressure in the presence of a catalyst. Biological fixation of $N_2$ faces the same energy barriers but operates in a more subtle fashion at ambient temperature and a $pN_2$ of 0.78 atm. by invoking the combined action of dinitrogenase and dinitrogenase reductase. The action of these enzymes is supplemented by ferredoxin or flavodoxin as a reductant and MgATP as an energy source. When one provides optimal conditions for nitrogenase activity it still requires 16 MgATP for the reduction of $8\ H^+ + N_2 \rightarrow 2\ NH_3 + H_2$. Under natural conditions, 20 to 30 MgATP are used per $N_2$ reduced. Faced with this energy demand, there have been evolutionary pressures for organisms possessing nitrogenase to develop a means to turn off nitrogenase when fixed nitrogen is available to supply the organism's needs. These control systems vary among $N_2$-fixing organisms. Our interest has centered on the system that invokes covalent modification of one unit of the dimer of dinitrogenase reductase.

When $NH_4^+$ enriched in $^{15}N$ became available, the $^{15}N$ furnished a means for quickly establishing the initiation and cessation of $N_2$ fixation and the influence of fixed nitrogen on the process of biological $N_2$ fixation. When *Azotobacter vinelandii* was given the choice of $^{15}N_2$ and $NH_4^+$ or other nitrogenous compounds, Wilson et al. (1943) found that the organisms showed a clear preference for $NH_4^+$ over nitrate, nitrite and a variety of amino acids. When $NH_4^+$ at 121 or

315 ppm N was present, only 0.9% or 0.0%, respectively, of the cell nitrogen was derived from the $^{15}N_2$. Burris and Wilson (1946) expanded on these observations and followed the time course of utilization of $NH_4^+$ and nitrate by *A. vinelandii*. Although switch-off of $N_2$ fixation of *A. vinelandii* by $NH_4^+$ was demonstrated may years ago, the detailed mechanism of switch-off in *A. vinelandii* has not been elucidated.

In contrast to *A. vinelandii*, the control in *Rhodospirillum rubrum* has been worked out in considerable detail. Kamen and Gest (1949) and Gest and Kamen (1949) demonstrated that *Rhodospirillum rubrum* is capable of fixing $N_2$. Although Schneider et al. (1960) reported that they had prepared cell-free extracts from *R. rubrum* that fixed $N_2$, there was difficulty in getting consistent $N_2$ fixation with extracts from this organism. Munson and Burris (1969) found that the time course of reductions catalyzed by *R. rubrum* nitrogenase preparations was non-linear and exhibited a lag phase. The inconsistencies arose because *R. rubrum* has a method for quickly turning off its $N_2$ fixation system and then turning it on again. Ludden and Burris (1976) first reported the presence of an activating system for nitrogenase in *R. rubrum* chromatophores. The factor could be separated from the dinitrogenase and dinitrogenase reductase on a column of DEAE cellulose. ATP and $Mg^{2+}$ or $Mg^{2+}$ plus $Mn^{2+}$ were necessary for the activation of dinitrogenase reductase. The activating factor if preincubated with *R. rubrum* preparations abolished the lag. It was apparent that the action of the activating factor was centered on dinitrogenase reductase and had no influence on dinitrogenase.

Several students in the laboratory had problems obtaining active nitrogenase preparations from *R. rubrum* despite the initial success of Schneider et al. (1960). Paul Ludden took up the problem and made the crucial observation that an extract from *R. rubrum* chromatophores would activate dinitrogenase reductase from the organism. The responses suggested that an enzyme was activating the Fe protein of nitrogenase, and with incubation the activity increased. The agent was designated the 'activating factor' by Ludden and Burris (1976).

It is interesting that Nordlund et al. (1977) in Sweden observed the factor independently. They derived the activating agent from NaCl extracts of chromatophore membranes. They also reported that it was sensitive to oxygen, reacted specifically with the Fe protein of nitrogenase and that its activity required ATP and $Mg^{2+}$.

Ludden and Burris (1979) reported that when inactive dinitrogenase reductase was activated it released an 'adenine-like molecule'. ATP and a divalent metal ion were required for activation of the protein. Nordlund and Eriksson (1979) observed that preparations were not active in $H_2$ production until treated with the activating factor.

Ludden and Burris (1976) reported that the addition of $Mn^{2+}$ and $Mg^{2+}$ enhanced the activity of extracts from *R. rubrum*. Anaerobic collection of *R. rubrum* aided in recovering active extracts from *R. rubrum*. In 1980 Ludden could verify that the activating factor was oxygen labile and sensitive to trypsin. It could by stabilized, and it could be precipitated with 30% polyethylene glycol 4000. The extract could be purified further on a column of DEAE cellulose. The presence of $NH_4^+$ in the culture medium did not suppress the production of activating factor. The dinitrogenase reductase recovered from *R. rubrum* had its full complement of Fe and S whether it was active or inactive.

It proved possible to separate the inactive Fe protein from *R. rubrum* into two components on polyacrylamide gels (Ludden and Burris, 1978). The inactive Fe protein carried phosphate, ribose and an 'adenine-like' unit. With an active Fe protein available it was possible to purify both dinitrogenase and dinitrogenase reductase from *R. rubrum*, and the properties of both were similar to those of other nitrogenase proteins. The modifying group did not appear on dinitrogenase reductase isolated from other organisms.

Nordlund et al. (1977) had suggested that the 'activating factor' formed a stable complex with dinitrogenase reductase to restore its catalytic activity. Ludden (1980) in contrast presented evidence that activity was restored when the activating enzyme removed the adenine-ribose-phosphate unit whose addition had inactivated the enzyme. The process required the activating enzyme, a divalent metal ion ($Mg^{2+}$, $Mn^{2+}$,

Fe$^{2+}$) and ATP. He pointed out that upon activation the second band disappeared from a gel used to separate components of inactive dinitrogenase reductase. So the picture emerged that *R. rubrum* dinitrogenase reductase was inactivated in the dark or aerobically by adding a unit with phosphate, ribose and adenine, and that the dinitrogenase reductase could be reactivated by enzymatic removal of these blocking units.

From this point the story has developed primarily with observations by Ludden and colleagues, and the observations and conclusions have been summarized by Ludden (1980) and by Ludden and Roberts (1989). They depict their scheme for activation and inactivation of dinitrogenase reductase as shown in Figure 1. We will define their terms and then develop the rationale. Starting at the left, we have the active form of dinitrogenase reductase with its shared Fe/S center and with two equivalent protein subunits. As illustrated at the top of the figure the dinitrogenase reductase is altered by the enzyme DRAT (dinitrogenase reductase ADP-ribosyltransferase) which with NAD and MgADP, ADP-ribosylates one of the two equivalent protein subunits and thus inactivates dinitrogenase reductase. This is accompanied by the release of nicotinamide. The inactivation occurs by ADP-ribosylation of the arginine in position 101. Note that DRAT is coded for by *draT* and that the inactivation is triggered by NH$_4^+$ or darkness.

The inactive complex can be reactivated by DRAG (dinitrogenase reductase-activating glycohydrolase) acting with MgATP and Mn$^{2+}$. ADP ribose is removed from arginine-101 of the inactive dinitrogenase reductase and is released. The DRAG is coded for by *draG*. Dinitrogenase reductase is restored to its normal active state in which the two subunits are equivalent.

The story is well told by Ludden and Roberts (1989) of how the activation and inactivation of dinitrogenase reductase was resolved and how the various components in the system fit together. Ludden and Burris (1979) presented evidence that both adenine and ribose were covalently bound to inactive dinitrogenase reductase and that both were removed when the dinitrogenase reductase was reactivated. Pope et al. (1985a,b) purified the displaced adenine moiety and demonstrated its identity as ADP-ribose by NMR and mass spectrometric analysis. Jouanneau et al. (1988) found ADP ribosylation also functioned in the inactivation of the nitrogenase system in *Rhodobacter capsulatus*. The attachment of the ADP to ribose was established by Pope et al. (1986).

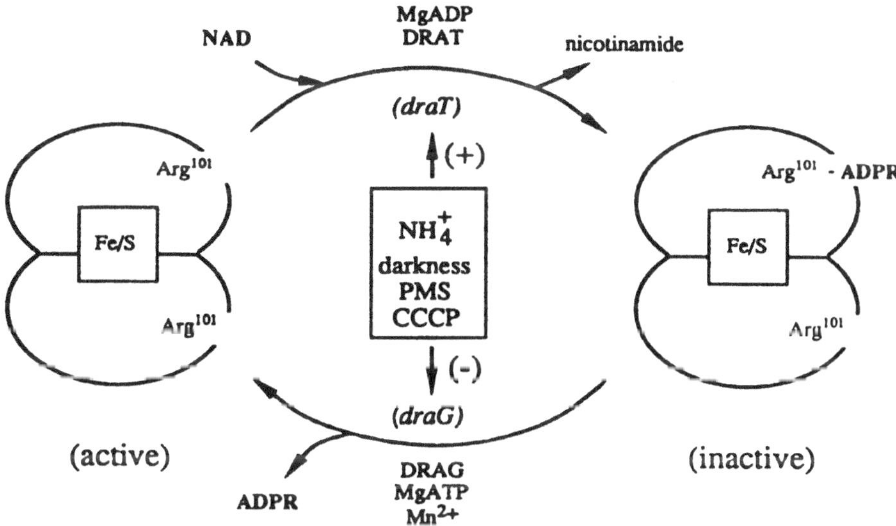



*Fig. 1.* Control of dinitrogenase reductase activity by ADP-ribosylation as depicted by Ludden and Roberts (1989). ADPR, ADP-ribose, PMS, phenazine methosulfate, CCCP, chlorocarbonylcyanide phenylhydrazone.

**Experimental**

We now shift to some of the specifics of our work and the work of others on the control of nitrogenase in *Azospirillum* sp. and related organisms. Under proper conditions *Azospirillum* species fix nitrogen vigorously. Characteristically they thrive at a low partial pressure of oxygen, and certain species have their $N_2$ fixation systems turned off readily by low concentrations of $NH_4^+$. *Azospirillum lipoferum* is the most studied species of the group.

*A. lipoferum* was first described by Beijerinck (1925) as *Spirillum lipoferum*, reexamined by Becking (1963), and brought into prominence when Döbereiner and Day (1976) reported that they had found it associated with the roots of grasses that did not exhibit symptoms of nitrogen deficiency whereas surrounding grass did. They described how to culture the organism and pointed out that it was microaerobic. Since then the organism has been renamed and studied extensively. Okon et al. (1976a) reported that *S. lipoferum* grew vigorously on malate, lactate, succinate and pyruvate and poorly on glucose. It grows aerobically when supplied $NH_4^+$, but when it is fixing $N_2$ the optimal $pO_2$ is a bit below 0.01 atm. (1 kPa).

Okon, Houchins, Albrecht and Burris (1977) reported that on $NH_4^+$ the doubling time of *S. lipoferum* was 1 h, whereas on $N_2$ it was 5.5 to 7 h. They found that its dinitrogenase reductase was activated by the activating factor from *R. rubrum*. Ludden et al. (1978) separated *S. lipoferum* extracts into three components: dinitrogenase, dinitrogenase reductase and an activating factor comparable to that recovered from *R. rubrum*. The efficacy of the activating factor was increased by added $Mn^{2+}$ and $Mg^{2+}$. The activation required ATP.

Hartmann and Burris (1987) reported that the nitrogenase activity of *A. brasilense* and *A. lipoferum* was completely inhibited by 2 kPa of oxygen; some activity was recovered by transferring to 0.2 kPa of oxygen. There was no covalent modification of dinitrogenase reductase comparable to that arising from switch-off by $NH_4^+$ as evidenced by Western-blotting and $^{32}P$-labeling experiments.

The question was posed by Ludden et al. (1982), 'Why is the Fe protein inactive as purified from *R. rubrum* grown on glutamate as its nitrogen source'? Although the inactive Fe protein proved capable of binding MgATP and undergoing a conformational shift upon binding MgATP, and it could undergo oxidation-reduction, it was incapable of transferring electrons to dinitrogenase at a substantial rate. Hence, it could not function in the reduction of $N_2$ or acetylene.

Apparently the only reported purification of the nitrogenase from the *Azospirillum* group is that of Song et al. (1985). Purification from *A. amazonense* was not unlike that for other nitrogen fixers and specific activities were comparable. They were unable to find an activating enzyme for *A. amazonense* so the organism differed in this respect from *A. lipoferum* and *A. brasilense*.

Vignais and coworkers (1987) reviewed information on regulation of $N_2$ fixation in the photosynthetic bacteria. They pointed out that it is generally accepted that $NH_4^+$ itself is not the real effector of switch-off of nitrogenase, and that glutamine synthetase may be involved directly or indirectly. However, they concluded that there is no direct evidence that glutamine is the in vivo effector of switch-off.

The switch-on-off process can be very rapid in $N_2$-fixing organisms. Sweet and Burris (1981) followed nitrogenase activity by measuring $H_2$ production with a hydrogen electrode. This permitted continuous measurement, and rapid changes could be observed. *R. rubrum* grown on $N_2$ on a glutamate medium turned nitrogenase off quickly when $NH_4^+$ was added and turned production of $H_2$ back on in seconds after exhaustion of the $NH_4^+$.

Cejudo et al. (1984) found that low levels of $NH_4^+$ turned off nitrogenase immediately in *Azotobacter chroococcum*, and that the activity returned when the $NH_4^+$ was exhausted. Inhibitors of $NH_4^+$ assimilation, such as methionine sulfoximine, prevented $NH_4^+$ switch-off.

Hartmann et al. (1986) reversibly inhibited the nitrogenase activity of *A. brasilense*, *A. lipoferum* and *A. amazonense*. The interesting observation was that methionine sulfoximine abolished switch-off in *A. lipoferum* and *A. brasilense* but not in *A. amazonense*. Obviously there was something different about the control of nitrogenase among species of the same genus. In *A. amazonense*, the in vitro nitrogenase activity of $NH_4^+$-treated cells was not decreased, and no evidence could be found for a modified Fe protein band on electrophoretic examination.

Hartmann et al. (1988) noted further differences among species and strains of the azospirilla. *A. brasilense* grew poorly on glutamate, aspartate, serine or histidine, and these amino acid even at concentrations of $10 \, \text{m}M$ showed little inhibition of $N_2$ fixation. In contrast, *A. lipoferum* and *A. amazonense* grew well on these amino acids as a source of carbon and nitrogen. *Azospirillum* spp. thus differ substantially in their metabolism of amino acids.

*Herbaspirillum seropedicae*, a $N_2$-fixing organism isolated and characterized by Döbereiner (1989), is microaerobic. Fu and Burris (1989) found its optimal $pO_2$ is from 0.04 to 0.2 kPa for $N_2$ fixation. $NH_4^+$ inhibited $N_2$ fixation only partially even at $20 \, \text{m}M$ concentration. Treatment with $NH_4^+$ gave no detectable change in the electrophoretic pattern of the nitrogenase components, so there is no evidence that ADP-ribosylation plays a part in turning off nitrogenase in this organism.

Control of nitrogenase by the DRAT and DRAG system is by no means universal among $N_2$-fixing microorganisms. The question arises whether one can transfer the system to other microorganisms devoid of it. Fu, Wirt et al. (1989) demonstrated that the DRAT system could be moved to *E. coli* and *Klebsiella pneumoniae*. When the DRAT was expressed in wild-type *K. pneumoniae* strains, they lost their ability to fix $N_2$. If the arginine 101 site on dinitrogenase reductase was eliminated, the dinitrogenase reductase could not be inactivated by ADP-ribosylation.

The mechanism of $NH_4^+$ switch-on-off in *A. brasilense* and *A. lipoferum* was investigated by Fu, Hartmann et al. (1989), and they established a correlation between in vivo regulation of nitrogenase activity by $NH_4^+$ or glutamine and the reversible covalent modification of dinitrogenase reductase. Dinitrogenase reductase ADP-ribosyltransferase (DRAT) activity was found in extracts of *A. brasilense* with NAD serving as the donor molecule. Dinitrogenase reductase-

activating glycohydrolase (DRAG) activity was present in extracts of both *A. lipoferum* and *R. rubrum*. The region homologous to *R. rubrum* *draT* and *draG* was identified in the genomic DNA of *A. brasilense* as a 12-kilobase *Eco*RI fragment and in *A. lipoferum* as a 7-kilobase *Eco*RI fragment. The authors concluded that a posttranslational regulatory system for nitrogenase activity is present in *A. brasilense* and *A. lipoferum*, and that it operates via ADP-ribosylation of dinitrogenase reductase as it does in *R. rubrum*.

The question of whether the regulatory system characteristic of *R. rubrum* could be transferred to a $N_2$-fixing organism lacking it and whether it would function there was posed by Fu, Burris and Roberts (1990). They found that they could transfer *draT* and *draG*, the genes coding for DRAT and DRAG, from *R. rubrum* or *A. lipoferum* to *K. pneumoniae* (an organism normally devoid of these genes) and demonstrate that they could function. The expressed *draT* and *draG* genes allowed *K. pneumoniae* to respond to $NH_4^+$ with a reversible regulation of nitrogenase activity that was correlated with the reversible ADP-ribosylation of dinitrogenase reductase. One can conclude that DRAT and DRAG in *K. pneumoniae* are sufficient to support $NH_4^+$-switch-off-on, and that ADP-ribosylation can serve as a regulatory mechanism for this organism.

Fu et al. (1990) cloned the *draTG* genes from *A. lipoferum* and demonstrated that they are located next to *nifH* (structural gene for dinitrogenase reductase) as in *R. rubrum*. The similarity in organization strongly suggests that these genes have been conserved in these two organisms during evolution. This is the first identification of this regional homology outside the photosynthetic bacteria.

*Herbaspirillum seropedicae* is an organism that has an unusually low optimal $pO_2$ for nitrogen fixation as measured by reduction of acetylene. As indicated in Figure 2, the optimal acetylene reduction occurred at about 0.1 kPa. *A. brasilense* had optimal acetylene reduction at about three times this $pO_2$.

*Acetobacter diazotrophicus* (formerly *Acetobacter nitrocaptans*) is a nitrogen-fixing organism with intriguing properties. This organism was

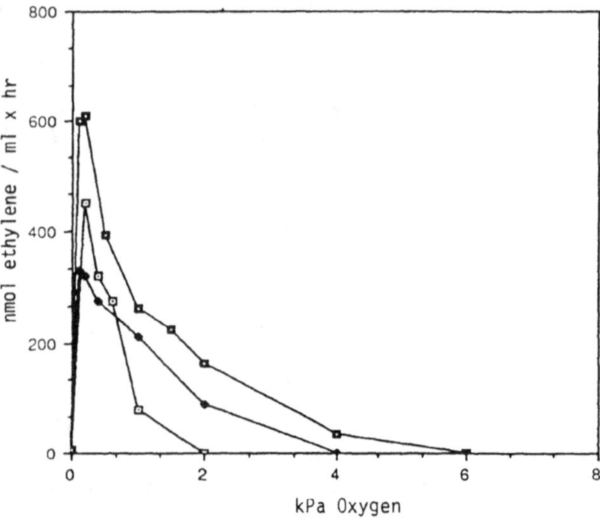

*Fig. 2.* Influence of pO$_2$ on acetylene reduction by *Acetobacter diazotrophicus* (-□-) in comparison with *Azospirillum brasilense* Sp7 (-·□·-) and *Herbaspirillum seropedicae* Z176 (-◆-). Samples (4 mL) of N$_2$-fixing cultures were transferred to an O$_2$-electrode chamber, C$_2$H$_2$ was added, and samples were removed for measurement of C$_2$H$_2$ reduction at different dissolved O$_2$ levels. The OD$_{580}$ of the culture was about 1.0 and contained about 500 μg protein mL$^{-1}$.

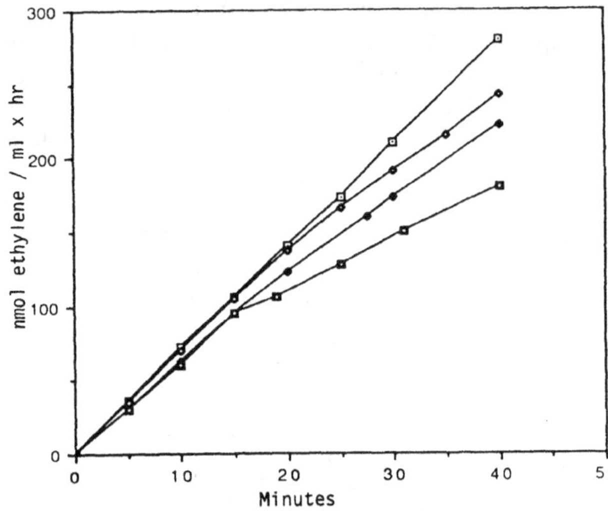

*Fig. 3.* Effect of amino acids and azaserine on nitrogenase activity of *Acetobacter diazotrophicus* PAL-5. At 15 min, 2 m$M$ azaserine (-□-), 5 m$M$ asparagine (-◆-), 5 m$M$ glutamic acid (-◇-) or water as control (-·□·-) was added. Reactions were run in an O$_2$-electrode chamber at 0.1–0.2 kPa O$_2$.

isolated from sugar cane and described by Boddey and Döbereiner (1988) and Döbereiner (1989). They have summarized recent studies on associative nitrogen fixation. Most free-living and associative nitrogen fixers achieve limited N$_2$ fixation because of the dearth of energy-yielding substrates available to them. However, *A. diazotrophicus* by growing in association with sugar cane has chosen a location rich in carbohydrate. It takes advantage of the available sucrose, grows actively in or on the cane stem and roots, and fixes an abundance of N$_2$. The physiology of the organism is particularly remarkable in two aspects: (a) it grows in high concentrations of sucrose (we grow it routinely in 10% sucrose and Döbereiner, 1989, reports that it grows in 30% sucrose), and (b) it fixes N$_2$ at a remarkably low pH (it grows and reduces acetylene below pH 3.0). There is no clear definition of the internal pH of the organism, but it grows on a nitrogen-free medium at pH 3.0 and lower. There are indications that *A. diazotrophicus* and other organisms that may be associated with it may fix over 200 kg N per hectare in a year.

The data in Figure 3 are from an experiment in which amino acids and azaserine were added to N$_2$-fixing cultures of *A. diazotrophicus*. At 15 min 2 m$M$ azaserine, 5 m$M$ glutamic acid and 5 m$M$ asparagine were added to the chamber. Glutamic acid and asparagine inhibited about 25% and azaserine about 50%.

As discussed, it has been shown that N$_2$ fixation can be turned off by NH$_4^+$ and that the mechanism in *R. rubrum*, *A. lipoferum* and *A. brasilense* is by ADP-ribosylation of arginine 101 of one of the protein subunits of dinitrogenase reductase. When this occurs, the two subunits can be separated by gel electrophoresis, because one has had its properties altered by adding the ADP-ribose at arginine 101. We found no evidence for such a switch-off-on process in *A. diazotrophicus*, so we examined the electrophoretic pattern of dinitrogenase reductase from this organism. Figure 4 indicates that *A. diazotrophicus* is not subject to control by covalent modification of its dinitrogenase reductase. The left lane shows the separation of two components from *A. brasilense* that had been treated with NH$_4^+$; this is an organism that undergoes switch-off by ADP-ribosylation of its dinitrogenase reductase. The normal subunit and the modified subunit from its dinitrogenase reductase are clearly separated. In contrast, only one band appears in extracts from *A. diazo-*

*trophicus.* Lane 2 is before and lane 3 after a 30 min treatment with 10 m$M$ NH$_4$Cl. Cells for lane 3 were held under 6 kPa O$_2$ for 30 min and those for lane 5 were anaeorobic for 30 min. Under no conditions was the dinitrogenase reductase altered by covalent modification as it is in *A. brasilense.* Figure 5 supports these conclusions and indicates that *A. diazotrophicus* lacks draT and draG genes.

Lane    1    2    3    4    5

*Fig. 4.* Immunoblots of crude extracts from *Acetobacter diazotrophicus* PAL-5 separated by SDS-PAGE with antiserum against dinitrogenase reductase of *Azotobacter vinelandii.* Lane 1, crude extract of *Azospirillum brasilense* treated with 5 m$M$ NH$_4$Cl for 30 min. Lanes 2, 3, 4, 5 from crude extracts of *A. diazotrophicus* PAL-5 before (lane 2) and after 30 min treatment with 10 m$M$ NH$_4$Cl (lane 3). Lane 4, 30 min at 6.0 kPa O$_2$; and lane 5, anaerobic for 30 min. The crude extracts were prepared by sonication (lanes 1 and 2) or by quick extraction (lanes 3, 4, 5).

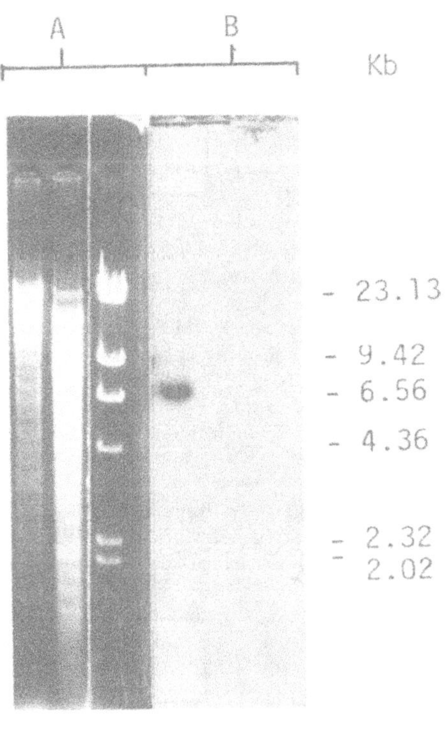


*Fig. 5.* Southern hybridization of genomic DNA of *Acetobacter diazotrophicus* PAL-5 with a *draTG* probe from *Azospirillum lipoferum.* Total DNAs from *A. lipoferum* (lane 1) and *A. diazotrophicus* (lane 2) were digested with EcoRI and hybridized to a $^{32}$P-labeled *draTG* probe from *A. lipoferum.* (A)Ethidium bromide-stained DNA on the agarose gel. (B) autoradiogram. Bacteriophage λ-DNA cut with Hind III was used as a size marker and control (lane 3).

## References


Becking J H 1963 Fixation of molecular nitrogen by an aerobic *Vibrio* or *Spirillum.* Antonie van Leeuwenhoeck 29, 326.

Beijerinck M W 1925 Über ein *Spirillum* welches freien Stickstoff binden kann? Zenthl. Bakt. Parasitkde. 63, 353–359.

Boddey R M and Döbereiner J 1988 Nitrogen fixation associated with grasses and cereals: Recent results and perspectives for future research. Plant and Soil 108, 53–65.

Burris R H and Wilson P W 1946 Ammonia as an intermediate in nitrogen fixation by *Azotobacter.* J. Bacteriol. 52, 505–512.

Cejudo F J, de la Torre A and Paneque A 1984 Short-term ammonium inhibition of nitrogen fixation in *Azotobacter.* Biochem. Biophys. Res. Commun. 123, 431–437.

Döbereiner J 1989 Isolation and identification of root-associated diazotrophs. *In* Nitrogen Fixation with Non-Legumes. Eds. F A Skinner et al. pp 103–108. Kluwer Academic Publishers, Dordrecht, The Netherlands.

Döbereiner J and Day J M 1976 Associative symbioses in tropical grasses: Characterization of microorganisms and dinitrogen-fixing sites. *In* Proc. 1st Internat. Symp. on Nitrogen Fixation. Eds. W E Newton and C J Nyman. pp 518–538. Washington State University Press, Pullman, WA.

Fu H and Burris R H 1989 Ammonium inhibition of nitrogenase activity in *Herbaspirillum seropedicae.* J. Bacteriol. 171, 3168–3175.

Fu H, Burris R H and Roberts G P 1990 Reversible ADP-ribosylation is demonstrated to be a regulatory mechanism in prokaryotes by heterologous expression. Proc. Natl. Acad. Sci. USA 87, 1720–1724.

Fu H, Fitzmaurice W P, Roberts G P and Burris R H 1990 Cloning and expression of *draTG* genes from *Azospirillum lipoferum.* Gene 86, 95–98.

Fu H, Hartmann A, Lowery R G, Fitzmaurice W P, Roberts G P and Burris R H 1989 Posttranslational regulatory system for nitrogenase activity in *Azospirillum* spp. J. Bacteriol. 171, 4679–4685.

Fu H-A, Wirt H J, Burris R H and Roberts G P 1989 Functional expression of a *Rhodospirillum rubrum* gene encoding dinitrogenase reductase ADP-ribosyltransferase in enteric bacteria. Gene 85, 153–160.

Gest H and Kamen M D 1949 Photoproduction of molecular hydrogen by *Rhodospirillum rubrum*. Science 109, 558–559.

Hartmann A and Burris R H 1987 Regulation of nitrogenase activity by oxygen in *Azospirillum brasilense* and *Azospirillum lipoferum*. J. Bacteriol. 169, 944–948.

Hartmann A, Fu H and Burris R H 1986 Regulation of nitrogenase activity by ammonium chloride in *Azospirillum* spp. J. Bacteriol. 165, 864–870.

Hartmann A, Fu H and Burris R H 1988 Influence of amino acids on nitrogen fixation ability and growth of *Azospirillum* spp. Appl. Environ. Microbiol. 54, 87–93.

Jouanneau Y, Meyer C and Vignais P M 1988 Regulation of nitrogenase activity in *Rhodobacter capsulatus*: ADP ribosylation of the Fe protein. *In* Nitrogen Fixation: Hundred Years After. Eds. H Bothe, F J deBruijn and W E Newton. p. 173. Gustav Fischer, Stuttgart.

Kamen M D and Gest H 1949 Evidence for a nitrogenase system in the photosynthetic bacterium *Rhodospirillum rubrum*. Science 109, 560.

Ludden P W 1980 Nitrogen fixation by photosynthetic bacteria: Properties and regulation of the enzyme system from *Rhodospirillum rubrum*. *In* Nitrogen Fixation. Eds. W E Newton and W H Orme-Johnson. pp 139–156. University Press, Baltimore, MD.

Ludden P W and Burris R H 1976 Activating factor for the iron protein of nitrogenase from *Rhodospirillum rubrum*. Science 194, 424–426.

Ludden P W and Burris R H 1978 Purification and properties of nitrogenase from *Rhodospirillum rubrum*, and evidence for phosphate, ribose and an adenine-like unit covalently bound to the iron protein. Biochem. J. 175, 251–259.

Ludden P W and Burris R H 1979 Removal of an adenine-like molecule during activation of dinitrogenase reductase from *Rhodospirillum rubrum*. Proc. Natl. Acad. Sci. USA 76, 6201–6205.

Ludden P W, Hageman R V, Orme-Johnson W H and Burris R H 1982 Properties and activities of 'inactive' Fe protein from *Rhodospirillum rubrum*. Biochim. Biophys. Acta 700, 213–216.

Ludden P W, Okon Y and Burris R H 1978 The nitrogenase system of *Spirillum lipoferum*. Biochem. J. 173, 1001–1003.

Ludden P W and Roberts G P 1989 Regulation of nitrogenase activity by reversible ADP ribosylation. *In* Current Topics in Cellular Regulation 30, 23–56. Academic Press, New York.

Nordlund S and Eriksson U 1979 Nitrogenase from *Rhodospirillum rubrum*. Relation between 'switch-off' effect and the membrane component. Hydrogen production and acetylene reduction with different nitrogenase component ratios. Biochim. Biophys. Acta 547, 429–437.

Nordlund S, Eriksson U and Baltscheffsky H 1977 Necessity of a membrane component for nitrogenase activity in *Rhodospirillum rubrum*. Biochim. Biophys. Acta 462, 187–195.

Okon Y, Albrecht S L and Burris R H 1976 Factors affecting growth and nitrogen fixation of *Spirillum lipoferum*. J. Bacteriol. 127, 1248–1254.

Okon Y, Houchins J P, Albrecht S L and Burris R H 1977 Growth of *Spirillum lipoferum* at constant partial pressures of oxygen, and the properties of its nitrogenase in cell-free extracts. J. Gen. Microbiol. 98, 87–93.

Pope M R, Murrell S A and Ludden P W 1985a Purification and properties of the heat-released nucleotide-modifying group from the inactive iron protein of nitrogenase from *Rhodospirillum rubrum*. Biochemistry 24, 2374–2380.

Pope M R, Murrell S A and Ludden P W 1985b Covalent modification of the iron protein of nitrogenase from *Rhodospirillum rubrum* by adenosine diphosphoribosylation of a specific arginine residue. Proc. Natl. Acad. Sci. USA 82, 3173–3177.

Pope M R, Saari L L and Ludden P W 1986 N-glycohydrolysis of adenosine diphosphoribosyl arginine linkages by dinitrogenase reductase activating glycohydrolase (activating enzyme) from *Rhodospirillum rubrum*. J. Biol. Chem. 261, 10104–10111.

Schneider K C, Bradbeer C, Singh R N, Wang L C, Wilson P W and Burris R H 1960 Nitrogen fixation by cell-free preparations from microorganisms. Proc. Natl. Acad. Sci. USA 46, 726–733.

Song S-D, Hartmann A and Burris R H 1985 Purification and properties of the nitrogenase of *Azospirillum amazonense*. J. Bacteriol. 164, 1271–1277.

Sweet W J and Burris R H 1981 Inhibition of nitrogenase activity by $NH_4^+$ in *Rhodospirillum rubrum*. J. Bacteriol. 145, 824–831.

Vignais P M, Willison J C, Allibert P, Ahombo G and Jouanneau Y 1987 Regulation of nitrogen fixation in photosynthetic bacteria. *In* Inorganic Nitrogen Metabolism. Eds. Q Ullrich et al. pp 154–159. Springer-Verlag, Berlin.

Wilson P W, Hull J F and Burris R H 1943 Competition between free and combined nitrogen in nutrition of *Azotobacter*. Proc. Natl. Acad. Sci. USA 29, 289–294.

# Biology of nitrogen-fixing Rhizobacteria

J. BOONJAWAT, P. CHAISIRI, J. LIMPANANONT, S. SOONTAROS, P. PONGSAWASDI, S. CHAOPONGPANG, S. PORNPATTKUL, B. WONGWAITAYAKUL and L. SANGDUAN
*Department of Biochemistry, Faculty of Science, Chulalongkorn University, Bangkok 10330, Thailand*

*Key words:* adhesion, glutamine synthetase, nitrogen fixation, rice lectin, rhizobacteria

## Abstract

In non-legumes associative nitrogen-fixing system, several genera of rhizobacteria have been reported. The object of this paper is to summarize the current understanding of how rhizobacteria adhere to the root surface of non-legumes especially rice and other cereal crops. Evidence for involvement of rice lectin in adhesion will be reviewed. An emphasis will be placed on the *Klebsiella* R15 ammonium assimilation system in free-living state and in associative state with rice seedlings. Nitrogenase and glutamine synthetase (GS) activities of associative *Klebsiella* increased significantly in the rhizosphere of rice comparing to the free-living state. In rice, the soluble form of GS specific activity appear to be slightly lower than in rice root in the absence of bacteria. These results suggest that nitrogen-fixing activity has been enhanced during association. The dinitrogen fixed should be changed to amino acids via GS-GOGAT pathway in bacteria. Transfer of fixed nitrogen and assimilation in the rice plant is the problem that needs to be solved in order to improve the efficiency of associative nitrogen fixation.

## Introduction

In non-legumes associative nitrogen-fixing system, diverse genera of diazotrophs can associate to the rhizoplane, or invade damaged cells and sometimes found distributed in the plant cells as endophyte without special form of compartmentation. *Azospirillum* spp., *Klebsiella* spp. and *Enterobacter* spp. have been reported as ubiquitous associative rhizobacteria in diversed habitats of grasses and cereal crops in both tropical and temperate regions (Table 1). Since rice is the most important food crop of the developing world, the object of this paper is to present recent advances in the biology of nitrogen-fixing rhizobacteria isolated from the rhizosphere of rice grown in wetland rice ecosystems.

### Adhesion of rhizobacteria to rice roots

Adhesion of *Klebsiella* spp. strain R15 and R17 on the root of rice cv. RD7 seedlings grown in sterile water have been demonstrated by fluorescent microscopy after labelling half of the bacterial inoculum ($10^8$ cells) with acridine orange (1:10,000), and by scanning electron microscopy as shown in Figure 1. Adhesion of bacteria on the root hairs and epidermal cells were observed after 2 h of inoculation, whereas *E. coli* were washed out and not detectable. After 36 h, enveloped structures entrapping bacteria were observed which were confirmed by transmission electron micrographs (Fig. 2) showing details of an electron dense layer surrounding the bacteria. These micronodule-like structures were sensitive to N-acetyl glucosaminidase, neuraminidase, glucosidase and trypsin, suggesting an involvement of exopolysaccharides (EPS) and glycoproteins in the binding between bacteria–bacteria, and plant-bacteria associations. Murty and Ladha (1987) reported that *Azospirillum lipoferum* strain 34H did not colonize the root tips of rice IR42 but embedded in the mucigel, whereas rice IR50 with less mucigel were colonized heavily on root hair primodia and root tips within

*Table 1.* Diversity of associative nitrogen-fixing rhizobacteria in non-legumes

| Bacterial species | Associative plants | Remarks |
| --- | --- | --- |
| *Azospirillum* spp. | Gramineae | Döbereiner et al., 1988 |
|   *A. lipoferum* | rice, sorghum, maize | |
|   *A. brasilense,* Sp7 | sugar cane, grasses | |
|   *A. amazonense* | | |
|   *A. halopraeferan* | Kallar grass | |
| *Herbaspirillum seropedicae* | | |
| *Bacillus azotofixans* | Grasses | |
|   *B. polymyxa* | Grasses and cereals | |
| *Acetobacter nitrocaptans* | Sugar cane | |
| *Enterobacter* spp. | Gramineae | |
|   *E. agglomerans* | Wheat, barley, ryegrasses | |
|   *E. cloacae* | Wetland-rice | |
|   *E. aerogenes* | | |
| *Enterobacter* and *Klebsiella* | Grasses and cereals | Korhonen et al., 1989 |
| *Klebsiella* spp. | Rice | You et al., 1986 |
|   *K. planticola* | Wetland rice | |
|   *K. oxytoca* | Rhizosphere of rice | Uozumi et al., 1984 |
| *Pseudomonas* spp. | Wetland rice | Watanabe et al., 1987 |
| *Alcaligenes faecalis* | Rice root | You et al., 1988 |

24 h. It is noted that Rhizobium surface polysaccharides were also shown to play a role in the infection process (Puhler et al., 1988), and involved in the cell–cell interaction of *Rhizobium*-legume symbiosis.

## Involvement of rice lectin in plant-bacterial association

The relationship between the agglutinated form of rhizobacteria and acetylene reduction activity (ARA) were tested in various rice cultivars, using various strains of associative rhizobacteria and free-living *Klebsiella pneumoniae* M5al as inoculants. Only those associative strains NG13, R15, R17 formed both micronodule-like structures and ARA, which fluctuated among rice cultivars. IR42 and IR58 seemed to be deficient in associative factor compared to RD7, because no ARA was detected, therefore the root exudate were compared for protein content and haemagglutinating activity of plant lectin. Table 2 shows that root exudate of RD7 contains significantly higher protein and lectin activity. Lectins were therefore purified from different tissues of RD7, and characterized as glycoproteins of similar molecular weight about 23KD, which

bind specifically to N-acetylglucosamine. By using $^{14}$C-labelled embryo lectin (EL) and Scatchard plots, lectin receptors on the surface of R15 and R17 were found to be 1.3–1.4 $\times 10^{-12}$ $\mu$mol cell$^{-1}$, and demonstrated by colloidal gold conjugated with root lectin (RL-Au), to be on the cell wall and glycocalyx (Boonjawat et al., 1988). By the same technique, distribution of lectin receptors were exhibited on the surface of root epidermal cells, especially in the extracellular slime.

The lectin content of rice roots were determined using indirect ELISA in 4 day old rice seedlings germinated in the dark. Variation of root lectins were observed in 8 rice cultivars (Table 3). Lectin content also decreased during development (4–7 days) especially total lectin in lighted condition (Table 4). Table 5 shows that lectin content decreased in the presence of exogenous ammonium chloride (20 mM).

Examination of lectin on the surface of leaf and root using immunofluorescent labelling showed that lectin specifically localized on the opening stoma and hydathode of 7 day leaf and was not detectable in the flowering stage, but the root lectin distributed all over the root surface and accumulated on root hair tips and mucigel of root tips until the flowering stage. The associa-

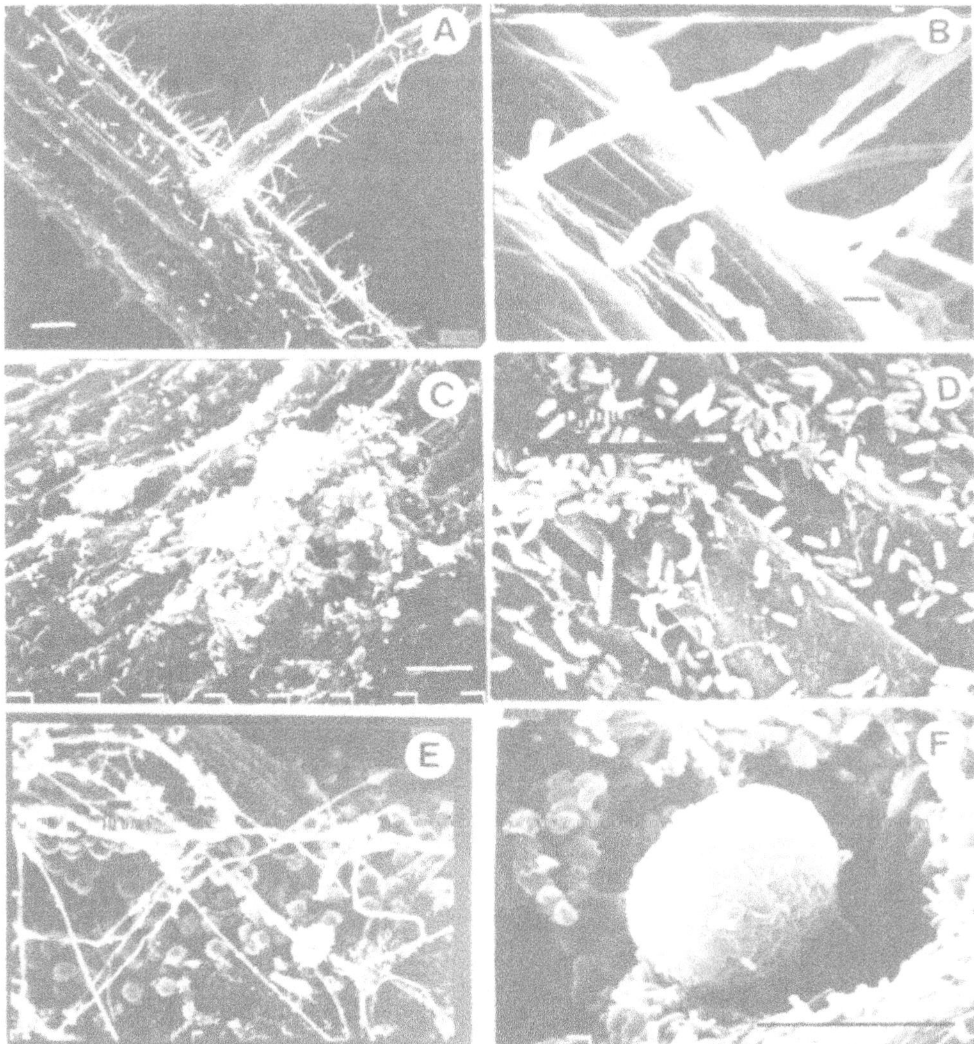

*Fig. 1.* Electron micrographs of washed rice root (RD7) inoculated with $10^8$ cells/100 $\mu$L PBS of *E. coli* (**A, B**), *Klebsiella* strain R15 (**C, D, E**) and R17 (**F**) at 36 h after inoculation. Bars equal to 100 $\mu$m in A and B; and 10 $\mu$m in C, D, E and F.

tion between secreted lectin in root exudate, root surface, lectin receptors and colonization pattern of *Klebsiella* spp. indicate the role of lectin together with bacterial and plant extracellular polysaccharides (EPS) in plant bacterial associations. Variation of lectin content with cultivar, developmental stage, and exogenous ammonium ion concentration indicate that both internal and external factors can affect plant-microbial interaction via lectin.

Many lectins that are produced in the root of leguminous plants are known as determinant of host-plant specificity in the *Rhizobium*-legume symbiosis. Dazzo et al. (1988) have shown that a lectin produced in clover root, trifoliin A can be detected in the root exudate, and bound specifically to the acidic capsular polysaccharide of *Rhizobium trifolii*. By introducing pea lectin gene into white clover roots using *Agrobacterium rhizogene* as a vector, the clover roots of transgenic plant become hairy and can be nodulated by a *Rhizobium* strain usually specific for pea (Diaz et al., 1989). The accumulation of lectin in peanut nodules was also found to be in relation to the symbiotic effectiveness of *Rhizobium* strains (Kishinevsky 1990). However lectins from rice, wheat and barley all have the same sugar specificity, and that might explain the non-

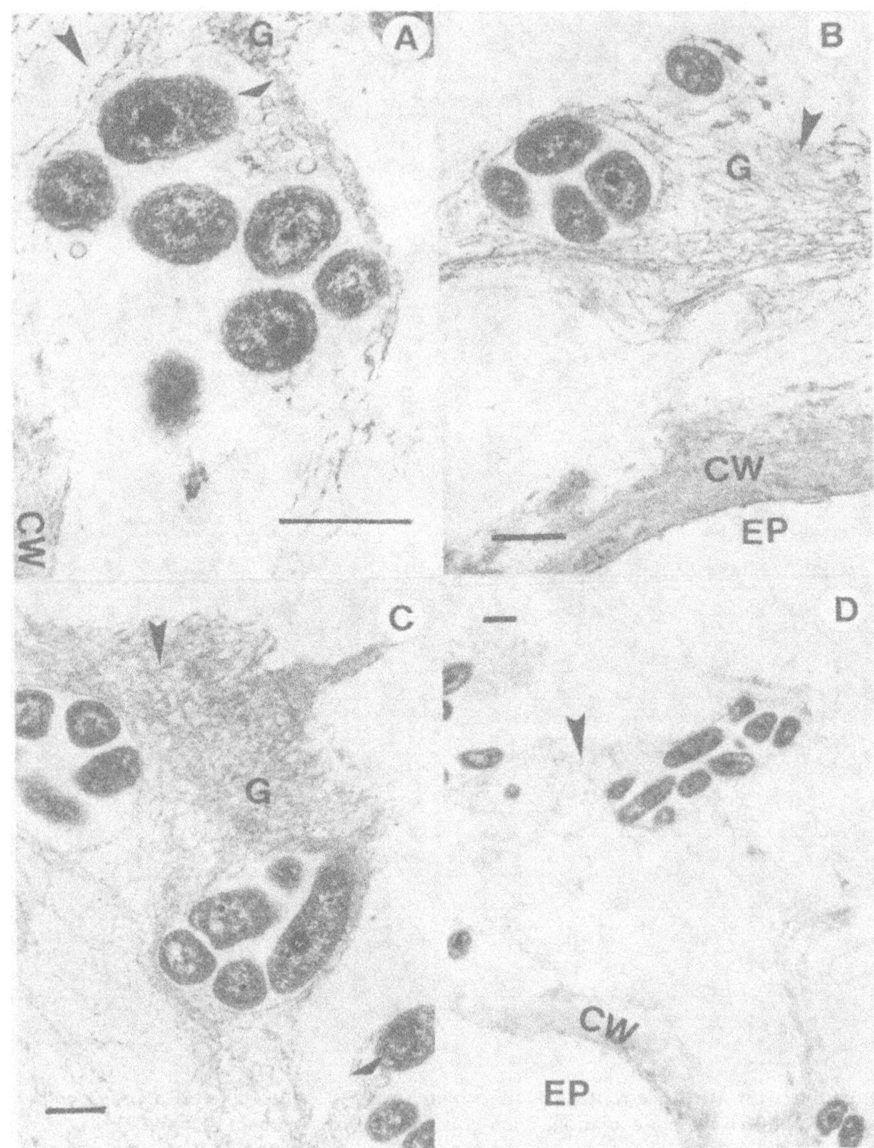

Fig. 2. Transmission electron micrographs of rice seedling roots in association with R17. Cross-section of micronodule showing adhesion site on the outer periphery of cell wall (CW). The bacteria are surrounded glycocalyx (6) which are composed of fibrous polysaccharide-like structure (long arrow head) and globular vescicle (short arrow head). Bars equal to 0.5 $\mu$m.

Table 2. Lectin in root exudate of rice. Root exudates were prepared from 7-day-old rice seedlings of rice RD7, IR42 and IR58

| Root exudate (105 plants) | Source of root exudate | | |
|---|---|---|---|
| | RD7 | IR42 | IR58 |
| Total dry matter (mg) | 107.2 | 3.0 | 5.3 |
| Per cent w/w protein | 43.0 | 80.4 | 28.0 |
| Lectin (HU/mg dry matter) | 64 | 8 | 8 |
| Total lectin (HU/105 plants) | 6,860 | 24 | 42 |

*Table 3.* Lectin content in root of 8 rice cultivars. All rice seeds were germinated in the dark for 4 days, and then roots were harvested and determined for lectin by indirect ELISA. Values are mean of 6 measurements

| Rice varieties | Lectin/Total protein (ng/mg protein) | Total lectin (ng/100 plants) |
| --- | --- | --- |
| NMS4 | 31 ± 7 | 92 ± 24 |
| RD7 | 210 ± 42 | 58 ± 6 |
| RD23 | 29 ± 2 | 34 ± 2 |
| SPBR60 | 13 ± 1 | 33 ± 3 |
| KTH17 | 8 ± 1 | 31 ± 2 |
| BMT470 | 19 ± 1 | 24 ± 1 |
| RD25 | 10 ± 1 | 17 ± 1 |
| KDML105 | 6 ± 0 | 5 ± 0 |
| Mean ± SD | 16 ± 9 | 36 ± 25 |

*Table 4.* Root lectin varies with developmental stage of rice seedling (cv. RD7)

| Time (day) | Lectin/Total Protein (ng/mg protein) | | Total lectin (ng/100 plants) | |
| --- | --- | --- | --- | --- |
| | Light | Dark | Light | Dark |
| 4 | 17 ± 1 | 15 ± 2 | 44 ± 1 | 58 ± 6 |
| 5 | 14 ± 1 | 19 ± 1 | 32 ± 2 | 38 ± 2 |
| 6 | 16 ± 1 | 11 ± 1 | 39 ± 2 | 15 ± 1 |
| 7 | 12 ± 1 | 12 ± 2 | 23 ± 2 | 28 ± 5 |

Values presented are mean of 6 measurements.

*Table 5.* Effect of ammonium chloride on root lectin. Rice seedlings (RD7) were grown in the dark in sterile water or supplemented with 2, 20 mM $NH_4Cl$ for 4 days, and root lectins were determined by indirect ELISA

| $NH_4Cl$ (mM) | Lectin/Total protein (ng/mg protein) | Total lectin (ng/100 plants) |
| --- | --- | --- |
| 0 | 31 ± 7[a] | 92 ± 20[a] |
| 2 | 24 ± 1[a] | 102 ± 4[a] |
| 20 | 13 ± 1[b] | 49 ± 2[b] |

Values presented are mean of 6 measurements significant different of lectin/total protein and total lectin between different letter (a, b) analysed by t-test at 95% confidence.

specific interaction of cereal crops and associative bacteria.

## Contribution of fixed nitrogen to plant nutrition

Substantial although very variable amounts of nitrogen fixed by associative rhizobacteria in paddy rice have been demonstrated with [15]N-dilution analysis in different cultivars of rice (Eskew et al., 1981). In other cereal crops, despite the high percentage of nitrogen derived from the atmosphere in some associations, significant nitrogen transfer was seen only in half of the associations and there was no correlation between nitrogen fixation and yield increases (Korhonen et al., 1988).

Plant growth promoting effects of associative *Azospirillum* spp. by increasing the root surface area and subsequently yield of associative crops have been reviewed by Okon (1988). Coinoculation of *Azospirillum* and *Rhizobium* can lead to increased legume forage and grain yield (Del Gallo and Fabbri, 1990). In rice inoculation of either *Klebsiella* strain R15, R17 or *Azospirillum* strain R25 always results in the net gain of plant vigor index, but the net gain in nitrogen fixation was detected only in some associations. In order to improve the efficiency of nitrogen transfer from associative rhizobacteria to rice plants, the biochemistry of regulatory enzymes in nitrogen metabolism have been studied.

## Ammonia assimilation in free-living and associative *Klebsiella* R15

Free-living *Klebsiella* sp. strain R15 contains glutamine synthetase (GS), which can be induced in nitrogen-free (NF) conditions or $N_2$-fixing cultures, and repressed in nitrogen-rich (NR) cultures containing 100 mM $NH_4Cl$ (Wongwaitayakul, 1988). Since the specific activity of glutamine dehydrogenase (GDH) was too low to be detected in both NF and NR conditions, whereas GOGAT activity was higher in $NH_4$ excess than in NF condition, indicating that GS-GOGAT are the key enzymes for ammonia assimilation in R15. Purified GS from R15 is an octamer (430 KD) of identical subunits (54 KD) as shown by molecular sieve chromatography on Sepharose 4B and denaturing gel electro phoresis. Also like symbiotic GS II of Rhizobium and Frankia CPI1, GS purified from free-living R15 in $N_2$-fixing condition is heat labile, but regulated by adenylation–deadenylation as classical GS (*glnA* gene) in other enteric bacteria. The activity of GS from *Klebsiella* R15 appears to be regulated by feedback inhibition mediated by a number of amino acids.

*Table 6.* Changes in the specific activity and total activity of glutamine synthetase (GS) in the free-living R15, and associative R15 comparing with free-living rice RD7 and rice RD7 in association with R15

| Source of GS | Day | Specific activity r GH/mg protein min | Total GS r GH/min | Ratio |
|---|---|---|---|---|
| *Bacterial fraction* | | | | |
| R15 | 0 | 0.28 | 8.00 | |
| R15 | 8 | 0.08 | 1.56 | |
| R15 + RD7 | 8 | 0.56 | 19.90 | +12.8 |
| *Root supernatant fraction (150 plants)* | | | | |
| RD7 | 8 | 1.00 | 77.28 | |
| RD7 + R15 | 8 | 0.84 | 57.00 | − 0.7 |
| *Root pelleted fraction (150 plants)* | | | | |
| RD7 | 8 | 0.19 | 2.61 | |
| RD7 + R15 | 8 | 0.31 | 5.60 | + 2.1 |

r GH: r Glutamyl hydroxamate

Comparative study of nitrogenase and GS activity in free-living R15 and in associative R15 after inoculation into aceptically grown rice seedings showed that associative nitrogen fixation, has increased from 2–3 nmol tube$^{-1}$ day$^{-1}$ in the absence in rice plant to >600 nmol tube$^{-1}$ day$^{-1}$ during 7 days of rice, RD7-R15 association. The specific activity and total activity of GS was determined on day 8 by transferase assay. Our results indicate an increase in GS specific activity, and total activity in the associative bacterial fraction. In the supernatant fraction of rice root, plants GS specificity slightly decreased, but the pellet fraction that might contain membrane bound GS tends to increase in associative condition (Table 6).

In the effective symbiotic nitrogen fixation system, the nitrogenase enzyme complex must be coupled with energy intensive processes of the host plant, and also protected from oxygen by leghemoglobin. The production of ammonia by nitrogenase in the symbiotic state is carried out by plant GS not via bacterial GS-GOGAT cycle as in free-living state. In rhizobia, three distinct GS enzymes and corresponding genes (*glnA*, *glnII*, *glnT*) have been identified, which are differentially regulated. In legumes, many nodule specific GS have been characterized which are induced by bacterial infection and responsible for ammonium assimilation. In the association between *Klebsiella*-rice the GS specific activities seem to be in the opposite direction, as the bacterial GS in associative form was higher than in free-living state, while the plant GS slightly decreased. The specific activity of GS in symbiont *Anabaena* with *Azolla caroliniana* is also lower than in free-living condition, and ammonia is assimilated by the fern GS (Orr and Haselkorn, 1982). The same phenomenon was reported in symbiotic *Nostoc-Anthoceros* (Joseph and Meeks, 1987). More understanding on the regulation of GS proteins and mRNA are required to trap the fixed nitrogen and increase the transfer efficiency in the *Klebsiella*-rice association.

## Conclusions

We have summarized here the basic biological approaches to study adhesion and interaction between rice and *Klebsiella* via lectin and EPS. We still have no firm conclusion on the ammonium assimilation in this associative system which appears to be inefficient, because the rice GS could not take over the bacterial GS for fixed-ammonium assimilation.

## References

Boonjawat J, Limpananont J and Horisburger M 1988 *In* Nitrogen Fixation: Hundred Years After. Eds. H Bothe, F J de Bruijn and W E Newton. p 792. Gustav Fischer, Stuttgart, New York.

Dazzo F B and Truchet G L 1983 Interactions of lectins and their saccharide receptors in the *Rhizobium*-legume symbiosis. J. Membrane Biol. 73, 1–16.

Del Gallo M and Fabbri P 1990 *In* Nitrogen Fixation: Achievements and Objectives. Eds. P M Gresshoff, L E Roth, G Stacey and W E Newton. p 654. Chapman and Hall, New York, London.

Diaz C L, Melcher L S, Hooykaas P J J, Lugtenberg B J J and Kijne J W 1989 Root lectin as determinant of host plant specificity in the *Rhizobium*-legume symbiosis. Nature 338, 579–581.

Döbereiner J, Reis V M and Lazarini A C 1988 *In* Nitrogen Fixation: Hundred Years After. Eds. H Bothe, F J de Bruijn and W E Newton. pp 717–722. Gustav Fischer, Stuttgart, New York.

Eskew D L, Eaglesham A R J and App A A 1981 Heterotrophic $^{15}N_2$ fixation and distribution of newly fixed nitrogen in a rice-flooded soil system. Plant Physiol. 68, 48–52.

Joseph C M and Meeks J C 1987 Regulation of expression of glutamine synthetase in a symbiotic *Nostoc* strain associated with *Anthoceros punctatus*. J. Bacteriol. 152, 626–635.

Kishinevsky B L, Nemas C, Friedman Y and Meromi G J 1990 *In* Nitrogen Fixation: Achievements and Objectives. Eds. P M Gresshoff, L E Roth, G Stacey and W E Newton. p 477. Chapman and Hall, New York, London.

Korhonen T K, Laakso T, Roenkkoe R, and Haahtela K 1989 *In* Recent Advances in Microbial Ecology. Eds. Hattori, Ishida, Maruyama, Morita and Uchida. pp 192–195. Japan Scientific Societies Press, Tokyo.

Murty M G and Ladha J K 1987 Differential colonization of *Azospirillum lipoferum* on roots of two varieties of rice (*Oryza sativa* L.). Biol. Fertil. Soils 4, 3–7.

Orr J and Haselkorn R 1982 Regulation of glutamine synthetase activity in free-living and symbiotic *Anabaena* spp. J. Bacteriol. 152, 626–635.

Okon Y, Sarig S and Blum A 1989 *In* Recent Advances in Microbial Ecology. Eds. Hattori, Ishida, Maruyama, Morita and Uchida. pp 196–200. Japan Scientific Societies Press, Tokyo.

Puhler A, Enenkel B, Hillemann A, Kapp D, Keller M, Muller P, Niehaus K, Priefer U B, Quandt J and Schmidt C 1988 *In* Nitrogen Fixation: Hundred Years After. Eds. H Bothe, F J de Bruijn and W E Newton. p 423. Gustav Fischer, Stuttgart, New York.

Watanabe I, So R, Ladha J K, Katayama-Fujimura Y and Kuraishi H 1987 A new nitrogen-fixing species of pseudomonad: *Pseudomonas diazotrophicus* sp. nov. Isolated from the Root of Wetland Rice. Can. J. Microbiol. 33, 670–678.

Wongwaitayakul B 1988 Purification and Characterization of Glutamine Synthetase in *Klebsiella* spp. R15. Thesis, Department of Biochemistry, Chulalongkorn University, Bangkok.

You I D, Fujii T, Sano Y, Komagata K, Yoneyama T, Iyama S and Hirota Y 1986 Dinitrogen fixation of rice-*Klebsiella* associations. Crop Science 26, 297–301.

You C B, Zhou F Y, Zhang D D, Wang H X and Yuan H L 1988 *In* Nitrogen Fixation: Hundred Years After. Eds. H Bothe, F J de Bruijn and W E Newton. p 802. Gustav Fischer, Stuttgart, New York.

# Biological nitrogen fixation associated with sugar cane

R.M. BODDEY, S. URQUIAGA, V. REIS and J. DÖBEREINER
*EMBRAPA, National Centre for Soil Biology Research (CNPBS), Km 47, Seropédica, 23851, Rio de Janeiro, Brazil*

*Key words:* Acetobacter diazotrophicus, Glomus clarum, N balance, $N_2$ fixation, [15]N isotope dilution, sugar cane, VA mycorrhiza

## Abstract

A recent [15]N dilution/N balance study confirmed that certain sugar cane varieties are capable of obtaining large contributions of nitrogen from plant-associated $N_2$ fixation. It was estimated that up to 60 to 80% of plant N could be derived from this source, and under good conditions of water and mineral nutrient supply, it may be possible to dispense with N fertilization of these varieties altogether. The recently discovered bacterium, *Acetobacter diazotrophicus*, apparently responsible for this $N_2$ fixation associated with the plants, has unique physiological properties for a diazotroph, such as tolerance to low pH, and high sugar and salt concentrations, lack of nitrate reductase, and nitrogenase activity which tolerates short-term exposure to ammonium. Furthermore, it also behaves as an endophyte, in that it is unable to infect sugar cane plants unless through damaged tissue or by means of VA mycorrhizae and is propagated via the planting material (stem pieces).

## Introduction

Four million motor vehicles in Brazil run on hydrated (95%) ethanol and all other cars run on gasohol containing 10 to 20% ethanol. At present this represents the largest biomass fuel program in the World, and this alcohol is produced entirely from sugar cane grown on approximately 4 million hectares, 8% of the total cropped area of the country. There are several advantages of the substitution of gasoline by ethanol: Firstly it reduces the dependence of Brazil (a country with limited petroleum reserves) on imported oil. Secondly it is less polluting than gasoline in terms of nitrogen oxides and carbon monoxide, contains no hexamethyl lead, and, most important for the rest of the planet, the growth of the crop assimilates as much (or more) carbon dioxide than is released during alcohol production and consumption. The main disadvantage of the program is that the overall cost of the ethanol produced is more expensive than gasoline (albeit

in national currency not in dollars for imported oil) until the price of a barrel of oil reaches approximately US$40. The program was planned in the early 1970s when this price seemed to be an imminent possibility.

The technology adopted for sugar cane production in Brazil is most favourable for a high positive energy balance. Most of the crop is cut by hand, and employs over half a million workers, thus economizing fuel for harvesters, and the fertilizer inputs, especially nitrogen, are modest. Breeding of sugar cane has been traditionally carried out under low nitrogen inputs and responses of these varieties to N fertilizer are small. In 135 NPK experiments carried out all over Brazil on the plant crop, only 19% of the studies showed significant yield increases (Azeredo et al., 1986). Average cane yield is between 65 to 70 t ha$^{-1}$, and N fertilizer inputs are rarely more than 60 and 120 kg N ha$^{-1}$ for the plant crop and ratoon crops, respectively. The crop typically accumulates between 100 and

200 kg N ha$^{-1}$ per· season (Orlando Filho et al., 1981; Sampaio et al., 1984) and virtually all of this nitrogen is removed from the field at harvest as the trash is almost always burned off before cutting. There are many areas of the country where sugar cane has been grown for decades or even centuries, and neither cane yields nor soil N reserves appear to fall with time despite this apparent deficit in N supply. These observations have led several researchers to suggest that this crop may benefit from plant-associated biological nitrogen fixation (BNF).

### Quantification of biological nitrogen fixation

Experiments using $^{15}$N-labelled N$_2$ gas performed at the Centro de Energia Nuclear na Agricultura (Piracicaba, São Paulo), showed that sugar cane incorporated some N from this source, but because of the difficulties of exposing plants grown in the field to controlled atmospheres, the agronomic significance of these N inputs could not be evaluated (Matsui et al., 1981; Ruschel et al., 1975). In a $^{15}$N-aided N balance experiment performed in pots containing 64 kg of soil, Lima et al. (1987) showed that the sugar cane variety CB 47-89 was able to obtain a large contribution of biologically fixed nitrogen, which they estimated to be in excess of 60% of the total N incorporated.

More recently our group at EMBRAPA-CNPBS have completed a three year $^{15}$N isotope dilution and N balance study on 10 sugar cane varieties grown in a concrete tank ($20 \times 6 \times 0.8$ m) filled with soil amended with $^{15}$N-labelled organic matter, and using *Brachiaria arrecta* as a non-N$_2$-fixing control plant (Urquiaga et al., 1991). The soil was fertilized with phosphorus, potassium and micronutrients and well irrigated throughout the experiment but no type of N fertilizer was added. In the first year yields of fresh cane of the commercial varieties were high, ranging from the equivalent of 175 to 230 t ha$^{-1}$, and in the varieties CB 45-3 and SP 70-1143 these high yields were maintained during the subsequent two ratoon crops. In these same varieties and the *Saccharum spontaneum* variety, Krakatau, the nitrogen accumulation also continued high and stable over the three years (Table 1).

However, other varieties (e.g. CB 47-89, NA 56-79, SP 71-799, Chunee) showed a decline in total N content after the first year as would be expected from the observed decline in the availability of soil N.

Over the whole three years the weighted mean $^{15}$N enrichments of all of the sugar cane varieties were much lower than that of the *B. arrecta* control, indicating large contributions of plant associated BNF (Table 2). At the second and third annual harvests (first and second ratoon

*Table 1.* Total N accumulation in aerial and root tissue of sugar cane and *Brachiaria arrecta* at three annual harvests (g N m$^{-2}$). Means of 4 replicates. After Urquiaga et al. (1991)

| Variety/ Species | Harvest | | | Total N all harvests + roots. |
|---|---|---|---|---|
| | 1987 | 1988 | 1989 (+ roots) | |
| CB 47-89 | 26.5ab[1] | 20.9bc | 14.0bc | 61.4bc |
| CB 45-3 | 27.6ab | 27.2b | 29.5a | 84.3ab |
| NA 56-79 | 24.6ab | 18.9bc | 14.4bc | 57.8c |
| IAC 52-150 | 27.1ab | 17.3bc | 15.2b | 59.6bc |
| P 70-1143 | 24.5abc | 21.6bc | 31.4a | 77.5bc |
| SP 71-799 | 24.4abc | 20.0bc | 12.5bcd | 56.9c |
| SP 79-2312 | 24.0abc | 27.7b | 11.8bcd | 63.6c |
| Chunee | 15.3c | 11.9cd | 5.8cde | 33.0d |
| Caiana | 5.6d | 2.8d | 4.3de | 12.8d |
| Krakatau | 29.4a | 45.2a | 28.2a | 102.8a |
| *B. arrecta* | 18.9bc | 4.3d | 1.7e | 24.9d |
| CV (%) | 28.9 | 36.3 | 36.9 | 25.0 |

[1] Means in the same column followed by the same letter are not significantly different at $p = 0.05$ (Duncan's multiple range test).

*Table 2.* $^{15}$N enrichment of aerial and root tissue of sugar cane and *Brachiaria arrecta* at three annual harvests (atom% $^{15}$N excess). Means of 4 replicates. After Urquiaga et al. (1991)

| Variety/ Species | Harvest | | | | Weighted mean all harvests + roots. |
| --- | --- | --- | --- | --- | --- |
| | 1987 | 1988 | 1989 | Roots | |
| CB 47-89 | 0.316bc[1] | 0.108de | 0.079 | 0.094 | 0.191bcd |
| CB 45-3 | 0.301bcd | 0.120bcd | 0.079 | 0.104 | 0.166cde |
| NA 56-79 | 0.310bc | 0.126abc | 0.081 | 0.132 | 0.198bc |
| IAC 52-150 | 0.292bcd | 0.115cd | 0.082 | 0.096 | 0.188bcd |
| Sp 70-1143 | 0.258cd | 0.109d | 0.079 | 0.086 | 0.146de |
| SP 71-799 | 0.292bcd | 0.109d | 0.076 | 0.104 | 0.183bcd |
| SP 79-2312 | 0.330b | 0.127abc | 0.085 | 0.090 | 0.198bc |
| Chunee | 0.341b | 0.128ab | 0.078 | 0.099 | 0.227b |
| Caiana | 0.310bc | 0.123abc | 0.079 | 0.097 | 0.190bcd |
| Krakatau | 0.240d | 0.097e | 0.076 | 0.090 | 0.133e |
| *B. arrecta* | 0.546a | 0.134a | 0.077 | 0.078 | 0.443a |
| CV (%) | 12.6*** | 6.3*** | 4.9ns | 24.0ns | 13.6*** |

[1] Means in the same column followed by the same letter are not significantly different at $p = 0.05$ (Duncan's multiple range test).
ns no significant differences between means at $p = 0.05$.
*** differences between means significant at $p < 0.001$.

crops) there were only small difference in the $^{15}$N enrichments between the different varieties and that of the control crop, which was due to the carry-over of labelled nitrogen from one harvest to the next in the stem bases and roots of cane varieties, which did not occur in the case of the *B. arrecta*. The interpretation of the $^{15}$N data was further complicated by the fact that the uptake of soil N by the *B arrecta* was almost certainly inhibited towards the end of each growing season due to shading of this crop by the tall sugar cane plants, and this probably resulted in a

*Table 3.* Total nitrogen accumulation of sugar cane and *Brachiaria arrecta* and estimates of nitrogen derived from BNF using N balance and $^{15}$N isotope dilution techniques (g N m$^{-1}$). means of 4 replicates. After Urquiaga et al. (1991)

| Variety/ Species | Final N content of soil | N accum. whole plant 3 years | Estimates of BNF contribution | | | |
| --- | --- | --- | --- | --- | --- | --- |
| | | | All three years | | Annual mean | |
| | | | N balance[1] | $^{15}$N[2] | N balance | $^{15}$N |
| CB 47-89 | 835 | 61.4bc | 39.7 | 34.8c | 13.2 | 11.6 |
| CB 45-3 | 864 | 84.3ab | 62.6 | 52.6b | 20.9 | 17.5 |
| NA 56-79 | 884 | 57.8c | 36.1 | 32.6c | 12.0 | 10.9 |
| IAC 52-150 | 924 | 59.6bc | 37.9 | 33.8c | 12.6 | 11.3 |
| SP 70-1143 | 852 | 77.5bc | 55.8 | 51.9b | 18.6 | 17.3 |
| SP 71-799 | 860 | 56.9c | 35.2 | 33.3c | 11.7 | 11.1 |
| SP 79-2312 | 845 | 63.6c | 41.9 | 35.4c | 14.0 | 11.8 |
| Chunee | 826 | 33.0d | 11.3 | 16.9d | 3.8 | 5.6 |
| Caiana | 837 | 11.6d | −10.1 | 6.7d | − 3.4 | 2.2 |
| Krakatau | 857 | 102.8a | 81.1 | 71.8a | 27.0 | 23.9 |
| *B. arrecta* | 830 | 24.9d | 3.2 | — | 1.1 | — |
| CV (%) | 5.1ns | 25.0*** | — | 29.2*** | — | 29.2 |

[1] N balance estimate of BNF contribution = total N accumulated by crop + mean total N content of soil in tank at emergence − mean total N content of soil in tank at final harvest. Mean change in soil N content from emergence until final harvest = 27.1 g N m$^{-2}$ with a standard error of the difference between the means of 22.0 g N m$^{-2}$. N balances greater than 37.3 g N m$^{-2}$ significantly greater than zero ($p = 0.05$, Student t test).
[2] $^{15}$N isotope dilution estimate of BNF contribution = (total N accumulated by the crop) × (1 − (weighted mean atom % $^{15}$N excess of sugar cane)/(weighted mean atom % $^{15}$N excess of *B. arrecta*).

somewhat higher·[15]N enrichment in the control crop than otherwise would have occurred. These difficulties are fully discussed in the original paper (Boddey et al., 1991), and because of them it was decided to perform a total N balance on the whole tank by the careful analysis of the N content of soil samples taken at plant emergence in the first year in comparison with samples taken at the final harvest. These data show that there were significantly ($p < 0.05$) positive N balances associated with the varieties CB 45-3, SP 70-1143, SP 79-2312 and Krakatau, and that there was a good agreement between the [15]N dilution and the total N balance estimates of the contributions of BNF to the sugar cane varieties (Table 3).

**Nitrogen fixing bacteria associated with sugar cane**

*Diazotrophs isolated from sugar cane*

In the 1950s Döbereiner (1959, 1961) found $N_2$-fixing bacteria of the genus Beijerinckia in high numbers in sugar cane fields, with selective enrichment in the rhizosphere and especially on the root surface. At the same time a new species of Beijerinckia was discovered (*B. fluminense*) associated with this crop (Döbereiner and Ruschel, 1958). Subsequently, other authors (Purchase, 1980; Graciolli and Ruschel, 1981; Graciolli et al., 1983) isolated a wide range of $N_2$-fixing bacteria from the roots, stems and even leaves of sugar cane including species of *Erwinia*, *Azotobacter*, *Derxia*, *Azospirillum* and *Enterobacter*. The species *Azospirillum amazonense*, the only member of the genus which utilizes sucrose, was isolated from sugar cane in Hawaii (Döbereiner, 1987, unpublished data). None of these bacteria seemed to occur in large enough numbers to account for the extremely high rates of $N_2$ fixation reported above.

Acetobacter diazotrophicus
More recently, a new species of $N_2$-fixing bacteria, *Acetobacter diazotrophicus*, was found to occur in large numbers in the roots and stems of sugar cane (Cavalcante and Döbereiner, 1988; Gillis et al., 1989). This most extraordinary diazotroph was originally isolated from semi-solid sugar cane juice inoculated with dilutions of sugar cane roots and stems which showed acetylene reduction (nitrogenase) activity in dilutions up to $10^{-6}$ to $10^{-7}$. Improved counting and isolation procedures were developed using N-free mineral medium containing 10% cane sugar and 0.5% cane juice acidified with acetic acid to pH 5.5.

The bacterium is a small, Gram-negative, aerobic rod showing pellicle formation (microaerobic aerotaxis) and acetylene reduction activity (ARA) also in N-free semi-solid medium with 10% sucrose but without cane juice, forming a thick surface pellicle after 5 days. Best growth occurs with high sucrose or glucose concentrations (10%) and strong acid production results in a final pH of 3.0 or less. Growth and $N_2$ fixation (more than 100 n moles $C_2H_2$. $ml^{-1} h^{-1}$) continues at this pH for several days (Fig. 1, Stephan et al., 1988). Ethanol is also used as a C source for growth and is oxidized to $CO_2$ and $H_2O$. Dark brown colonies form on potato agar with 10% sucrose, and dark orange colonies on N-poor (0.005% yeast extract) mineral agar medium with 10% sucrose and bromothymol blue. The bacterium possesses no nitrate reductase and $N_2$ fixation is not affected by high levels ($25\,mM$) of $NO_3$. Also $NH_4^+$ causes only partial inhibition of nitrogenase (Fu et al., 1988; Teixeira et al., 1987). This type of $NH_4^+$ regulation of nitrogenase, which was also observed in *Azospirillum amazonense* and *Herbaspirillum seropedicae*, suggests that there are at least two general categories of $NH_4^+$ inhibition of nitrogenase: a. via direct covalent modification of the Fe protein, and b. with no direct effect on the Fe protein where $NH_4^+$ may affect the electron donation pathway to nitrogenase (Fu and Burris, 1989).

Recent results of Reis et al. (1990) confirm that the nitrogenase activity (ARA) of *A. diazotrophicus* growing in 10% sucrose with a final pH of 2.3, is less inhibited by $5\,mM$ $NH_4Cl$ than when growing in medium with 1% sucrose (Fig. 2). Under these same conditions the $^{15}NH_4^+$ assimilation by whole cells was slower with 10% sucrose than in cells growing on 1% sucrose (Table 4). This could be interpreted as osmotolerance because cells growing on 10% sucrose were more tolerant to 1% NaCl in semi-solid medium (V. Reis, unpublished data).

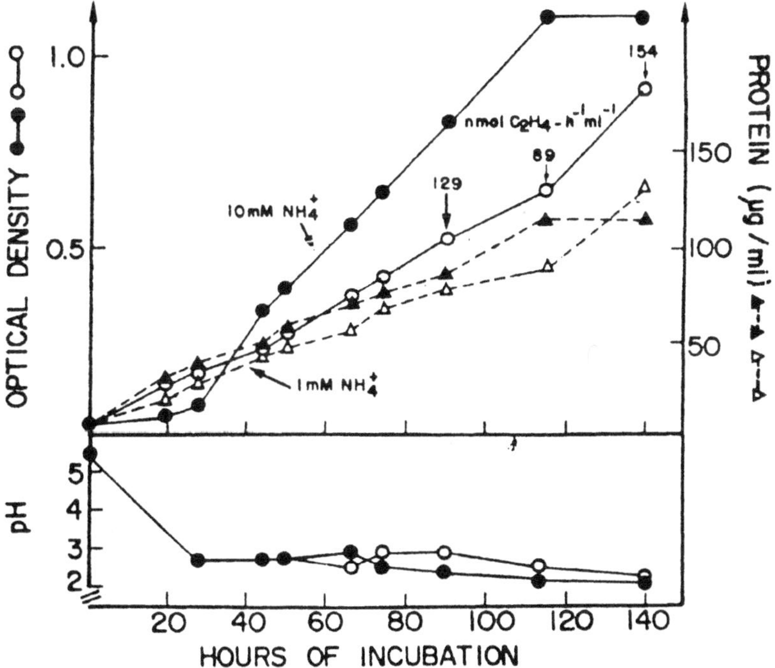



*Fig. 1.* Acid production and growth of *Acetobacter diazotrophicus* at two $NH_4^-$ concentrations in agitated liquid medium. With 1 m$M$ $NH_4^-$ the cultures continued growth after 24 h using $N_2$ as sole N source (no $NH_4^-$ was detected in the medium and $C_2H_2$ reduction activity was high. (after Teixeira et al., 1987)

Another interesting aspect is that *A. diazotrophicus* growing in 10% sucrose showed an optimum dissolved oxygen concentration for acetylene reduction in equilibrium with 0.2 kPa $O_2$ in the atmosphere, but continued to fix $N_2$ up to 4.0 kPa, showing a much higher $O_2$ tolerance than *Azospirillum* spp. (Reis et al., 1990).

The incomplete inhibition of $N_2$ fixation by $NH_4^+$ in these organisms, as well as the lack of nitrate reductase mentioned above, are of considerable ecological and agronomic importance because they may permit the complementation of plant-associated BNF with N fertilization.

Taxonomic studies based on DNA and rRNA

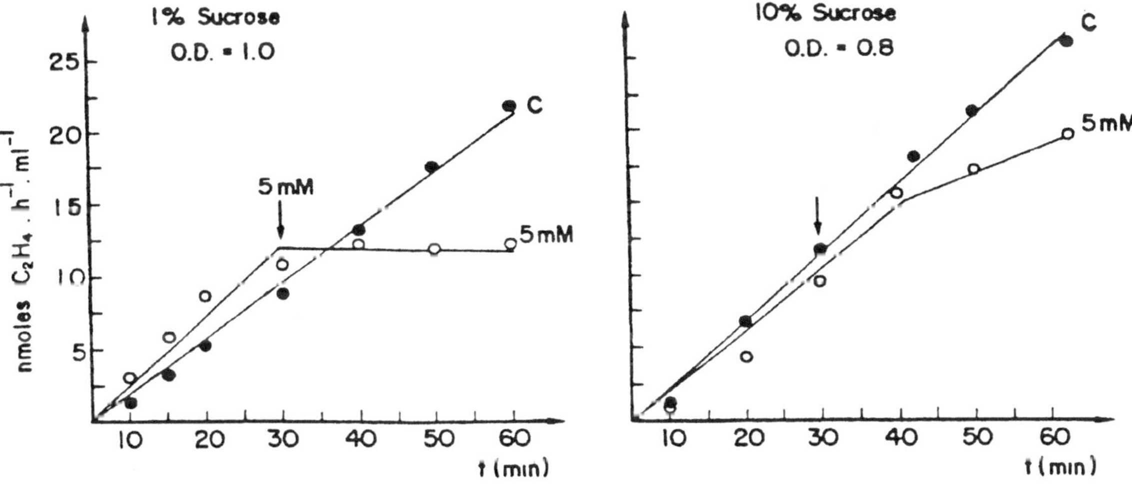



*Fig. 2.* Effect of 5 m$M$ $NH_4^+$ on the nitrogenase activity ($C_2H_2$ reduction) of cells of *Acetobacter diazotrophicus* (PAL-5) growing in LGI medium containing 1% or 10% sucrose.

Table 4. Incorporation of $^{15}N$ during 30 minute exposure to $(^{15}NH_4)SO_4$ (21.3 atom % $^{15}N$) by cells of *Acetobacter diazotrophicus* (PAL-5) grown for 48 h in LGI medium

| $(^{15}NH_4)SO_4$ | (Atom % $^{15}N$ excess) | |
|---|---|---|
| | 1% sucrose O.D. = 0.8 | 10% sucrose O.D. = 1.0 |
| 2 m$M$ | 0.760 | 0.265 |
| 5 m$M$ | 0.660 | 0.284 |
| 10 m$M$ | 0.667 | 0.289 |

analyses showed that the bacterium belongs to the *Acetobacter* rRNA cystron (Gillis et al., 1989) and is most closely related to *A. liquefaciens*. This latter species, however, does not fix $N_2$, does not form pigmented colonies on potato media and shows several other physiological differences. DNA/DNA binding experiments confirmed it to be a new species (Gillis et al., 1989), so that the name originally proposed, (*Saccarobacter nitrocaptans*, Cavalcante and Döbereiner, 1988) had to be changed to *Acetobacter diazotrophicus*.

The bacterium has been found in many sugar cane varieties in several regions of Brazil, and numbers were in the range of $10^3$ to $10^5$ in rhizosphere soil, $10^3$ to $10^7$ in washed roots, $10^3$ to $10^5$ in surface sterilized roots, $10^3$ to $10^6$ in basal and apical stems and $10^4$ to $10^7$ in sugar cane trash (Döbereiner et al., 1988). It was not found in soil between rows of sugar cane plants or roots from 12 different weed species taken from cane fields. It was also not found in grain or sugar sorghum, but was isolated from a few samples of washed roots and aerial parts of *Pennisetum purpureum* cv Cameroon, and from sweet potatoes (Döbereiner et al., 1988; Paula et al., 1989).

These observations indicate an $N_2$-fixing association very different from those known so far between plants and diazotrophs of the rhizosphere. The bacterium must be considered to be an endophyte which is propagated within stem cuttings. The propagation of $N_2$-fixing bacteria from stem cuttings into the developing sugar cane plant was first reported by Patriquin et al. (1980) although the microorganisms involved were not identified.

Sterile micropropagated sugar cane seedlings were not infected by *A. diazotrophicus* by tradi-

tional root inoculation methods, except in vitro in sugar-rich medium (V. Reis, unpublished) or in soil containing VA mycorrhizal fungi. In addition *Glomas clarum* spores containing *A. diazotrophicus* were most effective in introducing this diazotroph into roots and aerial parts of such sugar cane seedlings (Paula et al., 1991).

## Concluding remarks

The results of the $^{15}N/N$ balance study show that some sugar cane varieties can obtain large contributions of plant-associated BNF, ranging from 60 to 80% of total plant N, equivalent to over 200 kg N ha$^{-1}$ year$^{-1}$. The immediate practical application of these results in Brazil is to recommend the adoption, where possible, of the planting of these varieties and, in cases where water and P, K and micronutrient supply are optimized, it may be possible to dispense with N fertilization of these varieties altogether.

The bacterium apparently responsible for the plant-associated BNF has unique physiological properties for a diazotroph such as tolerance to low pH, and high sugar and salt concentrations, lack of nitrate reductase and nitrogenase activity which tolerates short term exposure to ammonium. Furthermore, it also behaves as an endophyte, in that it is unable to infect sugar cane plants unless the bacterium enters through damaged tissue or by means of VA mycorrhizae and is propagated via the planting material (stem pieces). Further studies of this fascinating association may not only lead to significant benefits for sugar cane and alcohol fuel production in Brazil and elsewhere, but also yield valuable information which may permit the development of viable $N_2$-fixing associations with other grasses or cereals.

## Acknowledgements

Financial support for this research was provided by the U.S. National Academy of Sciences, National Research Council, through a grant from the U.S. Agency for International Development.

# References


Azeredo D F, Bolsanello J, Weber M and Vieira J R 1986. Nitrogênio em cana-planta, doses e fracionamento. STAB 4, 26–32.

Boddey R M, Urquiaga S, Suhet A R, Peres J R and Neves M C P 1990 Quantification of the contribution of N₂ fixation to field-grown legumes: A strategy for the practical application of the ¹⁵N isotope dilution technique. Soil Biol. Biochem. 22, 649–655.

Cavalcante V A and Döbereiner J 1988 A new acid-tolerant nitrogen-fixing bacterium associated with sugar cane. Plant and Soil 108, 23–31.

Döbereiner J 1959 Influência da cana-de-açúcar na população de *Beijerinckia* do solo. R. Bras. Biol. 19, 251–258.

Döbereiner J 1961 Nitrogen-fixing bacteria of the genus *Beijerinckia* Derx in the rhizosphere of sugar cane. Plant and Soil 15, 211–216.

Döbereiner J, Reis V and Lazarine A C 1988 A new N₂ fixing bacteria in association with cereals and sugar cane. *In* Nitrogen Fixation: Hundred Years After. Eds. H Bothe, F J de Bruijn and W E Newton. pp. 717–722. Gustav Fischer, Stuttgart.

Döbereiner J and Ruschel A P 1958 Uma nova espécie de *Beijerinckia*. R. Biol. 1, 261–272.

Fu H-A and Burris R H 1989 Ammonium inhibition of nitrogenase activity in *Herbaspirillum seropedicae* J. Bacteriol. 171, 3168–3175.

Fu H-A, FitzMaurice W P, Lehman L J, Roberts G P and Burris R H 1988 Regulation of nitrogenase activity in azospirilla, herbaspirilla and acetobacter and cloning of drag and drat⁻ homologous genes of *A. lipoferum* Sp Br 17. *In* Nitrogen Fixation: Hundred Years After. Eds. H Bothe, F J de Bruijn and W E Newton. p 336. Gustav Fischer, Stuttgart.

Gillis M, Kerters B, Hoste D J, Kroppenstedt R M, Stephan M P, Teixeira K R S, Döbereiner J and De Ley J 1989 *Acetobacter diazotrophicus* sp. nov. a nitrogen fixing acetic acid bacterium associated with sugar cane. Int. J. Syst. Bacteriol. 39, 361–364.

Graciolli L A, Freitas J R de and Ruschel A P 1983 Bactérias fixadoras de nitrogênio nas raízes, caules e folhas de cana-de-açúcar (*Saccharum* sp.). R. Microbiol. 14, 191–196.

Graciolli L A and Ruschel A P 1981 Microorganisms in the phyllosphere and rhizosphere of sugar cane. *In* Associative N₂ Fixation. Eds. P B Vose and A P Ruschel. Vol. 2. pp 91–101. CRC Press, Boca Raton, FL.

Lima E, Boddey R M and Döbereiner J 1987 Quantification of biological nitrogen fixation associated with sugar cane using a ¹⁵N aided nitrogen balance. Soil Biol. Biochem. 19, 165–170.

Matsui E, Vose P B, Rodrigues N S R and Ruschel A P 1981 Use of ¹⁵N enriched gas to determine N₂-fixation by undis-turbed sugar cane plant in the field. *In* Associative N₂-Fixation. Eds. P B Vose and A P Ruschel. pp 153–161. CRC Press, Boca Raton FL.

Orlando Filho J, Haag H P and Zambello Jr. E 1980 Crescimento e absorção de macronutrientes pela cana-de-açúcar, variedade CB 41-76, em função da idade, em solos do Estado de São Paulo. Bol. Tec. Planalsucar 2, 1–128.

Patriquin D G, Ruschel A P and Graciolli L A 1980 Nitrogenase activity of sugar cane propagated from stem cuttings in sterile vermiculite. Soil Biol. Biochem 12, 413–417.

Paula M A de, Döbereiner J and Siqueira J O 1989 'Efeito da inoculação com fungo micorrízico VA e bactérias diazotróficas no crescimento e produção de batata-doce', in Congresso Brasileiro de Ciência do Solo, 22, Recife, SBCS, Programa e resumos, p. 109.

Paula M A, Reis V M and Döbereiner J 1991 Interactions of *Glomus clarum* with *Acetobacter diazotrophicus* in infection of sweet potato (*Ipomoea batatas*) sugar cane (*Saccharum* spp.) and sweet sorghum (*Sorghum vulgare*) Biol. Fert. Soils (*In press*).

Purchase B S 1980 Nitrogen fixation associated with sugarcane. Proc. S. Afr. Sugar Technol. Assoc., June, 173–176.

Reis V M, Zang Y and Burris R H 1990 Regulation of nitrogenase activity by ammonium and oxygen in *Acetobacter diazotrophicus*. An. Acad. Bras. Ci. 62, 317.

Ruschel A P, Henis Y and Salati E 1975 Nitrogen-15 tracing of N-fixation with soil-grown sugar cane seedlings. Soil Biol. Biochem. 7, 181–182.

Ruschel A P, Victoria R L, Salati E and Henis Y 1978 Nitrogen fixation in sugar cane (*Saccharum officinarum*). *In* Environmental Role of Nitrogen-Fixing Blue-green Algae and Asymbiotic Bacteria. Ed. V Grantall. pp 297–303. NFR. Uppsala (Ecological Bulletins, 26).

Sampaio E V S B, Salgado I H and Bettany J 1984 Dinâmica de nutrientes em cana-de acúcar. I. Eficiência da utilização de uréia (¹⁵N) em aplicação única ou parcelada. Pesq. Agropec. Bras. 19, 943–949.

Stephan M P, Teixeira K R S and Döbereiner J 1988 Nitrogen fixation physiology of *Acetobacter nitrocaptans*: Effect of oxygen, pH and carbon source on respiration and nitrogenase activity. *In* Nitrogen Fixation: Hundred Years After. Eds. H Bothe, F J de Bruÿn and W E Newton. p 207. Gustav Fischer, Stuttgart.

Teixeira K R S, Stephan M P and Döbereiner J 1987 Physiological studies of *Saccharobacter nitrocaptans* a new acid tolerant N₂-fixing bacterium. *In* Poster presented at 4th International Symposium on Nitrogen Fixation with non-legumes, Rio de Janeiro, Final program abstracts, p. 149.

Urquiaga S, Cruz K H S and Boddey R M 1990 Estimation of the contribution of biological nitrogen fixation to sugar cane using ¹⁵N and nitrogen balance techniques. Submitted. Soil Sci. Soc. Am. J.

# PHYSOLOGICAL PROPERTIES OF *AZOSPIRILLUM BRASILENSE* INVOLVED IN ROOT GROWTH PROMOTION


**YAACOV OKON, TAMI BAR, SARA TAL AND ELI ZAADY**
*Department of Plant Pathology and Microbiology,*
*Faculty of Agriculture,*
*The Hebrew University of Jerusalem,*
*P.O. Box 12, Rehovot 76100*
*Israel*


## INTRODUCTION

Free living *Azospirillum*, and other soil bacteria exert beneficial effects on plant growth and yield of many crops of agronomic importance, such as grain and forage grasses, legumes and tomatoes [16,39].

Plant growth promotion following *Azospirillum* inoculation is derived mainly from a general effect on root growth and function [40]. Crop yield has been enhanced by 10-30% in many field experiments. Associative $N_2$ fixation by *Azospirillum* is of less agronomical significance than expected [16]. The effects of *Azospirillum* on root morphology and physiology are summarized in Table 1.

The reader is also referred to recent reviews on physiology [27], genetics [17,19] and plant growth promotion [39,52] of *Azospirillum*.

In order to understand in greater detail the associative interaction of *Azospirillum* with roots it was proposed to develop genetically engineered mutants of *Azospirillum* lacking one property (Phenotype) [17,19,39]. This approach has been used with considerable success for a better understanding, of the basic biology of several plant bacterial interactions such as the *Rhizobium*-legume symbiosis [8] and plant pathogenic bacteria infection [34]. A list of proposed working hypotheses and desirable mutants and properties of *Azospirillum* are presented in Table 2.

This article summarizes results of our laboratory and others on the physiological properties of *Azospirillum brasilense* related to mechanisms involved in the *Azospirillum*-root interaction. Specifically we report on results on adsorption properties, the use of random transposon (Tn5) mutagenesis, poly-β–hdyroxybutyrate metabolism and indole-3-acctic acid metabolism.

## ADSORPTION PROPERTIES OF *A. BRASILENSE*

The bacterial surface plays an important role in establishment of beneficial as well as pathogenic bacterial-plant associations.

Lipopolysaccharides (LPS) from *A. brasilense* and *A. lipoferum* were found to contain, in all strains tested, rhamnose, glucose, glucoseamine and 3-deoxy-D-mannooctulosonic acid. LPS of some strains also contained fucose and galactose. 3-hydroxypalmitic acid and 3-hydroxymyristic acid predominated in the lipid fraction [9]. HPLC analysis of capsular (CPS) and exopolysaccharide (EPS) of *A. brasilense* and *A. lipoferum* showed the presence of N-acetylglucosamine, glucose, galactose and rhamnose. Amounts of CPS and EPS depended on the strain and the growth stage of the culture [14].

Table 1. Effects of *Azospirillum* on root morphology and physiology of grasses legumes and tomatoes.

|  | Reference |
|---|---|
| a. Promotes root diameter, density and length of root hairs | [26] |
| b. Promotes root surface area during early growth (3-4 weeks) | [21] |
| c. Promotes mineral and water uptake and dry matter accumulation in plants | [46] |
| d. Increases indole-3-acetic acid and indole butyric acid content in roots during early growth of maize (2-3 weeks) | [20] |
| e. Increases respiration and metabolic root enzyme activities | [22] |

Maximal effects of *Azospirillum* inoculation are obtained with optimal inoculum of $10^7$CFU/plant and depend on competitive rhizosphere population and organic matter content in soil

It was found that the capsule of *A. brasilense* selectively bound lectins which had specificity for N-acetylglucoseamine, N-acetylgalactoseamine, galactose, glucose and mannose [59], and N-acetylglucosamine with wheat germ agglutinin [15]. Calcofluor-binding polysaccharides were distributed between the CPS and EPS of *A. brasilense* and *A. lipoferum*. Two genes that affect EPS production in *A. brasilense* were identified by genetic complementation of *R. meliloti* exoB and exoC mutants [19]. ExoC mutants of *Rhizobium meliloti* fail to produce a periplasmic $\beta$–(1-2)glucan due to a deficiency in the enzyme phosphoglucomutase. ExoC mutants of *A. brasilense* still produces EPS but lacks a 110 KDa protein [19]. Cell aggregation and adsorption to roots in ExoB and ExoC mutants has not been studied, but mutants that are drastically impaired in attachment to roots of wheat and in aggregation (flocculation) were originally isolated as calcoflour (Cal⁻) mutants and are affected in EPS or CPS production. Four different recognized loci involved in the synthesis of surface polysaccharide in *A. brasilense* were located in the chromosome [19].

Table 2. Desirable mutants of *Azospirillum,* working hypotheses

| | Phenotype | Property relevant to rhizosphere |
|---|---|---|
| a. | Motility, chemotaxis, aerotaxis [19] | Motility towards roots and colonization |
| b. | Cell surface components LPS, EPS [19] | Adsorption to roots, colonization and elicitors of root response |
| c. | Production of indole-3-acetic acid  gibberellins, cytokinins {1,31] | Effect on root exudation, growth, metabolism of plant growth promoting substances in the root |
| d. | Carbon source, poly-β-hydroxy-butyrate metabolism | Colonization, survival and proliferation |
| e. | nif⁻, nitrogen metabolism [17] | Contribution of $N_2$-fixationn |
| f. | Production of siderophores, antibiotics, bacteriocins, chitinase | Survival and proliferation, competition antagonism against deleterious micro-organisms |
| | Pectinolytic activity | Root invasion |

Aggregation or autoagglutination of cells is a widespread phenomenon in asymbiotic $N_2$-fixing bacteria such as *Azospirillum, Klebsiella* and *Azotobacter* and may affect their dispersal, survival in soil and adsorption to plant roots [36,37,38,45]. Flocculation and floc formation in *A. brasilense* and *A. lipoferum* was related to β–linked exopolysaccharides, possibly cellulose [15,45].

Studies of aggregation and pellicle formation in *Azospirillum* by electron microscopy revealed the presence of an extracellular layer. Concavalin-A-ferritin, a lectin conjugated with ferritin was bound to the extracellular layer which was developed during the stationary phase but not exponential phase [37]. Aggregation of cells occurred towards the end of the exponential phase and during stationary phase [36]. Treatment of *A. brasilense* with EDTA resulted in loss of both aggregation capacity and the ability to adsorb to wheat roots. A glycoprotein (97 KD) purified from an EDTA extract of *A. brasilense* cells was suggested to be involved in aggregation of *Azospirillum*. Bacterial agglutination by this fraction was inhibited by D-glucose, melibiose and α-methylglucoside. No evidence as to the involvement of cellulose fibrils in aggregation of *A. brasilense* could be found [36].

The attachment to the roots and colonization by *Azospirillum* in the rhizosphere and its relationship to plant growth promotion is not well understood. Gafny et al. [24] found that bacterial adherence to roots increased with increasing bacterial concentrations, up to $10^9$ cell ml$^{-1}$ of the binding suspension mixture. Live metabolically active cells of *A. brasilense* are needed for adherence to plant roots. Electron microscopy observations showed polymeric fibrillar bridges in bacteria adsorbed to roots [24,41]. Bashan and Levanoni [5] suggested protein bridge in the adsorption of *Azospirillum* to sand. Production of these bridges was controlled by the amounts of nutrients available but the protein(s) were not characterized.

Recent studies in our laboratory [60] have demonstrated that when cells of *Azospirillum* have been treated with a maize root extract the number of bacteria adsorbed to the root of maize was increased, thus confirming the observation reported by Gafny et al. [24]. Sugars were tested for preventing aggregation of malate-grown *A. brasilense* cells. D-fructose (25–30 mM) strongly inhibited cell aggregation at various pH levels (5.0–8.4). Treatment of malate-grown cells with D-fructose significantly promoted adsorption of *Azospirillum* cells to maize roots. Highest adsorption was obtained in cells grown on D-fructose as sole carbon source and treated with maize root extracts. The main effect of D-fructose was on the bacteria and not on the roots. The results supported the idea that D-fructose molecules on the cell envelope bind to specific sites, thus preventing aggregation between cells but at the same time play a role in the adsorption to the roots [60]. This phenomenon may be related to $\beta$-(1-2) glycosidic linkages involved in binding of fructose with other fructose or other sugars.

It is possible that for preparation of *A. brasilense* inoculants by using fructose in the growth medium instead of malate, the adsorption of *Azospirillum* to the roots of plants will be increased. Thus with greater numbers of adsorbed bacteria it might be possible to obtain a more consistent effect on enhancement of root growth (unpublished data).

## RANDOM Tn5 MUTAGENESIS IN *AZOSPIRILLUM BRASILENSE Sp7*

The use of Tn5-Mob, permits transfer of plasmids to various recipients, but it is often not possible to identify phenotypes by this manner. Problems that may occur include incompatibility with indigenous plasmids in the recipient strain, formation of cointegrates with other plasmids, lack of expression of the plasmid-encoded trait in other genetic background [33] and Tn5 multiple insertions [2]. Lack of Tn5 polarity, or expression of genes within an operon downstream of the Tn5 site can be explained in a variety of ways [8].

There have been several reports on systems capable of generating random insertion mutants of a workable frequency in *A. brasilense* and in *A. lipoferum* [1,50].

In 1987 we started a search for random Tn5 mutation in *Azospirillum* lacking phenotypical traits related to the *Azospirillum*-plant association [18]. Tn5 insertions into the host genome (plasmids or chromosomal) enabled a screening for possible modifications of the resulting phenotype related to bacteria-plant interactions [7,57]. Properties of the strains and plasmids used are summarized in Table 3.

Complete medium for *E. coli* (37°C) and *Azospirillum* (30°C) was Luria broth (LB), pH 6.8. Minimal malate medium for *Azospirillum* with 0.8 or 0.15 gr/l $NH_4Cl$, pH 6.8, was prepared according to Tal and Okon [53].

*E. coli* strains S17-1 and HB101 were used to mobilize and donate the suicide plasmid pSUP5011, carrying the $Km^R$-transposon Tn5-Mob to *A. brasilense* Sp7. Bacterial overnight culture (150 strokes.min$^{-1}$) were concentrated by centrifugation (3,000 rpm for 10 min at 4°C).

100 µl of washed cells were plated onto solid nutrient agar (NA) with 1% yeast extract. Bacterial overnight growth was collected and plated onto solid medium containing both Str and Km to select for *A. brasilense* transconjugants.

For highest transconjugation frequency, the following were tested: *E. coli: A. brasilense* ratio, effect of donor-helper time period of sowing and mixing before mating, pre-mating mixture, effect of semi-solid mating medium, effect of $NH_4Cl$ content of medium, centrifugation and different growth media.

Highest transconjugation frequency was obtained with 1 donor to 10 recipients ratio. No previous contact between the two *E. coli* strains was necessary to promote conjugation. Previous adsorption of the *E. coli* suspension (4 to 6 h each after centrifugation) to the solid mating medium enhanced transconjugation frequency. Conjugation frequency was 5 times higher when using solid medium by the above procedure compared to semi solid mating medium.

*Azospirillum* cells grown on 0.8 g.l$^{-1}$ NH$_4$Cl (containing about 5% PHB) yielded higher transconjugation frequency than cells grown on 0.15 g.l$^{-1}$ NH$_4$Cl (containing about 40% PHB).

Highest transconjugation frequency was obtained when rich media (LB or NA) was used in all stages. From all the above experiments 3500 Km resistant mutants were screened for auxotrophy. Only 28 transconjugants were found to be stable auxotrophs and for only 8 of them the appropriate amino acid was found. This method was laborious and yielded only a few kinds of auxotrophic mutants, mainly His$^-$, Met$^-$ and Gln$^-$. No tryptophan auxotrophic mutant (Trp$^-$) was found.

Table 3. Bacterial strains and plasmids

| Strain or plasmid | Relevant Properties | Reference |
|---|---|---|
| ***E. coli* Strains** | | |
| S17-1(pSUP5011) | Tpa, recA, chromosomally RP4 derivatives | [49] |
| HB101(pRK2013) | F$^-$, recA13 | [6] |
| ***Azospirillum* Strain** | | |
| *A. brasilense* Sp7 | Spontaneous Str$^R$, ATCC29145 | [18] |
| **Plasmids** | | |
| pSUP5011 | ColE1 replicon Amp$^R$ Cm$^R$ Tc$^R$ Km$^R$ Tn5-Mob donor Tra$^-$ Mob$^+$ | [49] |
| pRK2013 | ColE1 replicon Km$^R$ helper plasmid Tra$^+$ Mob$^-$ | [23] |

Abbreviations and concentrations µg/ml: Ap - ampicillin, Cm - chloramphenicol (50), Km - kanamycin (20), Tc - tetracycline, Str - streptomycin (200), Tp - trimethoprim

Enrichment Experiments

The purpose of these experiments was to enrich auxotroph population after mating in order to find a Trp mutant.

The following toxic substances and antibiotics were tested for auxotroph enrichment: penicillin G, sodium cycloserine and sodium selenite.

The appropriate toxic substance concentration was found as follows.

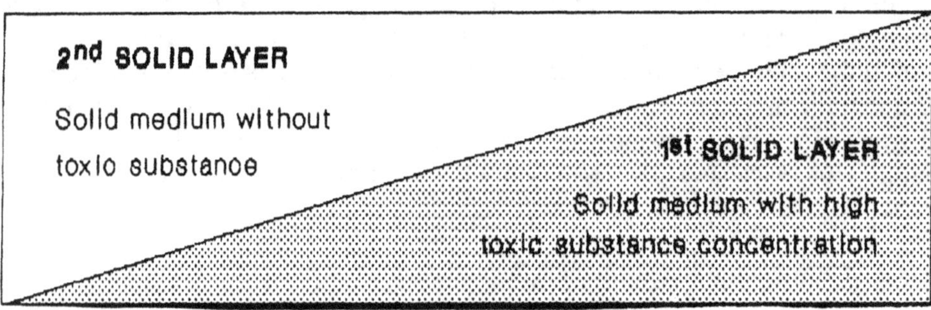


The gradient of toxic substance concentration was reached by diffusion of the toxic substance to the upper solid layer. Bacterial colonies developed only where the toxic substance concentration was below the lethal dose.

Bacterial suspensions from the mating plates were centrifuged and washed twice. Resuspended bacteria were transferred to minimal medium with a toxic substance. After 48 h shaking, 200 µl were plated on NA. Using cycloserine, 70% of the transconjugants were auxotrophs while using sodium selenite, 95% were auxotrophs. In these experiments more than 500 transconjugants per plate were obtained when $5 \times 10^6$ cells of *A. brasilense* were conjugated with $5 \times 10^5$ cells of each *E. coli* strains. More than 5000 transconjugants were tested and none of them appeared to be Trp⁻.

Second enrichment was done by transferring the bacteria from the toxic solution after 72 h shaking to minimal medium with Trp. Most of the developing colonies were not auxotrophs (98%), and the remaining auxotrophic colonies did not develop on minimal medium with Trp.

## INDOLE-3-ACETIC ACID PRODUCTION

The phytohormone indole-3-acetic acid (IAA) is produced by *Azospirillum* from its precursor tryptophan (Trp) [20]. It is utilized as an N source [56], followed by IAA production and excretion [13]. Addition of Trp at high concentrations was found to inhibit bacterial growth [3,25,56]. Trp conversion and immediate excretion of IAA was suggested to be a way to reduce toxic levels of Trp by *A. brasilense* [3]. Accumulation of IAA produced by *A. brasilense* Sp7, in the medium depends on culture age [61], Trp [56] and ammonia supplement, aerobic/anaerobic and static/shaking conditions [4]. IAA production induced by pulse injection of Trp was linear over time. IAA production was higher in young cultures and began less than 30 min after the Trp pulse [3,4].

The best-studied pathway from Trp to IAA involves conversion of Trp to IAA via indole-3-acetamide (IAM). In this pathway, tryptophan-2-monooxygenase (TMO) converts Trp to IAM, and then indole-3-acetamide hydrolase catalyzes the conversion of IAM to IAA. The genetic determinants for these enzymatic conversions are iaaM and iaaH, respectively [48,58]. This pathway is apparently found in *Azospirillum brasilense* Sp7. This observation is supported by strong toxicity of a-methyl Trp, as compared to 5-methyl Trp (which is typical of TMO) [35], induction of a protein with a molecular weight corresponding to the enzyme TMO and positive hybridization between iaaM and iaaH probes and *A. brasilense* total DNA [3].

Alternative pathway via tryptamine was suggested in *A. brasilense sp.* Cd. but not in *A. lipoferum* Sp RG20a [31] Four aromatic aminotransferases that can catalyze the transamination of Trp *in vitro* were found in *A. lipoferum* [44]. Three deaminated compounds of Trp, were identified in culture medium of *A. brasilense* 703Ebc [11].

The first step of Trp conversion to IAA via IAM requires oxygen [32]. Therefore, production of IAA under anaerobic conditions might indicate the existence of alternative pathway(s). Conversion of Trp and two deaminated Trp compounds, indole-3-lactic acid (ILA) and indole-3-pyruvic acid (IPyA), to IAA was analyzed in order to demonstrate alternative pathway(s) in *A. brasilense.* Conversion of ILA to IAA was reduced under anaerobic conditions, while IPyA conversion to IAA by *A. brasilense* Sp7 was higher under anaerobic than under aerobic conditions, in opposition to the spontaneous tendency.

Conversion of Trp and IPyA to IAA was reduced in the presence of ammonia.. Bacterial metabolism of Trp to IPyA, which is further converted to IAA is in agreement with the observed biosynthesis of IAA from Trp that proceeds via IPyA in most higher plants [10].

Trp conversion to IAA provides one molecule of ammonia per Trp molecule. Rapid feedback inhibition by ammonia on IAA production is therefore likely, considering that the optimal conditions for at least one pathway for IAA biosynthesis (via IPyA) are also the preferred conditions for nitrogen fixation [4].

Nitrogen fixation in *A. brasilense* Sp7 is known to be inhibited by oxygen [28], ammonia [29], Trp addition [4] and influenced by the addition of other amino acids to the medium [30].

Possible inhibition of nitrogenase activity by ammonia released during conversion of Trp to IAA was further indicated by the inhibition observed upon addition of Trp and by nitrogenase repression by Trp as was observed by Western blot against dinitrogenase reductase antibodies [4]. Under air and ammonia limitation, which are the most widespread conditions in soil, production of IAA and nitrogen fixation by *A. brasilense* is elevated and plant growth regulators may be provided by the bacterium. A summary of the suggested co-regulation by oxygen and ammonia of Trp conversion to IAA and nitrogenase activity is shown in Table 4.

To the best of our knowledge, it was not previously known that indole can serve as a precursor for IAA synthesis. Enzymatic conversion of indole to IAA was recently found in *A. brasilense* [4]. It may suggest that this pathway is a basis for enzymatic substitution of heterocyclic compounds in IAA biosynthesis by *A. brasilense* Sp7.

Table 4: Co-regulation of Trp conversion to IAA and nitrogen fixation by $NH_4^+$ and $O_2$ [4]

| | Effect of | |
| Metabolic Function | $NH_4^+$ | $O_2$ |
| --- | --- | --- |
| *Trp conversion to IAA via: | | |
| IAM or ILA | feedback inhibition | enhancement |
| IPyA | feedback inhibition | inhibition |
| $N_2$ fixation | feedback inhibition | inhibition |

*Trp conversion to IAA is followed by ammonia release.

## POLY-β-HYDROXYBUTYRATE METABOLISM (PHB)

Many bacterial species belonging to diverse taxa accumulate the reserve polymer PHB. PHB is the most important compound of the general group polyhydroxyalconates. Polymer accumulation is initiated under conditions of nutrient imbalance and serves as an electron and carbon sink. Under conditions of stress and starvation PHB functions as carbon and energy source [12].

The biosynthesis and degradation of PHB was studied in detail in *Azotobacter beijerinckii* [12]. Many similarities were found in studies with *Azospirillum brasilense* [53,54,55]. PHB synthesis in *A. brasilense* was favoured under oxygen limitation and under high C/N ratio. During a 7-day starvation bacteria containing 40% PHB survived, proliferated and respired at higher rates than cells containing 5% PHB [53]. Polymer rich cells of *Azospirillum* fixed $N_2$ in the absence of exogenous carbon and combined nitrogen. In the presence of stress factors such as ultraviolet irradiation, dessication and osmotic pressure, PHB-rich cells died less rapidly than PHB-poor cells [53]. Biosynthesis and degradation of PHB in *Azospirillum* and *Azotobacter beijerinckii* occurs via a cycle process in which six enzymes are involved (Figure 1) [12,53,55].

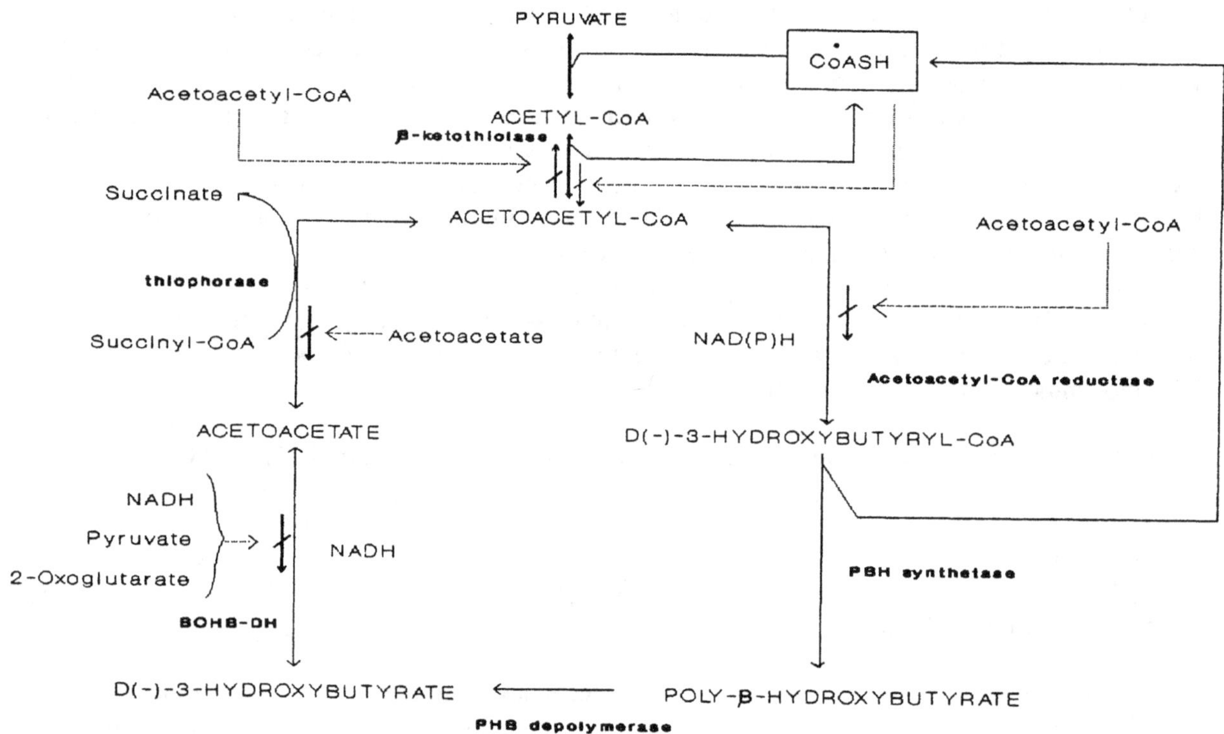


Figure 1. PHB cycle in *Azotobacter beijerincki*i [12] and *Azospirillum brasilense* [54,55].

In *Azospirillum brasilense* the enzymes of the PHB-cycle, of both synthesis and degradation processes, were more active in PHB rich cells than in PHB-poor cells. During 96 h of starvation PHB cycle enzymes were more active in PHB-rich cells. After 24 h of starvation there was a peak of activity in BOHB-DH, β-kethothiolase and thiophorase due to PHB degradation. Acetoacetyl-CoA reductase dropped to a minimum level because PHB could not be synthesized [55]. BOHB-DH of *A. brasilense* was purified 991-fold [54]. The enzyme is a tetramer, with identical subunits and a total molecular mass of 100 KDa. The enzyme is acidic and contains six disulphide bonds without free-SH groups. The purified enzyme is an $NAD^+$ oxidoreductase and is inhibited by NADPH, NADH, adenosine phosphates, pyruvate, acetyl-coenzyme A, oxaloacetate and 2-oxoglutarate [54].

PHB degradation and its utilization as sole carbon and energy source by *A. brasilense* could be involved in favouring establishment, proliferation, survival and competitions with other microorganisms in the rhizosphere [53]. Based on the above developments in the understanding of PHB metabolism in *A. brasilense* and the recent cloning of PHB-genes in other organisms [42,43,47,51] it is now more feasible to obtain mutants affected in synthesis and or degradation of PHB.

Acknowledgements:

We are greatly indebted to Prof. Dr. Claudine Elmerich for her valuable ideas and helpful techniques acquired during a sabbatical leave in 1987 of the first author (YO), at The Pasteur Institute, Paris.

Parts of this work were supported by a grant from the Friends of the Hebrew University of Jerusalem, Mexico and by the Research Foundation of The Society for the Protection of Nature in Israel.

## REFERENCES

1. Abdel-Salam, M.S. and Klingmuller, W. (1987) 'Transposon Tn5 mutagenesis in *Azospirillum lipoferum*: isolation of indole acetic acid mutants', Molecular General Genetics 210, 165-170.
2. Baker, C.J., Atkinson, M.M. and Collmer, A. (1987) 'Concurrent loss in Tn5 mutants of *Pseudomonas syringae* pv. *syringae* of the ability to induce the hypersensitive response and host plasma membrane $K^+/H^+$ exchange in Tobacco', Phytopathology 77, 1268-1272.
3. Bar, T. and Okon, Y. (1990) 'Indole-3-acetic acid synthesis in *Azospirillum brasilense* Sp7', (submitted).
4. Bar, T. and Okon, Y. (1990) 'Pathways for indole-3-acetic acid production by *Azospirillum brasilense* Sp7 and their possible co-regulation with nitrogen fixation,' (submitted).
5. Bashan, Y. and Levanony, H. (1988) 'Active attachment of *Azospirillum brasilense* Cd to quartz sand to a light-textured soil by protein bridging', J. General Microbiology 134, 2269-2279.
6. Boyer , H.W. and Roulland-Dussoix, D. (1969) 'A complementation analysis of the restriction and modification of DNA in *E. coli*', J. Molecular Biology 41, 459-468.

7. Bouzouklian, H., Fogher, C. and Elmerich, C. (1986) 'Cloning and characterization of the glnA gene of *Azospirillum brasilense* Sp7', Annals Institut Pasteur/Microbiologie 137B, 3-18.

8. de Bruijn, F.J. and Lupski, J. R. (1984) 'The use of transposon Tn5 mutagenesis in the rapid generation of correlated physical and genetic maps of DNA segments cloned into multicopy plasmids - a review', Gene 27, 131-139.

9. Choma, R.R., Russa, R., Mayer, H. and Lorkiewicz, L. (1987) 'Chemical analysis of *Azospirillum* lipopolysaccharides', Archives of Microbiology 146, 341-345.

10. Cohen, J.D. and Bialek, K. (1984) 'The biosynthesis of indole-3-acetic acid in higher plants', in A. Crozier and J.R. Hillman (eds.), The Biosynthesis and Metabolism of Plant Hormones (Soc. Exp. Biol. Sem. Ser. 23), Cambridge University Press, Cambridge, pp. 165-181.

11. Crozier, A., Arruda, P., Jasmin, J.M., Monterio, A.M. and Sandberg, G. (1988) 'Analysis of indole-3-acetic acid and related indoles in culture medium from *Azospirillum lipoferum* and *Azospirillum brasilense'*, Applied Environmental Microbiology 54, 2833-2837.

12. Dawes, E. A. (1986) 'Microbial energy reserve compounds', in E.A. Dawes (ed.), Microbial Energetics, Blackie, Glasgow and London, pp. 145-165.

13. De Francesco, R., Zanetti, G., Barbieri, P. and Galli, E. (1985) 'Auxin production by *Azospirillum brasilense* under different culture conditions', in W. Klingmuller, (ed.), Azospirillum III: Genetics, Physiology, Ecology, Springer-Verlag, Berlin, pp. 109-115.

14. Del Gallo, M. and Haegi, A. (1990) 'Characterization and quantification of exocellular polysaccharides in *Azospirillum brasilense* and *Azospirillum lipoferum'*, Symbiosis 8, (in press).

15. Del Gallo, M., Negi, M. and Neyra, C.A. (1989) 'Calcoflour and lectin-binding exocellular polysaccharides of *Azospirillum brasilense* and *Azospirillum lipoferum'*, J. Bacteriology 171, 3504-3510.

16. Dobereiner, J. and Pedrosa, F.O. (1987) 'Nitrogen-fixing bacteria in non-leguminous crop plants', Science Tech. Publishers/Springer-Verlag, Madison, WI.

17. Elmerich, C., Bouzouklian, H., Vieille, C., Fogher, C., Perroud, B., Perrin, A. and Vanderleyden, J. (1987) 'Azospirillum: Genetics of nitrogen fixation and interaction with plants', Phil. Transactions of the Royal Society London (B) 317, 183-192.

18. Elmerich, C. and Franche, C. (1982) 'Azospirillum genetics:plasmids, bacteriophages and chromosome mobilization', in W. Klingmuller (ed.), *Azospirillum* Genetics, Physiology and Ecology, Birkhauser, Basel, pp. 9-17.

19. Eyers, M., Michiels, K., Van Bastelaere, E., Croes, C., van Rhijn, P., van de Broek, A., Milcamps, A., DeMot, R. and Vanderleyden, J. (1990) *'Azospirillum* associations', in P.M. Gresshof, F. Roth, G. Stacey and W.E. Newton (eds.), Nitrogen Fixation: Achievements and Objectives, Chapman and Hall, New York, London, in press.

20. Fallik, E., Okon, Y., Epstein, E., Goldman, A. and Fischer, M. (1989) 'Identification and quantification of IAA and IBA in *Azospirillum brasilense* -inoculated maize roots', Soil Biology & Biochemistry 21, 147-153.

21. Fallik, E., Okon, Y. and Fischer, M. (1988) 'Growth response of maize to *Azospirillum* inoculation: effect of soil organic matter content, number of rhizosphere bacteria and timing of inoculation', Soil Biology & Biochemistry 20, 45-49.

22. Fallik, E., Okon, Y. and Fischer, M. (1988) 'The effect of *Azospirillum brasilense* inoculation on metabolic enzyme activity in maize root seedlings', Symbiosis 6, 17-28.

23. Figurski, D.H. and Helinski, D.R. (1979) 'Replication of an origin-containing derivative of plasmids RK2 dependent on a plasmid function provided in trans', Proceedings of the National Academy of Science, USA 76, 1648-1652.

24. Gafny, R., Okon, Y., Kapulnik, Y. and Fischer, M. (1986) 'Adsorption of *Azospirillum brasilense* to corn roots. Soil Biology & Biochemistry 18, 69-76.

25. Grossowicz, N. and Saheb, K. (1990) 'A specific inhibition of *Legionella pneumophila* growth by the auxins', in Annual Meeting of the Israel Society for Microbiology, Jerusalem, Hebrew University of Jerusalem Publication, p.69.

26. Hadas, R. and Okon, Y. (1987) 'Effect of *Azospirillum brasilense* inoculation on root morphology and respiration in tomato seedlings', Biology & Fertility of Soils 5, 241-247.

27. Hartmann, A. (1989) 'Ecophysiological aspects of growth and nitrogen fixation in *Azospirillum* spp.', in F.A. Skinner, R. Boddey and I. Fendrik (eds.), Nitrogen Fixation with Non-Legumes, Kluwer Academic Publishers, Dordrecht, pp. 123-136.

28. Hartmann, A. and Burris, R.H. (1987) 'Regulation of nitrogenase activity by oxygen in *Azospirillum brasilense* and *Azospirillum lipoferum*', J. Bacteriology 169, 944-948.

29. Hartmann, A., Fu, H. and Burris, R. (1986) 'Regulation of nitrogenase activity by ammonium chloride in *Azospirillum* spp.', J. General Microbiology 165, 864-870.

30. Hartmann, A., Fu, H. and Burris, R. (1988) 'Influence of amino acids on nitrogen fixation ability and growth of *Azospirillum* spp.', Applied and Environmental Microbiology 54, 87-93.

31. Hartmann, A., Singh, M. and Klingmuller, W. (1983) 'Isolation and characterization of *Azospirillum* mutants excreting high amounts of indoleacetic acid', Canadian Journal of Microbiology 29, 916-923.

32. Hutcheson, S.W. and Kosuge, T. (1985) 'Regulation of 3-indoleacetic acid production in *Pseudomonas syringae* pv. *savastanoi*. Purification and properties of trypthophan 2-monooxygenase', J. Biological Chemistry 260, 6281-6287.

33. Hynes, M.F., Quandt, J., O'Connell, M.P. and Puhler, A. (1989) 'Direct selection for curing and deletion of *Rhizobium* plasmids using transposons carrying the *Bacillus subtilis* sacB gene', Gene 78, 111-120.

34. Kerr, A. (1987) 'The impact of molecular genetics on plant pathology', Annual Review of Phytopathology 25, 87-110.

35. Kosuge, T. and Sanger, M. (1985) 'Indoleacetic acid, its synthesis and regulation: a basis for tumorigenicity in plant diseases', Recent Advances in Phytochemistry 20, 148-161.

36. Madi, L. and Henis, Y. (1989) 'Aggregation in *Azospirillum brasilense* Cd: conditions and factors involved in cell-to-cell adhesion', Plant and Soil 115, 89-98.

37. Madi, L., Kessel, M., Sadovnik, E. and Henis, Y. (1988) 'Electron microscopic studies of aggregation and pellicle formation in *Azospirillum* spp.', Plant and Soil 109, 115-121.

38. Nur, I., Steinitz, Y.L., Okon, Y. and Henis, Y. (1981) 'Carotenoid composition and function in nitrogen-fixing bacteria of the genus *Azospirillum*', J. General Microbiology 123, 27-32.

39. Okon, Y. (1985) '*Azospirillum* as a potential inoculant for agriculture', Trends in Biotechnology 3, 223-228.

40. Okon, Y., Fallik, E., Sarig, S., Yahalom, E. and Tal, S. (1988) 'Plant growth promoting effects of *Azospirillum*', in H. Bothe, F. de Bruijn and W.E. Newton (eds.), Nitrogen Fixation: Hundred Years After, Gustav Fischer, Stuttgart, New York, pp. 741-746.

41. Patriquin, D.G., Dobereiner, J. and Jain, D.K. (1983) 'Sites and processes of association between diazotrophs and grasses', Canadian Journal of Microbiology 29, 900-915.

42. Peoples, O.P. and Sinskey, A.J. (1989) 'Fine structural analysis of the *Zooglea ramigera* phbA-phbB locus encoding _β-kethothiolase and acetoacetyl CoA reductase nucleotide sequence of phbB', Molecular Microbiology 3, 349-357.

43. Peoples, O.P. and Sinskey, A.J. (1989) 'Poly-β-hydroxybutyrate biosynthesis in *Alicaligenes eutrophus* Hlb', J. Biological Chemistry 264, 5293- 5297.

44. Ruckdaschel, E., Kittell, B.L., Helinski, D.R. and Klingmuller, W. (1988) 'Aromatic amino acid aminotransferases of *Azospirillum lipoferum* and their possible involvement in IAA biosynthesis', in W. Klingmuller (ed.), Azospirillum IV: Genetics, Physiology, Ecology, Springer-Verlag, Berlin, pp. 49-53.

45. Sadisavan, L. and Neyra, C.A. (1985) 'Flocculation in *Azospirillum brasilense* and *Azospirillum lipoferum*: Exopolysaccharides cyst formation', J. Bacteriology 163, 716-723.

46. Sarig, S., Blum, A. and Okon, Y. (1988) 'Improvement of the water status and yield of field-grown sorghum (*Sorghum bicolor*) by inoculation with *Azospirillum brasilense*', J. Agricultural Science Cambridge 110, 271-277.

47. Schubert, P., Steinbuchel, A. and Schlegel, H.G. (1988) 'Cloning of the *Alcaligenes eutrophus* genes for synthesis of poly-β-hydroxybutyric acid (PHB) and synthesis of PHB in *Escherichia coli*', J. Bacteriology 170, 5837-5847.

48. Sekine, M., Watanabe, K. and Syono, K. (1989) 'Molecular cloning of a gene for indole-3-acetamide hydrolase from *Bradyrhizobium japonicum*', J. Bacteriology 171, 1718–1724.

49. Simon, R. (1984) 'High frequency mobilization of gram-negative bacterial replicons by the *in vitro* constructed Tn5-Mob transposon', Molecular and General Genetics 196, 413-420.

50. Singh, M. and Klingmuller, W. (1988) 'A Tn5 induced nifA like mutant of *Azospirillum brasilense*', in W. Klingmuller (ed.), Azospirillum IV: Genetics, Physiology, Ecology, Springer-Verlag, Berlin, pp. 26-31.

51. Slater, C.S., Voige, W.H. and Dennis, D.D. (1988) 'Cloning and expression in *Escherichia coli* of the *Alcaligenes eutrophus* H16 poly-β-hydroxybutyrate pathway', J. Bacteriology 170, 4431-4436.

52. Sumner, M.E. (1990) 'Crop responses to *Azospirillum* inoculation', Advances in Soil Science 12, 53-123.

53. Tal, S. and Okon, Y. (1985) 'Production of the reserve material poly-β-hydroxybutyrate and its function in *Azospirillum brasilense* Cd', Canadian Journal of Microbiology 31, 608-613.

54. Tal, S., Smirnoff, F. and Okon, Y. (1990) 'Purification and characterization of poly-β-hydroxybutyrate in *Azospirillum brasilense*', J. General Microbiology 136, 645-649.

55. Tal, S., Smirnoff, P. and Okon, Y. (1990) 'The regulation of poly-β-hydroxybutyrate metabolism in *Azospirillum brasilense* during balanced growth and starvation', J. General Microbiology 136, 1191-1196.

56. Tien, T.M., Gaskins, M.H. and Hubbell, D.M. (1979) 'Plant growth substances produced by *Azospirillum brasilense* and their effect on the growth of pearl millet (*Pennisetum americanum* L.)', Applied and Environmental Microbiology 37, 1016-1024.

57. Vieille, C., Onyeocha, I., Galimand, M. and Elmerich, C. (1989) 'Homology between plasmids of *Azospirillum brasilense* and *Azospirillum lipoferum'*, in F.A. Skinner, R. Boddey and I. Fendrik (eds.), Nitrogen Fixation with Non-Legumes, Kluwer Academic Publishers, Dordrecht, pp. 165-172.

58. Yamada, T., Palm, C. J., Brooks, B. and Kosuge, T. (1985) 'Nucleotide sequences of the *Pseudomonas savastanoi* indoleacetic acid genes show homology with *Agrobacterium tumefaciens* T-DNA', Proceedings of the National Academy of Sciences, USA 82, 6522-6526.

59. Yagoda-Shagam, J., Barton, L.L., Reed, W.P. and Chiovetti, R. (1988) 'Fluorescein isothiocyanate-labeled lectin analysis of the surface of nitrogen fixing bacterium *Azospirillum brasilense* by flow cytometry', Applied and Environmental Microbiology 54, 1831-1837.

60. Zaady, E. and Okon, Y. (1990) 'Cultural conditions affecting *Azospirillum brasilense* in cell aggregation and adsorption to maize roots', Soil Biology and Biochemistry (in press).

61. Zimmer, W. and Bothe, H. (1988) 'The phytohormonal interaction between *Azospirillum* and wheat', Plant and Soil 110, 239-247.

# REGULATION OF THE nifHDK GENES TRANSCRIPTION IN AZOSPIRILLUM BRASILENSE


C.Schipani, M.Bazzicalupo, A.Bussotti, R.Fani, E.Gallori, A.Grifoni, R.Pastorelli and M.Polsinelli

Department of Animal Biology and Genetics, University of Florence, Via Romana 17, 50125 Florence, Italy



ABSTRACT. The regulation of transcription of the structural genes for nitrogenase of Azospirillum brasilense was investigated. The function of two regions controlling the transcription of the nifH gene was studied. Data obtained suggested a mechanism of nif regulation in Azospirillum different from that described in Klebsiella pneumoniae . The effect of different nitrogen sources both on nitrogenase activity and on the transcription of the nifHDK cluster was also studied.


## 1. INTRODUCTION

The free living bacteria of the genus Azospirillum are able to fix nitrogen in association with the roots of several important grasses (Dobereiner and Day, 1976). As in other diazotrophs nitrogen fixation occurs only under limiting $NH_4^+$ and $O_2$ conditions. In A.brasilense and A.lipoferum the nitrogenase activity is regulated by a reversible covalent modification of the Fe-protein in response to the N-status of the cell, by a "switch-off" mechanism similar to that described in Rhodospirillum rubrum (Hartmann et al 1986). In addition to $NH_4^+$ and $O_2$, many other agents inhibit nitrogen fixation (Gallori and Bazzicalupo 1985) but up to now very little informations are known about their effect at nif transcriptional level. Several mutants altered in nitrogen metabolism were isolated in A.brasilense and their characterization (Pedrosa and Yates, 1984, Singh et al 1989) suggested the existance in Azospirillum of a nifA/ntrC -like system regulating nif genes transcription. This was confirmed by the analysis of the sequence upstream the nifH gene recently determined (Fani et al 1989; de Zamaroczy et al 1989). This region contains the -24, -12 GG-N10-GC sequence which is a characteristic feature of the NtrA-controlled promoters (Gussin et al 1986) and two upstream activator sequences (UAS) (Buck et al 1986) which in Klebsiella pneumoniae are thought to be the NifA binding site. Two transcription startpoints have been identified upstream the ATG of nifH ; moreover two classes of mRNAs of different length were found, the shortest (1.1 kb) corrensponding to the <u>nifH</u> gene and the longest (5.6 kb) to <u>nifHDK</u> genes (de Zamaroczy <u>et al</u> 1989).

In this work results of studies concerning with the effect of different nitrogen sources on nitrogenase activity and on <u>nifHDK</u> transcription are reported. The function of the region upstream the <u>nifH</u> gene and of a sequence inside the same gene in regulating transcription of <u>nifH</u> gene was also discussed.


## 2. MATERIALS AND METHODS

Bacterial strains used were <u>A.brasilense</u> SP7 and <u>E.coli</u> XL1-blue (Bullock <u>et al</u> 1987). Plasmids used were pAB1 (Quiviger <u>et al</u> 1982), pUC18 (Yanisch - Perron <u>et al</u> 1985), pBS-SK (Stratagene), pAFT1 and pAF330 (Fani <u>et al</u> 1989), pRK2013 (Figurski and Helinski 1979), pAF300 (Fani <u>et al</u> 1988). Plasmids pAF331, pAF332 and pAF333 were constructed by inserting different fragments from pAFT1 (Figure 1) into the PstI and/or the HindIII site of vector pAF300, after two subcloning steps in pUC18 and pBS-SK in order to provide the ends of each fragment with PstI and/or HindIII sites.

Nitrogen fixation "repression" conditions were obtained by growing <u>Azospirillum</u> at 35°C for 16-20 hours in 20 ml MSP (Bani <u>et al</u> 1980) supplemented with 20 mM $NH_4Cl$ in a 250 ml flask with shaking; the culture was then diluted into the same medium to an $OD_{550}$-0.2 and grown at 35°C up to an $OD_{550}$-0.8. For "derepression" 10 ml of "repressed" cells were centrifuged and resuspended to and $OD_{550}$-0.2 in MSP without $NH_4Cl$ in a flask where atmosphere was replaced by N plus 0.1% $O_2$; cells were then allowed to grow with gentle shaking for 8 hours at 35°. For nitrogenase assay 10% $C_2H_2$ was added to "derepressed" culture and $C_2H_4$ production was measured after 6 hours by gas-chromatography (Gallori and Bazzicalupo 1985). <u>E.coli</u> was grown at 37°C in LB. All the DNA manipulations have been extensively described (Maniatis <u>et al</u> 1989). Transformation of <u>E.coli</u> cells with plasmid DNA was carried out by the method of Hanahan (1983). Conjugal transfer of recombinant plasmids DNA from <u>E.coli</u> to <u>A.brasilense</u> was performed as described (Fani <u>et al</u> 1988). The activity of chloramphenicol acetyl transferase (CAT enzyme) was determined as described (Shaw 1975, Fani <u>et al</u> 1988).

For total RNA preparation cultures of <u>A.brasilense</u> were grown in nitrogen fixation "derepression" conditions. Each inhibitor was added at time 0 at a concentration of 10 mM, while $O_2$ at 20%. Cultures (80 ml) were stopped at appropriate times and total RNA extracted by boiling cells for 5 minutes in 3.3 ml 1%SDS, EDTA 2mM, TRIS HCl pH8 100 mM; 67 $\mu$l of KCl 2M were added and after centrifugation for 20 minutes at 7000 rpm, 3.8 g of CsCl plus 29 $\mu$l of mercaptoethanol were added to the supernatant and centrifuged at 33,000 rpm for 18 hours. The supernatant was removed and the RNA dissolved in 150 $\mu$l of TE. Labeling DNA with digoxigenin - 11 - UTP and detection was carried out according to the Boehringer Manual of DNA labeling and detection with non radioactive kit.

## 3. RESULTS AND DISCUSSION

### 3.1. Effect of different nitrogen compounds and oxygen on nifHDK transcription

The effect of some compounds on the nifHDK transcription was studied by dot blotting of RNA extracted from A.brasilense SP7 cells. Since two different classes of transcripts were found in A.brasilense (de Zamaroczy et al 1989), total mRNA extracted was hybridized with two probes, the former corresponding solely to nifH gene and the latter to an internal region of nifD gene (figure 1). The results of these dot blot experiments are shown in Table 1 together with data of nitrogenase activity of the same cultures used for RNA preparations. Data reported indicated that: i) the addition of ammonia or oxygen rapidly inhibited the transcription of nifHDK cluster; both agents blocked nitrogen fixation by inhibiting not only nitrogenase enzymatic activity ( "switch - off" effect) but also nifHDK transcription; ii) the repressive effect of ammonium chloride was faster than that of oxygen; iii) compounds inhibiting nitrogenase activity also blocked transcription of the structural genes for nitrogenase; however the addition of alanine and histidine inhibited transcription while still allowing nitrogen fixation.

Table 1. Dot blot hybridization of DNA probes H and D with RNA extracted from A.brasilense Sp7 cultures and their nitrogenase activity.

| Repression time(min) | Probe | $NH_4^+$ | $O_2$ | Glu | Ala | His | Gln | $KNO_3$ | $KNO_2$ |
|---|---|---|---|---|---|---|---|---|---|
| 0 | H | +++ | +++ | | | | | | |
| | D | + | + | | | | | | |
| 10 | H | + | ++ | | | | | | |
| | D | +- | + | | | | | | |
| 30 | H | − | + | +- | − | − | − | +- | − |
| | D | − | +- | − | − | − | − | − | − |
| 60 | H | − | − | | | | | | |
| | D | − | − | | | | | | |
| Nitrogenase activity | | 0 | 0 | 20 | 10 | 50 | 0 | 0 | 0 |

Abbreviations: Glu, Ala, His, Gln, $KNO_3$ and $KNO_2$ are glutamate, alanine, histidine, glutamine, potassium nitrate and potassium nitrite, respectively. Plus and minus represent the intensity of the hybridization spot; nitrogenase activity is expressed as percentage of the activity in nitrogen fixing conditions.

130

This result indicated that these aminoacids uncouple the effect on
transcription to that on the enzyme activity: iv) mRNA hybridized with
the H probe gave a signal more intense than with probe D confirmig the
presence of more copies of _nifH_ message in respect to _nifD_ (de
Zamaroczy _et al_ 1989). From all these data it is possible to
conclude that, in the experimental conditions tested, the trascription
of the two mRNA species is under te same control mechanism. Further
studies will need to establish whether the higher copy number of _nifH_
trancript is correlated to a higher concentration of NifH protein in
the cell.

The hypothesis that the presence of the 1.1 kb mRNA could be due to
the transcription of an extra copy of the _nifH_ gene onto the
chromosome was tested by hybridization experiments of plasmid pAB1
(containing an _A.brasilense_ DNA fragment including the _nifHDK_
genes) and the DNA extracted from _A.brasilense_ digested with
different restriction endonucleases. The hybridization pattern
obtained was that expected for a single copy of the _nifH_ gene in
contiguous to _nifDK_ (data not shown).

3.2. Identification of regions controlling _nifH_ transcription

Previous experiments concerning with the cloning of the _Azospirillum_
_nif_ upstream region in the promoter probe vector pAF300 (Fani _et al_
1989) suggested that the 2 Kb EcoRI-XhoI fragment (Figure 1) carries a
transcription promoter sequence and sequence(s) controlling the _nifH_
expression in response to nitrogen and oxygen availability.

In order to better identify the elements controlling transcription of
the _nifH_ gene in response to induction in _A.brasilense_ , a series
of subclones bearing different fragments of the region upstream the
_nifH_ gene was constructed in the vector pAF300 (Figure 1). The CAT
activity of cell extract of strain SP7 harboring the different
recombinant plasmids, grown either in nitrogen fixation derepressing
or repressing conditions, was tested.

Data obtained (Figure 1) showed that only the cells harboring the
plasmid pAF330 or pAF333 expressed the _cam_ gene according to the
availability of $NH_4Cl$. As expected for a NifA - dependent system, the
deletion of the region containing the UASs (plasmid pAF332) caused a
decrease in the transcription. However the residual level of
expression (about 20%), is high enough to suggest the existence of
some mechanism of control different from that described in
_K.pneumoniae_ . In fact it is known that in this bacterium the UAS, the
_nif_ promoter ( _pnif_ ) and NifA seemed to be sufficient conditions to
control _nif_ transcription. Preliminary results (not reported) of
experiments with the same recombinant plasmids, carried out in both
wild - type and _nifA_ mutant of _K.pneumoniae_ , confirmed this
hypothesis.

Another intriguing result concerns the effect of deletion of the
AluI-XhoI fragment inside the _nifH_ coding region: in plasmid pAF331
in fact transcription occurrs also in the presence of ammonium, while
plasmid pAF333, without the deletion, repressed transcription in
ammonium supplemented medium. The data could suggest that this

cis-acting region has an inhibitory effect on the NifA-mediated transcription at the UASs and _pnifH_ level.
In conclusion the results described in the present work showed that in _A.brasilense_ the control of _nif_ transcription is a complex mechanism involving aspects not yet described in other diazotrophs therefore deserving further investigations.

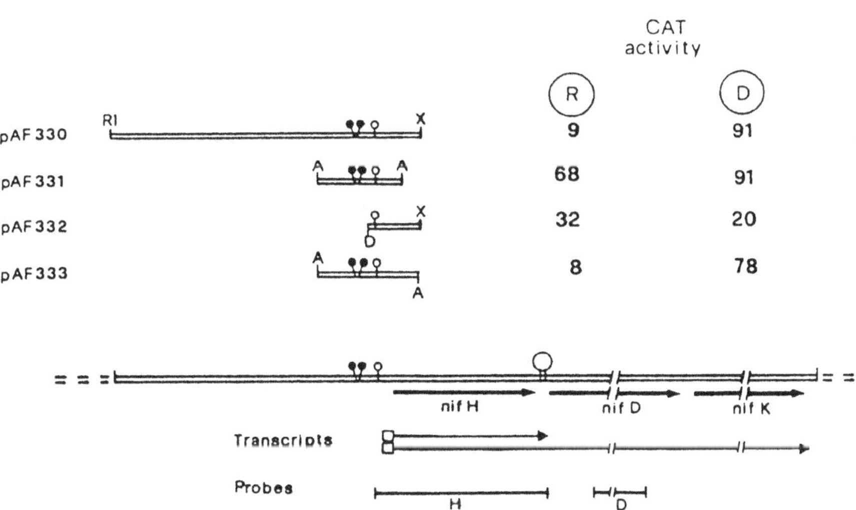


Figure 1. Map of the _nifHDK_ region of _A.brasilense_ and of the fragments cloned in the vector pAF300 and expression of their chloramphenicol acetyl transferase (CAT) activity in _A.brasilense_ . Symbols: open circle, putative _nifH_ promoter; close circles, putative upstream activator sequences (UAS); stem - loop, putative transcription terminator; the symbols of the restriction endonucleases are as follows: A, D, RI and X for AluI, DraI, EcoRI and XhoI. CAT activities are expressed as percentage of acetylated chloramphenicol in assays with extracts of _A.brasilense_ cells harboring the different plasmid grown in "derepressing" (D) or "repressing" (R) conditions.

## 4.AKNOWLEDGEMENTS

The technical contribution of Cristina Indorato is greatly appreciated. This work was supported by M.A.F. "Progetto Tecnologie Avanzate".


## 5.REFERENCES


Bani D, Barberio C, Bazzicalupo M, Favilli F, Gallori E, Polsinelli M (1980) Isolation and characterization of glutamate synthase mutants of _Azospirillum brasilense_ . J Gen Microbiol 119: 239-244
Buck M, Miller S, Drummond M, Dixon R (1986) Upstream activator sequences are present in the promoters of nitrogen fixation genes.

Nature 320: 374-378

Bullock WO, Fernandez JM, Short JM (1987) XL1-blue: a high efficiency plasmid transforming recA Escherichia coli strain with beta-galactosidase selection. Biotechniques 5,4: 376-379

Dobereiner J, Day JM (1976) Associative symbiosis in tropical grasses: characterization of microorganisms and dinitrogen - fixing sites. In: Newton WE, Nyman CY (eds) 1st International Symposium N2 - Fixation. Washington State University Press, Pullman, pp 518-536

Fani R, Bazzicalupo M, Ricci F, Schipani C, Polsinelli M (1988) A plasmid vector for the selection and study of transcription promoters in Azospirillum brasilense . FEMS Microbiol Lett 50: 271-276

Fani R, Allotta G, Bazzicalupo M, Ricci F, Schipani C, Polsinelli M (1989) Nucleotide sequence of the gene encoding the nitrogenase iron protein ( nifH ) of Azospirillum brasilense and identification of a region controlling nifH transcription. Mol Gen Genet 220: 81-87

Figurski D, Helinski DR (1979) Replication of an origin-containing derivative of plasmid RK2 dependent on a plasmid function provided in trans. Proc Nat Acad Sci USA 76: 1648-1652

Gallori E, Bazzicalupo M (1985) Effect of nitrogen compounds on nitrogenase activity in Azospirillum brasilense . FEMS Microbiol Lett 28: 35-38

Gussin GN, Ronson CW, Ausubel FM (1986) Regulation of nitrogen fixation genes. Ann Rev Genet 20: 567-591

Hanahan D (1983) Studies on transformation of Escherichia coli with plasmids. J Mol Biol 166: 557 - 580

Hartmann A, Fu H, Burris R (1986) Regulation of nitrogenase activity by ammonium chloride in Azospirillum spp. J Bacteriol 165: 864-870

Maniatis T, Fritsch EF, Sambrook J (1989) Molecular cloning II: a laboratory manual. Cold Spring Harbor Laboratory Press. Cold Spring Harbor, New York

Pedrosa FO, Yates MG (1984) Regulation of nitrogen fixation ( nif ) genes in Azospirillum brasilense by nifA and ntr ( gln ) type gene products. FEMS Microbiol Lett 23: 95-101

Quiviger B, Franche C, Luftalla G, Haselkorn R, Elmerich C (1982) Cloning of a nitrogen fixation ( nif ) gene cluster of Azospirillum brasilense . Biochimie 64: 495-502

Shaw WV (1975) Chloramphenicol acetyltransferase from chloramphenicol - resistant bacteria. Methods in Enzimology 43: 737-746

Singh M, Tripathi AK, Klingmuller W (1989) Identification of a regulatory nifA type gene and physical mapping of cloned new nif regions of Azospirillum brasilense . Mol Gen Genet 219: 235-240

Zamaroczy M de, Delorme F, Elmerich C (1989) Regulation of transcription and promoter mapping of the structural genes for nitrogenase ( nifHDK ) of Azospirillum brasilense Sp7. Mol Gene Genet 220: 88-94

Yanisch-Perron C, Vieira J, Messing J (1985) Improved M13 phage cloning vectors and hosts strains: nucleotide sequences of the M13mp18 pUC19 vectors. Gene 33: 103-119

CLONING OF THE nifA AND nifB GENES OF Azospirillum brasilense STRAIN FP2


M.A. KNOPIK, S. FUNAYAMA, L.U. RIGO, E.M. SOUZA, H.B. MACHADO
and F.O. PEDROSA
Dept. of Biochemistry, Universidade Federal do Paraná
C. Postal 19046
81504 Curitiba, PR, BRAZIL


ABSTRACT. A genomic library of Azospirillum brasilense was constructed
and used to isolate the nifA gene by complementation of a nifA$^-$ mutant
of A. brasilense (FP10). A recombinant plasmid, pMAK7, was isolated and
found to contain a functional nifA gene and to hybridize with nifA and
nifB probes from Herbaspirillum seropedicae.


## 1. INTRODUCTION

The nitrogen fixation (nif) genes of Azospirillum brasilense that have
been so far cloned or identified are nifHDK, nifE, nifUS, nifB and
fixABC (Gallimand et al., 1989). The nitrogenase structural genes of A.
brasilense, nifHDK, comprise a single operon disposed and transcribed in
the same order as that of Klebsiella pneumoniae (Perroud et al., 1985).
Regulation of nif gene expression in A. brasilense has been shown to
involve the nifA and ntrC genes acting in a cascade-type regulatory
mechanism (Pedrosa and Yates, 1984), similar to that described in K.
pneumoniae (Merrick, 1983; Ow and Ausubel, 1983).
   In this paper we report the cloning and characterization of the
nifA and nifB of A. brasilense strain FP2.


## 2. MATERIALS AND METHODS

### 2.1. Media and Bacterial Strains

A. brasilense strain FP2 (Sp7 Nal$^r$ Sm$^r$, w.t.) and its Nif$^-$ derivatives
(FP3, FP6, FP8, FP9 and FP10; Pedrosa and Yates, 1984) were grown in
NFbHP medium at 30°C. The semi-solid medium contained 0.15% agar. The
nitrogen source was ammonium chloride (20 mM), sodium glutamate (5 mM),
or N$_2$. Escherichia coli strains DH5 and MC1061 were grown in Luria broth
(LB) or agar (LA) (Maniatis et al., 1982). Antibiotics were added as
required.

## 2.2. Plasmids

The cosmid vector pLAFR3 (Tc$^r$, $\lambda$ cos, Inc. P1; Staskawicz et al., 1987) was used to construct a gene library of A. brasilense strain FP2. Plasmid pRK2013 (Km$^r$, Tra$^+$; Figurski and Helinski, 1979) was used as helper in triparental crosses. Plasmid R68.45 (Tra$^+$, Tc$^r$, Km$^r$, Ap$^r$, Inc. P1) was used in surinfection experiments to eliminate plasmids of the same incompatibility group (Haas and Holloway, 1976). Recombinant plasmids pMAK1 to 7 contain the A. brasilense nifA gene cloned into pLAFR3, as described herein.

## 2.3. Conjugation

Biparental and triparental matings were carried  out as described (Pedrosa and Yates, 1984).

## 2.4. Nitrogenase Activity

Nitrogenase activity (acetylene reduction) was assayed in N-free NFbHP semi-solid medium, as previously described (Pedrosa and Yates, 1984).

## 2.5. Construction of the Gene Library of A. brasilense

Total DNA of A. brasilense strain FP2 was purified, partially digested with Sau 3A. DNA fragments ranging from 20 to 30 kb were isolated using low melting point (LMP) agarose gel electrophoresis. Vector arms were prepared by digesting pLAFR3 separately with Eco RI or Hind III, followed  by dephosphorylation with alkaline phosphatase, and finally digestion with Bam HI. DNA fragments (2 $\mu$g) and vector arms (2 $\mu$g each) were ligated, and packaged into lambda particles. E. coli DH5 was infected with the phage suspension, and recombinant clones were isolated in LA medium containing tetracycline (15 $\mu$g/ml). Recombinant clones (3000) were collected and stores in 50% glycerol at $-20^o$C. Recombinant DNA techniques were carried out as described in Maniatis et al. (1982).

## 2.6. Hybridization

Specific DNA fragments, purified by LMP agarose gel electrophoresis, were labelled with digoxigenin-dUTP using the Genius Kit (Boehringer) according the manufacturer's instructions. Restriction fragments from plasmid pMAK7 or total DNA from A. brasilense were separated by electrophoresis and transferred to nylon membrane as described (Maniatis et al., 1982). Hybridizations were carried out under conditions of low stringency (60$^o$C for 18 hours and washed twice with 2X SSC at 65$^o$C). The nifA and nifB probes of Herbaspirillum seropedicae were a 0.4 kb Sal I/Bgl II and a 1.8 Pst I fragment from pEMS101 (Souza et al., submitted), respectively. The nifHDK probe of A. brasilense was the 6.7 Eco RI fragment from pAB35 (Jara et al., 1985).

3. RESULTS

## 3.1. Idetification of the nifA Gene of A. brasilense

In order to isolate the nifA gene of A. brasilense, the gene library was
conjugated with the nifA⁻ mutant of A. brasilense strain FP10, in a
triparental cross. Transconjugants were screened for $N_2$-dependent growth
on solid NFbHP medium under a nitrogen atmosphere containing 0.5-1.0% of
oxygen. Presumptive A. brasilense FP10 Nif⁺ transconjugants were
purified on NFbHP containing 20 mM ammonium chloride and tetracycline
(15 μg/ml), and then tested for nitrogenase activity. Seven
transconjugants were found to be capable of diazotrophic growth,
displayed acetylene reduction activity (Table 1), and contained
recombinant plasmids, as revealed by agarose gel electrophoresis.
Plasmid pMAK7 was chosen for further studies.

TABLE 1. Complementation of the nifA⁻ mutant (FP10) with
recombinant plasmids from a genomic library of A.
brasilense strain FP2

| STRAINS | nitrogenase activity $(nmol.C_2H_4.min^{-1}.mg\ protein^{-1})$ |
|---|---|
| FP2 | 18.2 |
| FP10 | 0.0 |
| FP10(pMAK1)[a] | 9.6 |
| FP10(R68.45)[b] | 0.0 |
| FP10(pMAK1)[c] | 9.5 |
| FP10(pMAK7)[a] | 9.2 |
| FP10(R68.45)[b] | 0.0 |
| FP10(pMAK7)[c] | 8.7 |

[a] A. brasilense FP10 transconjugants.
[b] Recombinant plasmids from the transconjugants were
eliminated by plasmid R68.45 (Inc. P1).
[c] Plasmids pMAK1 and pMAK7 were isolated from A.
brasilense FP10 transconjugants, transformed in E.
coli MC1061 and transferred back to FP10.

## 3.2. Other Physiological Studies

The above plasmids, named pMAK1 to pMAK7, were isolated, used to
transform E. coli MC1061, and finally transferred back to A. brasilense
FP10 by conjugation. All transformants tested showed nitrogenase
activity and diazotrophic growth. Elimination of the recombinant
plasmids (pMAK's) with plasmid R68.45, of the same incompatibility

group, restored the Nif⁻ phenotype (Table 1). A. brasilense
transconjugants harbouring pMAK7 failed to show acetylene reduction in
semi-solid medium supplemented with ammonium chloride (10 mM), but showed
wild-type nitrogenase switch-off by ammonium ions (0.1 mM). Expression
of nitrogenase activity at 37°C by FP10(pMAK7) transconjugants confirms
that the nifA gene product of A. brasilense is thermostable, as
previously suggested (Pedrosa and Yates, 1984).

3.3. Complementation of other Nif⁻ mutants of A. brasilense by pMAK7

Plasmid pMAK7 was unable to complement the regulatory FP8 and FP9
(ntrC⁻), and structural FP3 (nifHDK⁻) and FP6 (lacks MoFe protein)
mutants of A. brasilense, suggesting that this plasmid does not contain
functional homologous genes.

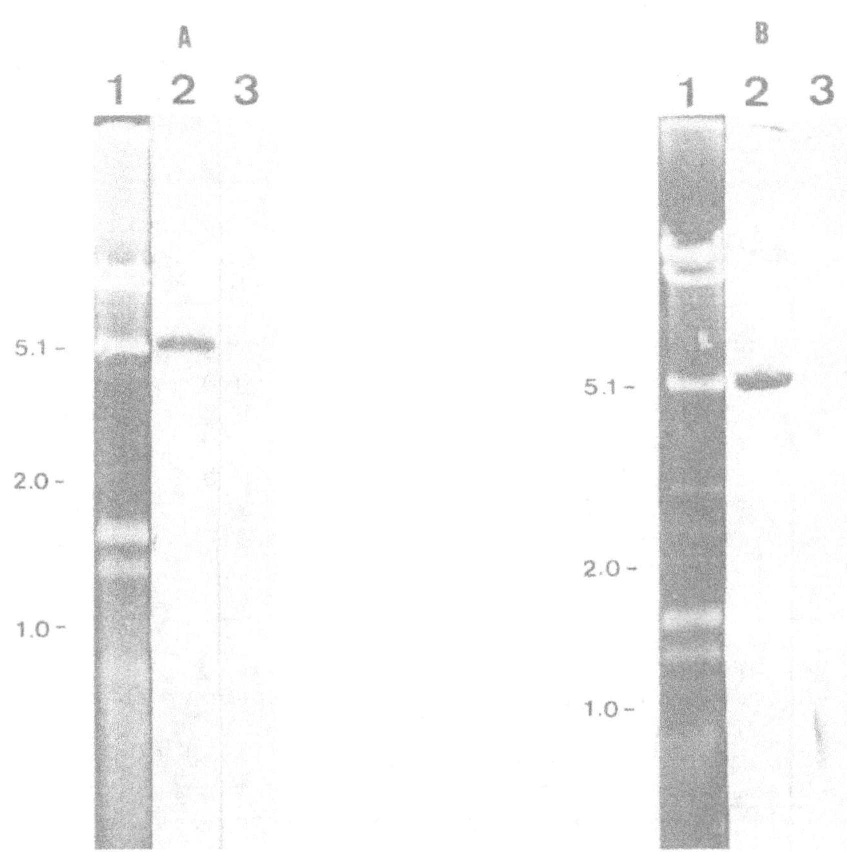

Figure 1. Hybridization of plasmid pMAK7 with nifA and nifB. A) Lane 1:
plasmid pMAK7 digested with Eco RI; lane 2: same as lane 1 hybridized
with the nifA probe; lane 3: total DNA from FP2 digested with Eco RI and
hybridized with the nifA probe. B) Lane 1: plasmid pMAK7 digested with
Eco RI; lane 2: same as lane 1 hybridized with the nifB probe; lane 3:
total DNA from FP2 digested with Eco RI and hybridized with the nifB
probe.  Numbers on the left of each panel refers to kilobase pairs.

3.4. The <u>nifA</u> gene of <u>A</u>. <u>brasilense</u> hybridizes to the <u>nifA</u> gene of
<u>H</u>. <u>seropedicae</u>

A strong hybridization band was detected on the 5.2 kb fragment from
plasmid pMAK7 and also from total DNA of FP2 digested with <u>Eco</u> RI, when
challenged with a probe carrying the <u>nifA</u> gene of <u>H</u>. <u>seropedicae</u>. This
same fragment was also found to hybridize with the <u>nifB</u> gene of <u>H</u>.
<u>seropedicae</u>. Plasmid pMAK7 failed, however, to hybridize with the <u>nifHDK</u>
genes of <u>A</u>. <u>brasilense</u>. These results indicate that the 5.2 kb <u>Eco</u> RI
fragment carries the nifA and nifB genes of A. brasilense (Fig. 1).


## 4. DISCUSSION

The results presented herein are: a)the <u>nifA</u> and <u>nifB</u> genes from <u>A</u>.
<u>brasilense</u> strain Sp7 has been cloned and partially characterized,
b)they are distant from the <u>nifHDK</u>  or <u>ntrC</u>  genes, c)the lack of
nitrogenase activity in <u>A</u>. <u>brasilense</u> FP10(pMAK7) transconjugants grown
in the presence of ammonium chloride suggests that the nifA gene is
expressed under control of an ammonium-regulated promoter, and
d)<u>A</u>. <u>brasilense</u> FP10(pMAK7), contrary to <u>A</u>. <u>brasilense</u> FP10(pCK3, carries
the <u>nifA</u> gene of <u>K</u>. <u>pneumoniae</u>) transconjugants, showed nitrogenase
activity at 37°C, indicating that the <u>nifA</u> gene product of <u>A</u>. <u>brasilense</u>
is thermostable at this temperature.


## 5. ACKNOWLEDGEMENTS

We thank CNPq, FINEP and VITAE for financial support.


## 6. REFERENCES

Figurski, D.H. and Helinski, D.R. (1979) 'Replication of an origin-
    containing derivative of plasmid RK2 dependent on a plasmid  function
    provided in <u>trans</u>', Proc. Natl. Acad. Sci. USA, 76, 1648-1652.

Galimand, M., Perroud, B., Delorme, F., Paquelin, A., Vieille, C.,
    Bozouklian, H. and Elmerich, C. (1989) 'Identification of DNA regions
    homologous to nitrogen fixation genes <u>nifE</u>, <u>nifUS</u> and <u>fixABC</u> in
    <u>Azospirillum brasilense</u> Sp7', J. Gen. Microbiol., 135, 1047-1059.

Haas, D. and Holloway, B.W. (1978) 'Chromosome mobilization by the
    plasmid R68.45: a tool in <u>Pseudomonas</u> genetics', Mol. Gen. Genet.,
    158, 229-237.

Jara, P., Quiviger, B., Laurent, P. and Elmerich, C. (1983) 'Isolation
    and genetic analysis of <u>Azospirillum brasilense</u> Nif ⁻ mutants',
    Can. J. Microbiol. 29, 968-972.

Maniatis, T., Fritsch, E.F. and Sambrook, J. (1982) Molecular cloning: a laboratory manual, Cold Spring Harbor Laboratory, Cold Spring Harbor, N.Y.

Merrick, M. (1983) 'Nitrogen control of the nif regulon in Klebsiella pneumoniae: involvement of the ntrA gene and analogies between ntrC and nifA', EMBO J., 2, 39-44.

Ow, D.W. and Ausubel, F.M. (1983) 'Regulation of nitrogen metabolism genes by nifA gene product in Klebsiella pneumoniae' Nature, 301, 307-313.

Pedrosa, F.O. and Yates, M.G. (1984) 'Regulation of nitrogen fixation (nif) genes of Azospirillum brasilense by nifA and ntrC (gln) type gene products', FEMS Microbiol. Lett., 23, 95-101.

Perroud, B., Bandhari, S.K. and Elmerich, C. (1985) 'The nifHDK operon of Azospirillum brasilense Sp7', in W. Klingmuller (ed.), Azospirillum III, Springer-Verlag, Berlin, pp.10-19.

Staskawicz, B., Dahlbeck, D., Keen, N. and Napoli, C. (1987) 'Molecular characterization of cloned avirulence gene from race 0 and race 1 of P. seryngae pv. glycina' J. Bacteriol., 169, 5789-5794.

REGULATION OF NITROGEN FIXATION GENES IN AZOSPIRILLUM:CHARACTERIZATION
OF A nif REGULATORY REGION


A.K.TRIPATHI[1], M.SINGH[2], W.KLINGMUELLER[*]
(1) School of Biotechnology, Faculty of Science
Banaras Hindu University, Varanasi-221005, INDIA.
(2) Microbiology Division, GBF, Mascheroder Weg 1,
D-3300 Braunschweig, F.R.Germany
*Department of Genetics, University of Bayreuth
8580 Bayreuth, F.R.Germany

ABSTRACT.   There  is  evidence  for  the  presence  of  ntrC  and  nifA  like
genes  and  their  involvement  in  the  expression  and  regulation  of
nitrogen  fixation  (nif)  genes  in  Azospirillum.   This  is  supported  by
the  isolation  of  our  group  of  Tn5  induced  mutant  of  A.brasilense  which
exhibited  NifA⁻  phenotype.   Tn5  mutagenized  nif  DNA  of  this  mutant
(Nif 27)  was,  earlier,  cloned  in  the  recombinant  plasmid  pMS188.   A
deletion  derivative  of  pMS188  has  been  created  to  remove  most  of  the
Tn5  DNA  present  on  a  HpaI  fragment.   The  wild  type  nif  regulatory  gene
seems  to  be  located  on  a  12kb  EcoRI  fragment.   Homology  of  this  nif
regulatory  region  has  been  detected  in  several  strains  of  A.brasilense
and  A.lipoferum.

INTRODUCTION

Genetic  evidence  for  the  regulation  of  nif  gene  expression  in
Azospirillum  brasilense  by  a  nifA  type  and  possibly  ntrC  type  system,
analogous  to  that  in  K.pneumoniae,  was  suggested  on  the  basis  of  the
ability  of  cloned  nifA  and  ntrC  genes  of  K.pneumoniae  to  complement
Nif⁻  mutants  of  A.brasilense  (Pedrosa  &  Yates,  1984).   Nucleotide
sequencing  data  of  A.brasilense  nifH  open  reading  frame  and  the  DNA
region  upstream  revealed  the  presence  of  -24/-12  consensus  $\sigma^{54}$ (NtrA)
dependent  promoter  element  and  two  upstream  activator  sequences  (UAS),
characterized  by  $_{54}$ TGT-N$_{10}$-ACA,  which  are  involved  in  NifA  mediated
activation  of  $\sigma^{54}$  dependent  nif  promoters  (Fani  et  al,  1989;  de
Zamaroczy  et  al,  1989;  Thoeny  &  Hennecke,  1989).
    In  our  experiments  with  Tn5  mutagenesis  of  A.brasilense,  three
different  types  of  Nif⁻  mutants  were  isolated  (Singh  &  Klingmueller,
1986).   Western  blots  of  proteins  from  derepressed  cultures  of  the
three  mutants  were  used  to  identify  nitrogenase  iron  protein.   The  two
forms  of  nitrogenase  iron  protein  were  completely  missing  in  Nif27
mutant.   This  mutant  was  complemented  by  pCK3  which  expressed
K.pneumoniae  nifA  gene  constitutively  but  not  by  pGE10  which  expressed
the  K.pneumoniae  ntrC  gene.   Further  characterization  of  the  mutant

Nif 27 revealed its inability to activate nifH lacZ fusion whereas wild type A brasilense activated the nifH-lacZ fusion quite strongly. All these evidences indicated for a regulatory NifA⁻ phenotype of the Tn5 induced mutant, Nif27 (Singh et al, 1989). The mutagenized nif region from Nif27 was cloned in a suicide vector pSUP202 and designated pMS188. By gene replacement technique, the mutagenized DNA region was recombined into wild type A.brasilense genome to obtain NifA⁻ phenotype. This confirmed that the NifA⁻ phenotype was conferred specifically by the cloned Tn5 mutagenized nif region. Further characterization of the nif regulatory region of A.brasilense is presented in this report.

## MATERIALS AND METHODS

Bacterial Strains and Plasmids: A.brasilense strains ATCC 29710, 29145 and A.lipoferum strains ATCC 29707, 29708, 29709, 29731 (Tarrand et al, 1978) were used. E.coli HB101, mutants of A. brasilense (Nif233, Nif232 and Nif27), and plasmids pMS188, pCK3 (used as K.pneumoniae nifA gene for hybridization purpose) are described in Singh & Klingmueller (1986) and Singh et al (1989).

Recombinant DNA techniques and Southern Hybridization: Total genomic DNA of Azospirillum strains was isolated and purified according to Singh & Klingmueller (1986). Minipreparations of plasmids was done according to Birnboim & Doly (1979). Restriction digestion of genomic/plasmid DNA, Southern transfer of DNA pattern on nitrocellulose paper, electroelution of probes, blunt end ligation, nick translation of probes with ($\alpha$-$^{32}$)dCTP, and hybridization was done by using established methods (Maniatis et al, 1982) and as described in Singh & Klingmueller (1986) and Singh et al, (1989).

## RESULTS

Hybridization of genomic DNA from A.brasilense and Tn5 induced nif⁻ mutants with Tn5 and pMS188

Hybridization of EcoRI digested total genomic DNA from wild type A.brasilense and three nif mutants Nif233, Nif232 and Nif27, with Tn5 probe (P1, Figure 1) revealed single hybridizing fragments in Nif233 Nif232 and Nif27 but none in wild type strain (Fig.2). When similar genomic pattern was hybridized with 17.9 kb insert of pMS188 (probe P2, Fig.1), an additional band of 12kb was observed in the lanes of wild type, Nif233 and Nif232 (Fig.2). This 12 kb EcoRI fragment represented the fragment of A.brasilense genomic DNA which has been mutagenized by Tn5 in the mutant Nif27. Homology of 17.9 kb EcoRI fragment in Nif27 with Tn5 probe indicated that the mutagenized fragment was not nifKDH region. As reported earlier (Quiviger et al, 1982) nifKDH in A.brasilense is located on a 6.7 kb EcoRI fragment. Since the size of Tn5 is 5.7 kb, had Tn5 integrated in nif KDH region, a 12.4 kb EcoRI fragment, instead, would have hybridized with Tn5.

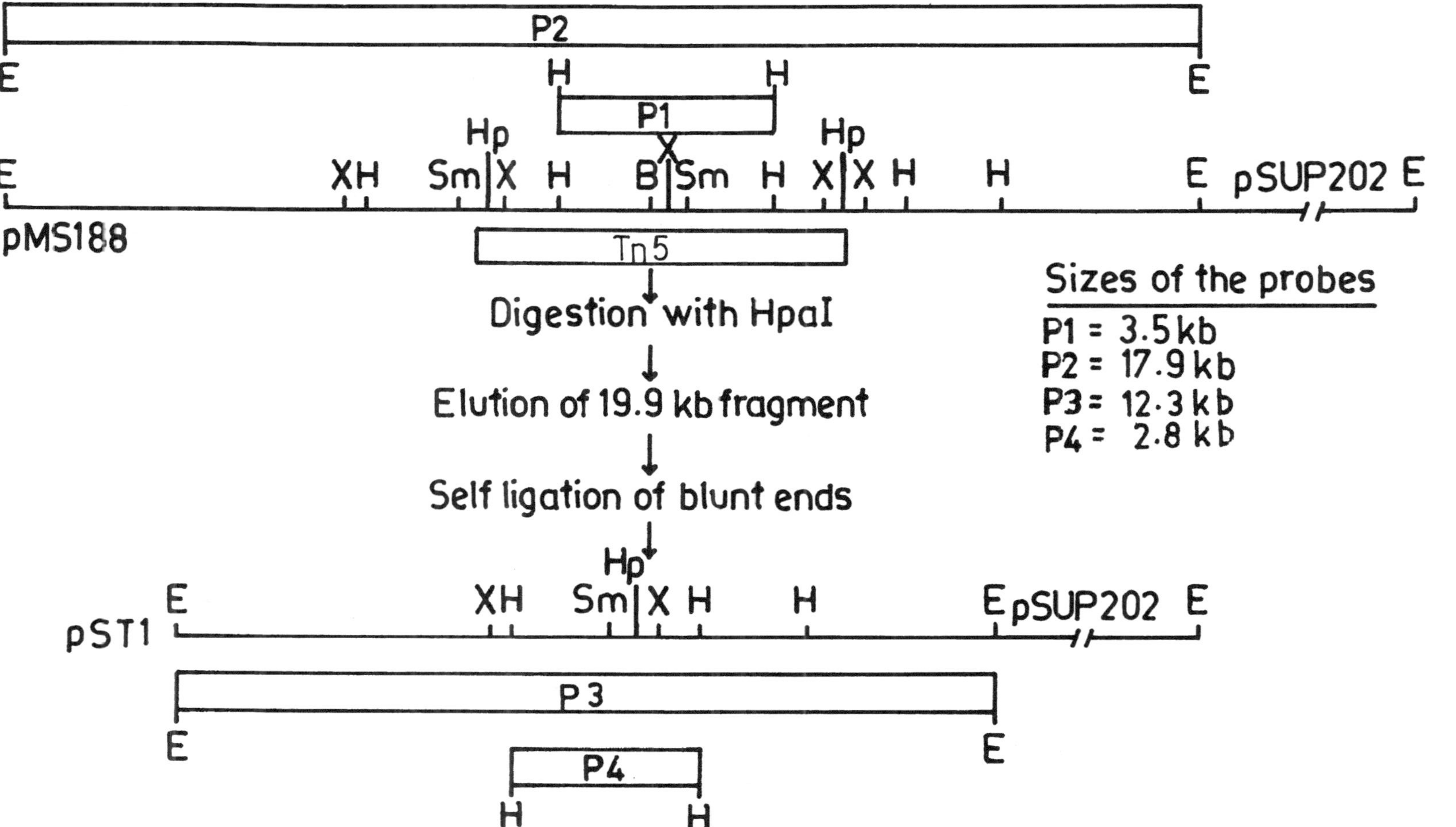



Figure 1. Physical map of the Tn5 mutagenized <u>nif</u> regulatory region of <u>A.brasilense</u>, and scheme for construction of pST1 (a deletion derivative of pMS188). Boxed regions P1, P2, P3 and P4 indicate restriction endonuclease fragments isolated for use as hybridization probes.

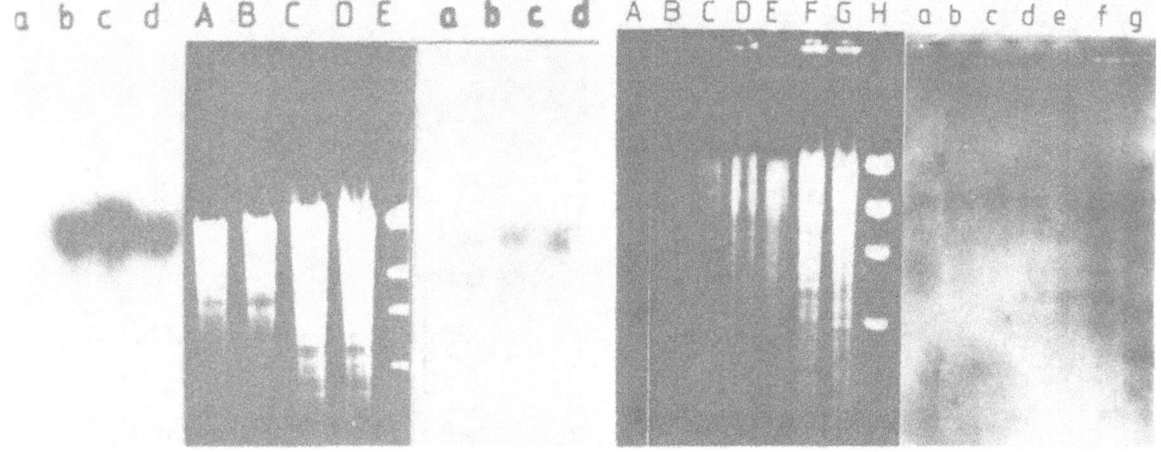



Figure 2                    Figure 3

Fig.2. Southern blot showing the hybridization of the EcoRI digested genomic DNA from the wild type (a) and mutants, 233(b), 232(c) 27(d) with the Tn5 probe (P1). The right side shows the same filter after removal of the probe and rehybridization with probe P2. In lanes A,B,C and D is the genomic DNA in agarose gel, respectively. The lane A shows the Lambda DNA digested with HindIII.

Fig.3. Conservation of the homology of the Tn5 mutagenized DNA from mutant 27 in Azospirillum. The EcoRI fragment of pMS188 (probe P2) was radioactively labelled and hybridized to EcoRI digested genomic DNA of A.brasilense 29710-Z8/35 (a,; nifHDK⁻ mutant), 29710(b), 29145(c), and the Alipoferum strains 29731(d), 29709(e), 29708(f), and 29707(g). Lane H shows the Lambda HindIII standard.

---

## Conservation of the homology of the Tn5 mutagenized DNA from mutant Nif27 in Azospirillum

Total genomic DNA from A.brasilense and A.lipoferum strains was hybridized with probe P2 (Fig.1) to study the inter- and intra- specific differences in the hybridization pattern. Figure 3 revealed that in A.brasilense strains same size of gragment (12kb) showed homology whereas in case of A.lipoferum the hybridization pattern was quite variable. Variability in the organization of nifKDH genes in two strains of A.lipoferum, i.e. ATCC29708 and ATCC29731, has also been reported (Fahsold et al, 1985).

## Construction of pST1, a deletion derivative of pMS188

Plasmid pMS188 when digested with HpaI, produced two fragments, one of ca 19.9 kb and the other of 5.4 kb. The 5.4 kb fragment was comparable to the Tn5 internal HpaI fragment. The 19.9 kb fragment was eluted and the blunt ends ligated. E.coli HB101 transformants, transformed with the ligation mixture, were screened for tetracycline resistance and Kanamycin sensitivity. The new plasmid construct (designated as pST1; Fig.1) was isolated from transformants and produced a 19.9 kb linearized HpaI fragment, and two fragments of 7.6 kb (vector DNA) and 12.3 kb (insert). Comparison of the restriction map of pMS188 with that of pST1 revealed a deletion of 5.4 kb fragment in pST1 insert.

Identification of pST1 homologous fragments in A.brasilense genome

Insert of plasmid pST1 was used as probe (P3; Fig.1) to identify homologous fragments in A.brasilense genomic DNA digested with EcoRI, Hind III and Sal I. The autoradiogram (Fig.4) revealed one band of 12 kb in EcoRl digest, three bands of 1.4 kb, 2.5 kb and ca 20 kb in Hind III digest, and three bands of 0.7 kb, 2.3 kb and 7.5 kb in SalI digest. The sizes of the fragments observed on autoradiogram matched very well with the restriction fragments present on pST1 (See Fig.1 for restriction map of pST1). In another experiment, a 2.8 kb Hind III fragment of pST1 (which contained the only HpaI site of pST1) was used as probe (P4) with the similar genomic pattern. In this case, only single bands of 12 kb in EcoRI digest, 7.5 kb in SalI digest and 2.5 kb in Hind III digest hybridized (autoradiogram not shown). these single bands in each lane represented the potential fragments suitable for cloning of the wild type regulatory nif gene which has been inactivated by Tn5 in mutant Nif 27.

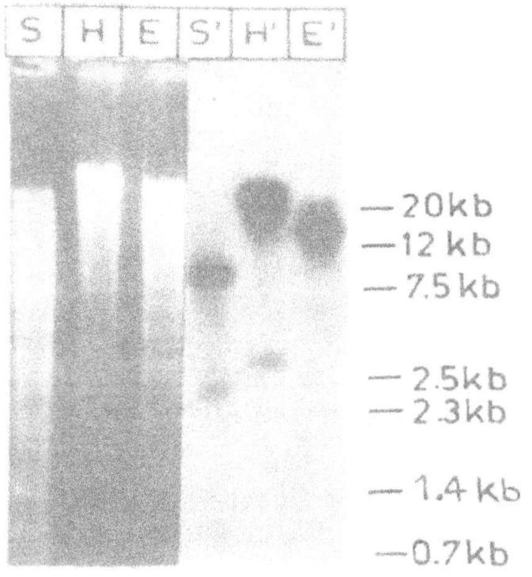


Fig.4
Agarose gel electropheresis of the A.brasilense genomic DNA digested with EcoRI(lane E), Hind III (lane H) and SalI(lane S). Southern blot of the electrophoretogram was hybridized with 12.3 kb insert of pST1 (probe P3). Autoradiogram of the Southern blot is represented on the right side in lanes E', H' and S'.

DISCUSSION

In the present report we have characterized the Tn5 mutagenized nif region of mutant Nif 27 by Southern hybridization technique. One paper in this volume (Elmerich et al) reports about identification of clones from A.brasilense gene bank which were homologous to Bradyrhizobium japonicum nifA. These clones have been shown to complement NifA⁻ mutant of A.brasilense. Bradyrhizobium japonicum nifA gene region has, earlier, been identified and cloned on the basis of its homology with K.pneumoniae nifA gene (Adams et al, 1984). The Tn5 mutagenized nif gene region of our clone (pMS188), however, failed to show any hybridization signal with K.pneumoniae nifA (results not shown) even under low stringency conditions.

In another report (Knopik et al, this volume), cloning of nifA gene from A.brasilense has also been achieved on the basis of complementation of FP10, a NifA⁻ type mutant (Pedrosa and Yates, 1984). Our mutant Nif27, too, has been characterized to exhibit NifA⁻ phenotype (Singh et al, 1989). Cloning of nifA from A.brasilense Sp7 by the two groups (described in this volume) report about a ca 5 kb EcoRI fragment which contained complete nifA gene. Our regulatory nif gene, on the other hand, is located on a ca 12 kb EcoRI fragment of A. brasilense Cd. The two strains of A.brasilense, i.e. Sp7 and Cd, are reported to be close relatives as both the strains harbour their nif structural genes on 6.7 kb EcoRI fragment (Wenzel et al, 1983).

Keeping abovementioned facts in view, it seems that the expression of nifA gene in A.brasilense may be regulated either by a cascade mechanism analogous to fix LJ genes of Rhizobium meliloti (David et al, 1988) or by a gene analogous to nfrX of Azotobacter vinelandii and A.chroococcum (Santero et al, 1988). Our Nif27 mutant of A.brasilense Cd, thus, appears to harbour defect in a gene which regulates the expression of nifA gene and exhibits a NifA⁻ phenotype.


ACKNOWLEDGEMENTS

This work was supported by the Bundesministerium fuer Forschung und Technologie. DAAD support to AKT is greatly acknowledged.

REFERENCES

Adams, T.H., McClung, C.R. and Chelm, B.K.(1984)'Physical organization of the Bradyrhizobium japonicum nitrogenase gene region.' J.Bacteriol. 159, 857–862.

Birnboim, H. and Doly, J.(1979) 'A rapid alkaline extraction procedure for screening recombinant plasmid DNA.' Nucleic Acids Res. 7, 1513–1525.

David, M., Deveran, M.L., Batut, J., Dedieu, A., Domergue, O., Ghai, J., Hertig, P., Boistard, P. and Kahn, D. (1988) 'Cascade regulation of nif gene expression in Rhizobium meliloti.' Cell.54, 671–683.

de Zamaroczy, M., Delorme, F. and Elmerich, C.(1989) 'Regulation of transcription and promoter mapping of the structural genes for nitrogenase (nif HDK) of Azospirillum brasilense Sp7.' Mol.Gen.Genet. 220, 88–94.

Fahsold, R., Singh, M. and Klingmueller, W. (1985) 'Cosmid cloning of nitrogenase structural genes of Azospirillum lipoferum.' In: Klingmueller, W.(ed) Azospirillum III: Genetics, Physiology, Ecology, Springer, Berlin, pp.30–40.

Fani, R., Allotta, G., Bazzicalupo, M., Ricci, F., Schipani, C. and Polsinelli, M. (1989) 'Nucleotide sequence of the gene encoding the nitrogenase iron protein (nifH) of *Azospirillum brasilense* and identification of a region controlling nifH transcription.' Mol.Gen.Genet.220, 81-87.

Maniatis, T., Fritsch, E.F. and Sambrook (1982) Molecular Cloning: A Laboratory Manual. Cold Spring Harbor Laboratory, Cold Spring Harbor, New York.

Pedrosa, F.O. and Yates, M.G. (1984) 'Regulation of nitrogen fixation (nif genes) of *Azospirillum brasilense* by nifA and ntr (gln) type gene products.' FEMS Microbiol. Lett. 23, 95-101.

Quiviger, B., Franche, C., Luftalla, G., Rice, D., Haselkorn, R and Elmerich, C.(1982) 'Cloning of a nitrogen fixation (nif) gene cluster of *Azospirillum brasilense*.' Biochimie 64, 495-502.

Santero, E., Toukdarian, A., Humphrey, R. and Kennedy, C.(1988) 'Identification and characterization of two nitrogen fixation regulatory regions, nifA and nfrX, in Azotobacter vinelandii and Azotobacter chroococcum.' Mol.Microbiol.2: 303-314.

Singh, M. and Klingmueller, W. (1986) 'Transposon mutagenesis in *Azospirillum brasilense*: Isolation of auxotrophic and NIF mutants and molecular cloning of the mutagenized nif DNA.' Mol.Gen.Genet.202, 136-142.

Singh, M., Tripathi, A.K. and Klingmueller, W. (1989) 'Identification of a regulatory nifA type gene and physical mapping of cloned new nif regions of *Azospirillum brasilense*.' Mol.Gen.Genet.219, 235-240.

Tarrand, J.J., Krieg, N.R. and Doebereiner, J. (1978) 'A taxonomic study of the *Spirillum lipoferum* group with descriptions of a new genus, *Azospirillum lipoferum* (Beijerinck) comb.nov. and *Azospirillum brasilense* sp.nov.' Can.J.Microbiol. 24, 967-980.

Thoeny, B. and Hennecke, H. (1989) 'The -24/-12 promoter comes of age.' FEMS Microbiol.Rev.63, 341-358.

Wenzel, W., Singh, M. and Klingmueller, W. (1983) 'Molecular cloning of nitrogen fixation genes from *Azospirillum*.' In Klingmueller, W. (ed.) Azospirillum II: Genetics, Physiology, Ecology. Birkhaeuser, Basel, pp.39-46 (Experientia supplementum, 48).

# *AZOSPIRILLUM*-PLANT INTERACTION: A BIOCHEMICAL APPROACH


ANITA HAEGI* AND MADDALENA DEL GALLO.
*Agrobiotechnology Dept. ENEA-Casaccia, Via Anguillarese 301. I-00060 Rome, Italy*
*(*present address: Experimental Plant Pathology Institute. MAF. Via Bertero 22, I-000156 Rome)*



SUMMARY. The composition of *Azospirillum lipoferum* Col 5 and *Azospirillum brasilense* Cd polysaccharides which surround the cell (capsular-polysaccharides, CPS) and the composition of polysaccharides loosely bound to the bacteria or released into the culture-medium (exo-polysaccharides, EPS) has been analysed - after hydrolysation - by HPLC and by chemical analyses.
Both fraction (CPS and EPS) are qualitatively similar in both bacteria, and are composed mainly by three sugars and by two organic acids. Relative amount of each monosaccharide unit, instead, varies independently among the two strains and during the culture growth. *A. lipoferum* Col5 CPS-fraction shows a peculiar pattern on glucose amount. which increases during the culture growth, more than other sugars.
This monosaccharide independent pattern suggests that more than one polysaccharide is present in *Azospirillum*-strains tested.


## Introduction

Previous observations have shown that initial step of plant root colonization by symbiotic or associated microorganisms starts generally from a chemotactic attraction of bacteria towards roots. Attractants are mainly component of root exudates (Gaworzewska and Carlile, 1982; Heinrich and Hess, 1984; Del Gallo et al. manuscript in preparation).
Once the bacteria reach the root, surface recognition events are likely to take place, similarly to other plant-microbe interaction, such as pathogens with the host plant (Halverson and Stacey, 1986). Actually, the specificity of plant-microorganism interactions depends on these initial mechanisms of recognition.
This type of interactions entail a complex series of differentiated events - till now not fully understood in many systems - which involves reciprocal stimuli and responses, mediated by specific molecules, both by the host and by the microorganism. Experimental evidences demonstrated that plant lectins - root surface glycoproteins - are implicated in such molecular exchanges (Püler et al., 1988).
The lectin hypothesis suggest that these host-plant glycoproteins, interact specifically with polysaccharide (PS) components of bacterial surface, initiating thus the infection process and determining specificity of interaction.
The involvement of lectin-PS interaction has been demonstrated for almost all Rhizobia and Bradyrhizobia symbioses. Molecular biology experiments carried out on mutated strains of *R. leguminosarum* show the determinant role of root lectins in *Rhizobium*-legume symbiosis

(Diaz *et al.*, 1989). It has been recently demonstrated also in *Bradyrhizobium*-soybean symbiosis, in which lectin role was still unclear, that lectin interaction, even though is not essential for nodulation, is important for competitivity (Bhagwat and Keister, 1990).
In general, in these type of interactions more than one exocellular-PS and lipo-PS (LPS) are involved, and side chain substituent play a determinant role in binding the lectin.
Besides, microorganism PS are important in the next step, when microorganisms, stimulated by the plant, start to produce fibrils which anchor them to the plant surface. This kind of attachment is already well understood in *Agrobacterium tumefaciens* (Matthysse, 1987) and in Rhizobium trifolii (the so called phase 2 adhesion; Dazzo *et al.* 1984).

The interaction of the diazotrophic, plant growth promoting rhizobacterium, *Azospirillum* spp with the host plant is still not well understood, particularly for what is concerning the first steps of interaction which lead to the establishment of the association.
A chemotactic attraction is likely involved at the beginning: Okon and Kapulnik (1986) observed that *Azospirillum* is attracted by a substance present in root exudates which is destroyed by trypsin, so likely to be of proteic nature, and Del Gallo et al. (manuscript in preparation) observed differences among several *Azospirillum* strains on chemotactic behaviour towards wheat and rice seedling roots.
On the basis of previous works on other plant-microorganism interactions, the involvement of plant lectins-bacterial PS has been suggested also for *Azospirillum* (Del Gallo et al. 1989). Yagoda-Shagam and co-workers (1988) showed that *Azospirillum* is able to bind to different FITC-labeled lectins which have different hapten specificity. These authors reported, also, that *A. brasilense* sp7 is able to bind to a lectin specific for 2-keto-3-deoxyoctonate, hypothesizing, thus, the involvement of LPS also in *Azospirillum*-lectin binding.
These results have been confirmed in our laboratory (Del Gallo and Haegi, 1990) when we observed, by fluorescence microscopy, the FITC-Concanavalin (Con A) and FITC-Wheat Germ Agglutinin (WGA) binding to *A. brasilense* Cd and to *A. lipoferum* Col 5. Both Con A and WGA bound CPS and EPS, but to a lower extent. LPS, instead, are not involved in the interaction with WGA, because, after solubilization of the PS-capsule, no fluorescence was observed any more. Besides, we have observed that lectin-binding depends on the culture age, with a maximum binding at the beginning of the stationary phase.

Flocculation and aggregation mechanisms are intensively studied in *Azospirillum*. (Sadasivan and Neyra, 1985; Del Gallo et al., 1989; Madi and Henis, 1989; Michiels et al., 1990). In fact, these processes can mimic the adhesion of bacteria to plant roots.
Flocculation, a morpho-physiological change which occurs in particular stress situations, involves a production of large amount of Calcofluor positive-PS (PS formed by 1-2 and 1-3 β-linked glucans). Recently it has been observed that an *Azospirillum brasilense* Sp7 calcofluor negative no-flocculating mutant, bind in a lower amount to root surface, even though, in a long term experiment, it can colonize the root and affect plant growth (Vanderleyden, comunication presented at the 8th International Congress on Nitrogen Fixation, Knoxville, TN, USA, 1990).
Umali-Garcia et al (1980) have shown that aggregates of *Azospirillum* are present on the surface of grass and wheat roots. Del Gallo et al. (1989) have shown that, in pure culture, aggregates of Azospirillum are entrapped in cellulose-like fibrillar material. However a correlation between flocculation and anchoring of the bacteria to plant is still ambiguous.

In our research we have focused on the possible involvement of PS-lectin binding in *Azospirillum*-plant interaction, thus exploring more in detail the composition of different exocellular-PS fractions produced by *A. brasilense* Cd and *A. lipoferum* Col 5 liquid cultures.

## Materials and Methods

*A. brasilense* Cd and *A. lipoferum* Col5, kindly supplied by J. Döbereiner, were used.
Inoculum preparation and culture media were performed as previously described (Del Gallo and Haegi, 1990).
Sample preparation for the HPLC analysis was performed following the scheme described by Del Gallo and Haegi (1990). Identification of neutral sugars was performed on a Bio-Rad Aminex HPX-87C. Polar sugars and organic acids were analyzed on a Bio-Rad Aminex HPX-87H column as described before (Del Gallo and Haegi, 1990) For quantification glucose, galactose and rhamnose were used as standard with the method of "the 2 point absolute titration curve" described on the Shimadzu Chromatopac C-R3A instruction manual.
Preliminary quantification of PS was performed by the Anthrone method using glucose as standard (Dische, 1962). Uronic acid quantification was performed following the method described by Blumenkrantz and Asboè-Hansen (1973).

## Results and Discussion

Exocellular PS have been divided in three different fractions: CPS tightly bound to the cell outside and EPS which are more loosely bound to the cell and easily detached by centrifugation, or free in the culture medium. EPS have been furtherly divided into EPS soluble and insoluble in ethanol. This fractionation was performed in order to separate different MW EPS fractions.
All fractions have been analyzed by HPLC. Results have shown that, in our cultural conditions, PS different fractions are mainly composed by glucose, galactose and rhamnose, the amount of each varying at different culture growth-stages and among strains.
In a previous work (Del Gallo and Haegi, 1990), we reported WGA-binding (correlated with the amount of CPS produced by the culture) to surface PS, which suggested the presence of N-acetyl-glucosamine (GlcNAc) in PS. Our HPLC analyses revealed, actually, the presence of this compound in the CPS, but this result has not been confirmed by preliminary NMR analysis carried out. Therefore, the presence of in *Azospirillum* CPS is still ambiguous.

### CAPSULAR-POLYSACCHARIDES.

The quantitative and qualitative composition of the CPS of both strains analyzed is shown in figure 1.
Qualitative composition (Figure 1, right) was similar between the two strains, but the amount of each monosaccharide unit varied independently with the culture age and among the two strains.
It can be noticed that in *A. lipoferum* peaks of total CPS at the third and fourth days were mainly due to the sharp increase of glucose. This independent pattern of monosaccharides unit amount during the culture growth, suggest the presence of different PS-chains.

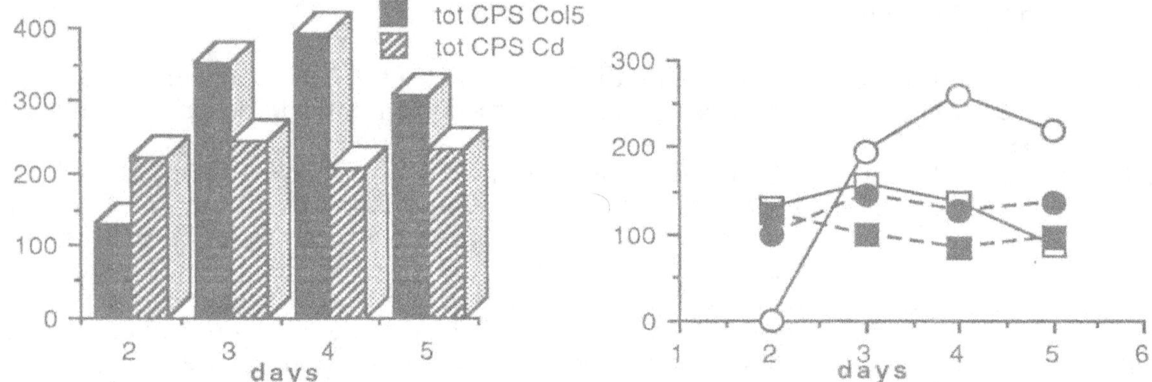

**Figure 1.** On the left: total CPS amount in both strain Col5 and Cd during a 5 days-growth. CPS value are expressed in μg/200 ml of culture. On the right: sugar composition of strain Col 5 (whole line) and of strain Cd (dashed line), μg/200 ml. (circle for glucose, square for galactose + rhamnose).

## EPS, ETHANOL-INSOLUBLE

HPLC analyses have shown that the monosaccharides composition of $EPS_i$ was similar to CPS, but ratios were different (figure 2).

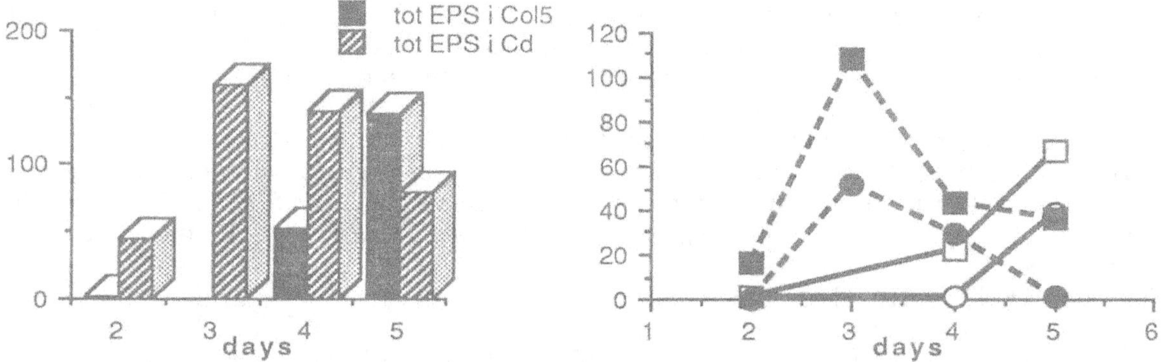

**Figure 2.** On the left: total $EPS_i$ amount in both strain Col5 and Cd during 5-days growth. $EPS_i$ value are expressed in μg/200 ml of culture. On the right: $EPS_i$ sugar composition of strain Col 5 (whole line) and of strain Cd (dashed line), μg/200 ml. (circle for glucose, square for galactose + rhamnose).

Strain Cd produced a larger amount of $EPS_i$ (just the opposite of CPS situation). In both strains the amount of each monosaccharide unit reflected total $EPS_i$ pattern. At the second day glucose was not present in Cd $EPS_i$. In the other cases glucose was present, but in undetectable amount (peak was present in the chromatogram, but not integrated because it was not discriminated from the baseline noise).

## EPS, ETHANOL SOLUBLE

Figure 3 shows, on the left, the pattern of total soluble EPS and on the right the pattern of each monosaccharide unit during culture growth. Glucose in Cd is always undetectable.

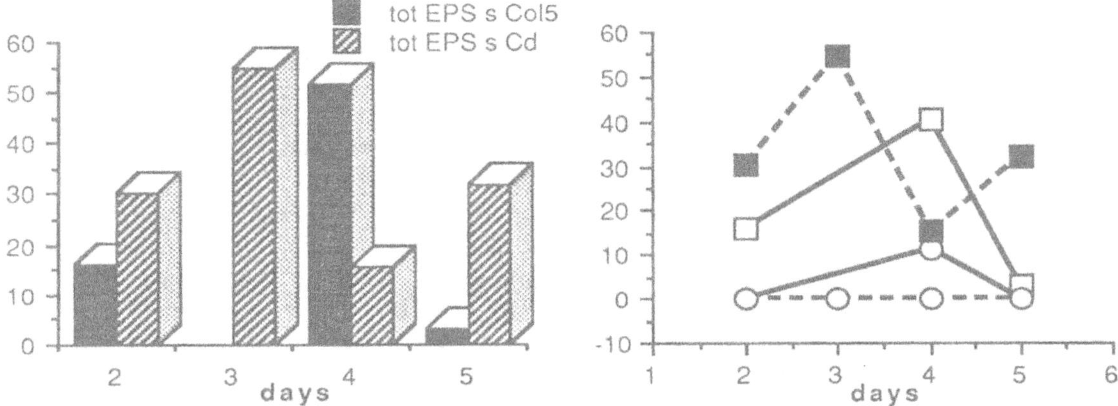

**Figure 3**. On the left: total EPS$_S$ amount in both strain Col5 and Cd during 5-days growth. EPS$_i$ value are expressed in μg/200 ml of culture. On the right: EPS$_S$ sugar composition of strain Col 5 (whole line) and of strain Cd (dashed line), μg/200 ml. (circle for glucose, square for galactose + rhamnose).

A comparison among all Azospirilla-PS shows that CPS are in higher amount than both EPS, and that EPS$_S$ were always recovered in a very low amount.

HPLC and chemical analyses have also shown the presence of uronic acids, mainly represented by galacturonic acid both in capsular and in exoPS. Preliminary studies carried out on organic acids in *Azospirillum*-PS suggest the presence of acetic and succinic acids. These could be present as substituent of the PS chains. This subject needs, however, further investigations because these components could play a detrminant role in lectin-PS interactions, as described in Rhizobia symbioses. Preliminary chemical analyses have also shown the

presence of a large contamination of β-hydroxybutyric acid in the capsular fraction. This is not surprising because *Azospirillum* is known to produce large amount of this compound.

However, β-hydroxybutyric acid can be easily removed by ethanol precipitation. Besides,

Yagoda-Shagam et al. (1988) observed that β-hydroxybutyric acid do not influence lectin binding.

Preliminary NMR studies have also shown a slight contamination with LPS. *Azospirillum* LPS composition have been studied by Choma et al. (1987). If we compare the LPS composition reported in that paper with the composition of PS we have analyzed, it is interesting to notice that LPS share common sugars with exo-PS, in particular, glucose and rhamnose. However, further investigation are needed in order to understand their eventual involvement in lectin binding.

Microscopical observations showed that, in our cultural conditions, both Cd and Col 5, did not bind Calcofluor, but bound lectins. This fact led to the conclusion that *Azospirillum* can produce - even in the same culture - different type of PS. Michiels et al. (1990) reached the same conclusion when observed that Calcofluor mutant strains of *A. brasilense* 7030 still produced EPS, but were affected in flocculation.

In conclusion, *Azospirillum*-PS which bind lectin, are composed by glucose, rhamnose and galactose as neutral sugars, galacturonic acid and putative acetic and succinic acid, each of these component vaying during the culture growth.

## Aknowledgement

We wish to thank Prof. Anna Laura Segre for carrying preliminary NMR analyses.

# References

Bhagwat, A.A. and Keister D.L. (1990) Competition Defective "Mutants of *Bradyrhizobium japonicum*" 8th Int. Congress on N$_2$-Fixation. Knoxville , TN, USA, B32.

Blumenkrants, N. Asboè-Hansen, G. (1973) "New Method for Quantitative Determination of Uronic Acids" Anal. Biochem. **54**, 484-489.

Choma, A., Russa, R., Mayer, H. and Lorkiewicz, Z. (1987) "Chemical analysis of *Azospirillum* lipopolysaccharides" Arch. Microbiol. **146**, 341-345.

Dazzo, F.B., Truchet, G.L., Sherwood, J.E., Hrabak, E.M., Abe, M. and Pankratz, S.H. (1984) "Specific Phases of Roor Hair Attachment in the *Rhizobium trifolii*-Clover" Symbiosis, Appl. Environ. Microbiol. **48**, 1140-1150.

Del Gallo, M and Haegi, A. (1990) "Characterization and quantification of exocellular polysaccharides in *Azospirillum brasilense* and *Azospirillum lipoferum*. Symbiosis, in the press.

Del Gallo, M., Negi, M. and Neyra, C.A. (1989) "Calcofluor- and Lectin-Binding Exocellular Polysaccharides of *Azospirillum brasilense* and *Azospirillum lipoferum*" J. Bacteriol. **172**, 3504-3510.

Del Gallo, M, Neyra, C.A., Waschutza, S., Hofman, N. and Fendrik, I. "First steps of the interaction between cereal and *Azospirillum*", manuscript in preaparation.

Diaz, C.L., Melchers, L.S., Hooykaas, P.J.J., Lugtenberg, B.J.J. and Kijne, JW. (1989) "Root lectin as determinant of host-plant specificity in the *Rhizobium*-legume symbiosis", Nature **338**, 579-581.

Dische, Z (1962) "General color reactions" Methods Carbohydr. Chem. **1**, 477-479.

Gaworzewska, E.T. and Carlile, M.J. (1982) "Positive chemotaxis of *Rhizobium leguminosarum* and other bacteria towars root exudates from legumes and other plants" J. Gen Microbiol. **128**, 1179-1188.

Halverson, L.J. and Stacey, G. (1986) "Signal Exchange in Plant-Microbe Interactions", Microbiol. Rev. **50**, 193-225.

Heinrich, D. and Hess, D. (1984) "Chemotactic attraction of *Azospirillum lipoferum* by wheat roots and characterization of some attractans" Can. J. Microbiol. **31**, 26-31.

Madi, L. and Henis, Y. (1989) "Aggregation in *Azospirillum brasilense* Cd: Conditions and factors involved in cell-to-cell adhesion" Plant Soil **115**, 89-98.

Matthysse, A.G. (1987) "Initial interactions of *Agrobacterium tumefaciens* with plant host cells" CRC Critical Rev. Microbiol. **13**, 281-307.

Michiels, K., Verreth, C. and Vanderleyden, J. (1990) "*Azospirilum lipoferum* and *Azospirillum brasilense* surface polysaccharides mutants that are affected in flocculation. J. Appl. Bacteriol., in the press.

Okon, Y. and Kapulnik, Y. (1986) "Development aand function of *Azospirillum*-inoculated roots. Plant Soil **90**: 3-16.

Püler, A., Enenkel, B., Hillemann, A., Kapp, D., Keller, M., Müller, P., Niehaus, K., Priefer, U.B., Quandt, J. and Schmidt, C. (1988) "*Rhizobium meliloti* and *Rhizobium leguminosarum* mutants defective in surface polysaccharide synthesis  and root nodule development" In "Nitrogen Fixation: Hundred Years After" ed. by H. Bothe, F.J. de Bruijin and W.E. Newton. pp. 431-436.

Sadasivan, L. and Neyra, C.A. (1985) "Flocculation in *Azospirillum brasilense* and *Azospirillum lipoferum*: Exopolysaccharides and Cyst Formation". J. Bacteriol. **163**: 716-723.

Umali-Garcia, M., Hubbel, D.H., Gaskins, M.H. and Dazzo, F.B.. (1980) "Association of *Azospirillum* with Grass Roots". Appl. Environ. Microbiol. **39**, 219-226.

Vanderleyden, J. (1990). "*Azospirillum* associations". 8th International Congress on Nitrogen Fixation, Knoxville, TN, USA

Yagoda-Shagam, J., Barton, L.L., Reed, W.P. and Chiovetti, R. (1988) "Fluorescein Isothiocyanate-Labeled Lectin Analysis of the Surface of the Nitrogen-Fixing Bacterium *Azospirillum brasilense* by Flow Cytometry" Appl. Environ. Microbiol. **54**, 1831-1837.

# Osmotolerance of diazotrophic rhizosphere bacteria

A. HARTMANN, S.R. PRABHU and E.A. GALINSKI
*GSF-Institut für Bodenökologie, Ingolstädter Landstr. 1, DW-8042 Neuherberg, Germany, Krishi Vigyan Kendra, Mitraniketan, P.O., Vellanad-695543, Trivandrum, Kerala, India and Institut für Mikrobiologie und Biotechnologie, Rhein. Friedrich Wilhelms-Universität, Meckenheimer Allee 168, DW-5300 Bonn, Germany*

*Key words:* Acetobacter, Azospirillum, compatible solutes, nitrogen fixation, osmotolerance, rhizosphere

## Abstract

In the genus *Azospirillum* tolerance towards high concentrations of sodium chloride, sucrose or polyethylene glycol increased in the order *A. amazonense A. lipoferum A. brasilense* and *A. halopraeferens*. In *A. brasilense* and *A. halopraeferens* the compatible solutes trehaloseglutamate and an unknown compound were identified. *A. halopraeferens* only could convert choline to the potent compatible solute glycine betaine. *Acetobacter diazotrophicus* tolerated high concentrations of *sucrose* and polyethylene glycol, but was very sensitive towards sodium chloride. In contrast to the more osmotolerant *Azospirillum* spp. amino acids such as glutamate, serine and histidine were efficiently utilized as carbon and nitrogen sources and betaine, choline and proline did not relieve osmotic stress.

New halotolerant bacteria (strains BE and TC) were isolated from the rhizosphere of rice growing in alkaline, saline soil in India. They were oxidase-positive, Gram-negative, very motile bacteria, which showed pleomorphic growth. In semisolid nitrogen free mineral medium they grew and fixed nitrogen microaerobically. These isolates required sodium ions for growth and they tolerated up to 2 $M$ sodium chloride in nitrogen containing mineral medium. At osmotic stress conditions the efficient compatible solute ectoine was synthesized.

## Introduction

When the water potential $\Psi$ in the cell environment increases, any microbial cell needs to regulate its turgor and intracellular composition to compensate water loss and to preserve metabolic functions. Many osmoregulatory responses involving synthesis and uptake of compatible solutes, changes in membrane composition or size regulation have evolved in the microbial world (Brown, 1990; Yancey et al., 1982). Diazotrophic rhizosphere bacteria face osmoregulatory challenges, when the plant lives in salt affected soil, the soil dries out or when they live inside the plant roots, where high organic solute concentrations may occur.

In *Azospirillum* spp. the tolerance towards sodium chloride increases in the order *A. amazonense, A. lipoferum, A. brasilense* and *A.*
*halopraeferens* (Hartmann, 1988a,b). The utilization of amino acids, choline and glycine betaine for growth is low in the more osmotolerant species (Hartmann et al., 1988). Proline and glycine betaine can only relieve osmotic stress in *A. brasilense* and *A. halopraeferens*. Here we report on further studies on osmoregulation of *Azospirillum* spp. and *Acetobacter diazotrophicus* and about new isolates of halotolerant diazotropic bacteria from the rhizosphere of rice.

## Material and methods

The *Azospirillum* strains were obtained from the Deutsche Sammlung für Mikroorganismen (DSM, Braunschweig). *Azospirillum halopraeferens* Au 4 was kindly provided by Dr. B. Reinhold (Hannover) and *Acetobacter diazo-*

156

*trophicus* PAL3· was a gift from Dr. J. Döbereiner (Rio de Janeiro). The *Azospirillum* and *Acetobacter* strains were grown in liquid mineral medium with ammonium or in semisolid nitrogen free medium as described previously (Gillis et al., 1989; Hartmann, 1988a; Reinhold et al., 1987). The strains TC and BE from the rhizosphere of rice were cultivated in *A. brasilense* mineral medium supplemented with 100 mM NaCl. The following minimal medium was used as osmotic stress medium to investigate the compatible solutes of TC and BE: NaCl (2 $M$), $K_2HPO_4$ (3 mM), $NH_4Cl$ (40 mM), $MgSO_4$ (0,4 mM), $FeSO_4$ (0.04 mM), $CaCl_2$ (0.5 mM), glucose (55 mM), yeast extract (0.01 g L$^{-1}$), pH 7.5–8.0. Nitrogen fixation activity was measured in semisolid media using the acetylene reduction technique (Hartmann et al., 1988). Bacterial growth was recorded by measuring the absorbance at 560 nm.

The new isolates TC and BE were obtained in semisolid malate medium (Nfb), supplemented with 50 mg L$^{-1}$ yeast extract according to Baldani and Döbereiner (1980). In addition 0.25% (w/v) NaCl was added. The enrichment cultures were transferred seven times to fresh Nfb medium and were finally streaked on malate and ammonium containing minimal medium agar plates. The oxidase and amino peptidase tests were performed using the reagent kit of Merck (Darmstadt).

For the determination of the compatible solutes freeze dried cells (70 mg) were processed using a modified Blye and Dyer-technique and the extracts were analyzed by the HPLC-method according to Galinski and Herzog (1990).

The conversion of choline to glycine betaine was investigated using $^{14}$C-choline purchased from Amersham. Logarithmically growing cells were incubated at 33 or 41°C with 35 $\mu M$ $^{14}$C-choline for 2 hours. Then 300 $\mu$L acetone was added to 200 $\mu$L incubation mixture. After drying in vacuum and redissolving in 20 $\mu$L water high voltage electrophoresis (800 V, 35 mA, 1 hour) was carried out using Whatman 3 mm paper and formic acid (0.75 $M$). Finally the dried paper was scanned for radioactivity. As reference substances $^{14}$C-choline and $^{14}$C-glycine betaine were used. $^{14}$C-glycine betaine was ob-tained from $^{14}$C-choline by enzymatic conversion with choline oxidase from Sigma.

The water potential of the different osmotic stress media were measured psychrometrically with the Wescor HR-33-T dewpoint microvolt-meter (Wescor Inc., USA). The following readings were obtained with minimal medium for *A. amazonense* and *Acetobacter diazotrophicus* (LGI with 1% sucrose): 3.1 bar; minimal medium for *A. lipoferum* and *A. brasilense*: 6.0 bar; minimal medium for *A. halopraeferens*: 8.8 bar; LGI-medium with 10% sucrose: 9.0 bar (1 bar = $10^5$ Pa).

## Results and discussion

### Comparison of the osmotolerance of Azospirillum *spp.* and Acetobacter diazotrophicus

Using sodium chloride, polyethylene glycol (PEG) 400 and sucrose as osmotic stress agents *A. halopraeferens* performed best among *Azospirillum* spp., followed by *A. brasilense*, *A. lipoferum* and *A. amazonese* (Fig. 1). *Acetobacter diazotrophicus* PAL3 was more tolerant towards osmotic stress by sucrose and PEG400 than *Azospirillum* spp.. However, PAL3 was very sensitive towards sodium chloride. In the presence of 0.1 $M$ NaCl the specific nitrogenase activity was reduced to 50% (in medium with 1% sucrose; Fig. 1) and to 20% in cultures with 10% sucrose (not shown). The nitrogen fixation activity of *Acetobacter diazotrophicus* was increased at 20% sucrose as compared to 1% (19 and 3 bars), demonstrating the adaptation of PAL3 to the high sugar concentrations in its natural habitat, the roots and stems of sugarcane.

In contrast to *A. brasilense* and *A. halopraeferens* glutamate, proline and glycine betaine could not relieve the osmotic stress induced by NaCl, PEG400 and sucrose in *Acetobacter diazotrophicus* PAL3. Glutamate, serine, alanine and histidine were efficiently used as carbon and nitrogen sources. Nitrogenase activity of PAL3 was inhibited to less than 10% in the presence of glutamate, which was also demonstrated for the osmosensitive species *A. amazonese* and *A. lipoferum* (Hartmann et al., 1988).

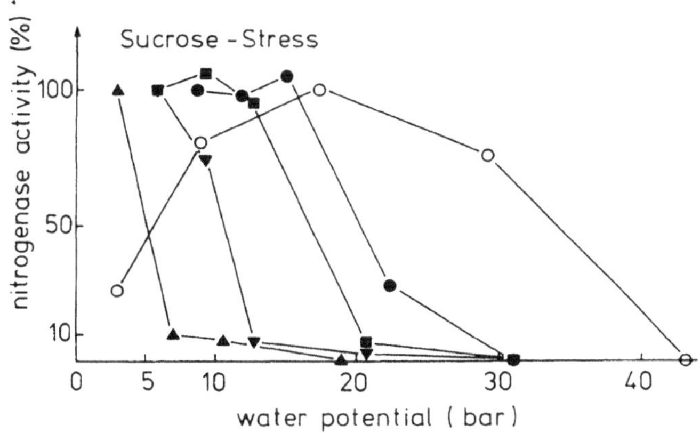

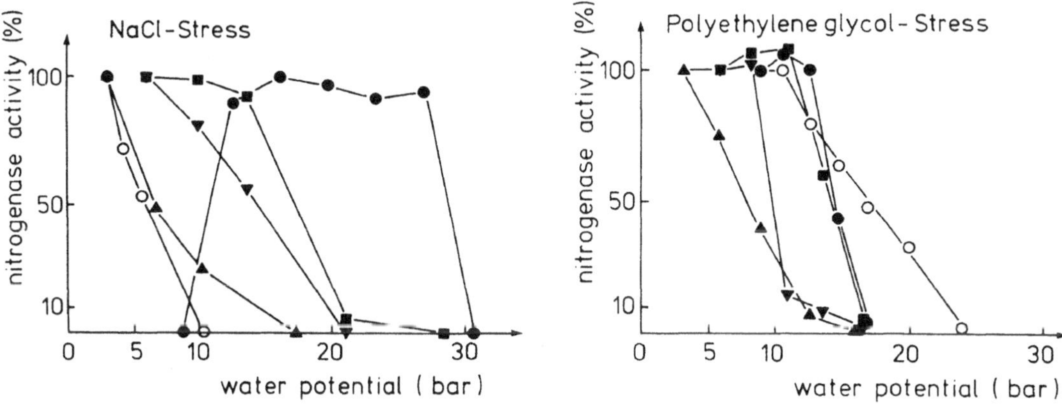

Fig. 1. Comparison of the impact of osmotic stress on nitrogenase activity of *A. amazonense* Y1 (▲), *A. lipoferum* Sp59b (▼), *A. brasilense* Sp7 (■), *A. halopraeferens* Au4 (●) and *Acetobacter diazotrophicus* PAL3 (○). The specific nitrogenase activity (nmol ethylene produced per 1 mL culture at $A_{560}$ of 1.0) was determined in semisolid media. For each strain the highest activity was set to 100%. The following concentrations of osmotic stress solutes were used: NaCl: 0, 0.05 (PAL3), 0.1, 0.2, 0.3 (Au4), 0.4, 0.5 and 0.6 M; polyethylene glycol 400: 0, 2.5, 5.0, 7.5, 10.0%(w/v); sucrose: 1, 5, 10, 20, 30% (w/v).

### Conversion of choline to glycine betaine in Azospirillum *spp.*

Choline and glycine betaine could relieve osmotic stress in *A. halopraeferens*, while only glycine betaine was active in *A. brasilense* (Hartmann, 1988b). The ability to convert choline to glycine betaine was tested with [14]C-labeled choline. The electropheretograms shown in Figure 2 clearly demonstrate, that only *A. halopraeferens* was able to convert choline to glycine betaine. Therefore *A. halopraeferens* harbours the choline-glycine betaine pathway similar to the *bet*-operon described for *E. coli* (Andresen

et al., 1988). It consists of a high affinity choline uptake (Hartmann, 1988b), choline dehydrogenase and betaine aldehyde oxidase.

### Isolation and osmoregulatory properties of newly isolated halotolerant $N_2$-fixing rhizobacteria

The strain TC was isolated from the roots of rice growing in salt affected, alkaline soil in Trichy, Tamil Nadu, India. The roots were washed several times with sterile water to get rid of loosely adhering soil. Root bits of 1 cm length were the starting material for isolation (see Material and

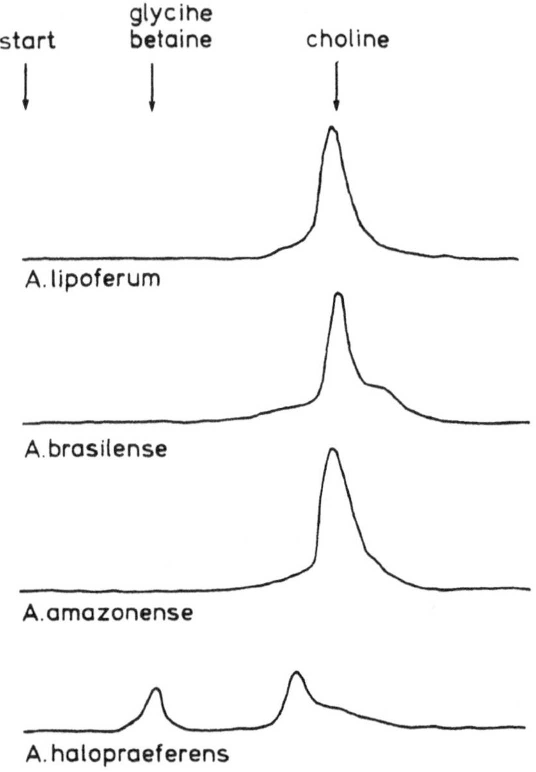



*Fig. 2.* High voltage paper electrophoresis of [$^{14}$C]-choline and its conversion products. *Azospirillum* cultures, growing at osmotic stress conditions, were incubated with 35 $\mu M$ choline for 2 hours. Electrophoresis was performed as described in Material and Methods. [$^{14}$C]-choline and [$^{14}$C]-glycine betaine were used as reference substances.

Methods). The strain BE was isolated from rhizosphere soil of a local rice variety grown at Cochin, Kerala in seawater affected paddy fields. Both isolates were oxidase-positive, amino peptidase-positive, Gram-negative and very motile bacteria. In nitrogen free semisolid medium they grew in a distinct subsurface pellicle. Both isolates showed pleomorphic growth (vibroid to long rods) and cyst formation.

The isolates grew very well in minimal medium with malate or glucose as carbon source. The maximal growth rate (66 min per generation) was obtained between 100 and 500 m$M$ sodium chloride (Fig. 3). Growth was sodium dependent with half maximal rates at 20 m$M$ NaCl. At salt stress conditions, lag phases occurred. At 2 $M$ sodium chloride the lag phase was about 48 hours long and was omitted from Figure 3.

*Nature of the compatible solutes*

*A. brasilense* Sp7 and *A. halopraeferens* Au4 showed a very similar pattern of compatible solutes. The main components were trehalose, glutamate and an yet unknown component (ratio 4:1:1). The occurrence and identity of trehalose and glutamate was corroborated by $^{13}$C-NMR analysis. In *Acetobacter diazotrophicus* PAL3 sucrose was the only organic solute which could

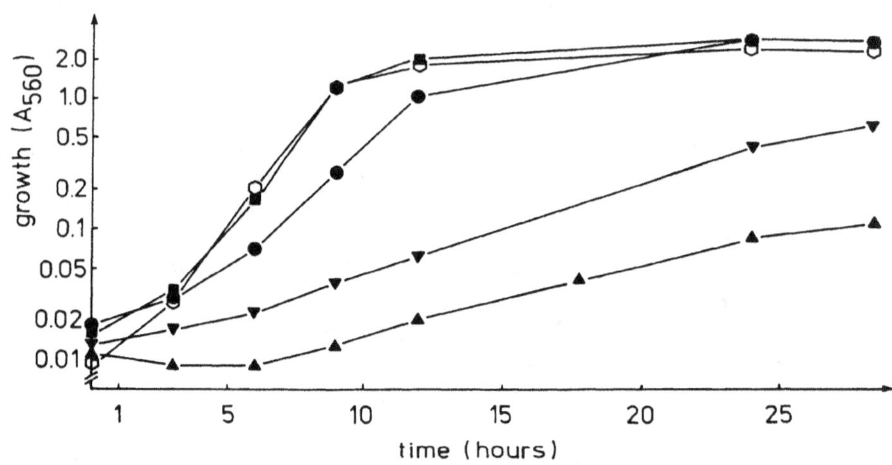



*Fig. 3.* Growth of the strain BE on mineral salt medium with malate as carbon and ammonium chloride (10 m$M$) as nitrogen source. Growth was performed in liquid shaking cultures (150 rpm, 33°C). Salt tolerance: 0.1 $M$ NaCl (○), 0.5 $M$ NaCl (■), 1.0 $M$ NaCl (●), 1.5 $M$ NaCl (▼) and 2.0 $M$ NaCl (▲).

be observed in high concentration. Since it was used as growth substrate the quantification of the intracellular part is difficult. The osmoregulation of *Acetobacter diazotrophicus* is clearly different from *Azospirillum* or *E. coli* (Larsen et al., 1987) and might resemble the situation of *Zymomonas mobilis*. These bacteria compensate high external glucose concentrations with high internal concentrations. This is achieved by high glucose import rates (Struch et al., 1990).

The halophilic isolates TC and BE contained the compatible solutes glycine betaine, ectoine and ectoine derivate Y (ratio 2:1:1 for TC and 2:1:2 for BE) when grown on complex medium. Ectoine is probably the main osmolyte in minimal medium (with glucose and $2\,M$ NaCl). The identity of ectoine needs to be proven by $^{13}$C-NMR analysis. Ectoine is a cyclic amino acid discovered for the first time in halophilic, phototrophic bacteria of the genus *Ectothiorhodospira* (Galinski et al., 1985). Meanwhile ectoine and related substances were found in other genera of halophilic bacteria too (Wohlfarth et al., 1990). Its biosynthesis starts with aspartate and proceeds in five enzymatic steps as described by Peters et al. (1990). The occurrence of the biosynthesis of this very efficient compatible solute in *Azospirillum* – like diazotrophic bacteria deserves further investigation.

## Acknowledgements

We are very grateful to Drs J Döbereiner and B Reinhold for providing bacterial strains. The contribution of AH was in part supported by the Deutsche Forschungsgemeinschaft (SFB 137), while he was at the Lehrstuhl für Mikrobiologie, Universität Bayreuth, Germany.

## References

Andresen A, Kaasen I, Styrvold OB, Boulnois G and Strom AR 1988 Molecular cloning, physical mapping and expression of the *bet* genes governing the osmoregulatory choline-glycine betaine pathway of *Escherichia coli*. J. Gen. Microbiol. 134, 1737–1746.

Baldani VLD and Döbereiner J 1980 Host-plant specificity in the infection of cereals with *Azospirillum* spp. Soil Biol. Biochem. 12, 433–439.

Brown AD 1990 Microbial Water Stress Physiology – Principles and Perspectives. Wiley, Chichester, UK.

Galinski EA and Herzog RM 1990 The role of trehalose as a substitute for nitrogen-containing compatible solutes (*Ectothiorhodospira halochloris*). Arch. Microbiol. 153, 607–613.

Galinski EA, Pfeiffer H-P and Trüper HG 1985 1,4,5,6-tetrahydro-2-methyl-4-pyrimidinecarboxylic acid: A novel cyclic amino acid from halophilic phototrophic bacteria of the genus *Ectothiorhodospira*. Eur. J. Biochem. 149, 135–139.

Gillis M, Kersters K, Hoste B, Janssens D, Kroppenstedt RM, Stephan MP, Teixeira KRS, Döbereiner J and De Ley J 1989 *Acetobacter diazotrophicus* sp. nov., a nitrogen-fixing acetic acid bacterium associated with sugarcane. Int. J. Syst. Bacteriol. 39, 361–364.

Hartmann A 1988a Osmoregulatory properties of *Azospirillum* spp. *In* Azospirillum IV, Genetics, Physiology, Ecology. Ed. W. Klingmuller. pp 122–130 Springer-Verlag, Berlin, New York.

Hartmann A 1988b Ecophysiological aspects of growth and nitrogen fixation in *Azospirillum* spp. Plant and Soil 110, 225 238.

Hartmann A, Fu H and Burris RH 1988 Influence of amino acids on nitrogen fixation ability and growth of *Azospirillum* spp. Appl. Environ. Microbiol. 54, 87–93.

Larsen PI, Sydnes LK, Landfald B and Strom AR 1987 Osmoregulation in *Escherichia coli* by accumulation of organic osmolytes: betaines, glutamic acid, and trehalose. Arch. Microbiol. 147, 1–7.

Peters P, Galinski EA and Trüper HG 1990 The biosynthesis of ectoine. FEMS Microbiol. Letters 71, 157–162.

Reinhold B, Hurek T, Fendrik J, Pot B, Gillis M, Kersters K, Thielemans S and De Ley J 1987 *Azospirillum halopraeferens* sp. nov., a nitrogen-fixing organism associated with roots of Kallar grass (*Leptochloa fusca* (L.) Kunth). Int. J. Syst. Bacteriol. 37, 43–51.

Struch T, Neuss B, Bringer-Meyer S and Sahm H 1990 Osmotic adjustment of *Zymomonas mobilis* to concentrated glucose and fructose solutions. Forum Mikrobiol. 1, 2–119.

Wohlfarth A, Severin J and Galinski EA 1990 The spectrum of compatible solutes in heterotrophic halophilic eubacteria of the family *Halomonadaceae*. J. Gen. Microbiol. 136, 705–712.

Yancey PH, Clark ME, Hand SC, Bowlus RD and Somero GN 1982 Living with water stress: Evolution of osmolyte systems. Science 217, 1214–1222.

# AZOSPIRILLUM-GRAMINEAE INTERACTION: EFFECT OF INDOLE-3-ACETIC ACID


P. BARBIERI[1], C. BAGGIO[2], M. BAZZICALUPO[3], E. GALLI[1], G. ZANETTI[4], M.P. NUTI[2]

1 Dip. di Genetica e di Biologia dei Microrganismi, Via Celoria 26, 20133 Milano
2 Dip. di Biotecnologie Agrarie, Via Gradenigo 6, 35131 Padova
3 Dip. di Biologia Animale e Genetica, Via Romana 19, 50145 Firenze
4 Dip. di Fisiologia e Biochimica Generali, Via Celoria 26, 20133 Milano



ABSTRACT. To evaluate the involvement of indole-3-acetic acid (IAA) in promoting plant response the effects of inoculation of wheat seeds with *Azospirillum brasilense* SpM7918, a very low IAA producer selected after Tn5 mutagenesis, have been compared with those caused by the inoculation with the wild type strain Sp6. In respect to Sp6, SpM7918 shows a reduced ability in promoting root system development as regards both the number and length of lateral roots and the distribution of root hairs. Similarly, inoculation with SpM7918 causes a decrease in mineral uptake, but less relevant than that observed for root system development.


## 1. INTRODUCTION

Several studies in the last years have pointed out that *Azospirillum* inoculation causes a beneficial effect on crop plants (1). Although the mechanisms underlying the plant responses are not fully understood so far, it is well known that *Azospirillum* inoculation promotes root development (2,3) and it has been proposed that the modifications induced by *Azospirillum* in the root system could be related to the observed improvement in mineral and water uptake by the inoculated plant (4,5). Besides nitrogen fixation, the ability of *Azospirillum* to excrete phytohormone-like substances as cytokinins, gibberellins, and auxins (6), could be involved in the stimulation of the root system development. Among natural phytohormones, indole-3-acetic acid (IAA) shows the greatest stimulating activity on plant growth (7 - 11).
In this work we report the isolation of an *Azospirillum* mutant excreting a low amount of indole-3-acetic acid (IAA) and its effects on roots development and on mineral uptake in durum wheat.

## 2. MATERIALS AND METHODS

### 2.1. Bacterial strains and growth conditions

The *Azospirillum* strains used in the present work are listed in Table 1.
Complete medium for *E.coli* (see Section 2.2) and *Azospirillum* was Luria broth (LB). Minimal medium (MSP) and growth conditions for *Azospirillum* strains were previously described (8). When required, antibiotics were added at the following concentration: rifampicin (Rif) 40 µg/ml; kanamycin (Km) 40 µg/ml.

Table 1. *Azospirillum* strains

| Strains | Relevant phenotype | Parent strain | Reference |
|---|---|---|---|
| A.*brasilense* Sp6 | wild type | - | (12) |
| A.*brasilense* SpF94 | Rif$^r$ | Sp6 | (13) |
| A.*brasilense* SpM7918 | Km$^r$ IAA$^-$ | SpF94 | This work |

## 2.2. Conjugation experiments

*E.coli* HB101 was used to donate the suicide plasmid pGS9 (14) carrying the Km resistance transposon Tn5 to *A.brasilense* SpF94. Matings were performed by the membrane filter method (14). Exconjugants were selected on LB agar supplemented with Km and Rif.

## 2.3. Determination of indole-3-acetic acid (IAA)

IAA excreted by bacteria was evidenced in cell-free culture broths by the colorimetric Salkowsky reaction (15) and measured photometrically at 530 nm using an IAA standard. When more exact measurements were required the cultural broths were extracted with ethylacetate and TLC and HPLC analyses were performed as previously described (8).

## 2.4. Nitrogenase assay

Nitrogenase activity was evaluated as follows. An exponentially growing culture of *Azospirillum* was washed and resuspended in MSP to an O.D. 550 nm of 0.2. The culture was incubated in a 0.1% $O_2$, $N_2$ atmosphere for 20 h. To start the assay, 10% acetylene was added. After 3 h of incubation, ethylene production was measured by gas chromatografy (16).

## 2.5. Inoculation experiments

Inoculation experiments on wheat seeds (*Triticum durum* var. *Appula*) were carried out in a hydroponic closed system as previously described (8). Bacterial inocula were grown in NFb medium (8) to give a final concentration in the test tubes of $10^6$ cfu/ml. After 14 days of incubation at 25°C in an illuminated room, the number and length of lateral and main roots were measured. Bacterial population, free in solution and associated to the roots, was determined by the plate count technique on MSP medium.

## 2.6. Scanning electron microscopy (SEM)

For microscopic examination 0.5 cm root segments were fixed in 3.5% glutaraldehyde buffered with 0.1M phosphate buffer pH 7.2. After two washings in phosphate buffer, root segments were postfixed with 1% $OsO_4$ in phosphate buffer for 2 h. Samples were washed twice and then, after gradual dehydration with ethanol, dried in a critical point dryer in $CO_2$ environment. The dry samples, coated with gold by sputtering, were examined under a Cambridge Stereoscan 250 MK2 SEM.

## 2.7. Mineral uptake determination

Plants for mineral uptake experiments were grown in modified Hoagland solution in a hydroponic system. $Fe^{2+}$ content was measured in inoculated and control plants grown for 8 days. After determination of fresh and dry weight, roots and leaves were treated at 550°C in muffle for 3 h. After cooling, 5 ml of 20% HCl were added to each sample and then diluted in 45 ml of distilled water. 10 ml of each sample were filtered and $Fe^{2+}$ content was determined by atomic adsorption.

For nitrate uptake plants at various incubation time were transferred in 100 ml of modified Hoagland solution at known $NO_3^-$ concentration. At regular intervals samples of 1 ml were drawn and treated with 4 ml of 5% perchloric acid. Nitrate concentration was measured photometrically at 210 nm using $KNO_3$ standards.

## 3. RESULTS

### 3.1. Transposon mutagenesis and isolation of mutants altered in IAA production

Transposon Tn5 was introduced in *A.brasilense* SpF94, a spontaneous Rif[r] derivative of Sp6, by conjugation. *E. coli* HB101 carrying the suicide vector pGS9 was used as donor. Km[r] exconjugants were selected, Rif was used to counterselect the donor strain.

About 2000 Km[r] Rif[r] exconjugants were analyzed for IAA production: each clone was inoculated in minimal medium MSP supplemented with tryptophan (100 µg/ml) as IAA precursor, and after 7 days of incubation the supernatants were collected and IAA was determined by Salkowsky reaction. This screening led to the isolation of two putative IAA overproducers and four putative IAA[-] mutants. The four IAA-negative clones were further analyzed: after growth in different cultural conditions (presence or absence of ammonia and tryptophan) the cultural broths were extracted and the IAA production was determined by TLC and HPLC. The strain SpM7918 was the lowest IAA producer. The presence of Tn5 in this strain was verified by Southern hybridization; a signal was found in an *Eco*RI fragment of about 17 Kbp (data not shown).

The growth in nitrogen-free medium and the IAA production in the presence of tryptophan of SpM7918 were compared with those of the wild type strain Sp6 and of the parent strain SpF94 (Fig. 1). The amount of IAA produced by the three strains at 3, 7, and 14 days is reported in Table 2; the IAA excreted by SpM 7918 was about 30% with respect to that produced by Sp6 and SpF94.

Table 2. IAA production ($\mu g/A_{670}$) at various times of growth

| Days<br>Strains | 3 | 7 | 14 |
|---|---|---|---|
| Sp6 | nd | 15.26 (100) | 42.68 (100) |
| SpF94 | 6.90 | 18.15 (118) | 37.40 (87) |
| SpM7918 | 3.14 | 5.32 (34) | 11.62 (27) |

Number in brackets are the percent in respect to Sp6.
nd = not determined

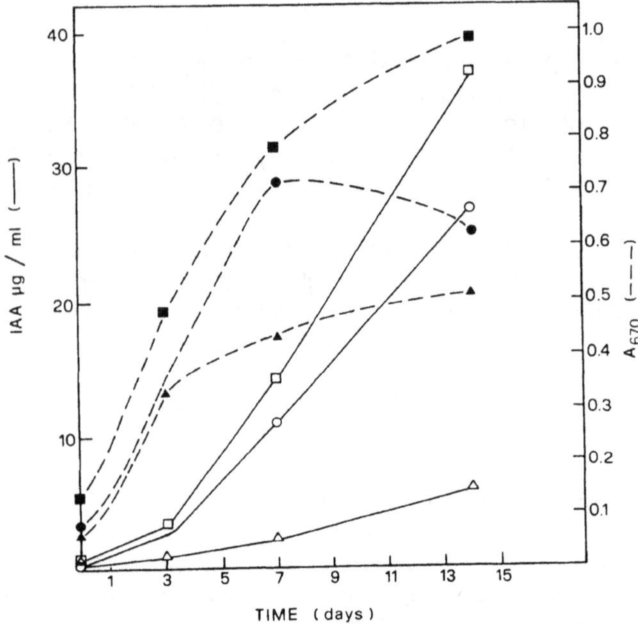

Fig. 1. Growth in N2 fixing conditions (close symbols) and IAA production (open symbols) of *A.brasilense* Sp6 (●,○); SpF94 (■,□); SpM7918 (▲,△).

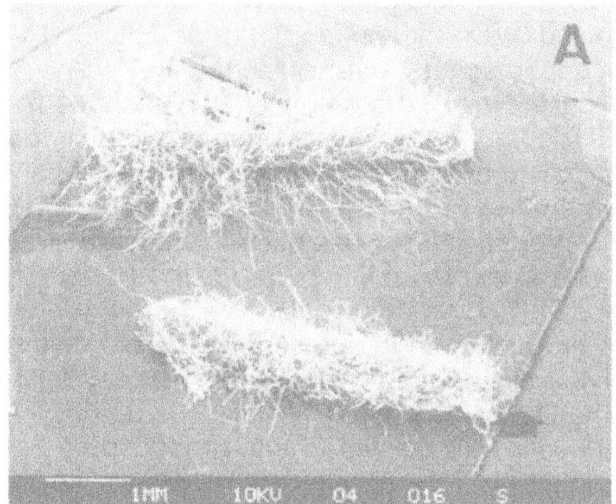

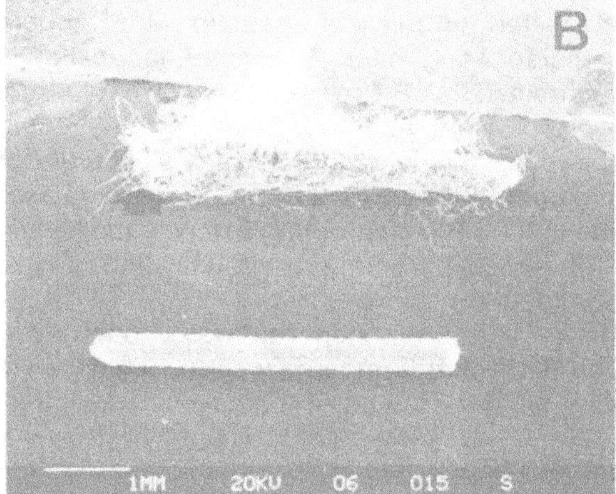

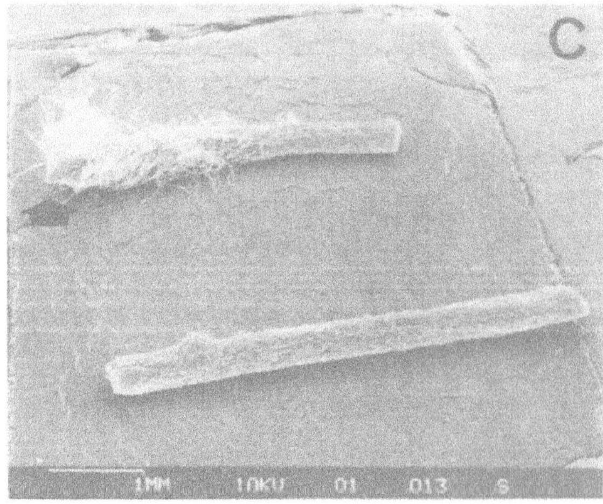

Fig. 2. Effect of inoculation with *A.brasilense* Sp6 (A) and with *A.brasilense* SpM7918 (B) on wheat root morphology. (C): Uninoculated control. The arrow indicates the root tip.

The nitrogenase activity of SpM7918 after 20 h of induction was evaluated (Table 3) and it was found to be comparable with that of the parent strain SpF94.

Table 3. Nitrogenase activity

| Strain | $C_2H_2$ reduction ($\mu$mol $C_2H_4$/hour/culture) |
| --- | --- |
| SpF94 | 0.869 (100%) |
| SpM7918 | 0.938 (108%) |

### 3.2. Effect of inoculation on wheat root development

The ability of Sp6 and SpM7918 to promote root growth was tested in *in vitro* inoculation experiments. Seedlings of *T. durum* var. *Appula* were inoculated with the strain to be tested, at a final concentration of $10^6$ cfu/ml, which was previously found to be optimal for root development (9). At the end of the incubation, plants were harvested and the number and length of lateral and main roots were measured (Table 4). In agreement with our previous experiments (8, 9) no significant differences were found between uninoculated control and inoculated plants in regard to number and length of main roots. On the contrary the inoculated groups significantly differ from the control group in the number and length of lateral roots; as far as these parameters are concerned, differences were also found between samples inoculated with Sp6 and with SpM7918. In particular, inoculation with SpM7918 led to a lateral root development lower than that observed in samples inoculated with Sp6.

Table 4. Effect of inoculation on the root system

| Inoculant strain | Lateral roots (number) | Lateral roots length (cm) | Main roots (number) | Main roots length (cm) |
| --- | --- | --- | --- | --- |
| None | 8.26 (4.16) | 5.48 (4.20) | 5.10 (0.47) | 44.41 (10.08) |
| Sp6 | 17.03 (7.25)[a] | 21.62 (18.99)[a] | 5.33 (0.74) | 38.11 ( 9.03) |
| SpM7918 | 12.50 (7.44)[ab] | 11.99 (12.30)[ab] | 5.06 (0.25) | 44.01 ( 9.69) |

Data represent means of 30 replicates. Standard deviation is given in paretheses.
a,b Values significantly different from: control (a); Sp6 (b). ($P \leq 0.05$).

As regards the counts of bacteria free in solution or associated to the roots, no remarkable differences were found between samples inoculated with Sp6 or with SpM7918, suggesting that the lower response elicited by SpM7918 is not due to a less viability of the strain in these conditions.

### 3.3. Changes in root hair distribution caused by inoculation

Root hair development was evaluated in uninoculated and inoculated roots by SEM. After few hours from inoculation, differences were observed independently from the strain used. Larger variations appeared only on prolonged incubation, i.e. 14 days. At this time, two 0.5 cm root segments, one including the root tip and the other excised at 3 cm from the root tip, were examined. As shown in Fig. 2, samples inoculated with Sp6 showed a great density of root hairs in both segments, while in the uninoculated control root hairs are present only in a narrow zone close to the root tip. Also in the samples inoculated with SpM7918, root hairs were present but only on the tip segment, although with a density greater than in the control.

## 3.4. Effect of inoculation on mineral uptake

$Fe^{2+}$ content varies according to the inoculant strain; in fact inoculation with Sp6 led to an increase of $Fe^{2+}$ in roots and leaves in respect to uninoculated plants, while the increase determined by the inoculation with SpM7918 is 6% less than that determined by Sp6 (Fig. 3). Inoculation with Sp6 led also to an increase in nitrate uptake, while the increase caused by SpM7918 was 7% less than that of Sp6 (17).

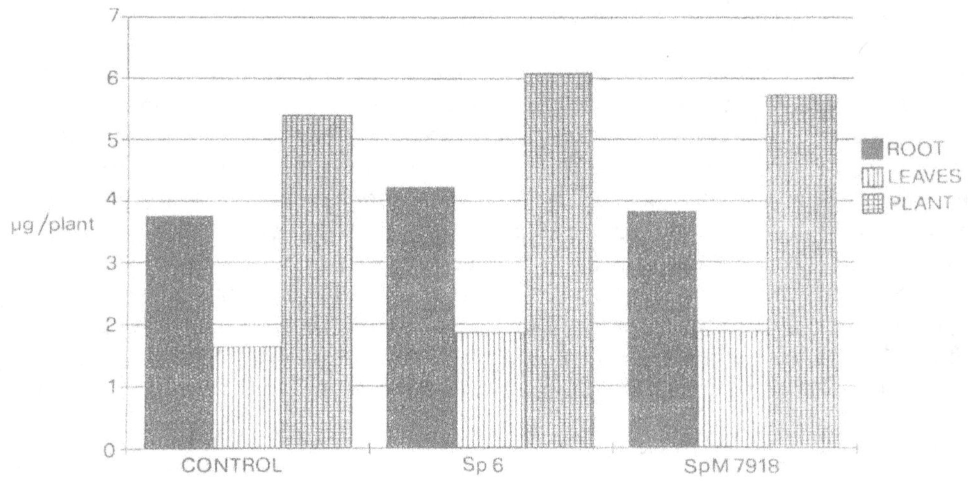


Fig. 3. Effect of inoculation on plant $Fe^{2+}$ content.

## 4. DISCUSSION

The enhancement of mineral and water uptakes in plants inoculated with *Azospirillum* could be related to the modification of the root system induced by the microorganism in the inoculated plant (1-7). As phytohormone-like substances excreted by *Azospirillum* could play an important role in promoting root development, we focused our attention on the selection of IAA⁻ mutants suitable for inoculation experiments. In the achievement of this objective two factors must be considered: i) The very low frequency of mutants totally unable to produce IAA could be due to the probable existence in *Azospirillum* of two biosynthetic pathways for IAA production (15), one via indole-3-pyruvic acid, the other via tryptamine. ii) The lack of a rapid method to evaluate IAA production greatly limits the number of clones that can be analyzed decreasing the efficiency of the screening.
By Tn5 mutagenesis we selected the low IAA producer strain SpM7918. *In vitro* inoculation of durum wheat indicates that SpM7918, although not totally uneffective, showed a reduced ability in promoting root system development in comparison with the wild type strain Sp6, that means a decrease of 27% and 45% as regards the number and length of lateral roots respectively. The observations on root hairs distribution seem to agree with these data; in fact roots inoculated with SpM7918 appear more similar to those of the uninoculated control than to those inoculated with Sp6. These data suggest a major involvement of IAA produced by *Azospirillum* in

promoting the development of root system. As far as mineral uptake is concerned, the decrease of $Fe^{2+}$ and $NO_3^-$ uptake observed in plants inoculated with the low IAA producer SpM7918 is less relevant than the decrease in the number and length of lateral roots. Although these data need to be further confirmed, however they seem to indicate that other factors, besides IAA production, may be involved in promoting a global beneficial effect in inoculated plants, as also assumed by other Authors (10, 11).


ACKNOWLEDGEMENTS

We thank Prof. G. Melone for his invaluable help in the electron microscopy experiments. This study was supported by the Consiglio Nazionale delle Ricerche, Italy, spacial grant I.P.R.A.


REFERENCES


1. Okon, Y. (1985) *Azospirillum* as a potential inoculant for agriculture. Trends Biotechnol. 3, 223-228.
2. Kapulnik, Y., Okon, Y., Henis, Y. (1985) Changes in root morphology of wheat caused by *Azospirillum* inoculation. Can. J. Microbiol. 31, 881-887.
3. Jain, D.K., Patriquin, D.G. (1984) Root hair deformation, bacterial attachment, and plant growth in wheat-*Azospirillum* associations. Appl. Environ. Microbiol. 48, 1208-1213.
4. Kapulnik, Y., Gafny, R., Okon, Y. (1985) Effect of *Azospirillum* spp. inoculation on root development and $NO_3^-$ uptake in wheat (*Triticum aestivum* cv. *Miriam*) in hydroponic system. Can. J. Bot. 63, 627-631.
5. Sarig, S., Blum, A., Okon, Y. (1988) Improvement of water status and yield of field-grown grain sorghum (*Sorghum bicolor*) by inoculation with *Azospirillum brasilense*. J. Agric. Sci. 110, 271-277.
6. Tien, T.M., Gaskins, M.H., Hubbell, D.H. (1979) Plant growth substances produced by *Azospirillum brasilense* and their effect on the growth of pearl millet (*Pennisetum americanum* L.). Appl. Environ. Microbiol. 37, 1016-1024.
7. Umali-Garcia, N., Hubbell, D.H., Gaskins, N.H., Dazzo, F.B. (1980) Association of *Azospirillum* with grass roots. Appl. Environ. Microbiol. 39, 219-226.
8. Barbieri, P., Zanelli, T., Galli, E., Zanetti, G., (1986) Wheat inoculation with *Azospirillum brasilense* Sp6 and some mutants altered in nitrogen fixation and indole-3-acetic acid production. FEMS Microbiol. Lett. 36, 87-90.
9. Barbieri, P., Bernardi, A., Galli, E., Zanetti, G. (1988) Effect of inoculation with different strains of *Azospirillum brasilense* on wheat root developments. In W. Klingmuller (ed.) "Azospirillum IV: genetics, phisiology, ecology", Springer-Verlag, Berlin Heidelber, pp. 181-188.
10. Morgenstern, E., Okon, Y. (1987) The effect of *Azospirillum brasilense* and auxin on root morphology in seedlings of *Sorghum bicolor* x *Sorghum sudanense*. Arid Soil Res. Rehabil. 1, 115-127.
11. Harari, A., Kigel, J., Okon, Y. (1988) Involvement of IAA in the interaction between *Azospirillum brasilense* and *Panicum miliaceum* roots. Plant Soil 110, 275-282.
12. Bani, D., Barberio, C., Bazzicalupo, M., Favilli, F., Gallori E., Polsinelli, M. (1980) Isolation and characterization of glutamate synthase mutants of *Azospirillum brasilense*. J. Gen. Microbiol. 119, 239-244.
13. Fani, R., Bazzicalupo, M., Ricci, F., Schipani, C., Polsinelli, M. (1988) A plasmid vector for the selection study of transcription promoter in *Azospirillum brasilense*. FEMS Microbiol. Lett. 50, 271-276.

14. Selvaraj, G., Iyer, V.N., (1983) Suicide plasmid vehicles for insertion mutagenesis in *Rhizobium meliloti* and related bacteria. J. Bacteriol. 156, 1292-1300.

15. Hartmann, A., Singh, M., Klingmuller, W. (1983) Isolation and characterization of *Azospirillum* mutants excreting high amounts of indoleacetic acid. Can. J. Microbiol. 29, 916-923.

16. Gallori, E., Bazzicalupo, M. (1985) Effect of nitrogen compounds on nitrogenase activity in *Azospirillum brasilense*. FEMS Microbiol. Lett. 28, 35-38.

17. Baggio, C., Antonello, F., Barbieri, P., Saccomanni, M., Nuti, M.P. (1990) $NO_3^-$ and $Fe^{2+}$ uptake in durum wheat is affected by *Azospirillum brasilense* inoculation. Abstract 5th Int. Symp. on Nitrogen Fixation with Non-Legumes. Florence. (In press).

# DNA RESTRICTION FINGERPRINT FOR THE IDENTIFICATION OF AZOSPIRILLA

S. VENTURA, L. GIOVANNETTI, A. GORI, S. MARUCA and P. MONTAINI
*Dip. di Scienze e Tecnologie Alimentari e Microbiologiche Univ. Firenze and Centro di Studio
dei Microrganismi Autotrofi CNR
Piazzale delle Cascine, 27
I-50144 FIRENZE*

ABSTRACT. DNA Restriction patterns of eighteen strains of Azospirillum were obtained with
SDS-PAGE and silver staining. Each strain was shown to have a unique, stable pattern that enables
its sure recognition. Strains were clustered with the UPGMA method into two main groups
corresponding to the species level.

## Introduction

Since some years azospirilla were proposed as  biological nitrogen fertilizers (Okon, 1985) and
experiments were performed to test the efficiency of selected strains.

A problem in this type of experiment is the accurate identification of the inoculant: indeed the
inoculated strain needs to be detected and traced in soil in order to evaluate its capability in root
colonization and its effect on plant yield.

The routine identification tests cannot give a unique fingerprint of each strain and do not let a single
strain to be surely recognized.

Looking for the solution of this problem we decided to consider some genomic features which could
help us to surely recognize single strains of bacteria: DNA base composition, genome size, nucleic
acids hybridization are  technique mainly used for the characterization and taxonomic positioning
of new species, and less used for quick identification purposes (Trüper and Krämer, 1981).

In recent years, restriction endonuclease analysis of the genome has proved a useful tool in the
differentiation of viruses and bacteria (Chowdhury *et al.*, 1986; Sorensen *et al.*, 1985).

Results obtained so far with diverse genera indicate that this technique is highly reliable and easy to
perform on a large number of isolates.

Most of the total DNA restriction pattern studies used agarose or polyacrilamide gels stained with
ethidium bromide: in this work a polyacrilamide gel was used to obtain the sharp separation of low
molecular weight fragments, coupled with the high sensitivity silver nitrate staining.

This paper describes the application of both this method and morpho-physiological routine tests to
the identification of several strains of the genus *Azospirillum*.

## Materials and Methods

BACTERIAL STRAINS

The strains used, and their origins, are listed in Table 1. Strains *Azospirillum* sp. were isolated during this work from the rizosphere of crop plants.

TABLE 1. *Azospirillum* strains used.

| Ref. | Species | Strain | Source | Obtained from |
|---|---|---|---|---|
| 14 | *A. brasilense* | Sp7 | *Digitaria* | DSM |
| 15 | *A. brasilense* | Cd | *Cynodon dactylon* | J.Döbereiner |
| 5 | *A. lipoferum* | Sp59b | Wheat | ATCC |
| 23 | *A. lipoferum* | SpRG6xx | Wheat | ATCC |
| 1 | *Azospirillum* sp. | TA2 | Tobacco | this study |
| 2 | *Azospirillum* sp. | TB2 | Tobacco | this study |
| 3 | *Azospirillum* sp. | TC2 | Tobacco | this study |
| 6 | *Azospirillum* sp. | PA1 | Pepper | this study |
| 7 | *Azospirillum* sp. | PB2 | Pepper | this study |
| 8 | *Azospirillum* sp. | PC1 | Pepper | this study |
| 9 | *Azospirillum* sp. | MCD1 | Maize | this study |
| 10 | *Azospirillum* sp. | MCE1 | Maize | this study |
| 11 | *Azospirillum* sp. | MCF2 | Maize | this study |
| 16 | *Azospirillum* sp. | MRA3c | Maize | this study |
| 17 | *Azospirillum* sp. | MRB2 | Maize | this study |
| 18 | *Azospirillum* sp. | MRC1 | Maize | this study |
| 12 | *Azospirillum* sp. | SA | Sorghum | this study |
| 13 | *Azospirillum* sp. | SB | Sorghum | this study |

IDENTIFICATION TESTS

Strains isolated in this work were attributed to the genus *Azospirillum* following Krieg and Döbereiner (1984) and Döbereiner and Pedrosa (1987).
All the eighteen strains employed were assayed for the species attribution using the following tests: growth with or without biotine, use of glucose, mannitol or α-ketoglutarate as sole carbon source, presence of pleomorphic cells, presence of pink, raised colonies and dissimilation of nitrite. All these tests were performed following the procedures described by Krieg and Döbereiner (1984).

DNA PURIFICATION AND SDS-PAGE

DNA was extracted, purified and digested as described by Giovannetti *et al.* (1990). DNA was completely digested with the endonuclease *Bgl*II which was chosen among ten restriction enzymes recognizing six base pairs sites because it gives a number and distribution of bands suitable for the computer analysis. DNA digests (10 μg) were loaded on a vertical discontinuous 7.5% (w/v) acrylamide gel and run at 20 mA for 17 h. 0.25 μg of DNA molecular weight marker VI (Boehringer Mannheim) were added to each sample as internal standard (Degli-Innocenti *et al.*, 1990). The gel was stained with silver nitrate and band patterns were recorded and rescaled according to Giovannetti *et al.* (1990) except that five bands of the molecular weight marker were used as internal standards to correct for the 'smiling' effect.

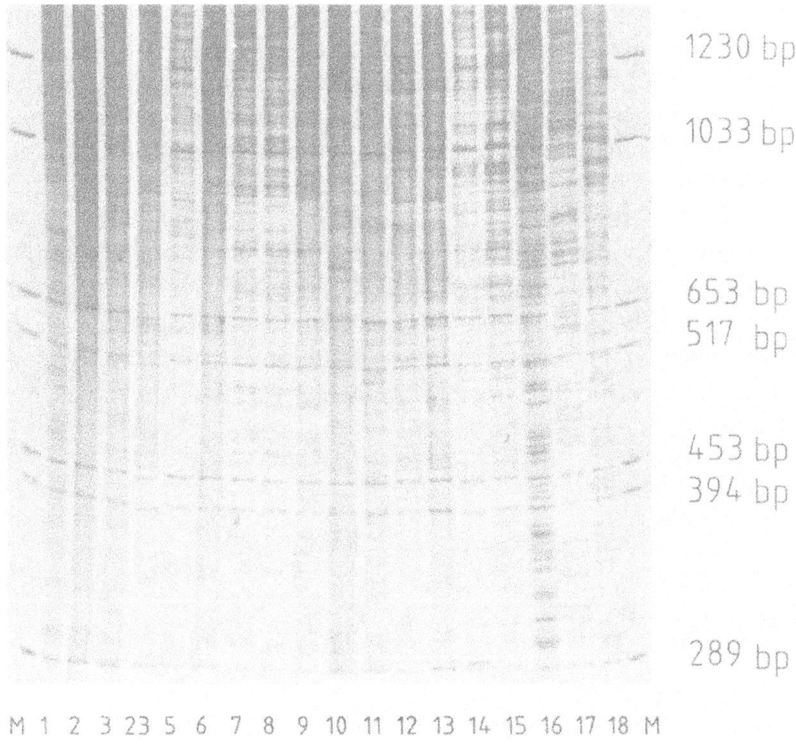


Figure 1. SDS-PAGE profiles of total DNAs digested with *Bgl*II. The strains are designated by the reference numbers listed in Table 1. Lanes M contain the molecular weight marker.

RESTRICTION PATTERN ANALYSIS

Only the range between 1200 and 300 base pairs was taken into account in the calculation, due to sticking together of bands in the higher part of the gel.

Similarity between all pairs of pattern was calculated with the Dice coefficient, $S_D$. Strains were clustered processing the matrix of Dice coefficients with the UPGMA method (Sneath and Sokal, 1973), using the SAS package.

## Results and Discussion

Isolation and growth conditions along with physiological characters excluded the attribution of our new isolates to *Azospirillum amazonense* and *Azospirillum halopraeferens*. Therefore only four reference strains of *Azospirillum brasilense* and *Azospirillum lipoferum* were included.

The restriction pattern of the eighteen strains is shown in Figure 1. It can be seen from this picture that each strain has a different pattern. Comparing patterns of strains Sp7 and Cd with the correspondent patterns obtained in a precedent work (Giovannetti *et al.*, 1990), it was found that the restriction pattern remains stable even after a prolonged interval of time. These observations demonstrate that the restriction pattern produces a fingerprint of the organism which enables a sure recognition of a single strain among stricly correlated others.

These patterns were used to try grouping microorganisms.

The first information supplied by the dendrogram shown in Figure 2 regards the relationship between plant and rizospheric organism: not any  group was constituted by azospirilla isolated from the same type of plant: the upper cluster in the picture houses isolates from tobacco, pepper and from two different cultivars of maize; the same situation is found in the remaining part of the tree.

Strains are grouped by UPGMA in two main clusters which merge at a similarity value of 0.53, plus one single member cluster much more different from the others, constituted by strain MRB2.

Cluster 1 comprises the two reference strains of *Azospirillum brasilense* which have a high degree of genomic similarity and six of our isolates; cluster 2 contains eight of our isolates and the reference strains of *Azospirillum lipoferum*. This distribution leds to the tentative assignement of our isolates to either one of the two species *brasilense* and *lipoferum* on the basis of their restriction pattern.

Apart from SA and SB, which come from two plants of sorghum, strains of the *lipoferum* group are not tightly related: this is not so surprising since DNA:DNA hybridization values among strain of *A. lipoferum* are low (Tarrand *et al.*, 1978).

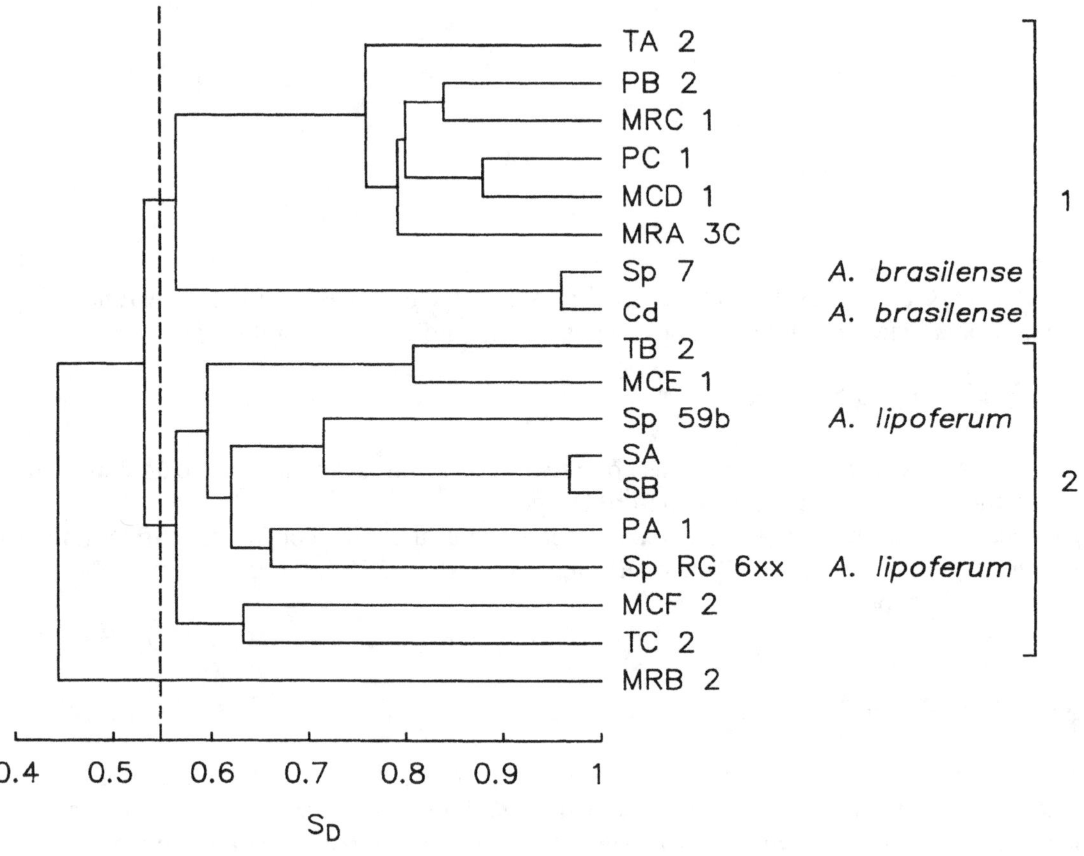


Figure 2. UPGMA clustering of the $S_D$ matrix obtained from *Bgl*II pattern analysis.

TABLE 2. Occurrence of pleomorphic cells in UPGMA clusters of strains.

| Cluster | Strain | Pleomorphic cells |
|---|---|---|
| 1<br>(A. brasilense) | TA2<br>PB2<br>MRC1<br>PC1<br>MCD1<br>MRA3C<br>Sp7<br>Cd | P |
| 2<br>(A. lipoferum) | TB2 | P |
| | MCE1 | |
| | Sp59b | P |
| | SA | P |
| | SB | P |
| | PA1 | P |
| | SpRG6xx | P |
| | MCF2 | |
| | TC2 | P |

The physiological properties listed under 'Materials and Methods' were employed for the phenotypic characterization of the strains.

Only strains SA and SB perfectely matched the response of *Azospirillum lipoferum* to the tests as reported in the Bergey's Manual, while the others showed variable degrees of discrepancy with the descriptions of both the two species, and a sure species attribution couldn't be done.

As regards the reference strains, some discrepancies were found, even in repeated experiments, in particular for biotine requirement.

Nevertheless when only the appearance of pleomorphic cells, which is a distinguishing character of *A. lipoferum*, was taken into account, a very high degree of congruence with the clusters obtained from the restriction pattern analysis occurred (Table 2.).

Data discussed in this paper confirm that the restriction pattern analysis separates very similar strains and therefore its use can be suggested for recognition of strains and identification of re-isolated strains from field experiments.

Since our isolates were divided by UPGMA clustering in two groups corresponding to *A. lipoferum* and *A. brasilense*, this method can be applied to separate correlated species; genus attribution should be previously obtained with phenotypic criteria because only very similar pattern can be compared.

## References

Chowdhury, S. I., Hammerschmidt, W., Ludwig, H., Thein, P. and Buhk, H. J. (1986) 'Rapid method for the identification and screening of Herpesviruses by DNA fingerprinting combined with blot hybridization', J. Virol. Meth. 14, 285-291.

Degli-Innocenti, F., Ferdani, E., Pesenti-Barilli, B., Dani, M., Giovannetti, L. and Ventura, S. (1990)

'Identification of microbial isolates by DNA fingerprinting: analysis of ATCC *Zymomonas* strains', J. Biotechnol. 13, 335-346.

Döbereiner, J. and Pedrosa, F. O. (1987) Nitrogen-fixing Bacteria in Nonleguminous Crop Plants, Science Tech Publishers, Madison.

Giovannetti, L., Ventura, S., Bazzicalupo, M., Fani, R. and Materassi, R. (1990) 'DNA restriction fingerprint analysis of the soil bacterium *Azospirillum*', J. Gen. Microbiol. 136, 1161-1166.

Krieg, N. R. and Döbereiner, J. (1984) 'Genus *Azospirillum* Tarrand Krieg and Döbereiner 1979, 79[AL] (effective publication: Tarrand, Krieg and Döbereiner 1978, 978)', in N. R. Krieg and J. G. Holt (eds.), Bergey's Manual of Systematic Bacteriology, Willliams and Wilkins, Baltimore, vol. 1, pp. 94-104.

Okon, Y. (1985) '*Azospirillum* as a potential inoculant for agriculture', Trends Biotechnol. 3, 223-229.

Sneath, P. H. A. and Sokal, R. R. (1973) Numerical Taxonomy. The Principles and Practices of Numerical Classification, W. H. Freeman, San Francisco.

Sorensen, B., Falk, E. S., Wisloff-Nilsen, E., Bjorvatn, B. and Kristiansen, B. E. (1985) 'Multivariate analysis of *Neisseria* DNA restriction endonuclease patterns', J. Gen. Microbiol. 131, 3099-3104.

Tarrand, J. J., Krieg, N. R. and Döbereiner, J. (1978) 'A taxonomic study of the Spirillun lipoferum group, with descriptions of a new genus, *Azospirillum* gen. nov. and two species, *Azospirillum* lipoferum (Beijerinck) comb. nov. and *Azospirillum* brasilense sp. nov.', Can. J. Microbiol. 24, 967-980.

Trüper, H. G. and Krämer, J. (1981) 'Principles of characterization and identification of prokaryotes', in M. P. Starr, H. Stolp, H. G. Trüper, A. Balows and H. G. Schlegel (eds.), The Prokaryotes, Springer-Verlag, Berlin, Heidelberg, New York, pp. 176-193.

# Alterations in membrane potential and in proton efflux in plant roots induced by *Azospirillum brasilense*

Y. BASHAN and H. LEVANONY
*Department of Microbiology, The Center of Biological Research (CIB), La Paz, P.O. Box 128, B.C.S., 23000, Mexico and Department of Plant Genetics, The Weizmann Institute of Science, Rehovot 76100, Israel*

*Key words:* Azospirillum, beneficial bacteria, membrane potential, plant-bacteria interaction, rhizosphere bacteria

## Abstract

Inoculation of soybean seedlings with *Azospirillum brasilense* Cd significantly reduced the membrane potential in every root part and was being maximal in the root elongation zone. Monitoring the proton efflux pattern of inoculated wheat roots by several *A. brasilense* strains and by *Pseudomonas* sp. for prolonged periods (up to 200 h) revealed a change from the bimodal pattern of proton efflux of non inoculated roots. This change was not related to root colonization ability but to bacterial capacity to induce changes in root surface area. Continuous perfusion of the plant nutrient solution with a fresh solution (from inoculation time), eliminated the enhancing effect of inoculation on proton efflux. We propose that *A. brasilense* inoculation influences membrane activity and subsequently proton efflux in roots, probably through the release of an as yet unidentified bacterial signal.

## Introduction

Changes in plant rhizosphere pH which result from an imbalance of ion uptake can be detected in the rhizosphere of several dicotyledonous plants. Proton extrusion through membranes of root cells, which result in acidification of the rhizosphere, is suggested as a major mechanism in mineral immobilization in plants (Spanswick, 1981).

The mode of action of the beneficial rhizosphere bacteria of the genus *Azospirillum* has been under a continuous debate for more than a decade (Bashan and Levanony, 1990). Among the several mechanisms proposed so far, increased mineral uptake by the plant as a result of inoculation has been suggested to play an essential role (Bashan et al., 1990; Murty and Ladha, 1988). However, no attempt has been made to relate mineral uptake to root membrane activity.

The aim of the present report is to demonstrate that inoculation with *A. brasilense* can affect membrane potential of root cells and release protons from these roots.

## Material and methods

*Bacterial strains and wheat cultivar, plants and bacteria growth conditions, hydroponic systems, nutrient solution for plants, bacterial inoculation, bacterial counts on roots, proton efflux and root surface area measurements, experimental design and statistical analysis*

These were done as previously described (Bashan, 1990; Bashan and Levanony, 1989; Bashan et al., 1989). Soybean seedlings (*Glycine max*, cv. Pella) were treated similarly to wheat seedlings.

*Electrophysiological measurements*

Intact soybean seedlings having a single root (63 h-old) were sampled and mounted horizontally on a plexiglass holder and washed for 3 h in an aerated solution composed of 1 m$M$ KCl, 1 m$M$ Ca(NO$_3$)$_2$, 0.25 m$M$ MgSO$_4$, and 66 m$M$ NaH$_2$PO$_4$, final pH 5.7 (1X) (Érsek et al., 1986). This rinsing was essential to equilibrate the cells in the perfusion medium.

Microcapillaries with glass micro-fibers (WP Instruments) were pulled to micro-electrodes using a vertical electrode puller. Micro-electrodes having a tip diameter of $0.6 \pm 2\ \mu$m, tip potential of $-2$ to $-16$ mV and tip resistance of 5 to 14 m$\Omega$ were used. Micro-electrodes were filled with 3 $M$ KCl, eliminating air bubbles trapped inside the micro-electrode. Each micro-electrode was microscopically tested for tip perfection. A reference salt-bridge, a 4 cm long piece of tube (2 mm inner diameter), was filled with 3 $M$ KCl in 2% agar. Both salt-bridges were connected through Ag/AgCl wire with an electrometer amplifier and a chart recorder. Micro-electrodes were inserted with a micromanipulator into the selected root site, continuously observed with a horizontally mounted stereoscopic microscope, illuminated with fiber optics. The plexiglass chamber containing the seedling and the holder (total volume of 7 mL) was perfused with 1 X medium at a flow rate of 8 mL min$^{-1}$ (Érsek et al., 1986; Findlay and Hope, 1976).

*Continuous perfusion of plant nutrient solution*

Continuous perfusion of the nutrient solution with fresh solution (4 mm/min) was carried out using a small pump. The overflow from the beaker was filtered through a 0.45 $\mu$m filter. This resulted in the bacteria and roots being continuously washed, but the bacteria were not removed from the root vicinity. This procedure allows replacement of a solution volume equivalent to the testing solution every 5 min. The replaced solution was collected, its volume was reduced by a roto-evaporation at $38 \pm 2$°C to its original volume, and the amount of released protons was determined.

**Results**

*The effect of inoculation by* A. brasilense *on the membrane potential of soybean cells in different root parts*

Non-inoculated soybean root parts exhibited similar E$_m$ values regardless of the root parts and within the range of $-146 \pm 6$ mV. Inoculation of roots with *A. brasilense* Cd significantly reduced the E$_m$ in every root part. The differences in E$_m$ reduction significantly varied between root parts, being maximal in the root elongation zone (Fig. 1).

*Changes in nutrient solution pH of wheat plants during prolonged periods by several rhizosphere bacteria*

Changes in pH of the nutrient solution were monitored continuously for 200 h. During this period a bimodal proton efflux activity was observed in non inoculated plants: two periods of proton efflux and two periods when an alkaline compound(s) was released. Inoculation of seedlings with *A. brasilense* Cd eliminated the fluctuation effect. Once the solution reached a pH of

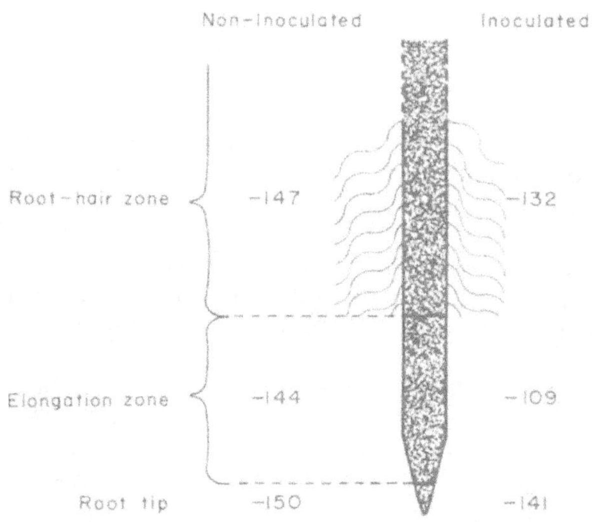


*Fig. 1* Membrane potential values (mV) for soybean cells at different sites of the root 16 h after inoculation with *A. brasilense* Cd.

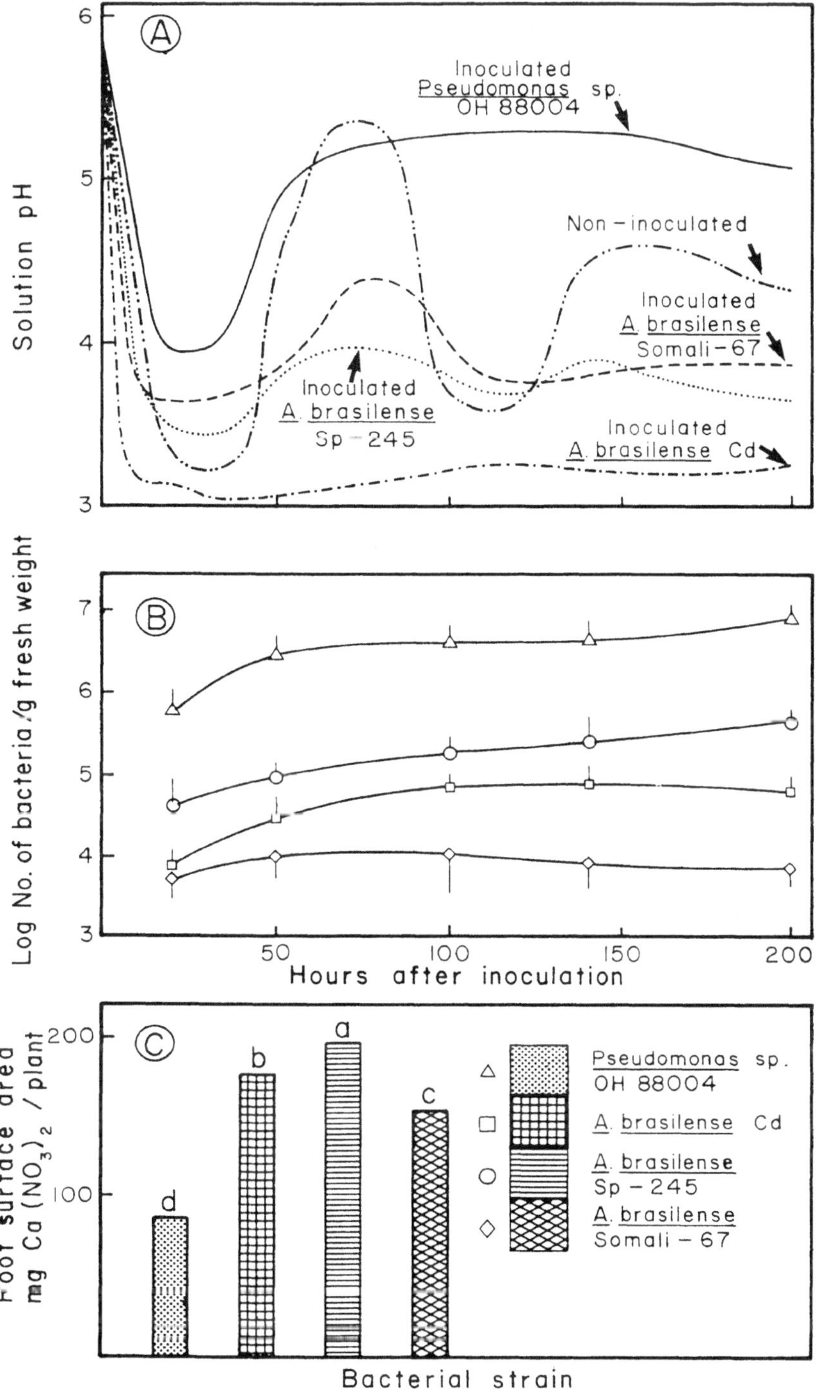


*Fig. 2.* Changes in nutrient solution pH (**A**) and in root colonization (**B**) caused by inoculation of wheat seedlings with several rhizosphere bacteria. All graphs were recorded automatically and drawn on a paper moving at a rate of 10 mm/h. (**C**) Changes in root surface area induced by inoculation with three rhizosphere bacteria.

3.1 ± 0.1 units, it remained at this low level throughout the entire period. Inoculation of wheat seedlings with *A. brasilense* strains Sp-245 and Somali-67 produced an intermediate result: the bimodal pattern of non inoculated plants was retained but at lower magnitudes. Inoculation with *Pseudomonas* sp. yielded a different pattern. After the initial pH decrease (up to 20 h from inoculation), the pH increased and remained at its highest level throughout the duration of the experiment (Fig. 2A).

Fluctuations in proton efflux were not related to colonization levels by the rhizosphere bacteria evaluated. All strains reached their maximal colonization level 50 h after inoculation. The best root colonizer (*Pseudomonas* sp.) produced the smallest effect. *A. brasilense* Cd which induced the most marked effects was a moderate colonizer (Fig. 2B).

Effect of inoculation, expressed as a change induced in root surface area, was not related to colonization levels. It was related to changes in proton efflux only in *A. brasilense* strains. The most marked effect on root surface area was induced by *A. brasilense* Sp-245, which affects proton efflux only moderately (Fig. 2C).

*Effect of continuous perfusion of the nutrient solution surrounding roots of inoculated wheat plants on the root proton efflux*

Inoculation with *A. brasilense* significantly increased proton extrusion over non-inoculated roots. Continuous perfusion of the plant nutrient solution with fresh solution did not affect proton efflux activity of non inoculated roots and the results resembled proton efflux in the still solution. When the nutrient solution of inoculated plants was continuously perfused (from inoculation time), the enhancing effect of inoculation on proton efflux was eliminated (Fig. 3A). However, the number of *A. brasilense* cells colonizing the wheat roots under these conditions was not different from the colonization numbers of roots in still nutrient solution (Fig. 3B, also compare to Fig. 2B).

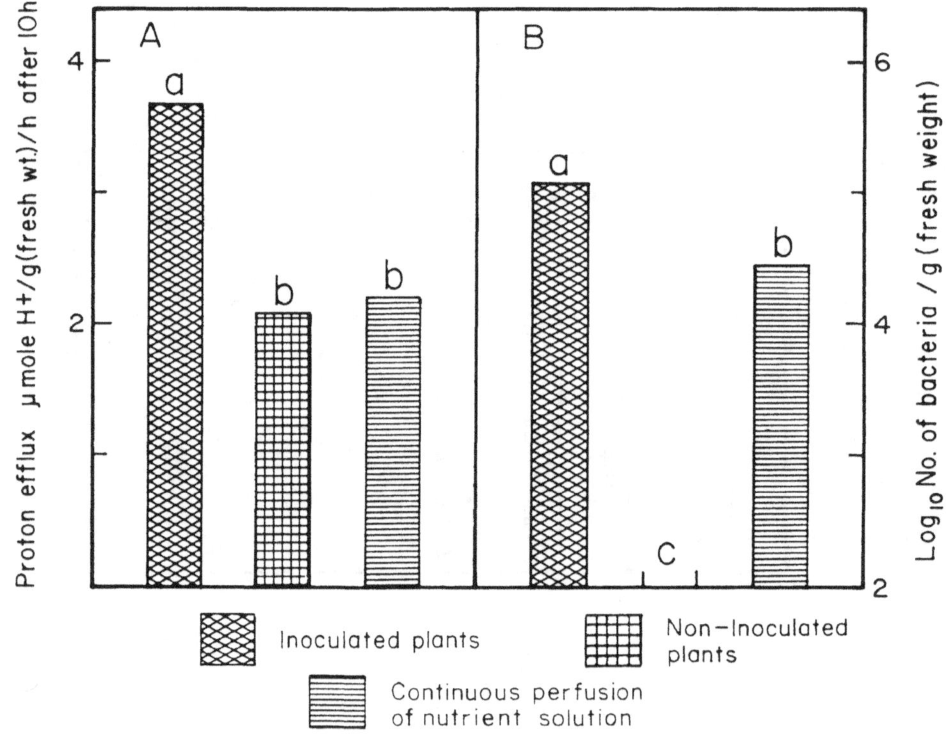


*Fig. 3.* Effect of continuous perfusion of the plant nutrient solution with fresh solution on: (**A**) proton efflux of wheat roots, and (**B**) the number of *A. brasilense* developed under these conditions.

## Discussion

The *Azospirillum* mode of enhancing plant growth is an open question. One of the mechanisms suggest thus far (Murty and Ladha, 1988), but not fully confirmed (Bashan et al., 1990), is enhancing the mineral uptake of plants induced by the inoculated bacteria. Changes in mineral uptake may be directly linked to root membrane activity and to its direct consequence, proton efflux from the roots. Significant changes in proton efflux activity of inoculated roots were recently demonstrated (Bashan, 1990; Bashan et al., 1989). This report adds evidence that concomitantly with proton efflux changes induced in the root by *Azospirillum* inoculation, membrane potential is also changed.

*A. brasilense* is known to have a preference to colonize the elongation zone of several plant species despite its ability to colonize every part of the root system. This report provides evidence that this preference affects membrane activity as well. Although changes in membrane potential were recorded in every part of the inoculated root, the greatest change was detected in the elongation zone of soybean roots.

The presence of *A. brasilense* in roots increased proton efflux. However, the inoculation did not increase the total capacity of root proton efflux. On the other hand, it did alter proton efflux pattern of non-inoculated roots.

Ultrastructural studies of the interaction between *A. brasilense* and root cells revealed that plant cell-walls prevent direct physical contact between the bacterium cell and the plasma membrane responsible for proton efflux (Levanony et al., 1989; Whallon et al., 1985). Thus, to alter membrane activity, the bacteria should release a diffusible signal(s) that has the ability to cross the plant cell-wall and be recognized by the membrane. This report provides preliminary evidence that *A. brasilense* may release such a signal(s). When the plant nutrient solution was continuously replaced by a fresh solution, even the presence of optimal bacterial numbers on the root surface did not enhance proton efflux.

In conclusion, this report shows that *A. brasilense* has the ability to reduce the membrane potential of inoculated roots and in particular of cells in the root elongation zone. Several strains of *A. brasilense* are capable of changing the pattern of proton efflux of wheat roots. The bacteria probably affect cell membranes through a release of bacterial signal(s).

## Acknowledgements

This study is dedicated to the memory of the late Mr. Avner Bashan from Israel. We thank Dr J Döbereiner, EMBRAPA, Brazil and Prof. F Favilli, University of Firenze, Italy for donating *A. brasilense* strains Sp-245 and Somali-67, respectively, and Mr Roy Bowers from the University of Baja California Sur for careful English corrections. All art-work was carried out by Mr Roberto Lomeli-Leos from the Center of Biological Research, La Paz, Mexico.

## References

Bashan Y 1990 Short exposure to *Azospirillum brasilense* inoculation enhanced proton efflux of intact wheat roots. Can. J. Microb. 36, 419–425.

Bashan Y, Harrison S K and Whitmoyer R E 1990 Enhanced growth of wheat and soybean plants inoculated with *Azospirillum brasilense* is not necessarily due to general enhancement of mineral uptake. Appl. Environ. Microb. 56, 769–775.

Bashan Y and Levanony H 1989 Effect of root environment on proton efflux in wheat roots. Plant and Soil 119, 191–197.

Bashan Y and Levanony H 1990 Current status of *Azospirillum* inoculation technology: *Azospirillum* as a challenge for agriculture. Can. J. Microb. 36, 591–608.

Bashan Y, Levanony H and Mitiku G 1989 Changes in proton efflux of intact wheat roots induced by *Azospirillum brasilense* Cd. Can. J. Microb. 35, 691–697.

Érsek T, Novacky A and Pueppke S G 1986 Compatible and incompatible rhizobia alter membrane potentials of soybean root cells. Plant Physiol. 82, 1115–1118.

Findlay G P and Hope A B 1976 Electrical properties of plant cells: methods and findings. *In* Encyclopedia of Plant Physiology, New Series, Vol 2 A. Eds. U Lüttge and M G Pittman. pp 53–92. Springer-Verlag, Berlin, New York.

Levanony H, Bashan Y, Romano B and Klein E 1989 Ultrastructural localization and identification of *Azospirillum brasilense* Cd on and within wheat root by immunogold labeling. Plant and Soil 117, 207–218.

Murty M G and Ladha J K 1988 Influence of *Azospirillum* inoculation on the mineral uptake and growth of rice under hydroponic conditions. Plant and Soil 108, 281–285.

Spanswick R M 1981 Electrogenic ion pumps. Annu. Rev. Plant Physiol. 32, 267–289.

Whallon J H, Acker G F and El Khawas H 1985 Electron microscopy of young wheat roots inoculated with *Azospirillum*. *In Azospirillum* III. Genetics, Physiology, Ecology. Ed. W Klingmüller. pp 223–229. Springer-Verlag, Berlin, New York.

# Effect of soil organic matter on chickpea inoculated with *Azospirillum brasilense* and *Rhizobium leguminosarum* bv. *ciceri*


MADDALENA DEL GALLO and PAOLA FABBRI
*Biotechnology Project, ENEA-Casaccia, Via Anguillarese 301, I-00060 Rome, Italy*



*Key words:*   Azospirillum, chickpea Rhizobium, *Cicer arietinum*, plant-growth promoting rhizobacteria, soil organic matter


## Abstract


*Azosprillum* inoculated with *Rhizobium* improved the nodulation of chickpea-*Cicer arietinum*. This interaction was further enhanced by organic matter present in the growth medium.


## Introduction

Scientific literature has reported already the positive effect that *Azospirillum* can have on *Rhizobium*-legumes symbioses. Sarig et al. (1986) reported a positive effect, under field condition, of *Azospirillum* inoculation on *Vicia sativa*, *Pisum sativum*, *Hedysarum coronarum* and *Cicer arietinum* spontaneously nodulated. Yahalom et al. (1987) reported the results obtained by an experiment carried out in in vitro conditions and concluded that a possible reason for increased susceptibility of the plant to *Rhizobium* infections, following *Azospirillum* application, may be due to the fact that it stimulates a formation of a larger number of epidermal cells that differentiate in infectable root hairs.

In our research we have focused on the effect of *Azosprillum* on chickpea-*Rhizobium* symbiosis and we have explored more in detail how soil organic matter can affect the interaction.

We have chosen chickpea for several reasons, among others because this plant respond very well to bacterial inoculation and because this system can be a good model of interaction, at root level, of plant and plant-growth promoting rhizobacteria (PGPR). Blocking nodulation (either increasing soil organic matter, or placing plants in a suboptimal light exposition), for instance, it is possible to study competition between *Azospirillum* and *Rhizobium* in the rhizosphere. In a previous paper (Del Gallo and Fabbri, 1990) we described the fact that *Azospirillum* can have a remarkable positive influence on chickpea-*Rhizobium* symbiosis, but we have also found that a strong competition between the two bacteria can be present at the root level, and that this competition can be affected by soil organic matter.

From field experiments previously conducted (unpublished results) we have observed that plants bred or selected to grow in an artificial environment (like our intensively exploited soils, continuously enriched by chemical fertilizers) respond to PGPR inoculation less than old variety or wild-type plants. This may be due to the fact that such a plant is pushed to utilize chemical fertilizers and to divert its energetics pool — carbon resources accumulated through the photosynthesis – to seed production instead of to the rhizosphere, to support a useless associated microflora. Moreover, in these particular conditions – such as a soil exploited for intensive agriculture – having all nutrients it needs concentrated, the plant develops a reduced root apparatus and has, thus, a reduced rhizosphere.

Chickpea, instead, is a plant generally cultivated in poor environments and has no need for large chemical fertilizers applications. Even recently bred cultivars are selected to grow in such a poor environment and association with its one

Rhizobium and, possibly, with an associated microflora.

## Material and methods

The bacterial strains utilized were: *Azospirillum brasilense* Cd and *Rhizobium leguminosarum* bv. *ciceri* USDA 3379 (Nitragin 27A8), kindly supplied by Peter Van Berkum. *Cicer arietinum* cv Califfo (an italian winter cultivar) seeds were kindly supplied by F. Calcagno. 3-days-old sterile chickpea seedlings were put aseptically in sterile pots (four plants per pot) and were inoculated with single or mixed cultures of bacteria (about $1 \times 10^5$ Rhizobia and $1 \times 10^6$ Azospirilla per plant) grown at the late exponential phase.

The soil utilized was a mixture of sand and agroperlite (2:1) enriched (or not) with peat in order to have 0, 8% or 22% of organic matter (o.m., Walkley–Black). In both 8% and 22% o.m. C/N ratio was 28. Nitrogen content was measured by Kjeldhal. Plants were grown in a Heraeus Heraphyt 2000 growth chamber with 16 hrs daylight; temperature was kept at 30°C during the light-period and at 25°C during the dark-period; humidity was kept constantly at 60%. Plants were watered twice a week alternatively with a Jensen N-free solution (Gibson, 1980) and tap water.

Plants grown at 22% o.m. were not nodulated – possibly because of the high N content of the substrate, 0.44%.

Controls not inoculated or inoculated with *Azospirillum* alone were carried out only at 22% o.m. level. At 0 and 8% o.m. plants without *Rhizobium*, or inoculated only with Azospirillum, did not survive up to seed supply (about 20 days). We inoculate Rhizobium either together with Azospirillum, when the seedling was planted in the pot, or 66 hours later. Twenty plants were considered for each treatment.

Data were elaborated statistically by Duncan's Multiple Range Test. In each group of data, different letter shows data statistically different from the control (a) for $p = 0.01^{**}$ or for $p = 0.05^{*}$.

## Results and discussion

The results obtained from the three experiments showed a positive effect of *Azospirillum* inoculation on *Rhizobium*-chickpea symbiosis.

### Azospirillum effect on roots and shoots

As described before (Del Gallo and Fabbri, 1990) the positive effect of *Azospirillum* inoculation was quite pronounced in the all experiments carried out, even more pronounced in plant inoculated with *Azospirillum* and later on with *Rhizobium* (Tables 1 and 2 and Figures 1 and 2).

Plants inoculated with both *Azospirillum* and *Rhizobium* had 2-fold dry weight with respect to plants inoculated with *Rhizobium* alone. This effect was more stressed at 22% o.m. level, when nodulation was inhibited: *Rhizobium* alone showed a slight positive effect on the plant, while *Azospirillum* increased 6-fold shoot dry weigh above controls. When the latter was inoculated together with *Rhizobium*, instead, its effect seemed to be inhibited, probably because of competition problems on root surface. However, the high level of organic matter by itself stimulated plant growth and precociousness. Plants grown at 22% o.m. were in a more advanced growth stage (at 52 days they were filling already

*Table 1.* Root dry weight. Root d.w. at 62 days, 22% o.m. was not determined because plants had already finished their life cycle.

| | Sand | | | Organic matter | | | | | | |
| | | | | 8% | | | 22% | | | |
| | R | R + A | R + A(66) | R | R + A | R + A(66) | C | R | A | R + A |
|---|---|---|---|---|---|---|---|---|---|---|
| 42 days | 0.39 a | 0.50 b* | 0.53 b* | 0.49 a | 0.66 b* | 0.80 c* | 2.02 a | 1.96 a | 2.87 b* | 2.17 ab |
| 52 days | 0.46 a | 0.51 b* | 0.61 c* | 0.52 a | 0.70 b* | 0.80 c* | 0.60 a | 1.40 b* | 4.75 c** | 1.76 b* |
| 62 days | 0.52 a | 0.63 b* | 0.88 c* | 0.68 a | 0.99 b** | 1.12 b** | n.d. | n.d. | n.d. | n.d. |

In each group of data, different letter shows data statistically different from the control (a) for $p = 0.01^{**}$ or for $p = 0.05^{*}$.

Table 2. Shoot dry weight. Shoot d.w. at 62 days, 22% o.m. was not determined because plants had already finished their life cycle.

| | Sand | | | Organic matter | | | | | | |
| | | | | 8% | | | 22% | | | |
| | R | R + A | R + A(66) | R | R + A | R + A(66) | C | R | A | R + A |
|---|---|---|---|---|---|---|---|---|---|---|
| 42 days | 0.44 a | 0.60ab** | 0.70 b** | 0.65 a | 0.76 b* | 0.85 c* | 2.54 a | 2.82 a | 3.55 b* | 3.06 ab |
| 52 days | 0.60 a | 0.70 b* | 0.82 c* | 0.83 a | 0.89 ab | 1.01 b* | 1.00 a | 1.50 a | 5.85 c** | 2.11 b* |
| 62 days | 0.81 a | 1.00 b* | 1.01 b* | 1.39 a | 2.13 b* | 1.85 b* | n.d. | n.d. | n.d. | n.d. |

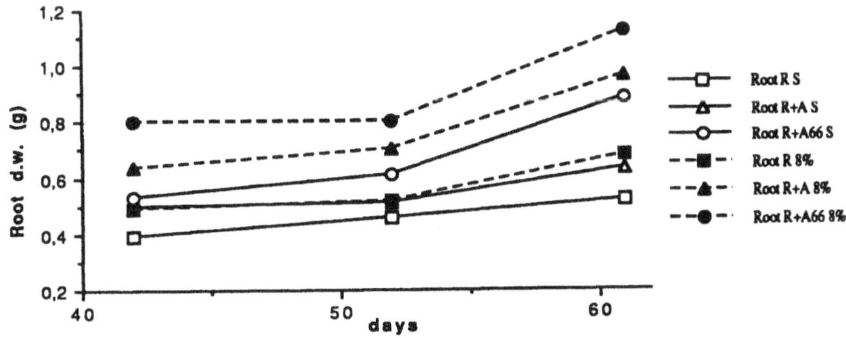


Fig. 1 Root dry weight of plant grown with 0 and with 8% organic matter. Statistic analysis of data is reported in table 1. R = rhizobium, A = azospirillum, R + A 66 = rhizobium was inoculated 66 h later than azospirillum, S = sand.

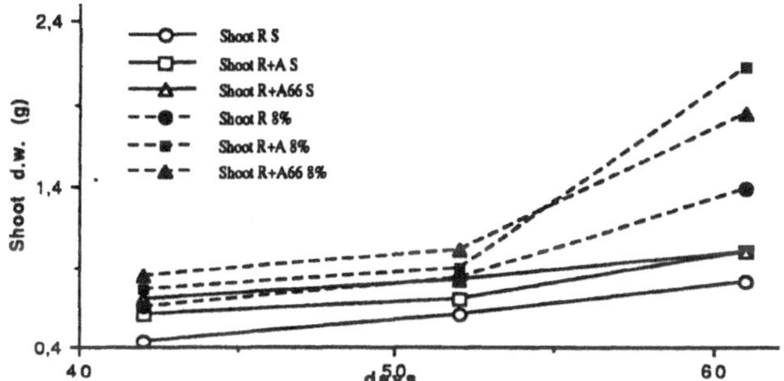


Fig. 2. Shoot dry weight of plant grown with 0 and with 8% of organic matter. Statistic analysis of differences are in Table 2.

pods, while plants grown at 0 or 8% o.m. did not start flowering yet) than plants grown at 0 and 8% o.m., so that a comparison between the two groups is impossible.

### Effect on nodule formation

As already described by Sarig et al. (1986) and Yahalom et al. (1987) *Azospirillum* affected chickpea nodulation both by anticipating nodule formation and increasing nodule number, size and dry weight (Tables 3 and 4, Fig. 3).

Table 3. Number of nodules per plant

| | Sand | | | 8% Organic matter | | |
| | Rhiz. | Rhi + Azo | R + A (66) | Rhiz. | Rhi + Azo | R + A (66) |
|---|---|---|---|---|---|---|
| 42 days | 6.00 a | 16.25 b* | 10.00 b* | 16.75 a | 27.66 b** | 27.08 b** |
| 52 days | 11.01 a | 15.37 a | 14.00 a | 19.81 a | 17.94 a | 20.19 a |
| 62 days | 15.00 a | 21.00 a | 18.00 a | 15.44 a | 26.37 b* | 23.13 b* |

*Table 4.* Nodules dry weight, g per plant

| | Sand | | | 8% Organic matter | | |
|---|---|---|---|---|---|---|
| | Rhiz. | Rhi + Azo | R + A (66) | Rhiz. | Rhi + Azo | R + A (66) |
| 42 days | 0.010 a | 0.030 b** | 0.049 c** | 0.031 a | 0.042 b* | 0.076 c*b** |
| 52 days | 0.028 a | 0.050 b* | 0.070 c*b** | 0.032 a | 0.051b* | 0.089 c*b** |
| 62 days | 0.030 a | 0.060 b** | 0.088 c** | 0.057 a | 0.168 b** | 0.189 b** |

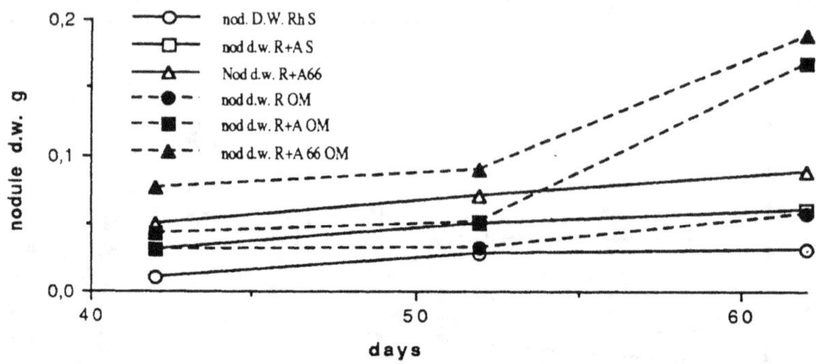


*Fig. 3.* Nodules dry weight.

## Effect on acetylene reduction activity

Results are shown in Figure 4. *Azospirillum* effect on acetylene reduction activity per plant was more accentuated in sand-grown plant, at 42 and 52 days. At the end of the experiment, instead, the activity decreased and reached the same value of plants inoculated with *Rhizobium* alone. At 0 and 8% o.m., *Azospirillum* effect was more pronounced, particularly when it was inoculated before *Rhizobium*. These values, however were similar, at the end of the experiment.

## Nitrogenase specific activity

The nitrogenase specific activity values are shown in Figure 5. Soil organic matter seemed to affect only *Azospirillum* inoculation, *Rhizobium*, instead, seemed not to be influenced at all: both groups of plant grown at 0 and at 8% o.m. showed the same pattern.

*Rhizobium*-inoculation 66 hours later than *Azospirillum* seemed not to affect nitrogenase activity in a different way than both bacteria inoculated together, but this was only on sand. When o.m. was present in the substrate, instead,

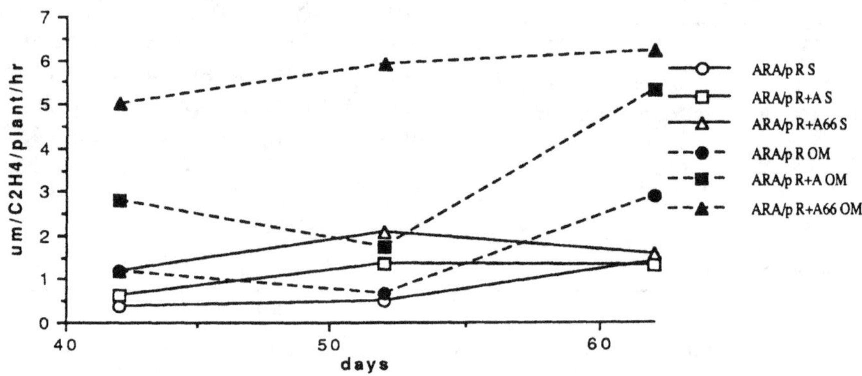


*Fig. 4.* Acetylene reduction activity/plant/h. All data were significantly different from the control (*Rhizobium* alone) except for plant inoculated with *Azospirillum* and *Rhizobium* together, without o.m. in sand, at 52 and at 62 days.

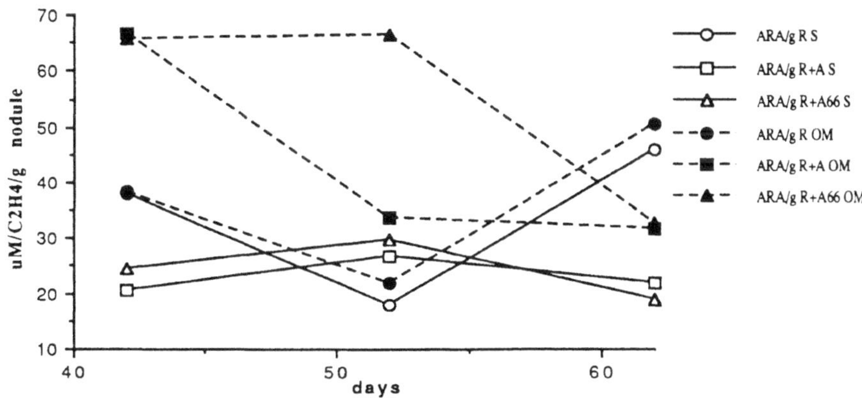

*Fig. 5.* Nitrogenase specific activity. Data were statistically different from controls (*Rhizobium* alone) only at o.m. 8%, in 42 and 52 days-old plants.

a pronounced effect was present until 52 days; 10 days later this difference disappeared. When both bacteria were inoculated together results showed a similar peak of nitrogenase specific activity, but for a shorter period. This may be due to the fact that the former group of plants was in a more advanced growth stage, and many senescent inactive nodules were present.

## Conclusions

In conclusion, *Azospirillum* inoculation, and PGPR in general (Burns et al., 1981; Grimes and Mount, 1984), can have a remarkable influence on Rhizobium-legumes symbiosis. Burns et al. (1981) reported that nodulation enhancement by *Azotobacter vinelandii* was probably caused by a non-excretable protein produced by this bacterium. Sarig et al. (1986) hypothesized that, following Azospirillum inoculation, more root hairs become susceptible to rhizobial infection.

However, it is difficult to explain, according to these hypotheses, the effect of Azospirillum on nitrogenase specific activity. An indirect effect of hydrogen recycling by Azospirillum hydrogenase, for instance, or a nitrogen fixation direct contribution inside nodules by this bacterium, can be present.

Further studies, however, are necessary to clarify the interaction.

## References


Burns T A, Bishop P E and Israel D W 1981 Enhanced nodulation of leguminous plant roots by mixed cultures of *Azotobacter vinelandii* and Rhizobium. Plant and Soil 62, 399–412.

Gibson A H 1980 Methods for legumes in glasshouses and controlled environment cabinets *In* Methods for Evaluating Biological Nitrogen Fixation. Ed. F J Bergersen. p 149. Wiley, Place.

Grimes H D and Mount M S 1984 Influence of *Pseudomonas putida* on nodulation of *Phaseolus vulgaris*. Soil Biol. Biochem. 16, 27–30.

Del Gallo M and Fabbri P 1990 Inoculation of *Azospirillum brasilense* Cd on chickpea (*Cicer arietinum* L.). Symbiosis 9, 283–287.

Plazinski J, Gärtner E, Mciver J, Jahnke and Rolfe B G 1984 Effect of *Azospirillum* strains on Rhizobium-legume symbiosis. *In* Advanced in Nitrogen Fixation Research. Eds. C Veeger and W E Newton. Nijhoff/Junk Publishers, Dordrecht, The Netherlands.

Plazinski J and Rolfe B G 1985 *Azospirillum-Rhizobium* interaction leading to a plant stimulation without nodule formation. Can J. Microbiol. 31, 1026-1030.

Sarig S, Kapulnik Y and Okon Y 1986 Effect of *Azospirillum* inoculation on nitrogen fixation and growth of several winter legumes. Plant and Soil 90, 335–342.

Yalom E, Okon Y and Dovrat A 1987 *Azospirillum* effect on susceptibility to *Rhizobium* nodulation and nitrogen fixation of several forage legumes. Can. J. Microbiol. 33, 510–514.

# Active attachment of *Azospirillum brasilense* to root surface of non-cereal plants and to sand particles

H. LEVANONY and Y. BASHAN[1]
*Department of Plant Genetics, The Weizmann Institute of Science. Rehovot 76100, Israel and Department of Microbiology, Division of Experimental Biology, The Center of Biological Research (CIB), La Paz, P.O. Box 128, B.C.S., 23000 Mexico ([1]Corresponding author)*

*Key words:* Azospirillum, bacterial adsorption, bacterial attachment, beneficial bacteria, plant growth-promoting rhizobacteria, rhizosphere bacteria

## Abstract

The rhizosphere bacterium *Azospirillum brasilense* Cd adsorbed strongly to light-textured and heavy-textured soils, but only slightly to quartz sand. Bacterial attachment to sand particles was mediated by a network made up of various sizes and shapes of fibrillar material. Inoculation of sand with an aggregate-deficient mutant resulted in no detectable fibrillar formation. Rinsing or agitating the sand, colonized by the wild-type and the mutant, had a greater effect on the mutant than on the parental strain. We propose that bacterial fibrils are essential for anchoring of *A. brasilense* to sand.

*A. brasilense* Cd was capable of efficiently colonizing the elongation and root-hair zones of tomato, pepper, cotton and soybean plants as well as of wheat plants. All inoculated plants demonstrated. (i) larger amounts of a mucigel-like substance on the root surface than non-inoculated plants, and (ii) fibrillar material which anchored the bacterial cells to the root surface. These fibrils established also connections between cells within bacterial aggregates. On non-water stressed soybean roots, most *A. brasilense* Cd cells occurred as vibroid forms. Whereas, those on roots of water-stressed plants (wilting) were cyst-like. A lower rhizosphere bacterial population was observed on water-stressed plants. When water stress conditions were eliminated, cells reverted to the vibroid form. A concomitant increase in the bacterial population was observed. We suggest that cyst-like formation is a natural response for *A. brasilense* Cd in the rhizosphere of water-stressed plants.

## Introduction

The beneficial rhizosphere bacterium *Azospirillum brasilense* Cd has been applied to numerous soil types world-wide in order to inoculate crop plant roots and to enhance plant yield (Bashan and Levanony, 1990; Michiels et al., 1989). Application of *A. brasilense* Cd cells to sand revealed that part of the bacterial population is actively attached to sand particles by a network of protein bridging between individual cells and between cells and sand particles (Bashan and Levanony, 1988b).

When inoculated onto cereal roots, *A. brasilense* Cd multiplied and formed small aggre-gates. The aggregates are mainly found in the root elongation and the root-hair zones (Bashan et al., 1986). The bacteria produced holdfast fibrillar material which anchored the cells to the root surface (Levanony et al., 1989).

The purposes of this study were: (i) To evaluate the importance of these bridges in the life cycle of the bacterium in the sand. This will establish whether or not they are essential for *A. brasilense* Cd attachment to sand. (ii) To evaluate the ability of *A. brasilense* Cd to colon-ize root surfaces of several non-cereal plants and to determine colonization sites. (iii) To examine the occurrence of, and conditions for pleomor-phic *A. brasilense* Cd in the rhizosphere. (iv) To evaluate whether· bacterial fibrillar connections to root surfaces are a general mode of attachment of *A. brasilense* Cd to plant root surfaces.

## Materials and methods

### Organisms

*Azospirillum brasilense* Cd (ATCC 29710) (aggregating strain, agg$^+$) and a non-aggregating mutant (agg$^-$) derived from a Tn5 mutant of strain Cd (strain 29710-10b) were used. The following plant species were used: tomato (*Lycopersicon esculentum*) cv. Na'ama; pepper (*Capsicum annuum*) cv. Ma'or; cotton (*Gossypium barbadense*) cv. Pima S-5; wheat (*Triticum aestivum*) cv. Deganit, and soybean (*Glycine max*) cv. Pella.

*Sand properties, bacterial and plant growth condition and inoculation, adsorption assays, bacterial counts from sand and roots, determination of percentage adsorption and experimental design*

These were previously described (Bashan and Levanony, 1988a,b; 1989a,b).

### Agitation and washing procedures

Strong agitation treatment of sand-bacteria mixtures after sand colonization (Vortex mixer) was performed at 180 rpm for 60 sec before the rinsing procedure, whereas low agitation was performed at 40 rpm for 90 sec. Inoculated sand was rinsed as previously described (Bashan and Levanony, 1988b). Two rinsing durations were used; 10 sec or 2 min.

### Induction of cysts in A. brasilense

Cyst-like forms of *A. brasilense* on root surfaces were induced by stopping irrigation of plants grown in vermiculite for 8 days, until plant leaves showed first signs of wilting. Then, irrigation was renewed.

### Definition of bacterial V- and C-forms

Bacterial cells observed on root surfaces were defined according to their size and the ratio of length to width. V−form bacteria had a length of $1.45 \pm 0.18\ \mu$m, width of $0.408 \pm 0.092\ \mu$m and average ratio (L/W) of 3.554:1 (mean of 228 measurements). C-form bacteria were defined as shorter and thicker cells with a length of $0.722 \pm 0.086\ \mu$m, width of $0.59 \pm 0.11\ \mu$m and average (L/W) ration of 1.224:1 (mean of 209 measurements).

## Results

### Description of active colonization of A. brasilense *of sand*

Random dispersal of single cells or very small microcolonies characterized the population density of inoculated sand or soil particles. Bacterial cell size was smaller compared to cells which were grown in liquid medium ($0.8$–$1.5\ \mu$m compared to $2$–$3\ \mu$m, respectively). Microcolonies were found in relatively small numbers and were located mainly in the shallow cavities of quartz particles. Many bacterial cells were attached to the surface of the sand particles by fibrillar ma-

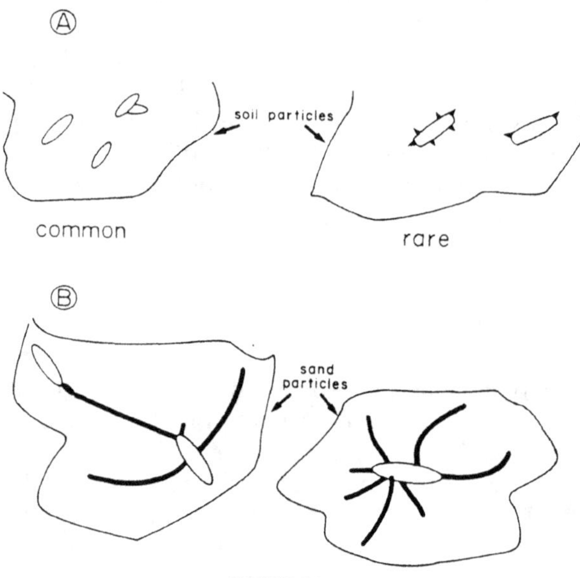


*Fig. 1.* Schematic presentation of *A. brasilense* agg⁺ cells colonizing sand. (**A**) Random dispersal of single cells on a soil particle. (**B**) fibrillar material connecting *A. brasilense* agg⁻ to sand.

terial (Fig. 1B). In soil, bacteria lacked any visible connection or had few very short connections to the soil (Fig. 1A). The bacterial fibrillar material was found to be either single-stranded or multistranded. It was located on any side of the bacterial cell (Fig. 1B).

*Attachment of* A. brasilense *agg$^+$ and agg$^-$ to sand following washing or agitation*

Bacterial attachment to sand by both strains, immediately after bacterial application, was negligible ($<0.001\%$). Population size of either agg$^+$ cells or agg$^-$ cells in sand was similar ($>10^7$ cfu mL$^{-1}$ after 48 h). However, the percentage of attachment of the two strains to the sand differed significantly, with the agg$^-$ mutant showing lower attachment rates (Fig. 2A). Slight rinsing of the sand (after sand colonization) decreased the adsorption of both strains, but there was a greater decrease in agg$^-$ cells. Increasing the washing time almost eliminated agg$^-$ cells from the sand, while significant number of agg$^+$ cells (approx. $10^6$ cfu g$^{-1}$ sand) remained attached to the sand. Despite a decrease in the total bacterial number of agg$^+$ cells caused by

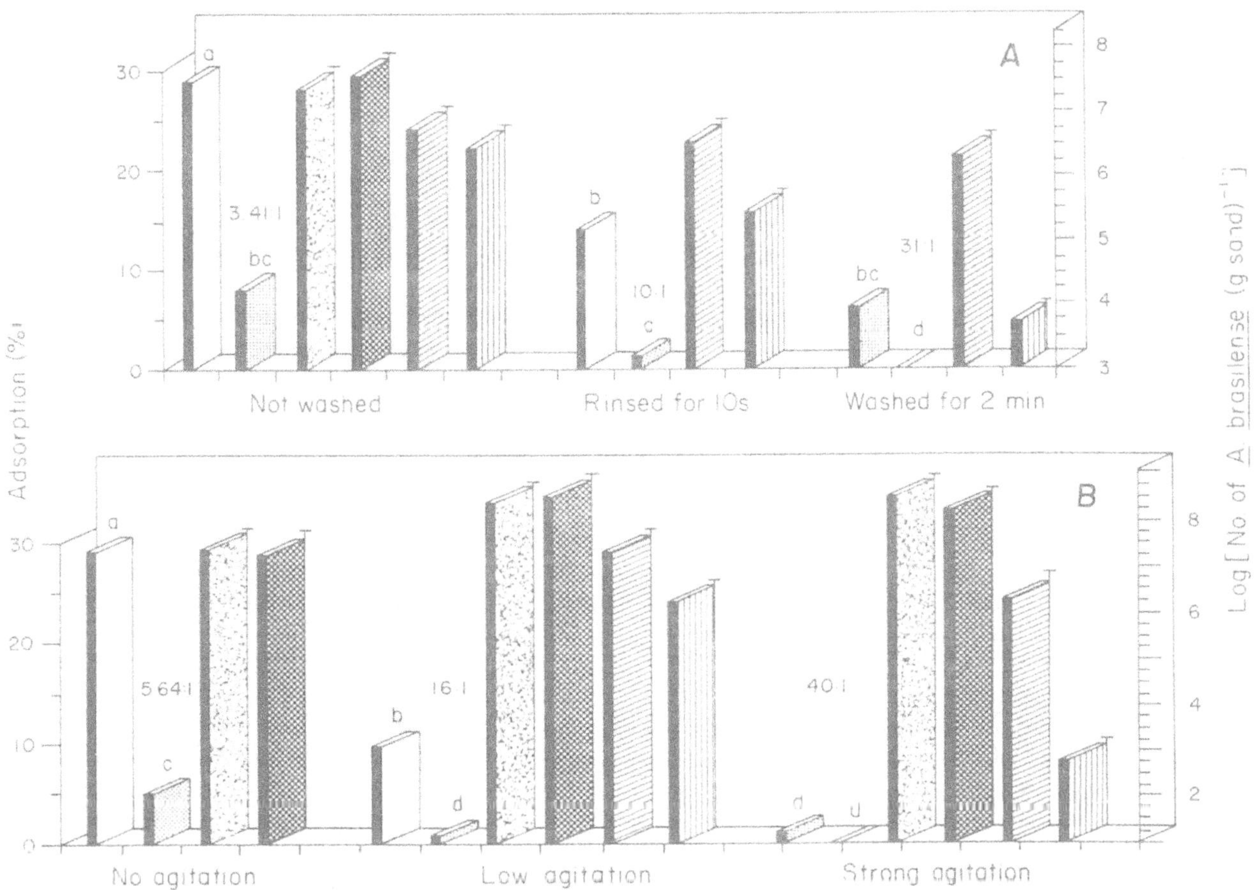


*Fig. 2.* (**A**) Percentage of adsorption and number of cells of *A. brasilense* (agg$^+$) and *A. brasilense* (agg$^-$) in sand before and following washing treatments. □ – adsorption of agg$^+$; ▦ – adsorption of agg$^-$; ▨ – number of cells of agg$^+$ adsorbed; ▥ – number of agg$^-$ cells adsorbed; total number of agg$^+$ cells (▨) and agg$^-$ cells (▩) in the sand mixture. (**B**) Percentage of adsorption and number of cells of *A. brasilense* (agg$^+$) and *A. brasilense* (agg$^-$) in sand before and after agitation. □ – adsorption of agg$^+$; ▦ – adsorption of agg$^-$; ▨ – number of cells of agg$^+$ adsorbed; ▥ – number of agg$^-$ cells adsorbed; total number of agg$^+$ cells (▨) and agg$^-$ cells (▩) in the sand mixture. Columns followed by a different letter (in each sub-figure separately) differ significantly at $p \le 0.05$. Bars represent SE. Number represent the adsorption ratio between agg$^+$ and agg$^-$ cells.

washing, the adsorption ratio between agg$^+$ and agg$^-$ strains increased with increasing the washing time about 10 times (Fig. 2A).

Generally, agitation decreased bacterial adsorption. Both strains developed large populations ($>10^8$ cfu g$^{-1}$ sand), and there was no significant difference between the strains. However, the adsorption ratio (between agg$^+$ and agg$^-$) increased as a result of agitation. When strong agitation was applied to sand, nearly all agg$^-$ cells were desorbed from the sand particles (840 cells out of $10^8$ cfu g$^{-1}$ sand remained) (Fig. 2B).

*Root colonization of tomato, pepper, cotton and wheat plants by* A. brasilense *Cd*

Schematic representations of root colonization by *A. brasilense* Cd is presented in Figure 3. Colonization patterns in the different plant species followed these schemes. Modifications in colonization regarded mainly the preferred colonization sites.

*Effect of watering regime on cell morphology of* A. brasilense *Cd colonizing the roots of soybean plants*

Under regular irrigation, the *A. brasilense* Cd population on the root surface increased exponentially reaching $10^6$ cfu cm$^{-1}$ root surface 20 days after sowing (Fig. 4A). Nearly all the bacterial cells observed on the root surface were single vibroid (V-form) forms. Stopping the irrigation of soybean seedlings affected both the population size and the bacterial cell shape. During the dry period, the number of *A. brasilense* Cd cells decreased to a low level, i.e., $10^4$ cfu cm$^{-1}$ root and the remaining cells were C-forms. Restarting irrigation of the plants resulted in the size of the bacterial population increasing 6 days later and the bacteria observed were V-form. Twenty six days after inoculation *A. brasilense* Cd cells were V-forms although a few cells kept their C-form throughout the experiment (Fig. 4C).

When no water was applied to the seeds until 8 days after sowing, C-forms appeared 3 days after sowing (Fig. 4D). In both irrigation regimes, the *A. brasilense* Cd population continued to increase for 2–3 days after the cessation of watering, and was composed mainly of V-form cells. Due to changes in the irrigation regimes, there were intermediate periods having V- and C-form *A. brasilense* Cd populations. In the absence of plants, the *A. brasilense* Cd population in the vermiculite decreased sharply regardless the irrigation regime. *A. brasilense* Cd population reached a low level, i.e., $10^3$ cfu g$^{-1}$ vermiculite; elimination of 99.9% of the original population (Fig. 4B).

## Discussion

When *Azospirillum* cells are applied to quartz sand, which holds them very loosely, they are readily detached by any water stream. Therefore, the bacterial cell should produce anchoring substance(s) to prevent this undesired washing below the root system.

*A. brasilense* Cd is known for forming large bacterial aggregates, both in liquid medium (Madi et al., 1988) and on root surfaces (Bashan et al., 1986).

Recently, we have shown that *A. brasilense* Cd is actively attached to sand particles by a network of proteinaceous bridges (Bashan and Levanony 1988b). Therefore, the focus of this study has been on the role of these bacterial fibrils in sand.

Generally, desorption of bacteria from soil particles by external mechanical forces such as washing and agitation, is difficult. Bacterial cells are strongly and permanently adsorbed by the soil particles (Bashan and Levanony, 1988a; Marshall, 1980). However, attachment of *A. brasilense* Cd to quartz sand particles is relatively weak. This study presents evidence demonstrating that single and multistranded fibrillar material are present and provide detachment resistance for the cells. Such phenomenon did not occur when a mutant incapable of producing these fibrils was inoculated into the sand. Furthermore, although agitation caused partial desorption of agg$^+$ cells, it had an even stronger effect on agg$^-$ cells, eliminating nearly all the bacterial cells from the sand. Therefore, it can be concluded that this network provides resistance against external physical forces applied to sand.

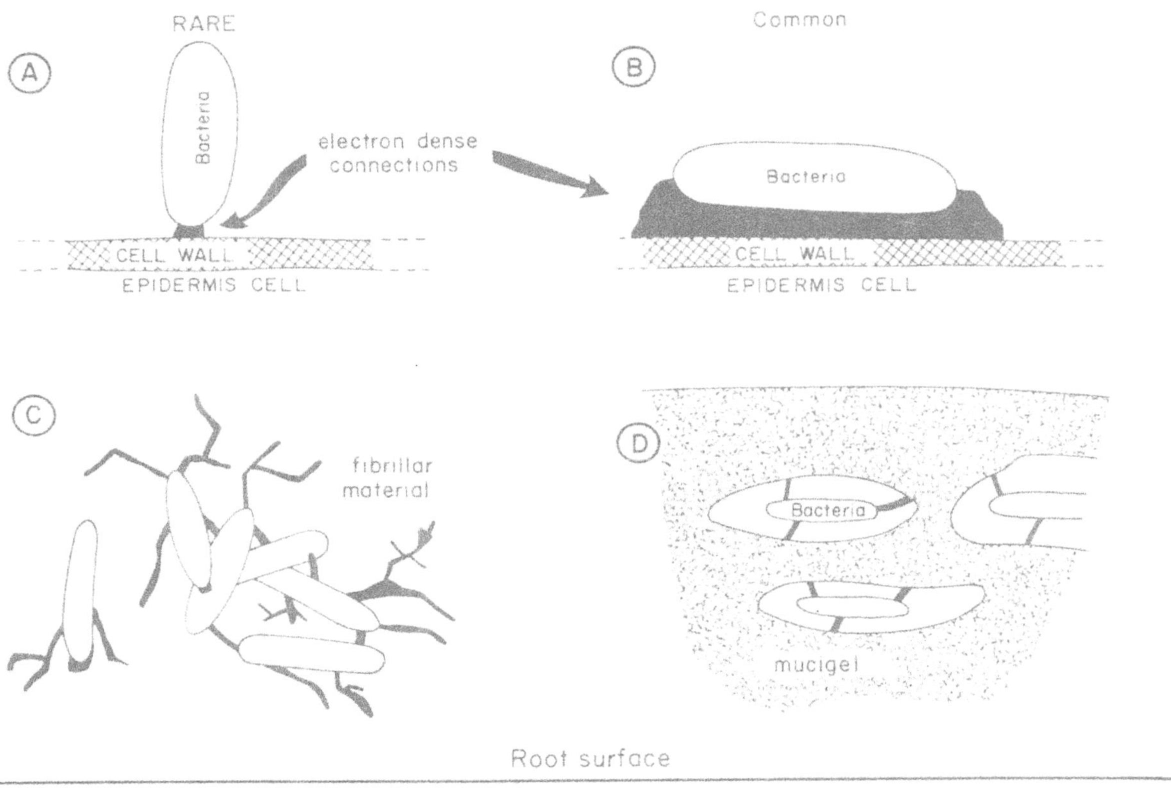


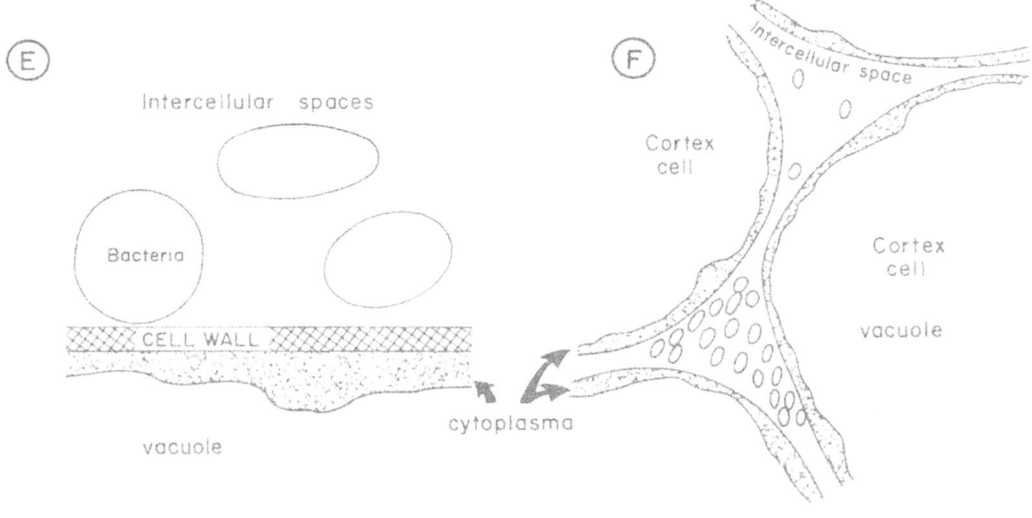


*Fig. 3.* Schematic representation of *A. brasilense* Cd colonization of: (**A–D**) root surface of wheat, tomato, pepper, cotton and soybean plants, showing intensive fibrillar material formation, and (**E–F**) colonization of the intercellular spaces of wheat roots showing no bacterial attachment to plant cell-walls via fibrillar connections.

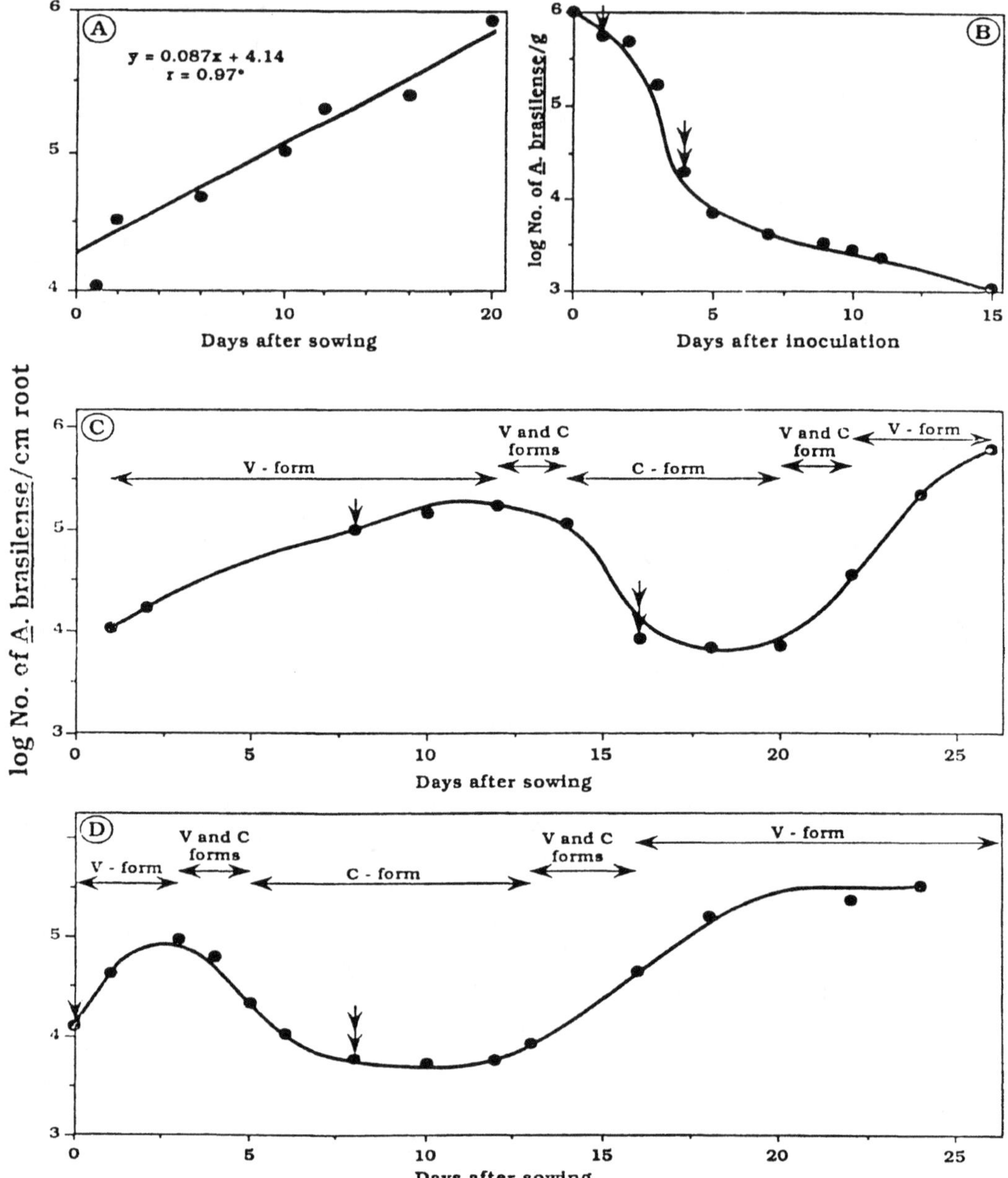

*Fig. 4.* Colonization of soybean roots (**A, C, D**) and vermiculite (**B**) by *A. brasilense* Cd under normal growth conditions without water limitation (**A**), and under water stress (**B–D**). ↓ – Daily irrigation was stopped; ↕ – irrigation renewed. Horizontal bars indicate the form of *A. brasilense* Cd cells found in this period of time (V-vibrio form; C-cyst form). *indicates significance of a correlation at $p \leq 0.05$. (Originally published in J. Gen. Microb. 137, 187–196, 1991.)

A unique feature of *Azospirillum* root colonization is the anchoring of bacterial cells to the plant surface by a network of fibrillar material. Fibrillar connections may play a role in the life cycle of *A. brasilense* Cd, whether it is on soil, sand (Bashan and Levanony, 1988a,b), wheat (Bashan et al., 1986; Levanony et al., 1990), or on several other non-cereal plant roots demonstrated in the present study.

Pleomorphism (vibroid or cyst-like forms of the bacterial cell) of *Azospirillum* in in vitro culture is well documented (Eskew et al., 1977;

Sadasivan and Neyra, 1987). The current study shows that this pleomorphism occurred in situ in the rhizosphere. Transition from the vegetative vibroid form to the cyst-like form and vice versa was observed following manipulation of water availability. The mechanism of this transition remains unknown.

In conclusion, our study suggests that fibrillar material plays an important role in attachment of *A. brasilense* Cd to root surface and to sand particles.

## Acknowledgements

This study is dedicated to the memory of the late Mr Avner Bashan from Israel. We thank Dr. M. Singh, GBF, Germany for donating *A. brasilense* strain 29710-10b and Mr Roy Bowers, University of Baja California Sur, La Paz, Mexico, for careful English corrections. All art-work was carried out by Mr Roberto Lomeli-Leos from the Center of Biological Research, La Paz, Mexico.

## References

Bashan Y and Levanony H 1988a Adsorption of the rhizosphere bacterium *Azospirillum brasilense* Cd to soil, sand and peat particles. J. Gen. Microb. 134, 1811–1820.

Bashan Y and Levanony H 1988b Active attachment of *Azospirillum brasilense* Cd to quartz sand and to light-textured soil by protein bridging. J. Gen. Microb. 134, 2269–2279.

Bashan Y and Levanony H 1989a Wheat root tips as a vector for passive vertical transfer of *Azospirillum brasilense* Cd. J. Gen. Microb. 135, 2899–2908.

Bashan Y and Levanony H 1989b Factors affecting adsorption of *Azospirillum brasilense* Cd to root hairs as compared with root surface of wheat. Can. J. Microb. 35, 936–944.

Bashan Y and Levanony H 1990 Current status of *Azospirillum* inoculation technology: *Azospirillum* as a challenge for agriculture. Can. J. Microb. 36, 591–608.

Bashan Y, Levanony H and Klein E 1986 Evidence for a weak active external adsorption of *Azospirillum brasilense* Cd to wheat roots. J. Gen. Microb. 132, 3069–3073.

Eskew D L, Focht D D and Ting I P 1977 Nitrogen fixation, denitrification, and pleomorphic growth in a highly pigmented *Spirillum lipoferum*. Appl. Environ. Microb. 34, 582–585.

Levanony H, Bashan Y, Romano B and Klein E 1989 Ultrastructural localization and identification of *Azospirillum brasilense* Cd on and within wheat roots by immunogold labelling. Plant and Soil 117, 207–218.

Madi L, Kessel M, Sadovnick E and Henis Y 1988 Electron microscopic studies of aggregation and pellicle formation in *Azospirillum* spp. Plant and Soil 109, 115–121.

Marshall K C 1980 Adsorption of microorganisms to soils and sediments *In* Adsorption of Microorganisms to Surfaces. Eds. G. Bitton and K C Marshall. pp 317–329. Wiley, New York.

Michiels K, Vanderleyden J and Van Gool A 1989 *Azospirillum*-plant root association: A review. Biol. Fert. Soils 8, 356–368.

Sadasivan L and Neyra C A 1987 Cyst production and brown pigment formation in aging cultures of *Azospirillum brasilense* ATCC 29145. J. Bacteriol. 169, 1670–1677.

# Association of *Alcaligenes faecalis* with wetland rice


C.B. YOU, W. SONG, H.X. WANG, J.P. LI, M. LIN and W.L. HAI
*Institute for Application of Atomic Energy, CAAS, P.O. Box 5109, Beijing 100094, People's
Republic of China*


*Key words:*    Alcaligenes faecalis, endosymbiont, wetland rise

## Abstract


*Alcaligenes faecalis* isolated from rice roots is widespread in paddy soil of China. It was found to be a
close association with rice. *A. faecalis* accumulate on the rice root surface, and part of them could enter
into the rice root. It can grow in the intercellular space, especially inside the root cells, and multiply and
fix dinitrogen there. *A. faecalis* could synthesize nitrogenase when it was grown in the medium
containing a high concentration of ammonia. The mechanisms of association are also discussed.


## Introduction

Rice is the staple food for the Chinese. The total
area of rice fields in China approximates to 26%
of the cultivated land, and the output of rice
grain accounts for 45% of the output of food
crops. Therefore the maintenance of nitrogen
fertility in paddy soils is very important (Yoshida
and Rinaudo, 1982). The conditions in paddy
soil may be aerobic or anaerobic, and almost all
major nitrogen-fixing groups can grow in this
ecosystem (Jia et al., 1989). The great majority
of bacteria associated with rice roots in China
belong to Enterobacteriaceae, *Pseudomonas* and
*Alcaligenes* (Jia et al., 1989), *Arthobacter, Azospirillum, Azotobacter, Bacillus, Derxia, Flavobacterium* etc. have also been reported as nitrogen-fixing inhabitants of rice root (Li and
You, 1991). Quite different from other reports
(Watanabe, 1986), *Alcaligenes faecalis* is widespread in paddy soil of China, and constitutes
the predominant diazotrophic strain isolated
from wetland rice roots (Jia et al., 1989). The
nitrogen-fixing activity of *A. faecalis* has not
been reported before (Qin et al., 1980). In this
paper we shall present the results obtained in the
association of nitrogen-fixers with rice plants
with emphasis on *A. faecalis*.

## Materials and methods

The strains of bacterium used in the experiments
were *Alcaligenes faecalis* A15 and newly selected
A15H1. Both *japonica* and *indica* types of rice
from different varieties were used in the pot and
field experiments. All analytical and examining
methods and techniques such as antibody preparation, FA staining of sample, protoplasts isolation, electromicroscopy and isotopic tracing etc.
were described previously (You and Zhou,
1989).

## Results

### Association with rice plants

Association of azospirilla and other diazotrophs
with cereals and forage grass have been described by many authors (Döbereiner and Pedrosa, 1987), but the nature of these association
and contribution of each partner have not so far
been ascertained.

### Chemotaxis of bacteria towards rice roots
Chemotatic reactions may be important in assisting motile bacteria to approach their host. The in

vitro assay determining the mobility of *A. faecalis* shows their chemotaxis in response to rice extract and exudate. Only a few bacteria are chemotatic to organic acids, which could be used as a carbon source for their growth (You et al., 1983). After a short-term incubation with *A. faecalis*, the bacteria were absorbed and randomly accumulated on the rice root surface to form a layer ranging from 50–100 $\mu$m. In a longer incubation, the bacteria colonized rice root epidermal surface with mucigel (You and Qin, 1982). However, when ammonia was added to the culture medium, the bacteria are removed from the root surface (You et al., 1983).

*Interaction between A. faecalis and rice plants*
The associative $N_2$-fixing activity of *A. faecalis* with rice was rather high. It could reach 3140 nmol $C_2H_4$ formed $g^{-1}$ dry root $h^{-1}$, and the *A. faecalis* excreted ammonia into the medium in the late log phase of its growth of $N_2$, and about 20–30% of the total amount of fixed nitrogen was excreted (You et al., 1983). About 1/3 of this could translocate rapidly into the leaves of rice plant in 69 h.

The exudates of *A. faecalis* (Table 1) and rice root have been examined. Rice exudate consisted of 2.25 mg of organic acids, 1.27 mg carbohydrate and 2.61 mg of amino acids per gram fresh root. In the exudate the organic acids from high to low concentration were in the order of citric acid, malic acid, succinic acid and lactic acid. Among the 15 kinds of amino acids the contents of the basic ones were higher than the acidic ones. The carbohydrates consisted of glucose, fructose and sucrose. As to phytohormones

only gibberell in-$GA_3$ was found in uninoculated rice root. The components and contents of rice root exudates were dependent on the variety of rice. Inoculation with *A. faecalis* stimulated the root exudation of rice and affected the components and contents of root exudates (Table 2). By using $^{14}CO_2$ tracer, 2.63% of photosynthate in uninoculated rice was excreted, while 3.52% was released from inoculated rice. Most of them were taken up by *A. faecalis* in the rhizosphere. The photosynthetic rate and excretory ability affected the nitrogenase activity in the rhizosphere. The higher the nitrogenase activity in the rhizosphere of the rice the greater the amount of root exudates.

*Non-nodular endorhizospheric nitrogen fixation*
*A. faecalis* not only accumulated on the rice root surface, but also could enter into the root itself. The bacteria were visible in the intercellular space of root epidermal cells, cortical parenchyma and inside the root cells. The immunofluorescence assay also confirmed the presence of these bacteria inside the root cells, while FA-stained bacteria were not observed in the non-inoculated control (You and Zhou, 1989). Additional evidence for bacteria endorhizocoenosis were obtained with B tracing, TEM, SEM etc. Protoplasts isolated from rice root and callus which had been inoculated with *A. faecalis* were estimated. The electromicrographs showed no bacteria on the protoplast surface. However, many bacteria could be seen inside the protoplasts broken by osmotic shock, and in the ultrathin section of protoplast (Fig. 1) (You and Zhou, 1989). *A. faecalis* could live symbioticaly

*Table 1.* Composition and content in exudate from *A. faecalis*

| Phytohormones ($\mu$g mL$^{-1}$) | | organic acids ($\mu$g mL$^{-1}$) | | Amino acids ($\mu$g mL$^{-1}$) | |
|---|---|---|---|---|---|
| IAA | 0.0097 | Malic acid | 2.152 | Glu | 0.80 |
| | | | | Ile | 0.12 |
| | | | | Ala | 0.98 |
| GA$_3$ | 0.0315 | Citric acid | 0.000 | Tyr | 0.22 |
| | | | | Phe | 0.17 |
| | | | | Leu | 0.22 |
| ABA | 0.0000 | Succinic acid | 0.000 | His | 0.24 |
| | | | | Lys | 0.86 |
| | | | | Agr | 0.44 |

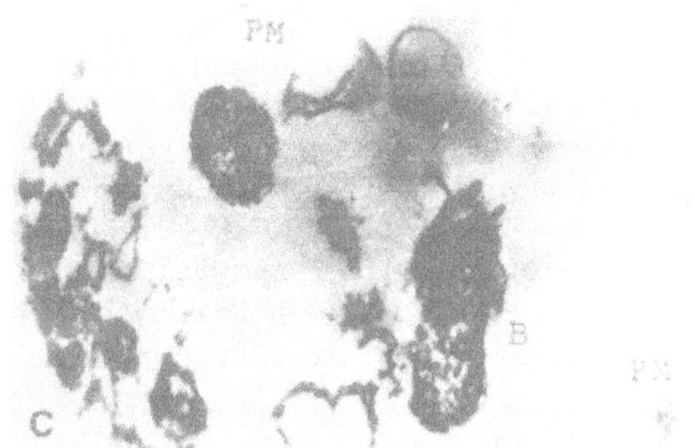

*Fig. 1.* Ultrathin section of protplast from rice root callus (TEM, 13000X). PM; proplast membrane; B: Bacterium.

with callus induced from rice root cells on N-free MS culture medium. This symbiont could fix dinitrogen. The rate ranged from 18.5 to 35.5 $\mu$g N fixed g$^{-1}$ dry weight callus day$^{-1}$. *A. faecalis* could grow and multiply in the rice root cell, as indicated by the fact that (i) the bacteria were visible in all callus cells which were grown from a single cell inoculated with this bacterium, and (ii). *A. faecalis* produce alkali during their growth and the pH value of medium rises along with the biomass, the presence of bromothyl blue in the medium indicates increases in pH by changing the colour from white to blue (You and Zhou, 1990).

More recently, using SEM and MPN-acetylene reducing assay (ARA), it has been shown that *A. faecalis* could grow and fix dinitrogen in the intercellular and intracellular spaces (Figs. 2, 3) in the rice plant (Table 3). Most of them grew in the centre part of the root. These results described above suggest that rice is in a very close association with *A. feacalis*.

## Nitrogen-fixing activity and its system

This is an integrated index of a complex parameter which can give the nitrogen fixation potential of a diazotroph and its usefulness as a biofertilizer. The maximum nitrogen-fixing activity of *A. faecalis* is 40 mg of N assimilated/g malic acid consumed.

*Table 2.* Content of organic acids and sugars in culture medium of rice non-inoculated and inoculated with *A. faecalis*

| Rice varieties | | Organic acids (mg g$^{-1}$ f.r) | | | | Carbohydrates (mg g$^{-1}$ f.r) | | |
|---|---|---|---|---|---|---|---|---|
| | | Lactate | Succin. | Citrate | Malage | Fruc. | Gluc. | Sucrose |
| Yuefu[a] | + | 0.31 | 0.00 | 0.05 | 0.31 | 5.56 | 0.00 | 9.03 |
| | − | 0.00 | 0.72 | 0.67 | 0.71 | 5.85 | 0.00 | 9.14 |
| Jingbai[a] | + | 0.78 | 0.00 | 0.22 | 0.00 | 2.97 | 0.00 | 0.00 |
| | − | 0.46 | 1.19 | 1.83 | 1.00 | 3.39 | 1.01 | 9.83 |
| Qinai[b] | + | 0.99 | 0.19 | 0.00 | 0.38 | 2.53 | 1.26 | 0.00 |
| | − | 0.53 | 0.00 | 0.64 | 0.44 | 8.13 | 1.01 | 1.01 |
| Rongjing | + | 0.12 | 0.00 | 0.10 | 0.84 | 2.71 | 0.34 | 0.00 |
| No.1[c] | − | 0.28 | 0.00 | 0.38 | 0.17 | 9.26 | 0.00 | 2.14 |

[a] *japonica* type; [b] Upland rice; [c] *indica* type. +: inoculated with *A. faecalis*, −: *uninoculated*.

*Table 3.* Comparison of acetylene reduction activit (ARA) at different parts of inoculated and noninoculated rice plants

| Treatments | ARA (nmol $C_2H_4$ formed g$^{-1}$ fresh wt h$^{-1}$) | | | |
|---|---|---|---|---|
| | Cut root | Crushed root | Cut stem[a] | Crushed stem[a] |
| Control | 27.93 | 153.34 | 30.93 | 58.78 |
| Inoculated with 10 mL culture[b] | 75.94 | 289.64 | 31.91 | 61.34 |
| Inoculated with 40 mL culture[b] | 146.93 | 689.28 | 33.06 | 170.78 |

[a] Stem part under the flooded water; [b] 1 mL $= 0.85 \times 10^8$ cells. Note: 40 rice plants were planted in a pot containing 2.5 kg sterilized soil.

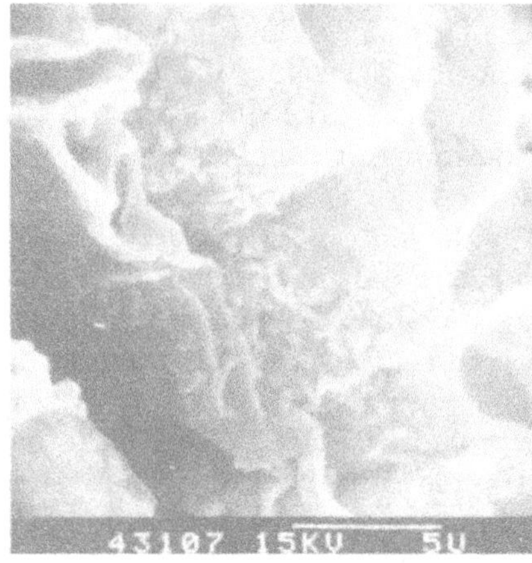

*Fig. 2.* SEM photo of *A. faecalis* in the intercellular space of rice root.

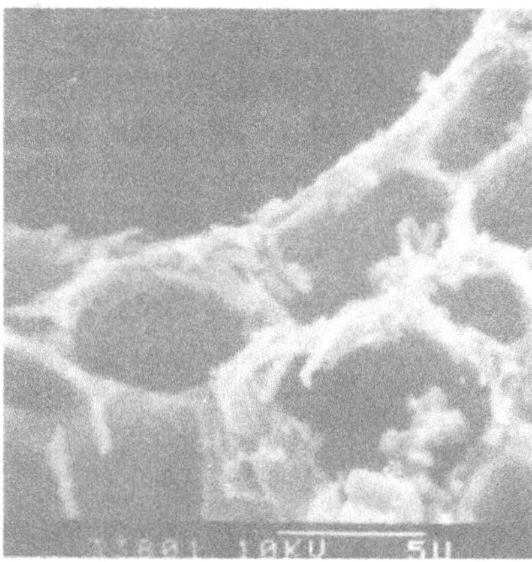

*Fig. 3.* SEM photo of *A. faecalis* in the inside of cells near xylem.

More interestingly, it was shown that nitrogenase was synthesized by the *A. faecalis* in the presence of a high concentration of ammonia (30 m$M$) in the culture medium, even though the nitrogen-fixing activity was absent (Zhang et al., 1989). This result was given by the SDS-PAGE, DEAE-cellulose chromatography, rocket and rocket-line immunoelectrophoresis (Zhang et al., 1989), and it was reconfirmed by the isolation and purification of nitrogenase protein. The MoFe protein from both $N_2$- and $NH_4^+$-grown cells of *A. faecalis* was purified to electrophoretic homogeneity. The physico-chemical properties such as amino acid composition, molecular weight, redox equivalents, metal contents etc.

were similar to each other, but they differed from one another only in circular dichroism (CD) spectra, suggesting their molecular structure might be different. By the aid of the antibiotic chloramphenicol, the further biosynthesis of nitrogenase was inhibited in $NH_4^+$-grown cells. The activity of nitrogenase synthesized during $NH_4^+$-grown period was depressed when the $NH_4^+$ was removed from the culture medium.

*Recent development of genetics*

*A. faecalis* was found to harbour a plasmid, but the *nif* gene was located on the chromosome. The result also shows that the *nif* genes in the $NH_4^+$-grown cells of *A. faecalis* were the same as in the $N_2$-grown cells. The *hup* genes are located on the same replicon in *A. faecalis*. A genomic library of total DNA from *A. faecalis* was constructed. A total number of $1.2 \times 10^6$ recombinants were obtained. The clones containing a consensus of *nif* HDK and *nif* A were screened, and the *nif* H subclone was isolated (Hai et al., 1990). The ammonia-resistant strains of *A. faecalis* which either harbour pCK3 or pCK5 plasmid or insertion of *nif* A in chromosome were performed. Recently, the rice plant was regenerated from a protoplast fused with an L-form of *A. faecalis*.

*Agronomic significance*

Using $^{15}N$ dilution techniques, dinitrogen fixation of diazotrophs was estimated in field experiments. The results indicate that the total N content in rice plants increased by up to 8% and total N fixed about $20-30 \, kg \, ha^{-1}$. So, as a potential nitrogen source for rice cultivation, associative nitrogen fixation deserves serious consideration.

## References


Döbereiner J and Pedrosa F O 1987 Nitrogen-fixing bacteria in nonleguminous crop plants. Sci. Tech. Publis. Springer-Verlag, Berlin, Heidelberg, New York. 350 p.

Hai W L, Zheng H G, Wang B and You C B 1990 Construction of a genomic library of *Alcaligenes faecalis* and screening of positive clone containing *nif* genes. *In* A Treatise on

Associative Nitrogen Fixation in Rice Rhizosphere. Ed. C B You. pp 213–219. Agric. Publ.

Jia X M, Mo W Y and Qian Z S 1989 Species and enumeration of nitrogen fixing bacteria in rice root system. Acta Agricul. Univ. Zhejangensis 15, 57–60.

Li J P and You C B 1991 Associative nitrogen fixation in wetland rice. *In* A Treatise on Associative Nitrogen Fixation in Rice Rhizosphere. Ed. C B You. pp 1–12. Agric. Publ.

Qiu Y S, Zhou S P, Mo X Z, You C B and Wang D S 1980 Investigation of $N_2$-fixation bacteria isolated from rice rhizosphere. J. Science 25, 383.

Watanabe I 1986 Nitrogen fixation by non-legumes in tropical agriculture with special reference to wetland rice. Plant and Soil 90, 343–357.

Xu Q 1981 Cropping system in relation to fertility of paddy soil in China. Ed. Inst. Soil Sci. Proc. Sym. Paddy Soil. Springer-Verlag, Berlin, Heidelberg, New York. 221 p.

Yoshida T and Rinaudo S 1982 Heterotrophic $N_2$-fixation in paddy soils. *In* Microbiology of Tropical Soils and Plant Productivity. Eds. Y R Dommergues and H G Diem. pp 75–107. Kluwer Academic Publishers, Dordrecht, The Netherlands.

You C B and Qiu Y S 1982 Nitrogen fixation of *Alcaligenes faecalis* in association with rice seedlings. Sci. Agricul. Sin. 15, 1–5.

You C B, Li X, Wang Y W, Qiu Y S, Mo X Z and Zhang Y L 1983 Associative $N_2$ fixation of *Alcaligenes faecalis* with rice plant. Biol. $N_2$ Fixation Newslett. Sydney 11, 92–103.

You C B and Zhou F Y 1989 Non-nodular endorhizospheric nitrogen fixation in wetland rice. Can. J. Microbiol. 35, 403–408.

Zhang D D, Zhou F Y, Li X and You C B 1989 Biosynthesis of nitrogenase in $NH_4^-$-grown cells of *Alcaligenes faecalis*. Acta Phytophysiol. Sin. 15, 35–40.

# Sunflower inoculation with Azospirillum and other plant growth promoting rhizobacteria

J. FAGES and J.F. ARSAC
*Pioneer France-Maïs, Chemin de l'enseigure, F-31840 Aussonne, France*

*Key words:* *Azospirillum lipoferum*, germination test, sunflower-growth promotion, *Xanthomonas maltophilia*

## Abstract

The bacterial microflora from sunflower rhizosphere (*Helianthus annuus* L.) and one strain of *Azospirillum lipoferum* of a different origin were screened for their ability to promote sunflower growth in a 6-day germination test and in pot experiments. Two *Azospirillum lipoferum* strains and one *Xanthomonas maltophilia* strain produced the best responses. These strains were chosen for field testing.

## Introduction

The beneficial effects of non symbiotic soil microorganisms living in the rhizosphere have been studied for several crop species (Okon and Hadar, 1987). Both direct beneficial microorganisms, like *Azospirillum* (Döbereiner and Pedrosa, 1987), and antagonists of plant pathogens, like *Pseudomonas* (Davison, 1988) or *Trichoderma* (Baker, 1989), can be classified as microbial yield enhancers. The rhizosphere of several crops have already been described but very little work has been done on rhizobacteria of sunflower. The sunflower is now, one of the most important oleaginous plants in the world and has an increasing economic and agronomic importance. Its potential for increases in yield is great.

The purpose of our study was to determine the potential plant growth-promoting effect of bacterial strains isolated from the rhizosphere of sunflower.

## Materials and methods

### Sunflower cultivar

Hybrid variety EMIL, (Pioneer Hi-Bred Int.)

### Bacterial strains

Forty-five gram negative bacterial strains isolated in our laboratory from the rhizosphere of sunflower and *Azospirillum lipoferum* CRT 1, strain isolated from maize roots (Fages and Mulard, 1988) were used in this study.

### Germination tests

A sheet of blotting paper was dipped in either a bacterial suspension at $10^7$ CFU mL$^{-1}$ or in a $KH_2PO_4$ solution at 21.25 mg L$^{-1}$ (Butterfield's solution) for the control. Three rows of 10 presterilized seeds were then arranged on this sheet. The seeds were then covered with another inoculated piece of paper. The sheets were rolled and placed vertically in a Becher flask to prevent cross-contamination. These flasks were then put into buckets. Each bucket contained one control roll and six (or less) inoculated rolls. The buckets were placed in the dark in a germination chamber at 18°C. The number of replicates (buckets) was 8. Two days after inoculation each roll was reinoculated with 10 mL of the bacterial suspension at $10^7$ CFU mL$^{-1}$. Control rolls received 10 mL of Butterfield's solution.

After 6 days the rolls were opened. Five seedlings (dead or abnormal) were eliminated and the fresh weight of the 25 remaining seedlings was measured.

### Pot experiments

Ten seeds were planted in a pot over a water tank containing 2.5 L of sterile pre-moistened sand. Inoculation was achieved by applying 1 mL of bacterial suspension at $10^7$ CFU mL$^{-1}$ on each seed. Control seeds received 1 mL of Butterfield's solution. Seeds were then covered with a sand layer 1 cm thick. Twelve replicates of each treatment were done. Pots were placed in a germination chamber for 14 days with a photoperiod of 14 h light and 10 h dark and a synchronous thermoperiod of 25/20°C. Plant nutritional needs were covered by adding the following presterilized solution: (mg L$^{-1}$) KNO$_3$, 384; K$_2$HPO$_4$, 54; KH$_2$PO$_4$, 109; NaNO$_3$, 17; NaCl, 12; NH$_4$NO$_3$, 16; Ca(NO$_3$)$_2$ 4H$_2$O, 732; MgSO$_4$ 7H$_2$O, 185; FeSO$_4$ 7H$_2$O, 1; ZnSO$_4$ H$_2$O, 1; H$_3$BO$_3$, 1; CuSO$_4$ 5H$_2$O, 0.03; MnSO$_4$ 4H$_2$O, 0.1; (NH$_4$)$_2$MoO$_4$, 0.03.

At emergence (after 4 days), thinning of seedlings was made leaving 8 out of 10 plantlets per pot. At harvest, after 14 days, the root and shoot dry matters were determined.

### Statistical analysis

Analyses of variance followed by Newman and Keuls test.

## Results and discussion

### Germination tests

In a first series of experiments we tested our isolates in order to select a group of promising bacterial isolates. Table 1 shows the result of four of these experiments.

According to the responses, the isolates can be classified in three groups:

(i) strains exhibiting a deleterious effect on the growth of sunflower seedlings. TSL 51 (Trial no. 2) and TSL 35 (Trial no. 3) strains for instance belong to this first group. These isolates were not selected for further tests.

(ii) strains exhibiting no significant differences compared with control. A majority of isolates belong to this group and were not selected for further tests. However, when the increase of fresh weight is over +3.0% the strains were used although the statistical threshold ($p = 0.05$) was not reached. Such is the case with CRT 1, TML 23 and TMS 11 which showed an increase of 4.1% (Trial no. 2), 4.1% and 3.9% respectively over control (results not shown in Table 1 for the last two strains).

(iii) strains exhibiting a significant increase on the fresh weight of sunflower seedlings. Strains TML 21 (Trial no. 1), TSL 32 (Trial no. 3) and TMS 17 (Trial no. 4) belong to this group.

In a second series of experiments we re-tested the pre-selected isolates to validate their potential before graduating to pot experiments.

---

*Table 1.* Germination tests. Effect of bacterial inoculation on fresh weight of 25 sunflower seedlings

**Fresh weight (g)**

| Trial no. 1 | | Trial no. 2 | | Trial no. 3 | | Trial no. 4 | |
|---|---|---|---|---|---|---|---|
| TML 21 | 10.99 a | CRT 1 | 12.37 a | TSL 32 | 11.97 a | TMS 17 | 13.78 a |
| TMS 11 | 10.71 ab | TSL 54 | 12.09 ab | TSL 34 | 11.64 ab | TSLp13 | 13.28 b |
| TSL 41 | 10.60 b | TSL 42 | 11.92 ab | TSL 15 | 11.61 ab | TSLp12 | 13.24 b |
| TSL 42 | 10.56 b | Control | 11.88 ab | Control | 11.33 b | TSL 22 | 13.20 b |
| TSL 53 | 10.47 b | TSL 55 | 11.79 ab | TSL 45 | 10.86 c | Control | 13.16 b |
| TSL 31 | 10.44 b | TSL 43 | 11.59 b | TSL 14 | 10.76 c | TSL 21 | 13.16 b |
| Control | 10.31 b | TSL 51 | 9.30 c | TSL 35 | 9.53 d | TSS 31 | 13.02 b |

Results followed by common letter within a trial are not significantly different at $p = 0.05$.
Weight variations between trials ar due to test duration which may vary by a few hours after 6 days.

All preselected strains were tested three times. Table 2 shows the results of two of these experiments.

In this series, only TSL 32 and CRT 1 strains exhibited a significant increase of 5.4% and 3.7% respectively on at least one occasion (Trial no. 2). They were selected for pot experiments. TMS 11 gave repeated non-significant negative response and was thus not used in the following pot experiments. The three other strains which gave a non-significant positive response on at least two occasions were selected.

Seedling fresh weight is a very simple and useful parameter. Since no photosynthesis occurs during the test, it expresses water absorption by the seedlings and is, therefore, directly correlated with the root surface. Seedling dry weight – which is only related to seed weight – and elongation measurements were more time consuming and less precise (data not shown).

It must be noted that despite the relatively small increases measured (due to very early measurements) the sensitivity and reliability of the test is such that the differences among bacterial treatments are often significant.

These results show that sunflower seedling root development was directly affected by bacterial inoculation. Thus, blotting paper germination tests can be a powerful tool for screening sunflower-bacterial associations using plant-growth promotion criterion. Application of this technique is an improvement over classical screening methods which utilize indirect criteria such as biological nitrogen fixation (Ela et al., 1982) or phytohormone production (Zimmer and Bothe, 1989).

*Pot experiments*

At a later stage, shoot and root dry matter measurements exhibited differences between bacterial treatments (Fig. 1).

A third parameter, the root surface area was measured using the method of Carley and Watson (1966). This method provided good results with other crops (Arsac et al., 1990; Okon and Kapulnik, 1986). No difference among treatments could be evidenced with this third parameter. Typically the root system of the sunflower is thin and dense and is poorly adapted to this type of measurement.

The effect on root dry matter is great and significant with the CRT 1 and TSL 32 strains while TML 21 and TML 23 gave weak and non-significant increases. No difference was noted with strain TMS 17. On shoot dry matter increases were less pronounced and non-significant at $p = 0.05$. However, at $p = 0.10$, TSL 32 and TML 23 gave a significant positive response.

The CRT1 strain belongs to the species *Azospirillum lipoferum* and has already been described as maize root-growth enhancer (Arsac et al., 1990). It shows that its ability is still effective on sunflower roots demonstrating that no specificity occurs in this case. The positive effect on shoots is very likely an induced effect explaining a weaker improvement and less significance with this parameter. This type of response was the same with maize, suggesting that the mechanism involved is identical although sunflower is a dicotyledonous plant.

The TML 21 and TML 23 strains were previously characterized as $N_2$-fixing strains of *Azospirillum lipoferum*. Unexpectedly, the sunflower responded differently to TML 21 and TML 23

*Table 2.* Germination tests. Effect of inoculation with preselected bacterial strains on fresh weight of 25 sunflower seedlings

| Fresh weight (g) | | | |
| --- | --- | --- | --- |
| Trial no 1 | | Trial no. 2 | |
| TMS 17 | 13.04 a | CRT 1 | 13.78 a |
| TSL 32 | 13.00 a | TSL 32 | 13.56 a |
| TML 21 | 12.84 ab | Control | 13.07 b |
| TML 23 | 12.83 ab | TMS 17 | 12.97 b |
| Control | 12.81 ab | | |
| TMS 11 | 12.38 b | | |

Results followed by common letter within a trial are not significantly different at $p = 0.05$.

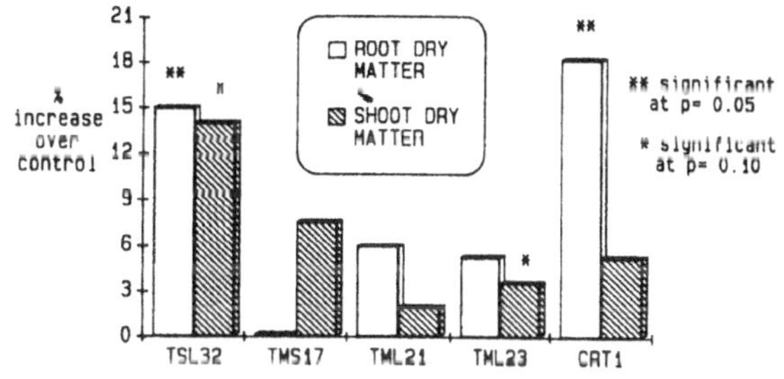


*Fig. 1.* Pot experiments. Effect of bacterial inoculation on shoot and root dry matter of 2 week old sunflower.

inoculation as compared with that of CRT1 which also belongs to the same bacterial species. Root-growth in particular, was not significantly improved either in pot experiments or in germination tests.

On the other hand, a positive effect on shoot development was observed with TML 23 strain. The mechanism involved was not clear.

The TSL 32 strain was characterized in another study as *Xanthomonas maltophilia*. It is the only strain giving positive and significant response on both parameters. Its positive effect appears to be greater after two weeks of growth than in the 6-day test. Although the biological variation was, at this development stage, more pronounced than in the germination test, a statistical significance was nevertheless obtained with this strain showing that the early increase was enhanced with time.

TMS 17 did not exhibit the same root-growth promotion effect obtained in the first test. Conversely with TSL 32 strain its early effect was no longer present at this later stage of development and therefore this strain was not selected for field testing.

The use of two complementary tests gives a first-hand understanding of the mechanisms involved and allows the screening of PGPR strains with agronomic potential on criteria directly linked with plant growth.

The results of this study led us to select, CRT 1, TSL 32 and TML 23 strains for field experiments.

## Acknowledgements

We thank Isabelle Penot and Joëlle Salaün for excellent technical assistance.

## References

Arsac J F, Lamothe C, Mulard D and Fages J 1990 Growth enhancement of maize (*Zea mays* L.) through *Azospirillum lipoferum* inoculation: Effect of plant genotype and bacterial concentration. Agronomie 10, 649–654.

Baker R 1989 Improved *Trichoderma* spp. for promoting crop productivity. Trends Biotechnol. 7, 34–38.

Carley H E and Watson R D 1966 A new gravimetric method for estimating root-surface area. Soil Sci. 102, 289–291.

Davison J 1988 Plant beneficial bacteria. Biotechnology 6, 282–286

Döbereiner J and Pedrosa F O 1987 Nitrogen-Fixing Bacteria in Nonleguminous Crop Plants. Springer-Verlag, Berlin.

Ela W S, Anderson M A and Brill J W 1982 Screening and selection of maize to enhance associative bacterial nitrogen fixation. Plant Physiol. 70, 1564–1567.

Fages J and Mulard D 1988 Isolement de bactéries rhizosphériques et effet de leur inoculation en pots chez *Zea mays*. Agronomie 8, 309–314.

Okon Y and Hadar Y 1987 Microbial inoculants as crop-yield enhancers. CRC Crit. Rev. Biotechnol. 6, 61–85.

Okon Y and Kapulnik Y 1986 Development and function of *Azospirillum*-inoculated roots. Plant and Soil 90, 3–16.

Zimmer W and Bothe H 1989 The phytohormonal interactions between *Azospirillum* and wheat. *In* Nitrogen Fixation with Non-Legumes. Eds. F A Skinner, R M Boddey and I Fendrik. pp 137–145. Kluwer Academic Publishers, Dordrecht, The Netherlands.

# Characterization of a pectinolytic activity in *Azospirillum irakense*

K. M. KHAMMAS[1] and P. KAISER
*Laboratoire de Microbiologie, Institut National Agronomique, 16 rue Claude Bernard, F-75231 Paris CEDEX 05, France. [1]Permanent address: Department of Biology, College of Science, Al-Mustansirya University, Bagdad, Iraq*

*Key words:*  Azospirillum, *A. irakense*, nitrogen fixation, pectate lyase, pectin methylesterase

## Abstract

A fifth and new *Azospirillum* species, *A. irakense*, a nitrogen fixing and pectinolytic bacterium was found associated with roots and rhizosphere of rice in the region of Diwaniya (Qadisya), Iraq. This species produces pectate lyase and pectin methylesterase activities and can fix nitrogen when pectin is the sole carbon source. The four other species of *Azospirillum* fail to show a pectinolytic activity.

## Introduction

*Azospirillum* spp. are widely distributed in the rhizosphere of grasses (Boddey and Döbereiner, 1982; Elmerich et al., 1991; Khammas et al., 1989). They were also reported to colonize the root cortex (Baldani et al., 1980). In the course of a survey of nitrogen-fixing bacteria living on the roots and in the rhizosphere of rice in the region of Diwaniya (Qadisya), Iraq, seven strains were isolated. Cells of the isolates are vibroid to S-shaped with one polar flagellum in liquid medium and additional lateral flagella on nutrient agar. They fix nitrogen under microaerobic conditions, grow on salts of organic acids, and can use sucrose, glucose, maltose, galactose, and arabinose, but not mannitol. The seven strains form a DNA-related group distinct from the four *Azospirillum* species already characterized and were recently identified as belonging to a new species, *A. irakense* (Khammas et al., 1989). The phenotypic characters were found to be very close to those of *A. amazonense*, with the following differences: growth occurred in the presence of 1 to 3% NaCl; the pII range was 5.5 to 8.5; myo-inositol was not utilized; pectin was slowly hydrolyzed and a pronounced swarming occurred on soft nutrient agar at 33°C. The purpose of our work was to further

examine the new species, to characterize its pectinolytic activity, and to look for its ability of using pectin as sole carbon source for nitrogen fixation.

## Materials and methods

### Bacterial strains

The following strains were used: *A. irakense* KBC1, KBC2, KA1, KA3, KAC4, KAC5, KAC6 (Khammas et al., 1989); *A. brasilense* Sp7 (ATCC 29145), Cd (ATCC 29710), G1, G2, Hr (isolated from rice rhizosphere in Iraq); *A. lipoferum* Sp59b (ATCC 29707), Br17 (ATCC 29709), Gr (isolated from rice rhizosphere in Iraq); *A. amazonense* Y1 (ATCC 35119), Y2 (ATCC 35120); *A. halopraeferens* Au4 (LMG 7108, DSM 3675); *Bacillus polymyxa* Pn$_2$ (isolated from rice rhizosphere in Iraq).

### Culture media

Media used were: the pectin plate agar medium of Plazinski and Rolfe (PR medium) (Plazinski and Rolfe, 1985); the same medium in which $(NH_4)_2SO_4$ was replaced by the same amount in weight of Casamino acids (PRC medium); the medium of Tien et al. (T medium) (Tien et al., 1981); the medium of Madjidi-Hervan (Madjidi-Hervan, 1982) devoid of $(NH_4)_2SO_4$ and supplemented by $5\,g\,L^{-1}$ of low methoxylated apple pectin instead of polygalacturonic acid (MH medium). PR and PRC media were used for detecting hydrolysis of pectin on plates. T and MH media were used for preparation of enzyme extracts. Inocula were from 1-day old shaked cultures in complete nutrient broth medium (Khammas et al., 1989). Residual pectin was measured by precipitation with ethanol (Kaiser, 1961).

*Pectinolytic enzyme preparation and assays*

Seven to 10-day old cultures at 30°C were centrifuged at 10,000 *g* for 10 minutes. Supernatants were used to determine enzyme activity. Cell crude extracts were prepared from bacteria harvested from 2 to 3-day old cultures, and disrupted by sonication in a 50 m*M* Tris-HCL buffer at pH 8.0.

Pectate lyase activity was determined by UV absorption at 235 nm using polygalacturonic acid as substrate (Rombouts, 1972). One unit of activity corresponded to the amount of enzyme that released 1 $\mu$mole of unsaturated galacturonic acid per minute. The thiobarbituric acid test (Sherwood, 1966) and the cup-plate technique (Dingle et al., 1953) were also used.

Pectin methylesterase activity was assayed by titrating the free carboxylic groups (Mehrotra et al., 1971). One unit of activity corresponded to the amount of enzyme that released 1 $\mu$mole of carboxyl group per minute. The protein content was determined by the Bio-Rad assay.

*Nitrogenase activity and efficiency of $N_2$ fixation*

Nitrogenase activity was determined in modified AAM semi-solid $N_2$ free medium (Khammas et al., 1989). When pectin $(5\,g\,L^{-1})$ was used as a carbon source, yeast extract $(20-100\,mg\,L^{-1})$ was added as starter to the medium. Serum bottles (25 mL) containing 10 mL of medium were incubated at 30°C for 5 to 9 days with pectin and 2 to 3 days with the other substrates. Nitrogenase activity was determined by the acetylene reduction test (Khammas et al., 1989).

For measuring the efficiency of $N_2$ fixation, experiments were performed in modified semi-solid AAM medium containing 0.05% sucrose or pectin as a carbon source. The total cell nitrogen content was determined by micro-Kjeldahl analysis at the end of growth.

## Results and discussion

*Pectinolytic activity*

In contrast to the four other *Azospirillum* species, the *A. irakense* strains were able to hydrolyze pectin and polygalacturonic acid. As shown in Fig. 1, a clear lysis zone was observed with *A. irakense*, whereas no lysis zone was observed with the other species. Residual pectin determination indicated that about 90% of pectin was degraded by *A. irakense* strains after 9 days (Fig. 2). The degradation was slow during the first four days and subsequently increased. The other *Azospirillum* species degraded only 5 to 7% of pectin during the same time, while *B. polymyxa* strain $Pn_2$ degraded about 80% of pectin in 2 days (Fig. 2).

All the strains of *A. irakense* exhibited pectate lyase activity in the supernatants of 7 to 9-day old cultures (Table 1), and strains from the four other species, including those isolated from Iraq,

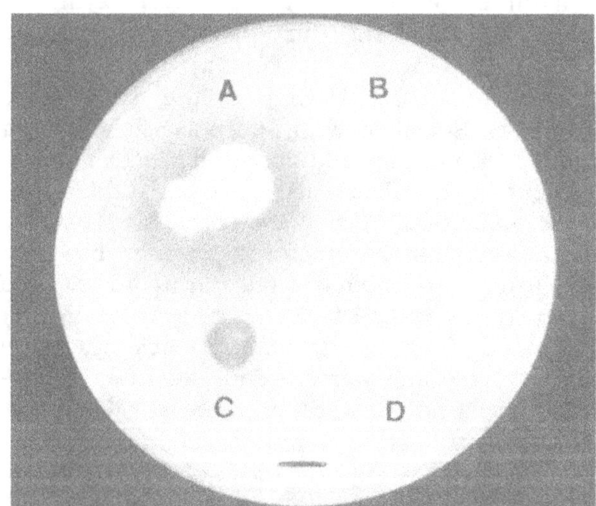


*Fig. 1.* Pectin hydrolysis on PRC plate after 9 days of incubation by *A. irakense* (A), *A. lipoferum* Sp59b (B), *A.brasilense* Cd (C) and Sp7 (D). *Bar* = 1 cm.

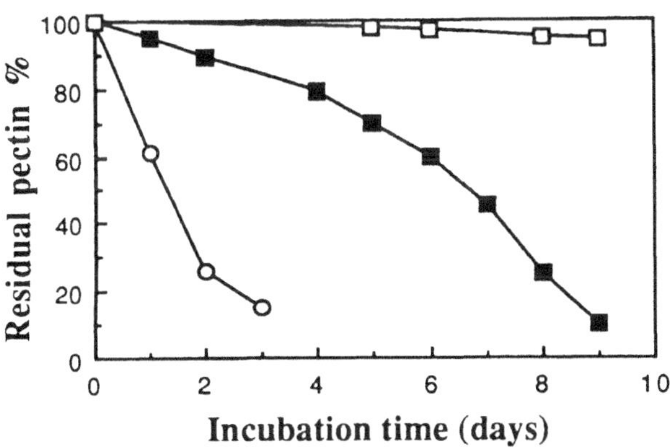

Fig. 2. Pectin degradation by *A. brasilense* Sp7 ( —□— ), *A. irakense* KBC1 ( —■— ), and *Bacillus polymyxa* Pn2 ( —○— ).

failed to show this activity. No pectate lyase activity was detected in supernatants during the first four days of incubation, but enzyme activity was found in crude extracts of cells from 2 to 3-day old cultures (data not shown). This suggests that the pectate lyase of *A. irakense* might be bound to the cell wall. Tien et al (1981) reported the same observation with some other strains of *Azospirillum*.

When pectin was replaced by glucose or sucrose, no pectate lyase activity was detected after 3 to 10 days of incubation even in cell crude extracts. Thus, pectate lyase of *A. irakense* seems to be inducible, as described for other bacteria (Ward and Fogarty, 1974). It required calcium for activity and the activity decreased sharply in the absence of calcium (data not shown) as reported for pectate lyase of other genera (Nagel and Wilson, 1970). Optimum pH was found in the same range as for other bacteria (Madjidi-Hervan, 1982; Nagel and Wilson, 1970), and was between 8.5 to 9.5 (data not shown). Six of the *A. irakense* strains tested showed pectin methylesterase activity (Table 1).

The four other species of *Azospirillum* failed to show pectinolytic activity and pectin could not support growth. These results are in contradiction with those of other authors (Plazinski and Rolfe, 1985; Rinaudo, 1982; Umali-Garcia et al., 1978), but are in agreement with those obtained by Rinaudo (1982), Myers and Hubble (1987), with *A. brasilense* Sp7.

Pectate lyase produced together with pectin methylesterase by *A. irakense* can hydrolyze methylated pectin, another important component of plant cell walls, suggesting the possibility of enhancing root invasion. In fact, several strains of this species were isolated from rice roots (Khammas et al., 1989). Compared with other pectinolytic bacteria such as *B. polymyxa*, *A. irakense* hydrolyzes pectin slowly, and the degradation time is 7 to 8-fold lower (Fig. 2). The specific activity of pectate lyase of *A. irakense* was also about 7-fold lower than that of *B. polymyxa* (Table 1).

*Table 1.* Pectate lyase and pectin methylesterase activities of *A. irakense*, *B. polymyxa*, and other species of *Azospirillum*[a].

| Strains | Pectate lyase units/mg protein | Pectin methylesterase units/mg protein |
|---|---|---|
| *A. irakense*[b] | | |
| KBC1 | 3530 | 30 |
| KBC2 | 3160 | 22 |
| KA1 | 3410 | 17 |
| KA3 | 2110 | 16 |
| KAC4 | 2420 | 15 |
| KAC5 | 1990 | 13 |
| KAC6 | 1900 | ND |
| *B. Polymyxa*[c] | | |
| Pn2 | 22950 | No activity |

[a] Results are the mean of 5 independent determinations.
[b] Activity from 9-day old culture supernatants.
[c] Activity from 2-day old supernatants.
ND not done.
No pectate lyase activity was detected with: *A brasilense*: Sp7, Cd, G1, G2, Hr; *A. lipoferum*: Sp596, Br17, Gr; *A. amazonense*: Y1, Y2; *A. halopraeferens*: Au4.

*Table 2.* Nitrogenase activity of *A. irakense* strains grown on pectin or other carbon substrates[a]

| Strain | Nitrogenase activity[b] | | |
| --- | --- | --- | --- |
| | Pectin[c] | Sucrose[d] | Na acetate[d] |
| KBC1 | 65 | 200 | 130 |
| KBC2 | 103 | 390 | 200 |
| KA1 | 75 | 230 | 150 |
| KA3 | 120 | 550 | 300 |
| KAC4 | 60 | 350 | + |
| KAC5 | 56 | 310 | + |
| KAC6 | 54 | 290 | + |

[a] Results are the mean of 3 independent determinations.
[b] Nanomoles of $C_2H_4$/h/ culture in modified semi-solid AAM medium.
[c] The culture was 9-day old.
[d] The culture was 2 to 3-day old.
+ activity not quantified.

## Nitrogenase activity and efficiency of $N_2$ fixation

All strains of *A. irakense* studied were shown to exhibit nitrogenase activity by the acetylene reduction test after 5 to 11 days of incubation in a medium containing pectin as sole carbon source and a small amount of fixed nitrogen as starter. Strains KA3 and KBC2 were more active than the others (Table 2). Nitrogenase activity was detected in the medium containing only pectin (without nitrogen starter) but the incubation time had to be extended up to 10–15 days. Other carbon substrates, sucrose, L-arabinose, acetate and galacturonate were used easily and appeared to be better substrates than pectin for nitrogenase activity. Sucrose was the best carbon source (Table 2). These results show that pectin is not a very good substrate for growth of *A. irakense.*

Strains of other *Azospirillum* species (except *A. halopraeferens*, which was not tested), inoculated in the same medium with pectin as carbon source, formed a very fine pellicle under the surface that disappeared after a few days and no nitrogenase activity was detected.

The rate of $N_2$ fixation in modified semi-solid AAM medium, measured by micro-Kjeldahl analysis, was 8 to 17 mg of nitrogen per g of sucrose, and 5 to 8 mg of nitrogen per g of pectin for strains KBC1 and KA3, respectively.

## Acknowledgements

We thank Dr C Elmerich for critical reading of the manuscript, R Vedel for plates photography, and Ms M Ferrand for typing the manuscript. K M K was the recipient of fellowships from the French Government and from the Iraqi Government. This work was supported by research funds from the Institut Pasteur.

## References

Baldani V L D, Baldani J L and Döbereiner J 1980 Host plant specificity in the infection of cereals with *Azospirillum* spp. Soil Biol. Biochem. 12, 433–439.

Boddey R M and Döbereiner J 1982 Association of *Azospirillum* and other diazotrophs with tropical graminae. *In* Non-symbiotic Nitrogen Fixation and Organic Matter in the Tropic. pp 28–47. Indian Society of Soil Science, New Delhi.

Dingle J, Solomons G L and Reid W W 1953 The enzymic degradation of pectin and other polysaccharides. J. Sc. Food Agric. 4, 149–155.

Elmerich C, Zimmer W and Vieille C 1990 Associative nitrogen fixing bacteria. *In* Biological Nitrogen Fixation. Eds. G Stacey, H J Evans and R H Burris. pp 211–257. Routledge, Chapman, and Hall, Inc., New York, London.

Kaiser P 1961 Etude de l'activité pectinolytique du sol et d'autres substrats naturels. Thèse de Doctorat, Université de Paris.

Khammas K M, Ageron E, Grimont P A D and Kaiser P 1989 *Azospirillum irakense* sp. nov., a nitrogen-fixing bacterium associated with rice roots and rhizosphere soil. Res. Microbiol. 140, 679–693.

Madjidi-Hervan E 1982 Les enzymes pectinolytiques de quelques *Erwinia* sp. et *Pseudomonas* sp. Thèse de Doctorat, Université Paris 11.

Mehrotra M D, Kaiser P and Reynaud C 1971 Recherche de l'activité pectinolytique in vivo et in vitro, d'un champignon phytopathogène *Gilbertella persicaria* (Eddy), Hesseltine. Ann. Inst. Pasteur, 120, 81–97.

Myers M L and Hubbell D 1987 Plant cell wall carbohydrates as substrates for *Azospirillum brasilense*. Appl. Environ. Microbiol. 53, 2745–2748.

Nagel C W and Wilson T M 1970 Pectic acid lyase of *Bacillus polymyxa*. Appl. Microbiol. 20, 374–383.

Plazinski J and Rolfe B G 1985 Analysis of pectolytic activity of *Rhizobium* and *Azospirillum* strains isolated from *Trifolium repens*. J. Plant Physiol. 120, 181–187.

Rinaudo G 1982 Fixation hétérotrophe de l'azote dans la rhizosphère du riz'. Thèse de Doctorat. Université Paris 11, France.

Rombouts F M 1972 Occurrence and properties of bacterial pectate lyases. PhD Thesis, Wageningen Agricultural University, Wageningen, The Netherlands.

Sherwood R T 1966 Pectin lyase and polygalacturonase production by *Rhizoctonia solani* and other fungi. Phytopathology 56, 279–286.

Tien T M, Diem H G, Gaskins M H and Hubbell D H 1981 Polygalacturonic acid transeliminase production by *Azospirillum* species. Can. J. Microbiol. 27, 426–431.

Umali-Garcia M, Hubbell D H and Gaskins M H 1978 Process of infection of *Panicum maximum* by *Spirillum lipoferum*: Environmental role of nitrogen-fixing blue-green algae and asymbiotic bacteria. Ecol. Bull. Stockholm 26, 373-379.

Ward O P and Fogarty W M 1974 Polygalacturonate lyase production by *Bacillus subtilis* and *Flavobacterium pectinovorum*. Appl. Microbiol. 27, 346–350.

# Associative N₂-fixation in plants growing in saline sodic soils and its relative quantification based on $^{15}$N natural abundance


K.A. MALIK, RAKHSHANDA BILAL[1], G. RASUL, K. MAHMOOD and M.I. SAJJAD[1]
*Nuclear Institute for Agriculture & Biology, Faisalabad and [1]Pakistan Institute for Nuclear Science & Technology, Rawalpindi, Pakistan*


*Key words:* Atriplex, Azospirillum, biological nitrogen fixation, $^{13}$C, Cynodon, Desmostachya, Enterobacter, Kallar grass, Kochia, Klebsiella, $^{15}$N natural abundance, Polypogon, saline soil

## Abstract


Saline-sodic soils are characterized by a very low nitrogen and organic matter content and thus are practically non fertile. However under these conditions, certain plants have been found to grow luxuriantly. One of such plants, *Leptochloa fusca* (Kallar grass) has exhibited nitrogenase activity associated with its roots as determined by acetylene reduction assay (ARA). Quantification of such nitrogen fixation was also carried out using $^{15}$N isotope dilution technique.

In addition to Kallar grass, other plant species growing in saline sodic soils namely *Atriplex amnicola*, *A. lentiformis*, *Sporobolus* sp., *Kochia indica*, *Desmostachya bipinnata*, *Cynodon dactylon*, *Suaeda fruiticosa* and *Polypogon monspilensis* have been screened for the presence of root associated nitrogenase activity. Some of the plant species tested showed high excised root acetylene reduction activity (ERARA). Isolation of diazotrophs from various fractions of the rhizosphere has also been carried out. *Azospirillum* was the dominant organism in niches closer to the roots, whereas there was a preponderance of the members of the family Enterobacteriaceae in general.

In order to have a relative estimate of the nitrogen fixing ability of different plant species screened, the delta $^{15}$N values of plant tops were estimated and were correlated with their ARA values. The delta $^{13}$C values of these plants were also determined which indicated that all the plants tested except *P. monspilensis* had the C-4 photosynthetic pathway.


## Introduction

Associative nitrogen fixation in the roots of non-legumes has been recognized as a possible significant component of the N cycle in a range of ecosystems including several extreme environments (Dart, 1986). Saline sodic soils are characterized by a very low nitrogen and organic matter content and are practically non fertile. However under these conditions certain plants have been found to grow quite well. Since the development of acetylene reduction methodology for detection of nitrogenase activity, many plants have been shown to harbor N₂-fixing bacteria in and around their roots (Jagnow 1983; Patriquin and Döbereiner, 1978). Using such techniques it has been shown that N₂-fixation in the rhizosphere contributes significantly to the N nutrition of plants growing in highly saline sodic low fertility soils (Malik et al., 1988).

A large potentially arable area in Pakistan is afflicted with salinity and sodicity. Extensive studies have been carried out in order to make economic use of these soils by growing salt tolerant plants. One of such plants, Kallar grass (*Leptochloa fusca*) has been highly successful in colonising these soils (Malik et al., 1986). Extensive studies on nitrogen fixation associated with its roots and quantification of such fixation using $^{15}$N isotope dilution techniques have previously been carried out (Malik et al., 1987; Malik and Bilal, 1989).

In addition to Kallar grass various other plant species are known to colonize salt affected soils. All such plants have been screened for possible associative nitrogen fixation using acetylene reduction technique, the results of which are being reported here.

Relative quantification of $N_2$-fixation has been made by estimating the $^{15}N$ natural abundance of the plant tops. This method is based on the observation that soil N is usually more abundant in $^{15}N$ than is atmospheric $N_2$ (Marioti, 1982; Shearer et al., 1978). As a result of this variation, non $N_2$-fixing plants whose primary source of N is soil derived N, would be expected to be more abundant in $^{15}N$ than $N_2$-fixing plants which take $N_2$ from the atmosphere as well as from the soil. Thus $N_2$-fixing plants tend to have values of $^{15}N$ nearer to that of atmospheric $N_2$.

Since the majority of the grasses in which associative nitrogen fixation was reported possessed the C-4 photosynthetic pathway, it led Döbereiner et al. (1972) to propose a relationship between C-4 plants and $N_2$-fixation associated with their roots. However, there have been some exceptions to this relationship but no systematic study in this regard has been carried out. With this objective, all the plants screened for associative nitrogen fixation were also analysed for their photosynthetic pathway based on the $^{13}/^{12}C$ ratios. This method is based on the observation that plants discriminate against $^{13}C$ during photosynthesis in ways which reflect plant metabolism and environment (Benedict, 1978; O'Leary, 1981).

## Materials and methods

### Study site

The study site was located between longitude 74–75°E and latitude 31–32°N at Biosaline Research Station (BSRS) near Lahore. Average rainfall in this area is about 500 mm. The mean annual air temperature range between 13°–25°C for minimum and 20°–40°C for maximum. Summers and winters are severe as air temperatures may be as high as 47°C in June and as low as 0°C in January. The soils of study plots belong to Khurerianwala soil series and are sandy clay loam in texture. These are calcareous, and highly saline sodic soils having pH 8.5–9.5, EC (of saturation extract) 10–40 mS $cm^{-1}$, total N is 0.02% and organic carbon is 0.2%.

### Acetylene reduction assay (ARA)

A large soil core of ca. 30 cm dia containing a plant was dug up to ca. 20 cm. Three such plant-soil samples for each species were collected from three different locations and assayed for excised root acetylene reduction (ERARA). The roots were separated and soil adhering to them was removed gently. The roots were then subjected to thorough washing with distilled sterile water. Approximately 5 g portions of washed roots were randomly taken in 30 ml capacity vials, sealed with serum caps and 10% v/v $C_2H_2$ atmosphere was provided. For preincubation, the atmosphere in the vials containing roots was replaced with nitrogen gas and incubated for 24 hours before providing 10% $C_2H_2$ atmosphere. Ten replicate samples were taken for both direct and preincubated roots for each plant and incubated at 30°C. The gas samples were analysed for $C_2H_4$ at various intervals using a gas chromatograph (Carlo Erba Fractovap Series 2150) fitted with 0.75 m $\times$ 2 mm stainless steel column packed with Porapak N (80–100 mesh, Water Associates Inc. USA), using flame ionization detector (FID). Gas sample (usually 100 $\mu$l) was injected by gas tight syringe (Hamilton, USA). The nitrogenase activity was expressed as nmol $C_2H_4 \, g^{-1}$ root dry wt. 30 mL capacity serum capped vials with 10% v/v $C_2H_2$ without any roots were used as control.

### Isolation of diazotrophic bacteria

Excess soil was removed by placing the rhizosphere under a gentle stream of water. When the roots were free of adhering soil, these were thoroughly washed in several changes of sterile distilled water. For the isolation of bacteria from the root interior, the roots were immersed in 5% NaOCl for 30 min, followed by washing in several changes of sterile distilled water. The roots were excised to 2 cm small pieces and were inoculated in sterile semi-solid nitrogen-free malate medium and combined carbon medium of

Rennie (1981). After 3 to 4 days of incubation at 30°C, a loopful of bacterial growth was transferred into a second vial of nitrogen free medium and was incubated further. The vials were observed daily for growth. The screw caps were replaced by serum stoppers and 10% v/v acetylene was added. The cultures were incubated at 30°C for 1 h, 100 $\mu$l of the gas sample was removed and analysed for ethylene by gas chromatography (Bilal and Malik, 1987). For isolation of diazotrophs the cultures yielding more than 100 nmol ethylene/h/vial were streaked on nitrogen-free medium plates supplemented with 0.01% yeast extract. Individual colonies were picked and reinoculated in semi-solid nitrogen-free medium for determining the nitrogenase activity. Various purified cultures giving positive acetylene reduction were retained, after being checked for purity on potato-dextrose or nutrient agar plates.

*Estimation of $^{15}N$ natural abundance*

The plant material was collected from the experimental site in the month of February. The fresh material was dried at 60°C for 2 days and was then ground in a Wiley mill to pass a 20 mesh screen. Total N was determined by the Kjeldahl method including steam distillation of the $NH_4^+$ into boric acid. Distillates were collected and concentrated for $^{15}N$ analysis. Samples were analysed by the Rittenburg method (Fiedler and Proksch, 1975) on a mass spec-

trometer fitted with a double inlet system (Varian Mat GD150). Sodium hypobromite was used for releasing $^{15}N$.

*Estimation of carbon isotope ratios*

Leaf tissue was collected from the field and dried in a forced air oven at 80°C for 24 hrs. The dried tissue (5–10 mg) was combusted at 750°C in an excess of oxygen and isotopic ratio ($^{13}C/^{12}C$) of the $CO_2$ evolved was measured on a mass spectrometer as described by Osmond et al. (1978). Atmospheric $CO_2$ contains about 1.1% of the heavier isotope $^{13}C$ and 98.9% of the lighter isotope $^{12}C$. The discrimination of $^{13}C$ in favour of $^{12}C$ has been highly correlated with the $C_3$ and $C_4$ pathways of photosynthetic metabolism. This characteristic when considered in relation to leaf anatomy, provides the most reliable criterion for distinguishing these two photosynthetic pathways (Smith and Brown, 1973).

**Results**

List of plants surveyed for root associated nitrogenase activity is presented in Table 1. Most of the plants screened belonged to the Graminae family while the rest were from the family Chenopodiacae. Out of the 14 plants species screened only five exhibited appreciable nitrogenase activity as determined by ARA whereas two species showed moderate activity while the

*Table 1.* List of plants surveyed for root associated nitrogenase activity by excised root acetylene reduction assay (ERARA)

| Plants species | Location | Family | ERARA* | Remarks |
| --- | --- | --- | --- | --- |
| *Leptochloa fusca* (L.) Kunth | BSRS | Graminae | + + | Introduced |
| *Cynodon dactylon* (L.) Pers. | BSRS | Graminae | + + + | Natural |
| *Desmostachya bipinnata* (L.) | BSRS | Graminae | + + | Natural |
| *Sporobolus arabicus* Bioss. | BSRS | Graminae | − | Natural |
| *Suaeda fruticosa* (L.) Forssk. | BSRS | Chenopod | − | Natural |
| *Kochia indica* Wight | BSRS | Chenopod | − | Natural |
| *Atriplex amnicola* P.G. Wilson | BSRS | Chenopod | + + | Introduced |
| *A. lentiformis* (Torr.) Wats. | BSRS | Chenopod | + + + | Introduced |
| *Panicum distichum* L. | BSRS | Graminae | + | Natural |
| *Andropogon gayana* | NIAB | Graminae | + | Introduced |
| *Cenchrus ciliaris* L. | NIAB | Graminae | − | Natural |
| *Panicum maximum* Jacq. | NIAB | Graminae | + | Introduced |
| *Triticum aestivum* L. (wheat) | NIAB | Graminae | + + + | Cultivated |
| *Oryza sativa* L. (rice) | NIAB | Graminae | + + + | Cultivated |

*, + + +, Activity in $\mu$mol; + +, more than 500 nmol; +, around 100 nmol g$^{-1}$ dry root h$^{-1}$.

remaining had either marginal or no activity. Root associated nitrogenase activity in *C. dactylon*, *A. lentiformis* and *D. bipinnata* was further studied from different locations at the experimental site (BSRS). The results of ARA of washed excised roots incubated directly or after preincubation with $N_2$ for 24 hrs are presented in Table 2. The rates of acetylene reduction varied with different plant species and locations. Maximum nitrogenase activity was exhibited by *A. lentiformis* both in case of direct and preincubation. In the case of *D. bipinnata*, plants sampled from location C showed relatively high nitrogenase activity. The root samples of *C. dactylon* were collected from 4 different locations. Out of these appreciable activity was detected only at one site.

The diazotrophs associated with the roots of grasses were isolated from the root surface (RP) or the root interior (HP). Most of the bacteria were however, isolated from the root surface as presented in Table 3. A total of 57 isolates were obtained using different media. The identification of these isolates were carried out using QTS-20 miniaturized identification system (DESTO Laboratories, Karachi, Pakistan) based on which an identification key was formulated. Some of the organisms which could be identified are listed in Table 4. *Enterobacters* have been shown to be dominant on the root surface whereas *Azospirilla* were exclusively isolated from root interior (HP) of Kallar grass and *Atriplex*. In addition to these *Citrobacter freundi* was also isolated from number of plant species.

All the different plant species growing at BSRS, Lahore were analysed for the $\delta$ $^{15}$N values. The respective rhizospheric soil (0–9 cms) was also analysed for $\delta$ $^{15}$N values. The results are summarised in Table 5. The $\delta$ $^{15}$N varied with plant species. Among the grasses other than Kallar grass, it ranged from +2.48 to +17.99. In case of Kallar grass the values ranged from −3.32 to +9.13. Among the Chenopods, the $\delta$ values ranged from +4.05 to +24.43. *Kochia* showed the maximum value whereas *A. amnicola* exhibited the lowest value. In addition, analysis of *Casuarina* and some legumes growing in the same area was carried out. Comparison

*Table 2.* Comparison of nitrogenase activity (ARA) of excised washed roots of different plants incubated directly or preincubated with $N_2$ for 24 hours. Activities are described as nmol $C_2H_4$ $g^{-1}$ dry roots

| Plants screened | Location | Direct without $N_2$ | | Preincubated with $N_2$ | |
|---|---|---|---|---|---|
| | | 21 h | 29 h | 2 h | 11 h |
| *Cynodon dactylon* | A | 0–105 (42) | 0–262 (87) | 132–2376 (1568) | 89–11550 (6512) |
| | B | 0–315 (116) | – – | 138–582 (402) | 209–4950 (1991) |
| | C | 00 – | 00 – | 38–700 (202) | 77–1320 (484) |
| | D | 567–7770 (3591) | 609–8845 (4060) | 4–1146 (556) | 44–5709 (3245) |
| *Atriplex* sp. | A | 798–4095 (1890) | 1392–4872 (2813) | 4–2700 (1094) | 88–13519 (5577) |
| | B | 1911–7665 (4326) | 12813–10498 (5394) | 284–2266 (1010) | 1078–6138 (3447) |
| *Desmostachya bipinnata* | A | 21–105 (63) | 29–149 (87) | 4–1174 (160) | 11–5060 (2156) |
| | B | 21–189 (63) | 29–174 (73) | 0–914 (248) | 0–3157 (1248) |
| | C | 21–2940 (987) | 29–6119 (1469) | 14–1654 (313) | 0–17974 (4829) |

The values in brackets are averages of ten replicates. h indicates hours of incubation.

*Table 3.* List of isolated diazotrophs from roots of plants of saline soils

| Plant origin | Root | Media | Isolate code | Total |
|---|---|---|---|---|
| *Leptochloa fusca* | RP | NFM | K4, K5, K6, K7, K8, K9, K10, K11, K12, K13, K14 | |
| | | CCM | K2, K3, KC11, K1 | 17 |
| | HP | NFM | KY1 | |
| | | CCM | K2HC2 | |
| *Atriplex* | RP | NFM | AX6, AX7, AX8, AX9, AX10, AX12, AX13, AX11, AX15 | 17 |
| | | CCM | AX1, AX2, AX3, AX4, AX5, AX14 | |
| | HP | NFM | AH1, AH2 | |
| *Triticum aestivum* | HP | CCM | QH7, ZH2b, AH6 | 3 |
| *Cynodon* | RP | NFM | Cd1, Cd2, Cd3, Cd4, Cd5, Cd6 | |
| | HP | NFM | CH1, CH2, CH4, CH6, CH7 | 13 |
| | | CCM | CH3, CH5 | |
| *Sporobolus* | RP | CCM | SP1 | 1 |
| *Andropogon* | RP | CCM | AP1, AP2 | 2 |
| *Kochia* | RP | CCM | KO–1 | 1 |
| *Desmostachya* | RP | CCM | DS1, DS2, DS3 | 3 |
| | | | Total | 57 |

*Table 4.* List of diazotrophs isolated from roots of plants growing at Biosaline Research Station, Lahore, Pakistan

| Plant origin | Identified organisms |
|---|---|
| *Cynodon dactylon* | *Enterobacter cloacae*, *E. agglomerans* |
| *Desmostachya bipinnata* | *Citrobacter freundi*, *E. agglomerans* |
| *Sporobolus arabicus* | *E. agglomerans* |
| *Kochia indica* | *C. freundi* |
| *Andropogon gayana* | *C. freundi* |
| *Atriplex* sp. | *E. agglomerans*, *Klebsiella pneumoniae* *E. cloacae E. intermedium* |
| *Triticum aestivum* | *E. agglomerans* |
| Kallar grass | *Azospirillum brasilense*, *Azotobacter* sp. *Enterobacter* sp. *Zoogloae* sp. |

was also made with nodulating Chickpea and *Phaseolus* which were grown on N free medium in the growth room. The delta values of *Melilotus* and *Sesbanea formosa* were +1.59 and +6.19 respectively.

The results of the soil $\delta$ $^{15}$N% are also presented in Table 5. The values ranged from +5 to +7 and did not show much variation.

The results of $\delta$ $^{13}$C% are presented in Table 6. All the plants except Polypogon had values between −16 to −14 which fall well within the range of C-4 plants. *Polypogon* however had $^{13}$C $\delta$ value of 30.49 and thus has a C-3 photosynthetic pathway.

## Discussion

Saline-sodic soils constitute an extreme environment in which plants are subjected to number of stresses. However, the plant growth itself exerts beneficial effects on the soil physical and chemical properties thus paving the way for other plant species to colonize (Sandhu and Malik, 1975). Among the plants screened, *D. bipinnata* and *S. fruticosa* were the two dominant species of the experimental site. Kallar grass was introduced and after 3–4 years of its cultivation, other plant species colonized (Mahmood et al., 1989). These essentially include *C. dactylon*, *K. indica*, *Poly-*

*Table 5.* Delta $^{15}$N of rhizospheric soil and leaf/shoot tissues of plants growing at Biosaline Research Station, Lahore, Pakistan

| Plant species | Delta $^{15}$N(%) | | Remarks |
| --- | --- | --- | --- |
| | Soil | Leaf/shoot | |
| **Kallar grass** | | | |
| 6 months old | +6.89(0.52) | +3.76(0.18) | green |
| 1 year old | +6.27(0.47) | +2.45(0.25) | green |
| 2 years old | +5.25(0.48) | −3.32(0.16) | green |
| 3 years old | +5.86(0.81) | +4.18(0.85) | green |
| 4 years old | +3.83(0.19) | +9.13(0.15) | green |
| 5 years old | nd | +2.89(0.12) | very young |
| **Other grasses** | | | |
| *Desmostachya bipinnata* | +5.09(0.61) | +2.48(0.19) | young |
| *Cynodon dactylon* | +5.02(0.29) | +6.30(1.51) | mature |
| *Polypogon monspilensis* | +6.86(0.70) | +17.99(0.67) | young |
| *Sporobolus* sp. | +7.95(0.88) | +14.49(0.44) | green |
| **Chenopodiaceae** | | | |
| *Sueda fruiticosa* | +5.84(2.35) | +9.05(0.42) | green |
| *Kochia indica* | +7.33(0.81) | +24.43(0.31) | green |
| *Atriplex lentiformis* | +5.61(0.71) | +4.89(0.84) | |
| *A. amnicola* | +5.09(1.87) | +4.05(1.33) | |
| *Casuarina* sp. | +6.51(0.07) | −1.85 to +5.74 | 2 yrs old |
| **Legumes** | | | |
| Chickpea | | +0.48(0.09) | −N medium |
| *Phaseolus vulgaris* | | −1.33(0.17) | −N medium |
| *Melilotus* sp. | nd | +1.49(0.54) | young |
| *Sesbanea formosa* | nd | +6.19(0.61) | young branches |
| *Acacia* Acc. 15771 | +4.43(0.19) | +14.12(2.21) | -do- |
| *Acacia* Acc. 15762 | +5.08(1.24) | +15.49(2.29) | -do- |

Figures in parenthesis are standard error.

*Table 6.* Delta $^{13}$C values of some plants growing at Biosaline Research Station, Lahore, Pakistan

| Plant species | Delta $^{13}$C (%) |
| --- | --- |
| *Leptochloa fusca* (Kallar grass) | −15.29 |
| *Atriplex amnicola* | −14.88 |
| *A. lentiformis* | −16.09 |
| *Sporobolus* sp. | −14.06 |
| *Kochia indica* | −12.63 |
| *Desmostachya bipinnata* | −14.16 |
| *Cynodon dactylon* | −14.22 |
| *Suaeda fruiticosa* | −14.20 |
| *Polypogon monspilensis* | −30.49 |

*pogon* sp., *Sporobolus* sp. In addition, *Atriplex* species have also been introduced from Australia.

All these plant species were subjected to ERARA for associative nitrogen fixation. *D. bipinnata*, *C. dactylon* and *Atriplex* spp gave high acetylene reduction values. Some species showed marginal or no activity. Excised root assay with pre-incubation has been criticised by some workers (Van Berkum 1980; Lethbridge et al., 1982). The ARA of excised roots was therefore performed without preincubation with $N_2$. There was a large variation in the ARA values which is not uncommon in such studies (Rao and Rao, 1984) and it is usually due to uneven distribution of bacteria on the root surfaces and to the difference between old and young roots (Capone and Buding, 1982).

Several types of diazotrophs can be isolated from the same root depending on the medium used. The use of nitrogen free malate medium ensures the isolation of *Azospirillum* (Rao and Rao, 1984; Reinhold et al., 1986). However, using combined carbon medium (Rennie, 1981) other diazotrophic species were also identified (Bilal et al., 1990). All the isolates obtained from the rhizoplane (root surface) belonged to family Enterobacteriaceae, predominantly *E. agglomerans*, followed by *E. cloacae*, *E. intermedium* and *K. pneumoniae*. In addition, *Citrobacter freundi* was also identified. *Azospirilla* were isolated from histoplane fractions of Kallar grass and *Atriplex* roots (Bilal et al., 1990).

The significance of nitrogen fixation associated with roots of grasses can only be demonstrated if it is properly quantified. $^{15}$N isotope dilution methodologies have quite extensively been used in the case of legumes (Chalk, 1985) where enriched $^{15}$N fertilizer sources is applied to both fixing and non fixing reference plants. Application of this methodology to quantify nitrogen fixation in grasses has also been made (Malik and Bilal 1989; Urquiaga et al., 1989). However, the problem of finding a good reference plant for saline environments has always made such experiments difficult.

In this study $\delta$ $^{15}$N values of the plants growing in saline soils have been determined. No efforts has been made to quantify nitrogen fixation but to have a relative picture as to the extent of fixation and correlate it with ARA values. Shearer and Kohl (1986) have reviewed the application of $^{15}$N natural abundance methodology to various ecosystems. Based on these methods, the plant using all the nitrogen through fixation should have zero $\delta$ $^{15}$N. Hence it is possible to grade various plants for their ability to support associative nitrogen fixation on the basis of their $\delta$ $^{15}$N. *D. bipinnata* and *Atriplex* sp. gave very high ARA whereas these two also had relatively low $\delta$ $^{15}$N values.

The data regarding Kallar grass showed variation with respect to the age and location. This is a perennial grass and is being continuously cut. However, the low delta values indicate the extent of nitrogen fixation. These values confirm the estimates of nitrogen fixation obtained earlier by using $^{15}$N isotope dilution technique (Malik et al., 1988). In the same area *Melilotus* sp. was also sampled and had $\delta$ $^{15}$N value of +1.49.

Shearer and Kohl (1986) reported $\delta$ $^{15}$N values of leaf tissue of number of legumes and non legumes. The mean values for Papilionoideae were +1.80; for Prosopis +8.90; for Acacias +10.60 and for non legumes it was 9.30. The $\delta$

$^{15}$N of non legumes which show no ERARA are similar to the one reported by Shearer and Kohl (1986) whereas the plants showing high ERARA have values nearer to the legume values.

The studies reported here have indicated possibilities of using $\delta$ $^{15}$N values of plants growing in an ecosystem, as an indicator for the extent of associative nitrogen fixation or the sources of its nitrogen nutrition.

## Acknowledgements

Financial support of this research was partly provided by the US National Academy of Sciences by means of a grant from USAID.

## References

Benedict C R 1978 Nature of obligate photoautotrophy. Annu. Rev. Plant Physiol. 29, 67–93.

Bilal R 1988 Associative nitrogen fixation in plants growing in saline environments. Ph.D Thesis. Punjab University Lahore, Pakistan.

Bilal R and Malik K A 1987 Isolation and identification of a $N_2$-fixing zoogloea-forming bacterium from Kallar grass histoplane. J. Appl. Bacteriol. 62, 289–294.

Bilal R, Rasul G, Mahmood K and Malik K A 1990 Nitrogenase activity and nitrogen fixing bacteria associated with the roots of *Atriplex* spp. growing in saline-sodic soils of Pakistan. Biol. Fert. Soil 9, 315–320.

Bilal R, Rasul G, Qureshi J A and Malik K A 1990 Characterization of *Azospirillum* and related diazotrophs associated with roots of plants growing in saline soils. World J. Microbiol. Biotechnol. 6, 46–52.

Capone D G and Budin J M 1982 Nitrogen fixation associated with roots and rhizomes of eelgrass *Zostera marina*. Plant Physiol. 70, 1601–1604.

Chalk P M 1985 Estimation of $N_2$-fixation by isotope dilution: An appraisal of techniques involving $^{15}$N environment and their application. Soil Biol. Biochem. 17, 389–410.

Dart P J 1986 Nitrogen fixation associated with non legumes in agriculture. Plant and Soil 90, 303–334.

Döbereiner J, Day J M and Dart P J 1972 Nitrogenase activity in the rhizosphere of sugarcane and some other tropical grasses. Plant and Soil 37, 191–196.

Jagnow G 1983 Nitrogenase activity in roots of non-cultivated and cereal plants: Influence of nitrogen fertilizer on population and activity of nitrogen fixing bacteria. Z. Pflanzenernaehr. Bodenkd. 146, 217–227.

Lethbridge G, Davidson M S and Sparling G P 1982 Critical evaluation of the acetylene reduction test for estimating the activity of nitrogen fixing bacteria associated with the roots of wheat and barley. Soil Biol. Biochem. 14, 27–35.

Mahmood K, Malik K·A, Sheikh K H and Lodhi M L K 1989 Allelopathy in saline agricultural land: Vegetation, successional changes and patch dynamics. J. Chem. Ecol. 15, 565–579.

Malik K A, Zafar Y Bilal R and Azam F 1987 Use of $^{15}$N isotope dilution for quantification of $N_2$-fixation associated with roots of Kallar grass. Biol. Fertil. Soils 4, 103–109.

Malik K A, Bilal R, Azam F and Sajjad M I 1988 Quantification of $N_2$-fixation and survival of inoculated diazotrophs associated with roots of Kallar grass. Plant and Soil 108, 43–51.

Malik K A and Bilal R 1989 Survival and colonization of inoculated bacteria in Kallar grass rhizosphere and quantification of $N_2$-fixation. Plant and Soil 110, 329–338.

O'Leary M H 1981 Carbon isotope fractionation in plants. Phytochemistry 20, 533–567.

Osmond C B, Zieglies H, Stichler W and Trimbon P 1975 Carbon isotope discrimination in alpine succulent plants supposed to be capable of crassulacean acid metabolism. Oecologia (Berl.) 18, 209–217.

Patriquin D G and Döbereiner J 1978 Light microscope observation of tetrazolium reducing bacteria in the endorhizosphere of maize and other grasses in Brazil. Can. J. Microbiol. 24, 734–742.

Rao R and Rao J L N 1984 Nitrogen fixation in soil samples from rhizosphere of rice grown under alternate flooded and non flooded conditions. Plant and Soil 81, 111–118.

Reinhold B, Hurek T, Niemann E G and Fendrik I 1986 Close association of *Azospirillum* and diazotrophic rods with different root zones of Kallar grass. Appl. Environ. Microbiol. 52, 520–526.

Sandhu G R and Malik K A 1975 Plant succession – A key to utilization of saline soils. Nucleus 12, 35–38.

Shearer G, Kohl D and Harper J E 1980 The nitrogen-15 abundance in a wide variety of soils. Soil Sci. Soc. Am. J. 42, 899–902.

Shearer G and Kohl D H 1986 $N_2$-fixation in field settings: Estimations based on natural $^{15}$N abundance. Aust. J. Pl. Physiol. 13, 699–756.

Smith B N and Brown N V 1973 The Kranz syndrome in the gramineae as indicated by carbon isotope ratios. Am. J. Bot. 60, 505–513.

Urquiaga S, Botteon P B L and Boddey R M 1989 Selection of sugarcane cultivars for associated biological nitrogen fixation using $^{15}$N labelled soil. *In* Nitrogen Fixation with Non Legumes. Eds. F A Skinner, R M Boddey and I Fendrik. pp 311–319. Kluwer Academic Publishers, Dordrecht, The Netherlands.

Van Berkum P and Sloger C 1984 A critical evaluation of the characteristics of associative nitrogen fixation in grasses. *In* Nitrogen and the Environment. Eds. K A Malik, S H M Naqvi and M I H Aleem. pp 139–150. NIAB, Faisalabad, Pakistan.

# GENETIC ANALYSIS OF THE AZOSPIRILLUM PLANT-ROOT INTERACTION

K. MICHIELS, A. VANDE BROEK, M. EYERS, C. CROES, A. MILCAMPS, P. DE TROCH, E. VAN BASTELAERE and J. VANDERLEYDEN.
F.A. Janssens Memorial Laboratory for Genetics, Catholic University of Leuven, W. De Croylaan 42, 3001 Heverlee, Belgium.

## 1. Introduction

*Azospirillum* spp are diazotrophic soil bacteria that associate with the roots of a variety of plants, mainly grasses and cereals. Inoculation of crop plants with *Azospirillum* has resulted in significant yield increases under certain conditions [for reviews, see 1, 2, 3]. We wish to better understand the mechanism of this plant root colonization and plant growth stimulation. Therefore, we are studying several bacterial determinants that could be involved in the plant interaction. Also, we are using the *uidA* (b-glucuronidase) reporter gene in *Azospirillum* for histochemical studies of colonized plant roots and to monitor gene expression during the interaction.

## 2. Results and discussion

### 2.1. *AZOSPIRILLUM BRASILENSE* EXTRACELLULAR POLYSACCHARIDES

Given the important role of bacterial extracellular polysaccharides (EPS) in diverse pathogenic plant-bacteria interactions, we initiated a search for *A. brasilense* Sp7 EPS genes to study the role of these components in *Azospirillum* rhizocoenoses.

Using random Tn5-mutagenesis, we isolated *A. brasilense* mutants showing reduced or increased intensity of fluorescence with Calcofluor (Cal mutants)[4]. So far, five different genetic loci affecting fluorescence have been identified on the chromosome. Non-fluorescent mutants still produce normal amounts of EPS in the culture supernatant, but have lost the ability to flocculate, which is characteristic for wild-type *A. brasilense*. In addition, we found that these mutants are drastically impaired in attachment to roots of wheat seedlings in an *in vitro* assay (Table 1). Fluorescence Microscopy of wild-type cells revealed the presence of Calcofluor-binding material thightly associated with the cells, and sometimes present as fibrillar material between the cells. Cal⁻ mutants had no or very few fluorescent material around the cells (result not shown).

We are now evaluating the relevance of the *in vitro* attachment data to the *in vivo* situation by measuring root colonization of sand-grown seedlings.

TABLE 1. Characteristics of some *A. brasilense* Cal mutants

| Strain | Cal phenotype (fluorescence) | Flocculation | % attachment to wheat seedlings |
|---|---|---|---|
| 7030 and Sp7 | Cal$^+$ | + | 9 |
| 7030TN5-11 | Cal$^d$ | + | 2 |
| 7030TN5-22 | Cal$^-$ | - | 0.6 |
| 7030TN5-23 | Cal$^-$ | - | 0.9 |
| 7030TN5-102 | Cal$^{++}$ | + | 10 |

By genetic complementation, we identified clones from an *A. brasilense* Sp7 clone bank that could correct the deficiency in EPS synthesis and restored the Fix$^+$ phenotype of *Rhizobium meliloti exo* mutants (Table 2). It should be noticed that cosmid pCal112 corrects the *exoC* mutant only for the Calcofluor phenotype but not for the Fix$^-$ phenotype.

TABLE 2. Correction of *R. meliloti exo* mutants with *A. brasilense* Sp7 DNA

| *R. meliloti exo* mutants | | *R. meliloti* transconjugants | |
|---|---|---|---|
| genotype | phenotype | recombinant cosmid | phenotype |
| *exoB* | Cal$^-$Fix$^-$ | pCal102 [5] | Cal$^+$Fix$^+$ |
| *exoC* | Cal$^-$Fix$^-$ | pCal112 [5] | Cal$^+$Fix$^-$ |
| *exoC* | Cal$^-$Fix$^-$ | pCal134 | Cal$^+$Fix$^+$ |
| *exoG* | Cal$^d$Fix$^{+/-}$ | pCal301 | Cal$^+$Fix$^+$ |
| *exoK* | Cal$^d$Fix$^+$ | pCal201 | Cal$^+$Fix$^+$ |
| *exoM* | Cal$^-$Fix$^-$ | pCal102 | Cal$^+$Fix$^+$ |
| *exoN* | Cal$^d$Fix$^+$ | pCal601 | Cal$^+$Fix$^+$ |
| *exoP* | Cal$^-$Fix$^-$ | pCal503 | Cal$^+$Fix$^+$ |

It was shown before that a mutation in the *exoC* locus in *R. meliloti* and *Agrobacterium tumefaciens* causes a pleiotrophic phenotype [6]. In addition to the lack of synthesis of exopolysaccharide, *exoC* mutants fail to produce a periplasmic b(1-2) glucan. It was recently elucidated that this is due to a deficiency in the activity of the enzyme phosphoglucomutase [7]. Proton-NMR analysis of the EPS isolated of the *R. meliloti exoC* mutant containing pCal112 revealed that the EPS produced is quite different from the wild-type EPS, particularly in the region of the non-carbohydrate substituents. Moreover this transconjugant is devoid of b(1-2) glucan, which also explains its Fix- phenotype. The *exoC* correcting locus on pCal134, in contrast, fully complements the *R. meliloti exoC* mutant (Fix$^+$, b(1,2)glucan synthesis, wild-type EPS). *A. brasilense* mutants in the *exoB* and *exoC* (pCal112) correcting loci were constructed previously, but still produce EPS [5].

We started the structural analysis of *A. brasilense* EPS. The glycosyl composition of Sp7 and Sp245 EPS fractions obtained after gel filtration chromatography was determined by alditol acetate derivatization and subsequent GLC-MS analysis. We found that the high $M_r$ EPS components have a different composition in Sp7 and Sp245, and that the high $M_r$ and low $M_r$ EPS components from Sp7 also have a different composition [poster abstract by De Troch et al., this volume].

## 2.2. PLASMID ANALYSIS

We showed by hybridization on blotted cell lysates prepared according to Kado and Liu (1981) that the *exoM* locus, similar to the *exoB* and *exoC* (pCal112) loci [8] is located on the 90-MDa plasmid (p90) in *A. brasilense* Sp7. This plasmid also carries two loci affecting general chemotaxis behaviour [9] and a locus homologous to *R. meliloti nodPQ* [10]. These findings suggest the potential involvement of plasmids, particularly p90, in the *A. brasilense*-plant interaction. We started a further search for plasmid-borne function op p90 and simultaneously on p115, the 115-MDa Sp7 plasmid. A physical map of p90 was constructed (result not shown). Several attempts by us and by Elmerich and coworkers (personal communication) to cure p90 have been unsuccessful, indicating that p90 may carry essential information for survival.

The p115 plasmid was tagged with Tn5 and Tn5-Mob by site-directed mutagenesis of a cloned p115 DNA fragment which was subsequently crossed back in the *A. brasilense* genome by homogenotization. Using this p115-tagged strain we observed that p115 curing increases with increasing growth temperature (Table 3). We observed that the cured fraction does not increase upon further subculturing, but is constant for a given growth temperature. Even at normal growth temperature (28°C), a considerable fraction of cells in an Sp7 population has lost p115 (0.25%).

TABLE 3. p115 plasmid curing as a function of growth temperature.

| Growth temperature | % cured cells |
|---|---|
| 28°C | 0.25 |
| 30°C | 2 |
| 37°C | 15 |
| 40°C | 19 |

We could not detect conjugative transfer of p115 Tn5 to a p115-free Sp7 recipient strain with a stable Tc$^r$ marker in the chromosome for selection, indicating that p115 is not self-transmissible. The Tn5-Mob containing p115 plasmid, however, could be mobilized with low frequency ($10^{-9}$ per recipient) to this recipient with pRI1J1 as a helper plasmid. We compared Sp7 and AB7200 (p115 cured derivative of Sp7) for a lot of properties including carbohydrate, amino acid and organic acids utilization (using API test systems), polysaccharide production, $NO_3^-$ reduction, chemotaxis, siderophore production, growth rate, $N_2$-fixation. The following differences were observed:
1)   AB7200 is no longer chemotactically attracted to valine in semi-solid swarm plates.
2)   AB7200 has a slightly reduced growth rate in minimal growth medium.
3)   AB7200 has only 50% of the acetylene reduction activity (ARA)of Sp7 in semisolid N-free medium. This may, however, be a consequence of the reduced growth rate, which could in turn be caused by the inability to synthesize a vitamin. An analogous situation exists in *R. meliloti*, where thiamin synthesis is encoded by the pSym2 plasmid.

## 2.3. GENES INVOLVED IN NITROGEN FIXATION

The organization of five regions, *nifHDK*, *nifE*, *nifUS*, *fixABC*, and an unidentified open reading frame between *nifE* and *nifUS* has been presented [11]. Recently, transcription of the *nifHDK* operon of *A. brasilense* has been analyzed [13] and the nucleotide sequence of *nifH* has been determined [12,13]. The 5' region contains an NtrA-dependent promotor and upstream activator sequences that show striking homology with upstream activator sequences for NifA-mediated activation in other diazotrophs [13]. The occurrence of a *nifA* type gene was already suggested by Pedrosa and Yates [14] and Singh et al [15].

From a collection of random *A. brasilense* Tn5 insertion mutants, we isolated potential $N_2$ fixation mutants as non-growing or poorly growing colonies on N-free minimal medium under a 2% $O_2$ atmosphere. Three classes of mutants could be distinguished on the basis of their residual ARA compared to Sp7: class 1 (0%), class 2 (3%), class 3 (15%). By site-directed mutagenesis with Tn3HoHo, we constructed transcriptional fusions of *lacZ* to the *nifH* and *nifD* genes cloned on pRK290 [16], and we conjugated these fusions into the above isolated mutants. *nifHDK* expression was visually detected by the formation of a blue pellicle in N-free semisolid medium containig X-Gal. Three class 1 mutants formed only a very pale blue pellicle, indicating that *nifHDK* expression was not induced. These mutants are likely to be regulatory and will be further analyzed.

## 2.4. NO$_3^-$ REDUCTASE GENES

*A. brasilense* can grow aerobically on $NO_3^-$ as a sole source of N ($NO_3^-$ assimilation). In addition, some *A. brasilense* strains also have $NO_3^-$ dissimilatory capacity. Boddey et al. [17] found that a majority of *A. brasilense* strains isolated from surface-sterilized wheat roots were able to dissimilate $NO_3^-$, as opposed to only a few strains isolated from non-sterilized roots and rhizoshere soil. Also, a $NO_3^-$ reductase deficient mutant of *A. brasilense* Sp245 lost the wild-type ability to infect the inner root of wheat [17]. These results suggest a role for the dissimilatory $NO_3^-$ reductase in the colonization of the inner root. Further, Zimmer et al. [18] demonstrated that $NO_2^-$ (the product of dissimilatory $NO_3^-$ reduction) can act as a growth factor stimulating root growth.

We have cloned *A. brasilense* genes involved in dissimilatory $NO_3^-$ reduction, and we are constructing the corresponding mutants These will help us to clarify the role of dissimilatory $NO_3^-$ reductase in the plant interaction.

## 2.5. IDENTIFICATION OF PLANT-INDUCED GENES

We have described previously the application of two-dimensional protein analysis for *A. brasilense* strain fingerprinting [19]. Here, we have used this technique to look for the specific induction of *A. brasilense* proteins by wheat root exudate. We found that, 6-15 hours after addition of exudate to log phase cultures of *A. brasilense* Sp7, 3 to 5 new spots appeared on the fingerprint, and 2 or 3 disappeared (results not shown). These changes were independent of the growth medium used for the *A. brasilense* cultures, indicating that they do not merely represent a switch to a different nutrient source. The most prominent new spot corresponds to an acidic protein of approx. 40 kDa. This protein is being purified for amino acid acid sequence analysis. Synthetic oligonucleotides based on this amino acid sequnce will be used to isolate the gene encoding this protein by hybridization on an *A. brasilense* Sp7 gene bank. With this strategy we hope to identify new *A. brasilense* genes that play a role in the interaction with the plant.

## 2.6. MONITORING THE *AZOSPIRILLUM*-PLANT INTERACTION

We are using the *E. coli uidA* gene, which encodes b-glucuronidase, as a reporter gene in *A. brasilense*. Firstly, by random mutagenesis with Tn5-*uidA*, we isolated *A. brasilense* Sp7 and Sp245 mutants that constitutively expressed high levels of b-glucuronidase. Attachment of these mutants to wheat seedling roots in the *in vitro* assay mentioned above, could be quantitatively and reproducibly measured by determining b-glucuronidase activity on whole roots with attached bacteria. We are now constructing a constitutively expressed *uidA* gene on a broad host range vector. This construct, when introduced into various mutants (p.e. Cal⁻ mutants, see above), will provide a rapid and reliable method to measure root attachment both *in vitro* and *in vivo*. Further, *A. brasilense* expressing *uidA* can be histochemically located on microscopic sections.

Secondly, we are constructing *uidA* fusions to the *A. brasilense nif, exo,* Cal, and *nar* genes, to study the expression of these genes during plant interaction.


## Acknowledgements

This work was supported by research funds from the Onderzoeksraad KU Leuven and the N.F.W.O.


## References

[1] ELMERICH, C. (1984).  Molecular biology and ecology of diazotrophs associated with non-leguminous plants.  Biotechnology **2**, 967-978.

[2] MICHIELS, K., VANDERLEYDEN J. and VAN GOOL, A. (1989).  *Azospirillum*-plant root associations: A review.  Biol. Fertil. Soils **8**, 356-368.

[3] OKON, Y. (1985).  *Azospirillum* as a potential inoculant for agriculture.  Trends Biotechnol. **3**, 223-228.

[4] MICHIELS, K., VERRETH, C. and VANDERLEYDEN, J. (1990).  *Azospirillum lipoferum* and *Azospirillum brasilense* surface polysaccharide mutants that are affected in flocculation.  J. Appl. Bacteriol. **69,** in press.

[5] MICHIELS, K., VANDERLEYDEN, J., VAN GOOL, A.P. and SIGNER, E.R. (1988).  Isolation and characterization of *Azospirillum brasilense* loci that correct *Rhizobium meliloti exoB* and *exoC* mutants.  J. Bacteriol. **170**, 5401-5404.

[6] LEIGH, J.A. and CHI CHANG LEE (1988).  Characterization of polysaccharides of *Rhizobium meliloti exo* mutants that form ineffective nodules.  J. Bacteriol.**170**, 3327-3332.

[7] UTTARO, A.D., CANGELOSI, G.A., GEREMIA, R.A., NESTER E.W. and UGALDE, R.A. (1990).  Biochemical characterization of avirulent *exoC* mutants of *Agrobacterium tumefaciens*. J. Bacteriol. **172**, 1640-1646.

[8] MICHIELS, K., DE TROCH, P., ONYEOCHA, I., VAN GOOL, A., ELMERICH, C., VANDERLEYDEN , J. (1989).  Plasmid localization and mapping of two *Azospirillum brasilense* loci that affect exopolysaccharide synthesis.  Plasmid **21**, 142-146.

[9] VAN RHIJN, P., VANSTROCKEM, M., VANDERLEYDEN, J., DE MOT, R. (1990).  Isolation of behavioral mutants of *Azospirillum brasilense* by using Tn5 *lacZ*.  Appl. Environ. Microbiol. **56**, 990-996.

[10] VIEILLE C. and ELMERICH, C. (1990).  Characterization of two *Azospirillum brasilense* Sp7 plasmid genes homologous to *Rhizobium meliloti nodPQ*.  Mol. Plant Microbe Interact., in press.

[11] GALIMAND, M., PERROUD, B., DELORME, F., PAQUELIN, A., VIEILLE, C., BOZOUKLIAN, H., ELMERICH, C. (1989). Identification of DNA regions homologous to nitrogen fixation genes *nifE*, *nifUS* and *fixABC* in *Azospirillum brasilense* Sp7. J. Gen. Microbiol. **135**, 1047-1059.

[12] FANI, R., ALLOTTA, G., BAZZICALUPO, M., RICCI, F., SCHIPANI, C. and POLSINELLI, M. (1989). Nucleotide sequence of the gene encoding the nitrogenase iron protein (*nifH*) of *Azospirillum brasilense* and identification of a region controlling *nifH* transcription. Mol. Gen. Genet. **220**, 81-87.

[13] de ZAMAROCZY, M., DELORME, F. and ELMERICH, C. (1989). Regulation and promotor mapping of the structural genes for nitrogenase (*nifHDK*) of *Azospirillum brasilense*. Mol. Gen. Genet. **220**, 88-94.

[14] PEDROSA, F.O., YATES, M.G. (1984). Regulation of nitrogen fixation (*nif*) genes of *Azospirillum brasilense* by *nifA* and *ntr(gln)* type gene products. FEMS Microbiol. Lett. **23**, 95-101.

[15] SINGH, M., TRIPATHI, A. and KLINGMÜLLER, W. (1989). Identification of a regulatory *nifA* type gene and physical mapping of cloned new *nif* regions of *Azospirillum brasilense*. Mol. Gen. Genet. **219**, 235-240.

[16] MILCAMPS, A. and VANDERLEYDEN, J. (1990). *In vitro* construction of gene fusions to the *nifHDK* operon of *Azospirillum brasilense*. FEMS Microbiol. Lett., in press.

[17] BODDEY, R.M., BALDANI, V.L.D., BALDANI, J.I., DÖBEREINER, J. (1986). Effect of inoculation of *Azospirillum* spp. on nitrogen accumulation by field-grown wheat. Plant and Soil **95**, 109-121.

[18] ZIMMER, W., ROEBEN, K., BOTHE, H. (1988). An alternative explanation for plant growth promotion by bacteria of the genus *Azospirillum*. Planta **176**, 333-342.

[19] DE MOT, R., VANDERLEYDEN, J. (1990). Application of two-dimensional protein analysis for strain fingerprinting and mutant analysis of *Azospirillum* species. Can. J. Microbiol. **35**, 960-967.

# Dinitrogen fixation and infection of grass leaves by *Pseudomonas rubrisubalbicans* and *Herbaspirillum seropedicae*


J.P. PIMENTEL[1], F. OLIVARES, R.M. PITARD, S. URQUIAGA, F. AKIBA[1] and J. DÖBEREINER
*EMBRAPA, National Centre for Soil Biology Research (CNPBS/EMBRAPA), Km 47, Seropédica, 23851, Rio de Janeiro, Brazil ([1]Universidade Federal Rural do Rio de Janeiro)*


*Key words:* Herbaspirillum, nitrogen fixation, napier grass, *Pseudomonas rubrisubalbicans*, sorghum, sugar cane

## Abstract


Bacteria causing mottled stripe disease in sugar cane, known as *Pseudomonas rubrisubalbicans*, were shown to be able to fix molecular $N_2$ and to grow on it. The root associated diazotroph known as *Herbaspirillum seropedicae*, after artificial inoculation caused mottled stripe disease symptoms on sorghum and Napier grass but not on sugar cane. Both bacteria could be reisolated from leaves even 60 days after. Sugar cane leaves contained large numbers of these bacteria even in the uninoculated controls. Additional physiological characteristics of six strains of *P. rubrisubalbicans* were compared with those of two *H. seropedicae* strains and were shown to be very similar.


## Introduction

The diazotroph *Herbaspirillum seropedicae* was described by Baldani et al. (1986) based on more than 100 isolates obtained from the rhizosphere and from surface sterilized roots of maize, rice and sorghum grown at three different sites. The organism, initially thought to be an additional *Azospirillum* species (Baldani et al., 1984), is smaller, shows a much broader pH tolerance (optimal growth from pH 5.0 to 8.0) and nitrogenase activity is more tolerant to higher $pO_2$. Nitrate is not dissimilated and recent data from Fu and Burris (1989) showed that $NH_4^+$ only partially inhibits nitrogenase activity.

The plant pathogen listed in Bergey's Manual (Palleroni, 1984) as *Pseudomonas rubrisubalbicans* is known to cause the mottled stripe disease in sugar cane. In Brazil however this disease is hardly ever observed and most cane cultivars are considered resistant. The classification of this organism as a *Pseudomonas* has been questioned (Gillis, person. com., Goor et al., 1986) on the basis of DNA:rRNA hybridizations with 35 strains. Recent DNA:rRNA analyses confirmed this and showed that *Pseudomonas rubrisubalbicans* is closely related to *H. seropedicae*.

In view of the apparent relationship between these two microorganisms the present paper compares dinitrogen fixation and some additional physiological characteristics as well as pathogenicity on various host plants.

## Materials and methods

The microorganisms used in this work were two *H. seropedicae* strains isolated from surface sterilized rice and sorghum roots respectively (Baldani et al., 1986). The *P. rubrisubalbicans* strains were obtained through courtesy of M. Gillis from the Belgian type culture collection and from the Brazilian collection in IAC, Campinas SP. The origins of all strains are summarized in Table 1.

The basic medium used for $N_2$-dependent growth and laboratory experiments was the N-

Table 1. Origin of bacterial strains

| Strain | Origin |
| --- | --- |
| *P. rubrisubalbicans* from the Belgian type culture collection | |
| LMG 1286 (ATCC 19308 T) | Sugar cane (*Saccharum officinarum*) leaves, USA |
| LGM 1278 | Sugar cane leaves, Mauritius |
| LMG 6415 | Sugar cane leaves, Reunion |
| LMG 6420 | Sugar cane, Jamaica |
| LMG 2284 | Sugar cane, Australia |
| LMG 2285 | Sugar cane, Australia |
| *P. rubrisubalbicans* from the Brazilian type culture collection | |
| IBSBF 175 | Sugar cane, Mauritius |
| IBSBF 198 | Sugar cane, Mauritius |
| *H. seropedicae* from the CNPBS/EMBRAPA collection | |
| Z 67 (ATCC 35892) | Surface sterilized rice roots, Brazil |
| Z 78 (ATCC 35893) | Surface sterilized sorghum roots, Brazil |

free, semi-solid malate medium (NFb) (Tarrand et al., 1978) used for *Azospirillum* and *Herbaspirillum* isolation. For growth with combined N, 0.005% yeast extract was added. Acid production was evaluated by the change in colour of the bromothymol blue in the same medium where malate was replaced by the various carbon substrates. Nitrogenase activity was evaluated by the $C_2H_2$ reduction assay in the same vials with semi-solid NFb medium, with the addition of various sucrose concentrations. Two plant experiments were carried out in pots with 1.5 kg red-yellow podzolic soil (pH 5.2) fertilized with 100 ppm P, 50 ppm K and 10 ppm N and planted with cuttings of Napier grass, sugar cane cv B4362 or with sorghum seeds (cv BR303). The plants were cut at the growing point to 10 cm leaf length and inoculated by infiltration with 1 ml of a suspension from nutrient agar slopes containing approximately $10^6$ CFU, 5 cm below the last internode.

Isolation of the inoculated bacteria from inoculated leaves was made by placing 2 to 4 mm large leaf pieces into the N-free semi-solid medium, or by streaking out leaf macerates on potato agar and replicating individual colonies with the typical *Herbaspirillum* appearance (small whitish smooth colonies with darker centers) into NFb medium for confirmation.

## Results and discussion

The type strain of *P. rubrisubalbicans* isolated from sugar cane in the USA as well as 5 additional strains of this species, all isolated from sugar cane in Reunion, Jamaica and Mauritius respectively were found to grow abundantly in N-free semi-solid medium showing growth pattern and nitrogenase activity (ARA) rates as high, or higher, than *H. seropedicae* (Table 2). All other *Pseudomonas* spp. found so far to fix $N_2$, as for example *P. diazotrophicus* (Watanabe et al., 1987), are unable to grow on $N_2$ as sole N source and ARA is less than 10% of that reported here. Two strains, also classified as *P. rubrisubalbicans*, were unable to grow with $N_2$ as sole N source, even though they grew well in the same medium supplied with yeast extract, forming the same growth pattern with an aerotactic veil, growing out into a subsurface pellicle. ARA in such cultures with these two strains was negligible (Table 2).

In addition to $N_2$-fixation the *P. rubrisubalbicans* strains showed a number of physiological characteristics identical to those of *H. seropedicae*. Acid was formed from arabinose but only variable (among strains, independent of origin) weak acid production was observed from glucose, galactose and mannitol. No acid was

*Table 2.* N$_2$-dependent growth and nitrogenase activity of *Pseudomonas rubrisubalbicans* and *Herbaspirillum seropedicae* in semi-solid malate medium (NFb) with 0.01 and 10% sucrose after 44 h growth

| Bacterial strain | n moles C$_2$H$_4$ h$^{-1}$ vial$^{-1}$ | | | |
| --- | --- | --- | --- | --- |
| | N-free | | 0.005% yeast extract | |
| | 0.01% | 10% | 0.01% | 10% |
| *P. rubrisubalbicans* group I[a] | | | | |
| LGM 2286 | 385 | 34 | 754 | 295 |
| LGM 1278 | 754 | 21 | 967 | 43 |
| LGM 6415 | 95 | 21 | 730 | 295 |
| LGM 6420 | 426 | 12 | 1066 | 139 |
| IBSBF 175 | 34 | 41 | 90 | 10 |
| IBSBF 198 | 266 | 25 | 1016 | 72 |
| *P. rubrisubalbicans* group II[b] | | | | |
| LGM 2284 | 0 | 0 | 0 | 0 |
| LGM 2285 | 0 | 0 | 5 | 0 |
| *H. seropedicae* | | | | |
| Z 67 | 92 | 13 | 111 | 66 |
| Z 78 | 34 | 28 | 142 | 25 |

[a] Group I strains grow in N-free NFb medium with an aerotactic pellicle.
[b] Group II strains do not grow in N-free NFb medium but form an aerotactic pellicle in NFb with yeast extract.

formed from glycerol. All strains grew very well in semi-solid media with these carbon sources. There was N$_2$ dependent growth and ARA at 37°C but at 40°C growth occurred only with yeast extract as N source (data not shown). The swarming on soft nutrient agar observed for *H. seropedicae* (Baldani et al., 1986) was also shown with all *P. rubrisubalbicans* strains tested. Tolerance to high sugar concentrations would be expected from isolates of sugar cane as also observed with *Acetobacter diazotrophics* (Gillis et al., 1989). Table 2 shows that all strains of *P. rubrisubalbicans*, as well as those of *H. seropedicae*, grow and fix N$_2$ with 10% sucrose even though this sugar is not being used.

When inoculated into leaves of Napier grass and sorghum by the traditional phytopathological methods, *H. seropedicae* caused the same symptoms as the *P. rubrisubalbicans* strains (Table 3).

*H. seropedicae* was pathogenic to sorghum which reacted with symptoms of red stripes on leaves and stems. Symptoms on Napier grass were mottled stripes and water soaking on the inoculation point. The sugar cane cv B 4362, a Brazilian cultivar succeptible to mottled stripe disease, did not develop symptoms when inoculated with *H. seropedicae*. *P. rubrisubalbicans* isolates from various origins (see Table 1) showed very similar symptoms of mottled stripe disease on sorghum which reacted by red mottled stripes, and on Napier grass which showed mainly water soaking on the inoculation points and mottled stripes. Sugar cane reacted to some of the isolates (LMG 2286, the type strain, and IBSBF 175) with red stripes and mottling as well as with stalk rot. The symptoms observed here on sorghum and Napier grass caused by *H. seropedicae* and those caused by the *P. rubrisubalbicans* strains on the three plants are the same as those reported in the literature (Egerton, 1955; Galli et al., 1980; Hale and Wilke, 1972a,b).

The inoculated bacteria could be reisolated from the leaves, 3 cm above the inoculation mark (Table 3) from all inoculated plants. All attempts to isolate these organisms from control plants failed except those from sugar cane which all contained N$_2$-fixing *H. seropedicae* (until dilution $10^{-5}$). Twenty six days after inoculation reisolation was also successful 10 cm below the inoculation mark, and 60 days after, from the

228

*Table 3.* Symptoms of mottled stripe disease (SM) and recovery of bacteria (RC) 8 days after inoculation

| Bacterial strain | *Saccharum* sp (cv B4362) | | *Pennisetum purpureum* | | *Sorghum bicolor* | |
|---|---|---|---|---|---|---|
| | SM | RC | SM | RC | SM | RC |
| *P. rubrisubalbicans* group I[a] | | | | | | |
| LGM 2286 | 2[b] | 3[c] | 1 | 3 | 2 | 2 |
| LGM 1278 | – | – | 1 | 3 | 2 | 2 |
| LGM 6415 | – | – | 1 | 2 | 2 | 2 |
| LGM 6420 | – | – | 1 | 2 | 1 | 1 |
| IBSBF 175 | 2 | 3 | 1 | 3 | 3 | 3 |
| IBSBF 198 | – | – | 0 | 1 | 0 | 2 |
| *P. rubrisubalbicans* group II[a] | | | | | | |
| LMG 2284 | – | – | 2 | 3 | 0 | 3 |
| LGM 2285 | – | – | 2 | 3 | 2 | 3 |
| *H. seropedicae* | | | | | | |
| Z 67 | 1 | 3 | 2 | 3 | 3 | 2 |
| Z 78 | – | – | 1 | 2 | 0 | 2 |
| Uninoc. control | 0 | 3 | 0 | 0 | 0 | 0 |

[a] Group I strains capable of $N_2$ dependent growth. Group II strains show variable low ARA after growth on yeast extract only.
[b] Intensity of symptoms 0 to 3.
[c] Number of NFb vials out of 3 with typical pellicle. Strains of group II were streaked out on potato agar and individual colonies testes in NFb medium with yeast extract.

3rd to 5th leaf above the inoculated leaf, indicating translocation of the bacteria within the plant (data not shown).

The confirmation, in this paper, of $N_2$ fixation of plant pathogens may have great economic implications, especially in a crop like sugar cane which in Brazil occupies now more than 4 million ha of land. With the exception of the recent report on *Agrobacterium tumefaciens* (Kanvinde and Sastry, 1990), no plant pathogen has as yet been shown to be able to fix $N_2$. Even though the extent of pathogenicity of our organisms will have to be confirmed with additional plant genotypes, the fact that diazotrophs inhabit grass leaves opens entirely new approaches to the investigation of plant-pathogen-diazotroph interactions.

## References


Baldani J I, Baldani V L D, Sampaio M J A M and Döbereiner J 1984 A fourth *Azospirillum* species from cereal roots. An. Acad. Bras. Cien. 56, 365.

Baldani J I, Baldani V L D Seldin L and Döbereiner J 1986 Characterization of *Herbaspirillum seropedicae* gen. nov., sp. nov., a root associated nitrogen-fixing bacterium. Inter. J. Syst. Bacteriol. 36, 86–93.

Edgerton C W 1955 Sugar Cane and its Diseases. Louisiana State University Press, Baton Rouge, LA.

Fu A H and Burris R H 1989 Ammonium inhibition of nitrogenase activity in *Herbaspirillum seropedicae*. J. Bacteriol. 171, 3168–3175.

Galli F, Carvalho P C T, Tokeshi H, Balmer F, Kimati H, Cardoso C O, Salgado C L, Krugner T L, Cardoso E J B N and Bergamin F A 1980 Manual de Fitopatologia: Doenças das plantas cultivadas. Ed. Agronômica Ceres, 2ª ed, vol 2. São Paulo, Brazil.

Goor M, Falsen E, Pot B, Gillis M, Kersters K and De-Ley J 1986 Taxonomic position of the phytopathogen *Pseudomonas rubrisubalbicans* and related clinical isolates. XIV Int. Congr. Microbiol. Wanchester, England (Abstract).

Hale C N and Wilke J P 1972a A comparative study of *Pseudomonas* species pathogenic to sorghum. N.Z.J. Agric. Res. 15, 448–456.

Hale C N and Wilke J P 1972b Bacterial leaf stripe of sorghum in New Zealand, N.Z. J. Agric. Res. 15, 457–460.

Kanvinde U and Sastry G R R 1990 *Agrobacterium tumefaciens* is a diazotrophic bacterium. Appl. Environ. Microbiol. 56, 2087–2092.

Palleroni J N 1984 Family I. Pseudomonadaceae. *In* Bergey's

Manual of Systematic Bacteriology. vol 1. Eds. N R Krieg and J G Holt. pp 141–199. Williams & Wilkins Co., Baltimore. MD.

Tarrand J J, Krieg N R and Döbereiner J 1978 A taxonomic study of the *Spirillum lipoferum* group, with descriptions of a new genus, *Azospirillum lipoferum* (Beijerinck) comb. nov. and *Azospirillum brasilense* sp. nov. Can. J. Microbiol. 24, 967–980.

Watanabe I, So R, Ladha J K, Katayama-Fujimura Y and Kuraish H 1987 A new nitrogen-fixing species of Pseudomonas: *Pseudomonas diazotrophicus* sp. nov. isolated from the root of wetland rice Can. J. Microbiol. 33, 670–678.

# GROWTH PARAMETERS OF MICROAEROBIC DIAZOTROPHIC RHIZOBACTERIA DETERMINED IN CONTINUOUS CULTURE


C.FRITZSCHE, J.UECKERT, and E.-G.NIEMANN
Institute of Biophysics
University of Hanover
Herrenhaeuser Str.2
3000 Hannover 21, FRG


## Abstract


Maximum growth rate, yield, and carbon source affinity of two *Azospirillum brasilense* strains and a not identified isolate were determined. Results are compared to data taken from literature.


## Introduction

Since the discovery that some diazotrophes will fix nitrogen only under microaerobic conditions several new strains and species have been isolated and described. Enrichment and consequent isolation from one rhizosphere sample usually contain more than one microaerobic diazotroph. Because rhizosphere microorganisms are believed to be carbon limited in their natural habitat, the substrate affinity to the carbon source will play an important role in the establishment of each strain. Continuous culture allows to grow the organisms under carbon limitation and the substrate affinity can be determined. In our laboratory continuous culture of diazotrophes under miroaerobic conditions was performed for a number of years (e.g. Kloss, 1983, Hurek, 1987, Fritzsche 1990). We report here the growth parameters of three strains isolated from rice and wheat.

## Materials and Methods

### BACTERIAL CULTURES

Isolate MB 35 is a gram-negative diazotrophic strain isolated from brasilian rice soil. *Azospirillum brasilense* Sp 245 and *Azospirillum brasilense* Sp 246 were kindly provided by J.Döbereiner. They were isolated from wheat roots (Döbereiner, pers.comm.).

### MEDIA AND GROWTH CONDITIONS

Isolate MB35 was grown in the following medium $(g \cdot l^{-1})$:$K_2HPO_4$ 0.2; $KH_2PO_4$ 0.3;

$MgSO_4 \cdot 7 H_2O$ 0.2; NaCl 0.1; $CaCl_2$ 0.02; $MnSO_4 \cdot H_2O$ 0.01; $Na_2MoO_4 \cdot H_2O$ 0.002;

$FeSO_4 \cdot 7H_2O$ 0.01; glucose 0.36 at 30°C, 6 µM $O_2$, and pH 6.4.

Sp 245 and Sp 246 were grown in ($g \cdot l^{-1}$): MOPS-buffer 0.5; $Na_5O_{10}P_3$ 0.05; $MgSO_4 \cdot 7\,H_2O$ 0.2; NaCl 0.1; $CaCl_2$ 0.3; $MnSO_4 \cdot H_2O$ 0.01; $Na_2MoO_4 \cdot H_2O$ 0.002; $FeSO_4 \cdot 7H_2O$ 0.01; L-malic acid (neutralized with KOH) 1.0 at 35°C, 6 µM $O_2$, and pH 6.8.

## CONTINUOUS CULTURE AND SAMPLING PROCEDURE

The strains were grown in continuous culture (Fritzsche 1989, Kloss 1983). Dissolved oxygen was measured with a Clark type electrode and regulated to constant concentrations by mixing a constant flow of $N_2$ and a regulated flow of artifical air (80% of $N_2$ and 20% of $O_2$). Steady states were sampled twice. The remaining substrate concentrations were determined by enzyme assays (L-malic acid assay, Boehringer GmbH, Mannheim, FRG or glucose assay, Sigma, Deisenhofen, FRG). The protein content was measured by the micro-Goa method with Benedict's reagent (Bergersen, 1980).

Results

The results of the determination of growth rates and substrate affinity are given in table 1. The results of the investigated strains are compared to data taken from literature. The yield at growth rates of approxamately 0.09 $h^{-1}$ were 100, 140, and 170 mg protein $\cdot g^{-1}$ carbon source for isolate MB 35, *A. brasilense* Sp 245, and *A. brasilense* Sp 246, respectively. From Fritzsche (1990) the yields of *A. brasilense* Sp 7 and Isolate R7 were calculated to be 100 and 90 mg protein $\cdot g^{-1}$ carbon source, respectively.

Table 1 Maximum growth rates ($\mu_{max}$) and $K_s$ values for the limiting carbon source of microaerobic diazotrophes investigated in this paper and of several other strains (data taken from literatur).

| Organism | $\mu_{max}$ [$h^{-1}$] | $K_s$ [µM] | Reference |
|---|---|---|---|
| *Azospirillum brasilense* Sp 245 | 0.20 | <2 | |
| *Azospirillum brasilense* Sp 246 | 0.21 | <2 | |
| Isolate MB 35 | 0.11 | <2 | |
| *Azospirillum brasilense* Sp 7 | 0.15 | 72 | Kloss, 1983 |
| *Azospirillum brasilense* Cd [b] | 0.09 | 80 | Cacciari, 1986 |
| *Azospirillum lipoferum* Br 17 [a] | 0.13 | | Volpon, 1981 |
| *Azospirillum halopraeferens* Au 4 [a] | 0.17 | | Reinhold, 1985 |
| *Xanthobacter autotrophicus* GZ 29 [a] | 0.14 | | Berndt, 1976 |
| *Arthrobacter giacomelloi* [b] | 0.08 | 83 | Cacciari, 1986 |
| Isolate R7 | 0.17 | <2 | Fritzsche, 1989 |

[a] determined in batch culture
[b] not determined under optimum $O_2$ concentration

## Discussion

The maximum growth rates of the microaerobic diazotrophes are within a range of
$0.11\ h^{-1}$ to $0.21\ h^{-1}$ (Table 1). The data of Cacciari are not taken into concideration
here, because the organisms were not grown at optimum oxygen concentrations
(Cacciari, 1986). Some influence of the temperature used for the investigations can
be seen. The slowest isolate MB 35 was investigated at 30°C, while the faster ones
were tested at temperatures of 35°C and above.

It has been found that during nitrogen fixation in continuous culture most of the
cellular N is in the form of protein and no substancial amounts of combined N is
released by the cells (Hurek 1987, Fritzsche 1989, Fritzsche 1990). Therefore, protein
content and protein yield serve as a simple measure for quantification of
nitrogenase activity. The yields of the strains varied between 90 and 170 mg
protein $\cdot g^{-1}$ carbon source.

Both the maximum growth rate and the yield  show only a two fold increase from
lowest to highest numbers. This might be attributed to similar abilities of the
nitrogenase of the investigated organisms under microaerobic conditions and to
similar physiological activities of the heterotrophic microaerobic diazotrophes. It is
important to realize, that these data were taken under optimum conditions in the
laboratory. Under natural conditions other mechanisms (e.g. oxygen tolerance,
Hurek, 1987) might be of greater importance for the establishment of
microorganisms in the rhizosphere.

The $K_s$-values for the limiting carbon source do not show a similar small range. The
$K_s$-values of <2 µM were estimated, because no carbon source was detected in the
supernatant. The detection limits of the assays were used for calculation. At least a
40 fold difference from lowest to highest numbers was observed. Under carbon
source limiting conditions the high affinity (low $K_s$-value) strains will have a
competitive advantage over the other strains.

It has been shown that in mixed culture with a non–diazotrophic strain isolate
MB_35 could succesfully compete in the presence and absence of $NH_4Cl$. The
competitive advantage was attributed to higher affinity to the limiting carbon
source. These experiments were performed in a multistage continuous culture system
(gradostat). The data stress the importance of the carbon source affinity for
competition (Fritzsche, in preparation).

The growth parameters of *A. brasilense* Sp 7 and *A. brasilense* Cd differ
significantly from those of *A. brasilense* Sp 245, and *A. brasilense* Sp 246. This
demonstrates the variability within one species.

The comparison of the growth parameters of microaerobic diazotrophes indicates
that maximum growth rate and yield of these strains are less variable factors. The
substrate affinity seems to be rather important for the evaluation of rhizosphere
organisms.

# References


Bergersen,F.J. (1980) 'Measurement of nitrogen fixation by direct means', in F.J.Bergerson (ed.), 'Methods for evaluating biological nitrogen fixation', John Wiley & Sons, Chichester, pp. 65–110.

Berndt,H., Ostwal,K.-P., Lalucat,J., Schuman,C., Mayer,F., and Schlegel,H.G. (1976) 'Identification and physiological characterization of the nitrogen fixing bacterium *Corynebacterium autotrophicum* GZ 29',Arch.Microbiol. 108,17–26.

Cacciari,I., Del Gallo,M., Ippoliti,S., Pietrosanti,T., and Pietrosanti,W. (1986) 'Growth and survival of *Azospirillum brasilense* and *Arthrobacter giacomelloi* in binary continuous culture', Plant and Soil 90,107–116.

Fritzsche,C., and Niemann,E.-G. (1989) 'Investigations on growth, nitrogen fixation and ammonium assimilation by a bacterium isolated from rice', in F.A.Skinner, R.B.Boddey, and I.Fendrik (eds.), 'Nitrogen fixation with non-legumes', Kluwer Academic Publishers, Dordrecht, pp.147–152.

Fritzsche,C. and Niemann,E.-G. (1990) 'Nitrogen fixation in continuous culture with $NH_4Cl$-containing media', Appl.Environ.Microbiol. 56,1160–1161.

Hurek,T., Reinhold,B., Fendrik,I., and Niemann,E.-G. (1987) 'Root-zone-specific oxygen tolerance of *Azospirillum* spp. and diazotrophic rods closely associated with kallar grass', Appl.Environ.Microbiol. 53,163–169.

Kloss,M., Iwannek,K.-H., and Fendrik,I. (1983) 'Physiological properties of *Azospirillum brasilense* Sp 7 in a malate limited chemostat' J.Gen.Appl.Microbiol. 29,447–457.

Reinhold,B., Hurek,T., and Fendrik,I. (1985) '*Azospirillum halopraeferans* sp.nov., a diazotroph associated with roots of *Leptochloa fusca* (Linn.) Kunth' in H.J.Evans, P.J.Bottomley, and W.E.Newton (eds) 'Nitrogen fixation research progress' Martinus Nijhoff Publishers, Dordrecht, p.427.

Volpon,A.G.T., De-Polli,H., and Döbereiner,J. (1981) 'Physiology of nitrogen fixation in *Azospirillum lipoferum* Br 17 (ATCC 29709)', Arch.Microbiol. 128,371–375.

# INFECTION OF INTACT ROOTS OF KALLAR GRASS AND RICE SEEDLINGS BY *AZOARCUS*


T. HUREK*, B. REINHOLD-HUREK, M. VAN MONTAGU AND
E. KELLENBERGER*
Laboratorium voor Genetica, Rijksuniversiteit Gent, K. L.
Ledeganckstraat 35, B-9000 Gent, * Biozentrum der Uni-
versität, Abteilung Mikrobiologie, Klingelbergstr. 70,
CH-4056 Basel.



ABSTRACT. A new group of nitrogen-fixing bacteria (proposed name
*Azoarcus*) occuring in high numbers in the root interior of field-grown
Kallar grass were monitored in roots of insitu-grown and reinoculated
plants. Bacteria were followed making use of the reporter gene *gusA*
introduced into *Azoarcus* BH72 and of antibodies specific for this
bacterium and dinitrogenase reductase from *Rhodospirillum rubrum*. High
spatial resolution was obtained by light- and electron microscopy.
Bacteria were shown to colonize rice and Kallar grass roots inter- and
intracellularly. Colonization was sometimes accompanied by appearance of
an extracellular matrix of at least partial bacterial origin. Possible
sites for $N_2$-fixation are shown.


## INTRODUCTION

Diazotrophic bacteria belonging to *Azoarcus* have been repeatedly found
in the rhizosphere of Kallar grass (*Leptochloa fusca* (L.) Kunth), growing on salt affected, low fertility soils in the Punjab of Pakistan
(Reinhold et al. 1986, own unpublished results). Applying now methods
with high spatial resolution we extend previous studies on colonization
of grass roots and include data on possible sites of $N_2$-fixation.

## MATERIALS AND METHODS

Reference bacteria were obtained from the Culture Collection Laboratorium Microbiologie, Gent, Belgium (LMG). Bacteria belonging to *Azoarcus*
and strain S7a2 were isolated by B. Reinhold-Hurek from a location in
the Punjab of Pakistan in 1984 and 1988. At the same time we took roots
of healthy looking Kallar grass from the same location and processed
them for embedding into LR White resin within three hours. We introduced
the gene for ß-glucuronidase into *Azoarcus* strain BH72 referred to as
strain D3 by random mutagenesis using a Tn5-*gusA* cassette. Procedures
of mutagenesis, staining and embedding are detailed elsewhere. A *gusA*
expressing mutant of *Azospirillum brasilense* 245 is a kind gift of Dr.

J. VanderLeyden, Leuwen, Belgium. Kallar grass and rice seedlings were grown gnotobiotically under light for one week in soft agar with L-proline as nitrogen source and 0.2 g D,L malate per liter before they were transferred into sterile dune sand (≤ 1 mg N/l) supplemented with 5 ppm urea-N. Plants were inoculated with *Azoarcus* strains BH72, D3 or as a control with the *gusA* expressing mutant of *Azospirillum brasilense* 245. SDS-soluble bacterial components were separated by SDS-PAGE (Laemm-

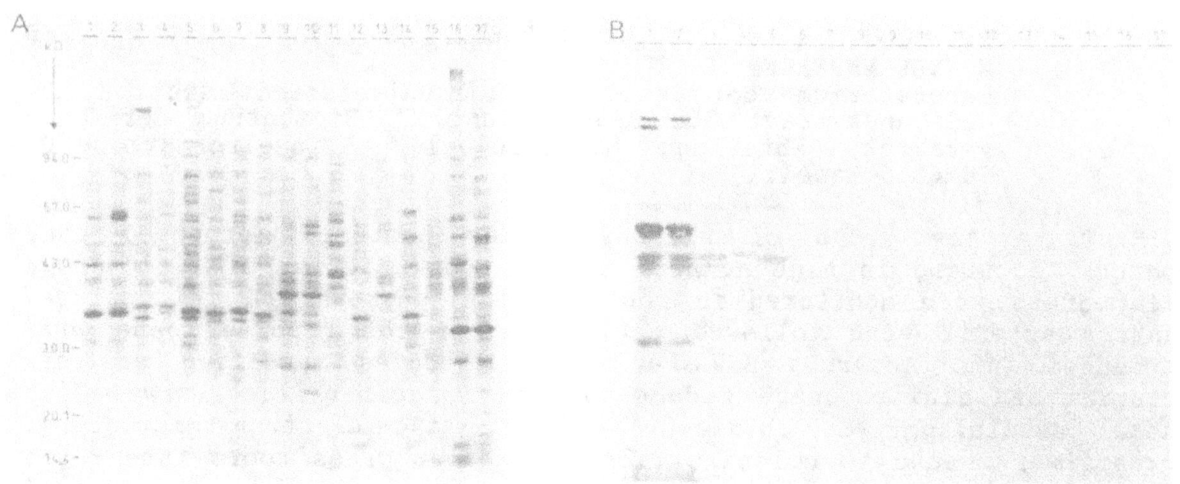

**Figure 1.** Immunological control of the polyclonal antibodies raised against whole cells of *Azoarcus* sp. BH72 by Western blotting. (A) Coomassie Blue stained protein profiles of whole-cell SDS-extraxts of strains belonging to the new genus *Azoarcus* (lanes 1-7) and reference strains (lanes 8-17). If not indicated otherwise, bacteria were grown on combined nitrogen. Lanes: 1, BH72; 2, BH72 grown on N₂; 3, SWuB3; 4, VB32; 5, S2; 6, 6a3; 7, S5b2; 8, S7a2; 9, *Herbaspirillum seropedicae* Z78; 10, *H. seropedicae* Z78 grown on N₂; 11, H6a2; 12, *Azospirillum halopraeferens* Au4; 13, *A. lipoferum* RP5; 14, *Alcaligenes piechaudii* NCMB 1051; 15, *Azorhizobium caulinodans* ORS571; 16, *Comomonas acidovorans* Stanier 14; 17, *C. terrigena* NCIB 8193. (B) Western blot of the same fractions, incubated with the whole-cell antiserum.

li, 1970), using a linear 8-15% polyacrylamide gradient. About 20 µg protein were loaded per lane. For Western blots, the protocol given by Towbin et al. (1979) was followed. Antibodies against whole cells of strain BH72 were raised in rabbits according to Reinhold et al. (1986). The antiserum was depleted with whole cell proteins of strain S7a2 by affinity chromatography. Polyclonal antibodies directed against the Fe-protein of nitrogenase-1 from *Rhodospirillum rubrum* are a kind gift of Dr. Ludden, Wisconsin, USA. For immunogold labeling both antisera were absorbed with roots of sterile, aseptically grown Kallar grass (10 mg dry weight/ml). As second antibody a goat anti rabbit gold conjugate (15 nm particles) was used. Immunogold silver staining for light microscopy was done as described previously (Reinhold and Hurek (1988). For electron microscopy, immunogold labeling was done on carbon coated

Formvar grids. Preparations were examined with a Zeiss EM 109 electron microscope operated at 80 kV or with an Olympus Vanox-S microscope equipped with modified Wollaston prisms and an analyzer to achieve differential interference contrast. Cytochemical controls for non-specific labeling regularly included omission of the primary antibody.

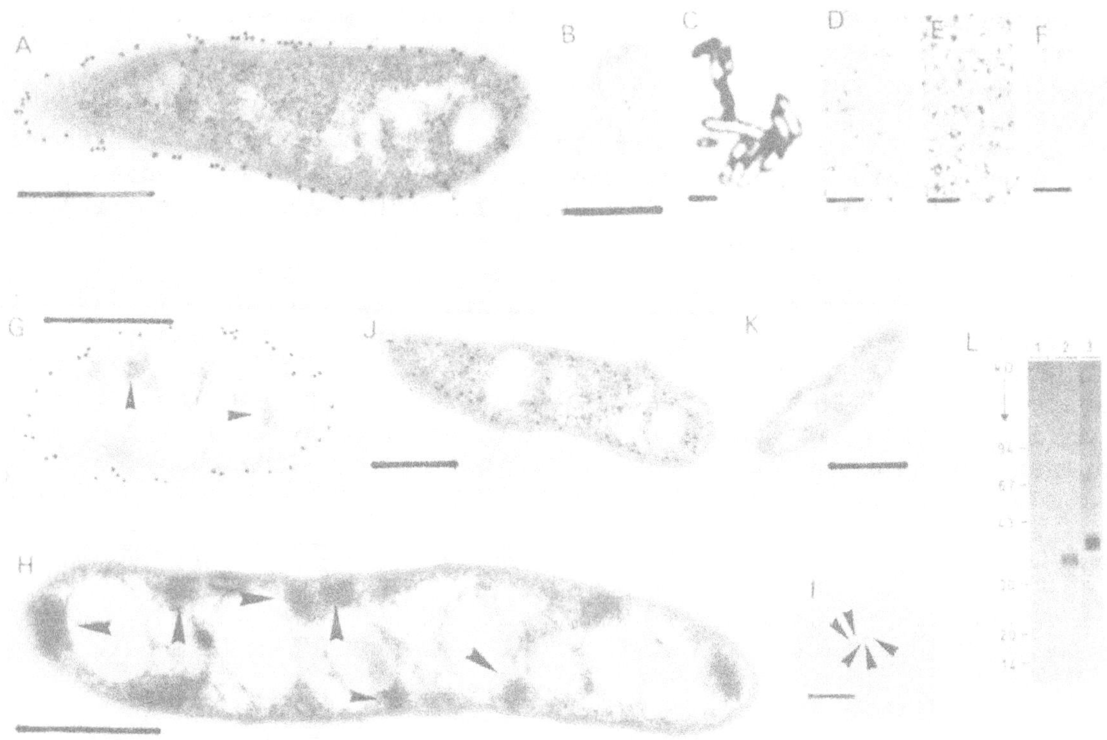

Figure 2. Further immunological controls by microscopy of immunogold labeled sections (A-K) and by Western blotting (L). Electron (A, B, G, H, J, K) and light microscopic (C, D, E, F, I) appearance of *Azoarcus* BH72 (A-E, J, K), D3 (G-I) and *Azospirillum halopraeferens* Au4 (F) grown on combined nitrogen (A, B, D-I, L) or on $N_2$ (C, J). Interference contrast (C), bright field (D, F, I), and phase contrast illuminations "under focus" (E) were used. Sections were incubated with the whole cell antiserum (A, C-G), with antiserum against dinitrogenase reductase (J, K), and with the whole cell antiserum depleted with cells of strain BH72 (B). (G-I) represent images of strain D3 stained with chromogenic substrate 5-Brom-4-chlor-3-indoxyl-β-D-glucuronide (X-Glu) prior to embedding. Note presence of cristals (arrows). Bars represent 0.5 µm (A-C, G, H, J and K) and 10 µm (D-F, I). (L), Western blots of whole cell SDS-extracts of strain BH72 grown on combined nitrogen (lane 1), or on $N_2$ (lane 2) and of 1 µg purified dinitrogenase reductase from *Klebsiella pneumoniae* (kind gift of Dr. J. Yates, England) after separation of components by SDS-PAGE. Preparations were incubated with antiserum against dinitrogenase reductase from *Rhodospirillum rubrum*.

238

## RESULTS AND DISCUSSSION

For detection of N$_2$-fixation and colonization sites of *Azoarcus* spp. in plants, polyclonal antibodies directed against the Fe-protein of nitrogenase-1 from *Rhodospirillum rubrum* and against whole cells of *Azoarcus* strain BH72 were used. Specificity of both antibodies was checked by (i) one dimensional SDS-PAGE, electroblotting and immunostaining of total SDS-extracts of reference bacteria (Fig. 1A, B; Fig. 2L) and protein preparations from roots of Kallar grass (not shown) with the respective antisera, (ii) by applying them on ultrathin sections (Fig. 2A, C-G, J,K), and (iii) by using an antiserum against strain BH72 depleted with whole cell extracts from the same bacterium (Fig. 2B). Antibodies proved to be specific at the level of Western blotting and of resin-embedded material. The whole-cell antiserum was "group specific" detecting no antigens in reference bacteria but only in SDS-extracts from members of *Azoarcus* (Fig. 1B, lanes 1-5, 7). Furthermore it did not significantly discriminate between repressed and derepressed N$_2$-fixing cells of strain BH72 (Fig. 1B, lanes 1 and 2) nor was it specific for nitrogenase (compare also Fig. 1B lanes 9 and 10). Application of antibodies directed against nitrogenase resulted in superior homogeneous labeling of the cytoplasma of nitrogenase-induced pure cultures of *Azoarcus* strain BH72 (Fig. 2J), whereas application of antibodies against whole cells led to

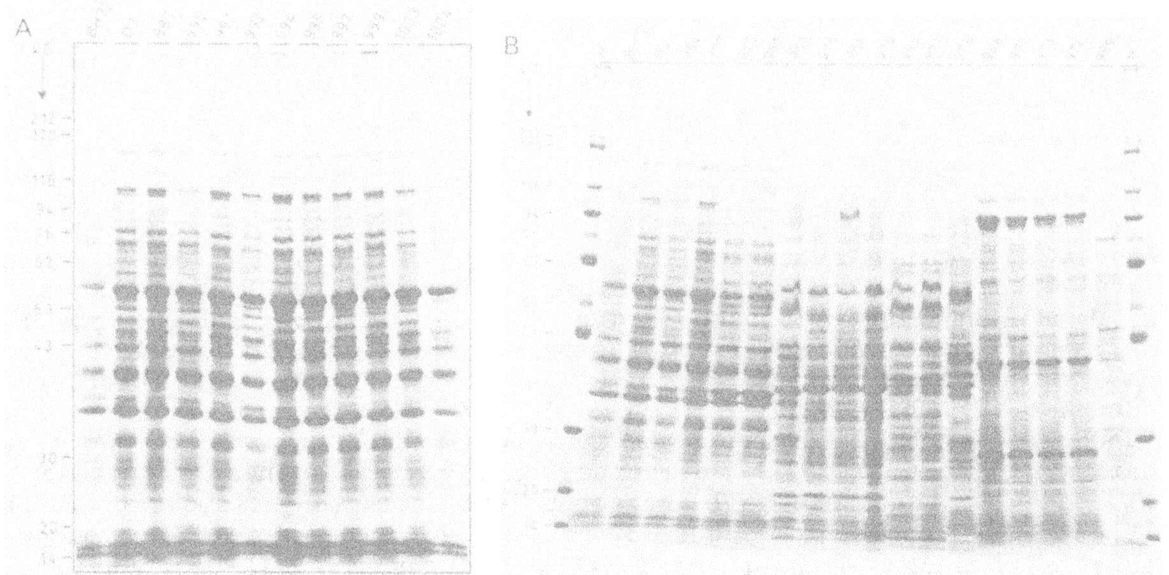

Figure 3. SDS-PAGE of reisolates from plants inoculated with *Azoarcus* sp. D3. Isolations were done after 1 week (A, gnotobiotic system) or after 2 months (B).

superior labeling  of the bacterial surface (Fig 2A, C,G) independent from the principal source of nitrogen used. The latter accounts also for

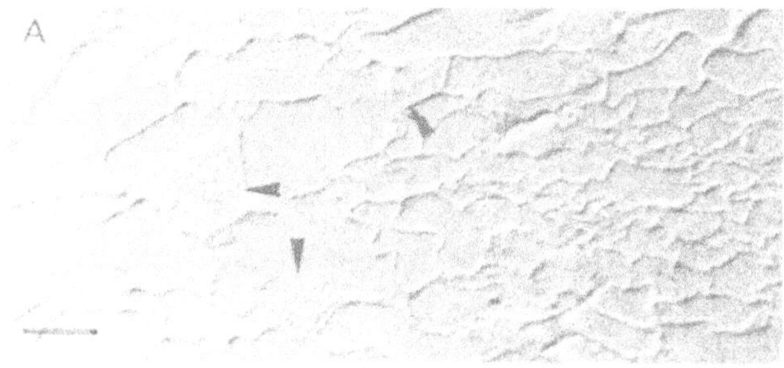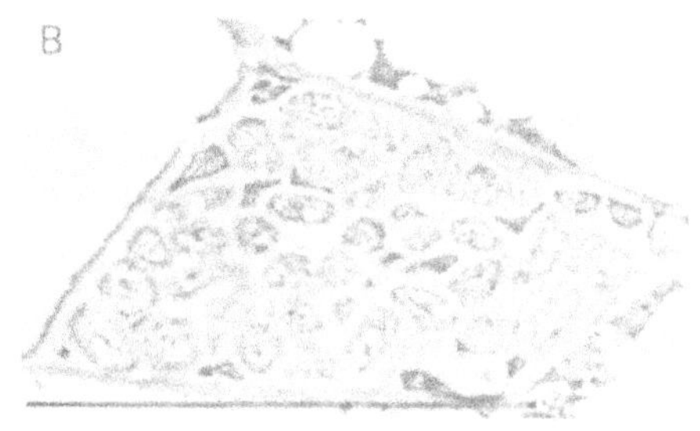

Figure 4. Colonization of Kallar grass by *Azoarcus* strain BH72 in gnotobiotic culture within one week. Interference contrast image of transverse section of heavily infected root (arrows) with absence of stele (A). Electron micrograph of subsequent sections from (A) showing colonization of the intercellular space (B). Note presence of mitochondria in adjacent cell. Bars represent 5 µm.


*Azoarcus* D3, stained with the chromogenic substrate X-Glu prior to embedding (Fig. 2G). Ideally, cristals in stained bacteria (Fig. 2H) appear blue in bright field and red in dark field (if dehydrated). Bacteria in resins can thus be targeted light microscopically and documented on black and white film (Fig. 2I) if embedded at low temperatures (Hurek, unpublished). In pure cultures of strain D3 all bacteria were stained after treatment with X-Glu.

As additional controls, SDS-PAGE of reisolates from plants inoculated with *Azoarcus* were performed. As expected, the almost identical protein patterns of the bacteria isolated after one week compared to the inoculum indicated absence of endogenous or exogenous contaminations(Fig.3A)whereas the observed heterogenicity among the protein patterns of the late isolates strongly indicates that for positive identification of *Azoarcus* in the open system the use of specific reporters is indispensable(Fig. 3B).

The following observations were made on Kallar grass and rice inoculated with *Azoarcus* strain BH72 or D3 (no difference observed, Hurek, unpublished) or on field-grown Kallar grass. Sections were examined light and electron microscopically. (i) Not all roots seemed to be penetrated (Fig. 5F). (iii) Penetration was commonly observed in the elongation zone (Fig. 4A, B) but not in the more mature parts of the root system; similar to other microorganisms closely interacting with plants no evidence for infection of root cap or meristem were obtained. (iv) Infection apparently terminated in the formation of large colonies intra- and intercellularly (Fig. 4A, Fig 5A-E). (v) Some evidence of overall host tissue distress was found especially in rice roots in form of altered host cell ultrastructure (not shown). (vi) The apparent occurence of *Azoarcus* in the stem base (Fig. 6A-C) of Kallar grass and in the stele of inoculated rice (Fig. 5D) may indicate that the stem base serves as a source of inoculum for new adventitious roots. (vii) The observation, that colonization of the root can be accompanied by the formation of an extracellular matrix (Fig. 5H) and by the occurence of infection-thread-like structures (Fig. 5B) may indicate that these

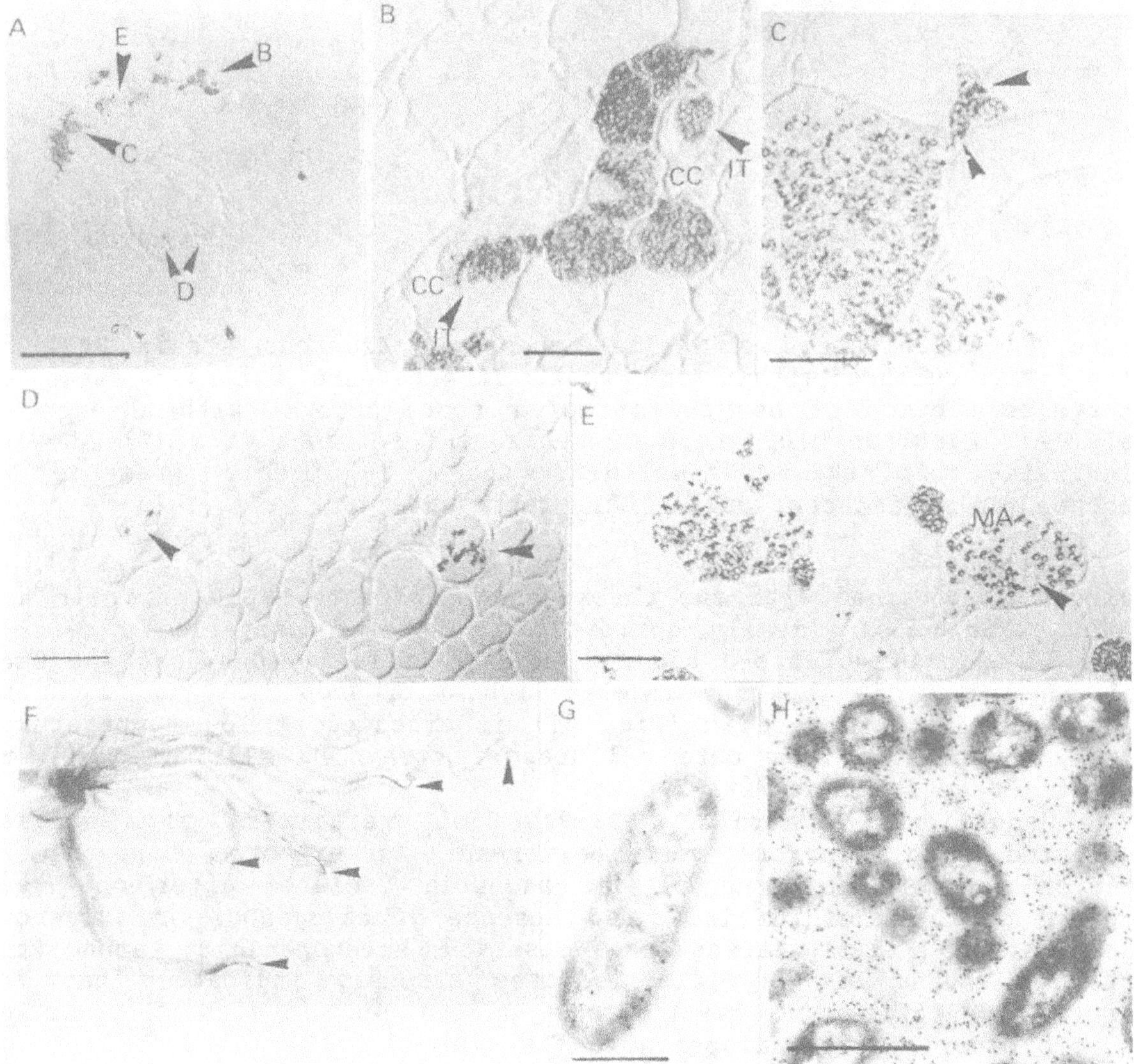

Figure 5. Colonization of rice by *Azoarcus* strain D3 in gnotobiotic culture within one week. Sections (A-E) and (H) were immunogold labeled for strain BH72. For light microscopy, signal was further amplified by silver staining (A-E). Interference contrast image of transverse section through rice root (A). Pictures (B-E) represent details of (A) at higher magnifications. *Azoarcus* (arrows) in cortex cell (CC) in an infection thread (IT) like structure (B). Spreading of *Azoarcus* apparently by moving through the primary wall (arrows) into the intercellular space (C). Colonization of the stele by *Azoarcus* (arrows) (D). Colonization of cortex cells by *Azoarcus* attended by formation of matrix (MA)(E) with (H) representing electron micrograph from *Azoarcus* embedded in matrix corresponding to area in (E, arrow). Whole intact root system of healthy rice plants stained for the presence of β-glucuronidase with X-Glu (F). Note staining of root tips (arrows). *Azoarcus* on the rhizoplane immunogold labeled for dinitrogenase reductase (G). Bars represent 100 μm (A), 10 μm (B-E), 0.5 μm (G, H).

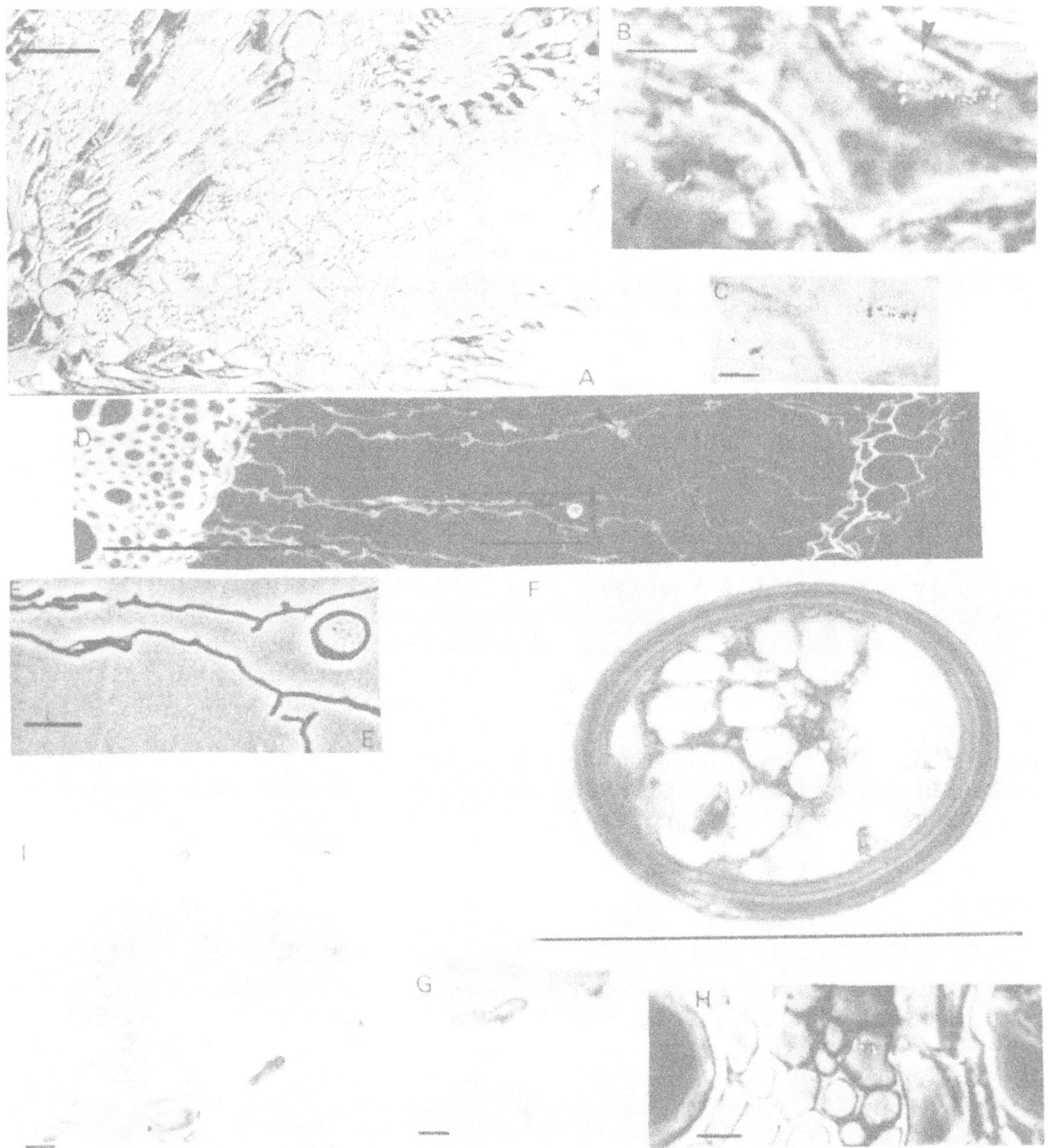

Figure 6. Colonization of stem base (A-C) and mature roots (D-H) of field grown Kallar grass and of two months old rice (I). Rice had been inoculated with *Azoarcus* strain D3. Sections (A-C, I) were immunogold labeled for strain BH72, sections (F-H) for dinitrogenase reductase. Transverse semithin section of lower stem base, where adventitious roots emerge shows complex vasculature with horizontically oriented xylem elements; differential interference-phase contrast illumination (A). Microcolonies of labeled bacteria (arrows) apparently inside stelar parenchymatic cells ; phase contrast image with "under focus" (B); (C), bright field illumination corresponding to a portion of (B). Both pic-

tures represent higher magnifications from (A, arrow). Autofluorescence image of transverse semithin section through mature root showing intercellular vesicles of probably VA-mycorrhizal fungi (arrows) (D). Phase contrast image of the boxed in area from (D) showing a vesicle rich in vacuoles (E). Electron micrograph of the other vesicle shown in (D) labeled with antibodies directed against dinitrogenase reductase (F). (G) Electron micrograph of bacteria in the aerenchyma and phase contrast image of immunogold-silver-stained bacteria in the stele (H) of mature roots of field grown Kallar grass labeled with the same antibodies. Fungus on the root surface of two-months-old rice probably invaded by *Azoarcus* showing dense gold-labeling over the cytoplasma (I). Bars represent 100 µm (A, D), 10 µm (B, C, E, F, H) and 0.5 µm (G, I).

bacteria are well adapted to the colonization of grass roots. (**viii**) We could not find any evidence, that the β-glucuronidase expressing derivative of *Azospirillum brasilense* 245 invaded rice roots to any significant extent. (**ix**) The rare occurence of bacteria in the rhizosphere (Fig. 5G, Fig. 6G, H) which apparently express proteins relevant for $N_2$-fixation makes it quite unlikely that the plant can efficiently exploit this special feature of these microbes in the system plant-diazotroph alone. Vigorous investigations on the embedded roots of the field-grown Kallar grass plants have shown that most of them contain mycorrhizal fungi some of them labeled with antiserum against whole cells of strain BH72 (Hurek et al., unpublished, see also Fig. 6I) and dinitrogenase reductase (Fig. 6F). Thus it seems to be possible that round bodies described earlier (Reinhold et al., 1987) are in fact the vesicles of VA-mycorrhizal fungi infected by *Azoarcus* sp. Whether the latter has an effect on nitrogen fixation is currently investigated.

## REFERENCES


Laemmli, U. K. (1970) 'Cleavage of structural proteins during the assembly of the head of bacteriophage T4' Nature (London) 227, 680-685.

Reinhold, B., Hurek, T., E.-G. Niemann, and Fendrik, I. (1986) 'Close association of *Azospirillum* and diazotrophic rods with different root zones of Kallar grass', Applied and Environmental Microbiology 52, 520-526.

Reinhold, B., Hurek, T., and Fendrik, I. (1987) 'Cross-reaction of predominant nitrogen-fixing bacteria with enveloped, round bodies in the root interior of Kallar grass' Applied and Environmental Microbiology 53, 889-891.

Reinhold, B. and Hurek, T. (1988) 'Location of diazotrophs in the root interior with special attention to the Kallar grass association', Plant and Soil 110, 259-268.

Towbin, H., Staehelin, T., and Gordon, J. (1979) 'Electrophoretic transfer of proteins from polyacrylamide gels to nitrocellulose sheets: Procedure and some applications', Proceedings of the National Academy of Science (USA) 76, 4350-5354.

# Strain-specific salt tolerance and chemotaxis of *Azospirillum brasilense* and their associative N-fixation with finger millet in saline calcareous soil

R. RAI

*Department of Microbiology, N.D. University of Agriculture and Technology, Kumarganj-224 229, (Faizabad) U.P., India*

*Key words:* *Azospirillum brasilense*, chemotaxis, *Eleusine coracana*, nitrogenase activity, root exudates, salt tolerant

## Abstract

Three salt-tolerant *Azospirillum brasilense* strains were isolated from the roots of finger millet grown in saline calcareous soil and characterized. The effect of various salts on growth and $N_2$ase activity of these strains was tested and strain STR1 was found more tolerant at higher concentrations of $Cl^-$, $SO_4^{2-}$ and $HCO_3^-$. Bicarbonate was found to be the most toxic. The content and concentrations of root exudates of finger millet genotypes were different and chemotaxis to sugars, amino acids, organic acids and root exudates was strain specific. Under salt stress, significant interactions between strains and genotypes of finger millet resulted in different responses of $N_2$ase activity, endo- and exorhizospheric population, dry weight of root, shoot and grain yield.

## Introduction

Salinity is the most widespread soil constraint in agriculture and the saline-affected area in India and elsewhere is increasing each year. Excess salts in soil adversely affected the survival, growth, nodulation, N-fixation and legume-*Rhizobium* symbiosis (Rai, 1987). *Azospirillum* is receiving much attention in view of its nitrogen fixing property and enhancing mineral uptake associated with a large number of grasses and grain crops (Döbereiner and Day, 1976; Rai, 1988). There is however, no information available on influence of various salts on growth, $N_2$ase activity in in vitro and associative nitrogen fixation in saline calcareous soil.

Associative nitrogen fixation and plant yield depend on, among other factors, the interaction of genotypes with strains (Rai, 1988). The mechanisms of interactions and factors determining the bacterial-host plant specificity for associative N-fixation are not well known and information concerning interstrain differences of *A. brasilense* is required especially for saline calcareous soil.

Since chemotaxis and its role in the establishment of the legume-*Rhizobium* symbiosis in saline soils is known (Rai, 1987), it is possible that chemotaxis may be an important factor for specific interaction. But the possible interrelationship between root exudates and strain specific chemotaxis and its role on associative N-fixation with finger millet in saline calcareous soil are not known. However, such a study was attempted and results are presented in this paper.

## Materials and methods

Thirty-five isolates of *A. brasilense* were isolated on TSA plates from roots of finger millet (*Eleusine coracana*) grown in saline calcareous soils as method described by Caceres (1982) and characterized using the *A. brasilense* strain Sp7 as a reference strain (Döbereiner and Day, 1976). After preliminary study of salt tolerance, only three strains STR1, STR2 and STR3 were selected for further study.

Effect of NaCl, Na$_2$SO$_4$ and NaHCO$_3$ (0 to 100 meq of Cl$^-$, SO$_4^{2-}$, and HCO$_3^-$) on growth (OD at 560 nm) and N$_2$ase activity of strains was assessed in nitrogen-free malate liquid and nitrogen-free malate with 0.175% agar media (Döbereiner and Day, 1976). For growth study, cultures were grown for 5 days at 32°C with respective salts and their concentrations using five replications at pH 8.5. For N$_2$ase activity, cultures were grown for 5 days at 32°C in 10 mL serum vials containing 7 mL medium with salts at different concentration and C$_2$H$_4$ was estimated (Rai, 1988).

*Chemotaxis*

These strains were grown in NFM liquid medium containing 80, 50 and 30 meq Cl$^-$, SO$_4^{2-}$ and HCO$_3^-$ for 72 h in microaerobic conditions at 32°C and pellet suspended in chemotaxis buffer to a density of 6 × 10 cells mL$^{-1}$. Chemotaxis was assessed by the blindwell chamber method (Rai, 1987). A chemotaxis ratio was obtained by dividing the number of bacteria passing into the test solution (2 m$M$ amino acids, sugars and organic acids) by the number passing into the control solution and the number of bacteria was counted with a haemocytometer.

*Root exudates and chemotaxis ratios*

Root exudates were obtained from 20-day-old seedlings of salt-tolerant genotypes RAU-7 and RAU-2 grown on moist wool over nutrient solution containing normal (control) and 100 meq Cl$^-$ in universal bottles. Cationic, neutral and anionic components of root exudates were determined (Rai, 1987). Amino acids were determined by automatic amino acid analyser. Sugars and organic acids were assessed and identified respectively by methods used by Rai (1987). Chemotaxis ratios for strains in response to root exudates were determined by methods described previously.

*Pot experiment*

To evaluate the response of strains in associative N-fixation, endo- and exorhizospheric populations, plants and root dry weight and grain yield of two genotypes, a pot culture experiment was conducted using 15 kg sterilised saline calcareous soil per pot with uninoculated control. Ten replicates were used for each treatment in a completely randomised design. Soil was sandy loam having pH 8.5; free CaCO$_3$, 31.9%; EC, 6.25 m mhos/cm; organic C, 0.21%, total soluble salts, 2.65 g kg$^{-1}$ soil. Fertilisers were applied at the rate of 20 kg N and 40 kg P$_2$O$_5$ ha$^{-1}$. Surface sterilised seeds were inoculated with each strain using 1 mL culture suspensions prepared in sterilised water and three plants were maintained in each pot. N$_2$ase activity (excised roots), endo- and exorhizospheric populations, root and plant dry weight were determined 35 days after sowing (Rai, 1988). Grain yield was recorded after maturity.

**Results**

Among the salts tested, bicarbonate was found to be the most toxic to strains, followed by sulphate and chloride (Table 1). Only strain STR1 showed reasonable growth and N$_2$ase activity at all concentrations of three salts. Differential response of strains to different concentrations of Cl$^-$, SO$_4^{2-}$ and HCO$_3^-$ was observed. However, greater C$_2$H$_2$ reduction was noted with strain STR1 followed by STR2 and STR3 at different concentrations of various salts.

Table 2 shows the amino acids, sugars and organic acids content of exudates of two genotypes with and without Cl. Analysis of root exudates demonstrated the presence of 8 amino acids, 4 sugars and 4 organic acids. Seedlings of RAU-7 produced a greater weight of amino acids, sugars and organic acids than RAU-2. Due to salt stress there is marked and significant increase in aspartic acid, fructose, malic and citric acids in the exudates of both the genotypes.

The chemotaxis response of three strains to amino acids, sugars, organic acids and root exudates of two genotypes is presented in Table 3. Results explain some of the differential response due to strains and genotypes. Strain STR1 showed positive and significant chemotaxis ratios to malic acid, citric acid (2 m$M$) and root exu-

*Table 1.* Effect of three salts on growth and $N_2$ase activity of salt-tolerant strains grown for 5 days

| Salt (meq) | | Strain | | | | | |
|---|---|---|---|---|---|---|---|
| | | STR1 | | STR2 | | STR3 | |
| | | 0 D | $C_2H_4$[a] | 0 D | $C_2H_4$[a] | 0 D | $C_2H_4$[a] |
| Cl | 0 | 0.35 | 95.2 | 0.30 | 90.9 | 0.25 | 80.3 |
| | 50 | 0.30 | 90.3 | 0.25 | 75.2 | 0.23 | 71.5 |
| | 75 | 0.28 | 85.6 | 0.22 | 60.8 | 0.18 | 56.6 |
| | 100 | 0.25 | 70.9 | 0.15 | 40.3 | 0.13 | 37.9 |
| $SO_4^{2-}$ | 0 | 0.35 | 95.2 | 0.30 | 90.9 | 0.25 | 80.3 |
| | 50 | 0.27 | 80.5 | 0.22 | 60.6 | 0.19 | 61.2 |
| | 75 | 0.23 | 70.3 | 0.15 | 45.2 | 0.13 | 43.5 |
| | 100 | 0.21 | 60.5 | 0.13 | 36.3 | 0.09 | 32.3 |
| $HCO_3$ | 0 | 0.35 | 95.2 | 0.30 | 90.9 | 0.25 | 80.3 |
| | 50 | 0.21 | 60.9 | 0.18 | 56.2 | 0.15 | 51.5 |
| | 75 | 0.18 | 51.6 | 0.13 | 40.0 | 0.10 | 35.7 |
| | 100 | 0.12 | 35.3 | 0.09 | 30.3 | 0.05 | 21.7 |
| LSD ($p = 0.05$) | | 0.019 | 3.051 | 0.013 | 2.615 | 0.011 | 2.069 |

[a] n moles $C_2H_4$ tube$^{-1}$ h$^{-1}$, O = NFM medium as control.

*Table 2.* Composition of root exudates ($\mu$g/plant) of finger millet genotypes grown in nutrient solution for 20 days

| Substance | With Cl$^-$ (100 meq) | | | Control | | |
|---|---|---|---|---|---|---|
| | RAU-7 | : | RAU-2 | RAU-7 | : | RAU-2 |
| Alanine | 2.9 | | 2.6 | 5.2 | | 4.8 |
| Aspartic acid | 8.9 | | 8.2 | 5.6 | | 4.7 |
| Glutamic acid | 4.9 | | 4.3 | 9.2 | | 8.7 |
| Clycine | 3.5 | | 2.9 | 5.6 | | 5.1 |
| Serine | 2.6 | | 3.1 | 4.5 | | 6.2 |
| Valine | 0.6 | | 0.8 | 0.9 | | 1.2 |
| Proline | 1.6 | | 2.1 | 3.5 | | 2.7 |
| Total | 25.1 | | 23.5 | 37.2 | | 35.9 |
| Glucose | 1.7 | | 1.3 | 0.5 | | 0.3 |
| Sucrose | 11.5 | | 10.9 | 12.2 | | 10.5 |
| Fructose | 13.2 | | 9.5 | 7.5 | | 6.0 |
| Maltose | – | | – | 0.9 | | 0.5 |
| Total | 26.4 | | 21.7 | 21.1 | | 17.3 |
| Citric acids | 22.5 | | 20.2 | 11.9 | | 12.6 |
| Succinic acid | 1.2 | | 1.5 | 0.6 | | 0.3 |
| Malic acid | 25.5 | | 23.6 | 18.3 | | 15.5 |
| Isocitrate | 3.9 | | 4.1 | 1.5 | | 1.7 |
| Total | 53.1 | | 49.4 | 32.3 | | 30.1 |
| LSD ($p = 0.05$) | 0.095 | | 0.062 | 0.039 | | 0.048 |

dates of RAU-7 and RAU-2 followed by STR2 and STR3.

The results for the effect of strains and genotypes on $N_2$ase activity, grain yield and other related attributes are presented in Table 4.

Significant interactions and striking differences among strains and genotypes were noted for all the studied characters. Strain STR1 showed more $N_2$ase activity, bacterial population and dry matter yields of roots, shoots and grain with

*Table 3.* Chemotaxis ratios for strains in response to selected amino acids, sugars, organic acids and root exudates of genotypes

| Substance (2 m$M$) | Chemotaxis ratios of strains | | |
|---|---|---|---|
| | STR1 | STR2 | STR3 |
| Aspartic acid | 3.7 | 2.9 | 2.5 |
| Glutamic acid | 3.2 | 3.0 | 2.8 |
| Sucrose | 4.7 | 5.2 | 3.8 |
| Fructose | 6.5 | 6.2 | 5.3 |
| Citric acid | 10.2 | 11.6 | 9.7 |
| Malic acid | 15.6 | 17.1 | 13.5 |
| RAU-7 root exudates | 9.6 | 7.8 | 5.5 |
| RAU-2 root exudates | 6.3 | 4.2 | 3.7 |
| LSD ($p = 0.05$) | 0.017 | 0.025 | 0.013 |

RAU-7 followed by RAU-2 than STR2 and STR3.

## Discussion

The present findings describe the isolation, and selection of salt-tolerant *A. brasilense* strains which are effectively growing and fixing nitrogen under stress conditions. The effect of various salts on growth and $N_2$ase activity of strains was variable and it appears that like most of the *Rhizobium* strains, *A. brasilense* strains are also sensitive to higher concentrations of salts (Rai, 1987). Genetic determinants in STR1 may be responsible for salt tolerance and resulted in more growth and $N_2$ase activity in batch cultures (under stress) or in association with finger millet genotypes in saline calcareous soil (Tables 1 and 4).

Excess production of malic acid, citric acid and aspartic acid in root exudates under salt stress indicates the primary protective mechanisms of finger millet genotypes under salt stress (Table 2). Chemotaxis was found to be strain specific and this was also related with content and concentrations of root exudates. It is thus possible that ability to respond to a wide range of readily diffusing compounds may facilitate the bacterial entry into rhizospheres and to the root surface. Interaction responses of strains and plant genotypes may be related to the chemotaxis because; (a) different strains have different chemotaxis ratios to various substances; (b) the two genotypes differ in their root exudates composition; and (c) the three strains differ in chemotaxis ratios between the finger millet genotypes. The present results also suggest that by selecting *A. brasilense* strains and finger millet genotypes under salt stress it is possible to increase the yield under saline calcareous soil at an economic dose of nitrogen.

*Table 4.* Effect of genotypes and strains on $N_2$ase activity, rhizospheric population, dry weight of plant, root (g/plant) and grain yield (g/pot)

| Treatment | $N_2$ase activity[a] | Endorhizo sphere[b] | Exorhizo sphere[b] | Root dry wt. | Plant dry wt. | Grain wt. |
|---|---|---|---|---|---|---|
| STR 1 + RAU-7 | 64.2 | 3.6 | 10.7 | 2.8 | 7.5 | 9.5 |
| STR 2 + RAU-7 | 55.5 | 2.8 | 8.8 | 2.3 | 6.2 | 7.2 |
| STR 3 + RAU-7 | 47.3 | 2.3 | 6.7 | 2.2 | 5.5 | 6.9 |
| STR 1 + RAU-2 | 57.5 | 3.1 | 9.9 | 2.4 | 6.9 | 8.7 |
| STR 2 + RAU-2 | 43.9 | 2.4 | 7.2 | 2.0 | 5.1 | 6.9 |
| STR 3 + RAU-2 | 39.3 | 1.8 | 5.4 | 1.7 | 4.6 | 6.4 |
| RAU-7 (Uninoculated) | – | – | – | 1.8 | 3.9 | 6.2 |
| RAU-2 (Uninoculated) | – | – | – | 1.5 | 3.5 | 5.7 |
| LSD ($p = 0.05$) | | | | | | |
| Strain | 1.513 | 0.135 | 0.146 | 0.056 | 0.113 | 0.606 |
| Genotype | 1.669 | 0.167 | 0.195 | 0.116 | 0.135 | 0.963 |
| Strain x genotype | 2.013 | 0.187 | 0.231 | 0.159 | 0.107 | 0.985 |

[a] mol $C_2H_4g^{-1}h^{-1}$.

[b] MPN of strain ($\times 10^4$).

## References

Caceres E A R 1982 Improved medium for isolation of *Azospirillum* spp. Appl. Environ. Microbiol. 44, 990–991.

Döbereiner J and Day J M 1976 Associative symbiosis in tropical grasses: Characterization of micro-organisms, and dinitrogen fixation sites. *In* Proceedings of the 1st International symposium on $N_2$-fixation. Eds. W E Newton and C J Nyman. Washington State University Press, Pullman, WA pp 518–538.

Rai R 1988 High-temperature-adapted *Azospirillum brasilense* strains: Growth and interaction response on associative nitrogen fixation, mineral uptake and yield of cheena (*Panicum miliaceum* L.) genotypes in calcareous soil. J. Agric. Sci. 100, 321–329.

Rai R 1987 Chemotaxis of salt-tolerant and sensitive *Rhizobium* strains to root exudates of lentil (*Lens culinaris* L.) genotypes and symbiotic N-fixation, proline content and grain yield in saline calcareous soil. J. Agric. Sci. 108, 25–37.

# NITRATE REDUCTASE ACTIVITY OF *AZOSPIRILLUM BRASILENSE* SP7 AND SP245 V – AND C – FORMS IN CONTINUOUS CULTURE.


J.UECKERT, J.DOEBEREINER[1], I.FENDRIK and E.-G.NIEMANN
Institute of Biophysics
University of Hannover
Herrenhaeuser Str.2
3000 Hannover 21, FRG

[1]EMBRAPA-CNPBS
Seropedica, Km 47
RJ 23851, Brazil



**Abstract.** In alginate beads immobilized c-forms of *Azospirillum brasilense* Sp7 and Sp245 exhibited high nitrate reductase activity in continuous culture under anaerobic conditions. Nitrite excretion of *Azospirillum* strain Sp245 was on a higher level than strain Sp7 by a factor of 3.


## Introduction

It has been observed, that under anaerobic conditions *Azospirillum* spp. are able to dissimilate nitrate and use it as an alternative electron acceptor (Neyra et al., 1977). Some strains of *A.brasilense* reduce nitrate to nitrite only, others further reduce nitrite to nitrogen gas (Scott et al., 1979). In addition nitrate or nitrite are used for assimilation in anerobiosis.
To elucidade the role of bacterial c-forms in infection processes and associations with graminee, c-forms have been induced by entrapment of *Azospirillum brasilense* strains Sp7 and Sp245 in alginate carriers. The nitrate reductase activity of c-forms was measured by determination of nitrite in the culture medium and compared to the data of v-forms in suspended culture. Strain Sp245 was of particular interest, since it has been shown to promote plant growth at certain nitrate levels (Ferreira et al., 1987).

## Materials and Methods

### MICROORGANISMS

*Azospirillum brasilense* Sp7 and Sp245 were both isolated from wheat roots, Sp7 from the ectorhizosphere and Sp245 (provided by Dr. Doebereiner) after root surface sterilization.

## MEDIA AND GROWTH CONDITIONS

The microorganisms were precultured on Nfb–medium with 0.5g/l $NH_4Cl$ and 0,1 g/l yeast extract added.

The media composition for the continuous culture, adapted to the requirement not to dissolve the beads, was as follows $(g \cdot l^{-1})$:  $MgSO_4 \cdot 7\ H_2O$  0.2; NaCl 0.1; $CaCl_2$  0.3; $MnSO_4 \cdot H_2O$  0,01; $Na_2MoO_4 \cdot H_2O$  0.002; $FeSO_4 \cdot 7\ H_2O$  0.01; $Na_5O_{10}P_3$  0.05; MOPS–buffer  0.5; L–malic acid (neutralized by KOH) 1.0; $KNO_3$ 1.0. The temperature was 35°C, oxygen 6 µM and pH 7.5.

Sodium alginate PROTANAL LF 20/60 was obtained from Protan GmbH, Hamburg, FRG.

## PREPARATION OF INOCULATED ALGINATE BEADS

Cells of the bacterial preculture were mixed with a sterile  3% (w/v) sodium–alginate solution and dropped into a 2.5% (w/v) $CaCl_2$–solution, which was gently stirred. Ionotropic gelation takes place immediately and forms beads with 1mm in diameter. After 1h in order to harden the gel the $CaCl_2$–solution was replaced by growth medium and added to the culture vessel.

The gel carriers could be dissolved by addition of 5% (w/v) tripolyphosphate.

## CONTINUOUS CULTURE AND SAMPLING

FIG. 1          Experimental Setup for Continuous Cultivation

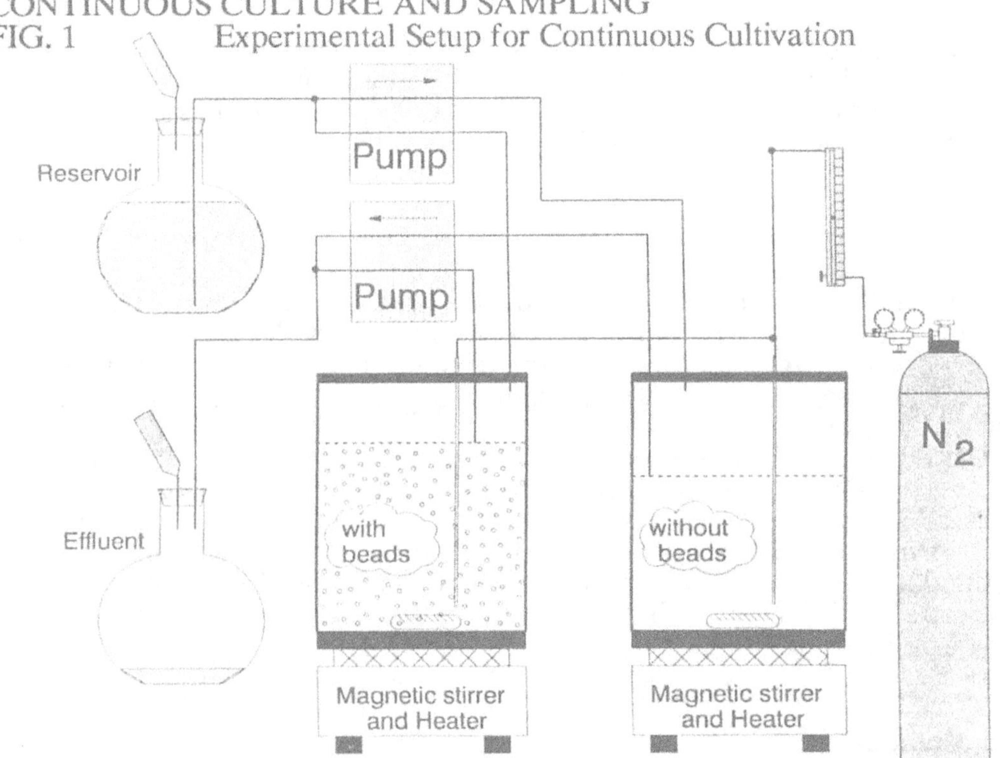


To establish anaerobic conditions in the culture vessels, the medium was flushed with nitrogen gas. The vessel with the immobilized and the one with the suspended culture (inoculated with the same microorganism) were connected to one medium flask and pump to obtain identical conditions. Sampling took place at dilution rates of $0.08h^{-1}$ at 4 volume changes, $0.22h^{-1}$ at 10, and (for washout conditions) of $0.48h^{-1}$ at 10–30 and $0.72h^{-1}$ at 30–38 volume changes. Nitrate was assayed by the method of Cataldo, (1975), nitrite by the test of Snell & Snell, (1949).

## Results

Immobilization of *Azospirillum brasilense* cells immediately induces the formation of sphere–like c–forms, which can be observed by light microscopy after dissolving the gel (Ueckert et al.,1990). Therefore the data display the comparison between entrapped c–forms and suspended v–forms.

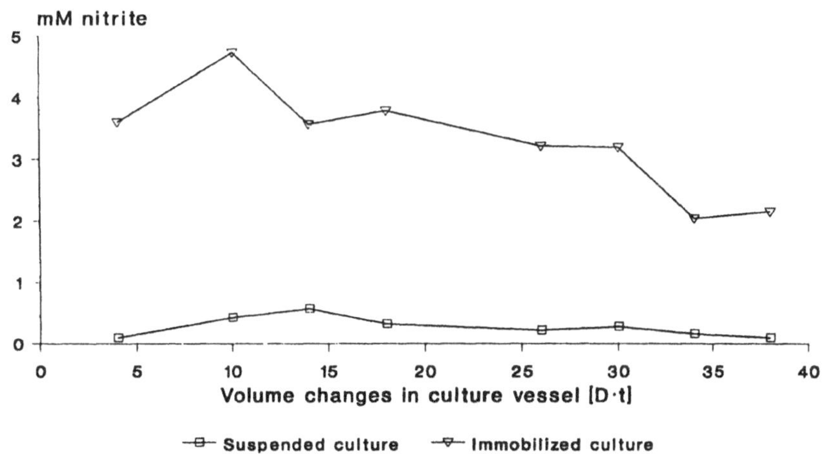



Espccially at a dilution rate of $0.08h^{-1}$ no washout of suspended cells occured and Fig. 2 indicates for strain Sp7 in immobilized culture high and in suspended low nitrite levels. This correlates with the protein concentrations of both cultures.

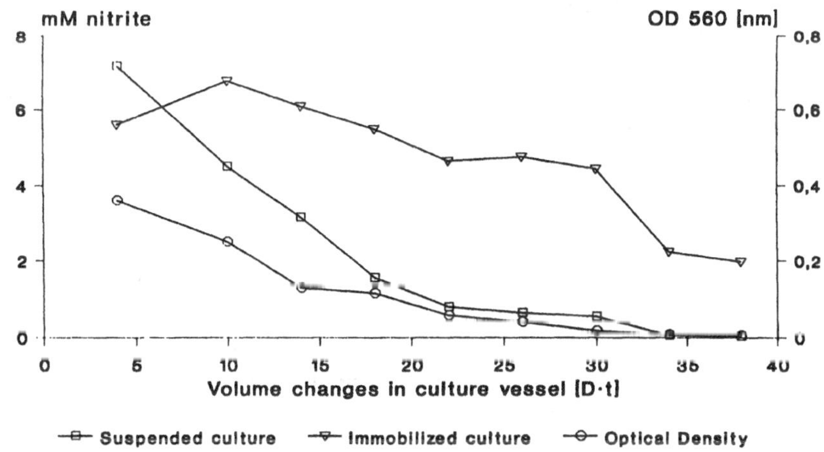

Entrapment allows a high protein load per ml. The values of immobilized and suspended strain Sp245 (Fig. 3) were roughly equal at about the same protein concentration, but on a higher level than Sp7. At dilution rates of $0.48h^{-1}$ and $0.72h^{-1}$ free suspended cells were washed out and the nitrite level in the medium decreased for both organisms, but with a slower rate for strain Sp245. The cysts entrapped in gel beads were resistant to washout conditions. This allows to evaluate their metabolic activity, since all products found in the medium originate from the c-forms.

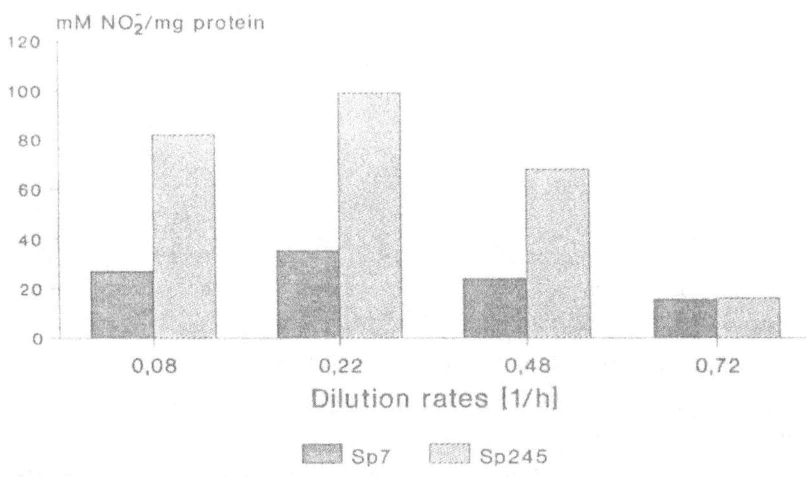


Figs. 2 and 3 show high rates of nitrite exretion, and consequently nitrate reductase activity of c-forms even after 38 volume changes in the culture vessel, especially for strain Sp245. The decline of nitrite concentrations at a dilution rate of $0.72 h^{-1}$ were the result of nitrite washout at constant production rate.

Fig. 4 displays the nitrite release per mg protein of both strains. Obviously the isolate from the endorhizosphere Sp245 excreted more nitrite by factor 3 than the ectorhizosphere isolate Sp7, except at $D= 0.72h^{-1}$. This indicates for strain Sp245 either the lack of a assimilatory nitrite reductase, a higher dissimilatory nitrate reductase activity or a higher induction level of both nitrite reductases.

## Discussion

*Azospirillum* c-forms were once considered resting stages without enzymatic activity (Eskew et al., 1977). In this study the proof of nitrate reductase activity of *Azospirillum brasilense* Sp7 and Sp245 c-forms under anaerobic conditions is given, as high as in vegetative cells. The c-forms seem to be metabolically activ non-motile cells, which may play a bigger role in the root infection process than expected. Due to encystation they probably survive unfavourable conditions better than vegetative cells. *Azospirillum* strain Sp245, isolated from surface sterilized roots, may support plant growth by excreting significant amounts of nitrite in presence of nitrate in the soil. It is postulated that nitrite reacts in the plant cell and forms an auxin-like substance (Zimmer et al. 1988).

## Acknowledgements

This work was funded by the Deutsche Forschungsgemeinschaft (Fe242/1-2) and supported under the Brazilian–German Scientific cooperation by BMFT.

## References

Cataldo, D.A. et al. (1975) Commun.Soil Sci.Plant Analysis 6, 71–80.

Eskew, M. et al. (1977) Appl.Env.Microbiol. 11, 582.85.

Ferreira, M.C.B. et al. (1987) Biol.Fert.Soils 4, 47–53.

Neyra, C.A. et al. (1977) Can.J.Microbiol. 29, 447–57.

Scott, D.B. et al. Arch.Microbiol. 121, 141–45.

Snell and Snell (1949) in: One test tube colorimetric methods of analysis, vol.3, van Nostraid, New York, pp. 804–805.

Ueckert, J. et al. (1990) in Gresshoff, Roth, Stacey and Newton (eds.), Nitrogen Fixation: Achievements and Objectives, Chapman and Hall, New York.

Zimmer, W. et al. (1988) Planta 176: 333–42.

MICROCALORIMETRIC STUDY OF THE GROWTH OF <u>AZOSPIRILLUM BRASILENSE</u>:
RELATION BETWEEN HEAT PRODUCTION AND OTHER GROWTH PARAMETERS

H. VANDENHOVE, R. MERCKX AND K. VLASSAK
Laboratory of Soil Fertility and Soil Biology
K. U. Leuven
Kardinaal Mercierlaan 92
3030 Heverlee
Belgium

ABSTRACT. We have grown <u>Azospirillum brasilense</u> sp. on a synthetic medium with either 5.0, 2.5 or 0.5 g malic acid/l to study the relation between heat output and other growth parameters. Most Pearson correlation coefficients between heat production, bacterial counts, $CO_2$ production, L-malic acid and $NH_4^+$ consumption were significant at the five percent level. Yield factors for cell dry weight production from total energy consumed (metabolized malic acid * heat of combustion of malic acid) were 42.26 µg/J in medium with 5.0 g malic acid/l, 27.97 µg/J in medium with 2.5 and 26.90 µg/J in medium with 0.5 g malic acid/l. Growth on 5.0 g malic acid/l resulted in the loss of 358.6 kJ/mole from the culture per mole of malic acid metabolized; on 2.5 g/l, this loss was 444.8 kJ/mole and on 0.5 g/l, 498.8 kJ/mole were lost per mole of malic acid metabolized. The energy efficiency was highest at the higher malic acid concentrations, ranging from 74 % with 5.0 g/l, over 65 % with 2.5 g/l to 62 % with 0.5 g malic acid/l.


## 1. INTRODUCTION

Apart from the obvious applications for the determination of thermodynamic functions (protein-protein interactions, enzyme-inhibitor reactions and thermal titrations), microcalorimetry has proved most important as an analytical technique. However, the specificity is low. This may restrict the practical use of the method, but for complex biochemical or microbiological processes it may sometimes be advantageous to use a nonspecific analytical method by which an overall value for the process or sum of processes can be obtained. Unknown or unexpected effects may thus be discovered which would not have been revealed by a more specific technique.

Furthermore, many biological and biochemical experiments can be designed in such a way that specific conclusions can be drawn from data obtained with this nonspecific method. For instance, the decrease in heat output when streptomycin or cycloheximide is added to a soil sample, gives an idea of the relative amount of respectively bacteria or fungi in the soil (Anderson and Domsch 1975; Ljungholm et al. 1979). Also calorimetric methods allow continuous registration of life processes; thus, it is possible to study the dynamics of bacterial growth.

Many microcalorimetric studies on microbial growth under anaerobic and aerobic conditions have been carried out.

The pioneering work has been rather qualitative: power-time curves of bacteria grown under growth-limiting conditions ('fingerprints') have been compared and used for bacterial classification (Monk and Wadsö 1975).

A lot of work has been done on the kinetic analyses of the power-time curves of bacterial cultures under anaerobic (Belaich et al. 1968; Belaich and Belaich 1976) and aerobic conditions (Dermoun and

Belaich 1979; Hashimoto and Takahashi 1982; Itoh and Takahashi 1984). Only a few investigators have studied the relation between the shape of the heat output profile and the underlying metabolic reactions (Dermoun and Belaich 1979).

Depending on the type of calorimeter, measurements can either be performed in ampoule mode or in flow-through mode. The ampoule mode is primarily used for calorimetric analysis of soil samples (Sparling 1981; Kimura and Takahashi 1985). Power-time curves, corresponding to the combined effect of several species are determined.

However, working in ampoule mode is very time-consuming (equilibration time) and has the additional disadvantages of possible $O_2$ depletion or $CO_2$ toxification (Mortensen et al. 1973; Ljungholm et al. 1979).

On the other hand, the flow-through mode is completely automatic and does not require any time for equilibration.Moreover, the problems of $O_2$ depletion and $CO_2$ toxification are eliminated (Mortensen et al. 1973; Ljungholm et al. 1979). Another advantage of the flow-through mode is the possibility for the reaction vessel to be outside the calorimeter. Thus, it is possible to measure the effect of adding growth factors or inhibitors to the bacterial culture. Simultaneously, other measurements on the reaction vessel can be performed (bacterial numbers, concentration of the carbon source, etc.). This is impossible in ampoule mode.

The purpose of this paper is to investigate the relation between heat output during aerobic growth of <u>Azospirillum brasilense</u> and growth parameters such as bacterial numbers, $CO_2$ production, L-malic acid and $NH_4^+$-N consumption.

After this characterization of the bacterial cultures, attempts will be made to standardize and optimize these cultures to use them in inoculation experiments. A standardized, reproducible inoculum is expected to decrease the experimental error in inoculation trials.


## 2. MATERIALS AND METHODS

### 2.1. Bacterial Strains

The bacteria used in all the experiments was <u>Azospirillum brasilense</u> a strain isolated from the rhizosphere of spring wheat by C. C. Weniger (Research Institute ITAL, Wageningen, Netherlands).

### 2.2. Medium

The bacteria were grown on a medium containing per liter: 0.6 g of $KH_2PO_4$, 2 g of $K_2HPO_4$, 0.2 g of $MgSO_4.7H_2O$, 0.1 g of NaCl, 4.9 g of KOH, 10 µg of $MnSO_4.H_2O$, 5 µg of $CaCl_2.H_2O$, 23.375 µg Fe.EDTA, 2 µg $Na_2MoO_4.2H_2O$ and either 0.5 g, 2.5 g or 5 g malic acid. The pH was adjusted to 6.8 with KOH.

### 2.3. Culture Conditions

Bacteria were grown in a 10 liter fermentor (Biolafitte) filled with 6 l of the medium. The pH was maintained between 6.75 and 6.85 through regular additions of $H_3PO_4$ (1.5 N). A sufficient $O_2$ supply was ensured by aerating with $CO_2$-free air at a flow rate of 210 l/h. The temperature was kept at 30 °C.

### 2.4. Microcalorimetric Measurements

The microcalorimetric measurements were performed with a flow microcalorimeter (Bioactivity Monitor, LKB 2277). This is a twin-conduction type calorimeter, the properties of which are described by Monk & Wadsö (1968, 1969).

The water bath temperature of the calorimeter was kept at 25 °C and the culture medium was led through the calorimeter by means of a peristaltic pump (LKB 2123 MicroPerpex pump) at a flow rate of

40 ml/h. At this flow rate the lag time between fermentor and measuring point is approximately 5 minutes. After measurement, the culture medium was led into a waste flask.

Before each run, the tubings were thoroughly washed with (1) CHOCl 10 % (20 minutes), (2) distilled water (5 minutes), (3) ethanol 95 % (20 minutes) and (4) distilled water (5 minutes) to avoid contamination. The flow rate was 40 ml/h.

The signal recorded from the microcalorimeter is in $\mu$Watts ($\mu$W) and results in a power-time curve. In order to obtain the heat evolution a graphical integration of the power-time curve had to be performed.

## 2.5. Bacterial Counts

The evolution in bacterial numbers was determined every hour by dilution plating of aliquots from the _Azospirillum_ cultures on malic acid medium. Spread plates were prepared by supplementing the media with 13 g agar (Difco) per liter.

## 2.6. Analytical Methods

2.6.1. $\underline{CO_2\ Production}$. Air samples from the headspace of the fermentor were taken every hour and the $CO_2$ concentration in the samples was measured by gas chromatography (Hewlett Packard 5700A, Porapack column length 2 m, internal diameter 2 mm). Helium was used as carrier gas. Oven and TCD detector temperatures were 60 $^\circ$C and 100 $^\circ$C, respectively.

2.6.2. $\underline{Malic\ Acid\ Concentration}$. Every hour, the malic acid concentration of the culture medium was measured with High Performance Ion Chromatography Exclusion (HPICE) using an Dionex Qic Ion Chromatograph with an HPICE-AS1 column, for the separation and determination of organic acid and a CMMS-1 suppressor. HCl (0.01 N) at a flow rate of 0.8 ml/min was used as eluens and TBAOH (0.05 N) at a flow rate of 2 ml/min as regenerans.

2.6.3. $\underline{NH_4^+\ Content}$. The $NH_4^+$ content in the culture medium was determined colorimetrically by a modified Berthelot reaction, using continuous flow analysis (Skalar 5100).

## 3. RESULTS

Results from the measurements of the heat evolution, the bacterial counts, the malic acid and $NH_4^+$ concentration, the $CO_2$ concentration in the headspace and the cell dry weight are shown in Tables 1 to 3.

For _Azospirillum brasilense_ grown on 5.0 g malic acid/l, heat production continued to increase to a value of 9.45 J/ml, when bacterial numbers were $1.44 * 10^9$/ml (Table 1) and cell dry weight was 1.40 mg/ml. Figure 4 shows the similar evolution of the heat production and the bacterial numbers. A substantial amount of malic acid was still present at that moment, implying that carbon was not the limiting growth factor. $CO_2$ concentration also increased continuously up to a value of 0.123 %. Only after 10 hours' growth there was a detectable uptake of $NH_4^+$; after 16 hours, the $NH_4^+$ concentration decreased to a value of 0.124 mg $NH_4^+$-N/ml.

For _Azospirillum brasilense_ grown on 2.5 g malic acid/l (Table 2), the total heat production (6.83 J/ml) was lower than for growth on 5.0 g/l. There was a final concentration of $4.6 * 10^8$ colony-forming units per ml (cfu/ml) with a total dry weight of 0.57 mg/ml. The concentration of malic acid still present after 16 hours' growth was 0.6952 g/l. The $CO_2$ concentration of the culture headspace increased during the first 12 hours of growth to 0.1355 %, after which it decreased slowly to a concentration of 0.1080 %. After 16 hours' growth, the $NH_4^+$ concentration was 0.173 mg $NH_4^+$-N/ml.

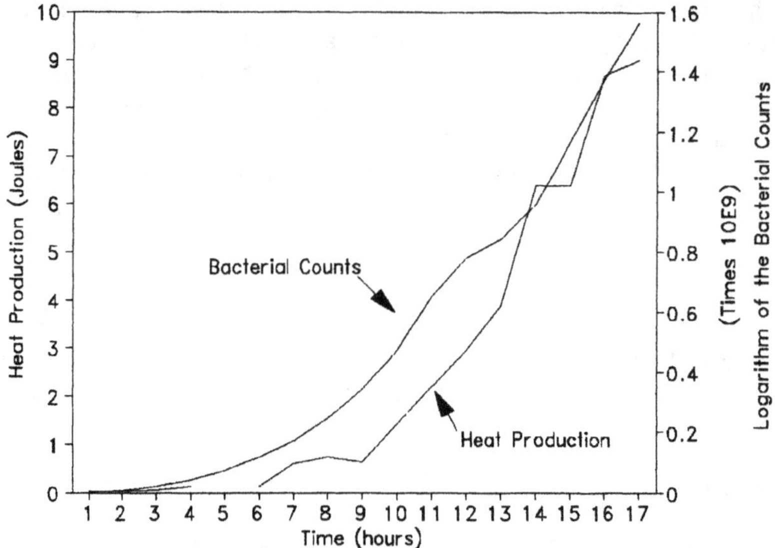


Figure 1.  Power-Time curve and corresponding Heat Production curve for <u>Azospirillum</u> <u>brasilense</u> grown on 5.0 g malic acid/l.

Table 1. Heat Production, Bacterial Counts, Malic Acid Concentration, $CO_2$ Concentration, $NH_4^+$-N Concentration and Cell Dry Weight for <u>Azospirillum</u> <u>brasilense</u> Grown on a Synthetic Medium with 5.0 g Malic Acid/l

| Time (hrs) | Heat Prod. (J) | Bact. Counts (cfu/ml) | Malic Acid (g/l) | $CO_2$ Conc. (V/V %) | $NH_4^+$-N (g/l) | Cell Dry Weight (g/l) |
|---|---|---|---|---|---|---|
| 1 | 0.010 | 4.00e6 | 4.5701 | 0.0150 | 0.252 | . |
| 2 | 0.049 | 4.40e6 | 4.2926 | 0.0161 | 0.242 | . |
| 3 | 0.122 | 8.90e6 | 4.6422 | 0.0175 | 0.210 | . |
| 4 | 0.250 | 1.89e7 | 4.6497 | 0.0219 | 0.236 | 0.000063 |
| 5 | 0.449 | . | . | . | . | . |
| 6 | 0.732 | 2.10e7 | 4.5760 | 0.0457 | 0.244 | 0.00020 |
| 7 | 1.076 | 9.70e7 | 4.5513 | 0.0627 | 0.228 | 0.00016 |
| 8 | 1.527 | 1.19e8 | 4.5298 | 0.0807 | 0.216 | 0.00018 |
| 9 | 2.133 | 1.03e8 | 4.3629 | 0.0856 | 0.228 | 0.00036 |
| 10 | 2.958 | 2.28e8 | 4.0867 | 0.0840 | 0.176 | 0.00041 |
| 11 | 4.080 | 3.50e8 | 3.3822 | 0.1223 | 0.180 | 0.00062 |
| 12 | 4.866 | 4.70e8 | 3.2883 | 0.1147 | 0.160 | 0.00074 |
| 13 | 5.258 | 6.20e8 | 2.6496 | 0.1189 | 0.150 | 0.00093 |
| 14 | 5.993 | 1.02e9 | 2.3051 | 0.1247 | 0.140 | 0.00112 |
| 15 | 7.290 | 1.02e9 | 2.1359 | 0.1100 | 0.122 | 0.00130 |
| 16 | 8.609 | 1.39e9 | 1.8196 | 0.1100 | 0.124 | 0.00142 |
| 17 | 9.485 | 1.44e9 | 1.5219 | 0.1100 | 0.124 | 0.00130 |

Table 2. Heat Production, Bacterial Counts, Malic Acid Concentration, $CO_2$ Concentration, $NH_4^+$-N Concentration and Cell Dry Weight for <u>Azospirillum brasilense</u> Grown on a Synthetic Medium with 2.5 g Malic Acid/l

| Time (hrs) | Heat Prod. (J) | Bact. Counts (cfu/ml) | Malic Acid (g/l) | $CO_2$ Conc. (V/V %) | $NH_4^+$-N (g/l) | Cell Dry Weight (g/l) |
|---|---|---|---|---|---|---|
| 1 | 0.019 | 2.00e6 | 2.6296 | 0.0156 | 0.244 | . |
| 2 | 0.046 | 4.80e6 | 2.6483 | 0.0176 | 0.252 | . |
| 3 | 0.086 | 1.04e7 | 2.6179 | 0.0182 | 0.262 | . |
| 4 | 0.155 | 1.28e7 | 2.6175 | 0.0234 | 0.264 | . |
| 5 | 0.269 | 2.20e7 | 2.6054 | 0.0375 | 0.258 | . |
| 6 | 0.443 | 2.50e7 | 2.5537 | 0.0547 | 0.254 | 0.00009 |
| 7 | 0.685 | . | 2.4478 | 0.0641 | 0.262 | 0.00014 |
| 8 | 1.017 | 7.40e7 | 2.3875 | 0.0891 | 0.258 | 0.00015 |
| 9 | 1.470 | . | 2.2675 | 0.0987 | 0.242 | 0.00020 |
| 10 | 2.086 | . | 2.1273 | 0.1102 | 0.226 | 0.00027 |
| 11 | 2.899 | 6.80e7 | 1.7688 | 0.1311 | 0.198 | 0.00037 |
| 12 | 3.945 | 1.90e8 | 1.3392 | 0.1355 | 0.181 | 0.00053 |
| 13 | 4.669 | 9.60e7 | 1.2260 | 0.1218 | 0.175 | 0.00052 |
| 14 | 5.355 | 2.70e8 | 1.0612 | 0.1162 | 0.168 | 0.00053 |
| 15 | 6.069 | 3.90e8 | 0.8496 | 0.1080 | 0.167 | 0.00056 |
| 16 | 6.833 | 4.60e8 | 0.6952 | 0.1089 | 0.173 | 0.00057 |

Table 3 shows growth parameter data for the 0.5 g malic acid/l medium. Heat production increased only to 2.032 J/ml. $CO_2$ concentration increased up to 0.0634 % after 9 hours' growth, after which it declined. All malic acid was metabolized after 16 hours. Hardly any $NH_4^+$-N was consumed (figures not shown).

Pearson correlation coefficients between the different growth parameters are shown in Tables 4 to 6: correlation coefficients of 0.8 or higher are significant ($p \leq .05$), and only the correlations between the $CO_2$ concentration in the culture atmosphere and the other growth parameters are too low to be significant at the 5 % level.

Table 3. Heat Production, Bacterial Counts, Malic Acid Concentration, $CO_2$ Concentration, $NH_4^+$-N Concentration and Cell Dry Weight for <u>Azospirillum brasilense</u> Grown on a Synthetic Medium with 0.5 g Malic Acid/l

| Time (hrs) | Heat Prod. (J) | Bact. Counts (cfu/ml) | Malic Acid (g/l) | $CO_2$ Conc. (V/V %) | Cell Dry Weight (g/l) |
|---|---|---|---|---|---|
| 1 | 0.008 | 5.80e6 | 0.5704 | 0.0101 | . |
| 2 | 0.026 | 3.80e6 | 0.5613 | 0.0106 | . |
| 3 | 0.059 | 4.80e6 | 0.5444 | 0.0145 | . |
| 4 | 0.112 | 8.70e6 | 0.5224 | 0.0205 | . |
| 5 | 0.198 | 1.45e7 | 0.5233 | 0.0253 | 0.000079 |
| 6 | 0.325 | 1.62e7 | 0.4941 | 0.0338 | 0.000097 |
| 7 | 0.506 | 1.90e7 | 0.4385 | 0.0434 | 0.000082 |
| 8 | 0.777 | 7.20e7 | 0.3529 | 0.0534 | 0.00012 |
| 9 | 1.118 | 7.90e7 | 0.2781 | 0.0634 | 0.000158 |
| 10 | 1.239 | 1.36e8 | 0.2467 | 0.0387 | 0.000147 |
| 11 | 1.346 | 1.14e8 | 0.2129 | 0.0418 | 0.000158 |
| 12 | 1.458 | 1.24e8 | 0.1813 | 0.0352 | 0.000187 |
| 13 | 1.583 | 1.42e8 | 0.1682 | 0.0313 | 0.000189 |
| 14 | 1.717 | 1.55e8 | 0.1238 | 0.0315 | 0.000197 |
| 15 | 1.864 | 1.50e8 | 0.0573 | 0.0315 | 0.000171 |
| 16 | 2.032 | 1.86e8 | 0.0000 | 0.0300 | . |

Table 4. Pearson Correlation Coefficients between the Different Growth Parameters for <u>Azospirillum</u> <u>brasilense</u> Grown on a Synthetic Medium with 5.0 g Malic Acid/l

| | Heat Prod | Bact. Counts | Malic Acid(1) | $CO_2$ | $NH_4^+$-N |
|---|---|---|---|---|---|
| Heat Prod. | 1 | | | | |
| Bact. Counts | 0.963 | 1 | | | |
| Malic Acid(1) | 0.970 | 0.978 | 1 | | |
| $CO_2$ | 0.862 | 0.720 | 0.771 | 1 | |
| $NH_4^+$-N(2) | 0.958 | 0.919 | 0.942 | 0.835 | 1 |

(1) Metabolized malic acid
(2) Metabolized $NH_4^+$-N

Table 5. Pearson Correlation Coefficients between the Different Growth Parameters for _Azospirillum_ _brasilense_ Grown on a Synthetic Medium with 2.5 g Malic Acid/l

|              | Heat Prod | Bact. Counts | Malic Acid(1) | $CO_2$ |
|--------------|-----------|--------------|---------------|--------|
| Heat Prod.   | 1         |              |               |        |
| Bact. Counts | 0.983     | 1            |               |        |
| Malic Acid(1)| 0.998     | 0.982        | 1             |        |
| $CO_2$       | 0.448     | 0.371        | 0.431         | 1      |

(1) Metabolized malic acid


Table 6. Pearson Correlation Coefficients between the Different Growth Parameters for _Azospirillum_ _brasilense_ Grown on a Synthetic Medium with 0.5 g Malic Acid/l

|                 | Heat Prod | Bact. Counts | Malic Acid(1) | $CO_2$ | $NH_4^+$-N |
|-----------------|-----------|--------------|---------------|--------|------------|
| Heat Prod.      | 1         |              |               |        |            |
| Bact. Counts    | 0.919     | 1            |               |        |            |
| Malic Acid(1)   | 0.997     | 0.904        | 1             |        |            |
| $CO_2$          | 0.782     | 0.638        | 0.786         | 1      |            |
| $NH_4^+$-N(2)   | 0.958     | 0.806        | 0.969         | 0.790  | 1          |

(1) Metabolized malic acid
(2) Metabolized $NH_4^+$-N


## 4. DISCUSSION

### 4.1. Shape of the Power-Time Curves

Figures 1, 2 and 3 show the power-time curves and the corresponding heat production curves for _Azospirillum_ _brasilense_ grown on 5.0 g/l, 2.5 g/l and 0.5 g malic acid/l, respectively.

For growth on 5.0 and on 2.5 g malic acid/l, the power-time curve is double-peaked; this is consistent with observations made by Dermoun and Belaich (1979) for growth without energy availability constraints.

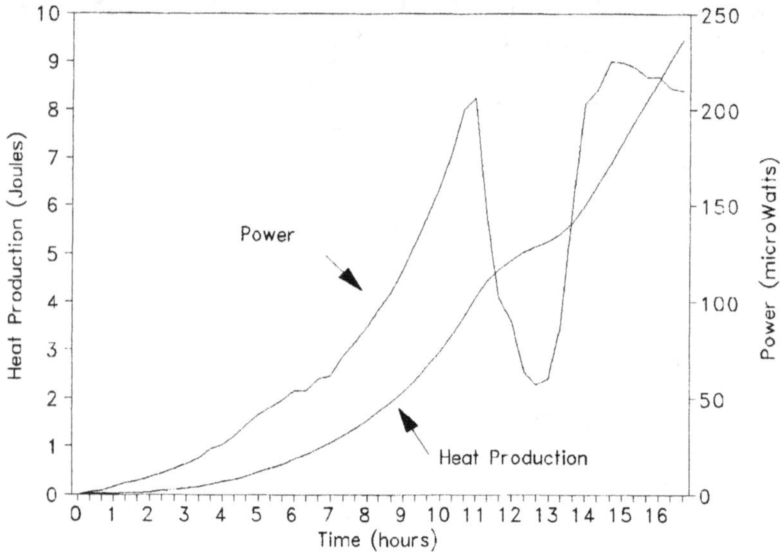


Figure 2.  Power-Time curve and corresponding Heat Production curve for _Azospirillum_ _brasilense_ grown on 2.5 g malic acid/l.

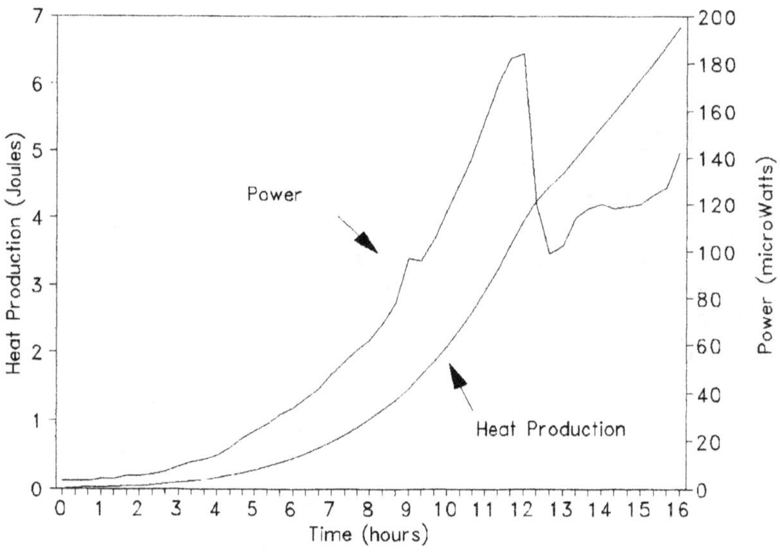


Figure 3.  Power-Time curve and corresponding Heat Production curve for _Azospirillum_ _brasilense_ grown on 0.5 g malic acid/l.

For <u>Azospirillum</u> <u>brasilense</u> grown on 0.5 g malic acid/l, the curve is single-peaked. A similarly shaped power-time curve for energy-limited growth has also been reported by Dermoun and Belaich (1979) for the aerobic growth of <u>Escherichia</u> <u>coli</u> on succinic acid medium.

The maximum of the power-time curve, followed by a decline, corresponds to the moment the bacteria start to multiply less fast; this is especially clear for the growth on 0.5 g malic acid/l.

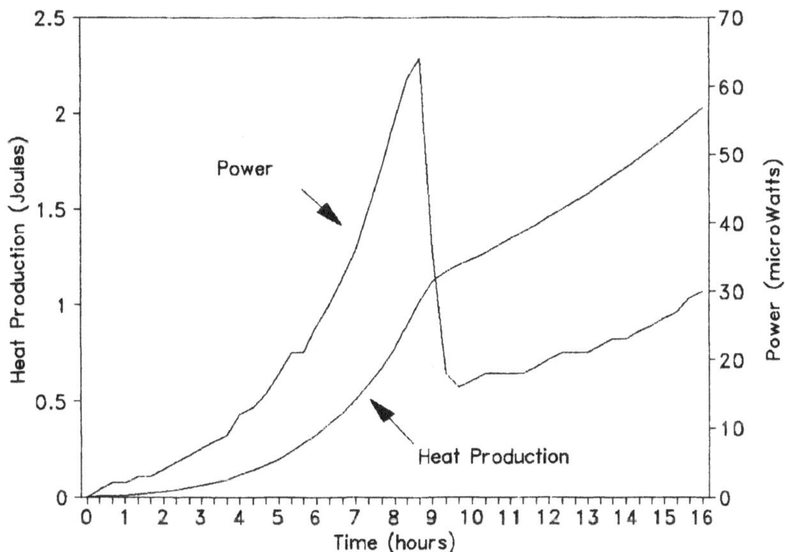


Figure 4.  Evolution of the bacterial counts and heat production for <u>Azospirillum</u> <u>brasilense</u> grown on 5.0 g malic acid/l.

## 4.2. Growth Kinetics

The specific growth rates, which were obtained after logarithmic transformation of the bacterial counts, were 0.396 per hour for growth on 5.0 g malic acid/l, 0.317 per hour for growth on 2.5 and 0.285 per hour for growth on 0.5 g malic acid/l.
The apparent growth rates could be derived from the exponential part of the heat output curves after logarithmic transformation and were 0.342, 0.364 and 0.457 per hour for the growth on 5.0 g/l, 2.5 g/l and 0.5 g malic acid/l, respectively.
The discrepancy between specific and apparent growth rate can be explained by that the apparent growth rate is related not only to the bacterial numbers but also to (1) the amount of malic acid metabolized and (2) the efficiency at which malic acid is metabolized.

## Dissipated Heat Accompanying Formation of a Single Cell and Yield Factors

The average heat production due to the formation of a single cell can be derived from the heat production against bacterial numbers plot. For growth on 5.0 g malic acid/l, the average dissipated heat accompanying formation of a single cell was $6.108 * 10^{-9}$ J/cell; this value was $1.540 * 10^{-8}$ for growth on 2.5 g malic acid/l and $1.077 * 10^{-8}$ J/cell for growth on 0.5 g malic acid/l. These results

264

are in agreement with the results obtained by Kimura and Takahashi (1985), who obtained a figure of $5.53 * 10^{-8}$ J/cell for soil microbes after addition of 12 % glucose to the soil.

Yield factors for cell dry weight production from total energy consumed (metabolized malic acid * heat of combustion of malic acid) were 42.26 µg/J in the medium with 5.0 g malic acid/l, 27.97 µg/J in medium with 2.5 and 26.90 µg/J in medium with 0.5 g malic acid/l. These values correspond fairly well with the average yield factor of 28.23 µg/J for aerobically grown heterotrophic bacteria (Payne 1970). The higher value for <u>Azospirillum brasilense</u> grown on 5.0 g malic acid/l indicates a higher growth efficiency (cell yield per Joule consumed - µg/J) at higher concentrations of malic acid.

### 4.3. $NH_4^+$-N and Malic Acid Assimilation and $CO_2$ Production

The peak in $CO_2$ concentration in the headspace corresponds with the maximum of the power-time curve (Tables 1, 2 and 3). For the cultures with an initial concentration of 0.5 g/l and 2.5 g malic acid/l we found a smaller $CO_2$ concentration after this peak (Tables 2 and 3, Figures 5 and 6), whereas at a concentration of 5.0 g malic acid/l the $CO_2$ concentration in the culture atmosphere remained constant (Table 1, Figure 4).

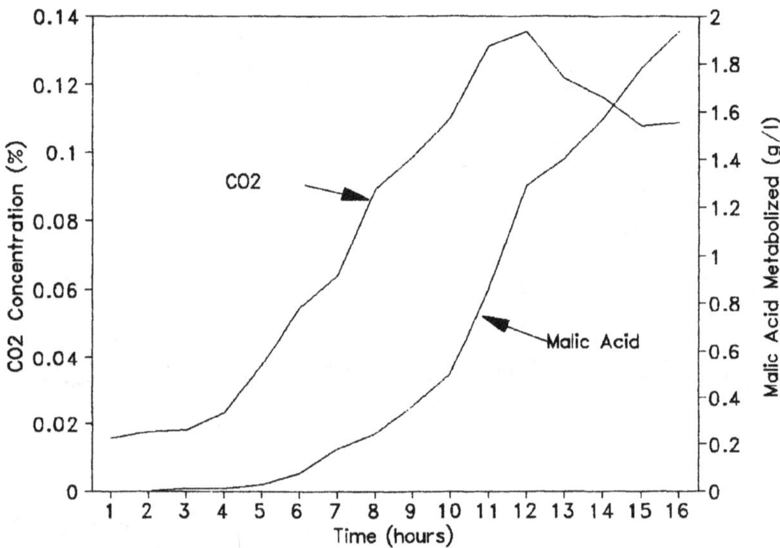


Figure 5. Evolution of the $CO_2$-concentration and the amount of malic acid metabolised for <u>Azospirillum brasilense</u> grown on 2.5 g malic acid/l.

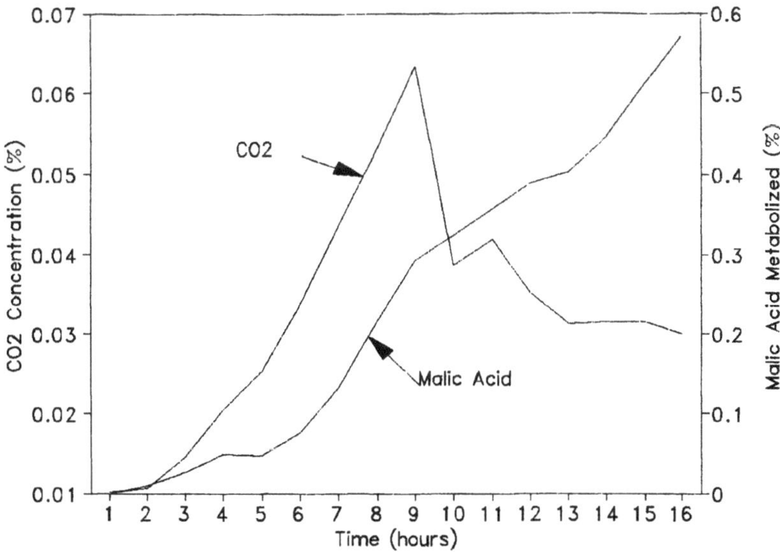



Figure 6. Evolution of the $CO_2$-concentration and the amount of malic acid metabolised for <u>Azospirillum brasilense</u> grown on 0.5 g malic acid/l.

A strong increase in both the amount of malic acid and $NH_4^+$ metabolized occurred after about 7 hours' growth, for the cultures grown on 2.5 g malic acid/l. This increase coincides with an increase in heat production. The simultaneous metabolization of malic acid and $NH_4^+$-N is not surprising, in view of the constant C/N ratio of the bacteria during growth.

The same trend occurs for the malic acid metabolization of bacteria grown on 0.5 g malic acid/l and for the $NH_4^+$-N assimilation for the medium with the highest malic acid content. In this medium, the malic acid metabolization started somewhat later than the increase in heat production.

## 4.4. Energy Efficiency

The average amount of energy dissipated per mole malic acid metabolized can be derived from a plot of heat production against the amount of metabolized malic acid. At a medium concentration of 5.0 g malic acid/l, 358.6 J/mole was lost; corresponding values were 448.8 J/mole for the 2.5 g malic acid/l medium and 498.8 J/mole at malic acid concentrations of 0.5 g malic acid/l.

If we assume that the microcalorimeter measures all the energy lost from the culture as heat and that the disappearance of malic acid from the medium implies biosynthesis, the energy efficiency of biosynthesis can be computed as (A - B) / A , with (A) the heat of combustion of malic acid times the amount of metabolized malic acid (this is a measure of all the energy available for the bacteria) and (B) the calorimetrically determined heat lost from the culture. The results for this equation are shown in Table 7. Irrespective of malic acid concentration, we find the same trend for the evolution of the energy efficiency: a high efficiency at the start of the culture is followed by a decrease and increases again at the end of the experiment to stagnate at a higher level. The lower efficiency when the bacteria are exponentially growing was also found by Westerhoff et al. (1983).

Again we observe a higher efficiency for bacteria growing on 5.0 g malic acid/l. After 16 hours' growth, an efficiency of about 74 % was obtained for bacteria grown on 5.0 g malic acid/l, whereas efficiencies of 65 % and 62 % were obtained for growth on 2.5 g/l and 0.5 g malic acid/l, respectively.

Table 7. Energy Efficiencies for the Growth of _Azospirillum_ _brasilense_ on either 5.0, 2.5 or 0.5 g malic acid/l

| Time (hrs) | 5.0 g/l | 2.5 g/l | 0.5 g/l |
|---|---|---|---|
| 1 | 0.992 | 0.896 | . |
| 2 | 0.987 | . | 0.706 |
| 3 | 0.751 | 0.709 | 0.767 |
| 4 | 0.401 | 0.483 | 0.760 |
| 5 | . | 0.355 | 0.567 |
| 6 | 0.355 | 0.519 | 0.562 |
| 7 | 0.217 | 0.649 | 0.606 |
| 8 | . | 0.599 | 0.633 |
| 9 | 0.335 | 0.603 | 0.607 |
| 10 | 0.498 | 0.588 | 0.607 |
| 11 | 0.680 | 0.661 | 0.613 |
| 12 | 0.644 | 0.690 | 0.614 |
| 13 | 0.735 | 0.663 | 0.595 |
| 14 | 0.742 | 0.653 | 0.604 |
| 15 | 0.707 | 0.653 | 0.627 |
| 16 | 0.692 | 0.640 | 0.634 |

Since air-flow measurements were rather inaccurate, it was impossible to  calculate the amount of $CO_2$ produced per hour; and hence, no conclusions can be drawn about the efficiency of the carbon metabolism. However, some indications about the efficiency of the carbon metabolism can be deduced from Figure 5 and Tables 1 to 3: a high relative difference between the amount of malic acid and the $CO_2$ concentration of the culture atmosphere coincides with a lower efficiency. For the 5.0 g malic acid/l as well as for the 2.5 g/l and 0.5 g/l, the evolution of carbon efficiency is comparable to the evolution of energy efficiency: high at the start, lower when the bacteria are exponentially growing, and again higher at the end  of the experiment.


## 5. CONCLUSIONS

It can be concluded that heat production is significantly ($P \leq .05$) correlated with bacterial numbers, malic acid and $NH_4^+$-N consumption. The correlations between $CO_2$ concentration in the culture atmosphere and the other growth parameters were rather low.

The values for the apparent growth rates, the dissipated heat accompanying formation of a single cell, the heat production per mole malic acid metabolized and the yield factor ($\mu$g/J substrate metabolized) were close to literature data.

The bacteria grown in a medium containing 5.0 g malic acid/l were growing fastest and most efficiently.

Hence, it may be possible to characterize a bacterial culture using microcalorimetrical data.

An advantage of microcalorimetry over other methods to characterize bacterial cultures is the possibility of estimating both the number and the activity of bacteria. Measurements of the ATP content of cells, $CO_2$ production and $O_2$ consumption only give an idea about the activity of the microorganisms; counting bacteria is very tedious and time-consuming. In addition, microcalorimetry is a continuous and nondestructive method. Furthermore, it is easy to determine the influence of different culture conditions (C/N ratio, carbon source, growth factors) on the heat output and hence on the physiological state of the bacteria. Microcalorimetry will facilitate the standardization of bacterial cultures and hence the production of reproducible inocula. With standardized, optimized and reproducible inocula, the variability in inoculation trials due to differences in the physiological conditions of the inocula can be eliminated.


ACKNOWLEDGEMENTS

This investigation was supported by a grant from the I.W.O.N.L. (Instituut voor de Aanmoediging van het Wetenschappelijk Onderzoek in Nijverheid en Landbouw).


REFERENCES

Anderson, J.P.E. and K. Domsch. 1975. Measurements of bacterial and fungal contributions to respiration of selected agricultural and forest soils. Can. J. Microbiol. 21: 314-322.

Belaich, J.P., J.C. Senez and M. Murgier. 1968. Microcalorimetric study of glucose permeation in microbial cells. J. Bacteriol. 95: 1750-1757.

Belaich, A. and J.P. Belaich. 1976. Microcalorimetric study of the anaerobic growth of Escherichia coli: Measurements of the affinity of whole cells for various energy substrates.
J. Bacteriol. 125: 19-24.

Belaich, J.P. 1980. Growth and metabolism in bacteria. In: Biological Microcalorimetry, pp. 1-42. Edited by Beezer, A.E. New York. Academic Press.

Dermoun, Z. and J.P. Belaich. 1979. Microcalorimetric study of Escherichia coli aerobic growth: Kinetic and experimental enthalpy associated with growth on succinic acid. J. Bacteriol. 140: 377-380.

Dermoun, Z. and J.P. Belaich. 1980. Microcalorimetric study of Escherichia coli aerobic growth: Theoretical aspects of growth on succinic acid. J. Bacteriol. 143: 742-746.

Itoh, S. and K. Takahashi. 1984. Calorimetric studies of microbial growth: Kinetic analysis of growth thermograms observed for bakery yeast at various temperatures. Agric. Biolog. Chem. 47: 1281-1288.

Kimura, J. and K. Takahashi. 1985. Calorimetric studies of soil microbes: Quantitative relation between heat evolution during microbial degradation of glucose and changes in microbial activity in soil. J. Gen. Bacteriol. 131: 3083-3089.

Ljungholm, K., B. Norén, R. Skold and I. Wadsö. 1979. Use of microcalorimetry for the characterization of microbial activity in soil. Oikos. 33: 15-23.

Monk, P. and I. Wadsö. 1975. The use of microcalorimetry for bacterial classification. J. Appl. Bacteriol, 38: 71-74.

Monk, P. and I.Wadsö. 1968. A flow micro reaction calorimeter. Acta Chem. Scand. 22: 1842-1852.

Monk, P. and I. Wadsö. 1969. Flow microcalorimetry as an analytical tool in biochemistry and related areas. Acta Chem. Scand. 23: 29-36.

Mortensen, U., B. Norén and I. Wadsö. 1973. Microcalorimetry in the study of the activity of microorganisms. Bull. Ecol. Res. Comm. (Stockholm), 17: 189-197

Payne, W.V. 1970. Energy yields and growth of heterotrophs. Ann. Rev. Microbiol. 24: 17-52.

Sparling, G.P. 1981. Microcalorimetry and other methods to assess biomass and activity in soil. Soil Biol. Biochem. 13: 93-98.

Westerhoff H. V., K. J. Hellingwerf and K. Van Dam. 1983. Thermodynamic efficiency of microbial growth is low but optimal for maximal growth rate. Proc. Natl. Acad. Sci. 80: 305-309.

# Nitrogen fixation in *para*-nodules of wheat roots by introduced free-living diazotrophs


Y.T. TCHAN[1], A.M.M. ZEMAN[2] and I.R. KENNEDY[2]
[1]*The University of Sydney, Department of Chemical Engineering and* [2]*Department of Agricultural Chemistry, Sydney, N.S.W. 2006, Australia*



*Key words:*   Azospirillum, nitrogen fixation, *para*-nodules, wheat, 2,4-D


## Abstract


Nitrogen-fixation ($C_2H_2$-reduction) was demonstrated in wheat root nodules (*p*-nodules) induced by 2,4-dichlorophenoxyacetate (2,4-D) and inoculated with *A. brasilense*. By lowering the $O_2$ tension it was possible to distinguish the nitrogenase activity of bacteria located within the *p*-nodule of the wheat root system from that in the rhizosphere. Using cytological evidence, nitrogenase activity was attributed mainly to be coming from the bacteria within the *p*-nodule. It was also shown that the host plant was able to supply the necessary substrate required for the bacterial $N_2$-fixation ($C_2H_2$-reduction) within the *p*-nodules.


## Introduction

The role of free living diazotrophs in the nitrogen economy has been discussed recently (Tchan, 1988). The difficulties in obtaining a genetically active *nif* gene in non-legumes prevented fast progress in this direction. Alternatively, introducing a free-living diazotroph with its fully equipped *nif* gene into non-legumes would avoid the difficulties involved in genetic manipulation. Giles and Whitehead (1977) reported the introduction of *Azotobacter* into the mycorrhizal system as a means of fixing $N_2$ for a woody plant (Pinus). Its validity, however, has been challenged by Terzachi and Christensen (1986).

Nie et al. (1980) reported that nodule like structures induced by 2,4-D can be inhabited by diazotrophs. However, using rhizobia and *Azotobacter*, Nie et al. did not provide convincing evidence of nitrogen fixation in the living nodule. This work was recently reviewed by Tchan and Kennedy (1989).

The term *p*-nodule (*para*-nodule) was introduced by Tchan and Kennedy to describe the chemically induced nodule since it differs from the naturally occurring legume nodule (Kennedy et al., 1990).

Although the formation of *p*-nodules using 2,4-D could be reproduced (Bender et al., 1990; Tchan and Kennedy, 1989) nitrogenase activity in non-legumes using rhizobia was not detected (Bender et al., 1990).

Similarly, Al-Mallah et al. (1989), Benson et al. (1990), and Cocking et al. (1990) were able to induce nodule formation at a low frequency using enzyme treatment in non-legumes, but like Jing et al. (1990) who used a mutant of *R. sesbania* to induce nodulation on rice roots, obtained only low nitrogenase activity in their systems.

Cellular manipulation to introduce a complete $N_2$-fixing system into non-legumes is of little value to the nitrogen economy (Patriquin, 1982) unless such introduced organisms can express their nitrogenase activity. Significant $N_2$-fixation had not yet been demonstrated in any previous work on induced nodules.

This paper describes positive $N_2$-fixation with *A. brasilense* in the *p*-nodules of the wheat root system.

## Materials and methods

### Bacterial culture, plant host and induction of p-nodules

Wheat seeds (cultivar Miskle) were surface sterilized using 0.5% $HgCl_2$ for 2.5 min. The seeds were germinated on potato-malate agar at 25°C for 2–5 days. Uncontaminated seedlings were transferred and grown in sterile hydroponic solution at 25°–30°C under continuous lighting. When the roots of the wheat plant were 5–7 cm in length, 0.1 mL of a 24 hour culture of *Azospirillum brasilense* containing $10^6$ to $10^7$ cells $mL^{-1}$ and a sufficient quantity of a sterile 2,4-D solution was added to a final concentration of 0.5–1 ppm (2,4-D). After 7 to 10 days, *p*-nodules were well formed on the wheat root system and ready for experimentation (see Fig. 1a).

### Nitrogen fixation ($C_2H_2$ reduction)

The acetylene reduction assay (ARA) was used to test for the $N_2$-fixing capacity of the *p*-nodules using 4–8 plants per flask. The flasks contained 10 mls of a 0.1% glucose solution or when the sugar was omitted, they were exposed to a light source. The $O_2$ tension in the flasks was reduced to that of 10% of air and they were then either shaken or not. A Shimadzu GC8A gas chromatograph fitted with a flame ionisation detector and a 1 metre column of Porapak T (Walters Millipore) was used for measurement of ethylene production.

### Cytological examination

Nodules formed on wheat roots by the 2,4-D/ *Azospirillum* system were incubated with iodonitrotetrazolium (INT) or 2,3,5 triphenyltetrazolium (TPT) to locate sites of strong reduction. Such sites were detached, crushed or sectioned and examined using light phase micro-scopy. Sites on the root that showed no visible reduction were similarly examined.

## Results and discussion

When the 2,4-D/*Azospirillum* wheat root system was first tested for ethylene production, only one plant per flask was used and no change in the flask's atmosphere was made, producing negative or erratic results. This led to two proposals: (1) that the normal $O_2$ tension of the air was too high for the nitrogenase of *A. brasilense* to operate efficiently and (2) that a single seedling may not provide adequate ethylene production for analysis. It was also necessary to distinguish ethylene produced by the azospirilla inside the plant root system including the *p*-nodule from the azospirilla located in the rhizosphere.

By shaking the flasks during the ARA incubation period and reducing the $O_2$ tension to 10% of air, the nitrogenase activity attributable to the azospirilla in the rhizosphere was greatly reduced. Yet under these conditions, the expression of nitrogenase inside the *p*-nodules was protected by the plant tissue. The number of plants tested in the ARA was also increased from one to 4 to 8 seedlings per flask, providing a more accurate estimation of ethylene production. The 0.1% glucose solution used in the flasks was to ensure that nitrogenase activity was not limited by the possible lack of energy supply during our assay. The suppression of nitrogenase activity in the shaken flask compared to the unshaken flask was demonstrated using a culture of *A. brasilense* (Fig. 2).

The shaken flask gave a value of 1.9 nmoles $h^{-1}$ of ethylene compared to the unshaken flask with 24.3 nmoles $h^{-1}$.

By comparing the ARA results of plant roots not inoculated with 2,4-D under shaken and unshaken conditions, a small amount of ethylene was produced by the azospirilla residing within the root system. In the presence of 2,4-D, however, the $N_2$-fixation capacity of the shaken flasks increased five fold indicating that only the azospirilla protected from oxygen within the plant cells were able to express their nitrogenase activity. Furthermore, the presence of 2,4-D

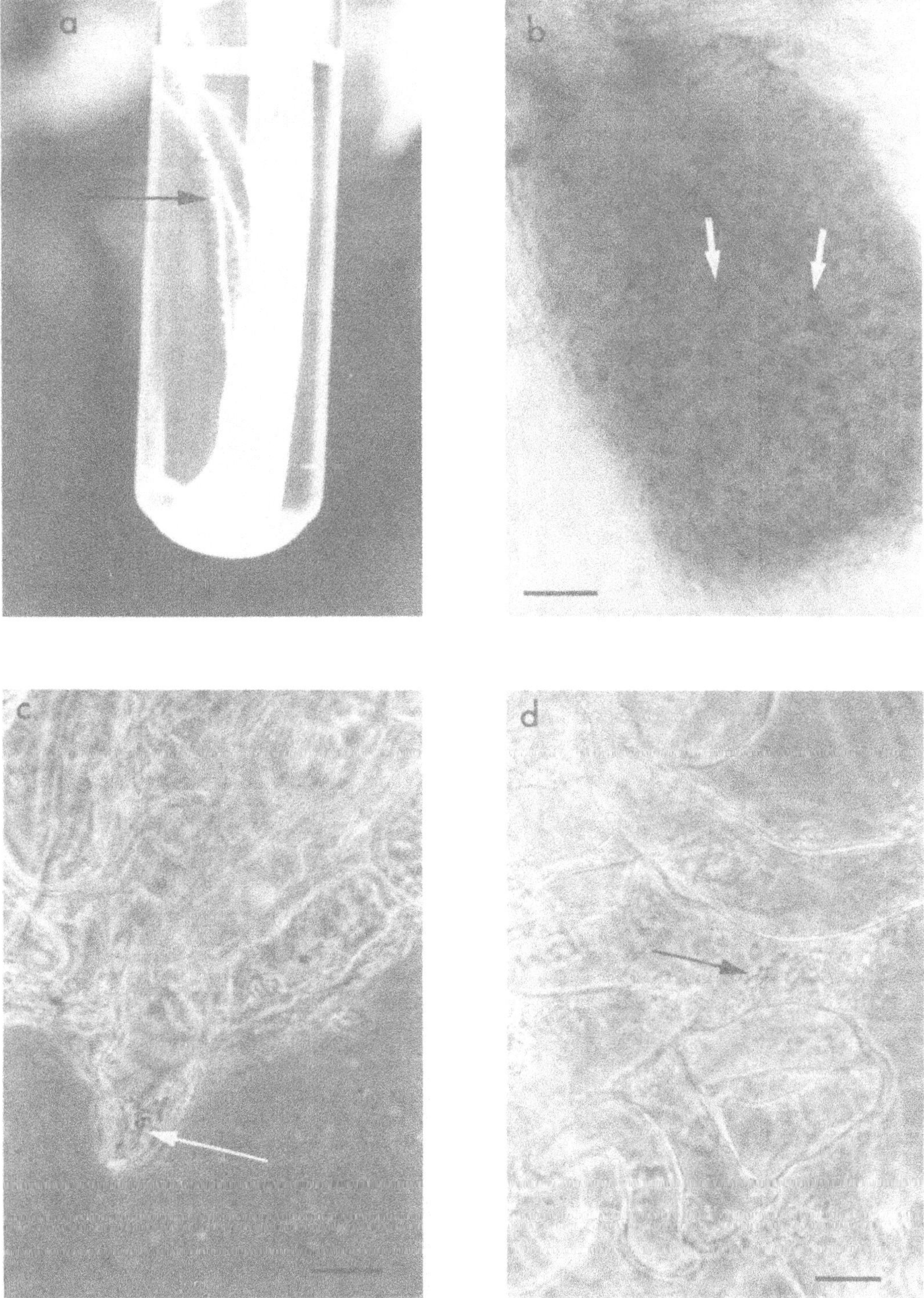

*Fig. 1.* **a**, *p*-nodules on wheat root induced by 2.4-D: **b**, *p*-nodule stained with INT showing reduction sites (arrow) within nodule cells (*bar* = 250 $\mu$m): **c**, Azospirilla within cells of nodule (*bar* = 20 $\mu$m) and **d**, Azospirilla located intercellularly (bar = 20 $\mu$m).

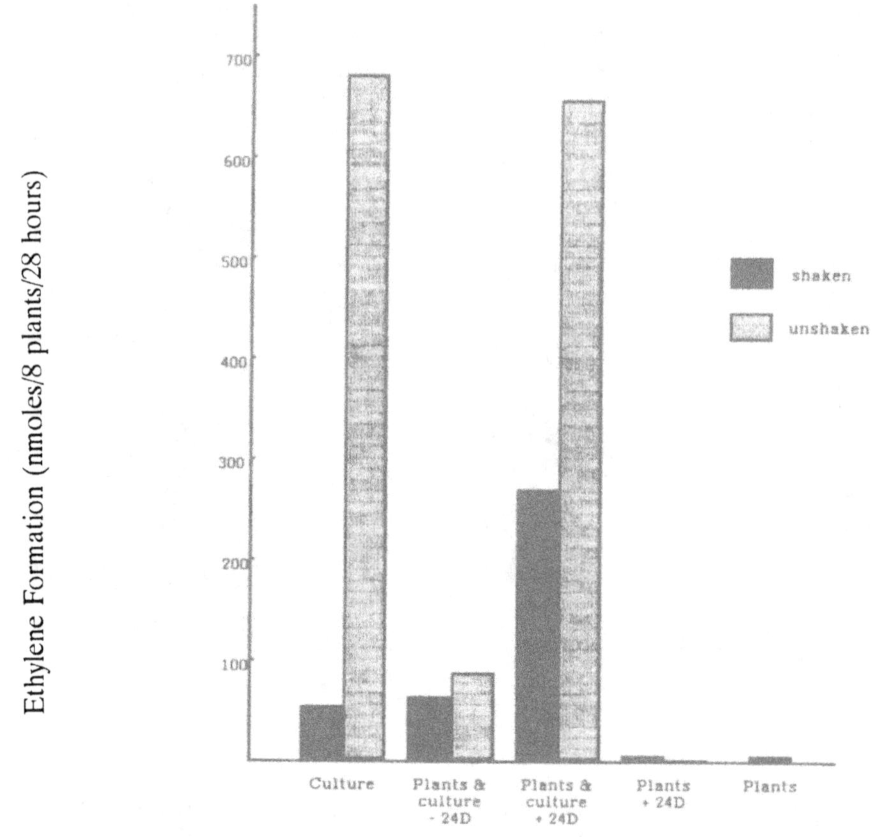



*Fig. 2.* N$_2$ fixation (C$_2$H$_2$-reduction) by *Azospirillum brasilense* in *p*-nodules of wheat plant roots.

considerably enhanced the nitrogenase activity of azospirilla within the rhizosphere.

Cytological investigations revealed that there were more azospirilla present within the *p*-nodule than in the root system. This indicates that the major part of ethylene production came from the azospirilla activity within the *p*-nodules (Fig. 1 b,c,d).

The data strongly suggested that *A. brasilense* in the *p*-nodule was potentially capable of fixing N$_2$ provided that adequate requirements and environmental conditions were fulfilled. Such potential for N$_2$-fixation from the 2,4-D treatment would have a greater significance if the required energy could be supplied by the host plant. This was tested by keeping the wheat seedlings in the dark for 18 hours prior to the ARA and detaching their seeds to reduce the supply of substrate to a minimum. The glucose in the solution was eliminated and the shaken flasks were exposed to a light source. Flasks containing plants not treated with 2,4-D were used as controls (Fig. 3).

The results showed that wheat plants with *p*-nodules (+2,4-D) produced 9.3 nmoles of ethylene per plant per hour compared to 2.3 nmoles of ethylene obtained in plants not treated with 2,4-D. This indicated that the host plant was capable of supplying the necessary substrate for N$_2$-fixation (C$_2$H$_2$-reduction) of the bacteria residing mainly in the *p*-nodule.

The demonstration of significant N$_2$-fixation in the wheat plant using the 2,4-D/*Azospirillum* system is an important development which may open new fields for both basic and applied aspects of N$_2$-fixation.

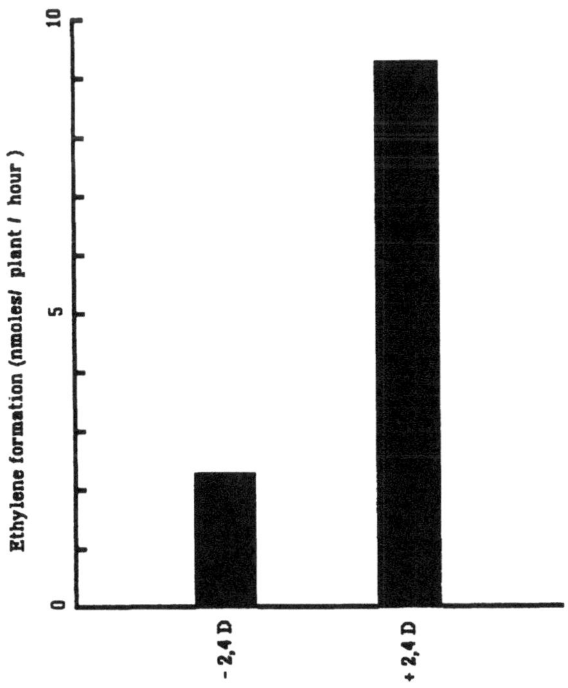

*Fig. 3.* N$_2$-fixation (C$_2$H$_2$-reduction) by *Azospirillum brasilense* in *p*-nodules of wheat plants using energy supplied by the host plant.

## Acknowledgements

We wish to acknowledge the financial assistance provided by the Rural Credits Development Fund (Reserve Bank of Australia) and the Wheat Research Council (Australia).

## References

Al-Mallah M K, Davey M R and Cocking E C 1989 Formation of nodular structures on rice seedlings by rhizobia. J. Exp. Bot. 40, 473–478.

Bender G L, Preston L, Bernard D and Rolfe B G 1990 Formation of nodule-like structures on the roots of the non-legumes rice and wheat. *In* Nitrogen Fixation: Achievements and Objectives. Eds. P M Gresshoff, E C Roth, G Stacey and W E Newton. Chapman and Hall, New York.

Benson E E, Al-Mallah M K, Davey M R and Cocking E C 1990 Enzyme-induced nodulation of *Brassica napus* by *Rhizobium. In* Nitrogen Fixation: Achievements and Objectives. Eds. P M Gresshoff, E C Roth, G Stacey and W E Newton. Chapman and Hall, New York.

Cocking E C, Al-Mallah M K, Benson F and Davey M R 1990 Nodulation of non-legumes by rhizobia. *In* Nitrogen Fixation: Achievements and Objectives. Eds. P M Gresshoff, E C Roth, G Stacey and W E Newton. Chapman and Hall, New York.

Giles K L and Whitehead H C M 1977 Re-association of a modified mycorrhiza with the host plant roots (*Pinus radiata*) and the transfer of acetylene reduction activity. Plant and Soil 48, 143–152.

Jing Y X, Li G S, Shan X Q and Li J G 1990 Rice root nodule with nitrogenase and hemoglobin. *In* Nitrogen Fixation: Achievements and Objectives. Eds. P M Gresshoff, E C Roth, G Stacey and W E Newton. Chapman and Hall, New York.

Kennedy J R, Zeman A, Tchan Y T, New P B, Sriskandarajah S and Nie Y F 1990 Biological nitrogen fixation and prospects for yield increases in wheat. Trans. 14th Int. Congr. Soil Sci. (Kyoto). III, 146.

Patiquin D G 1982 New developments in grass-bacteria associations. *In* Advances in Agricultural Microbiology. Ed. N S Subba Rao. pp 139–190. Butterworths Scientific, London.

Tchan Y T 1988 Some aspects of non-rhizobial diazotrophs: Their past and their future. *In* Microbiology in Action. Eds. W G Murrell and I R Kennedy. pp 193–208. Research Studies Press, John Wiley and Sons, Chichester.

Tchan Y T and Kennedy I R 1989 Possible N$_2$-fixing root nodule induced in non-legumes. Agric. Sci. (AIAS, Melbourne) 2, 57–59.

Terzachi B E and Christensen M J 1986 Should the transfer of nitrogen-fixing ability to an eukaryotic cell be reconsidered?: Evidence for non-basidiomycete fungal contamination. J. Plant Physiol. 122, 275–283.

# FIREFLY LUCIFERASE EXPRESSION ON NITROGEN FIXATION WITH NON-LEGUMES


A. J. PALOMARES, A. CEBOLLA, M. A. CAVIEDES, B. SANCHEZ, D. N. RODRIGUEZ, J. A. MUÑOZ, C. CORONADO AND F. RUIZ-BERRAQUERO
*Department of Microbiology and Parasitology*
*Faculty of Pharmacy. University of Seville.41012 Seville. Spain.*
*Ph. 34-54-628355. Fax. 34-54-233765.*



ABSTRACT. We have tested the firefly luciferase expression under *Rhizobium meliloti* symbiotic promoters control in nitrogen fixation bacteria non associated with legumes. Constitutive expression of *luc* gene was determined in the same bacteria using a fusion to $\lambda_{PR}$ promoter. Luciferase activity was found in all nitrogen fixer backgrowns but it depended of promoter activation conditions. Bioluminiscence was high enough in bacteria containing $\lambda_{PR}$-*luc* fusion to be observed by a dark adapted eye and photographed.


## 1. Introduction

The firefly luciferase gene isolated from a cDNA library from *Photinus pyralis* encodes an enzymatically active polypeptide with an apparent molecular mass of 62 kDa (1). The bioluminiscence is catalyzed by the enzyme luciferase which oxidizes a heterocyclic carboxilic acid (D(-)luciferin) in the presence of ATP, $Mg^{++}$ and oxygen. In the first step, luciferase (Enz) forms a tightly bound luciferyl-adenilate (Enz $LH_2$-AMP) which then reacts with molecular oxygen in a subsequent reaction. During the process, luciferin is decarboxylated and $CO_2$ is released. The product oxyluciferin is initially produced in an electronically excited state and a photon is emitted as it decays to the ground state. The quantum yield of the reaction is 0.88 with respect to $LH_2$, making this the most efficient bioluminiscent reaction known to man. The luciferase has specific requirement for ATP which makes it possible to perform assays in the presence of other nucleoside di-or triphosphates without any significant interference.

$$LH_2 + Enz + ATP \xrightarrow{\quad Mg^{++} \quad} Enz\ LH_2\text{-AMP} + PPi$$

$$Enz\ LH_2\text{-AMP} + O_2 \longrightarrow oxyluciferin^* + CO_2 + AMP$$

$$oxyluciferin^* \longrightarrow oxyluciferin + h\nu$$

Because of the emitted light can be measured with high sensitivity, luciferase activity has become a useful reporter for measurement gene expression. In addition, the assay is rapid, relatively inexpensive and requires no radioactive materials. Endogenous luminescence is undetectable in normal cells and moreover, detection can be achieved *in vivo* by several methods (2).

In this work we show the constitutive expression of the firefly luciferase gene driven by lambda rightward promoter in some nitrogen fixers bacteria non associated with legumes (*Azospirillum brasilense, Klebsiella pneumoniae* and *Azotobacter vinelandii* ). We have also estudied the *Rhizobium meliloti* simbiotic nitrogen fixation promoters expressed in free living diazotroph bacteria background using the firefly luciferase gene as a reporter marker. With this purpose we have constructed plasmids harbouring *nifA-luc, nifH-luc* and *fixA-luc* translational fusions, respectively.

## 2. Material and Methods

The following bacterial strains were employed: *Escherichia coli* HB101, *Azospirillum brasilense* Sp245, *A.brasilense* cd, *A. brasilense* Sp7, *Azotobacter vinelandii* UW and *Klebsiella pneumoniae* M5A1. Spontaneous nalidixic acid resistant mutants were isolated in all cases.The following plasmids were used: pRK293 (3), pKW101 (1), pMB210 (4), pMB211 (4), pRK2013 (5), pRK2073 (5) and pCHK57 (6). *E. coli, Azospirillum* and *Klebsiella* strains were routinely grown in LB medium (7). *A. vinelandii* was grown in C medium (0.8 m$M$ MgSO$_4$.7 H$_2$O, 1.7 m$M$ NaCl, 0.25% Yeast extract, 0.05% casaminoacids, 1% Mannitol). When required, antibiotics were used at the following concentrations (micrograms per milliliter): Tetracycline, 10; Kanamycin, 25; Nalidixic acid, 10; Spectinomycin, 80 and rifampicin, 25.

Restriction enzymes, T4 DNA ligase and Klenow polymerase, were purchased from Böehringer Mannheim Biochemicals and were used according to manufacturers instructions. Luciferin purified from *P. pyralis* was purchased from Sigma and Böehringer Mannheim Biochemicals.

Plasmid DNA was isolated by the alkaline-SDS method of Birnboim and Doly (8), and further purified by CsCl -ethidium bromide density gradients (9). Patch matings were performed according to Palomares *et al.*(10). Plasmids pRK2073 or pRK2013 were used as helper plasmids to mobilize transfer defective derivatives of RK2. Transconjugants were selected onto appropriate antibiotic selective medium. Luciferase expression was assayed by two methods: use of X-ray film to detect light emission by colonies, and the use a LKB luminometer equipped with a chart recorder for monitoring the time course of light emission by whole cells and extracts (10). Microaerobic and nitrogen starvation studies were performed as described elsewhere (6).

## 3. Results

### 3.1. CONSTRUCTION OF *RHIZOBIUM MELILOTI* SYMBIOTIC PROMOTER *-LUC* FUSIONS ON A BROAD HOST RANGE PLASMID VECTOR

A *Bam* HI fragment from pKW101 containing the firefly luciferase gene (*luc* ) was subcloned into the corresponding *Bam* HI site of pCHK57, pMB210 and pMB211, which contain in frame *lacZ* fusions to the 22th, 29th and 11th first codons of the *R. meliloti nifA, nifH* and *fixA*, respectively. *Hind*III fragments from the resultant plasmids containing out of frame *luc* fusions to these genes were isolated and cloned into the *Hind*III site of the broad-host-range plasmid pRK293; screening for insertionsal inactivation of kanamycin resistance was made.

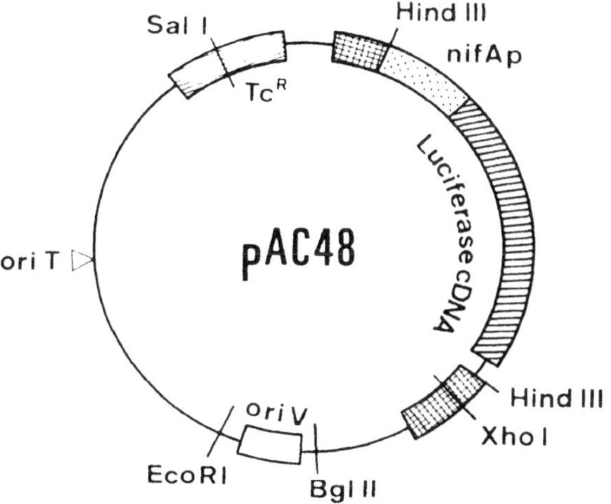

Figure 1. Structure of pAC48. The *luc* gene is indicated by the hatched box. *nifA* promoter is indicated with dots.

Since the DNA sequences of *nifA, nifH, fixA* and *luc* fragment and their 3' and 5' flanking regions were known (4,12, Wood, personal communication), digestion of the resultant plasmid with *Bam*HI followed by treatment with Klenow enzyme, and ligation of the blunt ends creates fusions to *luc* in the correct translational frame. Plasmid finally obtained and used for gene expression studies were pAC48, pAC87 and pAC88, which harbour in frame *nifA-luc, nifH-luc* and *fixA-luc* translational fusions, respectively. A scheme for the plasmid with a translational fusion between *nifA* promoter and *luc* sequence is presented in Figure 1. Accomplished fusions are schemed in Figure 2.

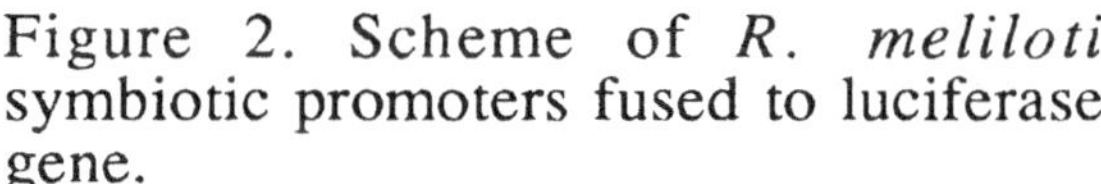

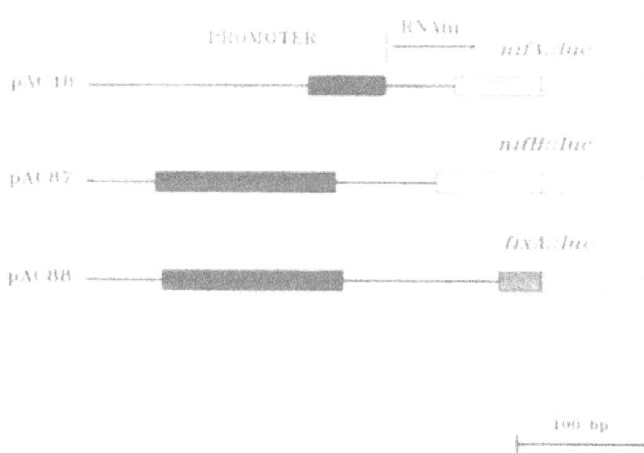

Figure 2. Scheme of *R. meliloti* symbiotic promoters fused to luciferase gene.

## 3.2. CONSTRUCTION OF λPR-LUCIFERASE ENCODING PLASMID

The *Hind*III λPR-*luc* DNA segment of pKW101 with the cI857 gene deleted, was inserted into the broad-host-range Inc P1 plasmid pRK293. The resulting plasmid, pAP2 (Fig. 3) contains, in addition to a functional *luc* gene the Inc P1 plasmid origin of replication, the origin of conjugal transfer and tetracycline resistance gene.

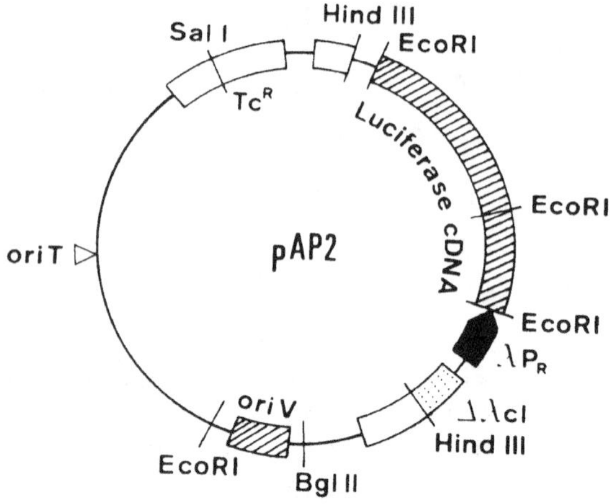

Figure 3. Structure of pAP2 plasmid. The *luc* gene is indicated by the hatched box. The location and direction of the $\lambda_{PR}$ promoter is indicated by a arrowed box.

## 3.3. *R.MELILOTI* SYMBIOTIC PROMOTERS LUCIFERASE EXPRESSION

Plasmids pAC48, pAC87 and pAC88 were established in *K. pneumoniae, A.vinelandii* and *A. brasilense.* Bacteria carrying plasmids were selected on the basis of tetracycline and rifampicin (*Klebsiella*) or nalidixic acid (*Azotobacter, Azospirillum*) resistances. Subsequently screening for luciferase activity by exposure to photographic film and luminometer measurements were effected. As shows in Table 1, expression of luciferase from *nifH* promoter was quite increased in nitrogen starvation condition in all bacteria except to *A. vinelandii.* The promoter of the exclusive symbiotic gene, *fixA*, only present activation in *k. pneumoniae* in nitrogen free medium as *R. meliloti.* The levels of luciferase expression from *nifA* promoter always were higher in nitrogen starvation than in rich medium. In contrast to *Rhizobium*, no significative activation was found under microaerobic conditions. Also, no luciferase expresion could be noted after photographic development in the cases of *Klebsiella, Azotobacter* and *Azospirillum* opposite to *R. meliloti*, where cells expresing luciferase were readily detected by 30 min of contact exposure to Ektachrome 200 film.

TABLE 1. Luciferase activity in relative light units/$OD_{600}$ of *R. meliloti* symbiotic promoters in nitrogen fixation bacteria*

| | *Klebsiella* | | | *Azotobacter* | | | *Azospirillum* | | | *Rhizobium* | | |
|---|---|---|---|---|---|---|---|---|---|---|---|---|
| | Aer. | Mic. | -NH$_4$ | Aer. | Mic.- | NH$_4$ | Aer. | Mic. | -NH$_4$ | Aer. | Mic. | -NH$_4$ |
| *ΦnifA-luc* | 104 | 77 | 218 | 8 | ND | 58 | 6 | ND | 83 | 920 | 6100 | 2600 |
| *ΦnifH-luc* | 281 | 316 | 5757 | 151 | ND | 328 | 10 | ND | 246 | 1380 | 24200 | 11400 |
| *ΦfixA-luc* | 5 | ND | 201 | 3 | ND | 0 | 5 | ND | 0 | 650 | 660 | 2880 |

* Average data of at least five experiments, referred to extracts determination. ND: not determined

## 3.4. EXPRESSION OF LUCIFERASE DRIVEN BY $\lambda_{PR}$ PROMOTER ON NITROGEN FIXATION BACTERIA ASSOCIATED WITH NON LEGUMES

Plasmid pAP2 was established in nitrogen fixation bacteria associated with non legumes by conjugal mating. As shows in Figure 4, colonies of *K. pneumoniae* (as representative) expressing luciferase are readily detected by a short exposure to X-ray film after blotting onto filter paper and soaking the blot with a luciferin solution. Luminometer measurements of luciferase activity were carried out with cell extracts prepared by sonication and intact cells (Table 2). It is clear that the level of luciferase activity is considerably higher in *K. pneumoniae* than in other nitrogen fixing bacteria examined.

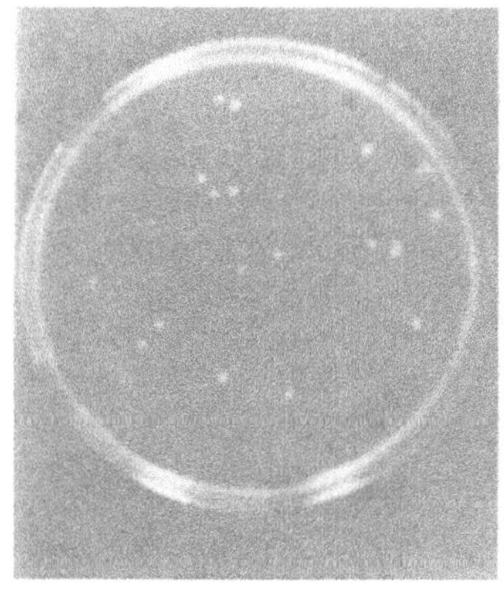 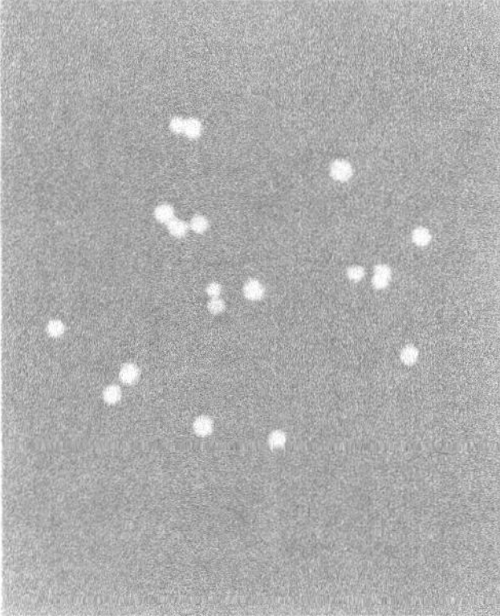

Figure 4. Luciferase activity in *K. pneumonia* colonies. Left, bacteria were grown in LB plates. Right, the same colonies were blotted onto Whatman paper, wetted with luciferin and exposed to X-ray film.

Surprisingly, data referred to luciferase activity in *A. vinelandii* were obviously different between intact cells and extract determinations. In *Azospirillum brasilense* the behaviour of pAP2 was quite similar in all of the strains examined, nevertheles the light units were clearly fewer than in *Klebsiella* and *Azotobacter*, but high enough to be detected with a naked eye (data not shown).

TABLE 2. pAP2 activity in nitrogen fixation bacteria non associated with legumes

| Bacteria | Activity (light units) / OD$_{600}$ | |
| --- | --- | --- |
| | Intact cells | Extracts* |
| *K. pneumoniae* M5A1 | 25.728 | 181.661 |
| *A. vinelandii* UW | 1.104 | 34.232 |
| *A. brasilense* Sp245 | 424 | 1.767 |
| *A. brasilense* cd | 480 | 900 |
| *A. brasilense* Sp7 | 600 | 590 |

* Extracts were prepared from mid-log cells and sonicated during 30 seconds, except for *A. vinelandii* (2 minutes)

## 4. Discussion

The firefly luciferase gene (*luc*) is a reporter gene wich can be used to visualize gene expression in procariotic, lower eucariotic, plant and mammalian organisms (2). The most appealing feature about firefly luciferase is that the gene product can be detected by non destructive methods, it does mean *in vivo*. These studies were undertaken to determine in nitrogen fixation bacteria, bearing the firefly luciferase gene linked to $\lambda_{PR}$ or *R. meliloti* symbiotic promoter, whether the pattern of light emitted was a reflection of the specific expression of procariotic promoters. The $\lambda_{PR}$-*luc* fusion, previously constructed, was used for measuring gene expression in a variety of nitrogen fixation bacteria non associated with legumes. It is of interest that $\lambda_{PR}$ promoter works in all nitrogen fixation bacteria tested, being outstanding in the case of *K. pneumoniae*.Significantly lower activity was observed in *Azospirillum* strains. These differences may indeed reflect a wide range of $\lambda_{PR}$ activity levels according to the host.

We also have examined the patterns of *luc* expression driven by the *R. meliloti nifA, nifH* and *fixA* promoters in *Klebsiella, Azospirillum* and *Azotobacter* backgrounds. Opposite to *Rhizobium* no significative oxigen tension influence was observed in bacteria studied (6, 11, 12).We were able to detect increase of luciferase activity when free nitrogen conditions were used. These data support the idea that the *R. meliloti nif* promoters are controlled by N-status, maybe *ntrC* mediated, in these bacteria (4, 13, 14).

## Acknowledgments

We thank  J. Dobereiner, Y. Okon and J. Casadesús for providing us with bacteria. This work was supported by the Spanish DGICYT ( Grant  PB 87-0918).

## References

1. de Wet, J.R., Wood, K.V., Helinski, D.R. and De Luca, M. (1985). 'Cloning of firefly luciferase cDNA and the expression of active luciferase in *E. coli* '. Proc. Natl. Acad. Sci. USA 82, 7870-7873.

2. Subramani,  S. and De Luca, M. (1988). 'Application of the firefly luciferase as a reporter  gene', in  J.K.Setlow (ed.), Genetic Engineering Principles and Methods, Plenum Press, New York and London, Vol. 10, pp. 75-89.

3. Ditta, G., Schmidhauser, T., Yakobson, E., Lu, P.,  Liang, X.-W., Finlay, D.R., Guiney, D. and Helinski, D.R. (1985). 'Plasmids related to the broad range vector, pRK290, useful for gene cloning and for monitoring gene expression'. Plasmid 13, 149-153.

4. Better, M., Ditta, G. and Helinski, D.R. (1985). 'Deletion analysis of *Rhizobium meliloti* symbiotic promoters'. EMBO J. 4, 2419-2424.

5. Ditta, G. (1986). 'Tn*5*  mapping of *Rhizobium*  nitrogen fixation genes'. Methods Enzymol. 118, 519-528.

6. Ditta, G., Virts, E., Palomares, A.J. and Kim, C.-H. (1987). 'The *nifA* gene of *Rhizobium meliloti*  is oxygen regulated'. J. Bacteriol. 169, 3217-3223.

7. Kahn, M., Kolter, R., Thomas, C., Figursky, D., Meyer, R., Remaut, D. and Helinski, D.R. (1979). 'Plasmid cloning vehicles derived from plasmid Col E1, F, R6K, RK2'. Methods Enzymol. 68, 268-280.

8. Birnboim, H.C. and Doly, J. (1979). 'A rapid alkaline extraction procedure for screening     recombinant plasmid DNA'. Nucleic Acids Res. 7, 1513-1523.

9. Meyer, R., Figurski, D. and Helinski, D.R. (1977). 'Physical and genetic studies with restriction endonucleases on the broad host range  plasmid RK2'. Mol. Gen. Genet. 152, 129-135.

10. Palomares, A.J., De Luca, M. and Helinski, D.R. (1989). 'Firefly luciferase as a reporter gene for measuring gene expression in vegetative and symbiotic *Rhizobium meliloti* and other Gram-negative bacteria'. Gene 81, 55-64.

11. Ditta, G., Virts, E. and Helinski, D.R. (1988). 'Oxygen regulation of *nif* genes in *Rhizobium meliloti* ', in R. Palacios and D.P.S. Verma (eds.), Molecular Genetics of Plant-Microbe Interactions, APS Press, St. Paul MN, pp. 109-110.

12. David, M., Daveran, M.-L., Batut, J., Dedieu, A., Domergue, O., Ghai, J., Hertig, C., Boistard, P. and Kahn, D. (1988). 'Cascade regulation of *nif* gene expression in *Rhizobium meliloti* '. Cell 54, 671-683.

13. Sundaresan, V., Ow, D. W. and Ausubel, F. M. (1983). 'Activation of *Klebsiella pneumoniae* and *Rhizobium meliloti* nitrogenase promoters by *gln* (*ntr*) regulatory proteins'. Proc. Natl. Acad. Sci. USA 80, 4030-4034.

14. Szeto, W.W., Nixon, B.T., Ronson, C.W. and Ausubel, F.M. (1987). 'Identification and characterization of the *Rhizobium meliloti ntrC* gene: *R. meliloti* has separated regulatory pathways for activation of nitrogen fixation genes in free-living and symbiotic cells'. J. Bacteriol. 169, 1423-1432.

Nucleotide   sequence   of   regulatory   regions   of   <u>Azospirillum</u>
<u>brasilense</u>.


Civardi L., Delledonne M., Marudelli M., Fogher C.
Istituto di Genetica vegetale
Università Cattolica, via E. Parmense, 84
Piacenza 29100 Italy.


Bacteria   of   the genus <u>Azospirillum</u> are diazotrophs associated with
the roots of grasses.  Considerable  progress   towards   understanding
the  molecular  biology of their association with plants is based on
the identification of plant-inducible genes. With this goal, we have
isolated and sequenced a set of  "constitutive"  promoter-containing
DNA  fragments  of <u>Azospirillum</u> <u>brasilense</u> Sp7 (1).  Transcriptional
fusions to CAT reporter gene of  plasmid  pKK232-8  (2)  allowed  us
identified  in  <u>E</u>. <u>coli</u>-constitutive  promoters at the frequence of
3.4%.  The first 17 sequences,  isolated at 20 µg/ml Cm,  have  been
classified  in  seven MIC classes of resistance from 20 to 280 µg/ml
of Cm (Fig. 1).


Table 1. Number of clones, for each MIC classes, and size (bp).

Concentration of Cm (µg/ml)

| | 20 | 80 | 100 | 120 | 140 | 200 | 280 |
|---|---|---|---|---|---|---|---|
| number of clones | 2 | 2 | 5 | 1 | 2 | 3 | 2 |
| size in bp | 2950 | – | 500 | 390 | 1070 | 590 | 467 |
| | 213 | 990 | 650 | | – | 1100 | 1300 |
| | | | 700 | | | 1600 | |
| | | | 1100 | | | | |
| | | | 2200 | | | | |


One fragment from each class has been sequenced and analysis of  the
general feature has,  for the moment,  revealed only a common TCCTTC
sequence.

This sequence is present also at position -150 from the start  codon of A. b. glnA (3), -232 of nifH (4) and -64 of nodG (5).
The  promoter  activity  of  these  fragments  will  be  tested  in Azospirillum after fusion with the reporter  genes  lacZ,  phoA  and luxAB.

1.  Civardi L.,  Fogher C.  1989.  XXXIII Conv.  S.I.G.A.,  Alghero, p.126.
2. Brosius J. 1984. Gene, 27:151-160.
3. Bozouklian M., Elmerich C. 1986. Biochimie 68:1181-1187.
4. Fani R. et al. 1989. Mol. Gen. Genet. 220:81-87.
5. Delledonne M. et al. Abstr. this Congress.

Nucleotide sequence and expression of an <u>Azospirillum</u> <u>nod</u> homologous region.


Delledonne M., Porcari R., Fogher C.
Istituto di Genetica vegetale
Università Cattolica, via E. Parmense, 84
Piacenza 29100 Italy


A 12 kb DNA segment of the <u>A</u>. <u>brasilense</u> Sp7 genome, that shows homology with an heterologous <u>R</u>. <u>meliloti</u> <u>nod</u> region has been cloned and partially sequenced (1). The construction of the physical map of the cosmid with this fragment and the DNA sequence analysis of some subclones has allowed the positioning and the identification of homology with <u>nod</u>G from <u>R</u>. <u>meliloti</u>. The nucleotide sequence of the putative <u>nod</u>G gene of <u>A</u>. <u>brasilense</u> shares 60% homology with nucleotide sequence of <u>R</u>. <u>meliloti</u> <u>nod</u>G (2). The sequence has a ORF of 765 bp coding for a protein of 255 aa residues and 27.3 kd MW. The predicted aminoacid sequence shows an overall 33% homology with <u>R</u>.<u>m</u>. <u>nod</u>G and two highly conserved regions (residue 69 to 93 and 133 to 190) with, rispectivaly, 86 and 88% similarity, (Fig 1).


Figure 1. Aminoacid sequence of the two highly conserved regions of <u>R</u>. <u>m</u>. <u>nod</u>G and <u>A</u>. <u>b</u>. <u>nod</u>G.

```
69                      KAMVAKIEADLGPVDILVNNAGITRD     nodG  A. b.
                        ::    . :::: :::::::::::::.:
68                      KALGQRAEADLEGVDILVNNAGITKD     nodG  R. m.


133  GRIINISSVNGVKGQAGQTNYSAAKAGVIGFTKALAAELATKGVTVNAIAPGYIGTDM
     :::::...:: :  :. ::::: : ::: :::.: :: :.:: .::: .:::.: . :
132  GRIINVTSVAGAIGNPGQTNYCASKAGMIGFSKSLAQEIATRNITVNCVAPGFIESAM
```


The nucleotide sequence of the <u>A</u>.<u>b</u>. <u>nod</u>G gene was compared against the nucleotide sequence data base. One sequence, for the <u>act</u>III gene of <u>Streptomyces</u> <u>coelicolor</u> homologous to known oxidoreductases, shows a 63% homology (3). The region upstream of the ATG start codon shows homology to the 16S rRNA of <u>E</u>. <u>coli</u>, and a large inverted repeat of 43 residues is present after the stop codon.

The presence of the ribosome binding site, the high G+C percent inside the ORF (65.5), and the G or C preferential presence as third base in codons (82%) give strong indications as coding frame of this DNA sequence (4). Preliminary results of Northern blot analysis suggests that the gene is costitutively expresss and that the signal is enhanced in presence of root extracts.


1. Antonelli M. et al. 1988. Abstr. XXXI Conv. S.I.G.A., Como.
2. Debellé F., Sherma S.B. 1986. Nucleic Acid Res. 14: 7453.
3. Hallam S.E. et al. 1988. Gene 74:305-320.
4. Gribskov M. et al. 1983. Nucleic Acid Res. 11:539.

# ISOLATION AND CHARACTERIZATION OF TN5 INDUCED AZOSPIRILLUM BRASILENSE MUTANTS ALTERATED IN CALCOFLUOR STRAINING


GIANNELLINI ILVA[a], GALLORI ENZO[b], BAZZICALUPO MARCO[b]
[a] M.A.E. Istituto Agronomico per l'Oltremare Firenze
[b] Dipartimento di Biologia Animale e Genetica Firenze


The significance of exopolysaccharides in the process of root colonization by bacteria of the genus Azospirillum is not known, although in has been demonstrated the lack of involvment in the attachment step (Michiels et al. 1989). On the other hand the production of extracellular polysaccharides has been reported to play a possible role in the encystement that permit the survival of these bacteria under stress conditions and in flocculation of Azospirillum brasilense and Azospirillum lipoferum (Sadasivan L. and Neyra C.A. 1985). In order to investigate the function of exopolysaccharides in the interaction of Azospirillum brasilense with plant roots we tryed to isolated mutants altered in the syntesis of this macromolecule. Following the work of Michiels et al. (1989) Azospirillum mutants, induced by transposon Tn5, were selected on plates with Calcofluor which binds to $\beta$1-4 and $\beta$1-3 linked glycans ( Wood P.J. and Fulcher R.G. 1978 ).

ISOLATION OF THE MUTANTS

In the crosses between E. coli HB101 (pGS9) and A. brasilense SPF94, Tn5 induced kanamycin resistent mutants were obtained with a freguency of $10^{-7}$ per recipient cell. 6000 exconjugant colonies obtained were assayed in plates with 0.02% Calcofluor: about 0.1% presented a different Calcofluor phenotype in comparison to parental strain.
Results of the Southern blot carried out with Tn5 fragment as probe, showed that the mutant strains SPE1, SPE3, SPE6, SPE17, SPE18, SPE19, SPE29, incorporated the trasposon Tn5 in the cromosomal DNA. Cutting total DNA with Eco RI gave hybridization bands in the range of 10,000-20,000 bp different for each mutant. The SPE6 strain showed two bands.

MICROSCOPE OBSERVATIONS

The bacterial mutants, referred to as SPE, were observed with the phase-contrast microscopie. All the mutants showed cellular size and shape similar to the motile spirillar form of the parental strain. Mutant SPE6 however showed spherycal, non motile cells similar to the cysts described by Sadasivan L. and Neyra C.A. 1985. Cysts of strain SPE6 were observed in MSP medium supplemented with succinate or malate, in LB and Antibiotic Medium 3 (Difco) in the exponential growth and in the stationary phase. This mutant strain in liquid colture presented, moreover, numerous clots.



<u>Azospirillum</u> mutants were observed under fluorescent microscopy in order to detect floc formation. Mutants SPE1, SPE17, SPE18, SPE19 and SPE29 showed floc formation similar to parental strain SPF94. Mutant SPE6 demostrated a higher level of flocs; while mutant SPE3 showed floc formation intermediate beetween parental and mutant SPE6.



Mutant SPE6, which demostrated a very unusual cellular form, was further characterized by nitrogenase assay and by thermal resistance. Nitrogenase activity either in liquid and semisolid medium was almost similar to that of parental strain SPF94. Microscopy observation of nitrogen fixing cells showed high percent of spirillum like forms, indicating that the shape of the cells is controlled by environmental factors. Experiments of thermal resistence showed that strain SPE6 is not resistant to highter temperature compared to the parental strain, suggesting that the spherical shape of this mutant is not a kind of resistance form.

REFERENCES

Michiels K., Vanderleyden J., Van Gool A. and Signer E.R. (1989) Isolation and properties of A. lipoferum and A. brasilense surface polysaccharides mutants, F.A. Skinner et al. (Eds) Nitrogen fixation with non-legumes, 189-195 by Kluwer Academic Publishers.
Sadasivan L. and Neyra C.A. (1985) Flocculation in A. brasilense and A. lipoferum: exopolysaccharides and Cyst formation, J. Bacteriol. 163:716-723
Wood P.J. and Fulcher R.G. (1978) Cereal. Chem. 55:952-966.

# IN VITRO CONSTRUCTION OF LACZ GENE FUSIONS WITH THE NIFHDK OPERON OF AZOSPIRILLUM BRASILENSE.


A. MILCAMPS
*F.A. Janssens Laboratory of Genetics*
*University of Leuven*
*Willem de Croylaan 42*
*3030 Heverlee, Belgium*


## 1. Introduction

In recent years, progress has been made on the identification and organization of genes involved in nitrogen fixation in the diazotroph *Azospirillum*. Hundreds of mutants impaired in nitrogen fixation have been isolated by various groups (1,2,3). The best known genes are *nif*HDK (4). Less well characterized are a number of genes that are presumably regulatory genes. A *ntr*C regulatory mutation and a *nif*A type locus have been reported (3,5). Recently, the nucleotide sequence of the *nif*H gene has been determined (6,7), revealing a *nif*-consensus promotertype that occurs in many other nitrogen fixing bacteria.

Studies on gene regulation are greatly facilitated by the use of reporter genes, as illustrated by numerous examples in prokaryotes and eukaryotes. Here we describe the use of an easy and reliable system to monitor differential gene expression under nitrogen fixing conditions.

## 2. Results and discussion.

Site-directed mutagenesis was carried out on cloned *A. brasilense* DNA in *E. coli*. A crusial element was the use of a suitable cloning vector in which no transcriptional readthrough from vector sequences into the insert DNA occurs. The combination of the broad host range pRK290 cloning vector and Tn3HoHo1 (8) as a promoter probe transposon was tested with the *nif*HDK operon of *A. brasilense* Sp7.

Four gene fusions of Tn3HoHo1 into the *nif*HDK operon, cloned in pRK290, were transferred to *A. brasilense* Sp7. Inoculated in semi-solid N-free medium, strains carrying a *nif*H and a *nif*D fusion showed ß-galactosidase activity in the presence of X-Gal after one night incubation, with the *nif*H fusion being more active than the *nif*D fusion. Tn3HoHo1 insertions in the antisense orientation or positioned 5' upstream of the promoter showed no activity.

Measurement of the ß-galactosidase activity (according to Miller) confirmed the results mentioned above.

To determine whether transcriptional or translational *lac*Z gene fusions had been formed, protein extracts were analysed by immunoblot with anti-ß-galactosidase serum. In this way, hybrid proteins with molecular weights larger than ß-galactosidase could be detected for both the *nif*H and *nif*D fusions, indicating that both are translational fusions. The estimated size of the hybrid proteins fits very well with the position of the Tn3HoHo1 insertion in the *nif*H and *nif*D open reading frame respectively. The higher β-galactosidase activity of the *nif*H fusion might be linked to the observation of de Zamaroczy *et al.* (6) that there are two transcripts for *nif*HDK (1,1 kb and 5,6 kb), the smaller one being more abundant than the larger one. Consequently *nif*H*lac*Z mRNA's would be more abundant in the cell than *nif*D*lac*Z mRNA's.

## References


1. Vanstockem, M., Michiels, K., Vanderleyden, J., Van Gool, A.P. (1987). Appl. Env. Microbiol. 53 (2): 410-415.
2. Singh, M.and Klingmüller, W. (1986). Mol. Gen. Genet. 202: 136-142.
3. Pedrosa, F.O. and Yates M.G. (1984). FEMS Microbiol. Lett. 23: 95-101.
4. Perroud, B., Bandhari, S.H., Elmerich, C., (1985). In Azospirillum III: Genetics, Physiology, Ecology (Klingmüller, ed.), Springer-Verlag, Berlin Heidelberg.
5. Singh, M., Tripathi, A.K., Klingmüller, M. (1989). Mol. Gen. Genet. 219: 235-240.
6. de Zamaroczy, M., Delorme, F., Elmerich, C. (1989). Mol. Gen. Genet. 220:88-94.
7. Fani, R., Allota, G., Bazzicalupo, M., Ricci, F., Schipani, C., Polsinelli, M. (1989). Mol. Gen. Genet. 220:81-87.
8. Stachel, S.E., An, A.G., Flores, C., Nester, E.W. (1985). EMBO J. vol.4:891-898.

# ISOLATION AND CHARACTERIZATION OF AZOSPIRILLUM BRASILENSE MUTANTS ALTERED IN THE PRODUCTION OF SIDEROPHORES


M.Bazzicalupo, R.Fani, O.Fantappiè, E.Gallori, E.Mori and M.Polsinelli

Department of Animal Biology and Genetics, University of Florence, Via Romana 17, 50125 Florence, Italy


## INTRODUCTION

Siderophores are low molecular weight compounds produced by microorganisms, able to bind Fe(III) from the environment where the metal is part of insoluble complexes. All the known siderophores are small peptides containing cathecol or hydroxammic groups which are responsible for the binding of the iron. The binding to the siderophore allows the transfer of the Fe(III) inside the cell enabling producing bacteria to compete for this otherwise unavailable elements (Neilands 1981). This competition is a very important factor for a possible utilization of these bacteria in agriculture as biopesticides (Lynch 1988).

Different bacteria species and different strains among the same species produce different types of siderophores; this variety has also been found in the genus Azospirillum (Hartmann 1988). The siderophore spirilobactin, produced by A.brasilense strain RG, the only one studied up to now, is a catechol containing 2-3 dihydroxybenzoic acid, ornithine and serine (Bachawat et al 1987). A.lipoferum strain D-2 also produces a catechol type siderophore (Saxena et al 1986).

## RESULTS AND DISCUSSION

Isolation of mutants . Mutants of A.brasilense strain SPF94 were isolated after transposon Tn5 mutagenesis; Tn5 was carried by the suicide plasmid pGS9 (Selvaraj and Iyer 1983) and transferred by conjugation from E.coli to A.brasilense . More than 7000 kanamycin resistant isolates were assayed for the siderophore production by the plate method based on the chrome azurol S dye ( Schwyn and Neiland 1987). Five mutants with reduced or absent halo in the indicator plates were selected and used for further characterization.

Characterization of the mutants . Mutants were first characterized by growing curves in minimal medium in iron limiting conditions as reported in Figure 1. Data indicated that iron starvation strongly affected the growth of mutant S5 and S6 in respect to that of the parental strain. Moreover the addition of FeCl (50 uM) after 8 hours to the iron deficient culture of mutant S6 rapidly restored the growth

of the cells (data not shown).
In order to better characterize the production of siderophores by the
_Azospirillum_ mutant strains, the presence of catechols in the
supernatants of the cultures was tested by the method of Arnow (1937).
In fact it has been proposed that the _Azospirillum_ siderophores
belong to the catechol type. The results indicated that catechol
compounds are excreted by growing cells in a sort of cyclic way with a
minimum after about 4-6 hours. Moreover results showed that mutants
produced more catechols than parental strain suggesting that the
mutation could affect the biosynthesis of the siderophore giving a
final product still positive to the catechol test, but unable to bind
Fe(III).

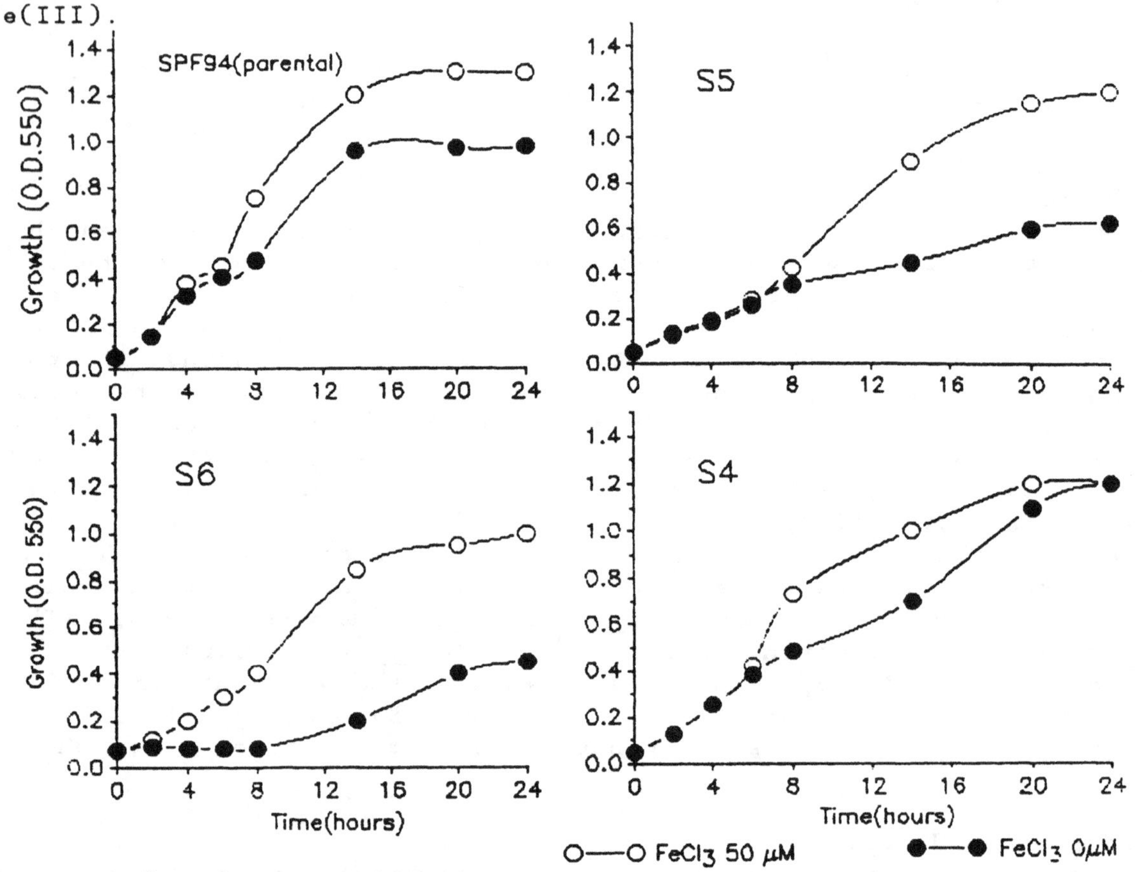


Figure 1.Growth curves of _A.brasilense_ strains in iron limiting conditions


REFERENCES
Arnow (1937) J Biol Chem  118 : 531-537
Bachawat et al (1987) J Gen Microbiol  133 : 1751-1758
Hartmann A  (1988) In Bothe _et al_ eds "Nitrogen fixation hundred years
    after", G.Fischer, Stuttgart
Lynch (1988) New Scientist, 28 April: 45-49
Neilands (1981) Ann Rev Biochem  50 : 715-731
Saxena et al (1986) J Gen Microbiol  132 : 2219-2224
Schwyn and Neiland (1987) Anal Biochem  160 : 47-56
Selvaraj and Iyer (1983) J Bacteriol  156 : 1292-1300

# TAXONOMIC RELATIONSHIPS BETWEEN [*PSEUDOMONAS*] *RUBRISUBALBICANS*, SOME CLINICAL ISOLATES (EF GROUP 1), *HERBASPIRILLUM SEROPEDICAE* AND [*AQUASPIRILLUM*] *AUTOTROPHICUM*


M. GILLIS*, J. DÖBEREINER, B. POT, M. GOOR, E. FALSEN, B. HOSTE, B. REINHOLD AND K. KERSTERS
*Laboratorium voor Microbiologie en microbiële Genetica*
*Rijksuniversiteit Gent*
*K.L. Ledeganckstraat 35*
*B-9000 Gent*
*Belgium*



ABSTRACT. By a polyphasic approach (DNA:rRNA and DNA:DNA hybridizations, and computer assisted analysis of auxanographic results) it was shown that *Herbaspirillum seropedicae*, EF group 1 strains and the generically misnamed [*Pseudomonas*] *rubrisubalbicans* and [*Aquaspirillum*] *autotrophicum* constitute one genus. Within this genus we delineated four groups that each deserve species rank. Species 1 contains EF group 1 strains and [*Pseudomonas*] *rubrisubalbicans* strains; species 2 is composed only of [*Pseudomonas*] *rubrisubalbicans* strains including the type strain. Species 3 contains *Herbaspirillum seropedicae* and one [*Pseudomonas*] *rubrisubalbicans* strain. Species 4 corresponds to [*Aquaspirillum*] *autotrophicum*.


## 1. Introduction

In previous research a phylogenetic relationship was found between the following groups of organisms: (i) [*Pseudomonas*] *rubrisubalbicans*, a generically misnamed group which is described to cause mottled stripe disease on sugar cane (Hayward, 1962); (ii) EF group 1 (Falsen, 1988), strains isolated from clinical material in Sweden; (iii) [*Aquaspirillum*] *autotrophicum*, a facultatively autotrophic hydrogen-oxidizing spirillum (Aragno and Schlegel, 1978).
As described here *Herbaspirillum seropedicae* (Baldani et al., 1986) and [*Pseudomonas*] *huttiensis* also belong to this group. The internal relationships between these taxa were further investigated.

## 2. Materials and Methods

The following strains were investigated: 13 [*Pseudomonas*] *rubrisubalbicans* strains, isolated from sugar cane, sorghum and maize in different countries; 19 EF group 1 strains, isolated from clinical material in Sweden; 2 [*Aquaspirillum*] *autotrophicum* strains isolated from an eutrophic lake in Switzerland; 3 representative *Herbaspirillum seropedicae* strains isolated from roots or rhizosphere soil preparations from maize, sorghum and rice plants in Brazil and 1 [*Pseudomonas*] *huttiensis* strain isolated from aq. dest. in New-Zealand. All strains were investigated in 147 auxanographic tests by using commercial API galleries,50 CH, 50 AO and 50 AA (API Systems

S.A., Montalieu-Vercieu, France). The results were clustered as described by Willems et al. (1989). DNA:rRNA hybridizations were performed as described (Willems et al., 1989). For DNA:DNA hybridizations, we used the initial renaturation rate method (De Ley et al., 1970). $N_2$-fixation was tested as described by Pimentel et al. (this volume).

## 3. Results and Discussion

All organisms investigated constitute a single rRNA subbranch (in rRNA superfamily III) that deserves generic rank.

The auxanographic results revealed 3 clusters (Phena I, II and III), two of them being divided in 2 subphena (a and b). The different clusters and subclusters can be differentiated phenotypically.

By DNA:DNA hybridizations we found four genotypic groups within this genus (D I, D II, D III and D IV). Within these groups the % of DNA binding is higher than 50%. Between the groups, no significant % binding was found. Group D I can be divided in 2 subgroups a and b (internal % DNA binding >85%).

Genotypic groups D Ia and D Ib correspond to subphenon Ia and Ib, respectively. Because both groups share more than 50% DNA binding, they can belong in a single species (1). It contains [*Pseudomonas*] *rubrisubalbicans* strains and EF group 1 strains.

The genotypic groups D II and D III correspond to subphenon IIa and IIb, respectively. Because there is no significant % DNA binding between D II and D III and because subphena IIa and IIb can be differentiated phenotypically, we propose to consider them as 2 separate species (2 and 3). Species 2 contains only [*Pseudomonas*] *rubrisubalbicans* strains while species 3 contains *Herbaspirillum seropedicae* and one [*Pseudomonas*] *rubrisubalbicans* strain. Plant associated $N_2$-fixers belong in species 2 and 3.

Group D IV corresponds with phenon IV and forms a fourth species containing only [*Aquaspirillum*] *autotrophicum*.

Nomenclatural proposals for the described genus and species will be made elsewhere (MS in preparation for the Int. J. Syst. Bacteriol.).

## 4. References


Aragno, M. and Schlegel, H. G. (1978) '*Aquaspirillum autotrophicum*, a new species of hydrogen oxidizing, facultatively autotrophic bacteria', Int. J. Syst. Bacteriol. 28, 112-116.

Baldani, G. I., Baldani, V. L. D., Seldin, L. and Döbereiner, J. (1986) 'Characterization of *Herbaspirillum seropedicae* gen. nov., sp. nov., a root-associated nitrogen-fixing bacterium', Int. J. Syst. Bacteriol. 36, 86-93.

De Ley, J., Cattoir, H. and Reynaerts, A. (1970) 'The quantitative measurement of DNA hybridization from renaturation rates', Eur. J. Biochem. 12, 133-142.

Falsen, E. (1988) 'Catalogue of strains CCUG', Culture Colection, 5th ed., University of Göteborg, Göteborg, Sweden

Hayward, A. C. (1962) 'Studies on bacterial pathogens on sugar cane', Mauritius Sugar Industry Research Institute, Occasional Paper N° 13, part 2, 13-30.

Willems, A., Busse, J., Goor, M., Pot, B., Falsen, E., Jantzen, E., Hoste, B., Gillis, M., Kersters, K., Auling, G. and De Ley, J. (1989) '*Hydrogenophaga*, a new genus of hydrogen-oxidizing bacteria that includes *Hydrogenophaga flava* comb. nov. (formerly *Pseudomonas flava*), *Hydrogenophaga palleronii* (formerly *Pseudomonas palleronii*), *Hydrogenophaga pseudoflava* (formerly *Pseudomonas pseudoflava* and "*Pseudomonas carboxydoflava*"), and *Hydrogenophaga taeniospiralis* (formerly *Pseudomonas taeniospiralis*)', Int. J. Syst. Bacteriol. 39, 319-333.

# STUDY ON THE BINDING OF AZOSPIRILLUM STRAINS TO PLANT TISSUES


GIANNELLINI ILVA[a], MANNELLI SOFIA[b], BAZZICALUPO MARCO[b]
[a] M.A.E. Istituto Agronomico per l'Oltremare Firenze
[b] Dipartimento Biologia Animale e Genetica Firenze


The aim of the present work is to study the mechanism of <u>Azospirillum</u> binding to plant tissue. The binding is supposed to be dependent on bacteria metabolism (Eyers et al. 1988); moreovar the involvment of bacterial products, like exopolysaccharides, in the binding to the roots has been hypothisized after the results obteined in Rhizobium (Finan et al. 1985; Michiels et al. 1988). Binding experiments described were carried out with different <u>Azospirillum</u> strains, including mutants obteined by trasposon Tn5 mutagenesis (referred to as SPE ) and selected for their altered production of exopolysaccharides tested by Calcofluor staining ( see poster B4). In particular we assayed the binding of <u>Azospirillum</u> to isolated cells from potato callus and to wheat roots.

BINDING OF LABELLED BACTERIA TO WHEAT ROOTS.

1) Preliminary experiments of binding of labelled bacteria to wheat roots were carried out with <u>Azospirillum</u> parental strain SPF 94 in order to establish optimal bacteria concentration. Labelled bacteria were used at different concentrations in the binding experiments as described in the methods section. Radioactivity bound to roots was measured together with radioactivity of total bacteria suspension. Results, show that the fraction of bacteria bound to the root decreases rapidly with the increase of the bacteria concentration. For further experiments the bacterial concentration of about $3.10^6$ was choosen as that resulting in a significant value of cpm bound to the roots together with an appreciable percentage respect to the total bacterial input.

2) In following experiments E. coli HB101 as negative control, <u>Azospirillum</u> brasilense parental strain SPF 94, <u>Azospirillum</u> <u>lipoferum</u> col 5 and Calcofluor-binding mutants were tested for their ability to bind wheat roots. Results reported in table 1 indicated that E. coli and heat killed <u>Azospirillum</u> did not bind to the roots, as already reported (Eyers et al. 1988) confirming the occurrence of an active and specific process. Moreover neither the Calcofluor negative nor the Calcofluor over binding mutants showed significant differences to the parental strain, as described (Michiels K. et al. 1989). These results suggested the independence of the binding to the root from the expopolysaccharides composition. This was also confirmed by the data obtained with A. <u>lipoferum</u> which showed binding similar to that of A. <u>brasilense</u>, although the exopolysaccharides composition was demostrated to be different (Del Gallo M. et al. 1989).

TABLE 1 - BINDING OF LABELLED AZOSPIRILLUM WITH WHEAT ROOTS

| BACTERIAL STRAINS | CALCOFLUOR FENOTYPE [a] | BACTERIA ADDED/ml | % cpm OF BACTERIA BOUND TO ROOTS [b] |
|---|---|---|---|
| HB 101  NEGATIVE CONTROL | - - | $9.10^6$ | 1,6 |
| SPF 94 | PARENTAL | $3.10^6$ | 11,1 |
| SPF 94 KILLED | | | 2,9 |
| SPE  1 | - - | $1.10^6$ | 13,0 |
| SPE  3 | + + | $3.10^6$ | 13,0 |
| SPE  6 | + + | $1.10^6$ | 15,7 |
| SPE 17 | - - | $2.10^6$ | 20,7 |
| SPE 18 | + | $9.10^6$ | 14,9 |
| SPE 19 | + | $4.10^6$ | 13,8 |
| Col   5 LIPOFERUM | | $1.10^6$ | 14,0 |

a) The - or + rapresents the minor or major of calcofluor binding activity in comparison
   to parental strain
b) Each value represents the average of 3 replicates 10 min. at 80°C

REFERENCES


Del Gallo M., Negi M. and Neyra C.A. (1989) Calcofluor - and Lectin - Binding Exocellular Polysaccharides of A. brasilense and A. lipoferum, J. of Bacteriology 171:3504-3510.
Eyers M., Vanderleyden J. and Van Gool A. (1988) Attachement of Azospirillum to isolated plant cells, FEMS Microbiol. Lett. 49: 435-439.
Finan T.M., Hirsch A.M., Leigh J.A., Johansen E., Kuldan G.A., Deegan S., Walker G.C., Signer E.R. (1985) Symbiotic Mutants of Rhizobium meliloti that uncouple Plant from Bacterial Differentiation, Cell, 40: 869-877.
Michiels K., Vanderleyden J., Van Gool A.P. and Signer E.R. (1988) Isolation and Characterization of A. brasilense Loci that correct R. meliloti exo B and exo C Mutations, J. Bacteriol. 170: 5401-5404.
Michiels K., Vanderleyden J., Van Gool A.P. and Signer E.R. (1989) Isolation and Properties of A. lipoferum and A. brasilense surface polysaccharides mutants, F.A. Skinner et al. Nitrogen fixation with non legumes 189-195 Kluwer Academic Publishers.

# EFFECT OF <u>Azospirillum lipoferum</u> AND GA$_3$ ON ROOT GROWTH IN CORN


M. FULCHIERI, C. LUCANGELI  AND R. BOTTINI
Departamento de Ciencias Naturales
Universidad Nacional de Río Cuarto
5800 Río Cuarto
Argentina



ABSTRACT. Pre-germinated corn seeds were aseptically grown with <u>Azospirillum</u> cultures or GA3. Both, inoculation with <u>Azospirillum</u> and <u>GA3</u> in concentrations similar to those produced by <u>Azospirillum</u> cultures, significatively increased root growth. Results sustain that GAs are effective in promoting root growth in very small quantities and the effect of <u>Azospirillum</u> might be mediated by GA production.


Inoculation of grasses with <u>Azospirillum</u> spp. often produces an enhancement of growth (1). Amongst the mechanisms proposed is production of plant hormones. <u>Azospirillum</u> produces auxins (3) and other hormone-like substances (6). We characterized Gibberellins $A_1$, $A_3$ , and iso-$A_3$ as endogenous from aseptic cultures of <u>A.lipoferum</u> str. op33 (2). Gibberellins increase root hairs and lateral roots (7). Such changes could enhance water and mineral uptake (5). Inoculation of wheat with <u>Azospirillum</u> increased $^{15}$N uptake, and this effect was mimicked by GA application (4). In this paper we examine the effects of inoculation with <u>A.lipoferum</u> on root growth of corn seedlings, and the possible involvement of GAs produced by the bacterium in the early events of seedling growth.

MATERIAL AND METHODS. Pre-germinated corn seedlings were submitted to the following tratments: control inoculated with sterile NFb culture medium; inoculation with <u>A.lipoferum</u> cultures (OD = 1.52 at 540 nm; $10^8$ cel.mL$^{-1}$) str WR 708, op33, and IAA$^-$-320; GA$_3$ $10^{-11}$, $10^{-12}$, $10^{-13}$, and $10^{-14}$ M. After incubation for 48 h  at 30°C and 100 % RH, the number of roots per plant, the lenght of roots, the root surface area (expressed as mL of NaOH), the root dry weight, and the hypocotile dry weight, were recorded. From roots of both treatments, inoculation with <u>A.lipoferum</u> str op33, and control, GA-like substances were analyzed by MeOH 80 % extraction (+ high sp. act. $^3$HGA$_5$ as internal standard) ——— RP C$_{18}$ chromatography ——— SiO$_2$ chromatography ——— enzymatic hydrolisis (for glucosyl-conjugates) ——— EtOAc partition (at pH 3.0) ——— RP C18 HPLC (10/73% MeOH) ——— microdrop (for free acid GAs) and inmersion (for glucosyl-conjugated GAs) dwarf rice cv. Tan-ginbozu bioassay.

298

RESULTS AND CONCLUSIONS

Table 1. Corn seedlings grown under different treatments.

| Treatment | Root Numb | Root Leng | Root Area | Root d wt | Hypoc d wt |
|---|---|---|---|---|---|
| WR 708 | 4.5000[a] | 10.2117[a] | 2.6958[a] | 27.8583[a] | 31.4333[a] |
| GA$_3$ 40 pg/mL | 4.5833[a] | 9.9683[a] | 2.7501[a] | 28.0500[a] | 29.5083[a] |
| Control | 4.3333[a] | 5.7150[b] | 1.8083[b] | 16.3083[b] | 24.5250[b] |
| op 33 | 4.4167[a] | 8.3408[a] | 2.9667[a] | 27.5250[a] | 32.3833[a] |
| GA$_3$ 40 pg/mL | 4.3333[a] | 6.5675[b] | 2.1083[b] | 20.8662[b] | 21.9667[b] |
| Control | 4.5000[a] | 3.8342[c] | 1.4583[c] | 16.5667[b] | 14.5583[c] |
| IAA$^-$ 320 | 4.1667[a] | 8.5442[a] | 3.1167[a] | 20.2750[b] | 17.2500[a] |
| GA$_3$ 40 pg/mL | 4.4167[a] | 9.8442[a] | 2.6983[a] | 28.0500[a] | 21.2583[a] |
| Control | 4.2500[a] | 6.0983[b] | 1.9633[b] | 16.3917[b] | 15.1917[a] |
| GA$_3$ 4 pg/mL | 4.0000[ab] | 6.9225[b] | 1.3000[c] | 15.4417[b] | 26.4583[ab] |
| GA$_3$ 40 " | 4.9167[a] | 9.2133[a] | 2.5167[a] | 25.3333[a] | 32.7917[a] |
| GA$_3$ 400 " | 4.6667[ab] | 10.6798[a] | 1.8666[b] | 25.8333[a] | 26.5750[ab] |
| GA$_3$ 4000 " | 4.4167[b] | 6.0967[b] | 1.1917[c] | 17.6833[b] | 18.5333[c] |
| Control | 3.6667[b] | 5.0450[b] | 0.9833[c] | 13.4083[b] | 19.4167[bc] |

Table 2. GA levels in ng of GA$_3$ per g f wt of root.

| Treatment | FA-GAs | FA(after hydr.) | GA-G/GE |
|---|---|---|---|
| op 33 | 7.5 ng | 3.8 ng | 3 ng |
| Control | 0.4 ng | 0.7 ng | no-detection |

All the 3 strains studied were effective in the promotion of root growth. Treatments with GA$_3$ 40 and 400 pg. mL$^{-1}$ were as effective as the inoculation with the different strains. However the effect could not be due only to GA-like substances, since inoculation with op 33 (an auxin over-producer) produced significant increases as compared with GA$_3$ treatments. Inocultion with _Azospirillum lipoferum_ op 33 enhanced the levels of GA-like substances (HPLC-Rt similar to those of $^3$H-GA$_{1/3}$, $^3$H-GA$_{5/20}$, and $^3$H-GA$_{4/36}$) in root extracts.

REFRENCES


1. Bashan Y, Ream Y, Levanony H, Sade A. 1988. Can J Bot.
2. Bottini R, Fulchieri M, Pearce D, Pharis R. 1989. Pl Physiol <u>90</u>, 45.
3. Crozier A, Arruda P, Jasmin J, Monteiro AM, Sandberg G. 1988. Appl Environ Microbiol <u>54</u>,
4. Kucey R. 1988. J Appl Bacteriol <u>64</u>, 187.
5. Lin W, Okon Y, Hardy F. 1983. Appl Environ Microbiol <u>45</u>, 1775.
6. Tien T, Gaskins M, Hubell D. 1979. Appl Environ Microbiol <u>37</u>, 1016.
7. Umali-García M, Hubell D, Gaskins M, Dazzo F. 1980. Appl Environ Microbiol <u>39</u>, 219.

**INTERACTIONS OF VA-MYCORRHIZAE WITH** *Acetobacter diazotrophicus* **INOCULATED INTO SWEET POTATOES (***Ipomoea batatas* **L. Lam.).**

M.A.PAULA, V.M.REIS, S.URQUIAGA, J.DÖBEREINER
*EMBRAPA, National Centre for Soil*
*Biology Research (CNPBS/EMBRAPA)*
*Km 47, Seropédica, 23851,*
*Rio de Janeiro, BRAZIL*

In ealier experiments effects of *Acetobacter diazotrophicus*, and of the VA-mycorrhizal fungus, *Glomus clarum,* on growth and N and P incorporation into sweet potatoes grown in fumigated soils were reported (Paula et al., 1989). *G. clarum* obtained from sweet potato spores grown in sterilized soil inoculated with this fungus and with an enrichment culture of *A. diazotrophicus* contained this bacterium besides several others, including a diazotroph *Klebsiella sp*. Inoculation of N-free LGIP medium with crushed surface sterized spores of *G. clarum* showed *A. diazotrophicus* growth with a typical dark yellow tick pellicle, wich was not observed when intact spores were inoculated. This shows that this new diazotroph, wich has been considered an endophyte can also infected a VAM fungus and remain viable even after two months storage of the spores at 4°C. The occurrence of *Azospirillum* on the surface or within the cytoplasm of VAM spores has been reported previously (Tilak et al., 1989).

Micropropagated sweet potatoes grown in fumigated and non fumigated soil, inoculated with *G. clarum* and *A. diazotrophicus*, showed increased spore formation in soil and within roots when compared to VAM inoculation alone. Plants inoculated with VAM spores containing the bacteria caused additional increases in number of spores formed within roots. Root colonization by *G. clarum* was also increased by *A. diazotrophicus*, this effect being even observed in non-fumigated soil. *A. diazotrophicus* infected aerial parts only when inoculated together with VAM or within VAM spores.

*G. clarum* and *A. diazotrophicus* inoculation, enhanced top and root growth in fumigated soil only. The best effects were observed with spores containing the bacteria.

The possibility of introducing diazotrophic microrganisms into plant roots by VAM fungal spores opens am entirely new field for biotechnologies wich envisage increasing replacement of N fertilizers in legumes and non-legumes.

TABLE 1: Colonization of micropropagated sweet potato plants by *Acetobacter diazotrophicus* (means of 4 replicates).

| Inoculant | number of *A. diazotrophicus*/$g^{-1}$ fresh wtx$10^4$ | | | | | |
| | Washed root | | Surface ster.roots | | stems | |
| | F* | NF* | F | NF | F | NF |
| Control | 0 | 0 | 0 | 0 | 0 | 0 |
| VAM spores | 0 | 0 | 0 | 0 | 0 | 0 |
| *A.diazotrophicus* | 9 | 3 | 2 | 10 | 0 | 0 |
| Mixed diazotroph* | 11 | 5 | 5 | 8 | 0 | 0 |
| VAM + *A. diazotrophicus* | 7 | 8 | 10 | 12 | 15 | 13 |
| VAM + mixed diazotrophs | 16 | 10 | 12 | 9 | 20 | 11 |
| VAM spores containing mixed diazotrophs | 10 | 8 | 13 | 7 | 30 | 21 |

+ enrichment culture in N-free LGIP medium obtained from sweet potato root macerate. * F-fumigated, NF-non fumigated.

REFERENCES
Paula, M.A.; Döbereiner, J.; Siqueira, J.O. (1989) Nutriçåo e produçåo de batata-doce micropropagada e inoculada com fungo micorrízico VA e bactérias diazotroficas. In: III Reuniåo Brasileira sobre Micorrizas, pp. 26. SMA/SA/USP, Såo Paulo.
Tilak, K.V.B.R.; Li, C.Y.; Ho, I. (1989) Occurrence of nitrogen-fixing *Azospirillum* in vesicular-arbuscular mycorrhizal fungi. Plant Soil, 116:286-288.

# GENETIC AND BIOCHEMICAL STUDIES ON THE PRODUCTION OF INDOLE-3-ACETIC ACID (IAA) IN AZOSPIRILLUM LIPOFERUM


E. RUCKDÄSCHEL, S. RAMSCHÜTZ, K. KRAFT AND W. KLINGMÜLLER
*Dept. of Genetics*
*University of Bayreuth*
*P.O. Box 101251*
*D-8580 Bayreuth, FRG*



ABSTRACT. In this study we examined *A. lipoferum* for the presence of different IAA biosynthetic pathways. The conversion of several indole compounds to IAA was measured by HPLC both after addition to a stationary growth phase culture and in an in-vitro test assay using crude cell extracts. The results indicate the presence of the transamination pathway only. In a second attempt genes involved in tryptophan synthesis and IAA production have been cloned.


## 1. Introduction

*Azospirillum* is a widely distributed soil bacterium living in close association with the roots of economically important plants. The study of this bacterium can help us to elucidate determinants playing a role in the complex plant-bacteria interaction.

The two most often discussed microbial factors influencing plant growth are the fixation of nitrogen and the production of plant growth substances by root associated bacteria. In the case of *Azospirillum* there are many reports on nitrogen fixation available , but little is known about the existing IAA pathways and the genes involved in IAA synthesis. In this report we therefore want to present some data about the biochemistry and genetics of IAA production in *Azospirillum lipoferum*.

## 2. Material and Methods

*Azospirillum* was grown in minimal medium (MAZ) for 24h in order to reach stationary phase. Then an indole compound was added at a concentration of 100 µg/ml and incubation was continued for another 12 hours.

For the preparation of cell extracts *Azospirillum* was grown in MAZ medium supplemented with 100 µg/ml tryptophan for 48 hours. Detection of aminotransferase activity was done as described earlier.

IAA production was measured by reversed phase HPLC analysis of an acidic ethylacetate extract of the reaction mixture. For the separation of indole compounds an ODS coloumn (250 X 5 mm; 5µm particle size) was used that was connected to a fluorimeter (excitation at 280 nm, emission at 350 nm). The mobile phase consisted of methanol:water: acetic acid (35:64:1,v/v/v) at a flow rate of 1 ml/ min.

The procedures for DNA isolation and manipulation were as described in Maniatis.

# 3. Results

## 3.1. BIOSYNTHESIS OF IAA IN AZOSPIRILLUM LIPOFERUM

In *Azospirillum lipoferum* IAA is produced via indole-3-pyruvic acid (transamination pathway). An additional pathway via indole-3-acetamide - that occurs in *Pseudomonas, Agrobacterium* and *Bradyrhizobium* - could not be detected:

a)   In supernatants of bacterial cultures grown for 24 hours in minimal medium supplemented with 100 μg/ml tryptophan indole-3- ethanol and indole-3-methanol were present in neutral ethylacetate extracts, whereas in acidic extracts indole-3-lactic acid and indoleacetic acid were identified by their retention times. Indole-3-ethanol and indolelactic acid are formed by the reduction of indoleacetaldehyde and indolepyruvic acid, respectively. Neither indoleacetamide nor tryptamine could be detected.

b)   Intermediates of other IAA pathways (i.e. tryptamine, indoleacetamide, indoleacetonitril) were not converted to IAA when added to a 24 hour old bacterial culture grown in minimal medium.

c)   In an in vitro assay (100 mM $KPO_4$ pH 8, 3 mM tryptophan, 0.1 mM pyridoxalphosphate, 3 mM NAD, 20 mM oxoglutaric acid, 1 mM DTT, 100 μg protein) no conversion of tryptophan to IAA was seen in the absence of oxoglutaric acid which is a necessary cofactor for the transamination reaction

## 3.2. CLONING OF GENES INVOLVED IN IAA BIOSYNTHESIS

*3.2.1. Cloning of the trp genes of A. lipoferum.* Since tryptophan is the precursor of IAA, the regulation of tryptophan synthesis is important for IAA production. The screening of a cosmid gene library by complementation of different *E. coli* trp⁻ mutants resulted in the isolation of two clones carrying the tryptophan biosynthesis genes of *A. lipoferum*. Both clones have an overlapping DNA region of about 15 kb in common. The <u>trp</u>E and the <u>trp</u>C gene were localized by complementation of different *E. coli* trp ⁻ mutants.

*3.2.2. Cloning of a gene encoding an aromatic amino acid aminotransferase.* Aminotransferases catalyze the first step in the transamination pathway. In *A. lipoferum* the transamination of tryptophan can be catalyzed by four different aminotransferases. The gene for one of these enzymes has been cloned.

*3.2.3. Complementation of a Tn5 induced IAA negative mutant.* A DNA fragment of the Tn5 induced IAA mutant Al37 containing the Tn5 region was cloned and used as a probe to identify the corresponding DNA region in a gene bank of *A. lipoferum* wildtype. The isolated clone can restore the mutant's IAA production to wildtype level.

AUTOREGULATION OF PH BY AZOSPIRILLUM SPP.


Martín R. Alonso and María Cristina Marzocca
Instituto de Suelos, INTA, Los Reseros y Las Cabañas
Castelar, Prov. Buenos Aires, REPUBLICA ARGENTINA


## Introduction

Becking (1961), showed that Beijerinckia grown on glucose has the ability of regulate pH
the culture medium by the production of acetic in alkaline medium, and an undetermined
alkaline sustance in an acid medium. Some workers have found Azospirillum growing in soils
where the pH seemed to be limiting. Charyulu and Rao (1980) reported Azospirillum in an
acid sulphate soil at pH 3,2. Marzocca (uniplished results) has found appreciable counts
of Azospirillum in alkaline soils at pH 9,5-10. The investigation of the tolerated and
optimum pH rang for Azospirillum has been done ap to now in media with malate as the carbon
source. The aim of the present work is to determine if Azospirillum has some capability,
like that of Beijerinckia, to regulate the pH of a culture medium with ethanol as the sole
carbon source, Alonso y Marzocca (1987). The ethanol presents two advantages over the malate to this pourpose. First, there seems to be only weak changes in the pH of the culture medium owing to the ethanol consumption, Alonso y Marzocca (1987). And second, the e-
thanol does not represents and additional buffer system to the medium like the malate.

## Material and Methods

The basal medium had the same composition of that used by Dobereiner and Baldani (1980).
The phosphate concentration in the  medium   was about the same 0,3 mM. The phosphate
and ethanol were added separately and the medium adjusted to different pH with KOH and
HCl. The ethanol and the malate were added at 0,5%. The inoculum used was equivalent to
three drops of an 72 hours old culture (O.D.= 0,8 at 475 nM) in the same medium, with
theadding of 0,25 g/l of $NH_4Cl$. The medium was dispensed in serum flasks, 4,5 ml per flasks.
Incubation temperature was 31-32°C. In order to mantain the pH 9,30 of the medium during
the incubation, it was found necessary to exclude the atmospheric $CO_2$. To this pourpose
the flasks were       incubated in  gas-tight larger flasks with 30 ml of 40% KOH. All observations were carried out by triplicate.

## Results

In the medium with ethanol at an initial pH of 4,30 Azospirillum changes the pH to the neutrality in the first 12 hours (figure 1). The modification of pH es produced during the lag-
phase and the growth curve is similar to that at pH 7. No growth was observed in the medium with malate at initial pH 4.10. It is propable that the bacterium could not modify
the pH of the medium owing to the resistance exerted by the malate buffer system. In the
medium with ethanol at initial pH 3,30 there were growth only in about 40% of the  inoculated flasks and the growth occurred at different times from the  inoculation (figure 2).
All the flasks showing growth had a pH 6,5-7,0. In the medium initial pH 9,30, about 65%
of the inoculated flasks presented growth with pH in the range 6,5-8,2.

## Conclusions

Azospirillum is capable of regulating pH in a medium with currently used buffer concentration and with ethanol as the sole carbon source. Depending on the initial pH, two events
can result: a)if initial pH is far removed from neutrality, the bacterium may or may not
grow; and b) if the bacterium grows, the pH of the medium tends to neutrality. Of course
Azospirillum can not modify the pH of the soil, but we hypothesize that it can change the
pH in the vicinity of small microhabitats in the soil and by this way it can endure adverse pH conditions, so that it can grow to some extent and fix nitrogen in soils that are
far from neutrality.

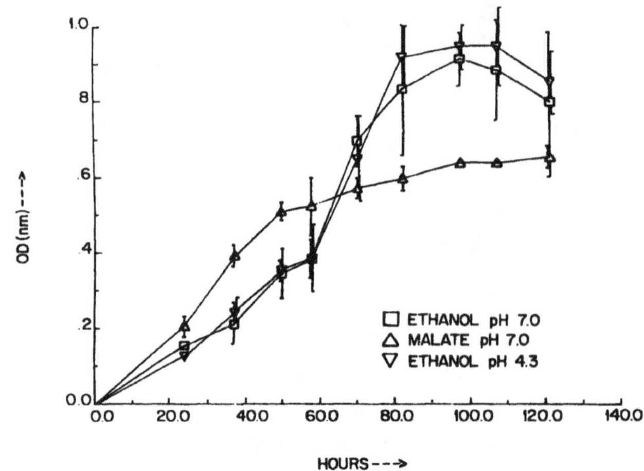



Figure 1. a) Growth (O.D. 540 nm) of the <u>Azospirillum</u> Strain ATCC 29145 in buffered with potasium phosphate (0,3 nM, about 0,5 g/l). Malate with initial pH 7. Ethanol with initial pH 7 and pH 4,3. B) pH evolution in the same media. Bars represents confidence intervals for 95% probability.

## REFERENCES


1) Alonso, M.R.; Marzocca, M.C. (1987). Culture of <u>Azospirillum</u> spp. with ethanol as the sole carbon source. IV International Symposium on N2. Fixation with Non-Legumes p. 152 Rio de Janeiro (Abstract).

2) Baldani, V.L.D.; Dobereiner, J. (1980). Host plant specifity in the infection of cereals with <u>Azospirillum</u> spp. Soil. Biochem. 12: 433 - 439.

3) Becking, J.H. (1961). Studies on nitrogen fixing bacteria of the Genus Beijerinckia. I Geographical and ecological distribution in soils. Plant and Soil XIV: 49 - 79.

4) Charyulu, P.B.B.N.; Rao, V.R. (1980) Influence of various soil factors on nitrogen fixation by <u>Azospirillum</u> spp. Soil Biol. Biochem. 12: 343 - 346.

# NITROGEN FIXING BACTERIA ISOLATED FROM MAIZE ROOT AND ANTAGONISTIC ACTIVITY AGAINST FUSARIUM SP.


L.HALDA [1], J.LEVIC [2], M.DENIC [2], V.PENCIC [2] AND C.A.NEYRA [3]

[1] *Lehrstuhl für Genetik, Universität Bayreuth*
*D-8580 Bayreuth*
[2] *Maize Research Institute "Zemun Polje"*
*Zemun, Yugoslavia*
[3] *Department of Biochemistry and Microbiology (C.A.N.)*
*Cook College, New Jersey Agricultural Experiment Station,*
*Rutgers University, New Brunswick, New Jersey 08903*



ABSTRACT. Enterobacter sp. is the major group of nitrogen fixing bacteria commonly found associated with maize roots grown in different Yugoslavian soils (chernozem, alluvial soil, pseudogley and smonitza). Nitrogen fixing ability of isolates was demonstrated with growth on N-free agar plates and acetylene reduction activity in N-free semisolid medium. Several of these nitrogen fixing bacteria, when tested in vitro, were shown to be able to restrict the growth of some Fusarium sp. causing maize root rot. These bacterial strains were found to suppress pigmentation and growth of Fusarium cultures. Culture discoloration was associated with poor growth of surface mycelia. Bacterial strain Enterobacter cloacae ZP 101 has a good nitrogen fixing potential and was the most promising in suppressing the growth of Fusarium sp. No differences in colony formation of Fusarium species were noticed in interaction with investigated non nitrogen-fixing bacterial strain. Introduction of biological control agents to soil born and seed born pathogens could potentially control maize root rot.


## 1. Introduction

Nitrogen fixing Enterobacter strains are associated with cereals such as wheat, maize and barley in tropical and temperate climates. Enterobacter cloacae, a nitrogen fixing bacterium is frequently found in the rhizosphere of plant species and may be a promising biological control agent of some diseases. Widespread and heavy economical diseases of maize (Zea mays L.) are Fusarium root and stalk rot. The majority of commercially grown maize genotypes carry little or no resistance to these diseases. Control of Fusarium root rot with chemicals has been limited by the domination of pathogens in the soil. In this case introduction of biological control agents to soil born and seed born pathogens could potentially control maize root rot.
The purpose of this study was to identify nitrogen fixing bacterial strains naturally living in maize root rhizosphere that could, also, have an inhibitory effect on Fusarium species, pathogens of maize root rot.

## 2. Material and Methods

Two maize inbred lines and their hybrid (Mo17, B73 and their ZP cross) growing on five different types of soil( chernozem, alluvial, brown forest, pseudogley and smonitza type of the soil) were collected. Root samples were prepared for excised root assay, enrichment culture, enumeration and isolation of bacteria. Healthy, medium, thick roots were selected for bacteria isolations and enumeration (Halda et al., 1988.).

Fungi/rhizobacteria interaction in vitro. The method described for <u>Pythium</u> inhibition assay by Gutterson et al.(2) was followed with some modifications.

All data were subjected to analyses of variance (2 factor factorial, RCBD). Treatment means were separated for every fungus/rhizobacterium interaction by LSD at 0.05 probability .

## 3. Results and Discussion

The nitrogen fixation by bacteria in associative symbiosis with maize roots apparently is dependent on the type of the soil and pH of the soil. Variations in nitrogenase activities were distinct between the same genotypes grown in different type of soil. For cross Mo17 X B73 nitrogenase activity expressed as nmol/h ranged from 14.49 nmol/h (on pseudogley) to 50.6 nmol/h (on chernozem) when non-sterilized roots are considered. For the same cross nitrogenase expressed no activity up to 17.48 nmol/h when sterilized roots are considered. Surface sterilization eliminate typical rhizosphere bacteria and severly inhibit nitrogenase activity. Little variation in nitrogenase activity between these genotypes was found on the same type of soil.

The majority of the nitrogen fixing bacteria were located in the rhizosphere ( $1.9 \times 10^5$ to $7.2 \times 10^7$ per gram of fresh root). Low numbers of nitrogen fixing bacteria were present on the sterilized roots ( $1.2 \times 10^4$ to $3.0 \times 10^4$ per gram of fresh root). The number of bacteria associated with maize roots, as well as nitrogenase activity, was higher in fertile, productive soil (chernozem, alluvial). The nitrogen fixing bacteria associated with the maize root showing appreciable acetylene reduction rates (higher than 10nmol/h/vial) were isolated and identified.

Three groups of nitrogen fixing bacteria were isolated depending on the type of soil. On fertile, productive soil (chernozem and alluvial) the main group of nitrogen fixing bacteria was <u>Enterobacter</u> sp. On pseudogley and smonitza the main group was a low nitrogen fixing unidentified yellow type of bacteria. On brown forest soil, beside <u>Enterobacter</u> sp., a low number of <u>Azospirillum</u> sp. was isolated.

Nitrogen fixing bacteria previously isolated on different types of soil were included in investigation of their antifungal activities. Characteristics of separated and dual culture of <u>Fusarium</u> species (<u>F.graminearum</u>, <u>F.verticillioides</u>, <u>F.oxysporum</u>, <u>F.solani</u>, <u>F.culmorum</u>, <u>F.tricinctum</u> and <u>F.sporotrichioides</u>) and bacterial strains (<u>E.cloacae</u> strains : ZP 101, ZP 102, ZP 103 and two unidentified strains) incubated over two, three, four, five and nine days on PDA and NBA at $25^0$C were studied. Some bacterial strains were found to suppress pigmentation and growth of <u>Fusarium</u> cultures. Culture discoloration was associated with poor growth of surface mycelia.

Nitrogen fixing bacteria which occur naturally in rhizosphere of maize root, when tested in vitro, were shown to be able to restrict the growth of some <u>Fusarium</u> sp. causing maize root rot. Bacterial strain of <u>E.cloacae</u> ZP 101 has a good nitrogen fixing potential and was the most promising in suppressing the growth of <u>Fusarium</u> sp.. Introduction of biological control agents to soil born and seed born pathogens could potentially control maize root rot.

## References

1. Halda et al.(1988). Mikrobiologija, Vol.25, No.1, 29-44

2. N.I.Gutterson et al.(1986). Journal of Bacteriology, Mar., 696-703

SOME CHARACTERS OF AZOSPIRILLUM SPP. ISOLATED FROM THE RICE FIELDS
NEAR HANOI, VIETNAM


Nguyen Ngoc Dung, Nguyen Thi Phuong Chi, Ly Kim Bang
Institute of Microbiology
National Centre of Scientific Research of Vietnam
Nghia do, Tu liem, Hanoi


## INTRODUCTION

In a recent paper it has been reported that Azospirilla occur
frequently in the rhizosphere of rice roots, near Hanoi (1).
Azospirilla have shown an effect on the growth of roots of rice
seedlings. Most of the following experiments have been carried out by
us at the department of genetics, University Bayreuth, Germany.

## MATERIALS AND METHODS

A. brasilense sp7, A. lipoferum sp59b and A. amazonense Y1 used as
reference have been obtained from Dr. A. Hartmann, Department of
Microbiology, University Bayreuth Germany; 29 strains have been
isolated from different fields, near Hanoi.
All media and methods used by us have been described by Tarrand et al.
(2, 3).

## RESULTS AND DISCUSSION

Morphological observation. Among 29 strains tested, there are 4
isolates, Arm 2-2, DA 10-1, Gl 1-0 and DA6 2-2 whose cells remain
mainly vibroid and motile, after they have been inoculated at 37°C for
36 hours in nitrogen-free semi-solid medium with malate as a C-source;
in medium containing glucose, their cells possess the enlarged vibroid
form. That is a morphological characteristics of A. brasilense. Under
the same conditions, the rest of isolates appear to be like A.
lipoferum, e.g. their cells become very large in nitrogen-free
semisolid malate medium and remain vibroid in the presence of glucose.
It is of interest that only the strains Arm 2-2, DA 10-1, Gl 1-0 and
DA6 2-2 are able to use sucrose and glucose, but not manitol and

citrate, for dinitrogen dependent growth. Figure 1 shows the $C_2H_2$ reduction by Arm 2-2 in nitrogen-free semi-solid medium with sucrose as a C-source. Glucose, fructose, manitol, ketaglutarate and citrate can be used by the other isolates for the dinitrogen dependent growth.

Effect of temperature. At about 30°C, the strains being able to use sucrose exhibit the best $C_2H_2$-reduction. For the rest of strains, it occurs the optimal temperature is between 25°C and 30°C. The $C_2H_2$-reduction in both groups appears to be still rather significant at 41°C.

Effect of oxygen. The sensitivity to the oxygen content in the gas phase of isolates is different, but most of strains show the best $C_2H_2$-reduction with 1%-4% of oxygen in gas phase, only DA6 2-2 and QO 2-2 possess a great sensitivity to oxygen.

Effect of pH. the effect of pH values on the reduction of acetylene has been established with two isolates, Arm 2-2 and Mat 2-1. Under slight alcaline condition, pH value about 7.1, both strains exhibit the best reduction of acetylene. At pH of 5.6 or 8.0 the reduction of acetylene is still rather significant.

Resistance to antibiotics. The resistance to some typical antibiotics of various species of Azospirillum has been described by many authors (2, 3). In comparison with the results obtained by those authors, we can say that the spectrum of resistance to antibiotics used of both groups, being able or not able to use sucrose, is similar to that of the known Azospirillum.


CONCLUSIONS


Though there are still many other tests about Azospirillum spp. living in the rhizosphere of roots of rice plants to be done, we can say that most of isolates (group 1) appear to be Azospirillum lipoferum. The rest (group 2) have characteristics of Azospirillum brasilense, except the property to use glucose and sucrose for the dinitrogen dependent growth. The optimal temperature for the reduction of acetylene of group 1 occurs at 25°C-30°C, but for group 2 it is at about 30°C. The sensitivity to the oxygen content in gas phase of both groups is similar to that known for Azospirillum. Under slight alcaline condition, pH 7.1, there is the best reduction of acetylene. They possess the same spectrum of resistance to antibiotics of Azospirillum brasilense and Azospirillum lipoferum.

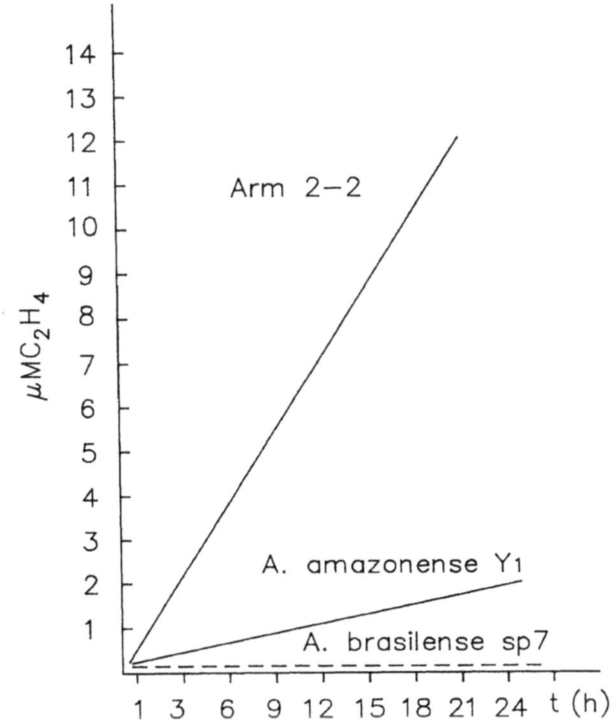


Fig.1 - $C_2H_2$-reduction by ARM 2-2 in comparison with the known Azospirillum in N-free semisolid medium with sucrose at 30°C, pH 6.89.

REFERENCES

1. Nguyen Ngoc Dung et al. (1987) in Klingmüller, W. (ed.) Azospirillum IV: Genetics Physiology Ecology. Springer-Verlag, p.242-246.
2. Tarrand J.I., Krieg N.R. and Döbereiner J. (1984) in Krieg, N.R. and Holt, J.G. (eds.) Bergey's Manual of Systematic Bacteriology, Vol.1. William and Wilkins, Baltimore.
3. Magalhäes I.M., Baldani J.I., Souto S.M., Kuykendall J.R. and Döbereiner J. (1983). An. Acad. Brasil. Cienc., 55(4) p.417-430.

ACKNOWLEDGEMENTS

We are grateful to Prof. Dr. W. Klingmüller, Genetics Department, Bayreuth University, Germany, for his effective help in all aspects; to DAAD for the financial aid to the stay in Germany and to Mrs. Phan Hong Lien for looking up and typewriting.

# MELANIN PRODUCTION BY *AZOSPIRILLUM LIPOFERUM* STRAINS

A. GIVAUDAN , A. EFFOSSE  and R. BALLY
*Laboratoire d'écologie microbienne CNRS  URA 697*
*Université Lyon I, Bat. 741, 43 Bd du 11 novembre 1918*
*69622 Villeurbanne, France*

Melanin production is a common property among fungi and streptomycetes , but it is less known among unicellular bacteria. Melanin formation from tyrosine is due to the catalytic activity of tyrosinase . Generally, this pigment is not essential for growth, but rather it  enhances the survival and competitive abilities of species in certain environments. Among nitrogen-fixing bacteria, *Rhizobium leguminosarum, R. meliloti* and *Sinorhizobium fredii*  produce a  dark-brown melanin pigment (4). One hypothesis concerning functions of melanin production in *Rhizobium* is that the tyrosinase functions in detoxifying polyphenolic compounds in senescing nodules (4). According to *Rhizobium* species, genetic studies have shown that *mel* genes are located either in the symbiotic plasmid or in cryptic plasmids (4). Among *Azospirillum*  spp., melanin formation has only been detected in *A.brasilense* Sp7 (8). However dark-brown pigment is only expressed in aging culture under influence of the cultural conditions conducive for encystement (8). In most *A.lipoferum* strains  tested , spontaneous non-motile forms appeared. *A. lipoferum*  4BpHT2 is a non-motile individual derived from *A.lipoferum* 4B, the parental strain isolated from the rhizosphere of rice (5).

In this work we reported that the non-motile *A. lipoferum*  4BpHT2 contrarly to the parental *A.lipoferum*  4B strain is melanin producer and we showed that most *A.lipoferum*  are melanin producers contrarly to *A.brasilense* strains. We located a homologous *mel*  region on a megaplasmid ( 270 Mda) of *A.lipoferum*  4B and  4BpHT2.


## Materials and Methods

*Bacterial strains*. Bacterial strains used in this study are *A. lipoferum*  (4B, 4BpHT2, 4T, Br10, Br17, B7C, 59b, SF50) and *A.brasilense* (Sp7, Sp245, RO7, A95, Cd)

*Detection of melanin production* . Bacteria were grown on TY(3) plate supplemented with

L-tyrosine (600 µg/ml or 1.5 mg/ml depending from  the experiments) and $CuSO_4$  (40 µg/ml). Grown colonies (5 days) were lysed with 20% sodium dodecyl sulfate. After a maximum of 48 h of incubation at room temperature, colonies that had produced a diffusible dark-brown pigment were scored as melanin producers (Mel +).

*Enzymatic assays. Azospirillum*  strains were cultivated in nutrient broth ( NB; Difco),  in nitrogen-free basal medium(NFb) (7) supplemented with malate (1g/l) and $NH_4Cl$ (0.5 g/l), in TY liquid medium . Media were supplemented with L-tyrosine and $CuSO_4$. Washed cells (24h old) were disrupted in potassium phosphate buffer 0.01M pH:7 by ultrasonic treatment. After centrifugation, the supernatant was used as intracellular enzyme extract. Phenol oxidase activity was measured using the Dopachrome assay procedure with L-dihydroxyphenylalanine (L-Dopa) as substrate (6).

312

## Results and Discussion

It was observed that old colonies of the non-motile *A.lipoferum* 4BpHT2 became dark- brown in colour. Thus we investigated whether other strains produced dark-brown pigment adding SDS to cultures grown (48h old) on TY plates supplemented with L-tyrosine and copper sulfate. Most *A.lipoferum* strains isolated from different origins were melanin-producers excepted Sp59b and 4B strains .This experiment confirmed that *A.lipoferum* 4BpHT2 produced melanin whereas no melanin production was detected in the wild strain 4B. This pigment seems to be a melanin-like pigment because no melanin production was detected on TY plate without addition of L-tyrosine. Although Sadivan and Neyra (1987) demonstrated that *A.brasilense* Sp7 produced a dark-brown pigment in aging cultures under carbon and nitrogen limitation conditions (conditions conduciving for encysting cells) (8), we found no melanin production among *A.brasilense* strains under cultural conditions tested.

Enzymatic extracts of *A.lipoferum* 4BpHT2 oxidized L-Dopa to Dopachrome. Thus apparently, *Azospirillum* cells synthesize melanin from L-Dopa by a polyphenol oxidase enzyme complex (tyrosinase and / or laccase). No activity was found in supernatant showing that enzyme was intracellular. *A.lipoferum* 4BpHT2 growing in several liquid media was tested for its enzyme activity. There was a higher level of enzyme activity when bacteria were cultivated in NFb medium or TY medium supplemented with L-tyrosine than in NB medium supplemented (activity 14-20 fold lower). In *Streptomyces glaucescens* tyrosinase expression is regulated by amino acids (2), According to our results, *Azospirillum* polyphenol oxidase activity was not enhanced when cells were cultivated in NFb medium supplemented with amino acids or with bactotryptone.

Enzymatic assays confirmed the lack of enzyme activity in *A.lipoferum* 4B wild strain. It is interesting to note that enzymatic activity levels were higher in non-motile *A.lipoferum* strains (4T, 4BpHT2) than in motile strains (B7C, Br17, 4B). Melanin production might be induced in *A.lipoferum* non-motile forms as well as in *A.brasilense* Sp7 cysts.

The function of melanin production in *Azospirillum* cells is unknown, but it could be associated to the survival of non-motile *A.lipoferum* strain in the rhizosphere. Further investigations are needed about the influence of polyphenolic root exudates on growth and polyphenol oxidase activity of *A.lipoferum*.

We are studying whether melanin production is plasmid-encoded in *A.lipoferum* 4BpHT2 strain. Plasmid patterns of wild strain 4B (Mel $^-$) and non-motile 4BpHT2 ( Mel$^+$) were similar and they harbored five cryptic plasmids (1). We showed that p270 plasmid DNA of 4BpHT2 hybridized with 1.55 kb fragment of pIJ702 plasmid containing *mel* genes of *Streptomyces antibioticus* . This homologous *mel* region was also located on p270 plasmid of the wild strain 4B (Mel $^-$) .To investigate whether p270 plasmid was responsible of phenol oxidase activity, curing experiments are undertaken. Using Tn5-Mob system, we obtained transconjugants of strain 4B carrying Tn5 insertion in p270 plasmid (1). We are currently attempting to cure this labelled Tn5-Mob plasmid using acridin orange treatments and heat curing.

## References


1. Bally, R. and A. Givaudan. (1988), Can. J. microbiol. 34, 1354-1357
2. Baumann, R. and H. P. Kocher (1976), In Genetics of industrial microorganisms, 2nd international symposium, Macdonald, K. D. (ed.)., London : Academic Press, pp 535-551
3. Beringer, J. E. (1974), J. Gen. Microbiol. 84, 188-198
4. Cubo, M. T., A. M. Buendia-Claveria , J. E. Beringer, and J. E. Ruiz-Sainz (1988), App. Env. Microbiol. 54, 1812-1817
5. Heulin , T., P. Weinhard and J. Balandreau. (1983), In W. Klingmüller (ed.), ,Azospirillum II Genetics, Physiology, Ecology, Experientia Supp.,48, pp 89-94
6. Katz, E., C. J. Thompson and D. A. Hopwood (1983), J. Gen. Microbiol. ,129, 2703-2714
7. Nelson, L. M. and R. Knowles (1978), Can. J. Microbiol. 24, 1395-1405
8. Sadivan, L. and C. Neyra (1987), J. Bacteriol. 169, 1670-1677

# EFFECT OF $pO_2$, $pCO_2$ AND AGITATION ON GROWTH KINETICS OF *Azospirillum lipoferum* UNDER FERMENTOR CONDITIONS .


1) E.PAUL, P.BLANC, A.PAREILLEUX, G.GOMA.
2) J.FAGES, D.MULARD.

1)*Centre de Transfert en Biotechnologie-Microbiologie, Dpt. Génie biochimique et alimentaire, UA-CNRS-544. INSAT Av. de Rangueil, 31077 Toulouse -Cedex, France.*
2) *Pioneer France-Maïs, laboratoire «Recherche Microbiologie Semences» chemin de l'enseigure, Borde Basse, 31840 Aussonne.*


The use of *Azospirillum lipoferum* as inoculant for agricultural crops requires a good control of cell production technology. In this way, it was interesting to define an optimal physiology state of the cells produced to ensure the best survival during storage and a good colonization of roots.

Only a few reports on the growth on organic acids or sugars of *A. lipoferum* in submerged cultures are available in literature. In this study, we report kinetic aspects of the growth of *A. lipoferum* crt1 isolated from corn roots by FAGES and MULARD (1).

The bacteria were grown in 2-L fermentors (SGI Toulouse, France) at 32°C, pH7. The dissolved oxygen (DO) concentration was measured by an Ingold amperometric probe. At the saturation, it was 7.2mg/l (32°C, $10^2$kPa).

## I) Influence of the Gazeous Environment on Growth.

The behaviour of *Azospirillum lipoferum* crt1 was first studied in batch cultures under standard conditions at 85% of saturation dissolved oxygen and 30g/l glucose concentrations. Then, the influence of various gazeous environments, i.e. various dissolved $O_2$ and $CO_2$ concentrations, were characterized using different conditions of agitation-aeration. We found a faster growth in the first stages of the culture at low dissolved oxygen fewer than 30% of saturation compared to the higher DO concentration used (up to 85% of saturation). The maximal biomass concentrations (dry weight and cell number) were however almost the same in all the experiments. In addition, as no evidence was found of an influence of shear stress on our results, the gazeous environments seemed to be a key parameter to control our growth kinetics.

Table 1. Growth characteristics of *Azospirillum lipoferum* for various agitation-aeration levels.

| operating conditions | | | growth characteristics | | |
|---|---|---|---|---|---|
| dissolved oxygen | initial agitation rate | initial $k_La$ | $\mu_{c1}$[a] | $\mu_{c2}$[b] | specific glucose consumption rate |
| (% of the saturation) | (rpm) | ($h^{-1}$) | ($h^{-1}$) | ($h^{-1}$) | (g/g.l) |
| 85 | 500 | 90 | 0.18 | 0.13 | 0.25 |
| 25 | 200 | 48 | 0.27 | 0.19 | 0.29 |
| 25 | 350 | 65 | 0.25 | 0.2 | 0.25 |
| 85 | 500 | 90 | 0.17 | 0.18 | 0.21 |
| 85 | 800 | 120 | 0.18 | 0.15 | 0.2 |
| 25 | 350 | - | 0.32 | 0.22 | 0.33 |
| 100 | 500 | - | 0.25 | 0.19 | 0.27 |
| 10 | 200 | - | 0.23 | 0.17 | 0.17 |
| 25 | 350 | - | 0.22 | 0.16 | 0.18 |
| 25 | 400 | - | 0.28 | 0.2 | 0.25 |
| 85 | 400 | - | 0.21 | 0.18 | 0.24 |

a: mean specific growth rate during the glucose non-consumption phase.
b: mean specific growth rate during the glucose consumption phase.

**II) Influence of the Dissolved $CO_2$ Concentration on Growth.**

A culture was carried out using $CO_2$ enriched air to regulate the dissolved $CO_2$ concentration in the medium at 30mbar (3.06%) during the first stages of the growth. This culture was compared to a reference culture carried out without $CO_2$ regulation. Since similar curves were obtained for the growths and the glucose consumptions, a possible influence of dissolved $CO_2$ concentration was discared.

Table 2. Growth characteristics of *Azospirillum lipoferum* at constant or non-regulated $CO_2$ partial pressure.

| operating conditions | | | growth characteristics | | |
|---|---|---|---|---|---|
| dissolved oxygen | initialagitation rate | dissolved $CO_2$ | $\mu_{c1}{}^{a}$ | $\mu_{c2}{}^{b}$ | specific glucose consumption rate |
| (% of the saturation) | (rpm) | (% initial) | (h$^{-1}$) | (h$^{-1}$) | (g/g.l) |
| 85 | 600 | 0.05 | 0.19 | 0.16 | 0.23 |
| 85 | 600 | 3.06 | 0.19 | 0.155 | 0.24 |

# III) Conclusion.

A detrimental effect of high dissolved oxygen levels and a favourable influence of low dissolved oxygen concentrations on growth were emphazised. Therefore, the gazeous environment needs to be considered as an operating parameter to ensure optimal growth in *Azospirillum* cultures.
Although *Azospirillum lipoferum* is described as a strict aerobe displaying an oxidative metabolism in presence of combined nitrogen our results demonstrate a sensitivity to oxygen, this specific behaviour being caracteristic of $N_2$-fixing organisms.

References:
(1) **FAGES, J., and D. MULARD.** 1988. Isolement de bactéries rhizosphériques et effet de leur inoculation en pots chez *Zea mays*. Agronomie 8:309-314.
(2) **PAUL, E., P. BLANC, G. GOMA, J. FAGES, D. MULARD and A. PAREILLEUX.** 1990. Effect of pO2, pCO2 and agitation on growth kinetics of *Azospirillum lipoferum* under fermentor conditions. Applied and Environmental Microbiology. (in press).

# SELECTION OF WETLAND RICE GENOTYPES FOR ASSOCIATED BIOLOGICAL NITROGEN FIXATION

S.Urquiaga, O.L.Silva, W.S.Trannin, H.F.Lopes,
F.O.Quintero and R.M.Boddey
*EMBRAPA/CNPBS, km 47, Seropédica, 23851,
Rio de Janeiro, BRAZIL.*

Experiments performed using $^{15}N$ labelled $N_2$ gas have demostrated that rice grown under wetland conditions is able to obtain a significant contribution from plant-associated biological nitrogen fixation (BNF) (Eskew et al., 1981; Yoshida and Yoneyama, 1981). N balance experiments performed in pots by App et al. (1986) showed that different genotypes differed in their capacity to benefit from this source. Wetland rice genotypes grown in Brazil exhibit large differences in their responses to fertilizer nitrogen, and in their capacity to accumulate nitrogen in low fertility soils when nitrogen is not applied, suggesting that some genotypes may obtain significant contributions from associated BNF.

In this study two experiments were performed, one with 20 genotypes in pots filled with 5.5 kg of soil (humic gley), pH 5.7, 0.06% N and fertilized with K, P and micronutrients and amended with 5.5 g/pot of compost labelled with $^{15}N$ (1.89% N, 5.25at.% $^{15}N$) (Urquiaga ct al., 1989). In the first experiment 4 successive crops were grown, each genotype replanted in the same pot. The other experiment was performed with 40 genotypes in a large concrete tank (120 $m^2$) filled with 62 tons of soil (Red-yellow podzolic) labelled with 0.316 g $^{15}N$ per $m^2$. The soil was flooded from 30 days after planting onwards.

The grain production, dry matter, N content and $^{15}N$ enrichment were evaluated.

The greatest differences in $^{15}N$ enrichments were found within the group of later maturing rice genotypes. This it is amongst the genotypes with the lowest enrichments within this group that genotypes may be found which possibly were able to obtain significant contributions of N from BNF. Among these genotypes are IR 22, De Abril, CNA 6750, BG 90-2, MG 444 and Metica 1. The genotypes which appeared to benefit least from BNF were IAC 4440 and Bluebelle.

Further studies using the same technique are now under way, in soil which now is more stable in $^{15}N$ enrichment and lower in N availability, which will improve the sensitivity of the technique.

REFERENCES

App, A.A.; Watanabe, I.; Ventura, T.S.; Bravo, M. & Jurey, C.D. (1986) The effect of cultivated and wild rice varieties on the nitrogen balance of flooded soil. Soil Sci. 141;448-452.

Eskew, D.L.; Eaglesham, A.R. & App, A.A. (1981) Heterotrophic $^{15}N_2$ fixation and distribution of newly fixed nitrogen in a rice-flooded soil system. Plant Physiol. 68:48-52.

Yoshida, T. & Yoneyama, T. (1980) Atmospheric dinitrogen fixation in the flooded rice rhizosphere as determined by the N-15 isotope technique. Soil Sci. Plant Nutr. 26:551-589.

Urquiaga, S.; Botteon, P.B.L. & Boddey, R.M. (1989) Selection of sugar cane cultivars for associated biological nitrogen fixation using $^{15}N$-labelled soil. In: Skinner et al. (Eds.). Nitrogen fixation with non-legumes. Ed. Kluwer Academic Publishers, p. 311-319.

ENRICHMENT OF TOMATO SAND CULTURES WITH COMPOSITE INOCULA OF
ASSOCIATIVE DINITROGEN FIXERS, P-DISSOLVING BACILLI AND VAM.


M. MONIB, M. SABER, A.M. GOMAA and N.A. HEGAZI
Faculty of Agriculture, Cairo University, Giza and
National Research Centre, Dokki, EGYPT



ABSTRACT. A pot experiment was designed to investigate effects of inoculation of tomato plants with a composite inocula of Azotobacter spp,
Azospirillum spp and either P-dissolving bacilli or VAM. Simultaneous
inoculation with Azotobacter spp + Azospirillum spp significantly increased number of leaves, root biomass, total dry matter and P- and N-
yields. Increases were magnified when VAM was included in the inocula.
A limited dose of NPK-fertilizers intensified response of plants to
inoculation to a magnitude comparable to those receiving full dose of
NPK-fertilizers. A significant cultivar effect was demonstrated.


## 1.1. Introduction

Sandy soils are almost devoid of functional populations of microorganisms related to biofertility. Therefore introduction of selected RMO to
the plant-soil system might be a workable approach to partially support
plant growth and to build up biofertility of such soils.  The present
work is an attempt to introduce mixed population of complementary
groups of RMO, e.g. associative dinitrogen-fixers and phosphate dissolving and extracting microorganisms. Tomato is used as a host plant.

## 1.2. Materials and Methods

The sandy soil used was of the following analysis: organic carbon,
0.04%; total-N, 0.01%; total phosphorous, 0.04%; W.H.C., 21.9%; pH,
7.3; E.C., 3.2 mmhos/cm; total microbial count, $7 \times 10^3$ c.f.u $g^{-1}$; total
mycorrhizal spores, 900-1200 spore $kg^{-1}$.
   A number of rhizopheric microorganisms (RMO) was isolated from
soils of Nile Valley.  They belonged to the $N_2$-fixing Azotobacter
chroococcum and Azospirillum lipoferum, and phosphate dissolving
Bacillus brevis (PDB).  Besides, enough mycorrhizal spores (VAM) were
collected.  Prior inoculation, liquid cultures were independently grown
at 30°C for 4-7 days. Seeds of 2 tomato cultivars, cv. Ace and cv. UC82
were germinated in trays filled with neutralized peat. Composite liquid
inocula were applied in the nursery with the following combinations: A.
chroococcum ($10^9$ cells mll) + A. lipoferum ($10^8$ cells $ml^{-1}$), A. chroococcum + A. lipoferum + B. brevis ($10^8$ cells $ml^{-1}$) and A. chroococcum +
A. lipoferum + VAM (500 spores $ml^{-1}$). Initially, each seedling received
ca. 5 ml broth consisting of equal portions of RMO tested, and after 10
days seedlings were thinned. Plastic pots containing 13.5 kg sandy soil
were prepared and 30 days-old seedlings were transplanted.  The experiment was carried out in a complete randomized block design with 5
replicates prepared for each of the following sets of pots: inoculated
plants, inoculated plants receiving as well full dose of NPK-fertilizers (180 kg N, 90 kg $P_2O_5$ and 90 kg $K_2O$ $h^{-1}$) and inoculated plants

receiving 1/3 rd of the full dose of NPK-fertilizers.  Pots were kept under ambient conditions of greenhouse and watered regularly.  At the end of flowering stage (80 days-old plants were uprooted, dried, weighed and powdered for analysis of total-N (Allen 1953) and total-P (Jackson 1960).  Data was computed using ANOVA 3 programme.

1.3. Results and Conclusions

Treating tomato plants distinguishably increased the percentage of infection of VAM up to >33-45% for either cultivars compared to non-treated plants (4-6%).  Mineral fertilization decreased to some extent such percentages.

In general, application of different combinations of composite inocula significantly increased total dry weight with various magnitudes (Table 1).  The percentage increase over control approximated 110% due to the application of formula composed of Azotobacter spp + Azospirillum spp.  Inclusion of BDB or VAM in the formula used resulted in additional increases, being 156% and 254% over control respectively. This could be attributed to the multibeneficial action of VAM in particular in the growth of tomato plants which was expressed in terms of high percentage of infection.  Application of a low dose of NPK-fertilizers intensified plant response to various formula of inocula. Highest increases were reported for fertilized plants simultaneously treated with composite formula of Azotobacter spp + Azospirillum spp + VAM.

Significant increases in dry weight of roots was attributed to inoculation as well. The cv. UC82 responded with a higher magnitude. Statistical significant differences were attributed to various degrees of interactions among inoculation, NPK-fertilization and cultivars.

Table 1: Response of tomato plant to inoculation and fertilization

| Treatments | Dry weight (gm/plant) | | Total yield (mg/plant) | |
| | plant | roots | N- | P- |
| --- | --- | --- | --- | --- |
| Control (+ 1/3 NPK) | 0.91 | 0.20 | 10.66 | 1.10 |
| Full dose of NPK | 2.62 | - | 69.31 | 5.21 |
| Azotobacter + Azosprillum + 1/3 NPK | 3.72 | 0.50 | 82.40 | 6.00 |
| Azotobacter + Azospirillum + Bacillus + 1/3 NPK | 3.95 | 0.75 | 81.53 | 5.84 |
| Azotobacter + Azospirillum + VAM + 1/3 NPK | 6.75 | 0.97 | 132.96 | 11.16 |
| LSD          0.05 | 1.18 | 0.19 | 23.47 | 2.33 |

N-yield was significantly increased in response to various combinations of composite inocula. Formula composed of <u>Azotobacter</u> spp + <u>Azospirillum</u> spp + VAM did result in distinguishable significant increases comparable to N-yield obtained with application of full doses of NPK-fertilizers. P-yield followed a similar trend as inoculation with various combinations of RMO - particularly in presence of low doses of NPK-fertilizers - did increase with significant levels the P-yield of plants. The inclusion of VAM, compared to BDB, in inocula tremendously increased P-yields of plants. Interaction among inoculation and cultivars was in favour of cv. UC82. Among other plant parameters which showed positive response to inoculation was the number of leaves.

1.4. References

Allen, O.N. (1953)  Experiments in Soil Bacteriology, Burgess
    Publishers, USA.
Jackson, M.L. (1960)  Soil Chemical Analysis, Prentice Hall Englewood
    Cliffs, N.J. USA.

PARTIAL PURIFICATION AND PROPERTIES OF TWO AROMATIC AMINO ACID
AMINOTRANSFERASES FROM THE A. brasilense UAP 14 STRAIN INDOLE-3-
ACETIC ACID PRODUCER.


Beatriz Eugenia Baca, L. Soto-Urzúa, and Y.G. Xochihua-
Corona.
Departamento de Investigaciones Microbiológicas, Instituto
de Ciencias, Universidad Autónoma de Puebla, México
P.O. Box 1622, 72000.


## INTRODUCTION

Several pathways are leading in higher plants and microorganisms from
tryptophan to IAA. The most widely spread pathway of IAA synthesis
seems to be the indole-3-pyruvic  acid (IPyA) pathway. The first step,
deamination of tryptophan leading IPyA could be cataluzed by an aminotransferase. We have detected two aromatic aminoacid aminotransferases
$AAT_1$ and $AAT_2$ from A. brasilense UAP 14 strain, as a first step in the
investigation of this hipothesis, attention has been directed towards
the isolation and characterization of $AAT_s$ which are present in extracts of Azospirillum. The present communication reports the isolation
of two $AAT_2$. Both enzymes may involved in the biosynthesis of phenylalanine and tyrosine, one of two would be involved in catabolism of his
todine. The results also suggested that $AAT_s$ contribute to IAA biosynthesis in Azospirillum brasilense.

## RESULTS

The following procedure was developed for the purification of two
aminotransferases which both convert tryptophan to IPyA. The purification steps includes $(NH_4)_2SO_4$ precipitation, DEAE-Sephacel Hydroxylapa
tite, and Gel-Filtration chromatographies. Since both enzymes utilize
tryptophan or tyrosine as an amino group donor, and IPyA of  hydroxy-
phenylpyruvate (pHPP) can be estimated easily. The purification of
enzymes were followed by determining the tryptophan and tyrosine amino
transforase activities of the various fractions.

DEAE-Sephacel chromatography.- The cells were grown and lysed by lysozyme. The cell free extract was precipitated by ammonium sulfate,
the precipitate was removed by centrifugation. Supernatant solution
was collected and dialysed. The clear solution from the previous step
was applied to a column (1.6 X 70 cm) of DEAE-Sephacel which has been
previously equilibrated. After washing the column with the same buffer,
protein was eluted at 12 ml./h, with a gradient of 0.45 M KCl in initial
buffer. Those fractions which eluted at 0.125 m KCl, and which possessed both enzymatic activities were pooled, and the protein precipitated by ammonium sulfate. The precipitate was suspended and dialysed
evernight.

Chromatography on hydroxylapatite.- The enzyme solution from the previous step was applied at the rate of 12 ml/H to a column (1X15 cm) of hydroxylapatite. The enzymes were eluted with a linear gradient formed from 2mM to 40mM potassium phosphate buffer. The combined active fractions were precipitated with ammonium sulfate. The precipitate was dissolved in a minimun amount of buffer and dialysed.

Sephadex G-100 Gel-Filtration chromatography.- The enzymes were applied to Sephadex G-100 column (2.6 X 100 cm). The molecular weight of native enzymes was estimated to be about 52,118. However when these fractions were pooled, and molecular weight determination of $AAT_s$ were made by gel electrophoresis under non desnaturating conditions, were estimated as 56,000 D. for $AAT_1$, and 44,000 D. for $AAT_2$. The enzymes show amino donor specificity for aromatic amino acid. Aminotransferase activity for other non aromatic amino acid was not detected. The enzymes showed a pH range, from 6.5 to 9 having their maximal activity at pH 8.5

## DISCUSSION

Our present sutyd demonstrated that IAA producer A. brasilense UAP 14 strain had two $AAT_2$, and these enzymes were partially purified from the extract of this microorganism. The enzymes were purified 105-fold with an appropiated recovery of   %. Nevertheless the purification procedure reported in this communication did not allow the separation and preparation in an homogenous state of two aromatic amino acid aminotransferases. In comparison with the molecular weight of $AAT_2$ present in E. coli (90,000 $\pm$ 4,000); in B. brevis (70,000); in R. leguminosarum those $AAT_s$ enzymes ranged from 59,000 to 61,000. While the enzymes from A. brasilense UAP 14 strain $AAT_1$ 56,000 and $AAT_2$ 44,000 have a lower molecular weught. Also the AAT of B. brevis showed a broad pH optima near 7.6, that one of E. coli has a sharp pH optima near 7.4. However $AAT_2$ from A. brasilense showed a broad pH optima near 8.5.

Pérez-Galdona et al (1989) have been described five aromatic amino acid aminotransferases in R. leguminosarum. Although the existance of more than one AAT ahd been demonstrated previously, the number of such enzymes and their biological significance remained uncertain. But is tempting to speculated that in Azospirillum $AAT_s$ could be involved in the terminal reaction in the biosynthesis of tyrosine and phenylalanine, and in IAA production.

## ACKNOWLEDGEMENTS

This work was partially supported by Consejo Estatal de Ciencia y Tecnología de Puebla. We thank Norma L. Arriaga by manuscript typing/

CHARACTERIZATION AND REGULATION OF TWO ENZYMATIC ACTIVITIES INVOLVED
IN BIOSYNTHESIS OF INDOLEACETIC ACID IN Azospirillum brasilense UAP 14
STRAIN.


B.E. BACA, L. SOTO-URZUA, Y.G. XOCHIHUA-CORONA AND A. CUERVO
GARCIA.
Departamento de Investigaciones Microbiológicas, Instituto
de Ciencias, Universidad Autónoma de Puebla Pue. México
P.O. Box 1622, 72000.


INTRODUCTION
Soil bacteria of genus Azospirillum live in association with the roots
zones of grasses and others plants, among which are the economical important culture plants corn, wheat, barley, rye, oat, and rice. There
are several reports indicating that Azospirillum can stimulate the
growth of the host plant by nitrogen fixation and production of plant
growth regulators as indoleacetic acid (IAA).

In this study, we showed that Azospirillum spp wild-type strains
produced IAA and we demonstrated the presence of two aromatic acid ami
notransferases associated with the production of IAA in A. brasilense
UAP 14 strain.

RESULTS
We studied the acumulation in the culture medium of IAA released by
Azospirillum wild-type strains, to understand the influence of nutrients
conditions on the synthesis of phytohormones by these microorganisms.
The IAA production was influenced by the nutrients contained in the
medium and the incubation time as is show in table 1. Auxin produced
by Azospirillum was released continously specially when the bacteria
were grown in medium supplemented by the amino acid precursor. IAA excreted in medium added with potassium nitrate was about 2 to 10-fold
(0.139-1.15 µg/ml) in A. lipoferum UAP 06, A. brasilense Az 30 INRA, and H 18
strains, but did not in A. brasilense. A very significative increase in
auxin production was observed, when the medium was supplemented with
tryptophan 50 to 100-fold higher (30-64 µg/ml) compared with $KNO_3$ and
$NH_4Cl$. Although the amino acid precursor was added at the begining, it
was also observed that IAA liberation was stimulated at 48h of growth.

In order to know if aromatic amino acid aminotransferase (AAT) could
be involved in biosynthesis of IAA we identified aminotransferase activity that react with tryptophan in crude extracts and ammonium sulfate
fractionation; from A. brasilense UAP 14 strain.

324

TABLE 1. IAA PRODUCTION BY Azospirillum spp. WILD-TYPE STRAINS.

| STRAINS | IAA µg/ml of the culture medium | | | | | | | | | | | | |
|---|---|---|---|---|---|---|---|---|---|---|---|---|---|
| | $NH_4Cl$ | | | $KNO_3$ | | | $NH_4Cl$ Trp | | | $KNO_3$ Trp | | | |
| | 48 | 72 | 96 | 48 | 72 | 96 | 48 | 72 | 96 | 48 | 72 | 96 | H |
| A. lipoferum UAP 06 | .02 | .07 | .15 | ND | .15 | 1.1 | .27 | 30 | 33 | 7 | 38 | 54 | |
| A. brasilense | | | | | | | | | | | | | |
| Az 30 INRA | .04 | .02 | .04 | ND | .2 | .88 | .04 | 5 | 29 | 34 | 39 | 45 | |
| H 18 | .03 | .05 | .27 | ND | .15 | .61 | 4 | 27 | 33 | 23 | 25 | 30 | |
| UAP 14 | .16 | .4 | .8 | ND | .10 | .13 | 15 | 48 | 63 | 5 | 25 | 64 | |

Specific activities in extracts from cultured without amino acid
ranged from 16 to 22 nmoles (IPyA)/min/mg of protein, whereas from cultured
with tyrosine, phenylalanine plus tyrosine or tryptophan ranged from 27
to 37 nmoles (IPyA)/min/mg of protein. When protein extracts were submitted to electrophoresis on non-denaturating gels and stained for AAT
activity, two bands (Fig. 1) were detected both converts tryptophan to
IPyA in the presence of 2-oxoglutaric acid. Omitting either tryptophan
or 2-oxoglutaric acid from the mixture reaction completely eliminated
the two bands. Substitution of phenylalanine or tyrosine for tryptophan,
reveal exactly the same protein pattern seen with tryptophan. Only the
larger band showed aminotransferase activity when histidine was used as
donor substrate. In conditions that may be expected to have an effect,
it was always observed the identical protein pattern. This result indicate that any of these two AATs are not regulated by the presence of
these amino acids.

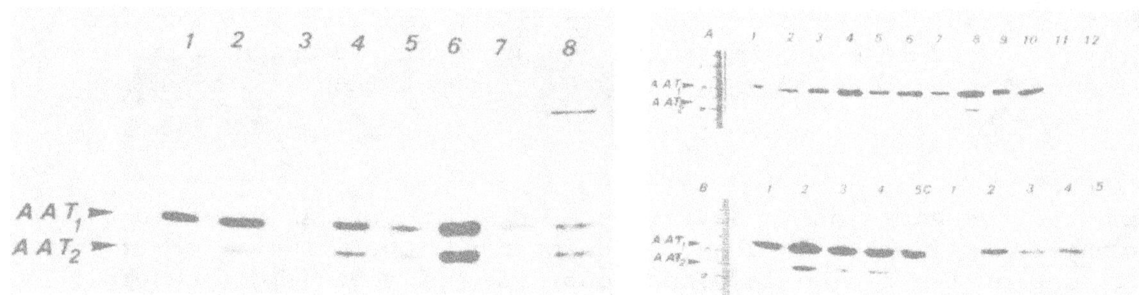

DISCUSSION

We quantified the IAA excreted by differents wild-type strains of
Azospirillum spp. The microorganisms produce IAA, during the late stationary growth phase and show a significative increase when tryptophan
is present in the medium. A. brasilense UAP 14 strain was grown under
different nutrient conditions, we observed the presence of two enzymes
$AAT_1$ and $AAT_2$. Characterization of the enzymatic activity and the electrophoretic mobility under non-denaturating conditions showed that the
se enzymes can be distinguished by their electrophoretic mobilities,
substrates specificities, and they are constitutively produced and
none of them are repressed by tyrosine.

Analysis of extracellular polysaccharides of *Azospirillum brasilense*

Paul De Troch, Saleela Philip-Hollingsworth, Guy Orgambide, Kris Michiels, Jos Vanderleyden and Frank B. Dazzo.
F. A. Janssens Memorial Laboratory for Genetics, W. Decroylaan 42, B 3030 Leuven
Dept. of Microbiology and Public Health, Giltner Hall, Michigan State University, East Lansing, Michigan 48824-1101

Bacteria of the genus *Azospirillum* live in association with the Graminae. *Azospirillum* has a growth-promoting effect due to nitrogen-fixation, phytohormone production and improved water household of the plant. *Azospirillum* is able to attach to plant roots (Umali-Garcia *et al.* 1980) and *A. brasilense* Sp245 can even invade into the root cortex. In plant bacteria interactions, signalling is an important factor. There are different signal molecules known in other microbe-plant interactions : $\beta$ 1-2 glucans in the interaction of *Agrobacterium* and *Rhizobium* for respectively tumor-formation and nodule-induction; hepta-saccharides which act as elicitors for the plant-defense mechanisms against plant pathogens; extracellular polysaccharides (EPS) necessary for the formation of nitrogen-fixing nodules on legumes by *Rhizobium*. The production of EPS by *Azospirillum* was shown by the fluorescent dye Calcofluor. In the F. A. Janssens Memorial Laboratory for Genetics, recent interest is focussed on the genetics of the synthesis of this EPS by *Azospirillum*. Tn5 induced *A. brasilense* mutants that are are affected in Calcofluor fluorescence have already been isolated (Michiels *et al.*, 1988a; 1990). Some *A. brasilense* genes are able to correct *R. meliloti exo* mutations. In this way, the *R. meliloti exoB, exoC, exoG, exoM, exoN, exoP* and *exoK* mutants have been genetically complemented by *A. brasilense* DNA. The *exoB, exoC1* and *exoM* correcting loci are located on the 90 MDa plasmid of *A. brasilense* Sp7 (Michiels *et al.*, 1988b; Marc Eyers, personal communication), whereas the *exoC2, exoG, exoK, exoN* and *exoP* correcting loci are located on the chromosome (Marc Eyers, personal communicaton). The *A. brasilense exoB* en *exoC1* mutants have been constructed by marker-exchange (Michiels *et al.* 1988b).

We have started the structure-analysis of EPS produced by *A. brasilense* Sp7 and Sp245 wild type strains. EPS were isolated from *Azospirilla*, grown on BIV agar at 30°C during 5 days. EPS were precipitated with ethanol, dissolved in and dialyzed against distilled water and lyophilized. The EPS were fractionated by gel permeation. The EPS were analyzed by $^1$H-NMR and by GC and GC-MS of alditol-acetate derivatives. The EPS isolated from *A. brasilense* Sp7 contains glucose (45%), galactose (26%), rhamnose (14%) and fucose (15%). The EPS, isolated from *A. brasilense* Sp245 contains glucose (60%), rhamnose (31%) and galactose (9%). The difference in EPS between the two strains could reflect a difference in host plant and in site of interaction : *A. brasilense* Sp245 has already been found in the endorhizosphere of wheat, while *A. brasilense* Sp7 was isolated from the surface of *Digitaria* (Döbereiner and Day, 1976; Baldani *et al.* 1986). This is the first evidence for differences in the surface of these two strains. The high glucose content in the HMW fractions of both strains could be due to the presence of a homoglucan, e.g. cellulose, in addition to an heteropolymer.

By microscopic observation of Calcofluor stained *A. brasilense*, fluorescence was detected on the cell surface as well on the surrounding material, which is fibrillar. The fibrillar material could be cellulose as suggested by other research groups (Del Gallo *et al.* 1989). The mutants were also microscopically observed for Calcofluor staining : some mutants lack the fluorescence of the cell surface, others lack the fluorescence of the surrounding material and some lack the fluorescence of both the cell surface and of the surrounding material. This means that the mutants we obtained are affected in the synthesis of different polysaccharides and probably in different steps of their biosynthesis. Calcofluor fluorescence-intensity was also measured quantitatively for these mutants. We also tested the attachment to wheat roots of these mutants, using radioactive labelled cells.

From these tests, it could be concluded that there is a correlation between floc formation, Calcofluor phenotype, presence of fibrillar material and attachment to wheat roots. It seems that cellulose or another Calcofluor-binding heteropolymer is important for floc formation and attachment to plant cells.


Baldani, V. L. D., Alvarez, M. A. de B., Baldani, J. I. and Döbereiner, J. 1986. Plant and Soil, 90, 35 - 46.

Delgallo, M., Negi, M. en Neyra, C. A. 1989. J. Bacteriol. **171**, 3504 - 3510.

Döbereiner, J. en Day, J. 1976. *In* Newton, E. W. en Nyman, C. J. (eds.). Proceedings of the 1st Intern. Symp. on $N_2$ Fixation. Washington State University Press, Pullman, 2, pp. 518 - 538.

Michiels, K., Vanderleyden, J., Van Gool, A. en Signer, E. R. 1988a. *In* Skinner, F., Boddey, R. M. en Fendrik, I. (eds.). Nitrogen fixation with non-legumes. Kluwer, The Netherlands, pp. 189 - 195.

Michiels, K., Vanderleyden, J., Van Gool, A. en Signer, E. R. 1988b. J. Bacteriol. **170**, 5401 - 5404.

Michiels, K., Vanderleyden, J. en Van Gool, A. 1989a. Biol. Fertil. Soils, **8**, 356 - 368.

Michiels, K., De Troch, P., Onyeocha, I., Van Gool, A., Elmerich, C. and Vanderleyden, J. 1989b. Plasmid **21**, 142 - 146.

Michiels, K., Verreth, C., and Vanderleyden, J. 1990. J. Appl. Bacteriol. 69, in press.

Umali-Garcia, M., Hubbell, D. H., Gaskins, M. H. and Dazzo, F. B. 1980. Appl. Environ. Microbiol. **39**, 219 - 226.



Acknowledgments
P. De Troch is the recipient of a fellowship of the IWONL. This work was sponsored by the National Fund for Scientific Research, Belgium, and the Onderzoekstoelage K.U.Leuven.

# $NO_3^-$ and $Fe^{++}$ uptake in durum wheat is affected by Azospirillum brasilense inoculation .


Baggio C., Antonello F., Barbieri P.*,Saccomani M., Nuti M.P.
Dip. di Biotecnologie Agrarie, Univ. di Padova via Gradenigo 6, 35131 Padova, Italy. * Dip. di Genetica e Biologia dei Microrganismi, Univ. di Milano, via Celoria 26, 20133 Milano, Italy.


## 1.Introduction

Several reports have suggested that inoculation of cereal plants with different species of the beneficial rhizosphere bacterium Azospirillum results in an enhanced plant growth (Levanony et al. 1989). The underlying mechanism by which Azospirillum affects plant growth are poorly understood. One mode of action could be linked to an enhancement of mineral uptake by the root system (Pacovsky 1990) and of root surface area (Levanony et al. 1989), although quantitative data are scanty. This communication provides evidence that A.brasilense promotes root and root hair growth along with the uptake of $NO_3^-$, and enhances $Fe^{++}$ accumulation in plant tissues.

## 2.Materials and Methods

Triticum durum cv Appulo was hydroponically grown in an illuminated chamber; the plants were aseptically or monoxenically maintained up to two weeks, and growth conditions were developed to minimize nitrogen losses through dissimilatory pathway. As inoculum the following strains were used:(1) wild-type A.brasilense Sp6, (2) mutants impaired in IAA production or nitrogen fixation, derived from the wild type by conventional mutagenesis or containing Tn5 as insertional gene inactivator (strains SpF103, SpF57, SpF7918 respectively).

## 3.Results

Uptake rate of NO3- by T.durum plants according to the inoculant strain. Two weeks after inoculation , 55 nmol $NO_3^-$x min-1xplant-1 are taken up in presence of Sp6, about 20% above control plants; after the same period, the increase of $NO_3^-$ uptake in the presence of SpF103 (IAA hyperproducer strain) is 59% above the control plants, while 7% decrease was significantly detected in the presence of SpF7918 (Tn5 mutant of Sp6, quasi-zero IAA

producer) (Fig.1).
$Fe^{++}$ content can increase up to 30% above control plants, following inoculation with Sp6; the increase is less, but still significant, in plants inoculated with SpF7918 (Fig.2).
Short term exposure of <u>T.durum</u> to <u>A.brasilense</u> Sp6 enhances root and root hair growth.

## 4.Conclusive remarks
Overall, this study extends and strenghtens previous findings and provides quantitatives data of nutrient uptake in plants inoculated with <u>Azospirillum</u>.
Although the syntesis of auxin (IAA) rapresents a relevant biochemical trait, in <u>Azospirillum</u>, affecting its relationship with the root system of <u>Triticum</u>, the presence of factor(s) other than IAA cannot be ruled out. Indeed mutants impaired in IAA synthesis exhibit stimulatory effects significantly higher than the controls, in both $NO_3^-$ uptake rate and $Fe^{++}$ accumulation by the plant.

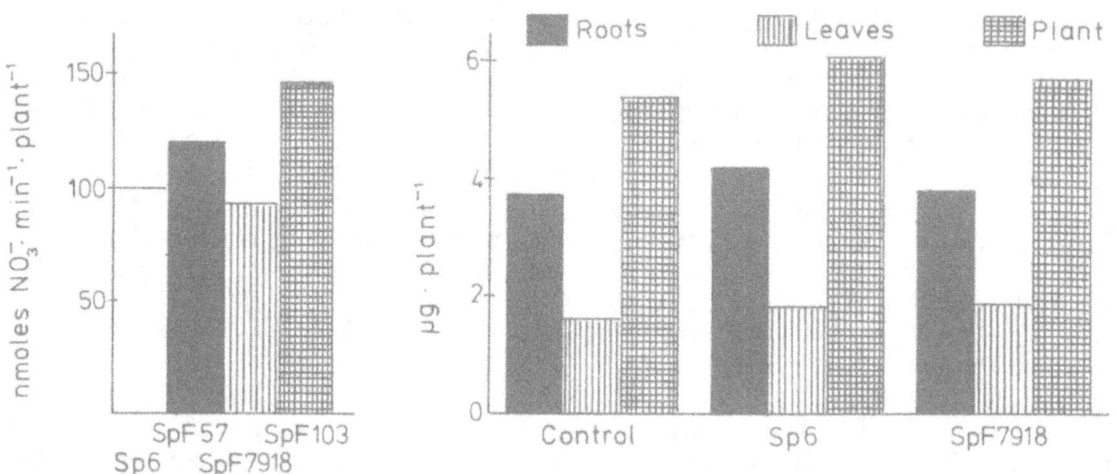


**Fig.1** Effect of <u>A.brasilense</u> strains on $NO_3^-$ uptake in respect to plants inoculated with wild type strain Sp6 (10th day from inoculation).

**Fig.2** Effect of <u>A.brasilense</u> strains on $Fe^{++}$ accumulation (8th day from inoculation).

## 5.References

Levanony, H., Bashan Y.(1989)' Enhancement of cell division in wheat root tips and growth of root elongation zone induced by <u>Azospirillum brasilense</u> Cd' Can.J.Bot. 67, 2213-2216.
Pacovsky, R.S.(1990) 'Development and growth effects in the <u>Sorghum</u>-<u>Azospirillum</u> association'J.Appl.Bacter. 68,555-563.

EFFICIENCY OF AZOTOBACTER STRAINS DEPENDING ON NITROGEN LEVEL AND  SUGAR
BEET GENOTYPES


B. KRSTIĆ, M. SARIĆ, ZORA SARIĆ*
Institute of Biology, Faculty of Sciences
Institute of Field and Vegetable Crops, Faculty of Agriculture*
21000 Novi Sad
Yugoslavia


ABSTRACT. The  six  sugar beet cultivars were grown in  sand  cultures,
supplied with different levels of nitrogen (1/8N,  1N,  2N).  The plants
were inoculated with different Azotobacter strains. Dry matter masses of
shoot as well as concentration and contents of nitrogen were determined.
Our  results show that the obtained effect considerably  dependent  upon
relationships between a certain genotype,  Azotobacter strain and  level
of nitrogen.

INTRODUCTION. Recently  attention has been focused on  the  relationship
between strains of nitrogen-fixing microoragnisms and plant species.  We
continued  our investigation of genetic specificity of plant  nutrition,
with plant genotypes and Azotobacter strains [1, 2, 3].
    The emphasis of this study is on the interaction between sugar  beet
genotypes and nitrogen level on the efficiency of different  Azotobacter
strains.

MATERIALS AND METHODS. We tested the effect of five Azotobacter  strains
(1S,  2S,  3S,  4S and 5S) on six sugar beet genotypes (1.  Ritmo, 2. KW
Lena,  3. KW Maja, 4. NS-Mo-1, 5. NS-H-92 and  6. NS-3xpZ).  Experimental
plants  were  grown in sand cultures with  Reid-York  nutritive  medium.
Three  nitrogen  level were used (1/8N,  1N and  2N).  The  plants  were
inoculated with 3 ml of the substrate containing $10^8$ Azotobacter  cells.
The control plant were grown without inoculation.  Tissue N was analysed
by  the  Kjeldahl  method.  All results were processed  by  analysis  of
variance (LSD test). The sum of positive or negative effect are given in
compare to control plants at LSD=5%.

RESULTS. The  effect of the association Azotobacter-sugar beet  depended
on 3 conditions:  sugar beet genotypes,  Azotobacter strain and nitrogen
content  in the nutritive medium. Considering the sugar  beet  genotypes
(Tab.  1) the inoculation was most effective for all parameters by NS-H-
92,  NS-3xpZ and NS-Mo-1 genotypes. The lowest effect of inoculation was
observed  in  cultivar Ritmo.  Regarding the effect  of  inoculation  on
shoots and roots the results show that strain affected the concentration
and content in shot specifically.

330

**TABLE 1. Sum of significant positive and negative effect caused by Azotobacter strain depending on sugar beet genotypes**

| Organs | | 1 | 2 | 3 | 4 | 5 | 6 | Total |
|---|---|---|---|---|---|---|---|---|
| | | | | **G e n o t y p e s** | | | | |
| | | | | **Dry matter mass** | | | | |
| Root | + | 5 | 5 | 2 | 5 | 5 | 4 | 26 |
| | − | 1 | 0 | 0 | 0 | 1 | 0 | 2 |
| Shoot | + | 4 | 3 | 3 | 5 | 4 | 3 | 22 |
| | − | 3 | 1 | 1 | 1 | 1 | 2 | 9 |
| | | | | **N-concentration** | | | | |
| Root | + | 1 | 5 | 4 | 3 | 5 | 9 | 27 |
| | − | 11 | 8 | 7 | 8 | 8 | 4 | 46 |
| Shoot | + | 7 | 9 | 8 | 13 | 13 | 12 | 62 |
| | − | 6 | 4 | 5 | 1 | 1 | 1 | 18 |
| | | | | **N-content** | | | | |
| Root | + | 8 | 9 | 7 | 9 | 10 | 8 | 51 |
| | − | 3 | 0 | 3 | 1 | 0 | 3 | 10 |
| Shoot | + | 9 | 12 | 10 | 10 | 12 | 10 | 63 |
| | − | 5 | 1 | 2 | 3 | 3 | 2 | 16 |
| Total | + | 34 | 43 | 34 | 45 | 49 | 46 | |
| | − | 29 | 14 | 17 | 14 | 14 | 12 | |

**TABLE 2. Sum of significant positive and negative effect on dry matter mass, N-concentration and content depending on Azotobacter strain**

| Organs | | 1S | 2S | 3S | 4S | 5S | Total |
|---|---|---|---|---|---|---|---|
| | | | | **S t r a i n s** | | | |
| | | | | **Dry matter mass** | | | |
| Root | + | 4 | 5 | 4 | 4 | 7 | 24 |
| | − | 1 | 0 | 1 | 1 | 0 | 3 |
| Shoot | + | 2 | 4 | 2 | 6 | 7 | 21 |
| | − | 3 | 1 | 3 | 1 | 0 | 8 |
| | | | | **N-concentration** | | | |
| Root | + | 4 | 5 | 4 | 6 | 6 | 25 |
| | − | 10 | 10 | 11 | 9 | 8 | 48 |
| Shoot | + | 13 | 12 | 12 | 12 | 13 | 62 |
| | − | 4 | 2 | 4 | 4 | 4 | 18 |
| | | | | **N-content** | | | |
| Root | + | 6 | 11 | 9 | 11 | 13 | 50 |
| | − | 2 | 3 | 2 | 3 | 1 | 11 |
| Shoot | + | 8 | 14 | 10 | 14 | 16 | 62 |
| | − | 8 | 1 | 6 | 1 | 1 | 17 |
| Total | + | 37 | 51 | 41 | 53 | 62 | |
| | − | 28 | 7 | 27 | 19 | 14 | |

Interaction between strains and genotypes (Tab. 2) stimulated to a various degree the dry mass, concentration and content of nitrogen. For example the Azotobacter strain 5S stimulated more strongly all investigated parameters when strain 1S was applied. In average, the poorest effect of inoculation upon dry matter, concentration and content of nitrogen was obtained in roots.

The effect of the Azotobacter strain on the dry matter mass, concentration and content of nitrogen slightly increased with increasing nitrogen dose in nutritive medium (Tab. 3)

TABLE 3. Sum of positive and negative effect on dry mass, concentration and content of nitrogen depending on N-level

| Nitrogen level | | Dry matter mass | | N-concentration | | N-content | | Total |
|---|---|---|---|---|---|---|---|---|
| | | Shoot | Root | Shoot | Root | Shoot | Root | |
| 1/8N | + | 2 | 2 | 8 | 19 | 13 | 14 | 58 |
| | − | 0 | 1 | 2 | 7 | 3 | 7 | 30 |
| 1N | + | 9 | 7 | 10 | 22 | 14 | 20 | 82 |
| | − | 2 | 9 | 18 | 4 | 7 | 8 | 48 |
| 2N | + | 14 | 10 | 8 | 20 | 22 | 28 | 102 |
| | − | 0 | 1 | 18 | 7 | 2 | 1 | 29 |
| Total | + | 25 | 17 | 26 | 61 | 49 | 62 | |
| | − | 2 | 11 | 48 | 18 | 12 | 16 | |

It may be assumed that the strains differed in their effect, also it was evident that specific relationships exist between certain Azotobacter strains and sugar beet genotypes.

## REFERENCES


1. Sarić Zora, Sarić M., Govedarica M., Stanković Ž. (1987) 'The effect of maize genotype and nitrogen level on the efficiency on different Azotobacter strains', Journal of plant nutrition 10(9-16), 1779-1786.
2. Sarić M., Sarić Zora, Govedarica M. (1988) 'Efficiency of strain combinations of different genera of nitrogen fixing bacteria on sunflower genotypes, Proceedings of 12th international conference, July 25-29, Novi Sad, Yugoslavia, pp. 187-192.
3. Milić Vera, Sarić M. (1988) 'Efficiency of Azotobacter in dependence of maize genotype and content of nitrogen in nutritive solution, Mikrobiologija 23, 1, 45-56.

# SPECIFIC RESPONSES OF AZOTOBACTER STRAINS AND SUGAR BEET GENOTYPES


SARIĆ M., SARIĆ ZORA*, KRSTIĆ B.
Institute of Biology, Faculty of Sciences
Institute of Field and Vegetable Crops, Faculty of Agriculture*
21000 Novi Sad
Yugoslavia



ABSTRACT. The ten gentopyes of sugar beet were grown in sand culture supplied with nutrient solution Reid-York without nitrogen. The plants were inoculated with 20 Azotobacter strains. After 30 days the dry matter mass of shoot of root, concentrations and contents of nitrogen were determined. The results show that the obtained effect considerably depended upon the relationships between a certain genotype and bacteria strain.


INTRODUCTION. In order to continue our investigations in to the problems of genetic aspect of mineral nutrition the interaction between the plant genotypes and strains of microorganisms was studied [1]. Our former results indicate the variability of effect nitrogen fixation in depending on genotype and diazotroph strains [2, 3]. These investigations dealt specifically with the relationships between Azotobacter strain, on one side and sugar beet genotypes on the other. The aim was to investigate the effect of numerous strains of different genera on genotypes of sugar beet plants in order to find the best strain/genotyp combination with respect to the efficiency of atmospheric nitrogen fixation.

MATERIAL AND METHOD. Seeds of ten genotypes of sugar beet (1.Ritmo, 2. KW Lena, 3. KW Maja, 4. MBO Ultra mono, 5. NS-Mo-1, 6. KWS Maja, 7. NS-1137 4n MM, 8. NS-701, 9. NS-H-92 and 10. NS-3xpZ) were grown in sand cultures on Reid-York solution without nitrogen. All plants were inoculated with 20 Azotobacter strains, without control plants. After 30 days, plant tissues were analysed on nitrogen by the Kjeldahl method. The sum of positive or negative effects are given in comparison to control plant at LSD 5%.

RESULTS. The achieved results have shown that sugar beet genotypes differently responded to definite strains of Azotobacter (Tab. 1). For example, positive cases regarding dry matter mass of shoots and roots were observed in genotypes NS-701, Ritmo, NS-H-92 and NS-1137 4n MM. The results show that in these genotypes had the lowest significant negative effect. Large significant positive effect in nitrogen content was indicated in NS-701, KW Maja and NS-3xpZ. In KW Lena, NS-701 and NS-H-92 gentotypes were found highly content of nitrogen.

TABLE 1. Sum of significant positive and negative effects caused of Azotobacter strain depending on sugar beet genotypes

| Organs | | 1 | 2 | 3 | 4 | 5 | 6 | 7 | 8 | 9 | 10 | Total |
|--------|---|---|---|---|---|---|---|---|---|---|----|-------|
| | | | | | | Genotypes | | | | | | |
| Dry matter mass | | | | | | | | | | | | |
| Root | + | 14 | 0 | 4 | 3 | 7 | 5 | 9 | 18 | 13 | 0 | 73 |
| | − | 1 | 19 | 4 | 12 | 0 | 4 | 0 | 2 | 0 | 10 | 52 |
| Shoot | + | 3 | 7 | 1 | 1 | 0 | 2 | 6 | 7 | 4 | 2 | 33 |
| | − | 6 | 0 | 6 | 4 | 7 | 2 | 1 | 1 | 9 | 10 | 46 |
| N-concentration | | | | | | | | | | | | |
| Root | + | 0 | 17 | 4 | 5 | 0 | 0 | 0 | 14 | 0 | 7 | 47 |
| | − | 18 | 0 | 4 | 4 | 15 | 18 | 17 | 1 | 9 | 2 | 88 |
| Shoot | + | 8 | 6 | 17 | 9 | 5 | 14 | 7 | 17 | 7 | 14 | 103 |
| | − | 2 | 3 | 1 | 0 | 5 | 0 | 0 | 0 | 3 | 2 | 16 |
| N-Content | | | | | | | | | | | | |
| Root | + | 6 | 2 | 9 | 0 | 1 | 0 | 1 | 0 | 19 | 0 | 38 |
| | − | 3 | 13 | 4 | 12 | 7 | 19 | 8 | 19 | 0 | 16 | 101 |
| Shoot | + | 5 | 17 | 8 | 8 | 2 | 15 | 16 | 17 | 7 | 5 | 100 |
| | − | 8 | 1 | 3 | 5 | 15 | 1 | 2 | 1 | 12 | 7 | 55 |
| Total | + | 36 | 49 | 43 | 26 | 15 | 36 | 39 | 73 | 50 | 28 | |
| | − | 38 | 36 | 22 | 37 | 49 | 44 | 28 | 24 | 33 | 47 | |

Also it may be assumed that on the average, for all genotypes, a higher positive effects was recorded, in all investigated parameters in shoot than in roots. According to the obtained parameters, the biggest number of positive effects were found in the following genotypes: NS-701, NS-H-92, KW Lena and KW Maja while the smallest positive effect was traced in genotypes NS-Mo-1.

As well as different strains showed different efficiency in the sense of synthesis of organic matter in various genotypes of sugar beet (Tab 2). According to the indicator of dry matter mass, strains number 11, 4 and 17 were the most effective, while the least effect was obtained in number 18. However, the concentration of nitrogen was the highest under the influence of strains number 18, 20, 19 and 14. The content of nitrogen was the highest under the influence of strains number 4, 6 and 10.

Taking into account the following parameters: dry matter mass, concentration and content of nitrogen strains number 4, 14, 11 and 6 were the most efficient on the average, for all the investigated genotypes. The highest negative effect was found in strain 3.

It may be concluded that relationship between plant genotype and Azotobacter strain depended on compatibility of partners metabolism in the system: plant genotype-microorganism strain.

TABLE 2. Sum of significant positive and negative effects on dry matter mass, N-concentration and content depending on Azotobacter strain

| Strain | Dry matter mass | | | | N-concentration | | | | N-content | | | | Total | |
| --- | --- | --- | --- | --- | --- | --- | --- | --- | --- | --- | --- | --- | --- | --- |
| | Root | | Shoot | | Root | | Shoot | | Root | | Shoot | | | |
| | + | − | + | − | + | − | + | − | + | − | + | − | + | − |
| 1. | 2 | 5 | 3 | 1 | 2 | 5 | 0 | 6 | 1 | 4 | 5 | 2 | 13 | 23 |
| 2. | 2 | 1 | 2 | 1 | 2 | 3 | 1 | 2 | 3 | 5 | 4 | 4 | 14 | 16 |
| 3. | 3 | 4 | 3 | 0 | 2 | 5 | 5 | 1 | 1 | 6 | 7 | 1 | 21 | 30 |
| 4. | 3 | 3 | 5 | 0 | 3 | 4 | 3 | 1 | 3 | 5 | 7 | 2 | 24 | 25 |
| 5. | 2 | 4 | 2 | 2 | 2 | 4 | 6 | 1 | 2 | 5 | 5 | 3 | 19 | 19 |
| 6. | 3 | 4 | 1 | 0 | 3 | 5 | 5 | 0 | 4 | 4 | 6 | 2 | 22 | 15 |
| 7. | 2 | 3 | 1 | 1 | 2 | 6 | 5 | 0 | 2 | 4 | 6 | 3 | 18 | 24 |
| 8. | 2 | 3 | 1 | 1 | 3 | 3 | 5 | 1 | 2 | 5 | 6 | 1 | 19 | 14 |
| 9. | 5 | 3 | 1 | 2 | 3 | 4 | 3 | 0 | 3 | 3 | 5 | 2 | 20 | 14 |
| 10. | 1 | 3 | 1 | 1 | 3 | 5 | 6 | 1 | 1 | 6 | 8 | 1 | 20 | 17 |
| 11. | 5 | 3 | 3 | 3 | 3 | 6 | 4 | 1 | 3 | 4 | 5 | 3 | 23 | 20 |
| 12. | 2 | 5 | 2 | 2 | 3 | 4 | 4 | 0 | 1 | 5 | 4 | 5 | 16 | 21 |
| 13. | 2 | 4 | 1 | 3 | 3 | 2 | 6 | 1 | 1 | 2 | 3 | 2 | 16 | 14 |
| 14. | 5 | 3 | 0 | 3 | 2 | 5 | 8 | 0 | 3 | 5 | 5 | 2 | 23 | 18 |
| 15. | 2 | 3 | 3 | 2 | 2 | 3 | 5 | 1 | 1 | 5 | 5 | 2 | 18 | 16 |
| 16. | 5 | 2 | 0 | 3 | 0 | 7 | 5 | 0 | 2 | 4 | 3 | 4 | 15 | 20 |
| 17. | 5 | 3 | 2 | 4 | 3 | 5 | 3 | 0 | 1 | 6 | 3 | 4 | 17 | 22 |
| 18. | 1 | 3 | 0 | 5 | 3 | 5 | 10 | 0 | 1 | 7 | 3 | 4 | 18 | 24 |
| 19. | 1 | 2 | 1 | 4 | 1 | 3 | 9 | 0 | 1 | 5 | 6 | 3 | 19 | 17 |
| 20. | 2 | 6 | 0 | 7 | 2 | 4 | 10 | 0 | 1 | 9 | 3 | 4 | 18 | 30 |
| Total | 55 | 67 | 32 | 45 | 47 | 88 | 103 | 16 | 37 | 98 | 99 | 54 | | |

## REFERENCES


1. Sarić M.R. (1983) 'Theoretical and practical approaches to the genetic specificity of mineral nutrition of plants', Plant and Soil 72, 137–150.
2. Sarić M.R., Sarić Zora, Govedarica M. (1987) 'Specific relations between some strains of diazotrophs and corn hybrid', Plant and Soil 99, 147–162.
3. Sarić M.R., Sarić Zora, Govedarica M. (1990) 'Variability of molecular nitrogen fixation in dependence on plant genotype and diazotroph strains', In N.H. El-Bassam, B.C. Loughman and M. Dambroth (eds), Genetic aspects of mineral nutrition, Martinus Nijhoff Publ., Dodrecht (in press)

ASSOCIATIVE DIAZOTHROPS OF DIFFERENT SYSTEMATIC GROUPS AND
THEIR EFFECT ON THE PRODUCTIVITY OF AGRICULTURAL CROPS.


L.F.VASYUK, T.A.POPOVA, V.K.CHEBOTAR, A.E.KHALCHITSKY,
N.S.IVANOV.
All-Union Agricultural Microbiology Research Institute
Leningrad,USSR


It was created the collection of the associative nitrogen-
fixing bacteria from genera Aquaspirillum, Azospirillum, Rhodospirillum, Arthrobacter, Flavobacterium, Bacillus, isolated
from rhizoplane of grain and fodder cereals, new fodder crops
from families of Apiaceae, Brassicaceae, Asteraceae.
The identification of isolates was done on the base of phenotype, chemotaxonomy and genosystematic tests. We revealed
the difference of bacteria associated with plant roots from
free-living ones.For instance Azospirillum brasilense strains
dominated in the soil, Azospirillum lipoferum on the plant
roots Rhodobacter sp. was found in the rhizoplane only.
Root diazotrophs are of great practical and ecological importance. We got positive results in greenhouse and field
trials when inoculated seeds of winter weat, oats, barley,
spring wheat, different types of fodder grasses, maral root,
beer's-breech, sorghum, potato with application of 40-60 N
kg/ha. Inoculation increased yield of the grain cereals by
300-500 kg/ha and fodder grasses by 1800-3000 kg/ha.
Mechanism of action of the associative diazotrophs is connected not only with activation of molecular nitrogen fixation. We found that these bacteria produced phytohormones,
have pectinase activity,enhanced mineral uptake (experiments
with 15N labelled manure). They also effect on pathogenic
microflora and yield quality: increase protein, lysin, ascor-
bic acid, carotene, phosphorus content.
Bioinoculants by trade mark "Myzorin","Flavobacterin" and
"Azorhizin" with specific action on different agricultural
crops were created on the base of some effective strains of
associative diazotrophs.

# CONSTRUCTION OF A GROWTH CHAMBER FOR LONG-TERMED AXENIC PLANT CULTIVATION


S. WASCHÜTZA, E.-G. NIEMANN, AND I. FENDRIK
University of Hannover,
Institute of Biophysics,
Herrenhaeuser Straße 2,
3000 Hannover 21, FRG



ABSTRACT. The construction of a growth chamber is described in which axenic rice plants were grown for 75 days until fruit-bearing stage.


## INTRODUCTION

Rice plants exude small organic molecules, which serve soil microorganisms in the rhizosphere as nourishment. For the plants' nutrition $N_2$-fixing microorganisms are considerable. In contrast to high yielding cultivated rice varieties, wild rice varieties grow without fertilizer. $N_2$-fixation might play a larger role with wild varieties. The objective was to obtain the whole spectrum of rice root exudates from Korean wild rice and cultivated rice varieties to compare their composition in regard to the plant/microorganism association. Sterile and specific inoculation experiments were carried out for a whole growth period of the plants. Therefore, long-termed prevention of contamination was required leading to the construction of the mentioned growth chamber.

## MATERIALS AND METHODS

PLANTS: Korean varieties of *Oryza sativa*(L.), namely the paddy cultivated varieties *Hwajinbyo, Dae-Chung, Sum-Jin, Nam-Young* and upland cultivated *Nong-Lim*, as well as one wild rice variety, all *Japonica* type.
GROWTH CONDITIONS: Surface sterilisation of the rice seeds was carried out according to Nilsson(1957). After sterilization the seeds were placed on nutrient agar plates. These were transferred through the lock into the growth chamber for germination. 3-5 days old seedlings were transplanted into flasks(4 per flask) and cultivated at 31°C / 26°C and 10/14 hours day/night cycle for up to 75 days. Some flasks contained sterile modified Hoagland nutrition solution(HNS) with either $KNO_3$ or $NH_4Cl$ (0,02 g N/l) and 0,4 mM phosphate buffer pH 7,0. Other flasks contained HNS together with rinsed and autoclaved sand or soil.
BACTERIA: Several strains of bacteria were isolated from Korean soil (wild rice and cultivated rice soil). These strains, which were fixing molecular Nitrogen as determined by acetylene reduction assay, were extensively characterized morphologically and physiologically and were used for inoculation experiments.
INOCULATION: In general, the plant nutrient solution was inoculated with a final concentration of 6 x $10^6$ cfu/ml.

GROWTH CHAMBER: A common glove-box was reconstructed into a growth chamber. For long-termed axenic plant cultivation it was necessary to comply with the plants' requirements and to vouch for sterility. Therefore, the top of the chamber was replaced by a pane of glass. Above the chamber, a 2000 (W) halogene bulb was installed. A vertical ventilator was used to reduce the lamp's heat and opalescent glass to spread the light equally in the chamber. Light intensity in the chamber was measured by a luxmeter. The highest light intensity was measured in the middle of the chamber with 30 (klux) on the ground up to 90 (klux) in 80 cm height. The lowest light intensity was measured at the flanks of the chamber with 10 to 15 (klux). The plants were placed in the middle of the chamber. It was important not to put up more than 18 flasks in the chamber, as the elder plants would be hindered in their gas exchange. In the chamber two UV-lamps were installed, another in the chamber's lock. All lamps were controlled electronically. As the heat, caused by the lamp, would have been too much for the plants (45°C) it was necessary to install a cooler to reduce the temperature by day to 31°C. At night 26°C were required, guaranteed by a temperable ventilator. It also was responsible for an equal air current in the chamber together with a buffer-sheet at the air-intake. The air was pumped continuously through sterile filters into the chamber causing an over-pressure to prevent contamination. On the opposite side of the air-intake, an air-exhaust was arranged, ending in a glycerine filled flask, also to prevent contamination. In the chamber a water-pool was placed causing an atmospheric moisture of (70 %). Condensed water at the cooler was directed through a flexible tube back to the water-pool. Before starting an experiment, the chamber and its lock were treated first by desinfectant, then by UV-light for one week. The formed ozon was removed afterwards by a water jet pump. Nutrient agar plates were used for sterility control frequently.

## RESULTS AND DISCUSSION:

In the flasks, filled with nutrient solution only, it was possible to observe sterile plant growth from germination to the fruit-bearing stage for 75 days. Corresponding to the results of Nowak et al. (1987) sterilization of soil by an autoclave was not efficient. 30 days after sterilization the soil containing flasks showed fungal growth. Further inoculation experiments were carried out in the described chamber with the isolates from Korean soil. Aliquotes of exudate containing solution were taken weekly to be examined by means of HPLC. There were less qualitative than quantitative differences in the exudates from wild/cultivated rice under sterile conditions. Besides several not yet identified sugars and organic acids, it was possible to detect the amino acids alanine, arginine, aspartic acid, glutamic acid, histidine, isoleucine, leucine, lysine, phenylalanine, serine, threonine, tyrosine and valine in the exudates of wild/cultivated rice. This special growth chamber is very useful for all long-termed plant experiments which require prevention of contamination.

ACKNOWLEDGEMENT: I would like to thank Mr. K.H. Iwannek and Mr. M. Hildebrandt for their technical assistance.

## REFERENCES:

Nilsson, P.E. (1957) 'Aseptic cultivation of higher plants', Arch. Microbiol. 26,285-301.
Nowak, A. and Wronskowska, H. (1987) 'On the efficiency of soil sterilization in autoclave', Zentralblatt Mikrobiologie 142, 521-525.

N2 FIXING BACTERIA IN THE RHIZOSPHERE OF NATURAL POPULATIONS OF ITALIAN RYEGRASS (<u>Lolium multiflorum</u> L.)


C. SCOTTI C. GARAU M. MOLINARI
Istituto Sperimentale per le Colture Foraggere
Viale Piacenza, 29
20075 Lodi
Italy

ABSTRACT. The aim of the present work is the research of coadaptated N2 fixing bacteria (NFB) and grass plants (<u>L. multiflorum</u>) in order to constitute performing meadows with reduced N input. Fifty-two NFB strains, isolated with the spermosphere model technique, are grouped in to 19 sets that are studied for acetylene reduction activity (ARA) in pure culture and in association with young and adult plants of L. <u>italicum</u>. The results of the strains till now examined show a decrease in ARA detected from the pure culture to the association with adult plants. In this last case strains tested affect positively, even not significantly, root D.M. while no effect is found on aerial traits D.M.Y. and number of leaves.

INTRODUCTION

The present work is aimed at the isolation and study of N2 fixing bacteria (NFB) in the rhizosphere of natural populations of L. <u>multiflorum</u> in order to select bacteria and plants coadaptated for the consitution of meadows, highly performing with reduced mineral nitrogen input. (1).

MATERIAL AND METHODS

NFB strains are isolated in the rhizosphere of <u>L. multiflorum</u> plants collected in permanent meadows 80-100 years old of the Lodi-Crema region in the Po Valley (Northern Italy). Isolation is carried out with the spermosphere model technique (2) using <u>L. multiflorum</u> cv. Crema. ARA is studied in: a) pure culture: on RCV medium with C sources; N = 0 or N = 7 mM nitrate. b) association with young plantlets (14-22 days old): RCV medium supplied or not with C sources; 3 strains tested; 5 replications/treatment. c) association with adult plant: microplots 5 cm 0 x 50 cm high, filled with sterile sand (10 plants/plot): 2 levels of N (2.50 and 5 mM nitrate) in absence or presence of C sources; 5 strains tested: 2 probable Enterobacteriaceae, 1 probable Azospirillum, 1 not determined, and A. brasilense SpCd (tester); 10 replications/treatment.

## RESULTS

Fifty-two N2 fixing Gram-, catalase+, non sporing rods are isolated and classified into 19 groups. a) <u>ARA in pure culture</u> raises from 5 to 500 nmol C2H4/h; 12% of the strains including SpCd, shows value between 350-500. b) <u>ARA in association with plantlets</u> is detectable (2-44 nmol C2H4/h/tube) only when C sources (C+) are present, while bacterial density is about the same in C- and C+. Addition of C sources in the C- tubes at the 19th day allows ARA detection at the 22nd day (3). c) <u>ARA in association with adult plants</u> (data of the 1st cut), measured on the intact plot (10 plants 2 months old), is detectable only in 3 strains (129, 141, SpCd) with values between 0.77 and 5 total micromol C2H4 after 20 hours of incubation under 20% acetylene. Strain 129 shows the greatest number of positive results (50% of plots) in C- and C+. Roots of all the 5 strains are found to reduce acetylene in C- treatments. No significant effect of inoculation is found neither in aerial, crown, roots D.M., nor in number of leaves. Nevertheless, in general, is higher and shoot/root ratio lower in inoculated plots than in testers. The addition of C sources (10 mM fructose + 10 nM Na-citrate) has a significant negative effect on the aerial and root D.M.; this effect is greater in the higher N level (5 mM nitrate).

## DISCUSSION

The introduction of plant as "environment" for bacteria, even in stable conditions of other environmental factors, influences bacterial population density and, to a greater extent, the process of N2 fixation. Our choice was to establish favourable conditions for the plant partner as for growth as for ARA measure (aerial part in open air). For a same strain, the measure of ARA in pure culture, in association with plantlets "in vitro", and finally in association with adult plants corresponds to a decrease of AR activity detected and to an increase of variability of results. Strains tested with adult plants seem to affect positively root D.M. rather than aerial D.M. This effect is found for all the strains studied.

## LITERATURE CITED

(1) Rotili P., Zannone L., Scotti C., Gnocchi G., Proietti S., (1987), "Agronomic and feeding value of <u>L. multiflorum</u> populations collected in the Po Valley", in INRA (ed), Natural variation and breeding for adaptation, Lusignan, 79-86.
(2) Thomas Bauzon D., Weinhard P., Villecourt P., Balandreau J., (1982), "The spermosphere model. I. Its use in growing, counting, and isolating N2 fixing bacteria from the rhizosphere of rice" Can. J. Microbiol., 28, 922-928.
(3) Martin P., Glatzle A., (1981), "Mutual influences of Azospirillum spp. and grass seedlings", in W. Klingmüller (ed), Azospirillum: genetics, physiology, ecology, Bayreuth, 108-120.

STUDIES ON FEW STRAINS OF AZOSPIRILLA COLONIZING MAIZE RHIZOSPHERE : THEIR ESTABLISHMENT ON ROOTS AND EFFECTS ON PLANT PHYSIOLOGY.


C. CASSINELLI, E.NORIS, E. SIGNORINI, D. TOLENTINO  AND G. PIRALI
Biotech. Dept. Research Institute G. Donegani, ENIMONT
Via Fauser 4, 28100 Novara, ITALY.


## Introduction


A few naturally occuring <u>Azospirilla</u> strains, characterized for IAA production and $N_2$-fixing ability both in "in vitro" tests and on roots, were further investigated for their effect on maize physiology and for their ability to colonize roots.
The changes in morphology and in nitrate absorption of inoculated roots, as well as leaf weight increase and resistance to dryness of treated plants are described.
Furthermore two strains, made resistant to kanamycin by the insertion of Tn5 and inoculated on maize seeds in greenhouse trials, have been monitored for their persistence on plant ectorhizosphere.
The results of <u>Azospirilla</u> growth and persistence along the root surface at different times of maize growth are reported.


## Physiology of maize


Maize seeds treated with <u>Azospirillum</u> were sown in soil or germinated and grown in hydroponics solutions.    The extent of root surface was evaluated by measuring the amount of hydrochloric acid absorbed to the roots in standard conditions.    The amount on nitrogen absorbed by plants was evaluated by analytical determination of the hydroponic solutions (as $KNO_3$ variations). When plants were growing in soil , they were exposed to water stress .
<u>Azospirillum</u> treatment of maize seeds increased (40%) the root surface of plant growing on sand and developed a greater number of side roots both in sand and soil.    Moreover plants grown in nitrogen free hydroponics cultures, or in soil under water stress, produce more biomass than untreated ones (120-150% increase fresh weight in first case, 30-33% increase in dry weight in water stressed plants).    Seed treatment with <u>Azospirillum</u> produces maize plants that absorb more nitrates ( + 12% ) than untreated seeds, in hydroponics solution.


## Survey of establishment and persistence of <u>Azospirillum</u>


<u>Generation of mutants</u>. For the evaluation of establishment and persistence of the <u>Azospirilla</u> strains in the maize rhizosphere mutants resistant to kanamycin by transposon mutagenesis ( Tn5 ) have been employed. The plasmid pGS9 in E.coli HB 101 was used to generate Tn5 insertion in <u>Azospirilla</u> isolated strains. In our conditions (2 x $10^8$ cfu E. coli and 2 x $10^9$ cfu Azospirillum on nitrocellulose membrane at 28° C for 16-18 h) we obtained frequencies of exoconjugant resistant to 100 µg/ml of kanamycin ( minimal salt medium ), ranging from 1 x $10^6$ to 3 x $10^8$.

344

The insertion of Tn5 was also verified with a biotynilated probe of a 3.4 Kb Hind III fragment of the pGS9, extracted from E. coli HB 101. A dot-blot of total DNA from the Azospirilum Kan$^{res}$ used in the colonization assay was performed with positive result.

<u>Greenhouse studies</u>. Maize seeds were treated with <u>Azospirilla</u>  to obtain $10^6$ cfu/seed, sown in pots and supported with 0-100-200  mg/kg of nitrogen  fertilization.  The population of the <u>Azospirilla</u> on seeds  and on ectorhizosphere  was monitored by  plate counts on minimal (MSP)  and  rich medium (NA) containing 50  μg/ml of kanamycin and 100  μg/ml of cycloeximide. The evaluation were made on samples collected every two weeks, on at least ten plants and three replicas.
The <u>Azospirillum</u> population at  15  and 30  days  after planting  was  higher  than  $10^6$ cfu/plant distributed 54% on seminal and 46 % on crown roots.


## Discussion

We verified the usefulness of the Tn5  mutagenesis for the study of the occurrence in the rhizosphere of <u>Azospirilla</u>.  The resistance to kanamycin,  coupled with the morphological characteristics  of the colonies on the used media,  has proven a reliable method  to evaluate an  <u>Azospirilla</u> population at the minimum level of $10^3$ cfu/plant.
The presence of a high population on seeds,  seedlings and roots in the first 15  days (30 days for a strain)  indicates a good level of persistence and colonization of our strains along  the root. After 45  days  bacteria  population became lower than  the recovery  limits,  probably due to uncontrolled environmental factors.  The population recovered was unaffected by the level of the N fertilizer (0 - 100 - 200 mg/kg).
The identity of the strains isolated from roots with those used for inoculum was confirmed by a total DNA "fingerprint" digested with restriction enzymes and run in PAGE.

The strains used in this study belong to the group that produces more IAA than the reference strains. The  effect  on  plant morphology  and physiology  found in  our  studies  are  probably  due  to the combination of the colonization ability,  the good level of persistence along the roots and the level of IAA produced.


## Acknowledgments

The present work was conducted within the contract " Programma Nazionale di Ricerca per la Chimica ", entrusted to Istituto Guido Donegani S.p.A. - Novara by the Ministro dell'Universita' e della Ricerca Scientifica e Tecnologica.

## References

1 Harris J. et al.,1989, Soil Biol.Biochem.,21, 59-64
2 Pederson,W.L. et al., 1978, Appl.Env.Microb.,35, 129-135
3 Hegazi,N.A., et al., 1983, Can.J.Microb., 29, 888-894
4 Vanstokem M., et al., 1987, Appl. Env. Microb., 53, 410-415
5 Abdel-Salam, M.S., Klingmuller,W., 1987, Mol.Gen.Genet., 210, 165-170
6 Carley, H.E., Watson, R.D., 1966, Soil Science, 102, 288-291
7 Lin, W. et al., 1983, Appl. Env. Microb.,45, 1775-1779

# ROOT-ASSOCIATED DIAZOTROPHS USED AS PLANT GROWTH STIMULATORS


V.Catska[1], S.Ruppel[2], I.Zaspel[2] and E.Gallori[3]
1) Institute of Microbiology, Praha, Czechoslovakia; 2) Research
Centre for Soil Fertility, Muncheberg, East Germany and
3) Department of Animal Biology and Genetics, Florence, Italy.


INTRODUCTION. Root-associated diazotrophs (RAD) are extensively
studied for thieir plant-growth promoting effect (1). This effect
include not only non-symbiotic N2- fixation (2) but also
phytohormone production (3) and the production of siderophores,
antibiotics or HCN (4). The possibility to use of these plant-growth
promoting bacteria for biological control of soil-born pathogens was
described (5). In this work we described the interactions among
individual components of the rhizosphere microflora and RAD .
Nevertheless the main aim is the possibility to use for inoculation
of plants beneficial microorganisms including RAD not only for
biological control of phytopathogen but also of saprophytic
phytotoxic micromycetes.

MATERIALS AND METHODS. For determination of diazotrophs was used
dilution plate method and for isolation many times washings of roots
- modified Harley and Waid method (6). For determination of
nitrogenase activity, acetylene reduction assay was used. For
determination of antibiotic activity - plate diffusion technique
(medium Czapek-Dox) was used. For determination of phytormones -
avena coleoptile section test and/or HPLC analyse was used. For
inoculation of seeds, roots or soil was used liquid fresh cultures
of microorganisms, or tomato roots infected with VAM fungi, or wheat
straw infected by Gaeumannomyces graminis var. tritici. For
determination of phytotoxic microorganisms algal test with Chlorella
vulgaris was used.

RESULTS AND DISCUSSION. With regard to decreasing environmental
quality and to necessity of obtaining healthy food products it must
be limited not only the chemical treatment of plants or soil with
pesticides, but also with anorganic nitrogen fertilizers as much as
possible. The investigated strains of RAD are able to fix $N_2$, but
more important it seems to be their production of phytohormone and
antibiotic metabolites. Especially cytokinine-like substances, IAA
and some other auxins were determinate in cultures of Klebsiella sp.
and Azospirillum brasilense. Antibiotic activity of these strains

including <u>Serratia rubidea</u> was also found against some phytopathogen and phytotoxic micromycetes . The previous laboratory results of good persistence of strain of <u>A.brasilense</u> Sp267 resistant to gentamycine especially in the first month of wheat growth showed that it is perspective to use some of RAD for inoculation of plants in pot and field experiments. It was found that it is possible to stimulate growth of some plants, and also the yield of them, directly or indirectly , by influence the other rhizoplane or rhizospere microflora. For inoculation of plants were used active strains of isolates obtained by many times root washings. The amount of RAD was lower in roots of wheat cultivated in monoculture but nitrogenase activity was 47% higher than of wheat cultivated in crop rotation. Also antibiotic activity <u>in vitro</u> was higher. It is also known that root's <u>Azospirillum</u> isolated strains are more competitive than the soil isolates (7). Already after two weeks we received in pot experiments  the stimulation of root and shoot dry matter and especially of root length by inoculation of winter wheat seeds with <u>Klebsiella</u> sp. Persistence of these bacteria on the roots was 8- and 10- times higher after seed inoculation with inoculum concentration $10^5$ and $10^7$ cells/ml respectively. In field experiments the average of 10% stimulation of growth (height of plants, dry matter of roots and shoots) and of corn yield in winter wheat or barley was received after the inoculation with some strains of RAD (<u>A.brasilense</u>, <u>S.rubidea</u>, <u>Klebsiella</u> sp.). Generally the inoculation of plants with plant growth-promoting rhizobacteria (<u>Agrobacterium radiobacter</u>, <u>Pseudomonas putida</u> , (8) or VAM fungi, influence positively also the occurrence of RAD. These beneficial microorganisms for plants usually decrease the amount of phytotoxic micromycetes and therefore the amount of RAD probably increase. On the other hand diazotrophic antagonists can also reduced the phytotoxic microflora. It was found that phytotoxic microorganisms inhibit nitrogenase activity of diazotrophs. From all results it can be assumed that for inoculation of plants it is possible to use not only RAD but also some other rhizobacteria  or VAM fungi, because these beneficial microflora can effect positively these RAD and thereforre the plant growth.

REFERENCES.
1) Michiels K. <u>et al</u>. (1989). Biol. Fertil. Soils 8, 379-407.
2) Dobereiner J. and Day J.M. (1975). In: Nitrogen Fixation by Free-living Microorganisms, W.D.P.Stewart ed., pp. 39-56. Cambridge University Press.
3) Tien T.M. <u>et al</u>. (1979). Appl. Environ. Microbiol. 37, 1016-1024.
4) Kloepper J.W. <u>et al</u>. (1989). Trends in Biotechnology 7, 39-44.
5) Weller D.M. (1988). Ann. Rev. Phytopathol. 26, 379-407.
6) Rennie R.J. (1981). Can. J. Microbiol. 27, 8-17.
7) Dobereiner J. (1988). In: Interrelationships between Microorganisms and Plants in Soil, V.Vancura and F.Kunc eds., pp. 229-242. Academia Prague and Elsevier Amsterdam.
8) Catska V. (1988). In: Interrelationships between Microorganisms and Plants in Soil, V.Vancura and F.Kunc eds., pp.463-468. Academia Prague and Elsevier Amsterdam.

# SCREENING PROGRAM FOR THE ISOLATION OF N2-FIXING BACTERIA OF THE GENUS AZOSPIRILLUM


ANDRIOLLO N., NORIS E., SIGNORINI E., TOLENTINO D. AND  PIRALI G.
Biotech. Dept. Research Institute G. Donegani, ENIMONT
Via Fauser 4, 28100 Novara, ITALY.


## Introduction

A screening program was started in order  to obtain a collection  of $N_2$-fixing  bacteria belonging to the  genus Azospirillum,  colonizing the rhizosphere of a wide spectrum of Gramineous  plants, to be tested in field trials as potentially useful soil inoculants.
More than six hundred samples of roots of wild and cultivated plants  were collected  all over Italy, and investigated for the presence of Azospirillum.
The isolation methods and the evaluation of the most  important aspects for their agronomic traits are described.
$N_2$-fixing ability (growth in  $N_2$-free medium,  $C_2H_4$  production under  low  $pO_2$), hormone production (colorimetric  estimation  and HPLC  identification  of  IAA  and  related  indoles)  and rhizosphere competence ($N_2$-fixation on homogenized maize roots) in laboratory assays have been evaluated .

## Isolation

Suspension of bacteria released by the roots samples were inoculated  in $N_2$  free selective semisolid medium for enrichement of Azospirilla like bacteria. After plating the mixed population on MSP medium containing Congo Red (1),  single colonies  were picked up and pure cultures estimated  in Acetilene Reduction Assay (ARA).  Characterization  with basic physiological  tests  indicated that 90% of the strains isolated were Azospirilla spp.
Using this procedure we collected  about 250  strains that were isolated from  about  the 30 % of samples collected.
Azospirilla were found in root samples of  every italian region,  and distributed on  every kind of crop, considering both nitrogen fertilized soil or uncultivated ones.

## Strain Evaluation

Bacterial  strains,  isolated for their ability  to grow in  the nitrogen free medium,  were further selected for some envisaged important characteristics for their field application.
Quantitative estimation  of  IAA production  and of  nitrogen  fixation  were carried out  on all the strains isolated.  The rhizosphere competence assay  (  $N_2$  fixation on homogeneized maize roots as a measure of presence of bacteria on roots) was set up for those with the higher $N_2$-fixing ability.

## NITROGEN FIXATION

The ARA on  liquid  culture in N-free medium  (MSP)  and $pO_2$  reduced (0.2%  $O_2$ in $N_2$) was applied on strain  isolated  and  results  reported  as  nmoles $C_2H_4$  $h^{-1}$  $mg^{-1}$  of total proteins (Method of Lowry)(2).

Three strains (ATCC 29710,  ATCC 29145,  Sp6),  have been used as reference;  the best producer (ATCC 29145)  evolved 714  nmoles $C_2H_4$   mg protein $^{-1}$  h $^{-1}$ (average of seventy tests). The 20 % of tested strains gave more than   1  $\mu$mole $C_2H_4$   mg protein $^{-1}$ h $^{-1}$, with therefore at least a 30 % higher $N_2$ fixing ability then those used as reference.

## IAA PRODUCTION

The  isolated  strains  grown  in  complete  medium  with  50  $\mu$g/ml  of  tryptophan  were  evaluated spectrophotometrically (530  nm) after Salkowsky colorimetric assay (3,4). The amount of IAA produced ranged from 0.5-18 $\mu$g/ml, the reference strain ATCC 29145 producing 13.9 $\mu$g/ml.

4 % of the strains tested had better performance than ATCC 29145 and their IAA production was further investigated by HPLC on acidic culture extracts (5, modified analytical conditions).

The colorimetric assay  gave  results consistent with the HPLC determination in almost every strain, even if it  overestimated the effective IAA production as expected.

## RHIZOSPHERE COMPETENCE

Cultures grown overnight in complete liquid medium,  centrifuged and washed, were used as inoculum on maize roots axenically growing in nitrogen free medium.

After  4  days the roots were  homogenized,  transferred on  N-free  semisolid medium  (MSP)  and the N-fixing ability was evaluated ( ARA method ).

A  linear correlation  was found between the cell  concentration present in  the homogenate  and  the ethilene measured.   33 % of the tested strains gave better results than the  ATCC 29145 strain used for comparison, indicating thus a potentially better establishment in the maize rhizosphere.

Combining data obtained by all tests we found that <u>Azospirilla</u> strains  growing better on maize roots belong  to the  group of high IAA producer,  suggesting therefore these two  parameters  as  important traits for agronomic application.

## Acknowledgments

The present work was conducted within the contract "Programma  Nazionale di Ricerca  per la Chimica", entrusted to  Istituto  Guido Donegani  S.p.A.  -  Novara,  by the Ministro dell'Universita'  e della Ricerca Scientifica e Tecnologica.

## References

1 Rodriguez Caceres, E.A., 1982 Appl.Env. Microb. 44 990-991

2 Turner, G.L., Gibson, A.H., 1980, in : " Methods for evaluating biological N- fixation ",
  ed. Bergenson F.J., 111-138.

3 Ehman A., 1977, J.Chromat., 132, 267-276.

4 Gordon, S.A., Weber, R.G., 1950, Pl. Physiol.,192-195

5 Tien, T.M., et al., 1979, Appl.Env.Microb., 37, 1016-1024.

# INTERACTIONS BETWEEN AZOSPIRILLUM AND SORGHUM RHIZOSPHERE ISOLATES UNDER DIFFERENT CULTURAL CONDITIONS


D. LIPPI, [*]I. CACCIARI, [*]P. QUATRINI and T. PIETROSANTI
Plant Radiobiochemistry and Ecophysiology Institute
C.N.R., Via Salaria Km 29- Monterotondo Scalo, Roma
[*]DABAC, University of Tuscia, Viterbo, Italy



Abstract.Three bacterial strains were isolated from sorghum rhizosphere and the growth was followed in pure and mixed cultures to study interactions with A.brasilense Cd.


## 1. Introduction

Plant growth promoting bacteria inoculated into plants compete for substrate with many indigenous microorganisms [1-2].If applied bacteria fail in this competition plant root colonization doesn't take place.To investigate interactions among Azospirillum and indigenous rhizosphere bacteria of sorghum plants,three isolates (S.I.) and a strain of A.brasilense Cd (Cd) were grown under different cultural conditions.

## 2. Organism and medium

Azospirillum brasilense Cd (ATCC 29710) and S.I. were grown at 30°C in succinate-sufficient batch cultures or in succi= nate-limited chemostat at D=0.025 and 0.11 h$^{-1}$. Medium [3] contained 5g/l and 0.56g/l of disodium succinate for batch and continuous culture, respectively,and 1g/l of $(NH_4)_2SO_4$, pH=6.8.

## 3. Results

Bacterial strains were isolated from the rhizosphere of 30d old sorghum plants grown in pots containing soil from a field (Inviolatella of Ist. Sper. Cerealicoltura, Roma-MAF) in which this plant was grown continuously for 4 y. Three bacteria were isolated following the procedure described in the Bergey's Manual for isolation of Azospirillum genus. The strains were aerobic, Gram-negative and the identification was started by means of the API 20B system. At present two bacteria were tentatively classified as a Pseudomonas sp.($C_2$) and a Spirillum sp. ($C_4$) while for the strain $C_6$, a not motile rod, the attribution was very doubtful. $\mu_{max}$ and $K_s$ of strains were determined by growing bacteria in batch and continuous cultures,respectively: Cd $\mu_{max}$ was 0.117h$^{-1}$ and $K_s$ was 1.8 $\mu$g/ml; values for S.I. varied from 0.026 to

350

$0.150h^{-1}$ ($\mu_{max}$) and from 1.2 to 2.5 $\mu g/ml$ ($K_S$). The popula= tion size of Cd was strongly reduced (of about 80%) in the presence of S.I. inoculated singly or in mixture, while it seemed to inhibit slightly the growth of $C_2$ and $C_6$.Interac= tions other than competition seemed to occur among the in= digenous strains. $C_2$ was stimulated by the presence of $C_4$ and $C_6$, the viable cell number increasing of about 170% as compared with pure culture. $C_2$ seemed to be the major re= sponsible for the repression of the growth of all the other components of the community.$C_6$,that in pure culture reached the highest values of viable cell number ($380 \times 10^7 \cdot ml^{-1}$) and biomass yield (Y = 38.5g d.w./moles of succinate consumed), was always depressed in presence of the other bacteria, Cd included. The isolated bacteria and Cd mixed cultures were grown also in chemostats at different D.The higher dilution rate seemed to stimulate growth and yield of the mixed cul- ture in presence of Cd. Total viable cell number were 115 x $10^6 \cdot ml^{-1}$, with Y=33, for D=0.11h$^{-1}$ and $46 \times 10^6 \cdot ml^{-1}$, with Y= 8.6,for D=0.025h$^{-1}$. In S.I. mixed culture, without Cd, only $C_4$ was greatly favoured by high growth rate,the cell number increasing of about 230%.Cd was very sensitive to S.I. pre- sence, mainly at the low D, where the cell number dropped from 47.5 to 3 x $10^6 \cdot ml^{-1}$.

## 4. Conclusions
Bacteria isolated from sorghum rhizosphere seem to overcome Cd either when carbon substrate is in excess or in limited supply. It has been stated [4] that in a mixed culture $\mu_{max}$ and $K_S$ regulate competition for the substrate,unless inter- actions occur. However in our work all the species can co- exist, although $\mu_{max}$ of $C_4$ was very close to the lowest D examined. The bacterial cell number ratio within the commu- nity shows that the different values of $\mu_{max}$ and $K_S$ account only partly for the results obtained. Interactions among indigenous bacteria and Cd appear complex and production of stimulatory and/or inhibitory substances can't be excluded.

## 5. References

1.Bennet,R.A. and Lynch,J.M. (1981) 'Bacterial growth and development in the rhizosphere of gnotobiotic cereal plants', J, Gen. Microbiol. 125, 93-102.
2.Fallik,E.,Okon,Y. and Fischer,M.(1988) 'Growth response of maize roots to Azospirillum inoculation: effect of soil organic matter content, number of rhizosphere bacteria and timing of inoculation', Soil Biol. Biochem. 20, 45-49.
3.Cacciari,I.,Del Gallo,M.,Ippoliti,S.,Lippi,D.,Pietrosan- ti,T. and Pietrosanti,W.(1986)'Growth and survival of Azo- spirillum brasilense and Arthrobacter giacomelloi in binary continuous culture', Plant Soil 90, 107-116.
4.Veldkamp,H.(1977)'Ecological studies with the chemostat', Advances in Microbial Ecology 1, 59-94.

# COMPETITION IN SORGHUM RHIZOSPHERE AMONG AZOSPIRILLUM AND INDIGENOUS BACTERIA


QUATRINI,P., CACCIARI,I.,*LIPPI,D.,*PIETROSANTI,T.
and **CARCUPINO, M.
DABAC , Universita' della Tuscia, Viterbo
*IREV-CNR, Via Salaria Km 29, Monterotondo, Roma
**Lab. Microscopia Elettronica,Facolta' SMF, Viterbo



ABSTRACT
Sorghum root colonization by Azospirillum is negatively
influenced by the presence of three different bacteria isolated from sorghum rhizosphere. Maximum growth rate of Azospirillum was lower than those of sorghum isolates.Interac-
tions other than competition also seem occur among strains.


## 1. Introduction
Seed or plant inoculation with Azospirillum under gnotobiotic conditions has been widely used in different bacte-
ria-root systems [1]. However, although many informations
on the factors involved in root colonization by Azospirillum are now available, little attention has been devoted to
bacterial interactions in the rhizosphere;the establishment
of inoculated bacteria can be greatly affected by the presence of indigenous populations. In the present work it was
studied how sorghum root colonization by A. brasilense Cd
can be influenced by the presence of three different bacteria isolated from the rhizosphere of sorghum.

## 2. Materials and Methods
Germinating sterile seeds of sorghum were planted in flasks
containing Fahreaus molten agar medium in which A.brasilen-
se Cd (ATCC 29710) and the sorghum isolates had been inoculated singly or in mixture.The three bacterial strains ($C_2$,
$C_4$ and $C_6$) were isolated from the rhizosphere of sorghum.$C_2$
and $C_4$ have been tentatively identified as a Pseudomonas sp
and a Spirillum sp., respectively. $C_6$ is a not yet identi
fied Gram-negative, not motile rod.The flasks were incubated in a growth-cabinet with a 16/8 h dark cycle at 27°C.The
maximum growth rate and final populations achieved by the
individual species in the rhizosphere of sorghum were calculated on the basis of the viable cell number per mg d. w.
root. The doubling time of bacteria was measured during logarithmic growth phase. For this purpose the bacterial populations on roots and adhering agar were counted by using

nutrient agar. To study bacterial competition and interaction in the rhizosphere of sorghum, _Azospirillum_ was co-inoculated with each of the bacterial isolates,singly or in mixture with the others.The final populations of each microorganism was determined after 2 and 5 days.

## 3. Results and discussion

Root colonization by _Azospirillum_ is known to be influenced by the rhizosphere microflora [2,3].The present study shows that growth rates can be partly responsible of the outcome of competition between _A.brasilense_ and sorghum isolates. The ecological success of a rhizosphere bacterium has been suggested to depend on its colonization potential and maximum growth rate [4,5].The maximum growth rate of _Azospirillum_ in the rhizosphere of sorghum was slow ($0.0128$ h$^{-1}$) as compared with those of the other strains ($0.0325-0.039$h$^{-1}$). However,the growth rate seems not to be the sole factor regulating bacterial interactions in sorghum rhizosphere. $C_4$ strain, when co-inoculated with _A.brasilense_ into the rhizosphere was stimulated and increased significantly,its final population reaching 4 times that when it was inoculated alone. On the other hand $C_6$, which exhibited the highest growth rate ($0.039$ h$^{-1}$) and colonization potential ($5.2 \times 10^7$ cells mg d.w.roots),was strongly repressed when coinoculated with _Azospirillum_ and at 5 days the population was only 22% of that when inoculated alone. Competition, antagonism and mutualistic interactions might occur at the same time within the microbial populations and the establishment of _Azospirillum_ into the rhizosphere was probably inflenced by many different interacting factor.

## 4. References


1-Bashan,Y. and Levanony,H.(1989) 'Factors affecting adsorption of _Azospirillum brasilense_ Cd to root hairs as compared with root surface of wheat',Can.J.Microbiol.35,936-944.
2-Bashan,Y. (1986) 'Enhancement of wheat root colonization and plant development by _Azospirillum brasilense_ Cd following temporary depression of rhizosphere microflora', Appl. Environ. Microbiol. 51, 1067-1071.
3-Fallik,E.,Okon,Y. and Fisher,M.(1988) 'Growth response of maize roots to _Azospirillum_ inoculation: effect of soil organic matter content, number of rhizosphere bacteria and timing of inoculation', Soil Biol. Biochem. 20, 45-49.
4-Bennett,R.A. and Lynch,J.M. (1981) 'Bacterial growth and development in the rhizosphere of gnotobiotic cereal plants', J. Gen. Microbiol. 125, 95-102.
5-De-Ming,Li and Alexander,M.(1986) 'Bacterial growth rates and competition affect nodulation and root colonization by _Rhizobium meliloti_', Appl. Environ. Microbiol.52, 807-811.

# NITROGENASE ACTIVITY IN THE TOBACCO RHIZOSPHERE


P. MONTAINI, S. FEDI, S. MARUCA and F. FAVILLI
Dipartimento di Scienze e Tecnologie Alimentari e
Microbiologiche, Università. P.le delle Cascine, 27 - 50144
Firenze, Italy



ABSTRACT. Root samples of six different tobacco cultivars were
submitted to the ARA assay. Nitrogenase activity occurred in 5 out of
the 6 tobacco cultivars assayed. The nitrogen fixing microflora
associated to the roots was mainly composed by Azospirilla but
<u>Bacillus</u> and <u>Pseudomonas</u> also occurred.


## 1. INTRODUCTION

With over 7000 hectares and an annual average yield of more than
150000 tons, the tobacco represents one of the most important
industrial cultivation in Italy (60% of the EC total production).In
spite of the fact that several tobacco cultivars are continuously
cultivated for 8-10 years in the same field without or with little
nitrogen fertilization, the average yield $(2.0-3.0 \ t \ ha^{-1})$ is
generally constant. The aim of the research was to investigate on the
nitrogenase activity and on the nitrogen fixing microflora associated
to the tobacco roots.

## 2. METHODS

Root samples of six different tobacco cultivars were submitted to the
ARA test (Burris, 1974). Root associated nitrogen fixing bacteria were
counted by MPN method (Hegazi, 1979) and isolated by successive
plating in Rennie solid medium (Rennie, 1981).

## 3. RESULTS

6 out of 5 tobacco cultivars showed very high nitrogenase activity
ranging from $502 \pm 35,1$ nmoles $C_2H_4 \ h^{-1} g^{-1}$ to $7200 \pm 2800$.
Rhizospheric nitrogenase activity was significantly correlated with

Table 1 – Nitrogenase activity of the roots of six different cultivars of tobacco.

| Cultivars | n moles $C_2H_4$ $H^{-1}$ $g^{-1}$ (d.w.) root | |
|---|---|---|
| | evaluated directly | evaluated in enrichment culture |
| Bright C60 | $4280 \pm 1450.2$ | $512200 \pm 57501$ |
| Bright BH11 | $7200 \pm 2800$ | $145330 \pm 12180$ |
| Maryland H609 | $1574 \pm 410.3$ | $61621 \pm 8650$ |
| Mac Nair H944 | $2562 \pm 820.4$ | $49884 \pm 6750$ |
| American Kentucky | $502 \pm 55.1$ | $7669 \pm 2550$ |
| Kentucky H151 x 104 | 0 | 0 |

data are means of three experiments each with six replicates $\pm$ standard deviation.

Table 2 – MPN of the root associated nitrogen fixing bacteria and occurrence of Azospirillum like-form.

| Cultivars | MPN $10^3$ $g^{-1}$ (d.w.) | Azospirillum like-form (a) |
|---|---|---|
| Bright C60 | 10 | +++ |
| Bright BH11 | 5.5 | +++ |
| Maryland H609 | 2.5 | +++ |
| Mac Nair H944 | 2.2 | +++ |
| American Kentucky | 0.8 | + |
| Kentucky H151 x 104 | 0 | 0 |

(a) +++ elevated occurrence of Azospirillum like-form

the number of the nitrogen fixing bacteria associated to the roots. The roots of the cultivars Kentucky H151 x 104 gave negative response either to the ARA test or to the count of rhizospheric nitrogen fixing bacteria. The rhizospheric nitrogen fixing microflora of the cultivars showing nitrogenase activity was mainly composed by Azospirilla, but _Bacillus_ and _Pseudomonas_ also occurred.

## 4. CONCLUSION

On the basis of the high nitrogenase activity and of the number of nitrogen fixing bacteria associated to the tobacco roots it is possible to affirm that the tobacco rhizosphere harbours an efficient nitrogen fixing system able to support the complete nitrogen requirement of the plants.

## 5. REFERENCES

Burris, R.H. (1974) Methodology, in "The biology of nitrogen fixation", North Holland Publ. Co., Amsterdam, pp.9-33.
Hegazi, N.A. et al. (1979) Soil Biol. Biochem. 11, 437-438.
Rennie, R. (1981) Can. J. Microbiol. 27, 8-14.

# EFFECT OF DIFFERENT PESTICIDES ON AZOSPIRILLUM BRASILENSE


E. Gallori, M. Bazzicalupo, E. Casalone, G. Di Biase, R. Fani and
M. Polsinelli
Department of Animal Biology and Genetics, University of Florence,
Italy.


## INTRODUCTION

The wide use of pesticides in modern agriculture has stimulated the study of their side-effects on non-target soil microorganisms, including the $N_2$-fixing bacterium Azospirillum sp., whose agronomic use has been recently adapted (1). Little information has been available on the detoxification processes of agrochemicals in the bacterial cell. Recent studies have suggested a possible involvement of glutathione (GSH), and its metabolyzing enzymes, in counteracting the toxicity of some pesticides (2).

The purpose of this work was to evaluate the effect of different pesticides (ten fungicides, six herbicides and four insecticides) on the growth of A.brasilense. The effect of the combined use of two fungicides (captan and thiram) and 1,8-naphthalic anhydride (NAF) antidote, on the growth, nitrogenase activity and GSH content was also investigated.

## MATERIAL AND METHODS

A.brasilense Cd (ATCC29710) was cultured in MSP medium (3). All pesticides used (listed in Table 1) were obtained from S.Ehrenstorfer, FGR; NAF from Aldrich Chem., USA. The cultures for nitrogenase assay were prepared and assayed as described (4). GSH assay was performed as described (2); bacterial cells for the assay were treated as reported by Daws et al. (5).

TABLE 1. Minimal inhibitory concentration of different pesticides on the growth of A.brasilense Cd.

| Compound | MIC ($\mu$g ml$^{-1}$) |
|---|---|
| **FUNGICIDES** | |
| Benomyl | >250 |
| Captafol | 45 |
| Captan | 10 |
| Carboxin | 110 |
| Dicloran | 150 |
| Folpet | 35 |
| Thiabendazole | 250 |
| Thiram | 12.5 |
| Zineb | 200 |
| Ziram | 20 |
| **HERBICIDES** | |
| 2,4 - D | >250 |
| Alachlor | " |
| Atrazine | " |
| Butylate | " |
| EPTC | " |
| Glyphosate | " |
| **INSECTICIDES** | |
| Dieldrin | >250 |
| Carbaryl | " |
| Malathion | " |
| Pyrethrine | " |

## RESULTS AND DISCUSSION

The MICs of different chemicals on <u>A.brasilense</u> growth are reported in Table 1. Results obtained indicated that dithiocarbamate (thiram and ziram) and phthalimide (captan,captafol and folpet) fungicides are the most toxic compounds tested. Captan and thiram, which showed the lowest MICs, have been further studied. The inhibition of bacterial growth by 10 ug/ml captan and 15 ug/ml thiram, was completely removed by 50 ug/ml cysteine or GSH (data not shown). The effect of the addition of 100 ug/ml NAF to cultures grown in the presence of sublethal concentrations of captan and thiram was also investigated. Results obtained showed that the combined use of NAF and each fungicide completely blocked the growth of <u>A.brasilense</u> (data not shown). Similar results were obtained for the nitrogenase activity of cultures in the presence of the same concentrations of chemicals (Table 2). To establish the GSH involvement in the mechanism of detoxification of fungicides, the GSH content of <u>A.brasilense</u> cells was determined for cultures grown in the absence and presence of sublethal concentrations of NAF, captan and thiram. Results obtained (Table 2) showed that 100 ug/ml NAF gave an increase of 50% of GSH also when NAF was used in combination with captan or thiram. This increase did not protect the cells against the toxic effects of fungicides.

## REFERENCES

1) Okon Y. and Hadar Y. (1987). CRC Crit.Rev.Biotec. 6: 61-85
2) Gallori E. <u>et al</u>. (1988). J.Gen.Microbiol. 134: 3173-3178
3) Bani E. <u>et al</u>. (1980). J.Gen.Microbiol. 119: 239-244
4) Gallori E. and Bazzicalupo M. (1985). FEMS Lett. 28: 35-38
5) Akerboom T.P.M. and Sies H. (1981). Methods Enzymol. 77: 373-382

**Table 2.** Acetylene reduction activity (ARA) and intracellular GSH content of <u>A.brasilense</u> Cd grown in the presence of different chemicals.

| Addition * | ARA ** | nmol GSH $(10^{10}$ cells$)^{-1}$ | % |
|---|---|---|---|
| None | 1.44 ± 0.25 | 20.6 ± 1.6 | 100 |
| Captan | 1.43 ± 0.24 | 19.3 ± 3.3 | 94 |
| Thiram | 1.41 ± 0.21 | 22.2 ± 4.1 | 108 |
| NAF | 1.45 ± 0.20 | 33.6 ± 0.3 | 163 |
| NAF + Captan | 0.38 ± 0.08 | 29.8 ± 1.3 | 144 |
| NAF + Thiram | 0.54 ± 0.10 | 32.3 ± 2.6 | 157 |

* The concentrations used in the experiments were: Captan 2 $\mu$g/ml$^{-1}$, Thiram 5 $\mu$g/ml$^{-1}$ and NAF 100 $\mu$g/ml$^{-1}$.
** Expressed as $\mu$mol acetylene reduced h$^{-1}$($\mu$g protein)$^{-1}$.

EXPERIMENTS OF <u>AZOSPIRILLUM</u> BACTERIZATION OF MAIZE SEEDS (ZEA MAIS L.)
IN OPEN FIELD



P. IACOPINI and A. CAPPELLINI*
Regione Emilia Romagna, Ufficio Agricolo di Zona, 29015
Castel San Giovanni (PC); * Azienda Agricola
Cappellini Dino & Figli, 29010 Pontenure (PC), Italia




ABSTRACT. The results of one year of bacterization field trial of
maize seeds with <u>Azospirillum</u> are reported and discussed.



INTRODUCTION

One of the main scope of the Agricultural Department of the Emilia
Romagna Region is to acquire, by field experiments, useful knowledge
on the use of $N_2$ fixing microorganisms as biofertilizers for cereals
in order to reduce chemical nitrogen fertilization and to controll the
environmental pollution.


MATERIALS AND METHODS

The field trials were carried out at Pontenure – Piacenza in the
Capellini Dino and sons farm, in the year 1985. Technical assistance
was given by the Dipartimento di Scienze e Tecnologie Alimentari e
Microbiologiche of the University of Firenze (Italia).
The experiments were run on plots of 300 $m^2$. Seeds of maize (CV. Sivam
600) were sown on 10th May 1985, with 62000 seeds $ha^{-1}$ equivalent to
4.5 plants $m^{-2}$ density. A strain of <u>Azospirillum lipoferum</u> M6 was used
as inoculum. A suspension of bacterial cells supported on a carrier
made with soil and farm yard manure (1:1) was applied when the plants
were 20 cm high. The soil was irrigated immediately after inoculation.
The control plots received 285 Kg N $ha^{-1}$ while the inoculated plots
received 35 Kg N $ha^{-1}$. A preemergency treatment with 1.5 Kg $ha^{-1}$ of
atrazina was made either to the control or to inoculated plots. The
harvest time was on 23rd September.

RESULTS

Table 1 - Effect of _Azospirillum_ inoculation on maize yield.

| Nitrogen (N) Kg ha$^{-1}$ | Grain yield (14% humidity) Kg ha$^{-1}$ | N unit employed for 100 Kg of dried grains | Cost (Lira) for 100 Kg of dried grains |
|---|---|---|---|
| 35 + Azospirillum | 8260 | 0.40 | 1975 |
| 285 | 8300 | 2.60 | 4481 |

The most interesting result has been obtained combining _Azospirillum_ inoculation with a very low level of nitrogen chemical fertilization. Indeed the productivity is equivalent to that obtained with the full nitrogen fertilization.


CONCLUSIONS

The results obtained showed that with _Azospirillum_ inoculation it is possible to maintain the same production level, to reduce either the cost of the nitrogen fertilization or the environmental pollution.


REFERENCES

Eid A.M., Hegazi N.A., Manib M. and Shokr E. (1984). Rev. Ecol. Biol. Sol., 21, 235-242.

Favilli F., Balloni W., Cappellini A., Granchi L. and Savoini G. (1987). Ann. Microbiol., 37, 169-181.

Kapulnik Y., Okon Y., Henis Y. (1987). Biol Fertil. Soil, 4, 27-37.

# NITROGEN FIXATION IN FILAMENTOUS CYANOBACTERIA

R. HASELKORN, M. BASCHE, H. BÖHME‡, D. BORTHAKUR*, P. B.
BORTHAKUR*, W. J. BUIKEMA, M. E. MULLIGAN+, AND D. NORRIS∞
*The University of Chicago*
*Department of Molecular Genetics and Cell Biology*
*920 E. 58th Street*
*Chicago, IL 60637*
*USA*

ABSTRACT. Filamentous cyanobacteria such as *Anabaena* differentiate cells specialized for nitrogen
fixation, called heterocysts. A large number of mutants of *Anabaena* 7120, defective in heterocyst
differentiation, were isolated. For some of these, the gene that complements the mutation has been
identified by conjugation with a cosmid library of wild-type DNA fragments. One early-acting gene has
been sequenced. This gene, *hetR*, is required to initiate differentiation, has no function in vegetative cells,
and when present in extra copies in wild-type cells, causes the formation of multiple heterocysts. The
organization of the genes for nitrogen fixation in *Anabaena* was also studied. The *nif* gene region of
vegetative cell DNA of *Anabaena* 7120 carries two interrupting elements: an 11-kb element inserted near the
carboxy terminus of the *nifD* gene and a 55-kb element inserted near the *nifS* gene. Both elements are
flanked by directly repeated sequences within which site-specific recombination occurs during heterocyst
differentiation. The result of these recombinations, in each case, is precise excision of the element and
fusion of the heterocyst chromosome. In the case of the 11-kb element, excision leaves behind the intact
*nifHDK* operon. The other rearrangement yields the *nifB fdxN nifSU* operon, where *nifBSU* have the same
significance as their *Klebsiella* counterparts.

## Introduction

Cyanobacteria have the same requirements for nitrogen fixation as other diazotrophs: synthesis of
the nitrogenase complex and its cofactors, a source of ATP and low potential electrons, and an
environment protected from oxygen. Unique among diazotrophs, cyanobacteria evolve oxygen in
the light. Filamentous strains such as *Anabaena* solve the problem of oxygen sensitivity of
nitrogenase by differentiating specialized cells called heterocysts at regular intervals along the
filament. In the heterocyst, the oxygen-evolving reaction of Photosystem II is turned off, $CO_2$
fixation stops, nitrogenase, cofactors and electron donor proteins are synthesized, and the oxidative
pentose pathway is activated. A double-layered envelope of polysaccharide and glycolipid is laid
down outside the existing cell wall. New connections between the heterocyst and the adjacent

---

*current address: Biotechnology Program, University of Hawaii, 3050 Maile Way, Honolulu, HI 96822

+current address: Dept. of Biochemistry, Memorial University, St. John's, Newfoundland A1B 3X9 Canada

‡current address: Botanical Institute of Bonn University, D-5300 Bonn 1 W. Germany

∞current address: Dept. of Chemistry, MIT, Cambridge, MA 02139

vegetative cells are established; these mediate the transport of carbohydrate and of fixed nitrogen. All of these changes require differential gene expression. The study of cyanobacterial nitrogen fixation and heterocyst differentiation has been enhanced by the development of methods for efficient conjugation of DNA from *E. coli* into *Anabaena* (20). Using these methods it has been possible to isolate developmentally significant genes by complementation of mutants, to inactivate genes by targetted insertional mutagenesis, to perform random mutagenesis with transposons, and to create fusions with reporter genes to measure gene expression at the single cell level. These results will be reviewed below. Most of our knowledge of the organization of the genes required for nitrogen fixation (*nif*) in *Anabaena* is based on physical mapping, relying on sequence similarity to known *nif* genes of *Klebsiella* and *Azotobacter* (15,19). Such studies additionally showed that several large DNA elements (11 kb and 55 kb) interrupt the *nif* genes in *Anabaena* vegetative cells and that these elements are excised during heterocyst differentiation (8,9,10). Those results will also be reviewed below.


## Methods

All of the experiments to be described from our laboratory use the strain called *Anabaena* sp PCC7120. The cells are maintained on medium BG-11 with nitrate or ammonia as N source and induced to differentiate heterocysts by transfer to BG-11 without N source. The various manipulations of DNA have been described extensively. We currently determine DNA sequences by making nested deletions with exonuclease III and nuclease S1 and then using the M13 universal primer in dideoxy synthesis. Mutagenesis involved treatment with diethyl sulfate followed by penicillin selection for cells unable to fix nitrogen (5). The construction of a cosmid shuttle vector for efficient conjugation from *E. coli* with *Anabaena* is described elsewhere, as is the procedure for complementation of mutants and the isolation of complementing DNA fragments (4,6,20).


## Results and Discussion

HETEROCYST DIFFERENTIATION

A new approach to the study of genes required for heterocyst differentiation is based on the conjugation system mentioned briefly in the Introduction. The system developed by Wolk and Elhai has three components: a shuttle vector built from a cryptic plasmid capable of replication in *Anabaena* fused to part of pBR322, a broad host range plasmid to make the apparatus for conjugal transfer of DNA, and a third plasmid that mobilizes the shuttle vector for conjugal transfer (6,20). The shuttle vector contains cloning sites for inserted fragments of *Anabaena* DNA. A recent improvement includes genes encoding methylase activities, capable of protecting DNA against the *Ava* restriction enzymes, in the *E. coli* strain in which the shuttle vector is maintained. Additionally, a low copy number shuttle vector with a *cos* site and a multiple cloning site flanked by termination sequences that block transcription into the cloned insert has been constructed (4). These principles have been applied in Wolk's laboratory to study the timing and the spatial distribution of transcripts of several genes expressed only in heterocysts (7).

In our laboratory, chemical mutagenesis and several rounds of penicillin selection were used to isolate aerobic Nif⁻ mutants of *Anabaena* 7120. Among these mutants were a number that displayed abnormal heterocyst morphology: thickened envelopes, weakened cell attachments, odd shape, or, in one case, no signs of differentiation at all (4). Each of these mutants was complemented *en masse* with the library of wild type DNA fragments, selecting for ability to fix $N_2$ and resistance to an antibiotic carried by the cosmid vector. From each complemented colony, plasmid DNA could be re-isolated and used to transform *E. coli*. Amplification of the complementing plasmid in *E. coli* provided the starting material for further analysis, i.e. trimming down the insert and repeating the complementation test, in order to define the minimum length of DNA needed for complementation. In this way, individual genes required for heterocyst differentiation have been cloned.

One mutant fails to initiate heterocyst differentiation when transferred to nitrogen-free medium (4). The wild type *Anabaena* gene (*hetR*) that complements the mutation in this strain is expressed at a low level in vegetative cells, although it is not required for vegetative growth. Its transcript increases in abundance within the first few hours of heterocyst induction. The transcript appears to be monocistronic. The gene's translation product is not similar in sequence to any known protein, including sigma factors and DNA-binding transcriptional activator proteins. The most interesting result obtained with this cloned gene to date is the following: when the wild type gene is introduced on a multicopy plasmid into either wild type *Anabaena* or the original mutant, on nitrogen-free medium the filaments produce clusters of heterocysts, two to five cells in a row, instead of the usual single heterocysts at approximately 10 cell intervals. Heterocysts are also formed on $N^+$ medium at a frequency expected for $N^-$ conditions. Thus the product of the *hetR* gene appears to play a key regulatory role in the decision to differentiate.

ORGANIZATION OF THE *NIF* GENES OF *ANABAENA*

The paradigm of *nif* gene organization is *Klebsiella pneumoniae*, in which all the known genes uniquely required for nitrogen fixation are clustered within 24 kb (17). In general, genes in the *nif* cluster that encode proteins with related functions are transcribed together. The *nifHDK* genes, which encode the components of the nitrogenase complex and the *nifEN* genes, which encode a structure required for synthesis of the Fe-Mo cofactor, comprise operons. The *nifUSV* genes are also cotranscribed. The functions of *nifU* and *nifS* are not yet known; *nifV* encodes homocitrate synthase and homocitrate is a part of the Fe-Mo cofactor (12). The *nifL* and *A* genes, both participating in regulation of transcription of the other *nif* genes, are cotranscribed. Finally, *nifB* and *Q*, both required for Fe-Mo cofactor synthesis, also comprise an operon (17).

The first studies of *Anabaena nif* genes indicated that their organization was different. The *nifD* and *K* genes were not contiguous and the *nifS* gene was 5' to *nifHD* rather than 3' as in *Klebsiella* (19). Moreover, these *nif* genes appeared to be interspersed with and surrounded by DNA that was not transcribed under nitrogen fixing conditions and therefore not likely to contain other *nif* genes. Subsequent work showed that although this description fits the situation in *Anabaena* vegetative cells, it is not the case in heterocysts. During heterocyst differentiation, two rearrangements occur in this region of the *Anabaena* chromosome. Both are excisions of DNA elements that interrupt the reading frames of *Anabaena nif* genes in vegetative cells. One excision results in the deletion of an 11-kb element from the *nifD* coding region, resulting in the formation of a *nifHDK* operon very similar to that of *Klebsiella* (10). The second excision removes a 55-kb element to the 5' side of the *nifS* gene that interrupts the reading frame of a gene called *fdxN* (8,9,14). This results in the formation of another operon, *nifB-fdxN-nifS-nifU* (15). Each excision is the result of recombination between directly repeated short sequences at the ends of the interrupting element. These short sequences differ for the two elements. Excision of the 11-kb element requires the product of a gene, *xisA*, located just inside the left end of the element (see Figure 1). Following the excision reaction, the excised 11-kb element is circular and it contains one copy of the short

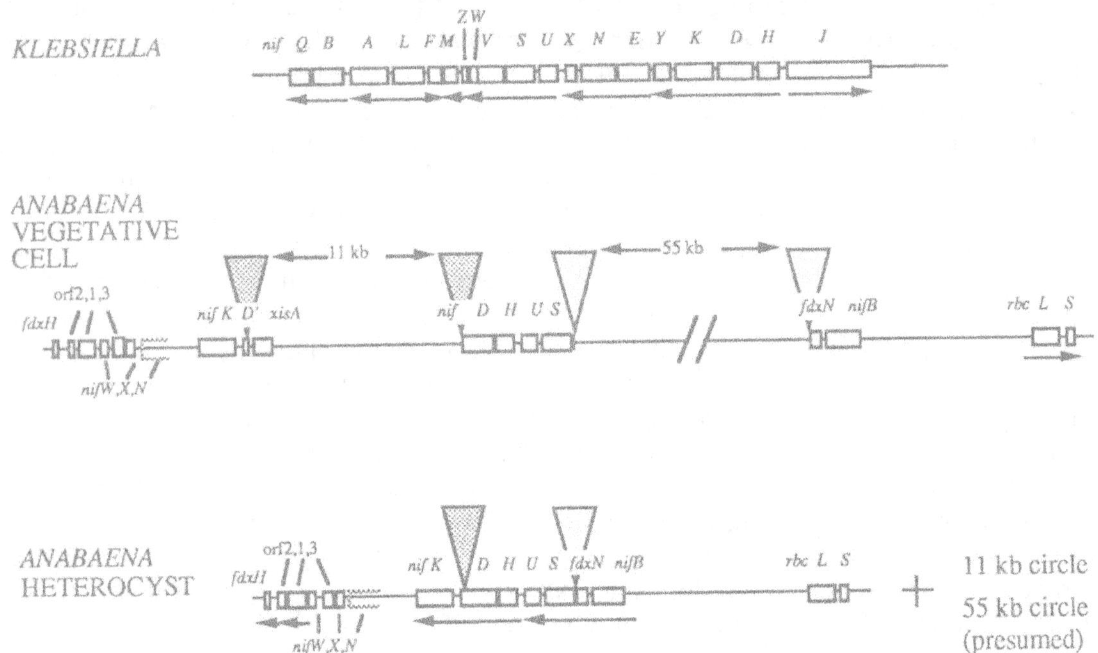

Figure 1. Organization of the *nif* gene region of *Klebsiella*, *Anabaena* vegetative cells, and *Anabaena* heterocysts after excision of the 11-kb and 55-kb elements. Shaded triangles indicate the location of short direct repeats within which the recombination events leading to excision occur. Horizontal arrows below each map indicate transcription units. Sizes of the transcripts produced by genes between *nifW* and *nifK* have not been determined. Sequencing of the *nifN-nifK* region has not been completed. (Reproduced by permission.)

terminal repeat. The other identical copy of the short repeat remains inside the *nifD* coding region. The *xisA* gene product catalyzes the precise excision reaction when the element is propagated as part of a plasmid in *E. coli* as well. The *xisA* gene has been cloned, sequenced and expressed in *E. coli* but the recombination reaction has not yet been observed *in vitro*. In terms of the sites and protein requirement for excision, the element resembles a lysogenic virus. As yet, no advantage has been found to be conferred on vegetative cells carrying the element.

The rearrangements are summarized in Figure 1, which also shows the *Klebsiella nif* gene map and the surrounding region of the *Anabaena* chromosome. The operon furthest to the right in the diagram is *rbcLS*, encoding the large and small subunits of RuBP carboxylase. These genes are transcribed in vegetative cells but not in heterocysts. The 10 kb between *rbcLS* and *nifB* have not been characterized in terms of sequence or transcripts yet.

The *nifB*, *nifS* and *nifU* genes were identified on the basis of sequence similarity to the corresponding genes of *Klebsiella*, *Azotobacter*, and several *Rhizobium* species (15). The *nifB* gene contains an open reading frame for a 53 kD protein whose sequence includes eight cysteine residues conserved among all *nifB* gene products sequenced to date. These cysteines are located in three highly conserved regions of the protein and are believed to function in the assembly of the Fe-Mo cofactor. The *nifS* gene encodes a 47 kD protein similar over its entire length to the corresponding *nifS* gene products of *Klebsiella* and *Azotobacter*. Three cysteines are conserved among the three proteins, one of which (cys 324) is embedded in a fully conserved stretch of 17 amino acids. The function of the *nifS* gene product is not known because *nifS* mutations in *Klebsiella*, for the most part, are polar on *nifV* and so are defective in Fe-Mo cofactor synthesis. The *nifU* gene of *Anabaena* is a bit different. The predicted size of the NifU protein is 32 kD and there is considerable sequence similarity with the corresponding proteins of *Klebsiella* and *Azotobacter*. However, this similarity is confined to the amino terminal half of the sequence and another short stretch from residues 240 to 272. Within the amino terminal half there are six conserved cysteines, widely spaced, while the 240-272 stretch contains a single conserved cys-X-X-cys. The conserved cysteines suggest that NifU, like NifS, has metal ligands and therefore might be involved in Fe-Mo cofactor synthesis. Neither of these putative proteins, however, has the correct size for the protein (20 kD) found associated with the dinitrogenase apoprotein of *Azotobacter* (18).

The *nifB* operon of *Anabaena* contains an additional open reading frame. Called *fdxN*, this gene encodes a 9 kD protein containing two domains, each with appropriately spaced cysteines for binding a 4Fe-4S center (14). Bacterial-type ferredoxins have not been seen in cyanobacteria previously except for the 9 kD Fe-S protein (product of the *psaC* gene) that forms part of the photosystem I reaction center.

The precise roles of these four genes in *Anabaena* nitrogen fixation are not known, in part because they have not yet been selectively mutated. Transcripts of the operon are seen only under conditions of nitrogenase induction. Three mRNAs have been detected: 4.4, 3.1 and 1.75 kb. These transcripts probably start at the same site 283 bp upstream of the *nifB* translation start. The shortest transcript terminates after *nifB*, the 3.1-kb RNA probably terminates after *nifS* and the longest RNA includes *nifU*. All of these RNAs appear to start at the same site, whose 5'-flanking sequence bears no resemblance to the typical *nif* gene promoters of *Klebsiella*. The operon contains an additional unusual feature: tandem repeats of the sequence CCCCAAT following *nifB* and *nifS*. Since these are approximately where the 1.75 and 3.1-kb transcripts terminate, it was considered that the tandem repeats are involved in transcription termination. However, a search of a cosmid library of *Anabaena* DNA fragments revealed many instances of these tandem repeats, one containing 25 copies, so they may not have a specific termination function.

Immediately following the *nifB* operon is the *nifHDK* operon. This operon produces three transcripts that correspond to *nifH*, *nifHD* and *nifHDK* in size. The start site for transcription of this operon was mapped by S1 nuclease protection and primer extension (S. J. Robinson, unpublished) and the 5'-flanking sequence of the transcript family was determined. This sequence is not easily related to the characteristic *nif* gene promoter of *Klebsiella* or *Rhizobium*. Both the *nifH* and the *nifB* 5'-flanking sequences contain CTA at -12 to -10 with respect to the start site and a few conserved residues in the -70 to -40 region, but there is no obvious "nif" consensus sequence. This situation is not improved by including the *fdxH* gene transcript (see below).

Following the *nifHDK* operon there are several kb of DNA for which we do not yet have sequence information. Beyond the unsequenced region we find, in the following order, *nifN*, *nifX*, orf 3, *nifW*, orf 1, orf 2 and *fdxH*. All of these except the last were assigned on the basis of similarity to sequences in *Klebsiella*, *Azotobacter* or both. *nifN* is required for the structure that assembles Fe-Mo cofactor. The *nifX* gene is in the same relative position in *Klebsiella* but its function is unknown. Orf 3 corresponds to a sequence (orf 3) in the *Azotobacter nif* gene region, not found in *Klebsiella*, of no known function. Orf 2 of *Anabaena* corresponds to orf 6 of *Azotobacter*. *nifW* is found in both *Klebsiella* and *Azotobacter* but its function is unknown also. Orf 1 of *Anabaena* encodes a 30 kD protein not recognized in any other previously sequenced *nif* gene region. However, sequence similarity to the *chlN* gene of *E. coli* (16) was discovered by R. Jones (personal communication) in a detailed search of sequence data bases. The *chlN* gene product is required for synthesis of molybdopterin, the Mo cofactor of nitrate reductase and other enzymes.

Interruption of the orf 1 (*chlN*) reading frame in *Anabaena* resulted in very slow growth on $N_2$, suggesting that the gene product participates in nitrogen fixation (2).

The last *Anabaena nif* gene shown in the Figure is *fdxH*, which encodes a 2Fe-2S ferredoxin found exclusively in heterocysts and shown previously to be an efficient donor of electrons to *Anabaena* nitrogenase *in vivo* (1). The *fdxH* gene is represented in two transcripts, 0.59 and 1.85 kb. The smaller transcript contains *fdxH* alone and initiates between *fdxH* and orf 2. The larger transcript initiates 42 bp before the ATG of orf 1 and contains orf 1, orf 2 and *fdxH*. Neither 5'-flanking sequence, *fdxH* or orf 1, contains information leading to an improved consensus for an *Anabaena nif* gene promoter.

Not all the genes required for nitrogen fixation in *Anabaena* are located in the region shown in the Figure. The transposon Tn5 can be introduced into *Anabaena* by conjugation from *E. coli*, if it is suitably methylated. The transposon can enter the chromosome at many sites, where it is detected by its ability to confer resistance to neomycin (3). Resistance provides a selectable marker for cloning the interrupted DNA fragment; in this way, a previously unrecognized gene required for nitrogen fixation was cloned. The DNA fragments on which the gene is located are not in the region shown in the Figure.

The genetic tools described in this paper, principally chemical and transposon mutagenesis for mutant isolation and cosmid complementation by conjugation for cloning genes, should make it possible to provide, eventually, a complete molecular description of the regulatory interactions required for heterocyst differentiation and nitrogen fixation in *Anabaena*.

## References


1. Böhme, H. and Haselkorn, R. (1988) 'Cloning and sequence determination of the gene encoding heterocyst ferredoxin from the cyanobacterium *Anabaena* 7120', Mol. Gen. Genet. 214, 278-285.
2. Borthakur, D., Basche, M., Buikema, W. J., Borthakur, P. B. and Haselkorn, R. (1990) 'Expression, nucleotide sequence and mutational analysis of two open reading frames in the *nif* gene region of *Anabaena* sp. strain PCC 7120', Mol. Gen. Genet. 221, 227-234.
3. Borthakur, D. and Haselkorn, R. (1989) 'Tn5 mutagenesis of *Anabaena* sp. strain PCC 7120: isolation of a new mutant unable to grow without combined nitrogen', J. Bacteriol. 171, 5759-5761.
4. Buikema, W. J. and Haselkorn, R. (1990) 'Isolation and complementation of nitrogen fixation mutants of *Anabaena* 7120: identification of a gene controlling heterocyst differentiation, submitted.
5. Chapman, J. S. and Meeks, J. C. (1987) 'Conditions for mutagenesis of the nitrogen-fixing cyanobacterium *Anabaena variabilis*', J. Gen. Microbiol. 133, 111-120.
6. Elhai, J. and Wolk, C. P. (1988) 'Conjugal transfer of DNA to cyanobacteria', Methods Enzymol. 167, 747-754.
7. Elhai, J. and Wolk, C. P. (1990) 'The genes for nitrogenase are developmentally regulated and exhibit a spatial pattern of expression in the cyanobacterium *Anabaena* under anaerobic as well as aerobic conditions', EMBO J. 9, in press.
8. Golden, J. W., Carrasco, C. D., Mulligan, M. E., Schneider, G. J. and Haselkorn, R. (1988) 'Deletion of a 55-kilobase-pair DNA element from the chromosome during heterocyst differentiation of *Anabaena* sp. Strain PCC 7120', J. Bacteriol. 170, 5034-5041.

9.  Golden, J. W., Mulligan, M. E. and Haselkorn, R. (1987) 'Two developmentally regulated genome rearrangements in *Anabaena* have different recombination site specificity', Nature 327, 526-529.

10. Golden, J. W., Robinson, S. J. and Haselkorn, R. (1985) 'Rearrangement of nitrogen fixation genes during heterocyst differentiation in the cyanobacterium *Anabaena*', Nature 314, 419-423.

11. Haselkorn, R. (1989) 'Excision of elements interrupting nitrogen fixation operons in cyanobacteria', in D. E. Berg and M. M. Howe (eds.) Mobile DNA, American Society for Microbiology, Washington, D. C., pp. 735-742.

12. Hoover, T. R., Imperial, J., Ludden, P. W. and Shah, V. K. (1989) 'Homocitrate is a component of the Fe-Mo cofactor of nitrogenase', Biochemistry 28, 2768-2771.

13. Lammers, P. J., Golden, J. W. and Haselkorn, R. (1986) 'Identification and sequence of a gene required for a developmentally regulated DNA excision in *Anabaena*', Cell 44, 905-911.

14. Mulligan, M. E., Buikema, W. J. and Haselkorn, R. (1988) 'Bacterial-type ferredoxin genes in the nitrogen fixation (*nif*) regions of the cyanobacterium *Anabaena* 7120 and *Rhizobium meliloti*', J. Bacteriol. 170, 4406-4410.

15. Mulligan, M. E. and Haselkorn, R. (1989) 'Nitrogen-fixation (*nif*) genes in the cyanobacterium *Anabaena* sp. strain PCC 7120: the *nifB-fdxN-nifS-nifU* operon', J. Biol. Chem. 264, 19200-19207.

16. Nohno, T., Kasai, Y. and Saito, T. (1988) 'Cloning and sequencing of the *E. coli chlEN* operon involved in molybdopterin biosynthesis', J. Bacteriol. 170, 4097-4102.

17. Orme-Johnson, W. H. (1985) 'Molecular basis of biological nitrogen fixation', Annu. Rev. Biophys. Biophys. Chem. 14, 419-459.

18. Paustian, T. D., Shah, V. K. and Roberts, G. P. (1990) 'Apodinitrogenase: purification, association with a 20-kDa protein, and activation by the Fe-Mo cofactor in the absence of dinitrogenase reductase', Biochemistry 29, 3515-3522.

19. Rice, D., Mazur, B. J. and Haselkorn, R. (1982), 'Isolation and physical mapping of nitrogen fixation genes from the cyanobacterium *Anabaena* 7120', J. Biol. Chem. 257, 13157-13163.

20. Wolk, C. P., Vonshak, A., Kehoe, P. and Elhai, J. (1984), 'Construction of shuttle vectors capable of conjugative transfer from *E. coli* to nitrogen-fixing filamentous cyanobacteria', Proc. Natl. Acad. Sci. U.S.A. 81, 1561-1565

RECENT  ASPECTS ON THE HYDROGENASE-NITROGENASE  RELATION-
SHIP IN CYANOBACTERIA


H. Bothe, T. Kentemich, Dai Heping

Botanisches Institut,  Universität zu Köln,  Gyrhofstr. 15, D-
5000 Köln 41, FRG


The  subject was reviewed several times in the past  (Benemann  and
Hallenbeck,  1978;  Bothe et al.  1978; Lambert and Smith 1981; Houchins
1984;  Papen et al.  1986;  Smith 1990).  This article will,  therefore,
concentrate on newer developments in the field.

I) Hydrogen metabolism in microorganisms

    In  microorganisms,  $H_2$  can be produced either by  nitrogenase  or
hydrogenase.  The  gas is formed by nitrogenase simultaneously with  the
reduction of the dinitrogen molecule following the equation:  $8 H^+ + 8 e^-$
$+ N_2 ---> 2 NH_3 + H_2$.  In addition,  $H_2$ is produced by nitrogenase in the
ATP-dependent reduction of $H^+$ when substrates other than $H^+$ are limiting
(Burns  and  Hardy 1975).  Therefore the  amount  of  $H_2$  produced  by
nitrogenase  is  variable both in vivo and in vitro,  depending  on  the
assay conditions employed. Hydrogen can also be produced by hydrogenases
in microorganisms.  This gas-production is generally ATP-independent and
affected  by  low  amounts of CO,  which allows  to  differentiate  the
formations catalyzed by hydrogenases and nitrogenases.  Newer data  show
that these enzymes can each be subdivided into three different classes.
    As  regard to nitrogenase,  the conventional Mo-enzyme is  expressed
under  $N_2$-fixing  conditions in the organisms when the  medium  contains
sufficient  amounts  of this element.  P.  Bishop and  coworkers  (1980)
discovered  that  Azotobacter vinelandii expresses  another  nitrogenase
when the medium is supplemented with vanadium but lacks molybdenum. This
alternative  nitrogenase has a V-Fe-S cofactor in the  prosthetic  group
and is encoded by the three structural genes vnf HDK which are different
from the nif HDK genes of the Mo-nitrogenase (Bishop et al.; 1989;  Pau
1989).  More  recently,  A. vinelandii was shown to have the gene  sets
(=anf  HDK)  for a third nitrogenase which probably contains  only  Fe-S
centres in the prosthetic group (Jacobsen et al. 1986; Pau et al. 1989).
    A. vinelandii  is the only organism shown so far to possess  three
nitrogenases.  The distribution of alternative nitrogenases among  organisms  has  not been investigated in much detail.  The closely  related
Azotobacter chroococcum possesses only a Mo- and a V-nitrogenase (Robson
et al.,  1989).  There  are indications that Clostridium pasteurianum

(Dilworth $et$ $al$., 1987), <u>Azorhizobium caulinodans</u>, <u>Pseudomonas</u> strains (Chan $et$ $al$. 1988) and two <u>Methanosarcina</u> strains (Scherer 1989) might also contain both Mo- and V-nitrogenase, although clear-cut evidence is missing. Another archaebacterium, <u>Methanosarcina barkeri</u> 227 may express even two different Mo-nitrogenases (Zinder and Lobo 1990) and the phototroph <u>Rhodobacter capsulatus</u> seems to possess Fe-nitrogenase in addition to the Mo-enzyme (Schneider and Müller 1990).

It has recently become clear that hydrogenases from the organisms can be subdivided into the three classes [Fe], [NiFe] and [NiFeSe] hydrogenases (Fauque $et$ $al$. 1988). The [Fe]-hydrogenase contains 2 (4Fe-4S) clusters and 1 (3Fe-xS) centre in the prosthetic group. Two different [Fe]-hydrogenases occur in <u>Clostridium pasteurianum</u> and the enzyme is also known in <u>Desulfovibrio</u> species. [Fe]-hydrogenases have the highest specific activities in the formation and uptake of $H_2$ and in the proton-deuterium exchange reaction. The [Fe]-hydrogenases are also characterized by their extreme sensitivity to carbon monoxide and nitrite.

The [NiFe]-hydrogenase is the most widely distributed enzyme of the three classes. It is usually membrane-bound and often orientated to the periplasmic space. It has a high affinity for $H_2$ and serves in $H_2$-utilization in most if not all organisms. The enzyme contains 2 (4Fe-4S) clusters, 1 (3Fe-xS) centre and additionally nickel in the prosthetic group. Ni is apparently involved in catalysis, although neither the oxidation states nor the role of it are well-defined currently. The [NiFeSe] hydrogenase has only been found in <u>Desulfovibrio</u> species (<u>D</u>. <u>baculatus</u>, <u>D</u>. <u>salexigenes</u>). Selenium occurs as selenocysteine and is encoded by TGA which is usually a stopcodon. The enzyme apparently has higher $H_2$-evolution and lower uptake activities than the [NiFe]-hydrogenase. It is possibly rewarding to search for such an enzyme in phototrophs which are still discussed in solar energy conversion programs (Kentemich $et$ $al$. 1990a).

## II. <u>NITROGENASES IN CYANOBACTERIA</u>

Physiological experiments recently indicated that <u>Anabaena variabilis</u> can express a V-nitrogenase (Kentemich $et$ $al$. 1988). This cyanobacterium can be grown in dependence on either Mo or V in the medium with essentially the same growth rate. Compared to Mo-grown cells, V-cultures form $C_2H_4$ with approximately 2/3 lower rates, produce more $H_2$ and reduce $C_2H_2$ partly to $C_2H_6$ (to approximately 3%) in light-dependent reactions. The expression of the V-nitrogenase could be detected after approximately 10d in a -Mo+V culture (= a culture growing without Mo but with V) growing in the exponential phase without a transient state of N-deficiency. Extensive purifications of the media to obtain Mo-deficiency are not necessary in the case of <u>A</u>. <u>variabilis</u>. Cyanobacteria apparently require more Mo than other organisms for growth (see Singh $et$ $al$. 1978). It is to be mentioned in this context that we could not demonstrate Mo-deficiency under our laboratory conditions with the non-photosynthetic <u>Azospirillum brasilense</u> Sp 7. Extensive purification of the medium might be necessary with this bacterium.

Recent findings indicate that <u>Anabaena variabilis</u> might contain also the third, Fe-nitrogenase (Kentemich $et$ $al$. 1990b). The original observation came from control cultures which had been kept without Mo and V for 130d. Such cultures showed heavy N-deficiency symptoms between days 30-50. They lost their blue-green colour and became green because they apparently degraded their phycobilins. The culture restarted growth after approx. 50d and regained the blue-greenish colour gradually. $C_2H_2$-reduction rates in such -Mo-V cultures were only 1/10 of that in Mo-grown cells similar as in _Azotobacter vinelandii_ (Pau _et al._, 1989). The Mo and V contents in the -Mo-V-cultures were less than $10^{-8}$M which was the detection limit of the determinations by atomic absorption spectrometry. It could be demonstrated that the expression of Mo-nitrogenase in such cultures required the addition of Mo to a final concentration of at least $10^{-7}$ M Mo. Characteristically, the addition of high concentrations of Mo to -Mo-V cultures caused a drastic increase in the $C_2H_6$- and $H_2$-formations within 2-4h. Such a rapid increase was also described for the Fe-nitrogenase from _Azotobacter vinelandii_ and was ascribed to improper incorporation of Mo into the cofactor of the Fe-nitrogenase (Pau _et al_. 1989). This increase was not observed when V-grown cells were supplemented with 0.5 M Mo indicating that V and -Mo-V cultures express two different alternative nitrogenases.

Southern hybridization experiments using probes for the structural gene H also indicated the existence of more than one nitrogenase H-gene in _A. variabilis_ (Kentemich _et al_. 1990b). Hybridizations with both _nif_ H (coding for the small subunit of Mo-nitrogenase from _Klebsiella pneumoniae_) and _anf_ H (same gene of the Fe-nitrogenase from _Azotobacter vinelandii_ gave two strong and one weak signals in each case. Two _nif_ H like sequences of unknown functions in _Anabaena_ 7120 (Rice _et al_. 1982) and even three in _Calothrix_ (Kallas _et al_. 1983) had been found earlier. The other structural gene, _nif_ K does not crosshybridize with the same gene from other organisms which is true also for _A. variabilis_ (Kentemich _et al_. 1990b), and _nif_ D generally gives weak hybridization signals.

## III. HYDROGENASES IN CYANOBACTERIA

It is now clear that cyanobacteria possess two different hydrogenases. This subject was controversial for some years, and our group believed that the so-called "reversible", bidirectional hydrogenase was an artifact of cell-free preparations resulting from the solubilization of the thylakoid-bound "uptake" hydrogenase from the membranes (Eisbrenner _et al_. 1981). The arguments were a) that the activity of the reversible hydrogenase increased the more, the more drastic the method was to break the cells and b) that no function was obvious for the reversible enzyme. Houchins and Burris (1981a,b) were able to separate both enzyme activities and to characterize them in detail. They also showed that the activity level of the reversible, but not of the uptake hydrogenase drastically increased by growing _Anabaena_ under anaerobic conditions (Houchins and Burris 1981b; Papen _et al_. 1986). Recently it as shown by immunogold labeling that the reversible hydrogenase resides at the cytoplasmic membrane (Kentemich _et al_. 1989). Earlier investigations had already indicated that the uptake hydrogenase is located on the thylakoid membranes. This particularly follows from the findings that $H_2$ supports photosystem I dependent $NADP^+$-reduction by thylakoid preparations (Eisbrenner et al. 1981). $H_2$ is utilized by the uptake hydrogenase in two different pathways which are the photosystem I and the respiration linked ones (Papen et al. 1986). The immediate electron acceptor for uptake hydrogenase probably is plastoquinone or a component close to plastoquinone. Both pathways of $H_2$-utilization share a common segment of electron carriers (the cytochrome $b_6f$ complex besides plastoquinone) before the electrons are allocated either to photosystem I or to the terminal cytochrome oxidase complex of respiration on the thylakoid membranes. The uptake hydrogenase is particularly expressed in the heterocysts and has no or minor activities only in vegetative cells (Lockau et al. 1988; Peterson and Wolk 1978; Eisbrenner and Bothe 1979; Eisbrenner et al. 1981). The experiments of Peschek (1979a,b, see also Eisbrenner et al. 1978), however, indicated that the enzyme is also present in the unicellular Anacystis nidulans. The role of the reversible hydrogenase and its electron acceptor at the cytoplasmic membrane are unknown up to date. The enzyme is present in heterocysts and vegetative cells, and activity levels increase in both cell-types under anaerobic growth conditions (Kentemich et al.1989).

Neither the uptake nor the reversible hydrogenase has been biochemically characterized in much detail, and the molecular biology of this field is still at infancy. This is probably due to the facts that both enzymes become unstable when purification progresses and that they are present in the cells only in low amounts. Cyanobacteria are packed with phycobilins (it is estimated to about 80% of the total protein), and a purification of any protein begins with throwing these 80% into the sink. The hydrogenases purified so far have molecular weights of 56.000 for the reversible enzyme from the non-fixing Spirulina maxima (Llama et al. 1979) and for the uptake protein from Anabaena 7120 (Houchins and Burris 1981b). The reversible hydrogenase from Anabaena 7120 was found to have the tendency to aggregate, and the monomer was presumably 55.000. These studies probably dismissed the smaller subunit of the hydrogenases. More recently Ewart et al.(1989) found molecular weights of 50.000 and 41.000 for the larger and smaller subunit of the reversible hydrogenase from Anabaena cylindrica. More interestingly, the larger subunit showed the $Na_2S_2O_4$ and methyl viologen dependent $H_2$-evolution activity, whereas the smaller subunit catalyzed the tritium exchange reaction. These observations, though very interesting, are so unusual for hydrogenases that it is tentative to assume that the 41 kDa protein is something else than the smaller subunit of the reversible hydrogenase. The 41 kDa protein does not share much sequence homologies with the hydrogenases from non-cyanobacteria. The reversible hydrogenase from the non-$N_2$-fixing Anacystis nidulans was found to consist of two subunits with the molecular weights 56 and 17 kDa, respectively. Both subunits, particularly the smaller one, were very susceptible to proteolysis. Western blot analysis showed that the enzyme is present both in heterocysts and vegetative cells (Kentemich et al. 1989).

## IV. ASPECTS OF THE REGULATION OF NITROGEN FIXATION AND HYDROGEN METABOLISM IN HETEROCYSTOUS CYANOBACTERIA

Except for few species, $N_2$-fixation proceeds in specialized cells (the heterocysts) under aerobic growth conditions and is strictly light-dependent (Bothe 1982). The requirement for light is explained by the fact that $N_2$-fixation requires both ATP and reductant. Heterocysts have a very active cyclic photophosphorylation to provide ATP. Photosystem I is also required to generate reductant (= reduced ferredoxin) to nitrogenase from electron donors like NADH or $H_2$ (Houchins and Hind 1982). As $H_2$-formation by nitrogenase is light-dependent in intact cells of autotrophically growing cyanobacteria (Bothe, 1982), also $H_2$-utilization by one or even both hydrogenases were to proceed in the light. The components involved in the light-activation of plant enzymes are the thioredoxins (Gleason and Holmgren 1988, Holmgren 1985, 1990, Buchanan 1980).

Cyanobacteria contain two different thioredoxins (Gleason and Holmgren, 1988). The one (thioredoxin m) shares strong sequence homologies with the thioredoxin from E. coli and with the thioredoxin m from chloroplasts. The cyanobacterial thioredoxin m activates $NADP^+$-dependent malate dehydrogenase from higher plants but not (Gleason and Holmgren, 1988) or poorly (Papen et al., 1983) malate dehydrogenase from cyanobacteria. Both NADH and NADPH-dependent malate dehydrogenases are rather active particularly in heterocysts (Papen et al., 1983) in contrast to other statements (Gleason and Holmgren, 1988) but they are apparently not controlled by the thioredoxin system. Thioredoxin, however, activates other enzymes in cyanobacteria like $NADP^+$-dependent isocitrate dehydrogenase (Papen et al., 1983) and glutamine synthetase at neutral pH-values (Papen and Bothe, 1984) and deactivates glucose-6-P dehydrogenase (Udvardy et al., 1984, Ip et al., 1984, Darling et al., 1986, Papen et al., 1983). The other protein (thioredoxin f) activates fructose-1-6-bisphophatase and some other enzymes from spinach but surprisingly not fructose 1-6-bisphosphatase from cyanobacteria (Whittaker and Gleason, 1984). No function could be assigned to thiroredoxin f in cyanobacteria as yet. It has a molecular weight of 25.5 kDa and is thus considerably larger than the spinach thioredoxin f (~12.000). In contrast to E. coli where the thioredoxin is apparently dispensable for growth, work with mutants from Anacystis R 2 recently showed that this photosynthetically growing cyanobacterium strictly requires the protein (Müller and Buchanan, 1989). Thioredoxin is probably reductant in the conversion of ribonucleotides to desoxyribonucleotides in Anacystis, whereas this function in E. coli might be fulfilled by glutaredoxin (Gleason and Holmgren, 1988).

If thioredoxin has a function in $N_2$-fixation and $H_2$-utilization by heterocysts, it should occur in these cells. The subject is controversial. Whereas Udvardy et al.(1984) reported an equal distribution of thioredoxin in both cell-types as determined by activity measurements, Rowell et al. (1985) found a very reduced level in heterocysts using the immunogold labeling technique. The latter authors postulated that glucose-6-P dehydrogenase as the key enzyme of carbohydrate catabolism is active also in the light due to the absence of thioredoxin. In vegetative cells, however, thioredoxin reduced by the photosynthetic electron transport deactivates glucose-6-P-dehydrogenase and thereby prevents carbohydrate degradation by the hexosemonophosphate shunt in the light.

The lab recently took up the subject (Dai Heping _et al._, in prep.). Both thioredoxins from _Anabaena variabilis_ were purified to homogeneity and polyclonal antibodies were raised against both proteins. Thioredoxin m levels were more or less the same in heterocysts and vegetative cells as determined by the ELISA-technique. Similarly, when thioredoxin m was enriched from heterocysts and vegetative cells by heat treatment and $(NH_4)_2SO_4$ precipitation, essentially the same protein amounts were required for maximal activation of $NADP^+$-dependent malate dehydrogenase from spinach. It could still be argued that the heterocyst preparations used were not pure enough to allow the determination of the thioredoxin content accurately. Immunogold labeling experiments were, therefore, performed which showed a more or less equal distribution of thioredoxin m in both cell-types of _A. variabilis_ and _A._ 7119. The same was found for thioredoxin f in _A. variabilis_, whereas the antibodies did not label heterocysts of _A._ 7119. Antibodies against thioredoxin m only poorly crossreacted with thioredoxin from _E. coli_ and the proteins from spinach. Surprisingly, they did not at all show crossreactivity with the protein from heterocysts or vegetative cells from _A. cylindrica_ in Western blot experiments. A similar observation was published earlier(Rowell _et al._, 1985).

The "uptake" hydrogenase but not the reversible protein was found to be activated by thioredoxin (Papen _et al._ 1986). This is consistent with the role of the uptake hydrogenase in reutilizing the $H_2$ evolved by nitrogenase(Fig. 1). A localisation of both nitrogenase and hydrogenase at or near the thylakoids is then to be expected. This, however, does not appear to be the case. As shown by immunogold labeling, nitrogenase is equally distributed inside the heterocysts (Bergman _et al._, 1985) and the same is found for thioredoxin m and f in both heterocysts and vegetative cells (Dai Heping _et al._, in prep.). The gold particles did not show any specific

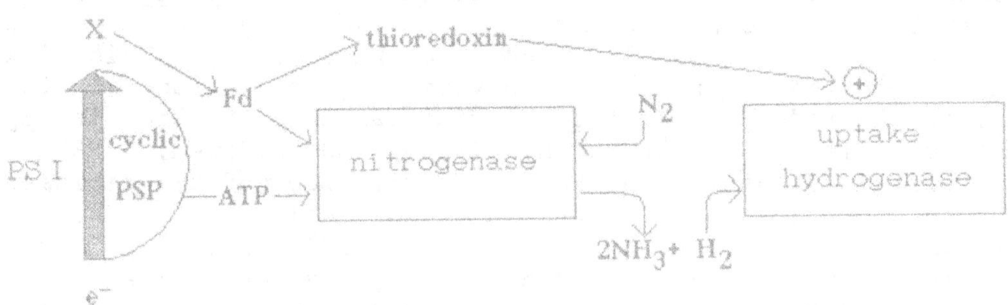


Fig.1 The activation of the uptake hydrogenase by thioredoxin

association with the thylakoids in the current experiments. The exchange of metabolites between thylakoids and cytoplasm inside a  cyanobacterial cell is possibly so rapid that a specific association of the  components mentioned is not required.

Acknowledgement:  This work was kindly supported by a grant  from  the BMFT (no 9342 A/ H2-01-04-90)

## V. REFERENCES

Benemann,  J.R.,  Hallenbeck,  P.C.  (1978) Basic and applied studies of hydrogenase in cyanobacteria In:  H.G.  Schlegel and  K.  Schneider (eds.),  Hydrogenases:  their cataly- tic activities, structure and function, Goltze, Göttin- gen, pp. 395-413.

Bergman,  B.,  Lindblad,  P.,  Petterson,  A., Renstrom, E.R., Tibers, E. (1985) Planta 329-334.

Bishop,  P.E.,  Jarlenski,  D.M.L.,  and Hetherington, D.R.  (1980) Proc. Natl. Acad. Sci. USA 77, 7342-7346.

Bishop,  P.E.  and Joerger, R.D. (1990) Ann. Rev. Plant   Physiol. Mol. Biol. 41, 109-125.

Bothe,  H.  (1982) Nitrogen fixation In:  N.G.  Carr and  B.C.  Whitton (eds.), The Biology of Cyanobacteria, Blackwell,  Oxford, 87-104.

Bothe, H.,Distler, E., Eisbrenner, G. (1978) Biochemie  60, 277-289.

Buchanan, B.B. (1980) Ann. Rev. Plant Physiol. 31, 341- 374.

Burns,  R.C.  and Hardy,  R.W.F. (1975) Nitrogen Fixation in   Bacteria and Higher Plants, Springer, Berlin-Heidelberg,   New York 1975

Chan,  Y.-K.,  Pau,  R.,  Robson, R.L., Miller, R.W. (1988) in H. Bothe, F.J.  de Bruijn and W.E.  Newton (eds) Gustav  Fischer, Stuttgart, New York, p. 88.

Darling,  A.J.,  Rowell,  P.,  Stewart, W.D.P. (1986) Bio chim. Biophys. Acta 850, 116-120.

Dilworth,  M.J.,  Eady,  R.R.,  Robson,  R.L.,  and Miller,  R.W. (1987) Nature (London) 327, 167-168.

Eisbrenner, G., Bothe, H. (1979) Arch. Microbiol. 123,  37-45.

Eisbrenner,  G.,  Roos,  P.,  Bothe,  H. (1981) J. Gen.  Microbiol. 125, 383- 390.

Eisbrenner,  G.,  Distler,  E.,  Floener,  L.,  Bothe,  H.  (1978) Arch. Microbiol. 118, 177-184.

Ewart,  G.D. Reed, K.C., Smith, G.,D. (1990) Eur. J.  Biochem. 187, 215- 223.

Fauque,  G., Peck, H.D.Jr., Moura, J.J.G., Huynh, B.H.,  Belier, Y., Der Vartanian,  D.V.,  Teixeira,  M.,  Przybyla,  A.E., Lespinat, P.A., Moura I., LeGall, J. (1988) FEMS Microbiol.  Reviews 54, 299 344.

Gleason, F.K., Holmgren, A. (1988) FEMS Microbiol. Rev.  54,  271-298.

Holmgren, A. (1985) Ann. Rev. Biochem. 54, 237-271.

Holmgren, A. (1990) J. Biol. Chem. 264, 13963-13966.

Houchins, J.P., Hind, G. (1982) Biochim. Biophys. Acta 682, 86-96.

Houchins, J.P. (1984) Biochim. Biophys. Acta 768, 227- 255.

Houchins J.P., Burris, R.H. (1981a) J. Bacteriol 146,  209-214.

Houchins, J.P., Burris, R.H. (1981b) J. Bacteriol 146,  215-221.

Ip., S.-M., Rowell, P., Aitken, A., Stewart, W.D.P. (1984) Eur. J. Biochem. 141, 497-504.

Jacobsen, M.R., Premakumar, R., and Bishop, P.E. (1986) J. Bacteriol. 170, 27-33.

Kallas, T., Rebière, M.-C., Rippka, R., Tandeau de Marsac  N. (1983) J. Bacteriol. 155, 427-431.

Kentemich, T., Danneberg, G., Hundeshagen, B., Bothe, H. (1988) FEMS Microbiol. 51, 19-24.

Kentemich, T., Bahnweg, M., Mayer, F., Bothe, H. (1989)  Z. Naturforsch. 44c, 384-391.

Kentemich, T., Haverkamp, G., Bothe, H. (1990a) Naturwiss. 77, 12-18.

Kentemich, T., Haverkamp, G., Bothe, H. (1990b) submitted for publication.

Lambert, G.R., Smith, G.D. (1981) Biol. Rev. 56, 589-660.

Llama, M.J., Serra, J.L., Rao, K.K., Hall D.O. (1979)  FEBS Lett. 98, 342-346.

Lockau, W., Peterson, R.B., Wolk, C.P., Burris, R.H. (1978) Biochim. Biophys. Acta 502, 298-308.

Papen, H., Neuer, G., Refaian, M., Bothe, H. (1983) Arch. Microbiol. 134, 73-79.

Papen, H., Bothe, H. (1988) FEMS Microbiol. Letters 23,  41-46.

Papen, H., Kentemich, T., Schmülling, T., Bothe, H. (1986) Biochemie 68, 121-132.

Pau, R.N. (1989) Trends in Biochem. Sciences 14, 183-186.

Pau, R.N., Mitchenall, L.A., and Robson, R.L. (1989) J. Bacteriol. 171, 124-129.

Peterson, R.B., and Wolk C.P. (1978) Plant Physiol. 61,  688-691.

Peschek, G.A. (1979a) Biochim. Biophys. Acta 548, 187- 202.

Peschek, G.A. (1979b) Biochim. Biophys. Acta 548, 203- 215.

Rice, D., Mazur, B.J., Haselkorn R. (1982) J. Biol. Chem.  257, 13157-13163.

Robson, R.L., Woodley, A.R., Pau, R.N., and Eady, R.R. (1989) Embo J. 8, 1217-1224.

Rowell, P., Kerby, N.G., Darling, A.J., Stewart, W.D.P. (1985) In: H.J. Evans, P.B. Bottomley, W.E. Newton (eds.) Nitrogen Fixation Research Progress, Martinus Nijhoff  Publisher, Dordrecht-Boston-Lancaster, pp. 387- 393.

Müller, E.G.D., and Buchanan, B.B. (1989) J. Biol. Chem. 264, 4008-4014.

Scherer, P. (1989) Arch. Microbiol. 151, 44-88.

Schneider, K., Müller, A. (1990), presented at the 8th  International Congress on Nitrogen Fixation, Knoxville,  TN, May 20-26, Abstract H-08.

Singh, H.N., Vaishampayan A., Singh, R.K. (1978) Biochem. Biophys. Res. Comm. 81, 67-74.

Smith, G.D. (1990) Hydrogen Metabolism in Cyanobacteria  In: H.D. Kumar (eds.) Phycotalk, Rastogi & Company, Subhash Bazar, Meerut-250 002 India, pp. 131-143.

Udvardy, J., Borbely, G., Juhasz, A., Farkas, G.L. (1984) J. Bacteriol. 157, 681-683.

Whittaker, M.M., Gleason, F.K. (1984) J. Biol. Chem. 259, 14088-14093.

Zinder, S.H., and Lobo, A.L. (1990), presented at the 8th International Congress on Nitrogen Fixation, Knoxville, TN, May 20-26, Abstract L-31.

# Azolla and other plant-cyanobacteria symbioses: Aspects of form and function

GERALD A. PETERS
*Department of Biology, Virginia Commonwealth University, Richmond, VA 23284-2012, USA*

*Key words:* Anthoceros, Azolla, cycads, Gunnera, $N_2$ fixation, symbiotic Nostoc/Anabaena

## Abstract

Nostoc, a genus of filamentous, heterocystous, cyanobacteria, is widely distributed in the free-living state. It is also the most common phycobiont in $N_2$-fixing lichens and occurs as the $N_2$-fixing symbiont in a small and diverse group of green plants. These include several bryophyte genera (e.g. Anthoceros and Blasia), a pteridophyte genus (Azolla; while the symbiont is referred to as *Anabaena azollae*, it may be a *Nostoc* spp.), a division of gymnosperms (the 10 cycad genera) and one angiosperm genus (Gunnera). In Gunnera the Nostoc apparently penetrates into the cells of the host. In the other associations Nostoc is extracellular but specific morphological modifications and/or structures of the host plant organs create an environment which fosters interaction and metabolite interchange.

The individual group of Nostoc-green plant symbioses other than Azolla are summarized in regard to the current understanding of their establishment, perpetuation, and host-symbiont interaction. This includes available information on recognition and specificity, mode(s) of infection if applicable, and a synopsis of morphological modifications of the partners. The symbiosis with *Azolla* is then addressed separately with a more indepth account of the foregoing areas. In addition, the concept of *Azolla* harboring a dominant, obiligately symbiotic Nostoc which has not been cultured as well as minor symbionts capable of free-living growth, the distinction between re-constituting and simply re-establishing the symbiosis, and current approaches to improving the symbiosis and to authenticating the establishment of new associations are considered.

## Introduction

Cyanobacteria (blue-green algae) comprise the largest group of photosynthetic procaryotes. They are ubiquitous, occurring in freshwater, marine and terrestrial environments around the globe, and morphologically diverse, ranging from unicellular through filamentous forms, the latter including unbranched, branched and multiseriate filaments. Filamentous forms are the most complex. Many of these, including the unbranched filamentous *Nostoc-Anabaena* group, are capable of differentiating two specialized cell types, heterocysts and akinetes (spores), from vegetative cells.

Although exhibiting cellular characteristics of Gram negative bacteria, cyanobacteria contain chlorophyll a and phycobiliproteins, possess a non-cyclic electron transport chain, and use water as an electron donor in photosynthesis. In addition to oxygenic photosynthesis, cyanobacteria have evolved the ability to carry out the seemingly incompatible cellular process of nitrogen fixation. The ability to fix atmospheric nitrogen is found in members of all the major cyanobacteria groups but fixation under aerobic conditions is restricted to a few unicellular species and to filamentous forms capable of differentiating heterocysts.

Symbiotic $N_2$-fixing associations occur between cyanobacteria and a diverse group of eukaryotes. In addition to algae and fungi, these include associations with bryophytes, a pteridophyte, a gymnosperm division and an angiosperm (Peters et al., 1986; Rai, 1990; Stewart et al., 1980). Although a broad segment of the plant kingdom is represented, only a very small number of plant genera are involved and such associations are, in fact, a rare occurrence in the plant kingdom. The cyanobacterium is almost invariably a *Nostoc* spp. Formation of hormogonia filaments by *Nostoc* spp. is a primary systematic criterion provisionally distinguishing *Nostoc* from *Anabaena* (Rippka et al., 1979) and nearly all symbiotically competent cyanobacteria produce these small-celled, highly motile filaments at some stage of their association with green plants (Campbell and Meeks, 1989; Stewart et al., 1980). The gliding motility of hormogonia or homogonia-like filaments, perhaps combined with a chemotactic response, appears to play a role in the establishment, or maintenance, of the functional symbioses. In contrast to the legume-Rhizobium and actinorhizal associations, where infection by the prokaryote alters development of plant tissue such that nodules are produced, the structures occupied by the symbiotic cyanobacteria are not a consequence of infection. Minor changes may occur in specific structures believed to facilitate metabolate interchange but the primary structures differentiate and develop in each host plant in the absence of the cyanobacterium. The structures occupied by the cyanobacterium contain a mucilagenous material and the host plants generally have an appreciable phenolic content but any role of the mucilage or phenolics remains to be determined. Salient features of the symbiotic cyanobacterium and the specific structures occupied in each plant group are addressed below in the descriptions of the functional symbioses.

**Bryophyte symbioses**

Bryophytes include the mosses (Musci), liverworts (Hepaticae) and hornworts (Anthocerotae). All are small, non-vascular, terrestrial plants with a dominant gametophyte generation which may exist in association with cyanobacteria. Associations with mosses are rare and epiphytic except for two species of *Sphagnum* which may contain Nostoc within their hayaline cells under certain growth conditions (Grandall and Holfsten, 1976). Associations are also rare in the liverworts. Of the 330 defined genera, only Blasia and Calvicularia are known to have endophytic associations; epiphytic associations have been reported for two other (Meeks, 1990). Associations with hornworts are more common. Four of the six hornwort genera, including Anthoceros and Phaeoceros, have known endophytic associations (Meeks, 1990).

In the bryophyte group the endophytic hornwort and liverwort symbioses, which are organizationally primitive but may be the most ancient of the plant-cyanobacterial associations (Sprent and Raven, 1985), have received the most attention. The symbiotic Nostoc occupies cavities about 0.5 mm in diameter on the ventral surface of the photosynthetic gametophyte thalli. In Anthoceros these are filled with mucus and termed 'slime' cavities. The symbioses have been reconstituted in Anthoceros (Enderlin and Meeks, 1983), Phaeoceros (Ridgway 1967) and Blasia (Rodgers and Stewart, 1977) from cultures of the separated partners. Pure cultures of Anthoceros can be adapted to submerged growth in liquid medium (Enderlin and Meeks, 1983). When symbiotically competent *Nostoc* spp. are co-cultured with the pure thallus in N-free medium, or incubated in such medium conditioned by Anthoceros growth, hormogonia filaments are formed at a high frequency (Campbell and Meeks, 1989). In nature, the hormogonia filaments presumably gain entry to the cavity through small openings called slime pores.

Earlier studies (Ridgway, 1967; Rodgers and Stewart, 1977) showed that the Nostoc of Blasia and several hornworts were interchangeable and that a variety of other Nostoc isolates, including those from cyads and Gunnera were symbiotically competent in Anthoceros. Subsequent studies using pure, liquid cultures confirmed and extended these findings; coculture of isolates from the other symbiotic associations, several pond and soil isolates and others of unknown origin were symbiotically competent with *Anthoceros punctatus* (Enderlin and Meeks, 1983; Meeks, 1990). However, a number of other strains classified as Nostoc or Anabaena were non-symbiotic in this system. A recent comparison of DNA digests from a free-living Nostoc and cultured isolates from lichens, liverworts, a cyad and Gunnera using probes to a single region of the genome revealed a diversity among the Nostoc symbionts which was not correlated with the phylogenetic status of the host (Leizerovitch et al., 1990). Competition experiments with competent Nostoc strains indicated that an Anthoceros cavity can be infected by more than one (Enderlin and Meeks, 1983). Therefore, in nature the cavities might contain more than one strain and/or different cavities on the same thallus may contain different strains (Meeks, 1990). The factors involved in symbiotic competence are a subject of current research. They are clearly limited to, but not a characteristic of all, Nostoc strains.

Cultured isolates of symbiotically competent Nostoc exhibit heterocyst frequencies of 3–5%. In symbiosis the filament morphology changes. Vegetative cells of the closely appressed filaments become larger and spherical. Heterocyst frequencies reach 30–50% (Duckett et al., 1977; Enderlin and Meeks, 1983; Rodgers and Stewart, 1977), although only half of these may be functional (Meeks, 1990). Connections between both cell types appear fragile (Enderlin and Meeks, 1983; Meeks, 1990), almost as if the cells were held in position by the surrounding mucilage. In field collected material all Nostoc cells appear to be nitrogen depleted and excised colonies show negligible $CO_2$ fixation (Duckett et al., 1977; Rodgers and Stewart, 1977). In contrast, Nostoc cells from the cultured Anthoceros association contain phycobiliproteins and cyanophycin granules along with other cellular inclusions (Meeks, 1990) and fix $CO_2$ at rates eightfold lower than in the free-living state. The lower rates are attributed to a posttranslational modification of the primary carboxylating enzyme rather than diminished synthesis (Steinberg and Meeks, 1989). Excised Nostoc colonies release fixed N as ammonia which is assimilated by the gametophyte (Meeks et al., 1985; Rodgers and Stewart, 1977) and transferred to the sporophyte when present (Rodgers and Stewart, 1977). Assimilation of ammonia by the gametophyte is via the GS-GOGAT pathway at low ammonia levels; at high levels assimilation occurs via this pathway and by GDH (Meeks et al., 1983). Symbiotic Nostoc also assimilates ammonia via GS-GOGAT but rates are again diminished compared to the free-living state and attributed to posttranslational control of GS rather than to diminished synthesis of the protein (Meeks, 1990; see also Lee et al., 1988; Joseph and Meeks, 1987). In the symbiosis, photosynthate is transferred from the host tissue to the Nostoc, possibly as sucrose (Stewart and Rodgers, 1977) which also appears as a constituent of the cavity mucilage (Ridgway, 1967). Interchange of metabolites between the partners is presumably facilitated by branched filamentous protrusions which proliferate on the wall of infected cavities and may exhibit transfer cell ultrastructure (Duckett et al., 1977; Rodgers and Stewart, 1977).

## Gymnosperm symbioses

Cycadophyta, a division of gymnosperms restricted to tropical and subtropical regions, includes the most primitive extant seed plants. There are three families with a total of ten genera and about 150 species. The genera include Cycas in the Cycadaceae, Stangeria in the Stangeriaceae, and Bowenia, Ceratozamia, Dioon, Encephalartos, Lepidozamia, Macrozamia, Microcycas and Zamia in the Zamiaceae. Except for an epiphytic species of Zamia, all are terrestrial (Lindblad and Bergman, 1990). In addition to normal roots, all examined cycads exhibit apogeotropic roots which include dichotomously branched coralloid roots. Although apogeotropic, these structures usually occur below the soil surface. They may, however, protrude above the soil in old plants and in potbound greenhouse specimens (Lindblad and Bergman, 1990). The coralloid roots of many species contain nitrogen-fixing cyanobacteria, almost always a *Nostoc* spp., as a symbiont.

In infected roots the cyanobacteria is localized in a distinct cell layer, or zone, which is readily discernible in sectioned material. The cell layer is present, but rather inconspicuous in uninfected roots (Milindasuta, 1975). In Macrozamia, the cells of this zone arise from the protoderm and the layer corresponds to a transformed epidermis enclosed by a persistent root cap which is usually interpreted as a secondary (outer) cortex (Milindasuta, 1975; Wittman et al., 1965). During development, cells of the transformed epidermis soon become separated from one another along their radial and transverse walls but the narrow tangential walls remain in contact with the normal (inner) and secondary (outer) cortical layers, creating intercellular spaces (Milindasuta, 1975; Wittman et al., 1965). Filaments of symbiotic cyanobacteria are usually confined to these intercellular spaces which contain mucilage of unknown origin/composition (Lindblad et al., 1985). The elongated host cells (Lindblad et al., 1985; Lindblad and Bergman, 1990) and protrusions on these cells earlier in their development (Grobellaar et al., 1984; Obukowicz, 1981) may function in metabolite interchange.

Several modes of entry have been postulated (Nathanielsz and Staff, 1975; Wittman et al., 1965) but descriptions of the infection process are limited to *Macrozamia communis*. In this species, cyanobacteria filaments enter the root via a break in the dermal layer and follow a continuous cortical channel of intercellular spaces to the cyanobacteria zone (Milindasuta, 1975; Nathanielsz and Staff, 1975). Since intercellular migration seems to involve digestion of the host cell wall middle lamellae, the possibility of intracellular migration has been noted (Nathanielsz and Staff, 1975). Although rare, intracellular filaments have been reported in *Cycas revoluta* and *Macrozamia communis* (Nathanielsz and Staff, 1975; Obukowicz et al., 1981). The cyanobacteria zone appears to be naturally free of bacteria (Grilli Cailoa, 1980; Wittman et al., 1965) and it has been suggested that phenolic constitutents of the host plant may be involved (Obukowicz et al., 1981).

Except for Calothrix in one *Encephalartos* species (Grobbelaar et al., 1984), the symbiotic cyanobacteria, including those referred to as Anabaena in Cycas (Obukowicz et al., 1981) and Macrozamia (Nathanielsz and Staff, 1975) are presumably *Nostoc* species (Stewart et al., 1980). Restriction fragment length polymorphism studies indicate that different *Nostoc* species/strains occur in the cycads and that while a particular strain can predominate in the root of one cycad, several different strains can infect a single cycad species (Lindblad et al., 1989). Isolated symbionts and symbiont-free cycads are readily cultured but Koch's postulate has not been accomplished via reestablishment of the symbiosis. Virtually nothing is known in regard to any recognition signals which lead to establishment of the symbiosis. The possibility that the production of hormogonia is triggered by a cycad constituent, with these motile filaments swarming to a site of entry, has not been excluded. In view of their occurrence in cycads (Obukowicz et al., 1981) and their involvement in signal transduction in other systems (Lynn and Chang, 1990), phenolics would certainly seem to merit attention.

The symbiotic cyanobacterium undergoes a basipetal gradient of differentiation within the corraloid roots. In the very apical portion of the coralloid root of twelve cycad species in four genera, cyanobacteria filaments were found to lack heterocysts and resemble hormogonia (Grilli Cailoa, 1980). Differentiation of heterocyst commenced immediately below this region resulting in a marked increase in heterocyst frequency in progressively older portions of the coralloid root. In the mid-portion, frequencies ranged from 15–30% and filaments began to dissociate in the older portion (Grilli Cailoa, 1980). Other studies on one of the same species, *C. revoluta*, and on *Z. skinneri*, made no mention of undifferentiated hormogonia-like filaments in the extreme apical portion. In these plants, cyanobacterial filaments near the root tip had 5–7% heterocysts which increased to about 45% in the older basal portion; multiple heterocysts accounted for an increasing percentage of the total (Lindblad and Bergman, 1986; Lindblad et al., 1985). In both species, nitrogenase activity was greatest near the tip and decreased with increasing heterocyst frequency. Vegetative cells of the Nostoc in subterranean corralloid roots are very similar to those of the free-living isolates in that they contain phycobilisomes, carboxysomes, and exhibit no distinct ultrastructural modifications (Lindblad and Bergman, 1989, 1990; Lindblad et al., 1985). Although extracts of the Nostoc removed from *C. revoluta* have carboxylase and phosphoribulokinase activities, intact filaments/cells do not show light-dependent $CO_2$ fixation (Lindblad et al., 1987) and the Nostoc in the subterranean structures are almost certainly adapted to heterotrophic carbon metabolism. In *C. circinalis*, where the corraloid root may have occurred near or at the soil surface, some photosynthetic $O_2$ is evolved by the cyanobacterium (Perraju et al., 1986). The symbiotic cyanobacterium may have to receive low levels of incident light to develop a functional photosynthetic apparatus. Earlier studies using $^{15}N_2$ not only established that N fixed by the symbiont in coralloid roots is distributed throughout the plant but also showed that light increased nitrogen fixation in coralloid roots growing near the soil surface in glasshouse cultivated Macrozamia but had no effect on fixation by the symbiont in roots excavated from beneath large plants growing in the field (Bergersen et al., 1965).

In symbiotic Nostoc from species of *Macrozamia*, *Cycas* and *Zamia* the levels of GS protein, its distribution between vegetative cells and heterocysts and its in vitro activity were effectively the same as in the free-living isolates (Lindblad and Bergman, 1986; 1990). The form in which fixed N is released by the symbiotic Nostoc is currently unknown. Analysis of xylem sap suggests two pathways of assimilation and transfer of the fixed nitrogen in the coralloid roots. The more common involves translocation of cittruline, glutamine, and some glutamate. In two genera, however, only glutamine and glutamate are translocated (Lindblad and Bergman, 1990; Pate et al., 1988).

**Angiosperm symbiosis**

Gunnera, a member of the water-milfoil family (Haloragaceae, syn. Gunneraceae), is the only angiosperm genus known to develop a symbiotic association with a cyanobacterium. The genus is restricted to warm moist climates and contains about 40 species of small to large herbaceous, frequently rhizomatous or stoloniferous, perennials. All species produce mucilage glands as a normal part of their development. These glands originate just below the shoot apex, occurring below the point of cotyledon attachment in seedlings and at the base of each true leaf in mature plants (Silvester and McNamara, 1976; Towata, 1985a). They are naturally infected by Nostoc and the symbiosis has been reconstituted with a variety of Nostoc isolates (Bonnett and Silvester, 1981; Silvester and McNamara, 1976). In young, healthy glands the Nostoc is considered to be intracellular, surrounded by a membrane of plant origin (Bonnett, 1990; Silvester and McNamara, 1976). The Nostoc is also encapsulated by host cell wall material, but there are differing reports on when, and the extent to which, this occurs (Bonnett, 1990; Silvester and McNamara, 1976; Towata, 1985b).

Mucilage produced by developing glands engulfs both the glands and the shoot apex (Silvester and McNamara, 1976; Towata, 1985a) and provides a matrix for hormogonia-like filaments of Nostoc (Bonnett, 1990). The mucilage is of unquestionable importance but its composition and precise function(s) are unknown. Colonization of each gland is an independent infection event, drawing on the apical Nostoc population (Bonnett and Silvester, 1981). The glands arise endogenously and initially consist of a mass of parenchyma cells between the epidermis and procambial area of the apex (Bonnett, 1990; Silvester and McNamara, 1976; Towata, 1985a). This mass of parenchyma becomes organized into discrete clusters of cells, or papillae (Silvester and McNamara, 1976), the outermost cells of which have tannin deposits (Towata, 1985a). The mucilage enveloped papillae extend outward, breaking through the epidermis and separate to form channels which lead to a cavity at their base (Bonnett, 1990; Towata, 1985a). Nostoc filaments in the mucilage covering the apex and developing glands migrate into the basal cavity via the mucilage filled channels where they penetrate the walls of meristematic host cells by an unknown mechanism and become intracellular (Bonnett, 1990; Silvester and McNamara, 1976; Towata, 1985a). As cyanobacteria are not known to produce cellulytic enzymes, an involvement of helper microbes in penetration of the cell wall has been suggested (Towata, 1985a). However, nothing is known about the capability of symbiotically-associated Nostoc strains to produce such enzymes and no microbes become intracellular with the Nostoc. While morphometric analysis of field collected material revealed a large proportion of degenerate Nostoc in recently colonized glands, suggesting that relatively few cells successfully establish themselves (Towata,

382

1985b), the possibility that such cells are from strains which lack symbiotic capability has not been excluded. The successful Nostoc cells initially multiply in the gland (Towata, 1985b). They then differentiate a high proportion of heterocysts and undergo other morphological and ultrastructural changes (Silvester, 1975; Silvester and McNamara, 1976). Concomitantly, they become devoid of phycocyanin and lose there capacity for $CO_2$ fixation, becoming dependent upon the plant for photosynthate (Silvester, 1975); see (Osborne, 1989) for a comparison of photosynthesis and productivity in a symbiotic and non-infected Gunnera. Symbiotic $N_2$ fixation can meet the total N requirements of the association (Silvester, 1975) and in nature Gunnera may rely upon the symbiotic Nostoc for nitrogen due to a diminished capacity to assimilate nitrate (Osborne, 1989).

**Pteridophyte symbiosis**

The heterosporous aquatic ferns comprising the genus *Azolla* Lam. are the only pteridophytes known to exist in symbiosis with a $N_2$ fixing cyanobacterium. They also are the only plant-cyanobacteria symbiosis used in agriculture and have been the subject of several recent, indepth reviews (Braun-Howland and Neirzwicki-Bauer, 1990, Nierzwicki-Bauer, 1990; Palzinski, 1990; Peters and Meeks, 1989; Shi and Hall, 1988). The genus is distributed in tropical and temperate regions. Currently, seven species are generally recognized; *A. caroliniana*, *A. filiculoides*, *A. mexicana*, *A. microphylla* and *A. rubra* in the section Euazolla and *A. pinnata* and *A. nilotica* in the section Rhizosperma. All species normally harbor a symbiotic cyanobacterium, known historically as *Anabaena azollae* Strass. The cyanobacterium is intimately associated with *Azolla* spp. throughout their life cycle. It grows in concert with the vegetative sporophytes, occupying cavities formed in the dorsal leaf lobes, and during sporogenesis it is packaged into sporocarps such that continuity of the symbiosis is maintained during sexual reproduction.

Azolla sporophytes have multibranched, prostrate floating stems which bear deeply bilobed, alternately arranged, overlapping leaves and ad-

ventitious roots (Peters and Calvert, 1983). The growth of the cyanobacterium (subsequently called symbiont) and fern is tightly coupled (Hill, 1975, 1989; Peters and Calvert, 1983). An ontogenetic sequence of leaf development occurs along every stem axis (Calvert and Peters, 1981; Peters and Calvert, 1983) and the symbiont undergoes a parallel pattern of differentiation and development (Hill, 1975, 1977; Kaplan et al., 1986; Kaplan and Peters, 1981). Actively dividing symbiont filaments reminiscent of hormogonia are associated with each stem apex (Hill, 1975, 1977; Peters and Calvert, 1983). As the cavity forms in each dorsal lobe, some of these filaments are partitioned into it by a rapidly differentiated plant trichome termed the primary branched hair (PBH) (Calvert and Peters, 1981). With further development, the PBH and associated filaments are completely engulfed within the cavity and the symbiont begins to form heterocysts and fix $N_2$ (Hill, 1977; Kaplan et al., 1986). Concomitantly, a second branched hair and up to 25 simple hairs form within the leaf cavity (Calvert and Peters, 1981; Peters and Calvert, 1983). In mature cavities where nitrogenase activity is maximal, cells of the symbiont are enlarged, cell division is greatly diminished, and heterocysts account for about 20–30% of the total cell population (Hill, 1975, 1977; Kaplan et al., 1986; Kaplan and Peters, 1981; Peters et al., 1980). The Fe-protein of nitrogenase is restricted to heterocyst (Braun-Howland et al., 1988). Both heterocysts and vegetative cells contain phycobiliproteins which capture light energy to drive photosynthesis and nitrogen fixation (Kaplan et al., 1986; Ray et al., 1979; Tyagi et al., 1981). During their ontogeny, both types of cavity hairs exhibit a basipetal development of transfer cell ultrastructure followed by senescence (Calvert et al., 1985). This profile parallels differentiation and development in the cyanobacterium and the cavity trichomes are assumed to mediate the interchange of nutrients and metabolites between the partners (Calvert et al., 1985). The center of each mature cavity is gaseous and the symbiont occurs around the periphery. It is not only closely associated with the cavity hairs but appears to be confined within a mucilagenous matrix by an inner (Nierzwicki-Bauer et al., 1989) as well as an outer (Peters,

1976; Peters et al.; 1978; Uheda, 1986) envelope. In old leaves the symbiont may form akinetes (Braun-Howland et al., 1988; Braun-Howland and Nierzwicki-Bauer, 1990; Peters, 1975) or simply exhibit moribund cells (Hill, 1975) and heterotrophic bacteria, especially Arthrobacter (Wallace and Gates, 1986), become more prevalent. Recently, however, an Agrobacterium was the only isolate obtained from the leaf cavities of an *A. filiculoides* strain (Plazinski et al., 1990).

$N_2$ fixation can provide the associations with their total N requirement (Peters and Mayne, 1974a; 1974b). The isolated symbiont releases fixed nitrogen as ammonium (Meeks et al., 1987; Peters, 1977). In the host, nitrogen fixed in mature cavities is translocated to the growing apices (Kaplan and Peters, 1981; Peters et al., 1985). The fern and the symbiont assimilate ammonium via the GS-GOGAT pathway (Meeks et al., 1985, 1987; Ray et al., 1978) but both the specific catalytic activity and the amount of GS protein are significantly diminished in the freshly isolate symbiont when compared to cultures of free-living Nostoc or Anabaena (Lee et al., 1988; Orr and Haselkorn, 1982; Ray et al., 1978; Stewart et al., 1980). Both partners can fix $CO_2$ (Ray et al., 1979) but within the association the symbiont receives fixed carbon, including sucrose, from the fern and is estimated to account for no more than 5% of the total $CO_2$ fixation (Kaplan and Peters, 1988). Recent analysis of the symbiont isolated from the apical region and progressively older cavities along the stem axis using western blots and polyclonal antibodies showed profiles for the Fe-protein of nitrogenase consistent with the earlier profiles of acetylene reduction in such segments (Hill, 1975; Kaplan et al., 1986; Kaplan and Peters, 1981; Peters et al., 1980) whereas apparent levels of GS and the large subunit of RuBisCO were greatest in the apical region and decreased with increasing heterocyst frequency (M. Garcia, personal communication).

Azolla is unique among the plant-$N_2$ fixing prokaryote symbioses: it maintains the symbiont during sporulation and sexual reproduction, precluding the need for a free-living growth state of the cyanobacterium. There is therefore no obvious need for postulated lectin-mediated recogni-

tion between *Azolla* species and free-living cultures of presumptive Azolla symbionts (Berliner and Fisher, 1987; Kobiler et al., 1981; Ladha and Watanabe, 1984; McCowen et al., 1987). A recent study indicates that the hemagglutinating agents extracted from the fern are polyphenols rather than lectins (Leizerovitch et al., 1988). Ten presumptive isolates of the symbiont have recently been characterized on the basis of their morphology, lectin binding and allozymes; six of the isolates were assigned to genus Anabaena and four to the genus Nostoc (Zimmerman et al., 1989). None of the presumptive symbiont isolates have been shown to reconstitute the symbiosis with a symbiont-free Azolla. Nevertheless, establishment of a functional association with Anthoceros was used to selectively isolate cyanobacteria with symbiotic capability from the total population of symbiotic cyanobacteria present in *A. caroliniana* (Meeks, 1990; Meeks et al., 1988). Based on the pattern of endonuclease restriction fragments hybridizing to several gene probes, the isolates obtained by the forced association with Anthoceros were distinct from the major symbiont of *A. caroliniana* and from other putative symbiotic cyanobacteria isolated from *Azolla* species (Meeks, 1990; Meeks et al., 1988). These results are in accord with an earlier restriction fragment length polymorphism study which showed a cultured isolate from *A. filiculoides* to be distinct from the freshly isolated symbiotic cyanobacterium from *A. filiculoides* and other species of *Azolla* (Franche and Cohen-Bazire, 1985). They indicate that *Azolla* species can contain minor symbionts capable of free-living growth (Franche and Cohen-Bazire, 1985; Meeks, 1990; Meeks et al., 1988). Such minor symbionts apparently occur in addition to the dominant strain, which may be an obligate symbiont. The occurrence of minor symbionts capable of free-living growth may also explain other differences noted in comparisons of cultured isolates and the freshly isolated symbiont (Gates et al., 1980; Ladha and Watanabe, 1982; Nierzewicki-Bauer and Haselkorn, 1986; Rosen et al., 1987; Zimmerman et al., 1989).

*Azolla* species are heterosporous, producing male and female sporocarps. The male microsporocarps are noticeably larger than the female megasporocarps at maturity (Calvert et al., 1983;

Konar and Kapoor, 1974). Both sporocarp types contain akinetes of the symbiont (Calvert et al., 1983; Campbell, 1893; Konar and Kapoor, 1974; Strasburger, 1873). The incorporation of the symbiont is reminiscent of the inoculation of the developing leaf cavity. At the very onset of sporulation a distinct set of branched epidermal trichomes develop near the sporangial initial on the apical meristem. These trichomes grow into the apical symbiont colony and filaments become entangled among them. The sporangial primordium along with the branched trichomes and entwined filaments of the symbiont are then displaced ventrally. Cells of the symbiont begin to differentiate into akinetes as they become enclosed in the developing sporocarp pair (S.k. Perkins, personal communication). At this stage each sporocarp contains a developing megasporangium. They may continue to develop as such, yielding a megasporocarp pair, or one or both may abort to continue development as a microsporocarp containing microsporangia (Konar and Kapoor, 1974). Only the symbiont cells enclosed in the megasporocarp are important to continuity of the symbiosis.

In the mature megasporocarps akinetes are packaged into the apical portion in a chamber near the indusial pore (Becking, 1987; Calvert et al., 1983; Campbell, 1893; Dunham and Fowler, 1987; Konar and Kapoor, 1974; Lin and Watanabe, 1988a; Strasburger, 1873). The megaspore germinates to produce the female gametophyte and the first archegonium formed is positioned such that if fertilized, the embryonic sporophyte grows upward into the cells of the symbiont, leading to reinoculation of the sporophyte. Several recent studies have addressed this reinoculation process (Becking, 1987; Dunham and Fowler, 1987; Lin and Watanabe, 1988a) but only one noted that the embryonic sporophyte differentiates a set of subapical branched trichomes (Dunham and Fowler, 1987). These trichomes are, however, the first cells of the sporophyte to encounter hormogonia-like filaments arising from the germination of akinetes in the apical chamber and they appear to play an essential role in the reinoculation process (Calvert et al., 1985; S.k. Perkins, personal communication).

Removal of the apical cap is a means of ob-taining symbiont-free Azolla (Lin and Watanabe, 1988b; Reddy and Fisher, 1988) and it has been shown that the symbiont in the cap of *A. microphylla* can reestablish the symbiosis when positioned on decapitated megasporocarps of *A. filiculoides* and vice versa (Lin et al., 1988). Initially, monoclonal antibodies reacting only with the symbiont from *A. filiculoides* were used to verify that the symbiont in the reestablished symbiosis was indeed the one present in the donor apical cap. Subsequently DNA probes specific for the symbiont in *A. filiculoides*, the symbiont in *A. microphylla*, and the symbiont in different strains of both species, were used to verify symbiont origin in nine such reestablished symbioses attained with various strains of these two *Azolla* species (Plazinski et al., 1989); it is not yet known whether crosses of this type can be accomplished with other species of Azolla. While eight of the nine contained the symbiont present in the donor cap, one contained the symbiont from the decapitated strain as well as that from the donor cap, suggesting incomplete removal of all akinetes during decapitation, but demonstrating that one Azolla strain can harbour two distinguishable populations, or strains, of the dominant symbiotic cyanobacterium (Plazinski et al., 1989). Whether the novel host-symbiont combinations resulting from such crosses have any enhanced agronomic potential remains to be determined. In the author's opinion, attainment of specifically desired agronomic attributes will more likely be acquired through controlled intraspecific, and perhaps interspecific, crosses of the host than by changes in the symbiont population.

As indicated above, use of DNA probes have revealed species and strain specific differences in the dominant symbiont of *Azolla*. Initially, analysis of sequence divergence based on *nif* gene hybridizations with the dominant symbionts from four species of *Azolla* in the Euazolla and five strains of *A. pinnata* (Rhizosperma) showed that while the dominant symbiont in all species arose from a common ancestor, the symbionts present in members of the Euazolla and Rhizosperma represented two divergent evolutionary lines (Franche and Cohen-Bazire, 1987). Subsequent hybridization studies concurred with, and extended, these findings. The use of eight genomic

DNA probes (six from the dominant symbiont of *A. caroliniana* and one each from the dominant symbiont of *A. microphylla* and *A. rubra*) to assess genetic variation in 21 isolates of the dominant symbiont occurring in the seven recognized *Azolla* spp. again indicate that the dominant symbionts in the Euazolla species and those in *A. pinnata* generally belong to divergent evolutionary lines (Plazinski et al., 1990). These studies further indicate, however, that the dominant symbiont in *A. nilotica* is distinct and appears to have genetically diverged from the other symbionts (Plazinski et al., 1990). In addition, these studies established 11 genotypes among the 21 isolates and concurred with others (Meeks et al., 1988; Peters and Meeks, 1989) in concluding that the dominant symbiont in association with *Azolla* spp. is more closely related to *Nostoc* spp. than to free-living *Anabaena* spp.

## Concluding remarks

Nitrogen-fixing cyanobacteria capable of entering into symbiotic associations with green plants are circumscribed by the genus Nostoc. Nostoc strains capable of forming functional symbiotic associations with plant genera other than Azolla are clearly normal constituents of the soil microflora capable of free-living growth. The overall manner in which such strains differ from Nostoc and other cyanobacteria which do not exhibit symbiotic capability is not yet resolved. The plant clearly has a greater impact on the cyanobacterium than the cyanobacterium has on the plant. The host plants cause symbiotically competent strains to modify their growth form prior to and after the establishment of the symbiosis, perhaps via specific chemical signals. Competent strains produce hormogonia-like filaments which lead to infection and then undergo alterations in their morphology and/or their biochemistry and physiology. The nature of the plant signals which govern such changes has not been elucidated. In Azolla, all species harbor the dominant symbiont during both vegetative and sexual propagation and this microbe may be an obligate symbiont. As with the other associations, the symbiont has no pronounced effect on plant structure but appears to alter its morphology, physiology and biochemistry in response to signals from the host.

## Acknowledgement

The author is indebted to Dr J C Meeks for his constructive comments.

## References

Becking J H 1987 Endophyte transmission and activity in the *Anabaena-Azolla* association. Plant and Soil 100, 183–212.

Bergersen F J, Kennedy G S and Wittman W 1965 Nitrogen fixation in the coralloid roots of *Macrozamia communis* L. Johnson. Aust. J. Biol. Sci. 18, 1135–1142.

Berliner M D and Fisher R W 1987 Surface lectin binding to *Anabaena variabilis* and to cultured and freshly isolated *Anabaena azollae*. Curr. Microbiol. 16, 149–152.

Bonnett H T 1990 The *Nostoc-Gunnera* association. *In* Handbook of Symbiotic Cyanobacteria. Ed. A N Rai. pp 161–171. CRC Press, Boca Raton, FL.

Bonnett H T and Silvester W B 1981 Specificity in the *Gunnera-Nostoc* endosymbiosis. New Phytol. 89, 121–128.

Braun-Howland E B, Lindblad P and Nierzwicki-Bauer S A, Bergman B 1988 Dinitrogenase reductase (Fe- protein) of nitrogenase in the cyanobacterial symbionts of three *Azolla*: Localization and sequence of appearance during heterocyst differentiation. Planta 176, 319–332.

Braun-Howland E B and Nierzwicki-Bauer S A 1990 *Azolla-Anabaena*, symbiosis: Biochemistry, ultrastructure, and molecular biology. *In* Handbook of Symbiotic Cyanobacteria. Ed. A N Rai. pp 65–117. CRC Press, Boca Raton, FL.

Calvert H E, Pence M K and Peters G A 1985 Ultrastructural ontogeny of leaf cavity trichomes in *Azolla* implies a functional role in metabolite exchange. Protoplasma 129, 10–27.

Calvert H E, Perkins S K and Peters G A 1983 Sporocarp structure in the heterosporous water fern *Azolla mexicana* Presl. Scanning Electron Microsc. 3, 1499–1510.

Calvert H E, Perkins S K and Peters G A 1985 Involvement of epidermal trichomes in the continuity of the *Azolla-Anabaena* symbiosis through the *Azolla* life cycle. Am. J. Bot. 72, 808 (Abstr.)

Calvert H E and Peters G A 1981 The *Azolla-Anabaena* relationship. IX. Morphological analysis of leaf cavity hair populations. New Phytol. 89, 327–335.

Campbell D H 1893 On the development of *Azolla filiculoides*, Lam. Ann. Bot. 26, 155–187.

Campbell E L and Meeks J C 1989 Characteristics of hormogonia formation by symbiotic *Nostoc* spp. in response to the presence of *Anthoceros punctatus* or its extracellular products. Appl. Environ. Microbiol. 55, 125–131.

Duckett J G, Prasad A K S K, Davies D A and Walker S 1977 A cytological analysis of the *Nostoc*-bryophyte relationship. New Phytol. 78, 349–362.

Dunham D G and Fowler K 1987 Megaspore germination, embryo development and maintenance of the symbiotic association in *Azolla filiculoides* Lam. Bot. J. Linnean Soc. 95, 43–53.

Enderlin C S and Meeks J C 1983 Pure culture and reconstitution of the *Anthoceros-Nostoc* symbiotic associations. Planta 158, 156–165.

Franche C and Cohen-Bazire G 1985 The structural *nif* genes of four symbiotic *Anabaena azollae* show a highly conserved physical arrangement. Plant Sci. 39, 125–131.

Franche C and Cohen-Bazire G 1987 Evolutionary divergence in the *nif* H.D.K. gene region among nine symbiotic *Anabaena azollae* and between *Anabaena azollae* and some free-living heterocystous cyanobacteria. Symbiosis 3, 159–178.

Gates J E and Fisher R W, Goggin T W, Azrolon N I 1980 Antigenetic differences between *Anabaena azollae* fresh from the *Azolla* fern leaf cavity and free-living cyanobacteria. Arch. Microbiol. 128, 126–129.

Grandall U and Holfsten A V 1976 Nitrogenase activity in relation to intracellular organisms in *Sphagnum* mosses. Physiol. Plant. 36, 88–94.

Grilli Cailoa M 1980 On the phycobionts of the cycad coralloid roots. New Phytol. 85, 537–544.

Grobbelaar N, Scott W E, Hattingh W and Marshall J 1987 The identification of the coralloid root endophytes of the southern African cycads and the ability of the isolates to fix dinitrogen. S. Afr. J. Bot. 53, 111–118.

Grobbelaar N, Small J G C, Marshall J and Hattingh W 1984 Metabolic studies of the coralloid roots of *Encephalartos transvenosus* and its endophyte. *In* Advances in Nitrogen Fixation Research. Eds. C Veeger and W E Newton. p. 54. Martinus Nijhoff, The Hague, The Netherlands.

Hill D J 1975 The pattern of development of *Anabaena* in the *Azolla-Anabaena* symbiosis. Planta 122, 179–184.

Hill D J 1977 The role of *Anabaena* in the *Azolla-Anabaena* symbiosis. New Phytol. 78, 611–616.

Hill D J 1989 The control of the cell cycle in microbial symbionts. New Phytol. 112, 175–184.

Joseph C M and Meeks J C 1987 Regulation of expression of glutamine synthetase in a symbiotic *Nostoc* strain associated with *Anthoceros punctatus*. J. Bacteriol. 169, 2471–2475.

Kaplan D, Calvert H E and Peters G A 1986 Nitrogenase activity and phycobiliproteins of the endophyte as a function of leaf age and cell type. Plant Physiol. 80, 884–890.

Kaplan D and Peters G A 1981 The *Azolla-Anabaena* relationship. X. $^{15}N_2$ fixation and transport in main stem axes. New Phytol. 89, 337–346.

Kaplan D and Peters G A 1988 Interaction of carbon metabolism in the *Azolla-Anabaena* symbiosis. Symbiosis 6, 53–68.

Kobiler D, Cohen-Sharon A and Tel-Or E 1981 Recognition between the $N_2$-fixing *Anabaena* and the water fern *Azolla*. FEBS Lett. 133, 157–160.

Konar R N and Kapoor R K 1974 Embryology of *Azolla pinnata*. Phytomorphology 22, 211–233.

Ladha J K and Watanabe I 1982 Antigenic similarity among *Anabaena azollae* separated from different species of *Azolla*. Biochem. Biophys. Res Comm. 109, 675–682.

Ladha J K and Watanabe I 1984 Antigenetic analysis of *Anabaena azollae* and the role of lectin in the *Azolla-Anabaena* symbiosis. New Phytol. 98, 295–300.

Lee K Y, Joseph C M and Meeks J C 1988 Glutamine synthetase specific activity and protein concentration in symbiotic *Anabaena* associated with *Azolla caroliniana*. Antonie van Leewenhoek 54, 345–355.

Leizerovitch L, Fleminger N, Kardisch N, Frensdorff A and Galun M 1988 Polyphenols, and not lectins, are responsible for haemagglutinating activity in extracts of Azolla filiculoides Lam. Symbiosis 5, 209–221.

Leizerovitch I, Kardish N and Galun M 1990 Comparison between eight symbiotic, cultured *Nostoc* isolates and a free-living *Nostoc* by Recombinant DNA. Symbiosis 8, 75–85.

Lin C and Watanabe I 1988 Study on the association between *Anabaena azollae* and *Azolla microphylla* during the germination of megasporocarps. Symbiosis 5, 199–208.

Lin C and Watanabe I 1988 A new method for obtaining *Anabaena*-free *Azolla*. New Phytol. 108, 341–344.

Lin C, Watanabe I, Liu C C, Zheng D-Y and Tang L F 1988 Reestablishment of symbiosis to *Anabaena*-free *Azolla*. *In* Nitrogen Fixation: Hundred Years After. Eds. H Bothe, N J de Bruijn and W E Newton. pp 223–227. Gustav Fischer, Stuttgart, Germany.

Lindblad P and Bergman B 1986 Glutamine synthetase: Activity and localization in cyanobacteria of the cycads *Cycas revoluta* and *Zamia skinneri*. Planta 169, 1–7.

Lindblad P and Bergman B 1989 Occurrence and localization of phycoerythrin in symbiotic *Nostoc* of *Cycas revoluta* and in the free-living isolated *Nostoc* 7422. Plant Physiol. 89, 783–785.

Lindblad P and Bergman B 1990 The cycad-cyanobacteria symbiosis. *In* Handbook of Symbiotic Cyanobacteria. Ed. A N Rai. pp 137–159. CRC Press, Boca Raton, FL.

Lindblad P, Bergman B, v. Hofsten A, Hallbom L and Nylund J E 1985 The cyanobacterium-*Zamia* symbiosis: An ultrastructural study. New Phytol. 101, 707–716.

Lindblad P, Hallbom L and Bergman B 1985 The cyanobacterium *Zamia* symbiosis: $C_2H_2$-reduction and heterocyst frequency. Symbiosis 1, 19–28.

Lindblad P, Haselkorn R, Bergman B and Nierzwicki-Bauer S A 1989 Comparison of DNA restriction fragment length polymorphisms of *Nostoc* strains in and from cycads. Arch. Microbiol. 152, 20–24.

Lindblad P, Rai A R and Bergman B 1987 The *Cycas revoluta-Nostoc* symbiosis: Enzyme activities of nitrogen and carbon metabolism in the cyanobiont. J. Gen. Microbiol. 133, 1695–1699.

Lynn D G and Chang M 1990 Phenolic signals in cohabitation: Implications for plant development. Annu. Rev. Plant Physiol. Plant Mol. Biol. 41, 497–526.

McCowen S W, MacArthur L and Gates J E 1987 *Azolla* fern lectins that specifically recognize endosymbiotic cyanobacteria. Curr. Microbiol. 14, 329–333.

Meeks J C 1990 Cyanobacterial-bryophyte associations. *In* Handbook of Symbiotic Cyanobacteria. Ed. A N Rai. pp 43–63. CRC Press, Boca Raton, FL.

Meeks J C, Enderlin C S, Joseph C M, Chapman J S and Lollar M W L 1985 Fixation of [$^{13}$N]$N_2$ and transfer of fixed nitrogen in the *Anthoceros-Nostoc* symbiotic association. Planta 164, 406–414.

Meeks J C, Enderlin C S, Wycoff K L, Chapman J S and Joseph C M 1983 Assimilation of $^{13}NH_4^+$ by *Anthoceros* grown with and without symbiotic *Nostoc*. Planta 158, 384–391.

Meeks J C, Joseph C M and Haselkorn R 1988 Organization of the *nif* genes in cyanobacteria in symbiotic association with *Azolla* and *Anthoceros*. Arch. Microbiol. 150, 61–71.

Meeks J C, Steinberg N A, Enderlin C S, Joseph C M and Peters G A 1987 *Azolla-Anabaena* relationship. XIII. Fixation of $[^{13}N]N_2$. Plant Physiol. 83, 883–886.

Meeks J C, Steinberg N S, Joseph C M, Enderlin C S, Jorgensen P A and Peters G A 1985 Assimilation of exogenous and dinitrogen derived $^{13}NH_4^+$ by *Anabaena azollae* separated from *Azolla caroliniana* Willd. Arch. Microbiol. 142, 229–233.

Milindasuta B-E 1975 Developmental anatomy of coralloid roots in cycads. Am. J. Bot. 62, 468–472.

Nathanielsz C P and Staff I A 1975 A mode of entry of blue-green algae into the apogeotropic roots of *Macrozamia communis*. Am. J. Bot. 62, 232–235.

Nierzwicki-Bauer S A 1990 *Azolla-Anabaena* symbiosis: Use in agriculture. *In* Handbook of Symbiotic Cyanobacteria. Ed. A N Rai. pp 119–136. CRC Press, Boca Raton, FL.

Nierzwicki-Bauer S A, Aulfinger H and Braun-Howland E B 1989 Ultrastructural characterization of an inner envelope that confines *Azolla* endosymbionts to the leaf cavity perifery. Can. J. Bot. 67, 2711–2719.

Nierzwicki-Bauer S A and Haselkorn R 1986 Difference in mRNA levels in *Anabaena* living freely or in symbiotic association with *Azolla*. EMBO J. 5, 29–35.

Obukowicz M, Schaller M and Kennedy G S 1981 Ultrastructure and phenolic histochemistry of the *Cycas revoluta-Anabaena* symbiosis. New Phytol. 87, 751–759.

Orr J and Haselkorn R 1982 Regulation of glutamine synthetase activity and synthesis in free-living and symbiotic *Anabaena* spp. J. Bacteriol. 152, 625–635.

Osborne B A 1989 Comparison of photosynthesis and productivity of *Gunnera tinctoria* Molina (Mirbel) with and without the phycobiont *Nostoc punctiforme* L. Plant Cell Environ. 12, 941–946.

Pate J S, Lindblad P and Atkins C A 1988 Pathways of assimilation and transfer of fixed nitrogen in coralloid roots of cycad-*Nostoc* symbioses. Planta 176, 461–471.

Perraju B T V V, Rai A N, Kumar A P and Singh H N 1986 *Cycas circinalis-Anabaena cycadeae* symbiosis: Photosynthesis and the enzymes of nitrogen and hydrogen metabolism in symbiotic and cultured *Anabaena cycadeae*. Symbiosis 1, 239–250.

Peters G A 1975 The *Azolla-Anabaena azollae* relationship. III. Studies on metabolic capabilities and a further characterization of the symbiont. Arch. Microbiol. 103, 113–122.

Peters G A 1976 Studies on the *Azolla-Anabaena azollae* symbiosis. *In* Proceedings of the 1st International Symposium on Nitrogen Fixation. Eds. W E Newton and C J Nyman. pp 592–610. Washington State University Press, Pullman, WA.

Peters G A 1977 The *Azolla-Anabaena* symbiosis. *In* Genetic Engineering and Nitrogen Fixation. Ed. A Hollaender. pp 231–258. Plenum, New York.

Peters G A and Calvert H E 1983 The *Azolla-Anabaena* symbiosis. *In* Algal Symbiosis. Ed. L J Goff. pp 109–145. Cambridge University Press, New York.

Peters G A, Kaplan D, Meeks J C, Buzby K M, Marsh B H and Corbin J L 1985 Aspects of nitrogen and carbon interchange in the *Azolla-Anabaena* symbiosis. *In* Nitrogen Fixation and $CO_2$ Metabolism. Eds. P W Ludden and J E Burris. pp 213–222. Elsevier, New York.

Peters G A and Mayne B C 1974a. the *Azolla, Anabaena azollae* relationship. I. Initial characterization of the association. Plant Physiol. 53, 813–819.

Peters G A and Mayne B C 1974b. The *Azolla, Anabaena azollae* relationship. II. Localization of nitrogenase activity as assayed by acetylene reduction. Plant Physiol. 53, 820–824.

Peters G A and Meeks J C 1989 The *Azolla-Anabaena* symbiosis: Basic biology. Annu. Rev. Plant Physiol. and Plant Mol. Biol. 40, 193–210.

Peters G A, Ray T B, Mayne B C and Toia R E Jr 1980 The *Azolla-Anabaena* association: Morphological and physiological Studies. In Nitrogen Fixation, Vol. II. Eds. W E Newton and W H Orme-Johnson. pp 293–309. University Park Press, Baltimore, MD.

Peters G A, Toia R E Jr, Calvert H E and Marsh B H 1986 Lichens to *Gunnera* – with emphasis on *Azolla*. Plant and Soil 90, 17–34.

Peters G A, Toia R E Jr, Raveed D and Levine N J 1978 The *Azolla-Anabaena azollae* relationship. VI. Morphological aspects of the association. New Phytol. 80, 583–593.

Plazinski J 1990 The *Azolla-Anabaena* symbiosis. *In* Molecular Biology of Symbiotic Nitrogen Fixation. Ed. P M Gresshoff. pp 51–75. CRC Press, Boca Raton, FL.

Plazinski J, Taylor R, Shaw W, Croft L, Rolfe B G and Gunning B E S 1990 Isolation of *Agrobacterium* sp. strain from the *Azolla* leaf cavity. FEMS Microbiol. Lett. 70, 55–60.

Plazinski J, Zheng Q, Taylor R, Croft L, Rolfe B G and Gunning B E S 1990 DNA probes show genetic variation in cyanobacterial symbionts of the *Azolla* fern and a closer relationship to free-living *Nostoc* strains than to free-living *Anabaena* strains. Appl. Environ. Microbiol. 56, 1263–1270.

Plazinski J, Zheng Q, Taylor R, Rolfe B G and Gunning B E S 1989 Use of DNA/DNA hybridization techniques to authenticate the production of new *Azolla-Anabaena* symbiotic associations. FEMS Microbiol. Lett. 65, 199–204.

Rai A N 1990 Handbook of Symbiotic Cyanobacteria. CRC Press, Boca Raton, FL. 253 p.

Ray T B, Mayne B C, Peters G A and Toia R E Jr 1979 *Azolla-Anabaena* relationship. VIII. Photosynthetic characterization of the association and individual partners. Plant Physiol. 64, 791–795.

Ray T B, Peters G A, Toia R E Jr and Mayne B C 1978 *Azolla-Anabaena* relationship. VII. Distribution of ammonia assimilating enzymes, protein and chlorophyll between host and symbiont. Plant Physiol. 63, 463–467.

Reddy P M and Fisher R W 1988 A new simple method to produce *Anabaena*-free *Azolla* by in vitro fertilization of micromanipulated megasporocarps. Plant Cell Rep. 7, 430–433.

Ridgway J E 1967 The biotic relationship of *Anthoceros* and *Phaeoceros* to certain cyanobacteria. Ann. Mo. Bot. Gard. 54, 95–102.

Rippka R, Deruelles J, Waterbury J B, Herdman M and Stanier R Y 1979 Generic assignments, strain histories and properties of pure cultures of cyanobacteria. J. Gen. Microbiol. 111, 1–61.

Rodgers G A and Stewart W D P 1977 The cyanophyte-hepatic symbiosis. I. Morphology and physiology. New Phytol. 78, 441–458.

Rosen B H, Johnson R C Jr, Fisher R W and Gates J E 1987 Ultrastructural localization of cell surface immunospecificity in *Anabaena azollae* using indirect fluorescent antibody staining. Am. J. Bot. 74, 1060–1064.

Shi D-J and Hall D O 1988 The *Azolla-Anabaena* association: Historical perspective, symbiosis and energy metabolism. Bot. Rev. 54, 353–386.

Silvester W B 1975 Endophyte adaptation in *Gunnera-Nostoc* symbiosis. *In* Symbiotic Nitrogen Fixation in Plants. Ed. P S Nutman. pp 521–538. Cambridge University Press, Cambridge.

Silvester W B and McNamara P J 1976 The infection process and ultrastructure of the *Gunnera-Nostoc* symbiosis. New Phytol. 77, 135–141.

Sprent J I and Raven J A 1985 Evolution of nitrogen-fixing symbioses. Proc. Roy. Soc. Edinburgh. 85B, 215–237.

Steinberg N A and Meeks J C 1989 Photosynthetic $CO_2$ fixation and ribulose bisphosphate carboxylase/oxygenase activity of *Nostoc* sp. strain UCD 7801 in symbiotic association with *Anthoceros punctatus*. J. Bacteriol. 171, 6227–6233.

Stewart W D P and Rodgers G A 1977 The cyanophyte-heptatic symbiosis. II. Nitrogen fixation and the interchange of nitrogen and carbon. New Phytol. 78, 459–471.

Stewart W D P, Rowell P and Rai A N 1980 Symbiotic nitrogen-fixing cyanobacteria. *In* Nitrogen Fixation. Eds. W D P Stewart and J R Gallon. pp 239–277. Academic Press, New York.

Strasburger E 1873 Über *Azolla*. Herman Davis Verlag, Jena. 86 p.

Towata E M 1985a Mucilage glands and cyanobacterial colonization in *Gunnera kaalensis* (Haloragaceae). Bot. Gaz. 146, 56–62.

Towata E M 1985b Morphometric and cytochemical ultrastructural analyses of the *Gunnera kaalensis / Nostoc* symbiosis. Bot. Gaz. 146, 293–301.

Tyagi V V S, Ray T B, Mayne B C and Peters G A 1981 The *Azolla-Anabaena* relationship. XI. Phycobiliproteins in the action spectrum for nitrogenase-catalyzed acetylene reduction. Plant Physiol. 68, 1479–1484.

Uheda E 1986 Isolation of empty packets from *Anabaena*-free *Azolla*. Plant Cell Physiol. 27, 1187–1190.

Wallace W H and Gates J E 1986 Identification of eubacteria isolated from leaf cavities of four species of the N-fixing *Azolla* ferns as *Arthrobacter* Conn and Dimick. Appl. Environ. Microbiol. 52, 425–429.

Wittman W, Bergersen F J and Kennedy G S 1965 The coralloid roots of *Macrozamia communis* Johnston. Aust. J. Biol. Sci. 18, 1129–1134.

Zimmerman W J, Rosen B H and Lumpkin T A 1989 Enzymatic, lectin, and morphological characterization and classification of presumptive cyanobionts from *Azolla* Lam. New Phytol. 113, 497–503.

STRUCTURE, METABOLISM AND NITROGENASE REGULATION IN THE *Azolla-Anabaena* ASSOCIATION


E. TEL-OR[1], E. BAR[1], C. WATAD[1], O. KLEIN[2] and C. FORNI[3]
*[1]Department of Agricultural Botany, The Hebrew University of Jerusalem, Rehovot 76100, Isarel; [2]Electron Microroscope Unit, The Weizmann Institute of Science, Rehovot 76100, Israel; [3]Department of Biology, II University of Rome, 00173 Rome, Italy*



ABSTRACT. Structural relations between the *Azolla* leaf cavity envelope, hair cells and the cyanobiont *Anabaena azollae* were tested with frozen hydrated preparations by scanning electron microscopy, exhibiting close interactions between the cyanobiont and the cavity envelope cells, and a mucilage layer covering parts of the cavity envelope. Hair cells contained starch granules and the cyanobiont exhibited amylase activity hence, both starch, fructose and sucrose can serve as carbon substrates of the cyanobiont. Light regulation of nitrogenase seems to involve the direct effect of light and the effect of oxygen as resolved by nitrogenase analysis and immuno-blot of dinitrogenase reductase. The respiratory activity of the cyanobiont is high and heterotrophic growth increased cytochrome oxidase activity.


## 1. Introduction

Structure-function relations in the *Azolla-Anabaena* association stimulate an increasing number of investigations on the anatomy, physiology, biochemistry, molecular biology and the implementation of the aquatic fern *Azolla* (Peters and Meeks 1989, Becking 1987). Our specific interest in metabolic relations and the intercellular signals in the *Anabaena azollae* association suggeted possible lectin receptor interactions and resolved the heterotrophic nature of the cyanobiont (Tel-Or et al. 1983, Rozen et al. 1986, 1988, Tel-Or et al. 1990). An unequal distribution of photosynthesis and nitrogen fixation along the stem axis in *Azolla* was recently reported (Kulasooriya et al. 1989). The regulation and control of nitrogen fixation by light, photosynthates and oxygen were investigated. Regulation of dinitrogenase reductase by light and $O_2$ was shown, and may explain the lack of synchronous response of $N_2$- fixation to optimal photosynthesis. The heterotrophic metabolism of *Anabaena azollae* was found to involve starch breakdown, and heterotrophic growth was facilitating $O_2$ removal by enhanced cytochrome oxidase activity.

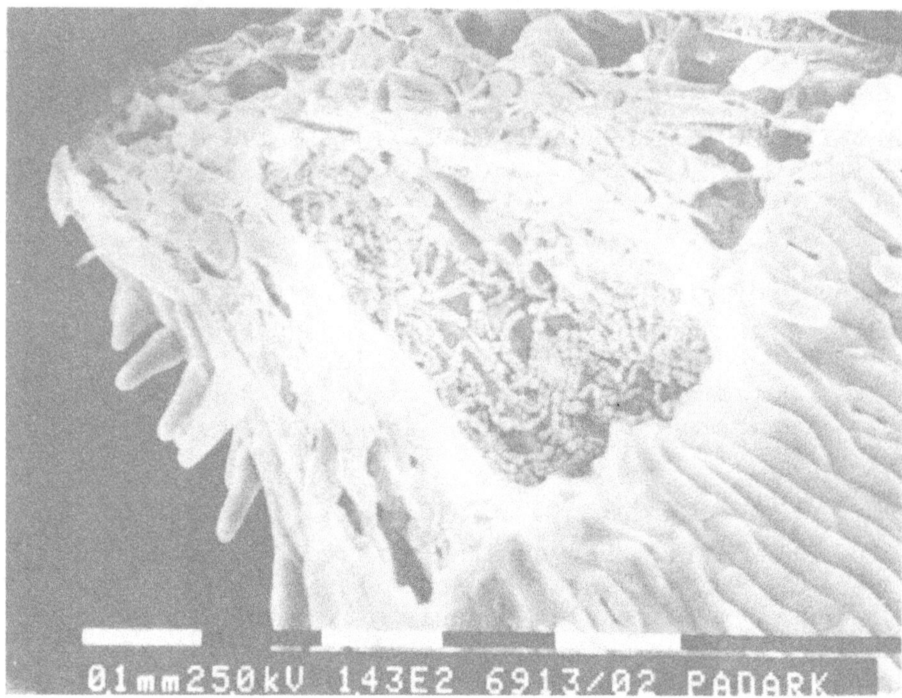

Figure 1.   Frozen hydrated leaf cavity envelope of *Azolla pinnata*

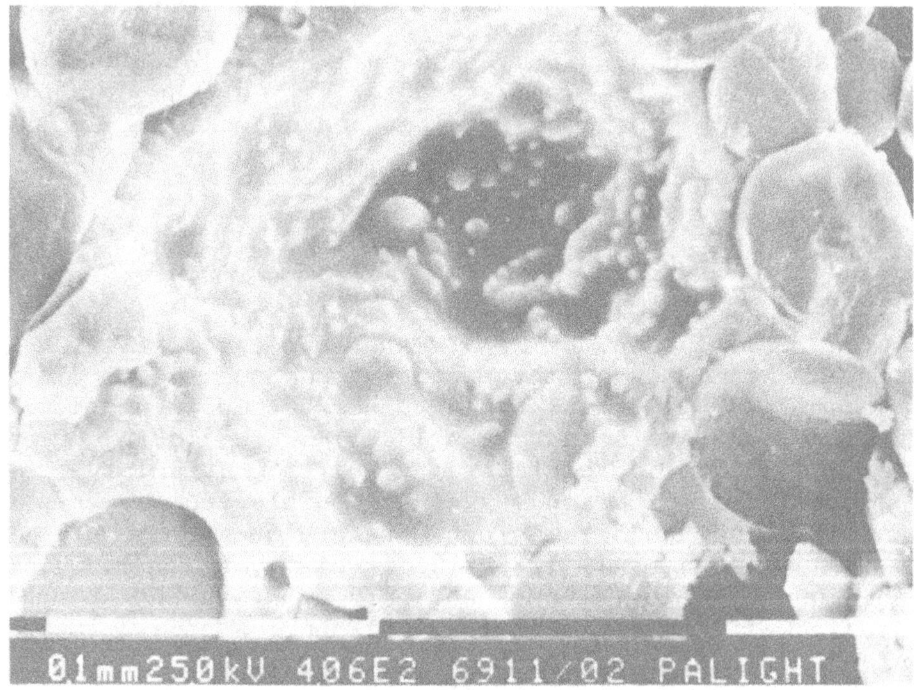

Figure 2.   Mucilage layer with immersed cyanobiont in *Azolla pinnata* leaf cavity

# 2. Results and Discussion

## 2.1 ANALYSIS OF LEAF CAVITY STRUCTURE

The structure and function relations in associations between
diazotrophic microorganisms and their hosts have been studied in the
legume root nodule and in the *Azolla* leaf cavity. More information is
obtained with the introduction of new fixation procedures and improved
resolution by electron microscopy. In *Azolla*, the leaf cavity was
shown to accomodate the cyanobiont *Anabaena azollae* that is closely
attached to the host epidermal hair cells and the cavity envelope
cells (Calvert et al. 1985, Nierzwicki-Bauer et al. 1989). The
intercellular exchange of metabolites is essential for successful and
efficient symbyiosis (Tel-Or et al. 1990).
We have recently introduced the frozen hydration fixation technique to
obtain preparations of improved quality for scaning electron
microscopy, where the structure of the *Azolla* leaf cavity envelope,
hair cells and *Anabaena azollae* cells were well preserved. Fig 1 shows
that the leaf cavity was intact and unharmed throughout its fixation
and opening in the cold stage, prior to the gold sputtering. The
cyanobiont cells are distributed uniformly along the envelope of the
leaf cavity, and did not aggregate, as shown in preparations obtained
by chemical gluteraldehyde fixation. Cells of *Anabaena azollae* in the
frozen hydrated samples, seem to interact more closely with the
envelope cells than with the hair cells. Higher resolution of these
preparations demonstrated more cymbiont cells interacting with the
cavity envelope and a minor fraction interacting with the few hair
cells in the cavity.
The frozen hydrated sample shown in Fig 2 demonstrates the mucilage
layer at the bottom of the leaf cavity. The mucilage is assumed to
contain carbohydrated, amino acids, peptides and other metabolites in
a viscous solution. Exchange and translocation of photosynthates and
fixed nitrogen products could be facilitated by surface interactions
between the host cavity and the cyanobiont cells. The mucilage may
provide suitable interface for metabolite exchange  and was found only
at the bottom of the cavity, while at the upper part filaments of the
cyanobiont were closely attached to the cavity envelope cells. This
attachment may involve lectin-receptor binding at the intercellular
surface. A fucose specific lectin of *Anabaena azollae*, detected by
heamagglutination was described (Kobiler et al. 1981) and
enrichment of fucose content in isolated leaf cavities of *Azolla* was
demonstrated (Tel-Or et al. 1983).

## 2.2 CONSUMPTION OF HOST PHOTOSYNTHATES BY *Anabaena azollae*

Previous analysis of soluble sugars in *Azolla* demonstrated high
content of sucrose and low fructose and glucose content (Peters et al.
1985), suggesting that sucrose is the major carbon source consumed by
the cyanobiont. A cultured isolate of *Azolla filiculoides* was shown to

take up fructose as the main substrate for growth in dark and light, and fructose uptake was faster than the uptake of glucose and sucrose (Rozen et al., 1986, 1988). Fructose supported the growth of *Anabaena azollae* isolated from *Azolla caroliniana* (Newton & Herman 1979). We have recently shown the effect of soluble host components on sucrose uptake by a fresh isolated cyanobiont from *Azolla filiculoides* (Tel-Or et al., 1990). In view of this information the content of soluble sugars in *Azolla filiculoides* was tested in fronds crushed gently by freezing and thawing. The homogenate was extracted in ethanol and separated by HPLC, provided with a differential refractometer. As shown in Fig 3, fructose and glucose content in *Azolla* was higher than

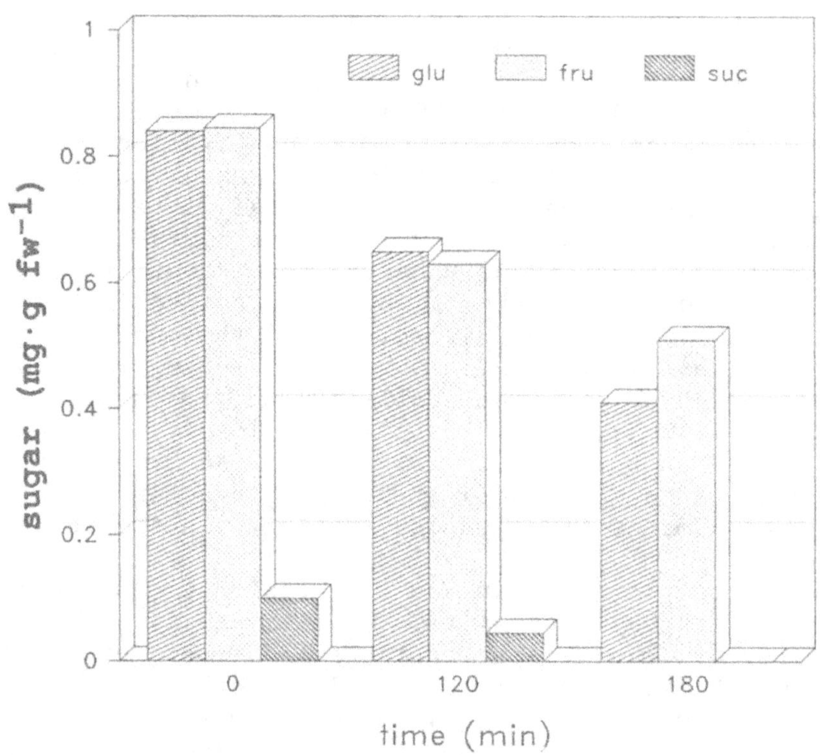


Figure 3.    HPLC analysis of soluble sugars in *Azolla* after addition of DCMU

that of sucrose. The  soluble sugars were tested in *Azolla* plants incubated with 10µM DCMU in the light for 2 and 3 h. It can be seen that a low content of sucrose was consumed within 2 h, while fructose and glucose were gradually consumed during 3 hours of the experiment. The decrease in sugar content in DCMU treated *Azolla*, reflects the consumption of these sugars for several metabolic activities: $N_2$-fixation, respiration and biosynthesis. The results shown are different than those reported by Peters et al. (1985), and it is unlikly that sucrose predominates fructose as carbon substrate for $N_2$-fixation by the cyanobiont.
*Azolla* leaves, fixed and sectioned for microscopic analysis, were stained for starch with iodine-potassium iodide and the stained hair cells suggested large starch granules in them. Fresh isolated

cyanobiont cells were tested for starch amylolytic activity, and cell extracts were electrophorized on starch containing polyacrylamide gels. The gel, stained with I-KI showed clear bands representing starch breakdown, suggested amylase activity. Similar amylolytic activity was observed with intact cells of the cyanobiont, suggesting the excresion of amylase by the cyanobiont. *In vitro* amylase activity in crude extarct of freshly isolated cyanobiont cells from *Azolla filiculoides* was tested as described in table 1. Activity of heat stable $\alpha$-amylse was higher than the heat sensitive $\beta$-amylase and specific activity of $\alpha$-amylase was between 2-3 $\mu$mol maltose·$\mu$g prot$^{-1}$·hr$^{-1}$. Amylase activity was observed in *Anabaena* sp. where $\alpha$-amylase activity was released frome lysosyme treated cells (Sarma et al., 1977). The demonstration of amylase activity in the symbiotic *Anabaena azollae* and the high content of starch in the host hair cells suggest that starch is employed as substrate for heterotrophic metabolism of the cyanobiont, and if so indicating that *A. azollae* does not depend solely on soluble sugars supply in the leaf cavity.

TABLE 1. *In vitro* amylase activity of the freshly isolated *Anabaena azollae*

| Enzyme preparation | pH | Amylase activity ($\mu$mol maltose·$\mu$g protein$^{-1}$·h$^{-1}$) |
|---|---|---|
| Cell extract | 6.2 | 2.05 |
| Preheated cell extract | 6.2 | 3.04 |
| Cell extract | 3.2 | 1.35 |
| Preheated cell extract | 3.2 | 0.22 |

The cyanobiont was isolated from crushed fronds of *Azolla filiculoides*, and purified by differential centrifugation. Preheating of cells extract was done for 10 min at 70°C. Cell soluble extract, obtained by sonication and centrifugation was incubated in 100 mM succinate buffer at the reported pH, with 0.2% starch and incubated for 8 h. Samples were then filtered and immidiately analysed for sugar content by HPLC.

## 2.3 LIGHT REGULATION OF NITROGENASE ACTIVITY IN *Anabaena azollae*

Nitrogenase activity of *Azolla filiculoides* and *Azolla pinnata* was not inhibited in the presence of 20$\mu$M DCMU. Furthermore, it was enhanced upon addition of DCMU, when tested in a specific device for multiple sampling and semi-continuous assay of acetylene reduction (Bar et al, 1988). DCMU enhanced nitrogenase activity may involve derepression of nitrogenase synthesis in the absence of photoevolved oxygen. Direct measurements of $O_2$ concentration in the leaf cavities of *Azolla filiculoides* in the dark demonstrated fast decrease of $O_2$ concentration (Grilli-Caiola et al. 1989). DCMU application may result in a similar drop of $O_2$ concentration, triggering nitrogenase activity. Kulasooriya et al. (1989) found unequal distribution of photosynthesis and nitrogenase activity in fragments along the stem axis of *Azolla*; i.e, higher nitrogenase in the mature and lower nitrogenase in the younger leaves, where photosynthetic activity was maximal. Bar et al. (1988) observed an anpredicted response of nitrogenase to light/dark transitions in *Azolla filiculoides* and *Azolla pinnata*. When light was turned off nitrogenase activity was switched off and dropped to zero. This drop was not accompanied by depletion of soluble carbohydrates within 10-20 min of dark intermission. Upon reillumination, nitrogenase activity was switched on within 10 min, which was too short to involve resynthesis of nitrogenase. Nevertheless, it reflects the light regulation of nitrogenase. These responses of nitrogenase activity to light were also reproduced with the freshly isolated cyanobiont cells, and were not observed in free living *Nostoc muscorum* 7119. An example of these phenomena is described in Fig 4. The acetylene reduction activity dropped to zero after 10 min in the dark, and was resumed 10-20 min after light was switched on. Light regulation of nitrogenase activity may involve specific modification of dinitrogenase reductase as reported for photosynthetic bacteria (Ludden et al. 1988). Reillumination of *Azolla* with low light intensity of $250\mu E \cdot m^{-2} \cdot sec^{-1}$, after the dark intermission, increased nitrogenase activity, above its control value, which stayed for 30 min and was then reduced to the level of the control activity (Fig 4). *Azolla*, reilluminated after the dark intermision with high intensity of $850\mu E \cdot m^{-2} \cdot sec^{-1}$, in the absence of DCMU, showed a different pattern. Nitrogenase activity was resumed for 20 min and dropped later, in the light, to zero. Nitrogenase activity of reilluminated *Azolla* with high light intensity in the presence of DCMU was not inhibited, suggesting involvement of $O_2$ in the light intensity effect. Photosynthetic activity of *Azolla*, measured by photoacoustics (Kulasooriya et al. 1989) or by infra red gas analysis of $CO_2$ fixation, was saturated at $850\mu E \cdot m^{-2} \cdot sec^{-1}$ and the concentration of photoevolved $O_2$ was much higher compared to $250\mu E \cdot m^{-2} \cdot sec^{-1}$. Nitrogenase activity reached saturation at $250\mu E \cdot m^{-2} \cdot sec^{-1}$ and was inhibited by $850\mu E \cdot m^{-2} \cdot sec^{-1}$. These differences in light saturation suggested that oxygen above certain concentration was harmful to nitrogenase in *Azolla* and the protection mechanisms in the heterocysts were no longer effective obove that critical threshold of $O_2$ concentration.

Analysis of dinitrogenase reductase polypeptides by immuno-blot confirmed that the 30KD polypeptide was detected in light grown *Azolla* (Fig 5, lane 1). Under high light intensity, the content of the 30KD polypeptide was reduced (lane 2). The additional band of 36KD polypeptide seems to represent active dinitrogenase reductase, which disappears under high light intensity. The same polypeptides of dinitrogenase reductase were detected in nitrogen fixing fresh isolated cynobiont from *Azolla filiculoides* (lane 4). The 36KD polypeptide was also resolved by immunoblot analysis reported by Braun-Howland et al. (1988).

The information obtained by the immunoblot analysis correlated with light and $O_2$ regulations of nitrogenase activity. The nature of the dinitrogenase reductase modification process has not been resolved.

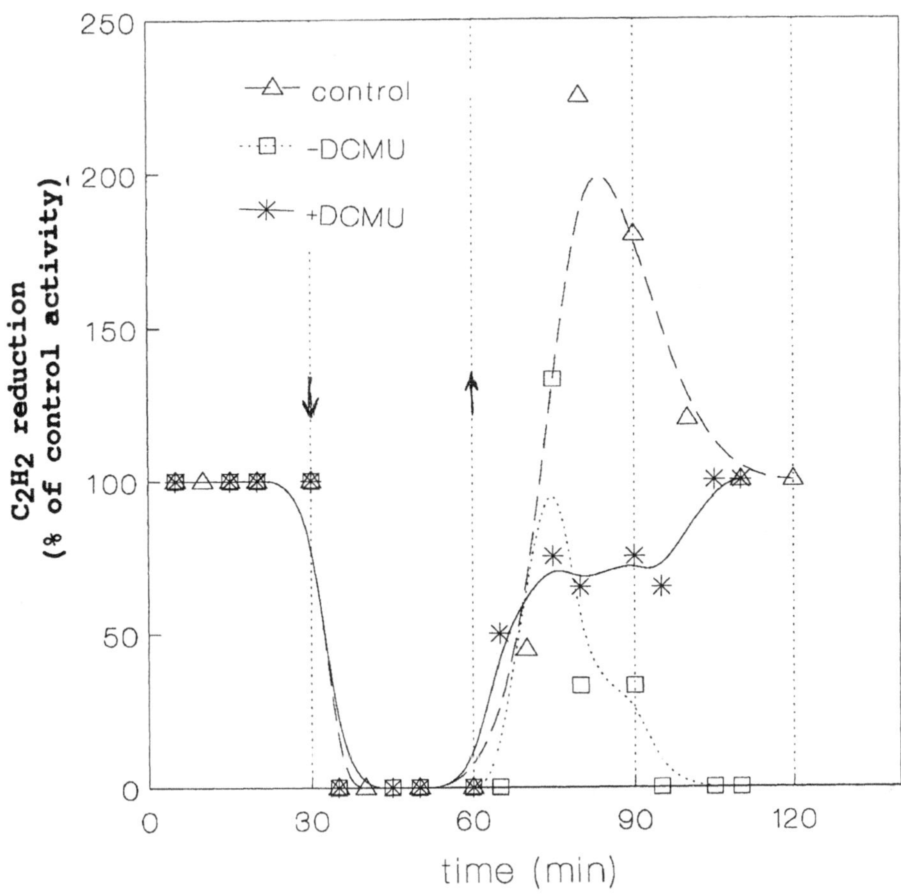



Figure 4.    Nitrogenase activity under intermittent light in *Azolla*. Control sample was illuminated with 250 $\mu E \cdot m^{-2} \cdot sec^{-1}$. DCMU samples were illuminated with 850 $\mu E \cdot m^{-2} \cdot sec^{-1}$. ↓- light off, ↑-light on.

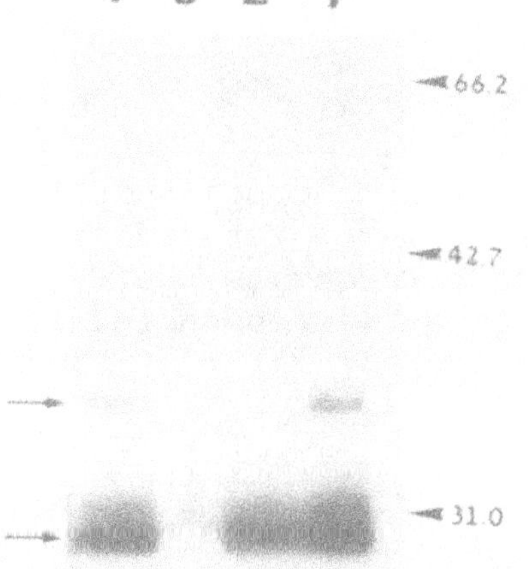



Figure 5.    Immuno-blot of dinitrogenase reductase with antibodies against Kp2. 1. Hyterocysts from Azolla in low light. 2. Heterocysts from Azolla in high light. 3. Vegetative cells. 4. Heterocysts from isolated cyanobiont grown in low light.

## 2.4 RESPIRATION AND $O_2$ PROTECTION IN *Anabaena azollae*

High rates of respiration activity and $O_2$ consumption of the cultured isolated cells of *Anabaena azollae* were reported by Rozen et al. (1988). Respiration was enhanced in cells grown heterotrophically either in the dark or light. The effect of heterotrophic growth on respiratory electron transport components, in the cultured isolate of *Anabaena azollae* and in the cyanobiont from *Azolla filiculoides* were studied. Membranes of vegetative cells exhibited high rate of cytochrome c reduction by NADH. This activity was sensitive to antimycin a, suggesting the participation of NADH oxidase. NADPH supported the reduction of cytochrome c which was insensitive to antimycin a, and could therefore involve ferredoxin–NADP reductase activity, as suggested by scherer et al. (1988). Heterocyst membranes from heterophically grown cells of cultured isolated *Anabaena azollae* and freshly isolated cyanobiont cells, had neither NADH nor NADPH supporting activity of cytochrome c reduction. These preparations showed enhanced $O_2$ uptake with reduced cytochrome c, and were enriched with cytochrome oxidase, as shown in table 2. Cytochrome oxidase activity was higher in heterocysts compared to vegetative cells, and the highest activity was observed in the fructose grown heterocysts of the cultured isolate. The affinity of cytochrome oxidase to $O_2$ shown in Table 2 was higher in heterocysts than in vegetative cells. The enrichemnt of cytochrome oxidase activity in the heterocysts, enables faster oxygen removal in these cells. The enhanced respiration in the heterotrophically grown cells of *Anabaena azollae in vivo*, and the enriched cytochrome oxidase activity in the heterocysts of the cyanobiont *in vitro*, demonstrates their capacity to modulate $O_2$ protection in the nitrogen fixing cells.

TABLE 2. Cytochrome oxidase activity in *Anabaena azollae*

| Cells | $V_{max}$ | | Apparent $K_m$ $[O_2]$ | |
|---|---|---|---|---|
| | Het. | Veg. | Het. | Veg. |
| Cultured isolated, autotrophically grown | 99 | 66 | 54 | 76 |
| Cultured isolated grown in light+10mM fructose | 629 | 85 | 49 | 73 |
| Fresh cyanobiont | 185 | 75 | 42 | 63 |

$V_{max}$ is expressed in nmol cytochrome c oxidized·mg prot$^{-1}$·min$^{-1}$. Apparent $K_m$ is calculated in µM $O_2$. Membranes were prepared from preparations of vegetative cells (Veg.) and heterocysts (Het.) isolated by the Yeda press procedure. The assay was conducted in 50mM Hepes pH 7.5, and followed with a spectrophotometer at 550nm with horse heart cytochrome c.

# 3. References


Bar, E., Kulasooriya, S.A., Tel-Or, E. (1988) Nitrogen fixation in *Azolla* is controlled by the availability of photosynthates to the cyanobiont, in Bothe, H., De Bruijn, F.J. and Newton, W.E. (eds.), Nitrogen fixation: Hundred Years After, p. 229.

Becking, J.H. (1987) Endophyte transmission and activity in the *Anabaena-Azolla* association. Plant and soil 100, 183-212.

Braun-Howland, E.b., Lindblad, P., Nierzwicki-Bauer, S.A., Bergman, B. (1988) Dinitrogenase reductase (Fe-protein) of nitrogenase in the cyanobacterial symbionts of three *Azolla* species: localization and sequence of appearance during heterocyst differentiation. Planta 176, 319-322.

Calvert, H.E., Pence, M.K., Peters, G.A. (1985) Ultrastructural ontogeny of leaf cavity trichomes in *Azolla* implies a functional role in metabolite exchange. Protoplasma 129, 10-27.

Grilli-Caiola, M., Canini, A., Moscone, D. (1989) Oxygen concentration, nitrogenase activity and heterocyst frequency in the leaf cavities of *Azolla filiculoides* Lam. FEMS Microbiol. Let. 59, 283-288.

Kobiler, D., Cohen-Sharon, A. and Tel-Or, E. (1981) Recognition between the $N_2$-fixing *Anabaena* and the water fern *Azolla*. FEBS Lett. 133, 157-160.

Kulasooriya, S.A., Arad, H., Canaani, O., Tel-Or, E., Malkin, S. (1989) Distribution of the $N_2$ fixation and photosynthetic activities in the *Azolla-Anabaena* symbiosis. Symbiosis 6, 117-128.

Ludden, P.W., Roberts, G.P., Lowery, R.G., Fitzmaurice, W.P., Saari, L.L., Lehman, L., Lies, D., Woehle, D., Wirt, H., Murrell, S.A., Pope, M.R., Kanemoto, R.H. (1988). Regulation of nitrogenase activity by reversible ADP-ribosylation of dinitrogenase reductase, in Bothe, H., De Bruijn, F.J. and Newton, W.E. (eds.), Nitrogen Fixation: Hundred Years After, pp. 157-162.

Newton, J.W. and Herman, A.I. (1979) Isolation of cyanobacteria from the aquatic fern, *Azolla*. Arch. Microbiol. 120, 161-165.

Nierzwicki-Bauer, S.A., Aulfinger, H., Braun-Howland, E.B. (1989) Ultrastructural characterization of inner envelope that confines *Azolla* endosymbionts to the leaf cavity periphery. Can. J. Botany 67, 2711-2719.

Peters, G.A., Kaplan, D., Meeks, J.C., Buzby, K.M., Marsh, B.H., Corbin, J.L. (1985) Aspects of nitrogen and carbon interchange in the *Azolla-Anabaena* symbiosis, in Ludden, P.W. and Burris, J.E. (eds.), Nitrogen Fixation and $CO_2$ Metabolism, pp. 213-222.

Peters, G.A., Meeks, J.C. (1989) The *Azolla-Anabaena* symbiosis. Basic biology. Ann. Rev. Plant physiol. Plant Mol. Biol. 40, 193-210.

Rozen, A., Arad, H., Schonfeld, M., Tel-Or, E. (1986) Fructose supports glycogen accumulation, Heterocysts differentiation, $N_2$ fixation and growth of the isolated cyanobiont *Anabaena azollae*. Arch Microbiol. 145, 187-190.

Rozen, A., Schonfeld, M., Tel-Or, E. (1988) Fructose-enhanced development and growth of the $N_2$-fixing cyanobiont *Anabaena azollae*. Z. Naturforsch. 43C, 408–412.

Sarama, T.A., Kanta, S., Ahuga, B.S., Kiran, U. (1977) Amylase activity in a blue green alga. Curr. Sci. 46 609–610

Scherer, S., Alpes, I., Sadowski, H. and Boger, p. (1988) Ferredoxin-$NADP^+$ oxidoreductase is the respiratory NADPH dehydrogenase of the cyanobacterium Anabaena variabilis. Arch. Biochem. Biophys. 267, 228–235.

Tel-Or, E. and Sandovsky, T., Kobiler, D., Arad, C. and Weinberg, R. (1983) The unique symbiotic properties of Anabaena in the water fern Azolla, in Papageorgiou G.C. and Packer, L. (eds.), Photosynthetic Prokaryotes: Cell Differentiation and Function. Elsevier Science, New York, pp. 303–304.

Tel-Or, E., Rozen, A., Ofir, Y., Kobiler, D., Schenfeld, M. (1990) Metabolic relations and intercellular signals in the *Anabaena-Azolla* association. Israel J. Botany, In press.

# REGULATION OF NITROGEN ASSIMILATION IN THE CYANOBACTERIUM *SYNECHOCOCCUS*


A. HERRERO, M. A. VEGA-PALAS, and E. FLORES
*Instituto de Bioquímica Vegetal y Fotosíntesis*
*Universidad de Sevilla-CSIC*
*Facultad de Biología*
*Apartado 1113, 41080-Sevilla*
*Spain*



ABSTRACT. *Synechococcus* is a unicellular cyanobacterium able to assimilate nitrate, nitrite or ammonium. Cellular elements involved in the assimilation of these compounds are subjected to repression by ammonium. Ammonium also promotes an immediate inhibition of the nitrate transport system. We have recently identified a gene (*ntcA*) whose product appears to be required for the expression of genes subjected to repression by ammonium. Mutants that constitutively express elements of the nitrate assimilation system have also been reported, suggesting that negative regulatory elements might also be involved in the control of nitrogen assimilation in *Synechococcus*.


## 1. Mechanisms for the assimilation of nitrate, nitrite and ammonium

The unicellular cyanobacteria of the genus *Synechococcus* are able to use, almost exclusively, nitrate, nitrite or ammonium as nitrogen sources for growth.

Nitrate assimilation involves nitrate entrance into the cell, nitrate reduction to the term of ammonium, and the incorporation of the resulting ammonium into organic material. Nitrate uptake takes place through an active transport system [8, 18] from which a component has been identified as a 48-kDa cytoplasmic-membrane protein [21, 25]. This transport system takes up nitrate with a high affinity, its $K_m$ being about 1 μM [8, 25, 31]. Nitrate reduction takes place in two successive steps. First, nitrate is reduced to nitrite in a two electron reaction catalyzed by nitrate reductase, which has been characterized as a 75-kDa single-polypeptide protein containing molybdenum [4]. Thereafter, the resulting nitrite is reduced to ammonium in a six electron reaction catalyzed by nitrite reductase, an enzyme constituted by a single polypeptide of about 50 kDa [22]. Electrons for nitrate and nitrite reduction in *Synechococcus* originate from the photolysis of water [5, 8, 28], reduced ferredoxin being the immediate electron donor for both reductases [23]. Nothing is known in *Synechococcus* about the pathways for molybdenum metabolism or for the synthesis of the molybdenum cofactor thought to be a component of nitrate reductase.

When nitrite is being utilized as the exogenous nitrogen source, it can enter the cyanobacterial cell both by passive diffusion as nitrous acid and by active transport [9]. Active transport of nitrite appears to take place through the same system that mediates nitrate transport [20]. With regard to the utilization of exogenous ammonium, an ammonium transport system has been identified in *Synechococcus* sp. PCC 7942 by means of studying methylammonium transport [2].

In *Synechococcus*, ammonium, either resulting from reduction of nitrate or nitrite or taken up from the outer medium, is almost exclusively incorporated into carbon skeletons through the action of the glutamine synthetase/glutamate synthase pathway (see [10]). Glutamine synthetase from these organisms exhibits the typical eubacterial structure [6].

## 2. Structural genes involved in nitrate assimilation

Through searching in expression gene libraries from *Synechococcus* sp. PCC 7942 for production of antigen reacting with antibodies prepared against the 48-kDa protein involved in nitrate transport, the gene encoding this protein (called *nrtA*) has been cloned [25] and afterwards sequenced [24]. Interruption of the *nrtA* gene with a kanamycin-resistance-encoding cassette allowed the isolation of a mutant impaired in nitrate transport whose characterization revealed that passive entrance of nitrate into *Synechococcus* cells can take place at external nitrate concentrations above 1 mM [25]. It is not known whether the *nrtA* gene product is the only component of the nitrate transport system in *Synechococcus*.

By means of transposon or NTG mutagenesis, a number of mutants have been isolated from *Synechococcus* sp. PCC 7942 that were impaired in nitrate reductase activity [16]. Complementation with a gene library from the wild type strain allowed the cloning of two different genes involved in nitrate reduction, *narA* and *narB* [16]. A third gene, *narC*, for nitrate reduction has been cloned by marker rescue of the ampicillin resistance encoded in the transposon [17]. These three *nar* genes appear to be distantly located from one another in the chromosome of *Synechococcus*. The precise function of none of these genes has been elucidated to date, although it has been suggested that *narB* represents the structural gene for nitrate reductase [1].

Mutants impaired in nitrite reduction have been isolated from *Synechococcus* spp. strains PCC 7002 [30] and PCC 7942 [32]. These mutants have not been genetically characterized. Nevertheless, closely upstream from the *nrtA* gene an open reading frame that may correspond to the nitrite reductase structural gene has been found [24].

# 3.   Regulation   of   nitrogen   assimilation

In *Synechococcus* sp., nitrate assimilation is subjected to control exerted mainly by ammonium, the end product of the reduction of nitrate. Ammonium acts as a repressor of the three proteins whose regulation has been studied to date, namely the 48-kDa nitrate transport element [21, 25, 29], nitrate reductase [13], and nitrite reductase [15]. The levels of all these proteins are much higher in the absence than in the presence of ammonium. An inducer, like nitrate or nitrite, is not required since considerable synthesis of those three proteins takes place in the absence of any nitrogen source. However, nitrate reductase stability has been shown to be higher in cultures supplemented with a nitrogen source than in nitrogen-free media [14]. In the case of nitrate reductase and nitrite reductase, it has been established that ammonium metabolism through glutamine synthetase is required for ammonium repression to be operative [13, 15].

Ammonium also regulates nitrate assimilation through inhibition of the nitrate/nitrite active transport system [7, 9, 18]. Again, ammonium is not the actual effector but operation of glutamine synthetase proved a requirement for ammonium-promoted control to be operative [7]. $CO_2$-fixation products also participate in the control of nitrate uptake in *Synechococcus*, though exerting a positive effect [11, 27]. The identity of the organic metabolites operating modulation of the nitrate transport activity has not been established. However, physiological experiments speak on a role of glutamine (or asparagine) as a negative effector and of α-ketoglutarate as a positive effector of nitrate transport in *Synechococcus* [26].

With respect to the assimilation of ammonium, both the ammonium (methylammonium) transport system [3] and glutamine synthetase [32] have been described as repressible by ammonium in *Synechococcus*. Levels of glutamine synthetase in nitrate-grown cultures are about twice those found in ammonium-grown cells [6, 32]. Finally, inhibition of purified glutamine synthetase from *Synechococcus* strain L-1402-1 by some amino acids, mainly alanine, has also been reported [6].

# 4.   Regulatory   genes   for   nitrogen   assimilation

From *Synechococcus* sp. PCC 7942, some pleiotropic mutants have been isolated that, as the result of single mutations, are simultaneously impaired in all the cellular elements known to be subjected to repression by ammonium [32]. These elements include the 48-kDa cytoplasmic-membrane protein involved in nitrate transport, nitrate reductase, nitrite reductase, methylammonium transport, and glutamine synthetase. Cellular levels of all of these elements in cultures supplemented with nitrate or lacking any added nitrogen source are similar to the levels found in ammonium-grown cells of the wild-type strain. The gene altered in those mutants, that has been named *ntcA*, seems to have a positive mode of action, its function being required for the expression of the structural genes subjected to ammonium repression in *Synechococcus* [32]. By means of phenotypic complementation with a gene library from wild-type *Synechococcus* sp. PCC 7942, *ntcA* has been cloned [32] and is currently being analyzed in detail at the molecular level. *ntcA* constitutes the first gene for nitrogen control identified in cyanobacteria. In these organisms, DNA sequences homologous to those of the enterobacterial *ntr* genes have not been found to occur. Nevertheless, a protein homologous to the *Escherichia coli glnB* gene product has recently been found in *Synechococcus* sp. PCC 6301 [12]. Whether this protein, which in *E. coli* participates in nitrogen control in combination with the *ntr* gene products, has any role in the regulation of nitrogen assimilation in *Synechococcus* remains to be established.

Two other types of mutants altered in the expression of the pathway for nitrate assimilation and exhibiting complex pleiotropic phenotypes have been isolated from *Synechococcus* sp. PCC 7942 [19]. Mutant FM16 has a low but significant level of nitrate reductase that remarkably appears to be freed from ammonium repression; it is also impaired in nitrate transport, but has increased levels of nitrite reductase that remain ammonium-repressible. On the other hand, FM2 is a transposon induced mutant that (as a consequence of a single mutation) lacks nitrite reductase but constitutively expresses both nitrate reductase [19] and the 48-kDa nitrate transport protein [21] as well as the ability to take up nitrate. The product of the gene altered in mutant FM2 might, in contrast to that of *ntcA*, have a negative action on the expression of nitrate assimilation genes in *Synechococcus*. Cloning of the gene altered in mutant FM2 is currently being attempted by means of marker rescue.

As it was mentioned above, it has been recently reported that the putative structural gene for nitrite reductase is closely linked to the gene for the 48-kDa nitrate transport element [24]. If a cluster of genes for nitrate assimilation actually occurred in the chromosome of *Synechococcus*, pleiotropic phenotypes could result, besides from mutation in a regulatory gene (as in the case of *ntcA*), from polar effects. An explanation at the molecular level of the alterations occurring in the pleiotropic mutants above described should prove most valuable for the understanding of the regulation of nitrogen assimilation in cyanobacteria.


## References


[1]     Andriesse, X., Van Arkel, G., and Weisbeek, P. (1990) 'Cloning of the nitrate reductase gene of *Synechococcus* PCC 7942', in Nitrate Assimilation: Molecular and Genetic Aspects (Third International Symposium), Bombannes (France), Abstracts p. 98.
[2]     Boussiba, S., Dilling, W., and Gibson, J. (1984) 'Methylammonium transport in *Anacystis nidulans* R2', J. Bacteriol. 160, 204-210.
[3]     Boussiba, S. and Gibson, J. (1987) 'Regulation of methylammonium/ammonium transport in the unicellular cyanobacterium *Synechococcus* R2 (PCC 7942)', FEMS Microbiol. Lett. 43, 289-293.
[4]     Candau, P. (1979) 'Purificación y propiedades de la ferredoxina-nitrato reductasa de la cianobacteria *Anacystis nidulans*', Ph. D. Thesis, Universidad de Sevilla.

[5]     Candau, P., Manzano, C., and Losada, M. (1976) 'Bioconversion of light energy into chemical energy through reduction with water of nitrate to ammonia', Nature 262, 715-717.

[6]     Florencio, F. J. and Ramos, J. L. (1985) 'Purification and characterization of glutamine synthetase from the unicellular cyanobacterium *Anacystis nidulans*', Biochim. Biophys. Acta 838, 39-48.

[7]     Flores, E., Guerrero, M. G., and Losada, M. (1980) 'Short-term ammonium inhibition of nitrate utilization by *Anacystis nidulans* and other cyanobacteria', Arch. Microbiol. 128, 137-144.

[8]     Flores, E., Guerrero, M. G., and Losada, M. (1983) 'Photosynthetic nature of nitrate uptake and reduction in the cyanobacterium *Anacystis nidulans*', Biochim. Biophys. Acta 722, 408-416.

[9]     Flores, E., Herrero, A., and Guerrero, M. G. (1987) 'Nitrite uptake and its regulation in the cyanobacterium *Anacystis nidulans*', Biochim. Biophys. Acta 896, 103-108.

[10]    Flores, E., Ramos, J. L., Herrero, A., and Guerrero, M. G. (1983) 'Nitrate assimilation by cyanobacteria', in G. C. Papageorgiou and L. Packer (eds.), Photosynthetic Prokaryotes: Cell Differentiation and Function, Elsevier Science Publishing Co., New York, pp. 363-387.

[11]    Flores, E., Romero, J. M., Guerrero, M. G., and Losada, M. (1983) 'Regulatory interaction of photosynthetic nitrate utilization and carbon dioxide fixation in the cyanobacterium *Anacystis nidulans*', Biochim. Biophys. Acta 725, 529-532.

[12]    Harrison, M. A., Keen, J. N., Findlay, J. B. C., and Allen, J. F. (1990) 'Modification of a glnB-like gene product by photosynthetic electron transport in the cyanobacterium *Synechococcus* 6301', FEBS Lett. 264, 25-28.

[13]    Herrero, A., Flores, E., and Guerrero, M. G. (1981) 'Regulation of nitrate reductase levels in the cyanobacteria *Anacystis nidulans*, *Anabaena* sp. strain 7119, and *Nostoc* sp. strain 6719', J. Bacteriol. 145, 175-180.

[14]    Herrero, A., Flores, E., and Guerrero, M. G. (1984) 'Regulation of the nitrate reductase level in *Anacystis nidulans*: activity decay under nitrogen stress', Arch. Biochem. Biophys. 234, 454-459.

[15]    Herrero, A. and Guerrero, M.G. (1986) 'Regulation of nitrite reductase in the cyanobacterium *Anacystis nidulans*', J. Gen. Microbiol. 132, 2463-2468.

[16]    Kuhlemeier, C., Logtenberg, T., Stoorvogel, W., Van Heugten, H. A. A., Borrias, W. E., and Van Arkel, G. A. (1984) 'Cloning of two nitrate reductase genes from the cyanobacterium *Anacystis nidulans*', J. Bacteriol. 159, 36-41.

[17]    Kuhlemeier, C. J., Teeuwsen, V. J. P., Janssen, M. J. T., and Van Arkel, G. A. (1984) 'Cloning of a third nitrate reductase gene from the cyanobacterium *Anacystis nidulans* R2 using a shuttle gene library', Gene 31, 109-116.

[18]    Lara, C., Romero, J. M., and Guerrero, M. G. (1987) 'Regulated nitrate transport in the cyanobacterium *Anacystis nidulans*', J. Bacteriol. 169, 4376-4378.

[19]    Madueño, F., Borrias, W. E., Van Arkel, G. A., and Guerrero, M. G. (1988) 'Isolation and characterization of *Anacystis nidulans* R2 mutants affected in nitrate assimilation: establishment of two new mutant types', Mol. Gen. Genet. 213, 223-228.

[20]    Madueño, F., Flores, E., and Guerrero, M. G. (1987) 'Competition between nitrate and nitrite uptake in the cyanobacterium *Anacystis nidulans*', Biochim. Biophys. Acta 896, 109-112.

[21]    Madueño, F., Vega-Palas, M. A., Flores, E., and Herrero, A. (1988) 'A cytoplasmic-membrane protein repressible by ammonium in *Synechococcus* R2: altered expression in nitrate-assimilation mutants', FEBS Lett. 239, 289-291.

[22]    Manzano, C. (1977) 'La reducción fotosintética del nitrato en el alga verde-azulada *Anacystis nidulans*', Ph. Thesis, Universidad de Sevilla.

[23]    Manzano, C., Candau, P., Gómez-Moreno, C., Relimpio, A., and Losada, M. (1976) 'Ferredoxin-dependent photosynthetic reduction of nitrate and nitrite by particles of *Anacystis nidulans*', Mol. Cell. Biochem. 10:161-169.

[24]    Omata, T. (1990) 'Cloning, nucleotide sequence and mutagenesis of the *nrtA* gene encoding the 45-kD protein involved in nitrate transport in the cyanobacterium *Synechococcus* PCC 7942', in Nitrate Assimilation: Molecular and Genetic Aspects (Third International Symposium), Bombannes (France), Abstracts pp. 54-57.

[25]    Omata, T., Ohmori, M., Arai, N., and Ogawa, T. (1989) 'Genetically engineered mutant of the cyanobacterium *Synechococcus* PCC 7942 defective in nitrate transport', Proc. Natl. Acad. Sci. USA 86, 6612-6616.

[26]    Romero, J. M., Flores, E., and Guerrero, M. G. (1985) 'Inhibition of nitrate utilization by amino acids in intact *Anacystis nidulans* cells', Arch. Microbiol. 142, 1-5.

[27]    Romero, J. M., Lara, C., and Guerrero, M. G. (1985) 'Dependence of nitrate utilization upon active $CO_2$ fixation in *Anacystis nidulans*: a regulatory aspect of the interaction between photosynthetic carbon and nitrogen metabolism', Arch. Biochem. Biophys. 237, 396-401.

[28]    Serrano, A. and Losada, M. (1988) 'Action spectra for nitrate and nitrite assimilation in blue-green algae', Plant Physiol. 86, 1116-1119.

[29]    Sivak, M. N., Lara, C., Romero, J. M., Rodríguez, R., and Guerrero, M. G. (1989) 'Relationship between a 47-kDa cytoplasmic membrane polypeptide and nitrate transport in *Anacystis nidulans*', Biochem. Biophys. Res. Commun. 159, 257-262.

[30]    Stevens, S. E. and Van Baalen, C. (1973) 'Characteristics of nitrate reduction in a mutant of the blue-green alba *Agmenellum quadruplicatum*', Plant Physiol. 51, 350-356.

[31]    Tischner, R. and Schmidt, A. (1984) 'Light mediated regulation of nitrate assimilation in *Synechococcus leopoliensis*', Arch. Microbiol. 137, 151-154.

[32]    Vega-Palas, M. A., Madueño, F., Herrero, A., and Flores, E. (1990) 'Identification and cloning of a regulatory gene for nitrogen assimilation in the cyanobacterium *Synechococcus* sp. strain PCC 7942', J. Bacteriol. 172, 643-647.

# CALCIUM, DINITROGEN FIXATION AND CALMODULIN IN A *NOSTOC*.


**L A Onek, P J Lea & R J Smith.**
Institute of Environmental & Biological Sciences, Division of Biology, Lancaster University, Bailrigg, Lancaster LA1 4YQ, U.K.


## Summary


Four inhibitors of Calmodulin and other calcium binding proteins, have been used to probe the calcium mediated effects upon dinitrogen fixation, heterocyst frequency and the intracellular calcium content of *Nostoc* 6720. The inhibitors both enhance and depress dinitrogen fixation depending upon their concentration, but do not significantly affect the heterocyst frequency or the intracellular calcium content. The results confirm the presence of two distinct calcium concentration dependent effects; a calcium dependent protection of nitrogenase from $O_2$ and a calcium mediated inhibition of $N_2$ fixation. The inhibitors provide tools to assist the identification of calcium binding proteins involved in the expression of the calcium mediated, physiological effects.


## Introduction.

The role of calcium as a 'second messenger' and calmodulin as a signal transducer protein in the regulation of diverse events in higher plant and animal cells is well established. Evidence for such calcium mediated regulation in cyanobacteria has come from a number of laboratories and involved such diverse studies as trichome motility, photosynthesis and heterocyst differentiation (1).

The effects of calcium upon dinitrogen fixation are complex. One aspect of the calcium requirement appears to involve the protection of the oxygen labile nitrogenase. In heterocystous species $N_2$ fixation declines as the calcium concentration of the growth medium is decreased below $10^{-5}$M calcium. The decline in $N_2$ fixation is $O_2$ dependent. The effect is much reduced or absent in anaerobic conditions (2). A similar $O_2$ dependent inhibition of $N_2$ fixation is observed on gross removal of calcium from cultures of *Gloeothece* by chelating agents such as EDTA and EGTA (3,4).

At low incident irradiance $N_2$ fixation is decreased as the calcium concentration is raised above $10^{-5}$ M; which indicates the presence of a second calcium dependent mechanism (5). This inhibition of $N_2$ fixation by calcium is confirmed by the action of the calcium ionophore A23187 and abscisic acid (ABA), both of which depress $N_2$ fixation (5,6). The activation and inhibition aspects of calcium are also seen in the concentration dependent effects of lanthanum (5).

The mechanism by which $N_2$ fixation is depressed by calcium at low irradiance is unknown, but might possibly involve the supply of reductant and energy from photosynthesis. The activity of Photosystem II from *Anacystis nidulans* has been shown to be calcium dependent both *in vivo* and *in vitro* (7). The activation of oxygen evolution in *Phormidium luridum* was attributed to a calcium dependent increase in the number of active photosynthetic units (8). The cellular calcium content as judged by $^{45}$Ca$^{2+}$-labelling experiments is inversely correlated with the incident irradiance, at least at low irradiance (9) and in light dark transitions (10). Thus the observed effects of calcium on PS II would suggest a mechanism whereby photosynthesis is activated to compensate for decreased energy flux. Recently a calcium binding protein of 56-60 kDa has been found in thylakoid extracts of *Nostoc* 6720 (11) and *A. nidulans* (12) which may be involved in such regulation.

In direct contrast to $N_2$ fixation the proportion of heterocysts (frequency) which differentiate increases as the calcium concentration increases. This effect is also confirmed by the actions of A23187, ABA and lanthanum (5,6). In calcium depleted cultures, grown over several mean generation times in the absence of additional calcium, heterocyst frequency is not decreased below a minimum of 5-6 %, but heterocyst differentiation is delayed. Proheterocysts take 2 to 3 times longer to form; though the maturation of $N_2$ fixing heterocysts then proceeds as normal.

In seeking a molecular mechanism in cyanobacteria through which a calcium mediated signal might be transposed, we have investigated calcium binding proteins in extracts of *Nostoc* 6720. In confirmation of the work of Pettersson and Bergman (13), one protein of 18 kDa specifically binds calcium and possesses the enzymological and immunological (14) characterisitics considered diagnostic of calmodulin.


## Materials and Methods

The heterocystous cyanobacterium *Nostoc* 6720 (ATCC 27895, PCC 6720 ), formerly known as *Anabaenopsis circularis* (G. S. West) Wolosz. et Miller (Rippka et al. ,1979), was grown on an 8 fold dilution of Allen & Arnon, (1955) medium supplemented with 20 mM $KNO_3$.

The procedures used in this study have been previously reported (5). They include-: induction of heterocyst differentiation by centrifugation / washing procedures and resuspension in nitrogen free medium, the estimation of heterocyst proportions by light miscroscopic analysis,  the estimation of nitrogenase activity by the acetylene reduction assay and the estimation of cellular calcium content and calcium influx by $^{45}$calcium labelling techniques.

The inhibitors; Trifluoroperazine (TFP) (17), Verapamil (18), Compound 48/80 (19) and $W_7$ (20) are potent inhibitors of calcium binding to calmodulin and other calcium binding proteins. In eukaryotic tissues the inhibitors have varied effects inhibiting both calmodulin dependent and other calcium mediated processes.

## Results

Dose response curves were constructed for the four inhibitors with respect to $N_2$ fixation, heterocyst frequency and the cellular content of calcium. The effects of all four inhibitors were very similar. At low concentrations of the inhibitors nitrogenase activity was enhanced as compared to the untreated control (Fig 1). As the concentration of inhibitor was increased the enhancement was reduced until, at concentrations exceeding $10^{-7}$ M, $N_2$ fixation was decreased below that of the control. Above $10^{-5}$ M the inhibitors were toxic; cellular calcium declined (Fig 3) and growth ceased.

In contrast to $N_2$ fixation, the proportion of heterocysts induced to form in the inhibitor treated cultures by removal of combined nitrogen was not significantly affected (Fig 2) even though in other treatments (5) both $N_2$ fixation and heterocyst frequency were affected.

As previously noted for trifluoroperazine (6) the total cell content of calcium did not vary in response to the inhibitors in the range of concentrations which affected $N_2$ fixation (Fig 3). Nor was the rate of influx of $^{45}Ca^{2+}$ affected by the inhibitors (data not shown).

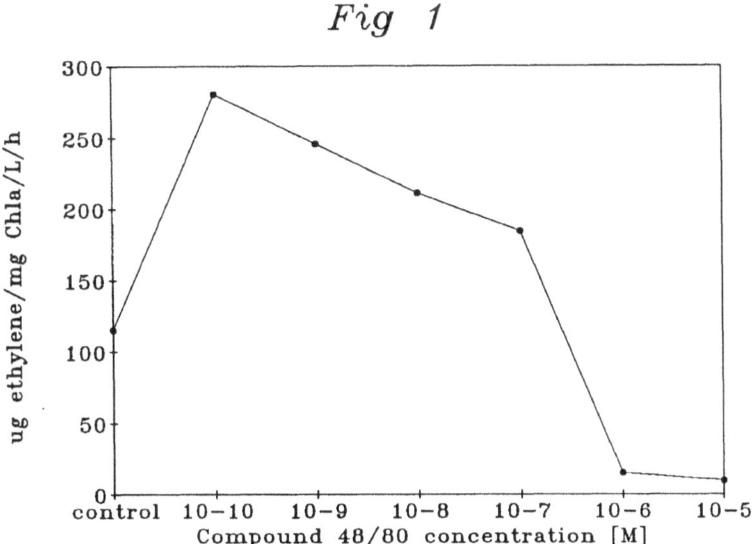


**Figure 1** The effect of inhibitor concentration on acetylene reduction in *Nostoc* 6720 in cultures caused to differentiate heterocysts by removal of nitrate in the presence of the inhibitor.

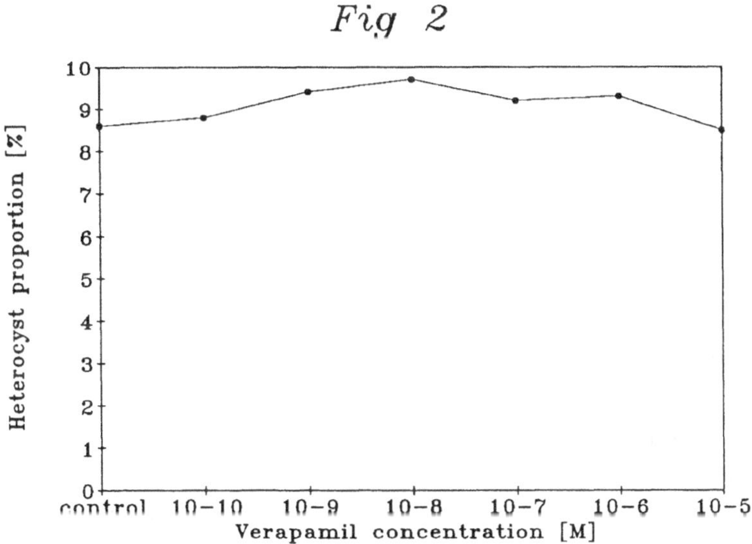


**Figure 2** The effect of verapamil concentration on the heterocyst frequency in *Nostoc* 6720 in a culture induced to form heterocysts by removal of nitrate in the presence of verapamil.

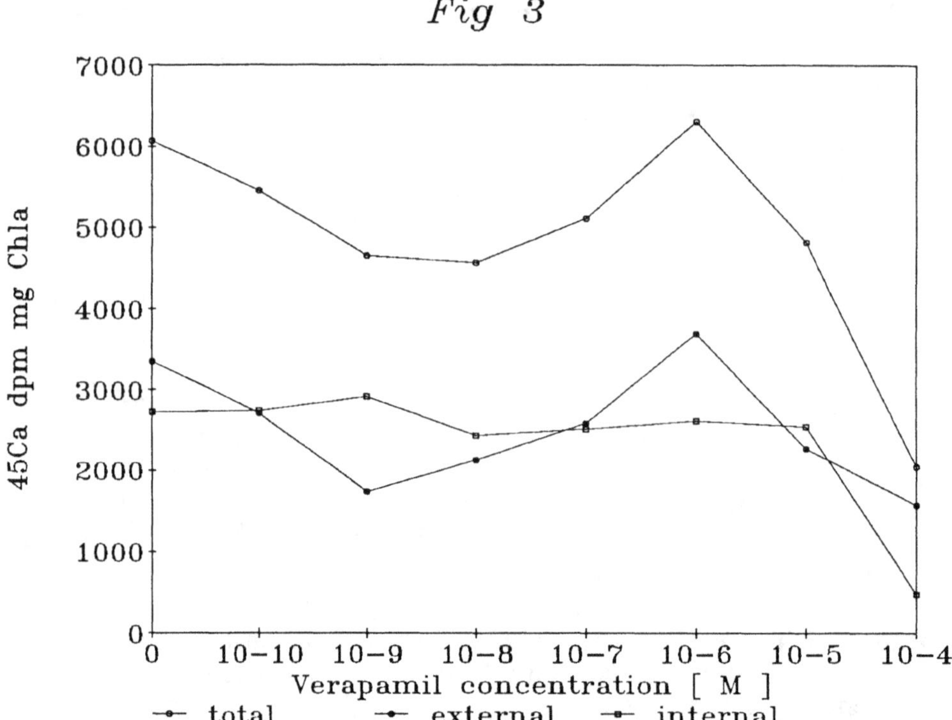



**Figure 3** The effect of verapamil concentration on the calcium associated with *Nostoc* 6720 cells.

## Discussion

The effect of calcium on dinitrogen fixation in cyanobacteria is likely to be indirect. As in calcium limited (2) or depleted (3) cyanobacterial cultures, the inhibition of nitrogenase activity at inhibitor concentrations exceeding $10^{-8}$ M does not occur in the absence of $O_2$ (unpublished data). The result provides further evidence of a calcium dependent mechanism(s) which maintains the anaerobic environment around the oxygen labile nitrogenase. The nature of the mechanism is not known, but considering the nature of the inhibitors, it probably involves calcium binding proteins.

The presence of the calcium ionophore A23187, ABA or growth in the presence of higher calcium concentrations decreases the rate of $N_2$ fixation (5,6). Low inhibitor concentrations enhance $N_2$ fixation (Fig 1). As with the effects upon heterocyst frequency (6) the inhibitors appear to prevent the suppression of $N_2$ fixation by A23187, ABA or calcium in the growth medium. A working hypothesis suggests that the inhibitors act by occluding calcium from a calcium binding protein, such as a cyanobacterial calmodulin, which otherwise acts to suppress $N_2$ fixation.

The proportion of heterocysts which are induced to form in a culture of *Nostoc* 6720 on removal of the combined nitrogen source is increased by the calcium ionophore A23187 or ABA or as the calcium concentration of the medium is increased from $10^{-5}$ to $10^{-3}$ M. The two effects of inhibitor action on $N_2$ fixation, enhancement followed by inhibition as inhibitor concentration increases, parallel the action of the inhibitor lanthanum (5). Unlike the action of lanthanum, the inhibitors had no significant effect on heterocyst proportion as compared to an untreated control. Both verapamil and TFP are able to counteract the increased proportion of heterocysts produced by ABA (6). Plausibly they inhibit the calcium mediated enhancement, but do not themselves affect heterocyst frequency.

The action of the inhibitors provides further evidence for calcium mediated regulation in cyanobacteria, but more significantly indicates the involvement of calcium binding proteins such as the cyanobacterial calmodulin. Activation of bovine cAMP phosphodiesterase and Pea NAD kinase by the putative cyanobacterial calmodulin we have isolated from *Nostoc* 6720 is completely inhibited by TFP, thus indicting that it could be involved. The inhibitors should prove useful in establishing the molecular mechanism between these calcium binding proteins and both the activation and inhibition of $N_2$ fixation since different concentrations of the inhibitors procure both effects.

## References

1. Smith, R. J. (1988) Review in Biochemistry of the Algae & Cyanobacteria. (Ed Rogers, L. J. & Gallon, J. R. ) Oxford Science Publications Clarendon Press London, pp185.

2. Rodriguez, H., Rivas, J. Guerrero, M. G. & Losada, M. (1990) Plant Physiol. 92, 886.

3. Gallon J. R. & Hamadi, A. F. (1984) J. Gen Microbiol. 130, 495.

4. Hamadi, A. F.& Gallon J. R.(1981) J. Gen. Microbiol. 125, 391.

5. Smith, R. J., Hobson, S. & Ellis, I. (1987) The New Phytologist 105, 531.

6. Smith, R. J., Hobson, S. & Ellis, I. (1987) The New Phytologist 105, 543.

7. Evans, E. H. ,England, R. R. & Manwaring, J. (1983) in Photosynthetic Prokaryotes: Cell differentiation and function. (ed. G. C. Papageorgiou & L. Packer) p. 175 Elsevier, New York.

8. Piccioni, R. G. & Mauzerall D. C. (1978) Biochem Biophys Acta 504, 384.

9. Smith R. J. & Wilkins A. (1988) The New Phytologist 109 157.

10. Becker, D. W. & Brand, J. J. (1985) Plant Physiology 79, 552.

11. Onec, L. A., Lea, P. J. & Smith R. J. (1990) First European Workshop, Molecular Biology of Cyanobacteria, Dourdan, France, p91.

12. McColl, S. M. & Evans, E. H. ibid p 60.

13. Pettersson, A. & Bergman, B. (1989) FEMS Microbiology Letts.

14 Van Eldik, L. J. & Wolchok, S. R. (1984) Biochem Biophys Res. Comm. 124, 752.

15. Rippka, R., Dueruelles,J., Waterbury, J. R., Herdman, M. & Stanier, R. Y. (1979) J. Gen. Microbiology 111, 1.

16. Allen M. B. & Arnon, D. I. (1955) Plant Physiol. 30, 266.

17. Levin, R. M. & Weiss, B. (1978) Biochem Biophys Acta 540, 197.

18. Schwartz, A. & Triggle, D. J. (1984) Annual Review of Medicine 35, 325.

19. Gietzen, K., Adamczyk-Engelmann, P., Wuthrich, A. Konstantinova, A. & Bader, H. Biochem Biophys Acta 736, 109.

20. Itoh, H. & Hidaka, H. (1984) J. Biochem 96, 1721.

# 'Respiratory protection' of the nitrogenase in dinitrogen-fixing cyanobacteria


G.A. PESCHEK, K. VILLGRATER and M. WASTYN
*Biophysical Chemistry Group, Institute of Physical Chemistry, University of Vienna, Währingerstrasse 42, A-1090 Vienna, Austria*



*Key words:* cyanobacteria, cytochrome oxidase, diazotrophy, mass spectrometry, nitrogenase, nitrogen fixation, respiratory protection


## Abstract


Several filamentous and unicellular cyanobacteria were grown photoautotrophically with nitrate or dinitrogen as N-sources, and some respiratory properties of the cells or isolated plasma (CM) and thylakoid (ICM) membranes were compared. Specific cytochrome c oxidase activities in membranes from dinitrogen-fixing cells were between 10- and 50-times higher than those in membranes from nitrate-grown cells, ICM of heterocysts but CM of unicells being mainly responsible for the stimulation. Whole cell respiration (oxygen uptake) of diazotrophic unicells paralleled increased cytochrome oxidase activities of the isolated membranes. Mass spectrometric measurements of the uptake of isotopically labeled oxygen revealed that (low) light inhibited respiration of diazotrophic unicells to a much lesser degree than that of nitrate-grown cells which indicates the prevailing (respiratory) role of CM in the former. Normalized growth yields of diazotrophic unicells grown in continuous light were significantly higher than those of cells grown in a 12/12 hrs light/dark cycle. Mass spectrometry showed that overall nitrogen uptake by the former was higher than by the latter; in particular, and in marked contrast to the time course of nitrogenase activity (acetylene reduction) there was no appreciable nitrogen uptake or protein synthesis during dark periods; likewise, there was no 14-CO$_2$ fixation, nor chlorophyll synthesis, nor cell division in the dark. By contrast, growth in continuous light gave sustained rates of nitrogen and carbon dioxide incorporation over the whole time range. Our results will be discussed in terms of 'respiratory protection' as an essential strategy of keeping apart nitrogenase and oxygen, either atmospheric or photosynthetically produced within the same cell.



*Abbreviations:* CM – cytoplasmic or plasma membrane, ICM – intracytoplasmic or thylakoid membrane, CCCP – carbonyl cyanide m-chlorophenylhydrazone, PCC – Pasteur Culture Collection


## Introduction

Cyanobacteria are oxygenic phototrophic prokaryotes uniquely combining the mechanisms of plant-type photosynthesis and cytochrome c oxidase-based respiration in one cell (Peschek, 1987; Stanier and Cohen-Bazire, 1977). In addition, many species – both filamentous and unicellular – are capable of diazotrophic growth (Gallon and Chaplin, 1988; Haselkorn, 1978; Kallas et al., 1983; Stewart, 1980), atmospheric nitrogen being assimilated via the nitrogenase enzyme complex. This enzyme as well as its biosynthesis are extremely oxygen labile regardless of the organism under investigation (Mortenson and Thorncley, 1979; Robson and Postgate, 1980). Aerobic dinitrogen fixers must therefore possess effective devices for avoiding contact between oxygen and nitrogenase. Such strategies are particularly imperative in the obli-

412

gately aerobic and phototrophic diazotrophic cyanobacteria performing both dinitrogen fixation and oxygen photoproduction. Three strategies are being considered for aerobic dinitrogen-fixing cyanobacteria:

(a) 'Spatial segregation' of nitrogenase and oxygen in special cell types of most filamentous diazotrophs. Aerobically, the nitrogenase is entirely confined to heterocysts which are devoid of photosystem II and Rubisco and, therefore, cannot produce oxygen in the light (Haselkorn, 1978; Stanier and Cohen-Bazire, 1977; Stewart, 1980).

(b) 'Temporal segregation' of nitrogenase (active only in the dark) and photo-synthesis (active only in the light) (Gallon and Chaplin, 1988; Mitsui et al., 1986; however, cf. Kallas et al., 1983; Pearson et al., 1979; Pearson and Howsley, 1980; and this article). The strategy of 'temporal segregation' would have to apply for nonheterocystous filamentous (Pearson et al., 1979; Stal and Krumbein, 1985) and unicellular (Gallon and Chaplin, 1988; Kallas et al., 1983; Mitsui et al., 1986) species.

(c) 'Respiratory protection' of the nitrogenase by effective scavenging of intracellular oxygen through sufficiently high rates of respiration, a mechanism which has been well established for obligately aerobic (nonphotosynthetic) soil bacteria of the Azotobacter group (Mortenson and Thorneley, 1979; Yates and Jones, 1974). Among cyanobacteria this mechanism appears to be operative in both heterocysts (Wastyn et al., 1988; Winkenbach and Wolk, 1973) and unicells (cf. this paper) where elevated levels of respiratory enzymes and/or electron transport components as well as respirable polyglucose reserve material (Ernst et al., 1984; Gallon and Chaplin, 1988; Mitsui et al., 1986) were reported. It is the only strategy that can explain sustained growth of unicellular diazotrophs in continuous light, and low growth rates would certainly be helpful in this context.

Slow penetration of (atmospheric) oxygen into thick-walled (heterocysts; e.g. Haselkorn, 1978; Lang and Fay, 1971; Stanier and Cohen-Bazire, 1977) or heavily ensheathed (unicells; Kallas et al., 1983) nitrogen-fixing cells might convey some additional protection. Certain nonheterocystous filamentous diazotrophs such as

the marine *Trichodesmium* sp. might effect some kind of 'pseudo-spatial segregation' through peculiar cell colonies (Carpenter and Price, 1976).

Our present investigation deals predominantly with unicellular diazotrophic cyanobacteria of the genera Gloeothece and Cyanothece. We hope that our results will contribute to clarifying the pivotal role of respiratory protection for the obligately aerobic and phototrophic growth of these organisms on dinitrogen as the sole N-source.

## Materials and methods

*Growth and harvest of the organisms*

Filamentous *Anabaena* PCC 7937 and 7120 and *Nostoc* PCC 8009 (*Nostoc* sp. strain Mac) were grown axenically in turbidostat cultures in modified medium D of Kratz and Myers (with or without nitrate) sparged with 1.5% carbon dioxide in air or oxygen-free nitrogen and illuminated with $100\,W\,m^{-2}$ warm white fluorescent light at 35°C as described (Wastyn et al., 1988). Axenic batch cultures of unicellular *Gloeothece* PCC 6501 and 6909, *Cyanothece* PCC 7822, and the sheathless mutant *Gloeothece* PCC 6909 She$^-$ were grown in medium BG-11 (with or without combined nitrogen) supplemented with $10\,mM$ bicarbonate and $1\,mM$ carbonate, sparged with 1.5% carbon dioxide in air or oxygen-free nitrogen, in 10-L flasks at 22°C and illuminated with $5\,Wm^{-2}$ warm white fluorescent light. Light intensities were measured with a YSI radiometer, type 65, at the surface of the vessels. *Anabaena* 7937 and *Nostoc* 8009 were gifts of Drs. C P Wolk and C van Baalen. All other species were obtained from the Pasteur Culture Collection (PCC), Paris, France, by courtesy of Mme. Rosmarie Rippka. For physiological measurements on isolated cells these were harvested from late logarithmic to linear cultures by centrifugation at room temperature (Peschek et al., 1988). The unicells and cell aggregates were surrounded by thick sheath layers irrespective of whether the N-source had been nitrate or dinitrogen; occasionally 'sheath-free' cells were obtained by applying a brief (3–5 s) burst of sonica-

*Table 1.* Respiratory oxygen uptake (nmol/min per mg chl) by unicellular cyanobacteria grown with nitrate or dinitrogen as N-sources

| | Nitrate-grown | | Dinitrogen-grown | |
|---|---|---|---|---|
| | Native | Sheath-free | Native | Sheath-free |
| *Cyanothece* 7822 | 46 | 53 | 78 | 370 |
| + 50 $\mu M$ CCCP | 86 | 95 | 89 | 405 |
| *Gloeothece* 6501 | 33 | 38 | 55 | 350 |
| + 50 $\mu M$ CCCP | 58 | 72 | 79 | 372 |
| *Gloeothece* 6909 | 29 | 34 | 67 | 285 |
| + 50 $\mu M$ CCCP | 61 | 83 | 75 | 313 |
| *Gloeothece* 6909 She⁻ | 26 | | 248 | |
| + 50 $\mu M$ CCCP | 59 | | 279 | |

Rates of oxygen uptake are mean values from three to five polarographic determinations on separately grown batches of cells, standard deviations never exceeding 12% of the corresponding mean.

tion at high power followed by low-speed centrifugation (Kallas et al., 1983); absence of spectroscopically detectable phycocyanin from the supernatants indicated that the cells had remained completely intact and the CM undamaged during the treatment (Table 1).

## Isolated membranes and cytochrome c oxidase

Cytoplasmic (CM) and thylakoid (ICM) membranes were isolated and purified from the cyanobacteria according to recently developed techniques (Murata and Omata, 1988; Peschek et al., 1988, 1989a,b). The activity of the cytochrome c oxidase (EC 1.9.3.1) in the membranes was measured by dual wavelength spectrophotometry using horse heart ferrocytochrome c as a substrate (Molitor and Peschek, 1986).

## Mass spectrometry

A quadrupole mass spectrometer (V.G. Gas Analysis (Winsford, U.K.) equipped with a Teflon membrane inlet system and a Faraday detector was used to determine the uptake (or evolution) of dinitrogen-28, dioxygen-32 and dioxygen-36 by intact *Gloeothece* 6909 She⁻ grown with nitrate or dinitrogen as N-sources and incubated in the dark or in the light in a so-called pseudo-open system (for technical details cf. Hoch et al., 1963; Radmer and Ollinger, 1980; Jensen and Cox, 1988; Dimon et al., 1988): A closed and translucent reaction vessel of 100 mL

contained 50 mL of the vigorously stirred cell suspension (equivalent to 35 $\mu$g chl mL$^{-1}$) in nitrate- and ammonia-free BG-11 buffered to pH 7.6 with 10 m$M$ K-phosphate at 22°C. In case of illumination the incident light intensity was 5 W m$^{-2}$. For oxygen exchange measurements 1.0 mL dioxygen-36 (99.2% isotopically pure) was introduced into the gas phase 30 min before the experiment was started, giving a ratio of oxygen-36/oxygen-32 partial pressures of 1:10 (cf. Fig. 2). In dinitrogen-28 fixation experiments argon-40 served as an internal reference (Dimon et al., 1988).

## Polarography

Rates of oxygen uptake (in the dark) by whole cells were also measured at 22°C and pH 7.6 using a Clark type oxygen electrode attached to a reaction chamber of 2.5 mL (YSI Oxygen Monitor, model 53) (cf. Table 1).

## Acetylene reduction

The reduction of acetylene to ethylene ('nitrogenase activity') was followed gas chromatographically according to standard procedures (e.g. Almon and Böger, 1988).

## Carbon dioxide fixation

Incorporation of C-14 from labeled bicarbonate into the acid-stable cell fraction was determined by liquid scintillation counting (Packard Tri-Carb 1500) according to conventional radiochemical techniques.

## Cell numbers

The number of cells in the suspensions was determined with a Bürker-Türk counting chamber. Cell numbers were averaged from at least 15 countings for each measuring point (cf. Fig. 3C).

## Chlorophyll and protein

Chlorophyll and protein were determined according to Mackinney, 1941, and Bradford, 1976, respectively. In order to ensure complete chlorophyll extraction the methanol was added to heavily sonicated (uni-) cells under nitrogen gas. Because of the massive and highly variable sheath layers around unicellular diazotrophs the chlorophyll content was assumed to be a more reliable measure of the cell mass than dry weight. protein or packed cell volume.

## Results and discussion

Specific rates of horse heart ferrocytochrome c oxidation by purified CM and ICM preparations from unicellular and filamentous cyanobacteria after growth on nitrate or dinitrogen are shown

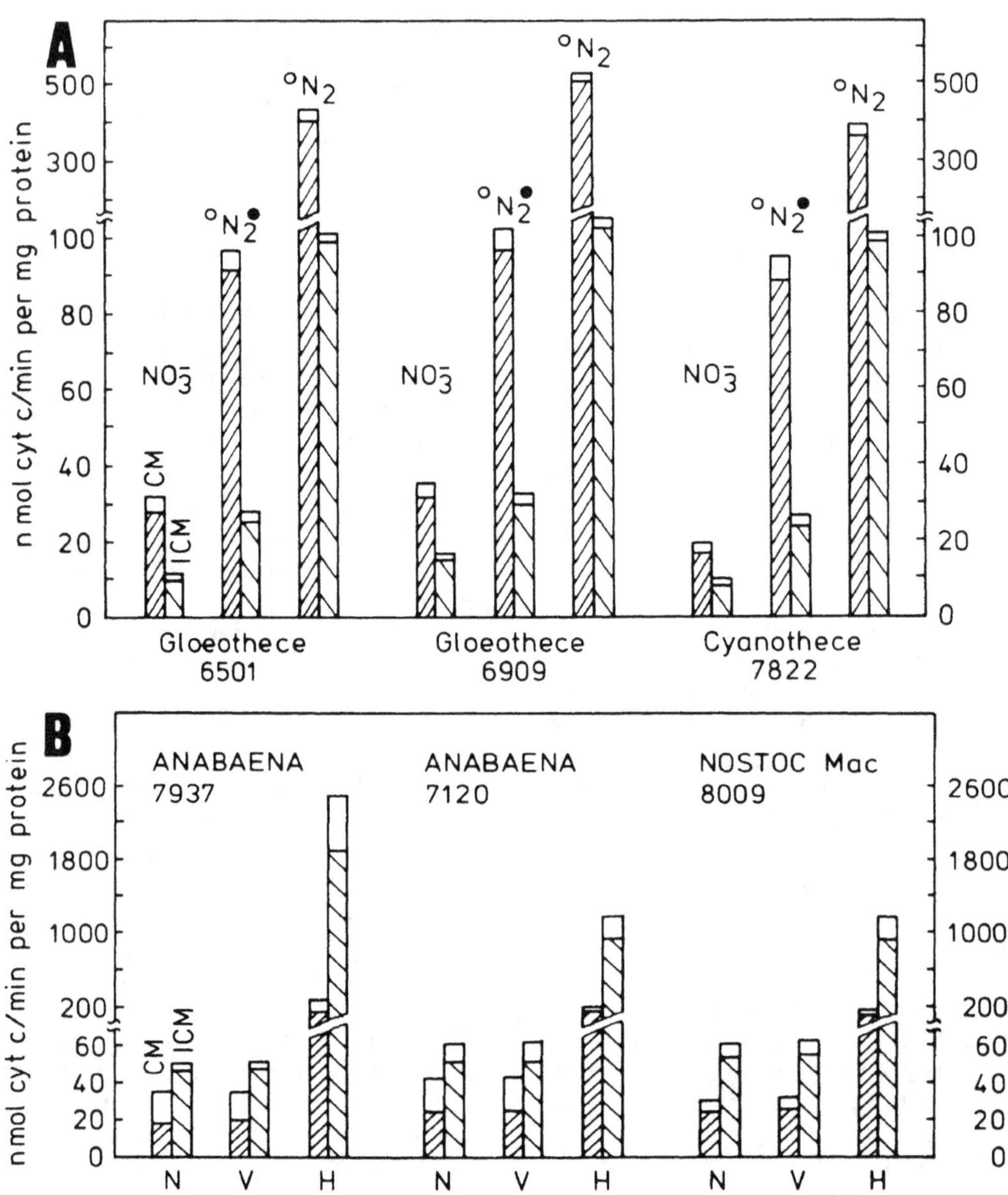


*Fig. 1.* Specific activities of the cytochrome c oxidase in isolated and purified plasma (CM) and thylakoid (ICM) membranes from unicellular (**A**) and filamentous (**B**) cyanobacteria grown with nitrate ($NO_3^-$; N) or dinitrogen ($N_2$; V; H) as N-sources. For further conditions of growth cf. Materials and methods. ○, diazotrophic growth in continuous light; ○●, diazotrophic growth in a 12/12 hours light-dark cycle. N, nitrate-grown filaments; V, vegetative cells of dinitrogen-grown filaments; H, heterocysts. Growth rates of the unicellular species were extremely low, doubling times ranging from about 35 to 50 hours at 22°C and 5 W m$^{-2}$ light intensity; yet, the strains did not tolerate more than 25°C and 10 W m$^{-2}$.

in Fig. 1A and B. respectively. Very high rates of cytochrome oxidase activity were measured with CM from unicellular diazotrophs and ICM from heterocysts. Vegetative cells of diazotrophic filaments exhibited cytochrome oxidase activities quantitatively similar to nitrate-grown filaments (Fig. 1B). Cytochrome oxidase activities in unicellular diazotrophs grown in a 12/12 hours light/dark cycle were intermediate between those in nitrate-grown cells and cells grown with dinitrogen in continuous light (Fig. 1A). Respiratory protection of nitrogenase in the former may therefore be less efficient than in the latter (cf. Fig. 3B and Table 2).

Elevated cytochrome oxidase activities in diazotrophic unicells (Fig. 1A) were paralleled by increased rates of whole cell respiration (Table 1), both being stimulated by factors of

*Table 2.* Cell yields ($\mu$g chl/mL culture) after nitrate-supported or diazotrophic growth under different light regimes

| | Continuous light | | 12/12 hrs light/dark | |
|---|---|---|---|---|
| | $NO_3^-$ | $N_2$ | $NO_3^-$ | $N_2$ |
| *Gloeothece* 6909 | 36 ± 5 | 34 ± 6 | 37 ± 4 | 27 ± 6 |
| *Gloeothece* 6909 She⁻ | 11 ± 2 | 12 ± 3 | 11 ± 2 | 7 ± 2 |
| *Cyanothece* 7822 | 44 ± 6 | 43 ± 7 | 45 ± 4 | 36 ± 7 |

Cultures were inoculated with cells equivalent to 1 $\mu$g chlorophyll and grown for ten or twenty days in continuous light (5 w.m⁻²) or a 12/12 hours light-dark cycle, respectively, at 22°C.

6–10 (if calculated on the basis of total cellular protein) and quantitatively compatible with each other. Table 1 also indicates that, with exceedingly high oxygen uptake rates the massive

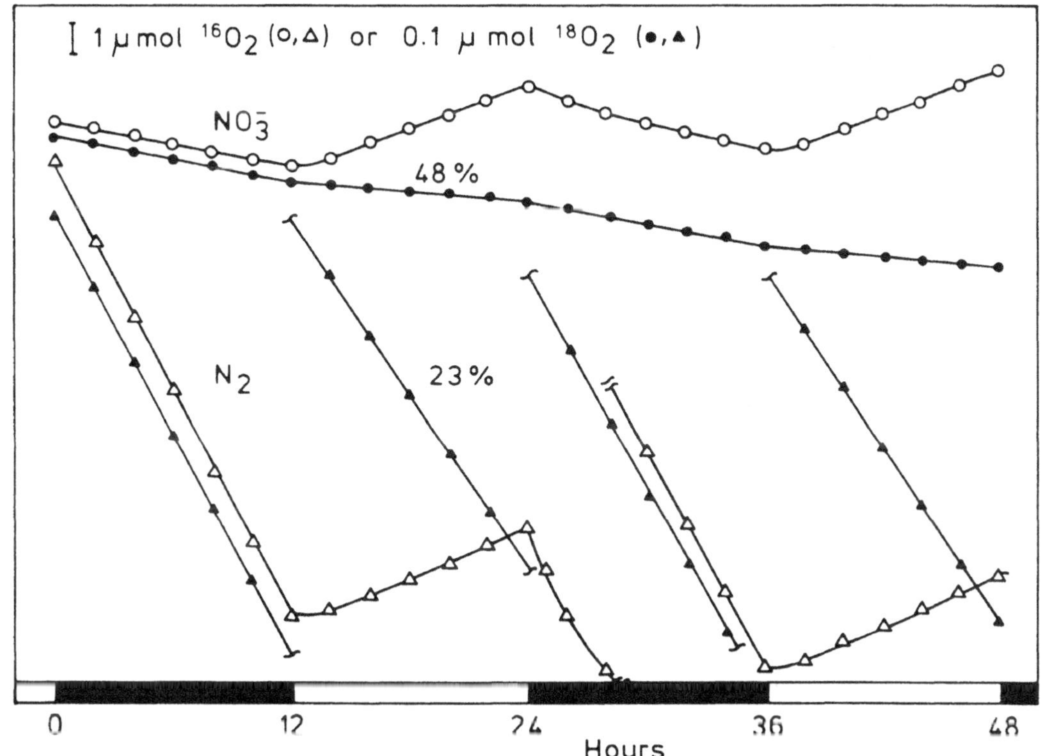


*Fig. 2.* Mass spectrometrically determined exchange of light (○, △) and heavy (●, ▲) oxygen by *Gloeothece* 6909 She⁻ grown with nitrate (○, ●) or dinitrogen (△, ▲) in continuous light, harvested and suspended in 10 m*M* K-Phosphate buffer (pH 7.6). During the experiment, the (nongrowing) cells were subjected to a light-dark cycle (light intensity was 5 W m⁻²; temperature was 22°C). The numbers of 48% and 23% indicate the degree of inhibition of net oxygen uptake (respiration) by illumination in nitrate and dinitrogen grown cells, respectively. The use of heavy oxygen in the gas phase permitted the discrimination between apparent photosynthetic oxygen evolution and specifically respiratory oxygen uptake in the light. For details cf. Materials and methods; also cf. the text.

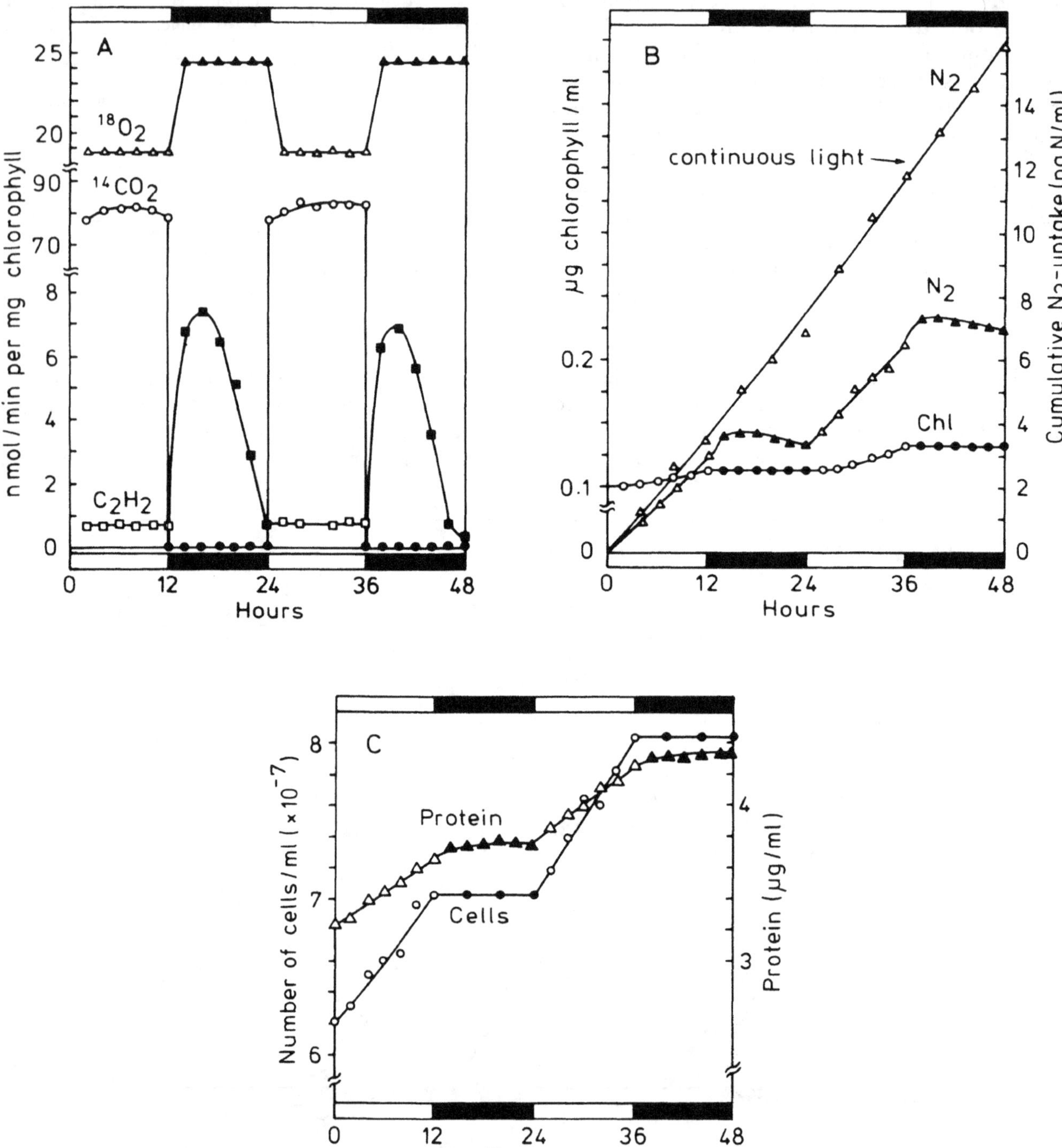

*Fig. 3.* (**A**) Rates of $^{18}O_2$ uptake (determined by mass spectrometry), $CO_2$ fixation (determined radiochemically), and acetylene reduction (determined by gas chromatography); (**B**) cumulative uptake of nitrogen (determined by mass spectometry) and chlorophyll synthesis; and (**C**) increase of cell numbers and protein, of diazotrophically growing cultures of *Gloeothece* 6909 She⁻ growing in a light-dark cycle of 12:12 hours. Experiments of (**B**) and (**C**) were directly performed on growing cell suspensions; for comparison, a culture growing in continuous light is also shown. Experiments (**A**) were performed on harvested and washed cells resuspended in 10 m*M* K-phosphate buffer (pH 7.6). Temperature was 22°C throughout; the light intensity was 5 W m⁻². The generation time of *Gloeothece* 6909 She⁻ under continuous light was about 65 hours.

sheath layers may become rate-limiting to the entry of atmospheric oxygen into the cells and, thus, to the overall process of respiration. Effects of the uncoupler CCCP on diazotrophic cells compared to nitrate-grown cells (Table 1) suggest that respiratory electron transport in the former is largely uncoupled from ATP synthesis like in *Azotobacter* sp. (Dalton and Postgate, 1969; Jones et al., 1973). This would conform to the notion that CM-respiration in cyanobacteria primarily serves trans-CM transport (and other energy-dependent) processes (e.g. proton-sodium antiport; Fry et al., 1986; Erber et al., 1986) rather than ATP synthesis. Alternatively though less probably the rate of respiratory electron flow in (the CM of) diazotrophic unicellular cyanobacteria might be limited by some component other than the ATPase.

That it is in fact the CM that gave rise for the major fraction of diazotrophically stimulated respiration (Fig. 1A) was clearly seen also in vivo with whole cells when photosynthetic evolution and respiratory uptake of oxygen in light and darkness were discriminated by mass spectrometric analysis of isotopically labeled oxygen in the gas phase (Fig. 2): The well-known and typically 'prokaryotic' inhibition of respiration in the light was much less pronounced with dinitrogen-grown cells than with nitrate-grown cells (inhibition by 23 and 48%, respectively). This indicates that a major fraction of respiratory electron transport is catalyzed by the CM which, due to the lack of chlorophyll (Murata and Omata, 1988; Peschek et al., 1989b) does not sense the light.

The lower efficiency of respiratory nitrogenase protection in unicellular diazotrophs grown under a 12/12 hours light/dark cycle compared to continuous light, which was concluded from correspondingly lower cytochrome oxidase activities (Fig. 1A), was reflected by a significantly lower growth yield under the former conditions (Table 2) and by similarly lower rates of nitrogen uptake determined by mass spectrometry (Fig. 3B). Fig. 3 also shows that, during the dark periods of cultures growing under a light-dark cycle there was no carbon dioxide fixation (A), chlorophyll synthesis (B) or cell division (C); likewise, nitrogen fixation (B) and protein synthesis (C) were negligible in the dark. Interest-

ingly, however, nitrogenase activity measured gas chromatographically with acetylene as artificial substrate (A) did not follow the pattern of true nitrogen uptake determined by mass spectrometry (C). In the present experiments, therefore, the rate-limiting step of nitrogen fixation might not be the activity of the nitrogenase per se but rather the biosynthesis of suitable carbon skeletons for incorporation of the ammonia-nitrogen. In obligately phototrophic (unicellular) cyanobacteria respiratory carbon metabolism as a whole (in contrast to respiratory electron transport; cf. Fig. 1) appears to be insufficient to provide these intracellular N-acceptor molecules. However, since diazotrophic growth in continuous light (with continuous photosynthetic oxygen evolution and carbon dioxide fixation but also with continuous, though very slow, dinitrogen fixation; cf. Fig. 3) is even more efficient than growth in a light-dark cycle (cf. Table 2) it must be concluded that 'respiratory protection' of the nitrogenase by way of greatly enhanced respiratory oxygen uptake is the predominant strategy of dinitrogen-fixing (unicellular) cyanobacteria.

## Acknowledgements

This work was supported by grants from the Austrian Science Foundation and the Kulturamt der Stadt Wien (to G.A.P.). We are grateful to Irene Steininger, Doris Zinner and Otto Kuntner for excellent technical assistance.

## References

Almon H and Böger P 1988 Nitrogen and hydrogen metabolism: Induction and measurement. Methods Enzymol. 167, 459–467.

Bradford M M 1976 A rapid and sensitive method for the quantitation of microgram quantities of protein utilizing the principle of protein-dye binding. Anal. Biochem. 72, 248–254.

Carpenter E J and Price C C 1976 Marine *Oscillatoria* (*Trichodesmium*): Explanation for aerobic nitrogen fixation without heterocysts. Science 191, 1278–1280.

Dalton H and Postgate J R 1969 Effect of oxygen on growth of *Azotobacter chroococcum* in batch and continuous cultures. J. Gen. Microbiol. 54, 463–473.

Dimon B, Gans P and Peltier G 1988 Mass spectrometric

measurement of photosynthetic and respiratory oxygen exchange. Methods Enzymol. 167, 686–691.

Erber W W A, Nitschmann W H, Muchl R and Peschek G A 1986 Endogenous energy supply to the plasma membrane of dark aerobic cyanobacterium *Anacystis nidulans*: ATPase-independent efflux of H⁺ and Na⁺ from respiring cells. Arch. Biochem. Biophys. 247, 28–39.

Ernst A, Kirschenlohr H, Diez J and Böger P 1984 Glycogen content and nitrogenase activity in *Anabaena variabilis*. Arch. Microbiol. 140, 120–125.

Fry I V, Huflejt M, Erber W W A, Peschek G A and Packer L 1986 The role of respiration during adaptation of the freshwater cyanobacterium Synechococcus 6311 to salinity. Arch. Biochem. Biophys. 244, 686–691.

Gallon J R and Chaplin A E 1988 Nitrogen fixation. *In* Biochemistry of the Algae and Cyanobacteria. Eds. L J Rogers and J R Gallon. pp 147–173. Oxford Science Publication, Clarendon Press, Oxford.

Haselkorn R 1978 Heterocysts. Annu. Rev. Plant Physiol. 29, 319–344.

Hoch G, Owens O H and Kok B 1963 Photosynthesis and respiration. Arch. Biochem. Biophys. 101, 171–180.

Jensen B B and Cox R P 1988 Measurement of hydrogen exchange and nitrogen uptake by mass spectrometry. Methods Enzymol. 167, 467–474.

Jones C W, Brice J M, Wright V and Ackrell B A C 1973 Respiratory protection of nitrogenase in *Azotobacter vinelandii*. FEBS Lett. 29, 77–81.

Kallas T, Rippka R, Coursin T, Revière M-C, Tandeau de Marsac N and Cohen-Bazire G 1983 Aerobic nitrogen fixation by nonheterocystous cyanobacteria. *In* Photosynthetic Prokaryotes – Cell Differentiation and Function. Eds. G C Papageorgiou and L Packer. pp 281–302. Elsevier Biomedical Publishers, New York.

Lang N J and Fay P 1971 The heterocysts of blue-green algae. II. Details of ultrastructure. Proc. Roy. Soc. Lond. B. 178, 193–203.

Mackinney G 1941 Absorption of light by chlorophyll solutions. J. Biol. Chem. 140, 315–322.

Mitsui A, Kumazawa S, Takahashi A, Ikemoto H, Cao S and Arai T 1986 Strategy by which nitrogen-fixing unicellular cyanobacteria grow photoautotrophically. Nature 323, 720–722.

Molitor V and Peschek G A 1986 Respiratory eletron transport in plasma and thylakoid membrane preparations from the cyanobacterium *Anacystis nidulans*. FEBS Lett. 195, 145–150.

Mortenson L E and Thorneley R N F 1979 Structure and function of nitrogenase. Annu. Rev. Biochem. 48, 387–418.

Murata N and Omata T 1988 Isolation of cyanobacterial plasma membranes. Methods Enzymol 167, 245–251.

Pearson H W and Howsley R 1980 Concomitant photoautotrophic growth and nitrogenase activity by cyanobacterium *Plectonema boryanum* in continuous culture. Nature 288, 263–265.

Pearson H W, Howsley R, Kjeldsen C K and Walsby A E 1979 Aerobic nitrogenase activity associated with a non-heterocystous filamentous cyanobacterium. FEMS Microbiol. Lett. 5, 163–167.

Peschek G A (1987) Respiratory electron transport. *In* The Cyanobacteria. Eds. P Fay and C Van Baalen. pp 119–161. Elsevier Biomedical Publishers, Amsterdam.

Peschek G A, Wastyn M, Trnka M, Molitor V, Fry I V and Packer L 1989 Charaterization of the cytochrome c oxidase in isolated and purified plasma membranes from the cyanobacterium *Anacystis nidulans*. Biochemistry 28, 3057–3063.

Peschek G A, Molitor V, Trnka M, Wastyn M and Erber W 1988 Characterization of cytochrome-c oxidase in isolated and purified plasma and thylakoid membranes from cyanobacteria. Methods Enzymol. 167, 437–449.

Peschek G A, Hinterstoisser B, Wastyn M, Kuntner O, Pineau B, Missbichler A and Lang J 1989 Chlorophyll precursors in the plasma membrane of a cyanobacterium, *Anacystis nidulans*. J. Biol. Chem. 264, 11827–11832.

Radmer R and Ollinger O 1980 Measurement of the oxygen cycle: the mass spectrometric analysis of gases dissolved in a liquid phase. Methods Enzymol. 69, 547–560.

Robson R L and Postgate J R 1980 Oxygen and hydrogen in biological nitrogen fixation. Annu. Rev. Microbiol. 34, 183–207.

Stal L J and Krumbein W E 1985 Oxygen protection of nitrogenase in the aerobically nitrogen-fixing, non-heterocystous cyanobacterium *Oscillatoria* sp. Arch. Microbiol. 143, 72–76.

Stanier R Y and Cohen-Bazire G 1977 Phototrophic prokaryotes: The cyanobacteria. Annu. Rev. Microbiol. 31, 225–274.

Stewart W D P 1980 Some aspects of structure and function in $N_2$-fixing cyanobacteria. Annu. Rev. Microbiol. 34, 497–536.

Wastyn M, Achatz A, Molitor V and Peschek G A 1988 Respiratory activities and $aa_3$-type cytochrome oxidase in plasma and thylakoid membranes from vegetative cells and heterocysts of the cyanobacterium Anabaena ATCC 29413. Biochim. Biophys. Acta 935, 217–224.

Winkenbach F and Wolk C P 1973 Activities of enzymes of the oxidative and the reductive pentose phosphate pathways in heterocysts of a blue-green alga. Plant Physiol. 52, 480–484.

Yates M G and Jones C W 1974 Respiration and nitrogen-fixation in Azotobacter. Adv. Microb. Physiol. 11, 97–135.

PHOTOSYNTHETIC OXYGEN EVOLUTION IN THE FILAMENTOUS CYANOBACTERIUM
OSCILLATORIA CHALYBEA: INTERRELATIONSHIP BETWEEN WATER SPLITTING,
HYDROGEN PEROXIDE DECOMPOSITION AND NITRATE METABOLISM


KLAUS P. BADER and GEORG H. SCHMID
Faculty of Biology, University of Bielefeld,
P.O. Box 8640
D-4800 Bielefeld 1
Germany



ABSTRACT.

Oscillatoria chalybea exhibits very unusual oxygen evolution patterns.
By mass spectrometric analysis we observed an $O_2$-uptake phenomenon leading to the formation of hydrogen peroxide with its subsequent and immediate decomposition upon flash or low light illumination. The phenomenon
is clearly photosystem II mediated and shows a strong oxygen dependency.


## INTRODUCTION

Photosynthetic oxygen evolution requires the successive accumulation
of four positive charges in the so-called S-state system (S-states=$S_i$;
i=0-4), before molecular oxygen is evolved. Under experimental conditions, single oxidation reactions can be obtained by illumination with
saturating but short light flashes in the microsecond-range. With such
assays flash sequences with no signal under the first flash, a maximum
under the third flash, an oscillatory periodicity of four and a more
or less pronounced damping are usually observed. These experiments are
the basis of the now generally accepted Kok-model. Still today, the chemical nature of the S-states or the structure, in which the charges are
stabilized, is unknown. A candidate in this context seems to be a manganese cluster able to perform differing valence transitions. By the
choice of the suitable organism, the preparation method and addition
of the appropriate chemicals it is possible, however, to study the oxygen-evolving complex under modified conditions such as with completely different redox- and S-state distributions.

## MATERIAL AND METHODS

For the above mentioned purposes we chose Oscillatoria chalybea, a
filamentous cyanobacterium performing oxygenic photosynthesis, for our
investigations. With this organism we observed a number of peculiarities
of the oxygen evolving system and electron transport. Oscillatoria chalybea was cultivated on clay plates in large Petri-dishes with culture
medium containing 1 g/l $NaNO_3$. Cultures from this medium were referred

420

to as 'nitrate-grown'. For the respective experiments, _Oscillatoria cha-
lybea_ was cultivated also on media in which nitrate had been substituted
by nitrogen-equivalent concentrations of ammonium sulfate, amino acids
or other nitrogen sources. Thylakoid preparations and thylakoid particle
preparations were prepared according to Bader and Schmid (1990).

_Oxygen Electrode Measurements_ were performed with a "Three Electrode-
System" conceived by Schmid and Thibault. A kinetic analysis of the signals due to photosynthetic oxygen evolution has been recently described by Schulder et al. (1990).

_Mass Spectrometric Analysis_ of the oxygen gas exchange was carried out
with a largely modified magnetic sector field mass spectrometer 'type
Delta' from Finnigan MAT, Bremen, which is a stable isotope ratio mass
spectrometer with a two-directional focusing device 'Nier type I'. The
essential laboratory-built adaptations to dynamic and extremely sensitive measurements of flash-induced oxygen and nitrogen gas exchanges
comprised a direct entry into the ion source by a specific short-cut
of the inlet system and the special geometry of the measuring cell. Details of the technique and the measuring device have been described
elsewhere (Bader et al. 1987, Renger et al. 1990).

RESULTS

When thylakoid preparations from the filamentous cyanobacterium _Oscilla-
toria chalybea_ grown on nitrate are exposed to a train of short saturating light flashes, a specific oxygen evolution pattern is observed
showing a variety of peculiarities in comparison to that usually described for _Chlorella_ or higher plant choroplasts (Fig.1).

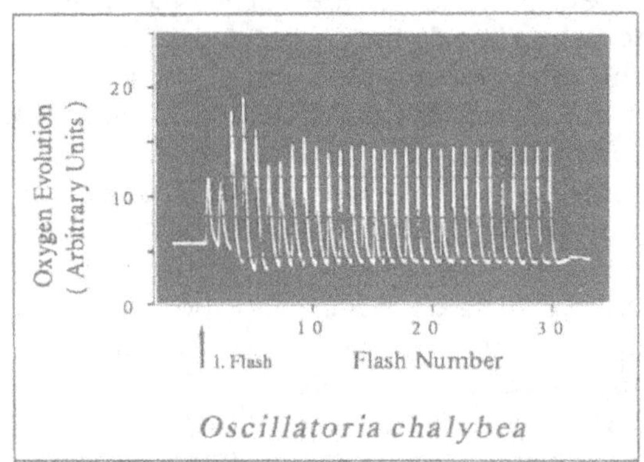

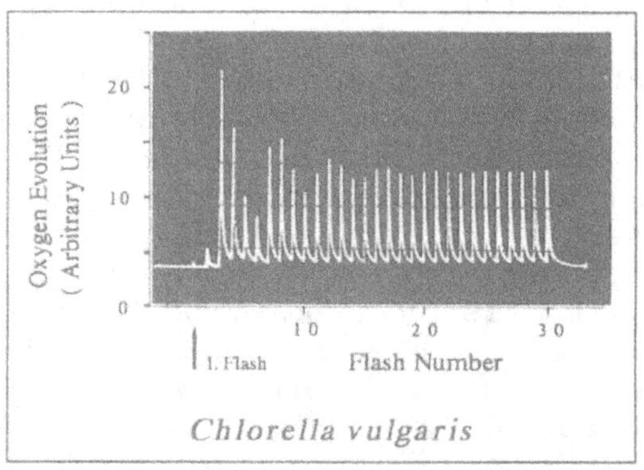


Figure 1. Polarographic recording of the oxygen gas exchange with thylakoids from _Oscillatoria chalybea_. The assay was exposed to a series
of saturating xenon flashes of a duration of 5 us (at half intensity)
and spaced 300 ms apart. For comparison a _Chlorella_-sequence is shown
recorded under identical conditions.

Characteristic for a typical _Oscillatoria_ pattern is the important signal under the very first flash of the sequence, generally followed by

the relatively **smaller** second one and with maximum oxygen yield under the fourth flash. Oscillation of <u>Oscillatoria</u>-patterns appears always strongly damped exhibiting nevertheless high steady-state amplitudes. Also characteristic for such sequences are relatively slow deactivation kinetics facilitating experiments which are otherwise difficult to perform due to the short life-times of the highly oxidized redox states $S_2$ and $S_3$ (Bader et al. 1987). Moreover, the flash pattern of <u>Oscillatoria</u> can be significantly influenced and altered by addition of alkylbenzyldimethylammonium chloride (ABDAC). This compound obviously interacts with the oxygen evolving complex of <u>Oscillatoria</u>. Upon addition of ABDAC the anyhow slow deactivation kinetics are further slowed down (hence stabilization of the S-states). Also, the oscillation of the sequence is strongly improved by reduction of the miss parameter (alpha) to about one half (Bader 1989). Table I shows a typical S-state distribution for <u>Oscillatoria chalybea</u> together with the transition probabilities alpha (misses) and gamma (double hits).

TABLE 1: S-State Distribution in Particle Preparations from *Oscillatoria chalybea*.. Values are derived from a mathematical fit according to the 4-state Kok model.

| Values for dark distribution (%) | | | | Transition probabilities | | relativ quadratic |
|---|---|---|---|---|---|---|
| $S_0$ | $S_1$ | $S_2$ | $S_3$ | misses ($\alpha$) | double hits ($\gamma$) | deviation $\Delta$ (%) |
| 46.2 | 42.0 | 4.3 | 7.5 | 30.3 | 1.8 | 0.96 |

From the fit it is clearly seen that a substantial percentage of meta-stable $S_3$ is present in this organism – a result which was also obtained when fitting the values in a 5-state Kok-model (results not shown). One of the possible interpretations for a signal under the first flash of a sequence, namely the light induced inhibition of an oxygen uptake process or interference between photosynthetic and respiratory electron transport could be clearly ruled out (Ref. 2 in Bader et al. 1987). In order to be able to further investigate this and other gas exchange phenomena we undertook mass spectrometric measurements. Under appropriate conditions, this technique represents the only direct approach for the independent **and** simultaneous quantitative determination of evolution and uptake phenomena within the same gas exchange. By means of this technique we were able to show that the electrochemically described signal under one single flash after dark adaptation really represents net oxygen (i.e. mass 32) evolution. Furthermore, we were able to show that metastable $S_3$ must be involved in this signal and that molecular water can be exchanged up to the highly oxidized $S_3$-state (Bader et al. 1987).

One of the most striking observations in context with oxygen gas exchange in <u>Oscillatoria chalybea</u> is the one that this organism evolves molecular oxygen of m/e=36 also after being supplemented with just $^{18}O_2$ in the gas phase over the assay. This phenomenon is strongly dependent on the oxygen partial pressure and might under appropriate conditions exceed $O_2$-evolution from water oxidation by far thus representing the bulk of the overall light-induced oxygen evolution in this organism (Fig.2).

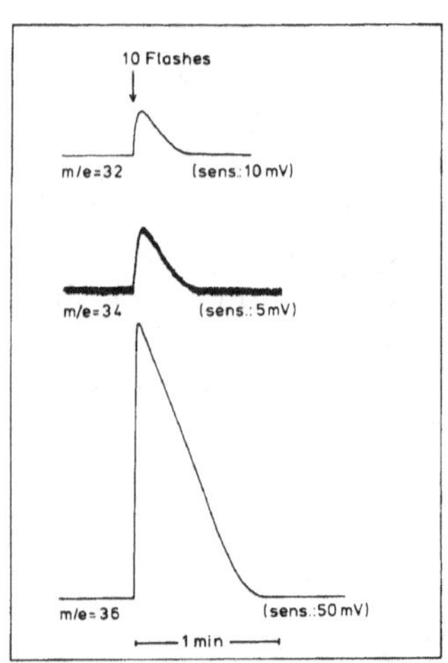

Figure 2. Case of an isotopic distribution of the dioxygen evolved in the presence of oxygen-18.

Such $^{18}O_2$-evolution signals can be observed with cultures which have been cultivated on nitrate, ammonium sulfate or various amino acids. It can even be observed with preparations from cultures that have been grown without any exogenous nitrogen source in the medium (Ref. 2 in Bader and Schmid, 1990). These results together with the observation of a period-2 oscillation of the respective flash patterns led us to the conclusion that <u>Oscillatoria chalybea</u> reacts in a very specific manner: addition of oxygen leads to a rapid oxygen uptake under formation of $H_2O_2$ followed by the immediate re-liberation of this oxygen upon subsequent illumination. Oscillation with a periodicity of 2 in this context hints at the participation of $S_2$ within this model of an $O_2/H_2O_2$-cycle. It could be imagined that in comparison to water splitting the reaction represents for certain metabolic purposes an energetic advantage to the organism, as the oxidation of $H_2O_2$ requires only two light quanta. The oxidation of $H_2O_2$ obviously does not include molecular water as an intermediate, but yields directly protons, electrons and $O_2$. This can be concluded from the negligibly small amount of mixed isotope $^{16}O=^{18}O$ (measured at m/e=34). It can be mathematically calculated that the oxidation of molecular water with both isotopes $H_2^{16}O$ and $H_2^{18}O$ involved at the appropriate concentrations must result in the preponderant oxygen evolution of mass 34. This can be seen from Fig.3.

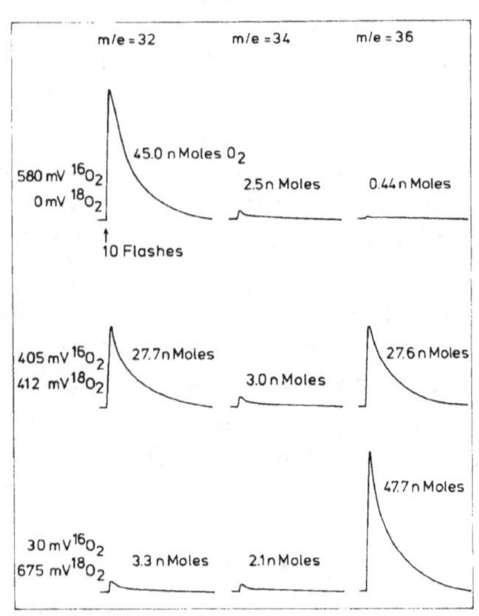


Figure 3. Photosynthetic oxygen evolution as the consequence of 10 light flashes in <u>Oscillatoria chalybea</u>. The assay contained 28% $H_2^{18}O$. After flushing with $N_2$, artificial gas atmosphere conditions were installed. The figure represents monoisotopic conditions with virtually 100% $^{18}O_2$ and 100% $^{16}O_2$, respectively, together with the exactly mixed condition of 50% $^{16}O_2$/50% $^{18}O_2$.
The signals of masses 32 ($^{16}O_2$), 34 ($^{16}O=^{18}O$) and 36 ($^{18}O_2$) were simultaneously recorded and the areas integrated to give the indicated absolute values.

In such assays we added $H_2^{18}O$ (to give a final concentration of 28%) to the liquid phase and investigated the isotopic distribution of the evolved oxygen at different background signals of the gas phase. Depending on whether the gas phase consisted of $^{16}O_2$ or $^{18}O_2$, the contribution of $H_2O_2$-oxidation to the overall gas exchange can be calculated, since the portion of water oxidation can easily be calculated from the expected isotope distribution and subtracted from the respective signal. As the oxygen uptake reaction leading to $H_2O_2$-formation works with each offered oxygen isotope, it appears trivial that we never observed a Kok-type sequence under normal air conditions with our mass spectrometer. Instead, we always observed sequences which could not be fitted in any model (Fig.4).

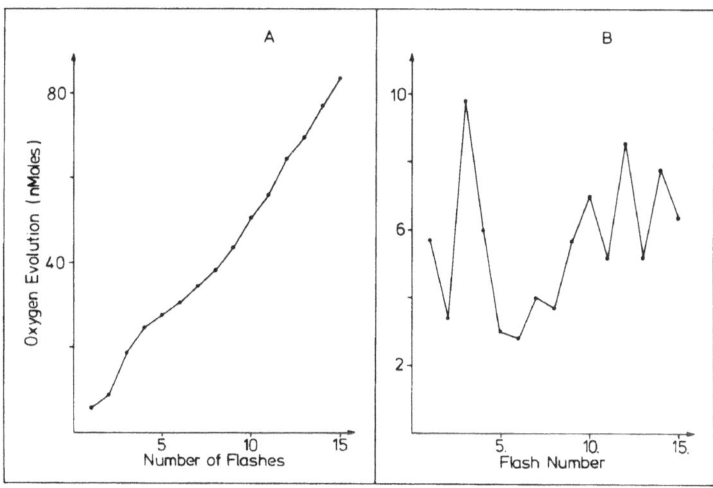

Figure 4. Mass spectrometric analysis of photosynthetic oxygen evolution in thylakoids from <u>Oscillatoria chalybea</u> by a sequence of short saturating light flashes. The oxygen yields were recorded at mass 32 and are depicted as the accumulated signals in dependence on the number of flashes as well as the single-flash resolutions of the respective signals.

Only the periodicity of 2 (<u>vide</u> <u>supra</u>) was repeatedly observed and can be improved by the appropriate assay conditions. A perfect Kok-sequence, however, is immediately obtained, if only the m/e=36 signal of an $H_2^{18}O$- but no $^{18}O_2$-containing assay is recorded, thus representing a condition that excludes interference with $O_2$-uptake and $H_2O_2$-oxidations.(Fig.5).

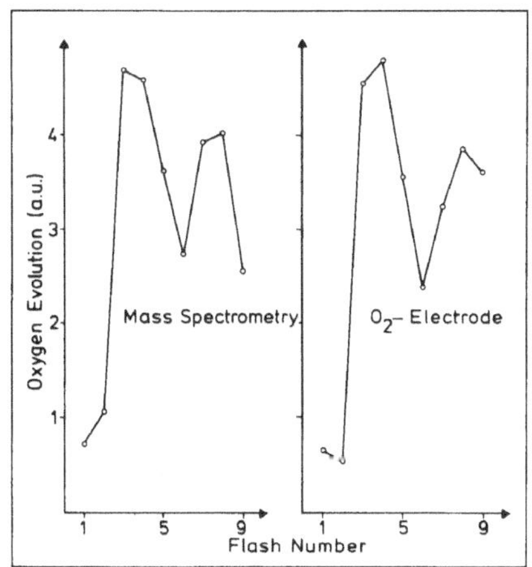


Figure 5.Comparison of the flash-induced oxygen signals in a thylakoid preparation of <u>Oscillatoria chalybea</u> measured by amperometry and mass spectrometry. The mass spectrometric assay contained $H_2^{18}O$ but no $^{18}O_2$ in the gas phase.

Comparison of this sequence to those recorded on our oxygen electrode shows an absolute identical picture. This might be explained by the assumption that the assay on the $O_2$-electrode due to the high polarization tends strongly towards anaerobiosis. As the $H_2O_2$-oxidation shows a much higher $O_2$-dependency, it appears rational that this process is

largely suppressed under such conditions. Thus, water oxidation clearly prevails so that the observed  signals can be (more or less exclusively) attributed to this reactions.
Recently, we investigated the effect of anilinothiophenes (Ant-2-p) on water oxidation as well as on hydrogen peroxide decomposition. (Fig.6)

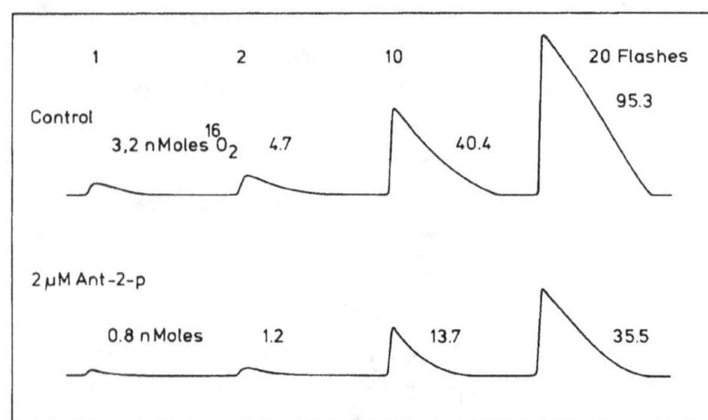


Figure 6. Effect of anilinodinit-rothiophene (ANT-2-p) on the oxygen evolution at m/e=32 in thylakoids from _Oscillatoria chalybea_. The assays were illuminated with 1, 2, 10 and 20 flashes, respectively.

Ant-2-p is one of the so-called ADRY-reagents which according to Renger accelerate the deactivation of the highly oxidized S-states. The results clearly show that the reagent affects both water oxidation and hydrogen peroxide decomposition. As we observed that nitrate reductase is induced in _Oscillatoria chalybea_ not only during growth on nitrate, but also on arginine or citrulline, but not on ammonium sulfate or other amino acids, further investigations concerning the interrelationship between photosynthetic, respiratory and nitrate reducing electron transport will be of particular interest.

<u>REFERENCES</u>


Bader, K.P. (1989) 'Alkylbenzyl dimethyl ammonium chloride, a stabilizer of the S-state system in the filamentous cyanobacterium _Oscillatoria chalybea_', Biochim. Biophys. Acta 975, 399-402.

Bader, K.P., Thibault, P. and Schmid, G.H. (1987) 'Study on the properties of the $S_3$-state by mass spectrometry in the filamentous cyanobacterium _Oscillatoria chalybea_' Biochim. Biophys. Acta 893, 564-571.

Bader, K.P. and Schmid, G.H. (1990) 'Quantitative determination of the distribution between two photosystem II-mediated oxidation reactions in _Oscillatoria chalybea_: A comparison between water oxidation and hydrogen peroxide decomposition', Z. Naturforsch. in press.

Renger, G., Bader, K.P. and Schmid, G.H. (1990) 'Mass spectroscopic analysis of $N_2$-formation by flash-induced oxidation of hydrazine and hydroxylamine in normal and Tris-treated tobacco chloroplasts', Biochim. Biophys. Acta 1015, 288-294.

Schulder, R., Bader, K.P. and Schmid, G.H. (1990) 'An amperometric study on the time constants of oxygen release in thylakoids of _Nicotiana tabacum_ and _Oscillatoria chalybea_', Z. Naturforsch. submitted.

# PHOTOSYNTHESIS AND RESPIRATION IN MIXOTROPHIC CULTURES OF THE CYANOBACTERIUM Anabaena variabilis.


E.FERNANDEZ VALIENTE; E.SANCHEZ MAESO; M.NIEVA
and M.C.AVENDAÑO.
Dpt.Biology. Autonomous University of Madrid.
Cantoblanco. 28049-Madrid.
Spain.



ABSTRACT.The cyanobacterium Anabaena variabilis take up and utilize exogenous fructose in the light simultaneously with $CO_2$. Fructose-supported growth was accompanied by an increase in nitrogenase activity. Oxidation of exogenous fructose was the main source of reductant for dinitrogen fixation. PSI activity was,however, required to obtain maximal nitrogenase activity. Respiration was enhanced by exogenous fructose both in the light and in the dark. The rate of respiration in the light was quite similar to that observed in the dark. The higher rate of $O_2$ consumption in the light seems to be the responsible for the decrease in the rate of net $O_2$ evolution, since photosynthetic electron transport, expressed per unit of chlorophyll, was also enhanced by exogenous fructose. The increase in PSI and PSII activities was probably a way to compensate the lower concentration of chlorophyll showed by fructose-grown cells.


## INTRODUCTION

Although photoautotrophy is the most common mode of growth of cyanobacteria, some strains are also known to assimilate organic compounds both in the light and in the dark (Smith 1982). Mixotrophy is a growth regime in which $CO_2$ and organic carbon are assimilated simultaneously and, hence, both respiratory and photosynthetic metabolism have to operate concurrently. This nutritional regime is distinct from photoheterotrophy in which growth on an organic compound in the light, take place in the absence of net $CO_2$ fixation.

Mixotrophic growth has been reported in a limited number of free-living and symbiotic cyanobacteria. Fructose,glucose and sucrose being the most commonly used substrates (Haury and Spiller 1981;Rozen et al.1988).

Fructose was shown to support growth and dinitrogen fixation in mixotrophic cultures of A.variabilis (Haury and Spiller 1981) and A.azollae (Rozen et al.1986;1988). Fructose-supported dinitrogen fixation suggests that respiration in the light must be enhanced by exogenous fructose. In fact, dark respiration was shown to be higher in mixotrophic cultures (Haury and Spiller 1981; Rozen et al.- 1988), however, no measurement of light respiration was reported in these studies.

Data on the magnitude of respiration in the light are controversial, although most reports state that respiration is strongly suppresed in the light. In a study with nine species of cyanobacteria, cultured under photoautotrophic conditions, Scherer and Böger (1982) showed that $CO_2$ evolved by respiration was immediately refixed in the light without being liberated. This fact means that single estimation of $CO_2$ can minimize the true rate of respiration in the light, which in some of the studied species, including A.variabilis, was considerable.

Information concerning to the photosynthetic activity of cyanobacteria under conditions of sugar supported growth is limited. It was reported in A.variabilis that net oxygen evolution decreased with fructose concentrations higher than 20 mM (Haury and Spiller 1981). Similarly in A.azollae the rate of net oxygen evolution decreased by a 40% in fructose-grown cells. However these cells showed an increase in PSI activity comparing to autotrophycally grown cells (Rozen et al.1988).

In this paper, we present data of a study conducted to investigate the relationships of photosynthesis and respiration in mixotrophic cultures of A.variabilis.

## MATERIALS AND METHODS

Organism and culture conditions. The cyanobacterium Anabaena variabilis ATCC 29413 was used for this study. Photoautotrophic cultures were grown under $N_2$-fixing conditions in the nitrogen-free medium reported previously (Mateo et al.1986). Mixotrophic cultures were grown in the same medium supplemented with 10 mM fructose. Batch cultures were gassed with air or with $CO_2$-enriched air, depending of the experiments, at 28ºC, under a constant light intensity of 90 $\mu E.m^{-2}.s^{-1}$.

Nitrogenase activity. Nitrogenase activity was determined by acetylene reduction as previously reported (Mateo et al.1986).

Oxygen exchange measurement. Oxygen evolution or consumption were monitored with a Clark-type $O_2$ electrode (Hach Chemical Company). Three ml aliquots of cell suspensions, with a cell density of 20 $\mu g$ Chl/ml, were placed in a temperature controlled cuvette and placed in the dark or illuminated with a quantum flux density of 300 $\mu E.m^{-2}.s^{-1}$. Respiratory $O_2$ consumption in the light was determined as the cyanide sensitive fraction of the $O_2$ consumed in the light by cells supplied with 10 $\mu M$ DCMU. KCN was added to a concentration of 2 mM.

<u>Photosynthesis.</u> Measurements of photosynthetic electron transport were made with spheroplasts prepared from auto-trophic and mixotrophic cultures by lysozime treatment. The methyl-viologen-catalysed Mehler reaction was followed as $O_2$ consumption in a reaction mixture containing 50 mM KCl, 0.13 mM methyl viologen, 2 mM KCN, 33 mM HEPES buffer, pH 8 and spheroplasts containing 30 µg Chl/ml. PSI and PSII activities were determined spectrophotometrically follow-ing the reduction of ferricyanide at 420 nm. The reaction mixture for PSI determination contained 5 mM sodium ascor-bate, 50µM 2,6,dichlorophenolindophenol (DPIP),1 mM FeCy, 10 µM DCMU, 25 mM HEPES buffer, pH 7.5 and spheroplasts containing 20 µg Chl/ml. The reaction mixture for PSII determination contained 50 µM DPIP,1 mM FeCy, 25 mM HEPES buffer, pH 7.5 and spheroplasts containing 20 µg Chl/ml. Chlorophyll concentration was estimated in 90% methanol ex-tracts as previously reported (Mateo et al.1986).

<u>Fructose uptake and oxidation.</u> Fructose incorporation was studied with U-$^{14}$C-fructose (290 mCi/mmol) as previously reported (Nieva et al.1990). Oxidation of fructose was estimated by measuring the conversion of U-$^{14}$C-fructose to $^{14}$CO$_2$, following the method previously reported (Nieva et al.1990).

## RESULTS AND DISCUSSION

<u>A.variabilis</u> shows a constitutive mechanism for the uptake of fructose, since non-adapted cells took up fructose without any lag period, with a rate of 280 nmol/mg dry weight x h, namely, a rate similar to that reported for other sugar-utilizing cyanobacteria (Smith 1982). A further activation of the uptake mechanism by its substrate is suggested by the fact that the rate of fructose uptake increased to 387 nmol/mg dry weight x h, in cells grown for 2 days in the presence of fructose. Similar results were reported in the symbiotic strain <u>A.azollae</u> (Rozen et al.-1988).

An increase in nitrogenase activity was observed when fructose was present in the growth medium (Fig.1). The increase was partially reduced by DCMU, but even in the presence of the inhibitor, the rate of nitrogenase activity was higher than in photoautotrophic cultures. These results indicated a very little contribution of photosynthesis in the supply of reducing power to nitrogenase activity in mixotrophic cultures. This contrast with the situation of of photoautotrophic cultures where nitrogenase activity was fully dependent of photosynthesis as indicated the total inhibition of dinitrogen fixation by DCMU. Nitrogenase activity in the dark in the presence of fructose showed an initial lag, and then increased to reach a maximum level at

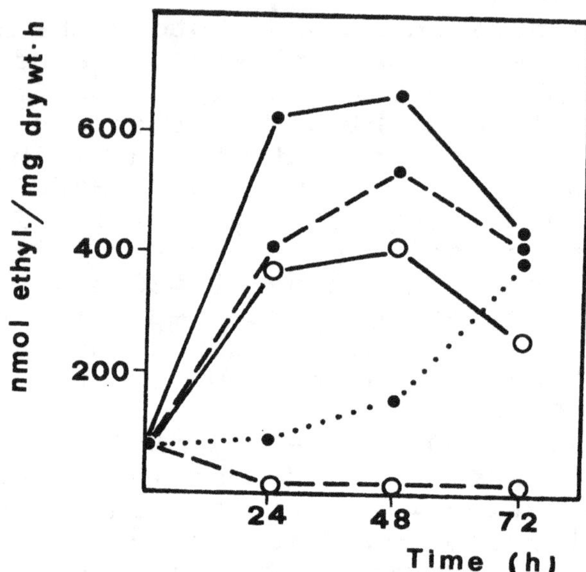

Figure 1. Effect of fructose on nitrogenase
activity of <u>A.variabilis</u>.● Cells grown with
fructose. ○ Cells grown without fructose.
(——)Light. (- - -)Light+DCMU. (·····) Dark.

the third day of culture. This maximum value was lower than
that reached by mixotrophic cultures indicating that a
functional PSI is required to reach maximal values of
nitrogenase activity.

As expected by the enhancement of nitrogenase activity,
exogenous fructose increased the rate of $O_2$-consumption
both in the dark and in the light (Table 1). Dark respiration of cells grown in the light in the presence of fructose was a 36% higher than that of autotrophically grown
cells. This increase was in the range to that reported in
<u>A.azollae</u> (Rozen et al.1988). $O_2$-consumption in the light
was measured in the presence of DCMU to avoid interference
with photosynthetic oxygen evolution. The data indicated
that the rate of $O_2$-consumption in the light was enhanced
by exogenous fructose. In photoautotrophic cultures about a
50% of the oxygen consumed in the light was sensitive to
cyanide and, hence, could be considered as true respiration. In mixotrophic cultures, this fraction sensitive to
cyanide was the only enhanced by exogenous fructose. The
fraction unsensitive to cyanide, which remained unchanged
by the presence of fructose, could be due to photorespira-
tion, since in cultures grown under high $CO_2$ levels this
fraction was not observed and cyanide fully inhibited the
consumption of oxygen in the light (not shown). These
results indicated that <u>A.variabilis</u> actively respire in the
light and that exogenous fructose enhanced respiration in
the dark and in the light.

The study of $CO_2$ evolved from radiolabelled fructose

TABLE 1. $O_2$ consumption (μmol/mg Chl x h) by <u>A.variabilis</u>

| Culture conditions | Dark | Light+DCMU | | |
|---|---|---|---|---|
| | | Total | CN-sensit. | CN-unsensit. |
| Autotrophic | 119 | 111 | 57 | 54 |
| Mixotrophic | 162 | 151 | 103 | 48 |

confirms the high rate of respiration in the light (Table 2). When photosynthesis was blocked by DCMU, cells adapted to growth on fructose evolved similar or even higher amounts of $CO_2$ in the light than in the dark. The amount of $CO_2$ evolved in the light significantly decreased when photosynthesis was active, indicating that most of the $CO_2$ evolved by respiration was reutilized in photosynthesis.

TABLE 2. Conversion of $^{14}C$-Fructose to $^{14}CO_2$ by cells of <u>A.variabilis</u> adapted to growth on fructose.

| Conditions | rate of $^{14}CO_2$ released (counts/mg dry wt x h) |
|---|---|
| Dark | 1443 |
| Light | 527 |
| Light+DCMU($10-_5$) | 1552 |

Net oxygen evolution in the light was shown to be partially inhibited by fructose (Haury and Spiller 1981,Rozen et al.1988). The data were interpreted as result of a decrease in the rate of photosynthetic oxygen evolution. These results, however, could be also discussed in terms of a possible increase in the rate of $O_2$-consumption in the light, which actually occurred, as we have shown above. Our data also indicated that net oxygen evolution was lowered by exogenous fructose (279 μmol/mg Chl x h in photoautotrophic cultures vs 227 μmol/mg Chl x h in mixotrophic cultures). However, if we correct the data taking into account the rate of oxygen consumed in the light (Table 1), we would obtain an estimation of the rate of gross photosynthetic oxygen evolution, which was practically the same in autotrophic (390 μmol/mg Chl x h) and mixotrophic (378 μmol/mg Chl x h) cultures.
From these considerations we can conclude that single estimation of oxygen exchanges could lead to misleading conclusions about the effect of exogenous fructose on photosynthesis. A more direct evidence can be obtained from the study of the photosynthetic electron transport.

TABLE 3. Rates of photosynthetic activities and chlorophyll concentration in <u>A.variabilis</u>. PSI and PSII activities and whole photosynthetic electron transport
were measured as described in Materials and Methods.
Whole electron transport is expressed as $\mu$mol $O_2$ con
sumed/mg Chlxh. PSI and PSII are expressed as nmol FeCy
reduced/mg Chlxh. Chlorophyll is expressed as $\mu$g/mg dry
wtxh.

| Culture conditions | Chl concentr. | Rates of electron transport | | |
|---|---|---|---|---|
| | | Whole | PSI | PSII |
| Autotrophic | 17.3 | 62 | 2.04 | 3.91 |
| Mixotrophic | 11.2 | 78 | 3.41 | 5.79 |

The activity of  the two photosystems, normalized to units
of chlorophyll,  were enhanced  by exogenous  fructose, and
accordingly with this,  the whole activity of the photosynthetic  electron  transport was  also  higher  in fructose-
grown cells (Table  3). This increase in the photosynthetic
electron transport  per unit of chlorophyll could  be a way
to compensate the lower concentration of chlorophyll showed
by cells grown on fructose, in such a way, that the rate of
photosynthesis could be maintained.

ACNOWLEDGEMENTS.
This work was supported by a grant from CICYT (PB86-0323).

REFERENCES
 Haury  JF  and  Spiller  H (1981)  Fructose uptake  and in-
     fluence  on  growth and  nitrogen  fixation  by <u>Anabaena</u>
     <u>variabilis</u>.J.Bacteriol.147,227-235
 Mateo P,Bonilla I,Fernandez-Valiente E and Sanchez Maeso E
     (1986) Essentiality of boron for dinitrogen fixation  in
     <u>Anabaena sp</u> PCC 7119.Plant Physiol.81,430-433.
 Nieva M,Avendaño MC,Sanchez-Maeso E and Fernandez-Valiente
     E.  (1990) Effect  of sodium  deficiency  on mixotrophic
     cultures of <u>A.variabilis</u>.J.Plant Physiol.In press.
 Rozen A,Schönfeld M and Tel-Or E  (1988) Fructose-enhanced
     development  and  growth  of  the  $N_2$-fixing  cyanobiont
     <u>Anabaena azollae</u>.Z.Naturforsch.43c,408-412.
 Scherer  S and  Böger P  (1982) Respiration  of blue-green
     algae in the light.Arch Microbiol.132,329-332.
 Smith AJ (1982) Modes of cyanobacterial carbon metabolism.
     In:Carr NG,Whitton BA(eds)The Biology of Cyanobacteria.
     Blackwell Sci.Publ.Oxford pp47-85.

# A POSSIBLE ROLE OF FLAVODOXIN IN NITROGEN FIXATION IN HETEROCYSTS FROM *Anabaena*..


M.C. Lázaro, M.F. Fillat, C.Gómez-Moreno and M.L.Peleato.
Departamento de Bioquímica y Biología Molecular y Celular.
Facultad de Ciencias. Universidad de Zaragoza. 50.009-
Zaragoza, Spain .



ABSTRACT. The possible role of flavodoxin in nitrogen fixation of *Anabaena* has been investigated, determining the levels of ferredoxin and flavodoxin in different nutritional conditions. It has been observed that ferredoxin decreases both in whole filaments and in heterocysts of cells grown in the presence and in the absence of a fixed nitrogen source, as a consequence of the decrease of the iron available. Consequently, flavodoxin increases in these conditions since it is supposed to replace the iron protein when it can not be synthetized. The amount of flavodoxin in  heterocysts of cells actively fixing nitrogen is, nevertheless, not affected by the level of iron in the culture. The enzyme ferredoxin-NADP$^+$reductase is also present in heterocysts of cells which are actively fixing nitrogen, indicating the possible role of this reductase in providing electrons to ferredoxin  (or flavodoxin) which will be ultimately used by nitrogenase.


## 1. Introduction

The requirement of iron in the culture medium for ferredoxin synthesis has been described in different microorganisms. It has also been  noticed that under iron deficient conditions,,  some of them synthetize a low molecular weight flavoprotein, named flavodoxin ( Smillie, 1965). It is proposed that flavodoxin replaces ferredoxin _in vivo_ in those reactions in which this iron protein participates (Yoch and Valentine, 1972). Several  _in vivo_ and _in vitro_ studies have been done on the effect of the substitution of ferredoxin by flavodoxin in the transfer of electrons between photosystem I and the enzyme ferredoxin NADP$^+$ reductase (FNR) (Trebst and Bothe, 1966, Fillat _et al_, 1988, Sandmann _et al_ 1990).  Flavodoxin  has also been shown to be very efficient _in vitro_ in providing electrons to nitrogenase in the acetylene reduction assay of this enzyme (Fillat _et al_, 1988). In some organisms such as *Klebsiella* and

*Azotobacter*, flavodoxin has been shown to be  the physiological electron donor to nitrogenase (Deistung and Thorneley, 1986),  in spite of the presence of ferredoxin which is probably  involved  in other functions (Bennett et al., 1987). Ferredoxin is accepted to be the immediate electron carrier to nitrogenase in cyanobacteria (Bothe et al. 1982, Scherer et al, 1988). This is supported by finding that  a specific ferredoxin isolated by Bohme and Schrautemeier (1987) from heterocysts of *Anabaena* ATCC 29413, which is different from that found in vegetative cells, is more effective  at reducing nitrogenase than the protein isolated from vegetative cell and also by the fact that it was found to be the product of  a gene which is expressed only under nitrogen-fixing conditions (Bohme and Haselkorn, 1988).
Our group has reported that flavodoxin replaces ferredoxin in cells that were grown under iron deficient conditions  (Sandmann et al, 1990). While the rate of photosynthesis was found to be equivalent in cells containing flavodoxin and in those with ferredoxin, the nitrogenase activity was significantly lower in cells grown in iron deficient media.
In order to study the substitution  of ferredoxin by flavodoxin in  heterocysts under iron deficient conditions and to determine the physiological role of both carriers, and that of  the enzyme FNR in cyanobacteria, heterocysts were isolated from cells cultured in different iron conditions.  Ferredoxin, flavodoxin and FNR were immunochemically assayed in these cells in order to determine the level of these proteins, and to ascertain their possible function in nitrogen fixation in heterocystous cyanobacterium.


## 2. Material and Methods


*Anabaena* PCC 7119  were grown as described (Sandmann and Boger, 1980) in media containing different iron concentrations, at 22ºC under light intensities of 4000 lx. Cells were cultured in the presence or absence of nitrogen in the media, according to the experimental purposes. Heterocysts  were isolated as described by Fay (1980), and prior to the heterocyst disruption, the cells were washed three times to avoid  contamination with egetative cells proteins.  The purity of heterocysts preparation was checked by light microscopy.

Antibodies against flavodoxin, ferredoxin and ferredoxin NADP+ reductase were obtained using purified proteins  isolated from whole filaments (Fillat et al, 1988). Proteins in crude extracts were quantified by the method of Laurell (1966) . At least three different determinations were performed for each data point. Chlorophyll was determined by spectroscopy using the Parsons and Strickland (1973) method.

## 3. Results and discussion.

In order to study the replacement of ferredoxin by flavodoxin in whole filaments of *Anabaena* as well as in heterocysts, aliquots of the cultures were grown in complete or iron depleted media. Crude extracts from 10 litre cultures at the exponential phase of growth were used to immunochemically determine the level of the different proteins that participate in the nitrogenase reduction pathway . Table 1 shows the data obtained for flavodoxin indicating the large increase that takes place in whole filaments when cells are grown in iron-deficient media, either in the presence of nitrogen or under nitrogen-fixing conditions. Surprisinglly, the results obtained with isolated heterocysts were quite different, since similar amounts of flavodoxin were found in heterocysts of cells grown under nitrogen fixing conditions, independently of the amount of iron in the culture medium.

Table 1.  Flavodoxin content in cells of *Anabaena* PCC 7119 grown in different conditions.  The results are expressed in nmol of flavodoxin by mg of chlorophyll.

| Cells grown in: | | Flavodoxin in: | |
| --- | --- | --- | --- |
| Nitrogen (mM) | Iron ($\mu$M) | whole filaments (nmol / mg cl) | heterocysts (nmol / mg cl) |
| ----- | 18 | 0.65 | 0.43 |
| ---- | 0.54 | 3.08 | 0.42 |
| 18.5 | 18 | 0.003 | 0.05 |
| 18.5 | 0.54 | 2.15 | 0.33 |

In the presence of a nitrogen source, that is, when nitrogenase is not active, an increase in the amount of flavodoxin found  in the  heterocysts is observed as the concentration of iron in the medium decreases.  The very small amount of heterocysts that can be isolated in the presence of nitrate raises some questions on the physiological meaning of these values.  It is quite clear, nevertheless, that heterocysts of *Anabaena* that are actively fixing nitrogen show a constitutive level of flavodoxin that is independent of  the amount  of iron in the culture.

The level of ferredoxin in whole filaments and heterocysts of cells grown in different nutritional conditions was also compared. Results in Table 2 indicate that the level of ferredoxin is always lower when iron is a limiting component of the culture. This occurs both in whole filaments and in heterocysts and also under nitrogen fixing conditions and when there is nitrate present in the culture. A significant increase in the amount of ferredoxin found in heterocysts of cells actively fixing nitrogen indicates  this protein  is involved in the transfer of electrons to nitrogenase.

**Table 2. Cross-reaction with ferredoxin antibodies of crude extracts from *Anabaena* cells grown in different conditions.**

| Cells grown in: | | Ferredoxin (nmol / mg chlorophyll) | |
| --- | --- | --- | --- |
| Nitrogen (mM) | Iron (µM) | whole filaments | Isolated heterocysts |
| – | 18 | 8 | 7.08 |
| -- | 0.54 | 1.23 | 2.38 |
| 18.5 | 18 | 3.74 | 1.13 |
| 18.5 | 0.54 | 0.75 | 0.77 |

It coult be argued  that the specific ferredoxin isolated from heterocysts is not detected by our immunochemical assay provided that it does not cross-react with the antibody against ferredoxin from vegetative cells. It should be mentioned that the decrease in the level of ferredoxin observed in heterocysts of cells grown in iron-limited media and under nitrogen-fixing conditions, does not correspond with the increase in the level of flavodoxin  mentioned above. This could explain the lower nitrogenase activity  detected <u>in vivo</u> in cells grown in an iron deficient medium. We can conclude then, that ferredoxin and flavodoxin behave in heterocysts of *Anabaena*  as alternative pathways for donating electrons to nitrogenase.

Table 3 shows data on the level of ferredoxin-NADP$^+$ reductase in whole filaments and isolated heterocysts of cells grown under nitrogen fixing conditions and also in the presence of nitrate both at saturating and limiting concentrations of iron in the culture medium. A slight decrease in the level of FNR is observed in cells grown in the media with low concentrations of iron.

 This difference could be interpreted as a consequence of the stress that is produced in the metabolism as a consequence of the limiting concentrations of iron. What seems to be more important  is the increased level of FNR found in heterocysts of cells actively fixing nitrogen. A two-fold difference was observed with respect to whole filaments and  a four-fold  difference with respect to the few heterocysts present when there is a nitrogen salt in the medium.

Several reductases have been proposed to be involved in the transfer of electrons to nitrogenase. Neuer and Bothe (1982) documented the presence of a pyruvate-ferredoxin oxidoreductase in heterocysts but very low nitrogenase activity was detected with pyruvate as the electron donor. Ferredoxin-NADP$^+$ reductase, which is present in all photosynthetic organisms could catalyze the

Table 3.   FNR or protein cross-reactive with FNR-antibody detected in crude extracts of cells grown in different nutritional conditions. The results are expressed in nmol / mg clorophyll. The cells were collected in the exponential phase, with FNR active.

| Cells grown in: | | FNR | |
| --- | --- | --- | --- |
| Nitrate | Iron | whole   filaments | Isolated heterocysts |
| (mM) | (µM) | (nmol / mg  chlorophyll) | (nmol / mg  chlorophyll) |
| - - - - - | 18 | 1.15 | 2.01 |
| - - - - - | 0.54 | 0.73 | 2.29 |
| 18.5 | 18 | 1.02 | 0.49 |
| 18.5 | 0.54 | 0.54 | 0.42 |

reduction of ferredoxin (or flavodoxin), with NADPH  which is generated via the pentose phosphate pathway. This activities have been shown to be operative in the dark in whole filaments (Schrautemeier _et al._,1984  ) but our data very strongly indicate that it could be a major pathway in providing reducing power to nitrogenase in heterocysts.

Acknnowledgements: this work was supported by grant 0397.E from  the EEC and grant BIO-8810413 from CICYT, Spain.

# 4.References

Bennett, L.T., Jacobson, M.R., and Dean, D.R. (1987). Isolation, sequencing, and mutagenesis of the _nif_  F gene encoding flavodoxin from _Azotobacter vinelandii_.  J.Biol.Chem. _263_, 1364-1369.

Bohme, H. and Schrautemeier,B. (1987). Comparative characterization of ferredoxins from heterocysts and vegetative cells of _Anabaena variabilis_. Biochim.Biophys. Acta, _891,_1-7.

Bohme, H. and Haselkorn, R. (1988). Molecular cloning and nucleotide sequence analysis of the gene coding for heterocysts ferredoxin from the cyanobacterium _Anabaena_  sp. strain PCC 7120. Mol.Gen.Genet. _214, _278-285.

Bothe, H.,Yates,M.G. and Cannon,F.C. (1982). Nitrogen fixation. In Lauchli,A. and Bieleski,R.(Eds) Encyclopedia of Plant  Physiology. New Series, Springer, Berlin.

Deistung, J., and Thorneley, R.N.F. (1986). Electron transfer to nitrogenase. Characterization of flavodoxin from *Azotobacter chroococum* and comparison of its redox potentials with those of flavodoxins from *Azotobacter vinelandii* and *Klebsiella pneumoniae* (*nif* F-gene product). Biochem.J. 239, 69-75.

Fay, P. (1980). Heterocyst isolation. Methods Enzymol. 69, 801-803.

Fillat, M.F., Sandmann, G. and Gómez-Moreno, C. (1988). Flavodoxin from the nitrogen-fixing cyanobacterium *Anabaena* PCC 7119. Arch.Microbiol. 150, 160-164.

Laurell, C.B. (1966). Quantitative estimation of proteins by electrophoresis in agarose gel containing antibodies. Anal.Biochem. 15 ,45-52.

Neuer,H. and Bothe,H. (1982). Electron transport to nitrogenase in heterocysts of cyanobacteria. BBA 716, 359-365.

Parsons, H.B. and Strickland, T.R. (1973). Pigment analysis. In: Stein,J.R. (ed.) Handbook of phycological methods: culture methods and growth measurements. Cambridge University Press. Cambridge,pp360-367.

Sandmann, G. and Böger, P. (1980). Cooper- induced exchange of plastocyanin and cytochrome c-553 in cultures of *Anabaena variabilis* and *Plectonema boryanun.* Plant Sci. Lett. 17, 417-424.

Sandmann,G. Peleato, M.L., Fillat, M.F. Lázaro, M.C. and Gómez-Moreno, C. (1990). Consequences of the iron dependent formation of ferredoxin and flavodoxin on photosynthesis and nitrogen fixation on *Anabaena* strains. Photosynthesis Res. in press.

Scherer ,S.,Almon,H. and Boger,P. ( 1988). Interaction of photosynthesis, respiration and nitrogen fixation in cyanobacteria. Photosynthesis Research, 15, 95-114.

Schrautemeier,B., Bohme,H. and Boger, P. (1984). In vitro studies on pathways and regulation of electron transport to nitrogenase with a cell free extract from heterocysts of *Anabena variabilis.* Arch. Microbiol. 137, 14-20.

Smillie,R.M. (1965) Isolation of two proteins with chloroplasts ferredoxin activity from a blue-green alga. Biochem. Biophys.Res.Commun. 20, 621-629.

Trebst, A. and Bothe, H. (1966). Zur rolle des phytoflavins im photosynthetischen electronentransport. Ber. Dtsch.Bot.Ges. 79, 44-47.

Yoch, D.C. and Valentine, R.C. (1972). Ferredoxins and flavodoxins of bacteria. Ann.Rev. Microbiol. 26, 41-51.

# PHOTOSYNTHESIS AND NITROGEN FIXATION IN THE UNICELLULAR CYANOBACTERIUM <u>GLOEOTHECE</u> PCC 6909


**L.J. STAL, K.S. MYINT and J.J. ORTEGA-CALVO**
*Laboratory for Microbiology,*
*University of Amsterdam,*
*Nieuwe Achtergracht 127,*
*NL-1018 WS Amsterdam, The Netherlands*



ABSTRACT. The unicellular, nitrogen-fixing cyanobacterium *Gloeothece* sp. PCC 6909 was grown diazotrophically in shaking batch cultures and in well-aerated continuous cultures. The cultures were grown under alternating light-dark cycles. The two types of cultures differed totally with respect to the cyclic pattern of nitrogenase activity. In batch culture it was found exclusively in the dark, whereas in continuous cultures nitrogenase activity occurred predominantly in the light, concomitantly with photosynthetic oxygen evolution. In contrast to what has been reported previously, *Gloeothece* does not necessarily separate nitrogenase activity from oxygenic photosynthesis.


## 1. Introduction

Many cyanobacteria are capable of fixing dinitrogen. These phototrophic organisms can easily meet the high demand of nitrogenase for energy (ATP) and reducing equivalents (reduced ferredoxin). On the other hand cyanobacteria produce oxygen which is paradoxal because nitrogenase is extremely oxygen-sensitive (Robson and Postgate, 1980). Nitrogen-fixing cyanobacteria possess mechanisms to protect nitrogenase from inactivation by atmospheric and photosynthetic oxygen. Regarding the latter, two different strategies are known. Certain cyanobacteria differentiate special cells - the heterocysts - in which nitrogenase is located. These cells have lost the capacity of oxygenic photosynthesis (Donze et al. 1972). Such organisms thus separate nitrogen fixation spatially from oxygenic photosynthesis.
An increasing number of non-heterocystous, nitrogen-fixing cyanobacteria, filamentous and uni-cellular, is discovered. Evidence is accumulating that these organisms have in common that nitrogen fixation is separated temporally from oxygenic photosynthesis (Mitsui et al. 1986). Under alternating light-dark cycles non-heterocystous organisms therefore

would fix nitrogen predominantly during the dark. It is not known how a temporal separation can be achieved, particularly under continuous illumination. Mitsui et al. (1986) assumed that nitrogen fixation and photosynthetic oxygen evolution were confined to different phases in the cell division cycle of *Synechococcus*. On the other hand Huang et al. (1990) presumed that a circadian clock would be involved in the same species.

The capacity of the unicellular cyanobacterium *Gloeothece* sp. (formerly *Gloeocapsa*) to fix nitrogen aerobically was reported more than 20 years ago (Wyatt and Silvey, 1969). Despite a lot of effort the precise mechanism, by which *Gloeothece* protects nitrogenase from oxygen, is unknown (Gallon and Chaplin, 1988). Likewise the interactions of nitrogenase and photosynthesis are obscure.

## 2. Materials and Methods

*Gloeothece* sp. PCC 6909 was kindly provided by Dr. R. Rippka (Institute Pasteur, Paris). The organism was grown in medium BG 11 (Rippka et al. 1979) with either nitrate or $N_2$ as the nitrogen source for growth. Batch cultures were grown in Erlenmeiers in an orbital shaking incubator. Continuous cultures were constructed as described in Van Liere et al. (1979). Nitrogenase activity was assayed using the acetylene reduction technique. Oxygen exchange was measured polarographically.

## 3. Results and Discussion

All non-heterocystous, aerobic nitrogen-fixing cyanobacteria known sofar, grow under continuous illumination. Nitrogen-fixing cultures of *Gloeothece* reduce acetylene and evolve oxygen concomitantly. This does however not necessarily implicate that oxygenic photosynthesis and nitrogen fixation occur simultaneously in one single cell. A possibility that has been suggested several times is that non-heterocystous nitrogen-fixing cyanobacteria separate nitrogen fixation temporally from photosynthesis (Huang et al., 1990, Mitsui et al., 1986, Stal and Krumbein, 1987).

If *Gloeothece* was grown under a cycle of alternating light and dark periods, nitrogenase was predominantly or exclusively found during the dark (e.g. for a 16-8 h light-dark culture: Fig. 1) (Mullineaux et al., 1981). Fig. 1 also shows that if this culture was shifted to continuous light, nitrogenase was induced practically at the same point of time. This phenomenon has been observed in other non-heterocystous cyanobacteria as well (Huang et al., 1990, Mitsui et al, 1986, Stal and Krumbein, 1987).

It appeared not possible to maintain the rhythmic pattern of acetylene reduction in continuous light for more than one cycle. This was already reported by Gallon et al. (1988),

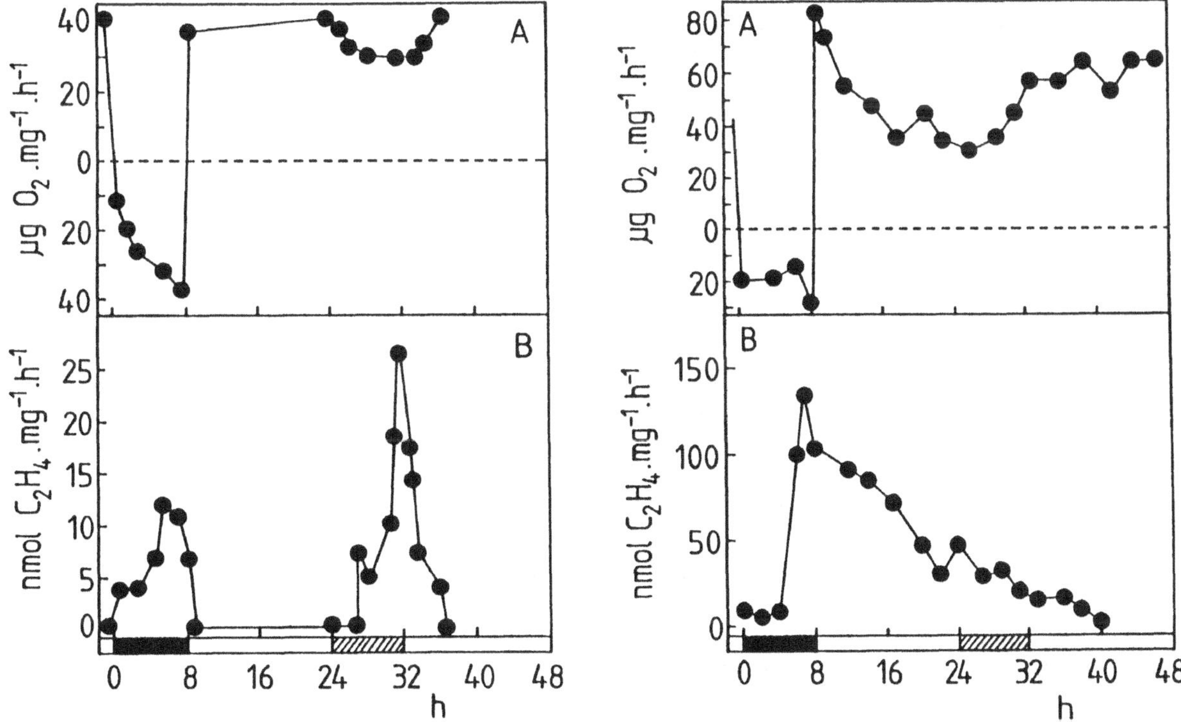

Fig. 1. Oxygen exchange (A) and acetylene reduction (B) in a batch culture of *Gloeothece,* grown under a 16-8 h light-dark cycle. Specific rates are expressed on a protein basis. The culture was shifted to continuous light and the measurements were continued (right part of figure). The position of the original dark phase is indicated by the shaded area.

Fig. 2. As Fig. 1. but using a continuous culture (D=0.005 $h^{-1}$) of *Gloeothece*, grown under a 16-8 h light-dark cycle.

but in contrast in *Synechococcus* RF-1 the rhythmic pattern of nitrogenase was retained for at least 4 days (Huang et al., 1990).
Fig. 1 also shows the rates of oxygen production or, where appropriate, - consumption. It was noticed that the rate of respiration increased with increasing nitrogenase activity. Another remarkable observation was that when the culture was shifted to continuous light, net photosynthetic oxygen evolution also decreased. In some cases, when nitrogenase activity was very high, it changed even into net oxygen uptake (Table 1). Experiments using the inhibitor of photosystem II, 3-(3,4-dichlorophenyl)-1,1-dimethylurea (DCMU) suggested that oxygenic photosynthesis may be (partly) switched off in cultures that were shifted from a light-dark cycle to continuous light (Table 1). In addition a high rate of oxygen uptake occurred.
The (partly) switch-off of oxygenic photosynthesis was also observed in nitrate-grown cultures of *Gloeothece* (lacking nitrogenase). Therefore it was concluded that the switch-off of photosynthesis was not necessarily associated with

Table 1. Photosynthesis and respiration in $NO_3^-$- and $N_2$-grown cultures of *Gloeothece* PCC 6909. A comparison is made between a culture grown under continuous light, a 12-12 h light-dark regime and a 12-12 h light-dark culture that was shifted to continuous light ("fooled" culture).

| Culture | Incubation | Oxygen exchange | |
|---|---|---|---|
| | | ($\mu$g $O_2$ . $mg^{-1}$ protein . $h^{-1}$) | |
| | | $NO_3^-$-grown | $N_2$-grown |
| continuous light | light | +11 | +22 |
| | light + DCMU | − 7 | − 4 |
| | dark | −15 | −25 |
| 12-12 h L-D | light | +18 | −26 |
| | light + DCMU | − 8 | −45 |
| | dark | −11 | −57 |
| "fooled" 12-12 h | light | + 9 | −21 |
| | light + DCMU | − 3 | −26 |
| | dark | − 9 | −36 |

The incident photon fluence rate was 40 $\mu$mol.$m^{-2}$.$s^{-1}$
The 12-12 h light-dark culture was sampled at 5 h after the light was switched off and nitrogenase was fully induced. The "fooled" cycle was sampled at the same point of time as the normal 12-12 h light-dark culture.

Table 2. Effect of light and photosynthetic - and atmospheric oxygen on the rate of acetylene reduction in *Gloeothece* PCC 6909, grown in continuous light or in a 12-12 h light-dark cycle.

| Culture | Incubation | Acetylene reduction | |
|---|---|---|---|
| | | (nmol $C_2H_2$ .$mg^{-1}$ protein.$h^{-1}$) | |
| | | oxic | anoxic |
| continuous light | light | 6.0 ± 0.5 | 44.7 ± 52 |
| | light + DCMU | 45.5 ± 6.4 | 163 ± 17 |
| | dark | 4.4 ± 0.9 | 5.6 ± 1.1 |
| 12-12 h L-D | light | 156 ± 25 | 638 ± 33 |
| | light + DCMU | 237 ± 5 | 536 ± 5 |
| | dark | 5.0 ± 1.5 | 38 ± 8 |

The incident photon fluence rate was 40 $\mu$mol. $m^{-2}$ . $s^{-1}$.
The 12-12 light-dark culture was sampled at 5 h after the light was switched off when nitrogenase was fully induced. Experiments were done in triplicate.

nitrogenase. More likely this phenomenon would be related to the imposed light-dark cycle. On the other hand the high rate of oxygen uptake was probably closely related to the capacity of nitrogen fixation. The hypothesis of a switch-off of oxygenic photosynthesis is also supported by the effect of DCMU on acetylene reduction (Table 2). If *Gloeothece* was grown under continuous light, the inhibition of the photosynthetic oxygen production by DCMU stimulated acetylene reduction several times. This was expected since in such cultures photosynthetic oxygen evolution and nitrogenase activity co-existed. However, in a light-dark cycle (e.g. 12-12 h light-dark; but it also holds for other light-dark cycles) the effect of DCMU was negligible, which was attributed to the switch-off of oxygenic photosynthesis and to the uptake of oxygen.

The stimulation of acetylene reduction by light is especially obvious from the data presented in table 2. In cultures that were grown under continuous light, the stimulation of acetylene reduction was particularly clear in the presence of DCMU. The stimulation of light is therefore explained as to cover the high demand of nitrogenase for ATP and/or low-potential reducing equivalents (reduced ferredoxin). This interpretation does not agree with the work of Maryan et al. (1986) who concluded that in *Clocothcce* respiration and not photosynthesis supports nitrogenase activity. However, it should be considered that the outcome of experiments may vary with the physiological state of the organism.

Therefore it was decided to investigate the characteristics of nitrogenase activity in *Gloeothece* in continuous culture The continuous cultures were well-aerated and it was assured that the culture liquid was air-saturated at all times. Continuous cultures have the advantage of reaching a steady state and that by adjusting the dilution rate, the growth rate of the organism is determined. The diurnal variations of acetylene reduction and oxygen exchange of a light-limited continuous culture, grown at a dilution rate of 0.005 $h^{-1}$ and a 16-8 light-dark cycle, are shown in Fig. 2. The results were very different from the batch culture experiments. The continuous culture possessed nitrogenase activity predominantly during the light, concomitantly with photosynthetic oxygen evolution. There was also no fixed endogenous rhythm, as was shown when the culture was shifted to continuous light.

The differences between the rhythmic patterns of acetylene reduction in batch culture and continuous culture are explained in terms of local oxygen concentrations. In batch culture in the light an accumulation of photosynthetic oxygen will occur that would inhibit nitrogenase activity.

On the other hand, oxygen depletion will allow nitrogenase activity, although a shortage of ATP and reduced ferredoxin in the dark would permit only low rates. In well-aerated continuous cultures the situation is different. There will be no supersaturation nor a depletion of oxygen. The provision with sufficient ATP and reducing equivalents is guaranteed in the light. Nitrogenase and respiratory electron transport compete for electrons and probably this will not favour nitrogenase activity in the dark.

# 4. References

Donze, M.; Haveman, J.; Schiereck, P. (1972) Absence of photosystem 2 in heterocysts of the blue-green alga *Anabaena*. Biochim. Biophys. Acta 256, 157-161.

Gallon, J.R.; Chaplin, A.E. (1988) Recent studies on $N_2$ fixation by non-heterocystous cyanobacteria. In: Bothe, H.; de Bruijn, F.J.; Newton, W.E. (Eds.) Nitrogen fixation: Hundred years after. Gustav Fischer, Stuttgart, pp. 183-188.

Gallon, J.R.; Perry, S.M.; Rajab, T.M.A.; Flayeh, K.A.M.; Yunes, J.S.; Chaplin, A.E. (1988) Metabolic changes associated with the diurnal pattern of $N_2$ fixation in *Gloeothece*. J. Gen. Microbiol. 134, 3079-3087.

Huang, T.C.; Tu, J.; Chow, T.J.; Chen, T.H. (1990) Circadian rhythm of the prokaryote *Synechococcus* sp. RF-1. Plant Physiol. 92, 531-533.

Maryan, P.S.; Eady, R.R.; Chaplin, A.E.; Gallon, J.R. (1986) Nitrogen fixation by *Gloeothece* sp. PCC 6909: respiration and not photosynthesis supports nitrogenase activity in the light. J. Gen. Microbiol. 132, 789-796.

Mitsui, A.; Kumazawa, S.; Takahashi, A.; Ikemoto, H.; Cao, S.; Arai, T. (1986) Strategy by which nitrogen-fixing unicellular cyanobacteria grow photoautotrophically. Nature 323, 720-722.

Mullineaux, P.M.; Gallon, J.R.; Chaplin, A.E. (1981) Acetylene reduction (nitrogen fixation) by cyanobacteria grown under alternating light-dark cycles. FEMS Microbiol. Lett. 10, 245-247.

Rippka, R.; Deruelles, J.; Waterbury, J.B.; Herdman, M.; Stanier, R.Y. (1979) Generic assignments, strain histories and properties of pure cultures of cyanobacteria. J. Gen. Microbiol. 111, 1-61.

Robson, R.L.; Postgate, J.R. (1980) Oxygen and hydrogen in biological nitrogen fixation. Ann. Rev. Microbiol. 34, 183-207.

Stal, L.J.; Krumbein, W.E. (1987) Temporal separation of nitrogen fixation and photosynthesis in the filamentous, non-heterocystous cyanobacterium *Oscillatoria* sp. Arch. Microbiol. 149, 76-80.

Van Liere, L.; Mur, L.R.; Gibson, C.E.; Herdman, M. (1979) Growth and physiology of *Oscillatoria agardhii* Gomont cultivated in continuous culture with a light-dark cycle. Arch. Microbiol. 123, 315-318.

Wyatt, J.T.; Silvey, J.K.G. (1969) Nitrogen fixation by *Gloeocapsa*. Science 165, 908-909.

# Diurnal and seasonal variations of nitrogen fixation and photosynthesis in cyanobacterial mats

M. VILLBRANDT, W.E. KRUMBEIN and L.J. STAL[1]
*Geomicrobiology Division/ICBM, University of Oldenburg, P.O. Box 2503, DW-2900 Oldenburg, Germany. [1]Laboratory for Microbiology, University of Amsterdam, Nieuwe Achtergracht 127, 1018 WS Amsterdam, The Netherlands*

*Key words:* cyanobacteria, microbial mats, nitrogen fixation, *Oscillatoria limosa*

## Abstract

In situ measurements of nitrogenase activity and photosynthesis were performed simultaneously in cyanobacterial mats of intertidal sand flats of the Southern North Sea. Two types of cyanobacterial mats, which differed in species composition and biomass content, were investigated. The measurements were done monthly during 3 years to detect seasonal variations of nitrogen fixation and photosynthesis. Diurnal variations were investigated as well. The results showed that (i) freshly colonized sediment with the cyanobacterium *Oscillatoria limosa* as the dominant organism revealed the highest specific nitrogenase activities (ii) nitrogenase activities were highest in spring and summer, when mat development was initiated and (iii) diurnal fluctuations of nitrogenase activity indicated that it occurred temporally separated from oxygenic photosynthesis.

## Introduction

Cyanobacteria are photoautotrophic, procaryotic microorganisms, that use water as electron donor and evolve oxygen in a plant-like type of photosynthesis. Cyanobacteria are found in a wide range of habitats, because they have a low demand of nutrients and show remarkable capacities to adapt to extreme environmental conditions. Environments that are low in combined nitrogen will select cyanobacteria with the capacity for the fixation of dinitrogen. The marine environment is such a nitrogen-depleted system.

Marine intertidal sediments are frequently inhabited by dense communities of cyanobacteria, generally known as microbial or cyano bacterial mats. Stal et al. (1985) described the structure of a microbial mat that developed on intertidal sand flats of the island Mellum which is part of the shallows of the North Sea. The importance of nitrogen fixation in these mats was documented by Stal et al. (1984). They noticed the absence of heterocystous cyanobacteria in the microbial mats. These organisms are especially well adapted, since they carry the oxygen-sensitive nitrogenase in differentiated cells (heterocysts) that have lost the capacity of oxygenic photosynthesis. This adaptation has been regarded as so essential, that the importance of nonheterocystous nitrogen-fixing cyanobacteria has been doubted for a long time. But recently, increasing numbers of non-heterocystous nitrogen-fixing cyanobacteria have been isolated. Evidence is accumulating that such organisms temporally separate the process of nitrogen fixation from oxygenic photosynthesis.

The North Sea microbial mats consist of a number of different cyanobacteria. *Microcoleus chthonoplastes* is the dominant organism in established mats. This organism did not fix nitrogen in laboratory culture. A second species,

*Oscillatoria limosa* very often is present in considerable numbers in the mature mats. The latter species was also identified as the pioneer organism which initially colonizes the barren sand. Stal and Krumbein (1981) isolated this organism and later showed (Stal and Krumbein, 1987) that nitrogen fixation and oxygenic photosynthesis were separated temporally. Most experiments that have been done on nitrogen fixation in non-heterocystous cyanobacteria used laboratory cultures. Extrapolations of such results to the field situation are not always justified, because culture conditions often are totally different from those in the natural environment. Therefore we started a program in which detailed measurements on nitrogen fixation in the North Sea microbial mats were carried out. In situ measurements of nitrogenase activity were carried out simultaneously with microprofile measurements of photosynthesis and oxygen.

## Materials and methods

### Area of investigation

Cyanobacterial mats that developed on the intertidal sand flats of the North Sea island of Mellum were investigated. The structure and development of these microbial mats have been described in detail by Stal et al. (1985). Two types of mats were recognized. Large areas of the sand flats were colonized by cyanobacteria of which the filamentous *Oscillatoria limosa* was the dominant species. In the upper intertidal areas established mats were found that consisted mainly of *Microcoleus chthonoplastes* but were often accompanied with *O. limosa*. In freshly colonized sediments *O. limosa* was loosely associated with the sand and occurred over a depth of up to 10 mm. *M. chthonoplastes* built tough and leathery structures in which the layer of cyanobacteria was often not more than 1 mm. The biomass of these established mats, however, was very high and was virtually devoid of sand grains. On an average the content of chlorophyll *a* in the mature mat was 4 times higher than in a mat consisting of *O. limosa* alone.

### Nitrogenase activity

Nitrogenase activity was assayed using the acetylene reduction technique. The measurements were done in situ using gas domes (Stal, 1988). Diurnal measurements were done by introducing a gas dome (50 mL) at random in an experimental area of 1 m$^2$, every 15 min. The gas domes were incubated for 2 hours after which the gas phase was sampled using Vacutainer-tubes (Becton and Dickinson). Acetylene and ethylene were determined gas-chromatographycally, using acetylene as an internal standard (Stal, 1988). The sediment which was enclosed by the gas domes was extracted with methanol and assayed for chlorophyll *a* using the method of Stal et al. (1984).

### Oxygen and photosynthesis

Microprofiles of oxygen and photosynthesis were measured using micro-electrodes (Revsbech et al., 1983). The measurements were done at hourly intervals during a 24 h period.

## Results and discussion

Nitrogenase is extremely sensitive to molecular oxygen and in addition to the availability of combined nitrogen, oxygen is one of the major factors that determines synthesis and activity of this enzyme. Phototrophic mats accumulate oxygen during the light. The degree of oxygen accumulation depends on a number of factors such as biomass content, incident light intensity and the diffusion characteristics of the sediment. During the dark, the cyanobacteria respire oxygen. If a large number of organisms is present, anoxic conditions may develop, especially if diffusion of air is limiting (e.g. if the sediment is water-saturated). With low biomass and a loose sediment air diffusion is sufficient to provide the organisms with oxygen and the cyanobacterial layer will remain oxic during the night. This gives rise to very typical patterns of concentration of dissolved oxygen during a diurnal cycle.

*Diurnal variation of nitrogen fixation*

In the young microbial mat community which consisted almost entirely of *O. limosa* no nitrogenase activity was detected during the day. The sediment stayed oxic during the night. Nitrogenase activity was detected during the dark period (7 hours) when a high average rate of acetylene reduction was measured (10 $\mu$mol $C_2H_2$ reduced mg Chl $a^{-1}$ $h^{-1}$). The highest rate of acetylene reduction measured in a young mat community which consisted almost entirely of the cyanobacterium *O. limosa* agreed with the maximum of nitrogenase activity measured in laboratory cultures of *O. limosa*. Other investigations on pure cultures of *M. chthonoplastes* (Malin and Pearson, 1988) and microbial communities in which *M. chthonoplastes* is the dominant species (Paling et al., 1989; Skyring et al., 1988) demonstrate this cyanobacterium as potential nitrogen fixing organism. In the mature mat which turned anaerobic up to the cyanobacterial layer soon after oxygenic photosynthesis stopped no nitrogenase activity was detected during the night. In this system nitrogenase activity was sometimes detected during the day with a negative correlation to oxygenic photosynthesis. In both systems peaks of nitrogenase activity were marked at sunrise. Bautista and Paerl (1985) also observed a peak of nitrogenase activity late at night or in the early morning. Figure 1 shows the results of a measurement carried out in a mature mat out in September 1988. It was a misty day and oxygenic photosynthesis was limited by light intensity which did not reach values above 70 klux. Temperatures varied between 15°C at night and 25°C by day. During the night (from 10 p.m. until 4 a.m.) no acetylene reduction occurred (data not shown). Our data confirmed with the assumption of Griffith et al. (1987) who assumed that differences in the diurnal pattern of acetylene reduction are due to environmental conditions rather than to differences between the strains of cyanobacteria present.

*Seasonal variation of nitrogen fixation*

In May only low nitrogenase activity was detected in the mature mat whereas in the freshly colonized sediment the nitrogenase activity was

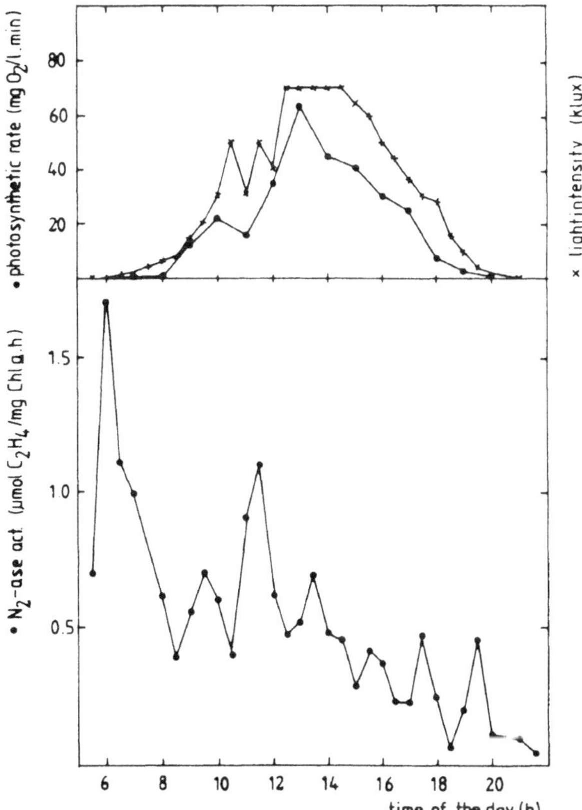

*Fig. 1.* Daily pattern of oxygenic photosynthesis, light intensity and acetylene reduction measured in September 1989 in a mature mat system.

below the limit of detection. In June and July the highest annual amounts of nitrogenase activity were observed in both systems. In autumn nitrogenase activity decreased. This may be due to seasonal variations in cyanobacterial intracellular carbohydrate reserves or seasonal variations of the cyanobacterial composition of the mat. Table 1 shows the average values of the acetylene reduction rates calculated from all performed measurements over the years from 1986 till 1989.

*In situ interactions with oxygen and photosynthesis*

A negative correlation of nitrogenase activity and oxygenic photosynthesis was observed in both systems. This observations agreed with that of Bebout et al. (1987) who found an inverse correlation of acetylene reduction and $CO_2$-fixation in a similar ecosystem.

*Table 1.* Average annual rates of acetylene reduction. Values were calculated of all measurements performed in the same months during the years from 1986 till 1989

| Month | Specific nitrogenase activities ($\mu$mol $C_2H_4$ mg Chl $a^{-1}$ $h^{-1}$) | |
| --- | --- | --- |
| | Mature mat | Freshly colonized sediment |
| May | 0.1 | 0.0 |
| June | 0.5 | 2.5 |
| July | 1.5 | 7.5 |
| August | 0.5 | n.d. |
| September | 0.3 | 1.0 |

## Conclusions

Nitrogen fixation in the cyanobacterial mat occurred temporally separated from oxygenic photosynthesis. In a diurnal cycle in which oxic conditions prevailed during the dark, nitrogenase activity was found predominantly during the night. If the mat turned anoxic during the night, nitrogenase activity was detected at dusk and dawn. Highest nitrogenase activities were always found at dawn. At that point of time oxygenic photosynthesis was not yet detectable.

It was also concluded that the nitrogen fixing cyanobacterium *O. limosa* was the major nitrogen-fixing organism in the mat, but chemolithotrophic bacteria might have contributed a little to the total nitrogen fixed. The development of the cyanobacterial mat depended greatly on the capacity of nitrogen fixed. This was obvious from two observations: (i) freshly colonized sediment usually showed highest specific nitrogenase activities, and (ii) rates of acetylene reduction were highest in spring or early summer, when mat development started.

## Acknowledgements

We thank the 'Mellumrat' for allowing us to carry out the fieldwork in the nature reserve. Financial support was granted from the 'Deutsche Forschungsgemeinschaft'.

## References

Bautista M F and Paerl H W 1985 Diel $N_2$ fixation in an intertidal marine cyanobacterial mat community. Mar. Chem. 16, 369–377.

Bebout B M, Paerl H W, Crocker K M and Prufert L E 1987 Diel interactions of oxygenic photosynthesis and $N_2$ fixation (acetylene reduction) in a marine microbial mat community. Appl. Environ. Microbiol. 53, 2353–2362.

Griffiths M S H, Gallon J R and Chaplin A E 1987 The diurnal pattern of dinitrogen fixation by cyanobacteria in situ. New Phytol. 107, 649–657.

Malin G and Pearson H W 1988 Aerobic nitrogen fixation in aggregate-forming cultures of the nonheterocystous cyanobacterium *Microcoleus chthonoplastes*. J. Gen. Microbiol. 134, 1755–1763.

Paling E I, McComb A J and Pate J S 1989 Nitrogen fixation (acetylene reduction) in nonheterocystous cyanobacterial mats from the Dampier Archipelago. Western Australia. Aust. J. Mar. Freshwater Res. 40, 147–153.

Revsbech N P, Jorgensen B B, Blackburn T H and Cohen Y 1983 Microelectrode studies of the photosynthesis and $O_2$, $H_2S$ and pH profiles of a microbial mat. Limnol. Oceanogr. 28, 1062–1074.

Skyring G W, Lynch R M and Smith G D 1988 Acetylene reduction and hydrogen metabolism by a cyanobacterial/sulfate-reducing bacterial mat ecosystem. Geomicrobiol. J. 6, 25–31.

Stal L J 1988 Nitrogen fixation in cyanobacterial mats. Meth. Enzymol. 167, 474–484.

Stal L J, Grossberger S and Krumbein W E 1984 Nitrogen fixation associated with the cyanobacterial mat of a marine laminated microbial ecosystem. Mar. Biol. 82, 217–224.

Stal L J and Krumbein W E 1981 Aerobic nitrogen fixation in pure cultures of a benthic marine *Oscillatoria* (cyanobacteria). Fems Microbiol. Lett. 11, 295–298.

Stal L J and Krumbein W E 1987 Temporal separation of nitrogen fixation and photosynthesis in the filamentous, non-heterocystous cyanobacterium *Oscillatoria* sp. Arch. Microbiol. 149, 76–80.

Stal L J, van Gemerden H and Krumbein W E 1984 The simultaneous assay of chlorophyll and bacteriochlorophyll in natural microbial communities. J. Microbiol. Meth. 2, 295–306.

Stal L J, van Gemerden H and Krumbein W E 1985 Structure and development of a benthic marine microbial mat. FEMS Microbiol. Ecol. 31, 111–125.

# SUPEROXIDE DISMUTASE DETECTION IN *ANABAENA AZOLLAE* STRASB. ISOLATED FROM *AZOLLA FILICULOIDES* LAM.


A. CANINI, F. GALIAZZO, G. ROTILIO and M. GRILLI CAIOLA
*Department of Biology,*
*II University of Rome "Tor Vergata"*
*V. E. Carnevale, 00173 Rome,*
*Italy*



ABSTRACT. Percoll and sucrose gradients were used to isolate respectively heterocysts and vegetative cells of *Anabaena azollae* symbiont of *Azolla filiculoides*. Both heterocysts and vegetative cells devoid of contaminating bacteria and *Azolla* debris, were used to assay the presence of the enzyme superoxide dismutase (SOD) by activity-stained polyacrylamide gel electrophoresis. Types of SOD were determined by different sensitivity to 2mM cyanide and 2mM hydrogen peroxide. Heterocysts contained only a faint band of both hybrid iron-manganese (Hy) and iron (Fe) superoxide dismutase. Vegetative cells possess Mn-SOD, Hy-SOD and Fe-SOD. Hy-SOD is detected for the first time in cyanobacteria. The higher activity was present as Fe-SOD in vegetative cells while total heterocysts SOD activity was only 15% with respect to vegetative cells.


## 1. Introduction

The fern *Azolla* is in symbiotic association with nitrogen-fixing *Anabaena azollae* Strasb. (Peters and Meeks (1989)) and bacteria (Grilli (1964)). Numerous physiological studies have shown that nitrogenase present in heterocysts of cyanobacterium is inactivated by oxygen (Goldberg et al. (1987); Hill (1988)). However, the precise mechanism by which oxygen inhibits nitrogenase is not known. Severe cellular damage can also be attributed to the reduced oxygen species, superoxide, hydrogen peroxide and hydroxyl radicals (Hill (1988); Becana and Rodrigues-Barrueco (1989)). In addition to other diverse nitrogenase protection mechanisms, Puppo and Rigaud (1986) have proposed the enzyme superoxide dismutase (SOD) . SOD catalyzes the dismutation of $O_2^-$ to $H_2O_2$ and $O_2$, thereby protecting cells from the effects of oxygen intermediates.

SOD activity in free living cyanobacteria was described in *Anabaena* PCC 7119 filaments by Garcia-Gonzales et al. (1986) and in *Anabaena cylindrica* by Mackey and Smith (1983) and Henry et al.(1978) while Fe-SOD and Mn-SOD were detected in *Plectonema boryanum*, *Anacystis nidulans* and *Anabaena variabilis* by Okada et al. (1979) and Asada et al. (1977).

Until now, no investigations have been carried out on the presence of SOD in heterocysts and vegetative cells of *Anabaena azollae* in symbiosis with *Azolla*. It is extremely difficult to isolate single components from this symbiosis, due to the great complexity of its organization. In fact, a single leaf cavity contains *Anabaena azollae*, bacteria, hair cells and a mucilaginous matrix delimited by an inner envelope (Grilli Caiola and Albertano (1988); Nierzwicki-Bauer et al. (1989)).
These difficulties however, are not present in free-living cyanobacteria. For this investigation it has

been important to isolate all cell types of cells with accuracy, avoiding contamination of bacteria and *Azolla* debris.

Many techniques have been adopted for the isolation of heterocysts. Peterson and Burris (1976) described the use of Ludox for the isolation of heterocysts from free- living *Anabaena*. Also Kaplan et al. (1986) used a procedure for isolating heterocysts but this did not allow the separation of bacteria. After several preliminary attempts we adopted Percoll density gradients that have for many years been used in the separation and purification of cells, viruses and subcellular particles. Percoll consists of colloidal silica particles coated with PVP that has many superior properties when compared with conventional Ludox (Pertoft et al. (1978)).

The present work reports the use of Percoll gradients in obtaining isolate heterocysts free from contaminating bacteria and *Azolla* debris.The distribution of SOD in heterocysts and vegetative cells of *Anabaena azollae* is also reported.


## 2. Materials and Methods

### 2.1. PLANT MATERIALS.

*Azolla filiculoides* Lam. plants were obtained from the Naples Botanical Garden and cultured for a month on Hoagland medium in the open air.

### 2.2. PREPARATION OF CELL EXTRACTS.

2.2.1. *Heterocysts. Azolla* plants collected after a month of growth were gently disrupted in a mortar. 0.1 M phosphate saline buffer (pH 7.4) (PBS) was added to the plants and the suspension was filtered 4 times on nylon mesh of 20 µm large to remove *Azolla* debris. The obtained filtrate containing bacteria and *Anabaena azollae* filaments, constituted of vegetative cells and heterocysts, was centrifuged for 10 min at 1500 g. The pellet was dissolved in spheroplasting medium containing 0.350 M mannitol, 0.01 M EDTA and 5 mg/ml Lysozyme in 0.1 M PBS to disrupt the bacterium and vegetative cell wall and incubated at 37 °C for 1 h. Spheroplasts were obtained and heterocysts collected by centrifugation at 3000 g and resuspended in 0.1 M PBS containing 0.5 M mannitol. After centrifugation at 1500 g spheroplasts were washed once with 0.1 M PBS and sonicated in an ethanol-ice bath at 20 W power for 90 sec at 30-s intervals with a MSE Soniprep 150. Heterocysts were separated from still present bacteria and disintegrated vegetative cells by means of Percoll gradients (Pharmacia) at density 1.080 g/ml containing sucrose 2.5 M. The gradients were centrifuged for 50 min at 70,000 g. The band containing whole heterocysts was removed and washed twice with 0.1 M PBS. The pellets obtained after centrifugation were stored at - 2O °C. Heterocysts obtained from 120 gradients were suspended in a volume of 0.2 ml of 0.1 M PBS and sonicated in an ethanol-ice bath at 20 W power for 11 min at 30-s intervals. The heterocyst disruption was controlled by light microscope.

2.2.2. *Vegetative cells.* The same procedure previously used to eliminate *Azolla* was followed. The pellets containing vegetative cells, heterocysts and bacteria were dissolved in PBS. Filaments were separated by discontinuous sucrose gradients according to Peters and Mayne (1974). Final pellets were stored at - 20 °C. Filaments obtained from 36 gradients were suspended in 0.2 ml of 0.1M PBS and sonicated in an ethanol-ice bath, at 20 W power for 120 sec.

For both types of extract preparations, cell debris was removed by centrifugation for 30 min at

30000 g. The cell extracts were used to assay enzyme activity and for protein determination. Protein was determined according to Lowry et al. (1951).

## 2.3. SUPEROXIDE DISMUTASE ASSAY

Superoxide dismutase isozymes were separated by electrophoresis on 7.5% or 10% polyacrylamide gels and stained for the activity according to Beauchamp and Fridovich (1971). The isozymes were quantified by densitometric scanning with a LKB Ultroscan XL laser densitometer (Hassan and Fridovich (1977)).
Types of SOD were determined according to their sensitivity to 2 mM cyanide and 2 mM hydrogen peroxide.

# 3. Results and Discussion

## 3.1. ISOLATION OF *ANABAENA AZOLLAE* CELL TYPES.

After disruption of *Azolla* fern, examination of the filtrate by light microscope evidenced filaments of *Anabaena azollae,* much *Azolla* debris and bacteria. The treatment with lysozyme before centrifugation allowed the disruption of bacteria and vegetative cells without damaging heterocysts which appeared intact having a thick envelope and polar bodies.Vegetative cells which had remained whole were disrupted by short sonication. Heterocyst isolation was performed by centrifugation on Percoll gradient. Percoll forms gradients within a density range 1.0-1.3 g/ml and is non-toxic to cells (Pertoft et al. 1977). Percoll stock solution was osmotically adjusted with sucrose 0.25 M, and diluted to density 1.08 g/ml. At this density it is possible to separate particles such as mitochondria, peroxisomes and microalgae.

Suspension containing heterocysts,fragments and other still intact cell types was layered on top of the gradient. Centrifugation was carried out in an angle-head rotor for 50 min at 70,000 g. After centrifugation, there appeared four different bands, one of which corresponded to the pellet (Fig. 1). The top and second band contained vegetative debris and bacteria. The primary lower band contained whole substantially purified heterocysts identifiable by their refractile polar bodies. Pellet contained *Azolla* debris.

Heterocysts were collected and Percoll was removed with a further centrifugation in 0.1 M PBS. Heterocyst purity was confirmed by means of observations made to a phase contrast microscope. Several heterocyst preparations were carried out to obtain a high enough quantity of protein to assay SOD activity by electrophoresis.

Attempts to isolate heterocysts using sucrose gradients were not successful because of the high contamination levels.

To isolate vegetative cells the following procedure was employed. Filaments of *Anabaena azollae* were separated from *Azolla* using sucrose gradients (Peters and Mayne, (1974); Leonardi et al. (1989)). Remaining bacteria were eliminated by repeated washing in 0.1 M PBS, followed by centrifugation. The purity of vegetative cells was controlled by means of a phase contrast microscope. Short sonication allows only the disruption of vegetative cells. Whole heterocysts were eliminated by centrifugation.

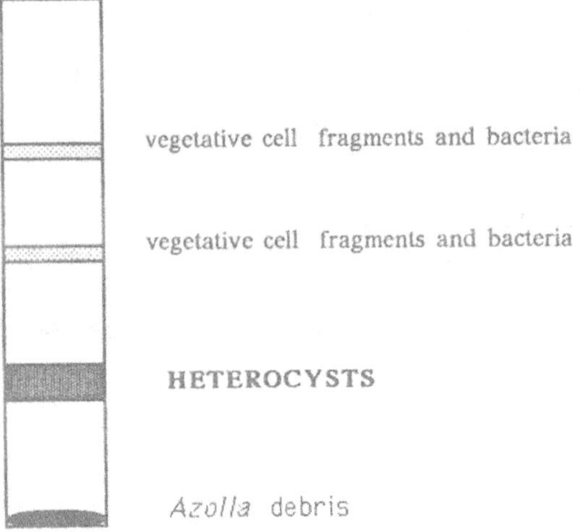



Figure 1. Separation of heterocysts by means of Percoll gradient at density 1.08 g/ml.

## 3.2. SUPEROXIDE DISMUTASE

Electrophoretic analysis of the extracts obtained from vegetative cells showed the presence of three different types of superoxide dismutase: Mn-SOD, Fe-SOD and Hy-SOD (Mn,Fe-SOD) (Tab. 1).

TABLE 1. Types and isoenzymatic forms of SOD present in vegetative cells and heterocysts

|                  | Mn-SOD | Hy -SOD | Fe-SOD   |
|------------------|--------|---------|----------|
| vegetative cells | 1[a]   | 1[a]    | 3[a]     |
| heterocysts      | 0      | 1[a] ( f) | 2 [a] (f) |

a: isoenzymatic forms; (f) : faint bands observed.

These forms corresponded to Mn-SOD, Hy-SOD and Fe-SOD respectively, on the basis of their sensitivity toward cyanide and hydrogen peroxide inhibitors. Fe-SOD was constituted of 3 isoenzymes as evidenced by the activity bands, while one band of activity was present for both Mn-SOD and Hy-SOD. The presence of Mn-SOD and Fe-SOD concorded with the enzymes found in free-living *Anabaena* by Okada et al. (1979).

Only two faint bands of activity were detected in heterocysts corresponding to Hy-SOD and Fe - SOD respectively (Tab. 1). Hy-SOD form was detected for the first time in cyanobacteria.

Tab. 2 shows the percentage of SOD types in heterocysts and vegetative cells with respect to the total SOD activity content in each cell types. The highest activity, 86%, is present as Fe-SOD in vegetative cells. Mn-SOD and Hy-SOD are very low and each corresponded to 7 % of the total. Heterocysts contained 75 % Fe-SOD and 25 % Mn-SOD.

TABLE 2. Percentage of Mn-SOD, Hy-SOD and Fe-SOD on total SOD activity present in heterocysts and vegetative cells.

| | Mn-SOD | Hy-SOD | Fe-SOD |
| --- | --- | --- | --- |
| | | ( % ) | |
| vegetative cells | 7 | 7 | 86 |
| heterocysts | - | 25 | 75 |

However, total SOD activity in heterocysts was 15 % compared to the total present vegetative cells.

The technique employed is useful in the detection of SOD in the different *Anabaena azollae* cells. Thus, further experiments have been planned to verify the involvement of SOD in nitrogenase protection.

## 4. References


Asada, K., Kanematsu, S., and Uchida, K. (1977) 'Superoxide dismutase in photosynthetic organism: absence of the cuprozinc enzyme in the eukaryotic algae', Arch.Biochem. Biopys. 179, 243-256.

Beauchamp, C. and Fridovich, I. (1971) 'Superoxide dismutase: improved assays and an assay applicable to acrylamide gels', Anal Biochem. 44, 276-287.

Becana, M. and Rodriguez-Barrueco, C. (1989) 'Protective mechanisms of nitrogenase against oxygen excess and partially-reduced oxygen intermediates', Physiol. Plant. 75,429-438.

Goldberg, I., Nadler, V., and Hochman, A. (1987) 'Mechanism of nitrogenase switch-off by oxygen', J. Bacteriol. 169, 874-879.

Garcia-Gonzales, M., Mateo, P., and Bonilla, I. (1988) 'Boron protection for $O_2$ diffusion in heterocysts of *Anabaena* sp. PCC 7119', Plant Physiol. 87, 785-789.

Grilli Caiola, M. and Albertano, P. (1988) 'Recognition mechanisms in the *Azolla-Anabaena* symbiosis ', in S. Scannerini et al. (eds.), Cell to cell Signals in Plant, animal and microbial symbiosis, NATO ASI Series H17,Springer-Verlag,Berlin Heidelberg, pp. 27-38.

Grilli , M. (1964) 'Infrastrutture di *Anabaena azollae* vivente nelle foglioline di *Azolla caroliniana',* Ann. Micr. *14, 69-90.*

Hassan, H.M. and Fridovich, I. (1977) 'Enzymatic defenses against the toxicity of oxygen and of streptonigrin in *Escherichia coli*', J. Bacteriol., 129, 1574-1583.

Henry, L.E.A., Gogotov, I.N., and Hall, D.O. (1978) 'Superoxide dismutase and catalase in the protection of the proton-donating systems of nitrogen fixation in the blue-green alga *Anabaena cylindrica*', Biochem. J. 174, 373-377.

Hill, S. (1988) 'How is nitrogenase regulated by oxygen ?', FEMS Microbiol. Rev. 54, 111-130.

Kaplan, D., Calvert, H.E., and Peters, G.A. (1986)'The *Azolla-Anabaena azollae* relationship. XII. Nitrogenase activity and phycobiliproteins of the endophyte as a function of leaf age and type', Plant Physiol. 80, 884-890.

Leonardi , D., Marini , S., Coletta, I., and Grilli Caiola,M. (1989) 'Anticorpi monoclonali contro *Anabaena azollae* Strasb.', Giorn. Bot. It. 123, 1-2 suppl. 1, 181.

Lowry, O.H., Rosebrough, N.J., Farr, A.L., and Randall, R.J. (1951) 'Protein measurements with the folin phenol reagent', J. Biol. Chem. 193, 265-275.

Mackey, E. J. and Smith, G.D. (1983)' Adaptation of the cyanobacterium *Anabaena cylindrica* to high oxygen tensions', FEBS Lett. 156,108-112.

Nierzwicki-Bauer, S.A., Aulfinger, H., and Braun-Howland E.B. (1989) 'Ultrastructural characterization of an inner envelope that confines *Azolla* endosymbionts to the leaf cavity periphery', Can. J. Bot. 67, 2711-2719.

Okada, S., Kanematsu, S., and Asada K. (1979) 'Intracellular distribution of manganese and ferric superoxide dismutases in blue-green algae', FEBS Lett. 103, 106-110.

Pertoft ,H., Rubin, K., Kjellen, L.,et al.. (1977) ' The viability of cells grown or centrifuged in a new density gradient medium, Percoll (TM)', Exp. Cell Res. 110,449-457.

Pertoft , H., Laurent, T.C., Lååb, T., et al. (1978) 'Density gradients prepared from colloidal silicia   particles coated by polyvinylpyrrolidone (Percoll)', Anal. Biochem. 88,271-282.

Peters, G.A. and Mayne, B. (1974) 'The *Azolla, Anabaena azollae* relationship. I. Initial characterization of the association', Plant Physiol. 53, 813-819.

Peters, G.A. and Meeks, J.C. (1989) ' The *Azolla-Anabaena* symbiosis: basic biology', Ann.Rev. Plant  Physiol. Plant Mol. Biol. 40,193-210.

Peterson , R. B. and Burris, R.H. (1976) ' Properties of heterocysts isolated with colloidal silica', Arch. Microbiol. 108,35-40.

Puppo, A . and Rigaud, J. (1986)' Superoxide dismutase : an essential role in the  protection of the  nitrogen fixation process?', FEBS Lett. 201, 187-189.

# Are bacteria the third partner of the *Azolla-Anabaena* symbiosis?

FRANCISCO CARRAPIÇO
*Departamento de Biologia Vegetal, Faculdade de Ciências de Lisboa, Campo Grande, Bloco C2, P-1700
Lisboa, Portugal*

*Key words:*   Azolla, bacteria, megasporocarps, symbiosis

## Abstract

The development of the prokaryotic colony in *Azolla filiculoides* indicates that *Anabaena azollae* is maintained through the life cycle of the fern and present in the leaves and megasporocarps. The same biological pattern is applied to the bacteria that are also present in these structures and seems to follow a development pattern identical to the cyanobacteria and probably can be considered the third partner of this symbiotic association.

## Introduction

The presence of bacteria in the *Azolla* leaves were previously described by Grilli (1964) and later referred by several authors, with some reserve, as an organism present in the leaf cavity of the fern (Forni and Grilli Caiola, 1986; Gates et al., 1980; Peters et al., 1978). In spite of this, organism having been identified by the first time by Bottomley in 1920 as *Pseudomonas*, the correct identification was done by Wallace and Gates (1986) and included in the genus *Arthrobacter*. Based on this identification Petro and Gates (1987) suggest that these bacteria may be the third component of the symbiosis. Using different methods Carrapiço and Tavares (1987, 1989) have shown, by electron microscopy studies, that these bacteria were always present as a permanent member of the prokaryotic colony at different stages of leaf development of *Azolla filiculoides* and based on these data they have also suggested that these bacteria were the third partner of the symbiosis. At the same time Forni et al. (1987, 1989) have identified by biochemical tests several species of *Arthrobacter* present in different species of *Azolla*, suggesting the same hypothesis. The first references of the presence of bacteria in sporocarps was made by

Peters and Calvert (1982) and by Tavares and Carrapiço (1988). In this work, further research was made on this symbiotic association, trying to integrate and develop the understanding of the relationship between the bacteria and the life cycle of the fern.

## Materials and methods

The *Azolla filiculoides* plants were collected in Alcochete (70 km from Lisbon) and maintained in cultivation chambers. The megasporocarps were collected in April of 1987 and 1988 directly from the wild and prepared for further studies. Light and electron microscopy research were made on dorsal lobe leaves and in megasporocarps. For electron microscopy studies, leaves were submitted to a double step fixation process by 4% glutaraldehyde in 0.05 $M$ cacodylate buffer (pH 7.2) and in 1% $O_sO_4$ in 0.05 $M$ cacodylate buffer (pH 7.4) with or without ruthenium red (Carrapiço and Tavares, 1989). The leaves were dehydrated in acetone series and embedded in Epon-Araldite (Mollenhauer, 1964). The megasporocarps were fixed in a fixative consisting of formaldehyde-glutaraldehyde (2%–2.5%) in 50 m$M$ PIPES buffer (pH 7.4) for

3 h at room temperature and postfixed in 1% O,O, in 50 mM cacodylate buffer (pH 6.8) for 2 h at room temperature. The formaldehyde was prepared fresh from paraformaldehyde immediately before use. The megasporocarps were dehydrated through an ethanol series and embedded for several days in LR White's resin. Thin sections (60–80 nm) of the specimens were cut with a glass knife and Porter-Blum MT-2 ultramicrotome and mounted on copper grids. The sections were post-stained in 2% (w/v) uranyl acetate and Reynold's lead citrate and the observations were made in a Hitachi-12 electron microscope operating at an accelerating voltage of 75 Kv. Some megasporocarps were prepared for scanning electron microscopy and observed at 10–15 Kv in a Jeol JSM-820 electron microscope. For the light microscope, semi-thin sections (800 nm) were made from the specimens prepared for the electron microscope and transferred to a small drop of distilled water on a microscope slide. The sections were stained with acqueous toluidine blue and photographed in a Dialux Leitz Wetzlar microscope.

## Results

Bacteria which show a coryneform or rod morphology are found in the dorsal lobe of the leaf cavity in close association with the transfer hairs and *Anabaena* cells (Fig. 1). The bacteria and the cyanobacteria are immersed in a fibrillar network, revealed by the ruthenium red, that fills the leaf cavity and establish connections between *Azolla* and the partners of the association (Fig. 2). Megasporocarps, with an oval morphology, are about 1 mm in length and 0.5 mm in breadth (Fig. 3). Inside this structure, below the indusium exists a cavity that contains *Anabaena* cells (akinetes) (Figs. 4, 5), that are maintained through the life cycle of the fern. This cavity observed in transmission electron microscopy shows the simultaneous presence of *Anabaena* akinetes and bacteria cells (Figs. 6, 7). These bacteria are morphologically similar to the bacteria found on the leaf cavity (Figs. 2, 7, double arrows).

## Discussion

The presence of bacteria in the *Azolla-Anabaena* symbiosis was referred for the first time by Grilli (1964) by means of transmission electron microscopy and observed in the leaf cavity of the fern. These bacteria show a coryneform or rod morphology typical of the genus *Arthrobacter* (Carrapiço and Tavares, 1989; Gates et al., 1980; Peters et al., 1978) and were taxonomically identified by Wallace and Gates (1986). Recently, Forni et al. (1989) provide information on the enzymatic activities and the nutritional characteristics of the bacteria isolated. These authors have isolated from five *Azolla* species (*A. caroliniana, A. filiculoides, A. mexicana, A. microphylla* and *A. pinnata*) different species of a bacterium of the genus *Arthrobacter*. The species isolated were *A. globiformis, A. nicotianae, A. aurescens* and *A. cristallopoietes* or *A. pascens*, suggesting that this prokaryotic organism may be the third member of the symbiotic association.

In our observations we have observed that the bacteria were always present in all stages of leaf development in close association with the primary branched hairs, simple hairs or *Azolla* epidermal cells, following a location identical to the cyanobacteria *Anabaena azollae* (Carrapiço and Tavares, 1989). To confirm the idea that *Arthrobacter* is the third member of the symbiosis it would be necessary to look for the sexual structures of the fern. The presence of the bacteria in these reproductive structures was previously referred by Peters and Calvert (1982) in the microsporocarps of *Azolla filiculoides* and afterwards, in preliminary observations, in the indusium cavity of the megasporocarps of the same species by Tavares and Carrapiço (1988). Further research by means of transmission electron microscopy was done in this structure, showing that the bacteria found in the megasporocarp were morphologically similar to those found in the different stages of leaf development and strongly suggesting that this prokaryotic organism was retained throughout the sexual cycle of the fern, as it happens with the *Anabaena* cells, and probably can be considered the third partner of this symbiosis. Further research is needed to determinate the exact role of the bacteria involved in the symbiosis and if the

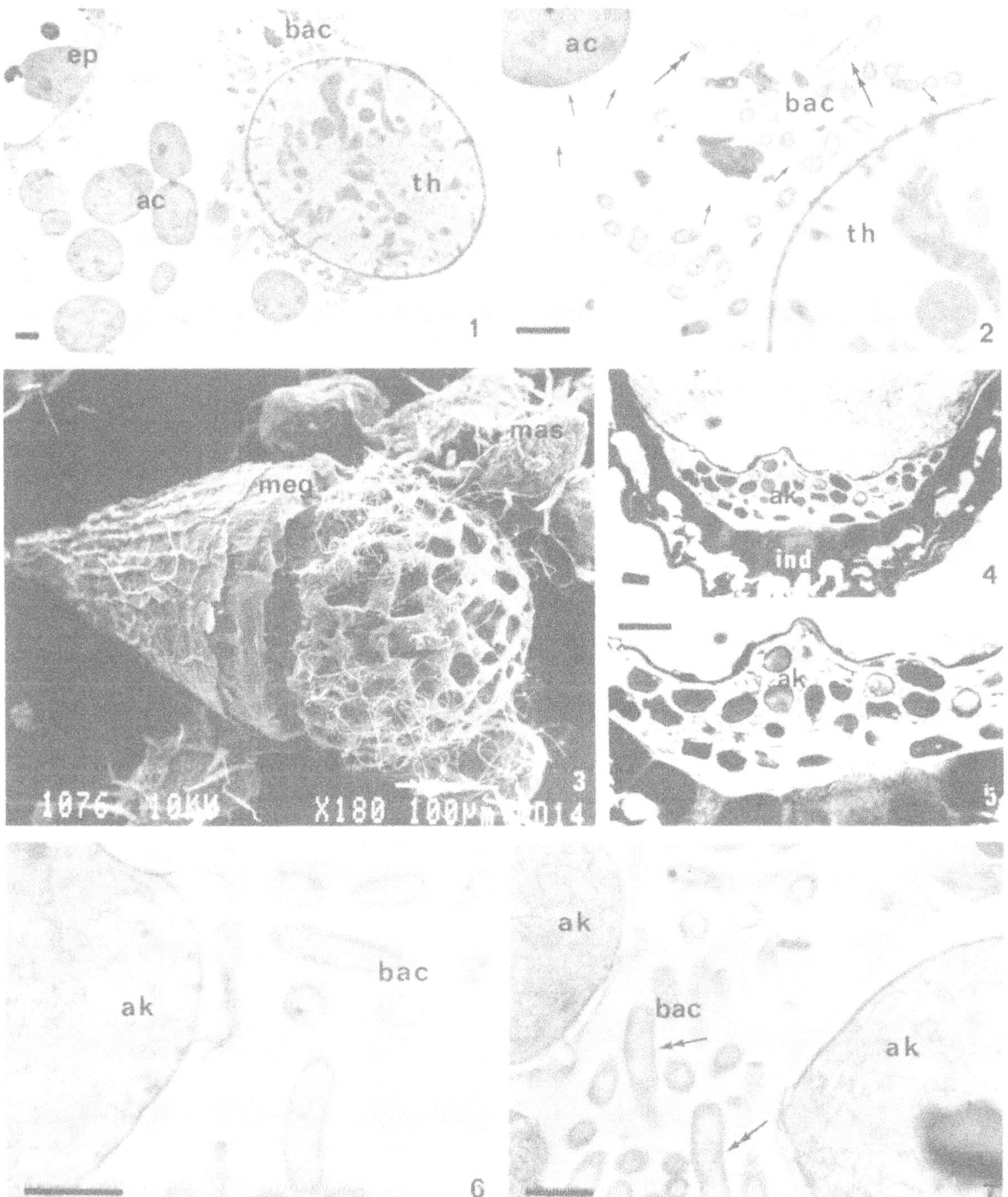

*Fig. 1.* Ultrastructural aspect of the leaf cavity. The transfer hair (th) is surrounded by a great number of bacteria (bac). *Anabaena* cells (ac) are close to these cells and *Azolla* epidermal cells (ep). *Bar* = 1 $\mu$m.

*Fig. 2.* Magnification of Figure 1, showing the fibrillar network (single arrows), where bacteria and *Anabaena* cells are immersed. *Bar* = 1 $\mu$m.

*Fig. 3.* A scanning electron microscopy micrograph of *Azolla filiculoides* megasporocarp (meg) surrounded by massulae (mas).

*Figs. 4. and 5.* Two images of the indusium zone of the megasporocarp, showing the *Anabaena* akinetes (ak) in the cavity below the indusium (ind). *Bar* = 10 $\mu$m.

*Figs. 6 and 7.* Micrographs of the indusium cavity showing *Anabaena* akinetes and bacteria cells. Some bacteria are morphologically similar to the bacteria found on the leaf cavity (double arrows, compare figure 2 with figure 7). *Bar* = 1 $\mu$m.

species found in the sporocarps are the same recently isolated from leaf symbiosis.

## Acknowledgement

The author thanks Mr R Tavares for the technical support in the work.

## References

Bottomley W B 1920 The effect of organic matter on the growth of various water plants in culture solution. Ann. Bot. 39, 353–365.

Carrapiço F and Tavares R 1987 New data on the *Azolla–Anabaena* symbiosis. I. Morphological and histochemical aspects. *In* Abstracts of the 4th International Symposium on Nitrogen Fixation with Non-Legumes. Rio de Janeiro, 23–28 August 1987.

Carrapiço F and Tavares R 1989 New data on the *Azolla–Anabaena* symbiosis. I. Morphological and histochemical aspects. *In* Nitrogen Fixation with Non-Legumes. Eds. F A Skinner et al. pp 89–94. Kluwer Academic Publishers, Dordrecht. The Netherlands.

Forni C and Grilli Caiola M 1986 Studio in culture dei batteri isolati dalle cavita fogliari di *Azolla* Giorn. Bot. Ital. 120, n. 1–2, suppl. 2, 100.

Forni C, Grilli Caiola M and Gentili S 1987 Bacteria in the *Azolla-Anabaena* symbiosis. *In* Abstracts of the 4th International Symposium on Nitrogen Fixation with Non-Legumes. Rio de Janeiro, 23–28 August 1987.

Forni C, Grilli Caiola M and Gentili S 1989 Bacteria in the *Azolla-Anabaena* symbiosis, In Nitrogen Fixation with Non-Legumes. Eds F A Skinner et al. pp 83–88. Kluwer Academic Publishers, Dordrecht. The Netherlands.

Gates J E, Fisher E W and Candler R A 1980 The occurrence of the coryneform bacteria in the leaf cavity of *Azolla*. Arch. Microbiol. 127, 163–165.

Grilli M 1964 Infrastrutture di *Anabaena azollae* vivente nelle foglioline di *Azolla caroliniana*. Ann. Microb. Enzim. XIV, 69–90.

Mollenhauer H H 1964 Plastic embedding for use in electron microscopy. Stain Technol. 39, 111–114.

Peters G A, Toia R E Jr, Raveed D and Levine N J 1978 The *Azolla-Anabaena azollae* relationship. VI. Morphological aspects of the association. New Phytol. 80, 583–593.

Peters G A and Calvert H E 1982 The *Azolla-Anabaena azollae* symbiosis. *In* Advances in Agricultural Microbiology. Ed. N S Subba Rao. pp 191–218. Oxford & IBH Publishing Company, New Delhi, India.

Petro M J and Gates J E 1987 Distribution of *Arthrobacter* sp. in the leaf cavities of four species of the N-fixing *Azolla* fern. Symbiosis 3, 41–48.

Tavares R and Carrapiço F 1988 Bacteria associated with the *Azolla-Anabaena* symbiosis. *In* Abstract of IBEREM 88. Lisbon. December 1988.

Wallace W H and Gates J E 1986 Identification of eubacteria isolated from leaf cavities of four species of the N-fixing *Azolla* fern as *Arthrobacter* Conn and Dimmick. Appl. Environ. Microbiol. 52, 425–429.

# Effects of antibiotic treatments on *Azolla-Anabaena* and *Arthrobacter*

C. FORNI[1], E. TEL-OR[2], E. BAR[2] and M. GRILLI CAIOLA[1]
[1]*Department of Biology, II University of Rome, Via E. Carnevale, I-00173 Rome, Italy and*
[2]*Department of Agricultural Botany, Hebrew University of Jerusalem, Faculty Agriculture, 76100 Rehovot, Israel*

*Key words:*  Anabaena, Arthrobacter, Azolla, erythromycin, novobiocin

## Abstract

The leaf cavities and the sporocarps of the water fern *Azolla* contain bacteria belonging to the genus *Arthrobacter* Conn and Dimmick. The number of these bacteria can be decreased by treating the plants with novobiocin and erythromycin. The latter antibiotic kills the *Anabaena*, so *Anabaena* free plants can be obtained. In these plants the bacteria disappear almost completely. The novobiocin does not affect the heterocyst differentiation, but it decreases the $N_2$-fixation activity by affecting the synthesis of the dinitrogenase reductase.

## Introduction

Bacteria, belonging to different species of the genus *Arthrobacter* Conn and Dimmick, are present in the leaf cavities of the fern *Azolla* Strasb. (Wallace and Gates, 1986; Forni et al., 1989), as well as in the sporocarps of *A. filiculoides* Lam. (Forni et al., 1990). The role played by these bacteria in the association is still unknown, but basing on the ubiquity and the constant bacterial population densities, Petro and Gates (1987) suggested the possibility of a controlled situation among *Azolla*, *Anabaena* and *Arthrobacter*. The study of the relationship among the three partners present problems, since sometimes it is difficult to determine the contributions of each component to the metabolism of the association. Therefore we wanted to test the possibility to affect one component of the association, i.e. the bacteria and to obtain bacteria-free *Azolla*. We determined the sensitivity to the antibiotics of the *Arthrobacter* spp., isolated from the leaf cavities and sporocarps of the fern. Successively we treated the association with the antibiotic to which the bacteria were sensitive.

This manuscript is a report of the effects of some antibiotic treatments on the bacteria as well as on the other two partners of the symbiosis.

## Materials and methods

### Antibiograms

*Arthrobacter globiformis* Conn and Dimmick (Forni et al., 1989) and *A. nicotianae* Giovannazi-Sermanni (Forni et al., 1990), isolated respectively from the leaf and microsporocarp of *Azolla filiculoides* Lam. were tested for their sensitivity to the antibiotics (Bacto-Unidisk, Difco). They were sensitive to erythromycin and novobiocin.

### Treatments of the association with the antibiotics

Plants of *A. filiculoides* were grown in the phytotron of the Department of Agricultural Botany at 27°C/22°C. under light/dark regime of 16/8 h respectively in the International Rice Research Institute (IRRI) medium (Watanabe et al., 1977).

For the treatment with the antibiotics, plants were grown in IRRI medium with added erythromycin or novobiocin (60 $\mu$g mL$^{-1}$). Growth rates of the plants were measured by doubling time using the formula of Moretti and Siniscalco (1988).

### Acetylene reduction

Plants were floated in 12 mL vials containing 2 mL growth medium, 10% acetylene in air and sealed with rubber stoppers. Samples were illuminated (90 $\mu$E m$^{-2}$ sec$^{-1}$) at 25°C for 1 h. Ethylene production was determined by a Gow-Mac model 69-100 gas chromatograph provided with a Poropack N column and flame ionization detector.

### Bacterial enumeration

The dorsal lobes of a plant (up to 15th leaf) were removed by manual dissection and surface sterilized in 5% NaClO for 30 sec. and rinsed in sterile water. The surface sterilized leaves were aseptically transferred to a glass homogenizer containing 2 mL of 0.9% NaCl. The leaves were homogenized and various dilutions of the homogenate were plated on TRN medium (Forni et al., 1989) in triplicate using 0.9% NaCl as diluent. Colonies were counted after three-four days at 30°C.

### Scanning electron microscope (SEM)

Samples were fixed in 5% glutaraldehyde in 0.1 $M$ phosphate buffer, pH 7.2 for 2 h at 4°C. Then the samples were rinsed twice in 0.1 $M$ phosphate buffer and dehydrated in a graded series of ethyl alcohol, critical point dried and coated with colloidal gold. The observation were made by a Jeol JSM-35 C microscope.

### SDS-page and immunobolt

*Azolla* plants were rapidly frozen in liquid nitrogen and the cyanobiont was isolated by gentle crushing of the *Azolla* fronds. Cell suspension was filtered through 8 layers of gauze and the *Anabaena* cells were separated by centrifugation (750 $g$ for 10 min.). The cell were resuspended in

1 mL of supernatant and 0.5% of lysozyme was added. The cell suspension was incubated on ice for 30 min. The addition of 1 mL SDS-lysis buffer (10%-$\beta$ mercaptoethanol, 10% SDS, 2 m$M$ EDTA, 100 m$M$ tris-base, 20% glycerol) was followed by 2 min vortex cell disruption in order to break the bacterial and vegetative cells. The suspension enriched with intact heterocysts was separated by centrifugation (2000 $g$ for 10 min.), resuspended in the running buffer (5%-$\beta$ mercaptoethanol, 2.5% SDS, 1 m$M$ EDTA, 10 m$M$ tris-base, 10% glycerol, 0.01% bromo-phenol-blue) and disrupted by harsh sonication (5 min. in 70% pulses). The suspension was boiled for 10 min and centrifuged in Eppendorf tubes for 5 min. in order to remove the broken cells. Samples with equal soluble protein content were loaded on 9–18% gradient polyacrilamide 0.1% SDS-gel and allowed to run for 12 h (250 V) in Bio-Rad Electrophoresis System at 4°C. Gels were electroblotted to a nitrocellulose membrane (Schleicher & Schuell BA 85, 0.45 $\mu$m). The membrane was transferred to blocking solution (PBS with 10% milk, adjusted to pH 7.5 with HCl) for at least 3 h followed by additional 3 h in the first antibodies solution (blocking solution containing anti dinitrogenase reductase polyclonal antibodies diluted 1:500, kindly provided by Prof. Ada Zamir, The Weizman Institute for Science, Israel) for at least 3 h. The membrane was washed three times, 20 min. each, with blocking solution with 0.5% Tween 20, and transferred to the second antibodies buffer (blocking solution with $^{125}$I-protein A diluted 1:1000) for 60 min. The membrane was washed three times, 20 min. each, drained by filter papers and autoradiographed at −70°C using Kodak X-OMAT film.

## Results

### Effects of the antibiotics on the numbers of bacteria

The bacterial strains tested were sensitive to both erythromycin and novobiocin. After 10 days of treatment of the plants with 60 $\mu$g mL$^{-1}$ of erythromycin or novobiocin, the number of

*Table 1.* Numbers of bacteria in plants of *Azolla filiculoides* Lam. treated with antibiotics

| Treatment | Bacteria/Plant |
| --- | --- |
| Control | 528.333 + 60.069 |
| Erythromycin | 3.666 + 2.032 |
| Novobiocin | 5.166 + 1.001 |

bacteria in the leaf cavities decreased respect to the control (Table 1).

## Effects of the antibiotics on the association

The presence of the antibiotic in the medium affected all the association. *Azolla* treated with antibiotics regressed and the doubling time of *Azolla* increased to 21 and to 27 days in the plants respectively treated with erythromycin and novobiocin in respect to the four days of the control. The nitrogenase activity decreased in the erythromycin treated plants within the first two days and almost completely disappeared after 8 days (Table 2). In the novobiocin treated plants, the nitrogenase activity was affected, but was still present after 8 days (Table 2). The heterocyst frequency was similar in the treated plants (15%) and in the control (16%).

## Effect of novobiocin on the dinitrogenase reductase of Anabaena

The novobiocin in the medium affected not only the nitrogenase activity, but also its synthesis was inhibited as it was shown by the dinitrogenase reductase (Fig. 1). The amount of 30 kD polypeptide decreased (Fig. 1) and almost completely disappeared after 9 days. We don't know if the novobiocin effect was also on other proteins, although the presence of the antibiotic had no significant effect on the general protein pattern

*Table 2.* Nitrogenase activity (ARA)* in plants treated with antibiotics

| Treatment | Days | | |
| --- | --- | --- | --- |
| | 2 | 5 | 8 |
| Control | 1626 + 124.4 | 1177 + 221.3 | 1137 + 138.2 |
| Erythromycin | 145 + 22.6 | 9 + 2.5 | 8 + 0.7 |
| Novobiocin | 1433 + 77 | 878 + 62.9 | 933 + 5.5 |

* ARA = nmol $C_2H_4$ $h^{-1}$ $g^{-1}$ fresh weight.

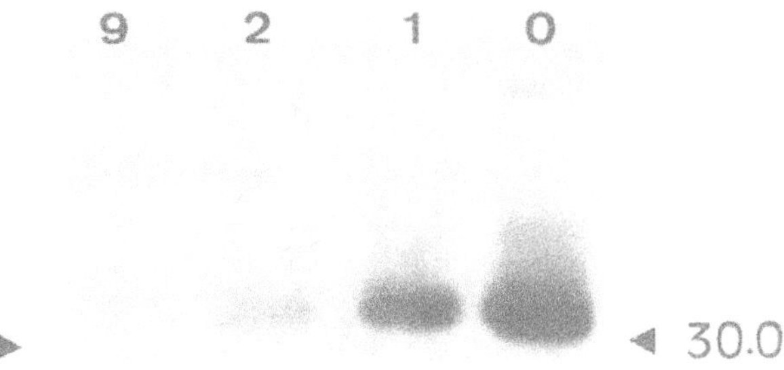


*Fig. 1.* Immunoblot of the dinitrogenase reductase extracted from the heterocysts of *Anabaena azollae* (0), after 1, 2 and 9 days of treatment with 60 $\mu$g mL$^{-1}$ of novobiocin.

of the heterocyst after Coomassie staining (data not shown).

## Azolla free of the symbiotic Anabaena

After 8-10 days of growth in the erythromycin medium the plants showed symptoms of senescence due to the effect of the drug on the cyanobacterium. After 15 days of treatment only the plants transferred to a medium supplemented with 0.5 gL$^{-1}$ NaNO$_3$ survived. In the leaf cavities of these plants the cyanobacterium disappeared and we obtained *Anabaena*- free plants (Fig. 2). In these plants the hair cells were still present (Fig. 2), while the bacteria were either mainly localized on the hair cells or completely absent. The *Anabaena*-free plants grew very well in IRRI medium supplemented with nitrate (0.5 g L$^{-1}$) (doubling time of 6 days vs 4 days of the control).

## Discussion

The novobiocin and erythromycin treatment affected to a different extent the *Anabaena* and the *Arthrobacter*. The erythromycin was the most effective on both partners, in particular the endophyte was so strongly affected to die and as a consequence of the treatment *Azolla* free of the symbiont has been obtained. Until now, the methods employed to achieve cyanobacterium-free *Azolla* involved treatments of a multitude of antibiotics (Peters and Mayne, 1974), which was most harmful to the fern or micromanipulation of megasporocarps (Lin and Watanabe, 1988), which degree of success depend upon avoiding

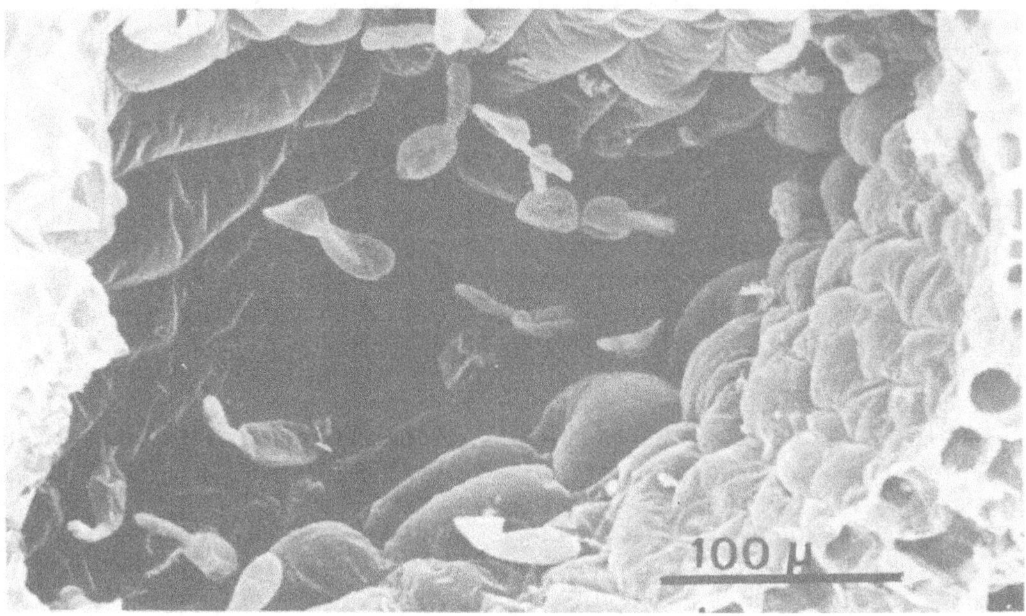



*Fig. 2.* Empty leaf cavity of *Azolla filiculoides* treated with erythromycin.

injury to megaspore apparatus. The erythromycin treatment is an easy and successful method and yields fronds which remained free of the symbiont and grow quite well in nitrate containing medium.

The disappearance of *Arthrobacter* from the leaf cavities of the *Anabaena* free plants suggests the dependence of the bacteria on the *Anabaena* for their growth. However the few bacterial cells surviving starvation, due to the absence of the endophyte, were localized on the hair cells surviving starvation, due to the absence of the endophyte, were localized on the hair cells of the fern. This may indicate the presence of viable hair cells. As for the novobiocin treatment, *Anabaena* survived the treatment, but the nitrogenase activity decreased with time, although the cell morphology and heterocyst frequency were similar to the control. However the activity was fully recovered upon the removal of the antibiotic from the medium (data not shown). It is likely that the decreasing of the activity may be due to the inhibition of the synthesis of dinitrogenase reductase as shown by immunoblot data.

The drug novobiocin is a specific inhibitor of DNA gyrase subunit B in bacteria (Drlica and Franco, 1988). According to Kranz and Haselkorn (1986) this drug completely blocks the activation of *Rhodopseudomonas capsulata nif* H D K transcription. Moreover the effect is specific on the *nif* H gene, i.e. a *nif* H:: *lac* Z fusion is not activated and the NifH, -D and K proteins are missing from the extracts of novobiocin treated cultures of the same microorganism (Kranz and Haselkorn, 1986).

As in *R. capsulata*, it is likely that in the *Azolla-Anabaena* treated cultures the novobiocin affected the cyanobacteria DNA gyrase and consequently the transcription of the *nif* H gene.

### Acknowledgements

We wish to thank Prof. Ada Zamir for providing us the anti Fe-protein antibodies and Mrs. Naomi Bahat for the precious help given for the SEM observation.

### References

Drlica K and Franco R J 1988 Inhibitors of DNA topoisomerases. Biochemistry 27, 2253–2259.

Forni C. Grilli Caiola M and Gentili S 1989 Bacteria in the *Azolla-Anabaena* symbiosis. *In* Nitrogen-Fixation with

Non-Legumes. Eds. F A Skinner et al. pp 83–88. Kluwer Academic Publisher, Dordrecht, The Netherlands.

Forni C, Gentili S, Van Hove C, and Grilli Caiola M 1990 Isolation and characterization of the bacteria living in the sporocarps of *A. filiculoides* Lam. Ann. Microbiol. 40, 235–243.

Kranz R G and Haselkorn R 1986 Anaerobic regulation of nitrogen-fixation genes in *Rhodopseudomonas capsulata*. PNAS 83, 6805–6809.

Lin C and Watanabe I 1988 A new method for obtaining *Anabaena*-free *Azolla*. New Phytol. 108, 341–344.

Moretti A and Siniscalco Gigliano G 1988 Influence of light and pH on growth and nitrogenase activity on temperate-grown *Azolla*. Biol. Fertil. Soils 6, 131–136.

Peters G A and Mayne B C 1974 The *Azolla, Anabaena azollae* relationship. I. Initial characterization of the association. Plant Physiol. 53, 813–819.

Petro M J and Gates J E 1987 Distribution of *Arthrobacter* sp. in the leaf cavities of four species of the N-fixing *Azolla* fern. Symbiosis 3, 41–48.

Wallace W H and Gates J E 1986 Identification of eubacteria isolated from leaf cavities of four species of the N-fixing *Azolla* fern as *Arthrobacter* Conn and Dimmick. Appl. Environ. Microbiol. 52, 425–429.

Watanabe I, Espinas C R, Berja N S, and Alimagno B A 1977 Utilization of the *Azolla-Anabaena* complex as nitrogen fertilizer for rice. IRRI research Paper Series 11, 15.

# The *Anabaena-Azolla* symbiosis: Diversity and relatedness of neotropical host taxa


W.J. ZIMMERMAN[1], I. WATANABE[2] and T.A. LUMPKIN[3]
[1]*Department of Natural Sciences, University of Michigan-Dearborn, Dearborn, MI 48128, USA,* [2]*Soil Microbiology Department, International Rice Research Institute, P.O. Box 933, Manila, Philippines and* [3]*Department of Agronomy and Soils, Washington State University, Pullman, WA 99164, USA*



*Key words:* azolla, DNA polymorphisms, isoenzymes, taxonomy


## Abstract


The *Anabaena-Azolla* association has proved to be an effective biofertilizer in tropical regions of wetland rice production. Three neotropical host species, *A. microphylla*, *A. caroliniana*, and *A. mexicana*, are similar in vegetative morphology (growth habits, frond dimensions, trichome cell number) and ecophysiology (relative heat tolerance). They were observed during our investigation to also be genetically alike and distinct from other taxa.


## Introduction

A major constraint to *Anabaena-Azolla* utilization as a rice biofertilizer is the inability to distinguish strains and even some species of the host. There are seven designated species of *Azolla*, an aquatic fern, divided by secondary reproductive features into two sections. Species traits in Section Azolla have been historically confined to anatomical characters and are of little value among neotropical taxa (three of five species). Varietal separation has never been achieved.

The morphological similarity of cultivated accessions has always been a major obstacle to the proper care of germplasm. Maintenance of live biomass is necessary since *Azolla* accessions rarely sporulate in culture collections. Fingerprinting assists the selection of isolates for field use, provides a means to monitor cross-contamination and somaclonal variation, and facilitates identification of some species. In this paper, we review our isoenzyme and DNA fingerprinting results thus far with azolla.

**Species code**

| | | | |
|---|---|---|---|
| *A. filiculoides* | FI | *A. pinnata* | P″ |
| *A. rubra* | RU | *A. pinnata* v. *pinnata* | PP |
| *A. caroliniana* | CA | *A. pinnata* v. *imbricata* | PI |
| *A. microphylla* | MI | *A. nilotica* | NI |
| *A. mexicana* | ME | | |

## Materials and methods

Isoenzyme comparisons by gel electrophoresis were first completed using known protocols (Zimmerman et al., 1989a,b). For DNA comparisons, genomic DNA digestions, fractionation, blotting, and Southern hybridizations were completed by standard techniques (Maniatis et al., 1983). Numerous gene probes were obtained for hybridizations. Thirty heterologous probes from maize, pea, lentil, squash, rice, soybean, *Clarkia*, barley, tomato, lettuce and the bacteriophage M13 were tested. A genomic DNA library (fragment sizes from 0.5–2 kb) was also made from an ME strain and partially screened. Another homologous probe, a 2.3 kb *Hind* III insert in pBS, from MI 431 (FAAS) was donated courtesy of J. Plazinski (Plazinski et al., 1989).

## Results and discussion

### *Isozymes*

Two initial objectives were achieved (Van Cat et al., 1989; Zimmerman et al., 1989a): 1) zymograms of healthy vs stressed leaf tissue, leaf vs root tissue, and leaf tissue with or without the cyanobiont were compared for discrepancies, and 2) the potential of this methodology for hybrid identification in *Azolla* was tested. These objectives were followed by a mass screening of accessions. Banding patterns for New World accessions were scored. The fingerprinting results were then interpreted from a perspective of biochemical taxonomy within Section Azolla (Zimmerman et al., 1989b).

Healthy leaves yielded more extensive information in their zymograms than did stressed leaves. Phosphorus deficiency was chosen since it is a common field stress and the most important limiting nutrient for *Azolla*. A reduced staining intensity of enzymes or the lack of visualization of loci was often observed in the zymograms of P-deficient fronds. The presence of the cyanobiont in leaves did not influence banding patterns (comparing FI + *Anabaena azollae*). Root zymograms of PI were found to be unlike those of leaves for two of four tested enzymes. Additionally, the presence of phenolics and other root compounds interfered with the resolution of zymograms.

Hybrids of FI × MI were successfully confirmed by observing the patterns of certain multi-loci enzymes such as PGM or SKDH. New hybrids (including those from crosses of ME × MI) continue to be confirmed at IRRI by enzyme electrophoresis.

The subsequent mass screening of accessions within this section was combined with data garnered from trichome morphology prior to final species analyses. Twelve enzymes were stained. Nine contained 17 polymorphic loci and 126 allelomorphic characters; three (ALD, G6PDH, F1, 6DP) appeared to be monomorphic. At least two loci each were present in SKDK and 6PGD.

FI accessions were enzymatically distinctive (e.g., IDH) and also identified by prominent one-celled leaf trichomes. True RU accessions were found only in Australia and New Zealand, and possessed barely discernable trichomes. The Japanese RU was concluded to actually be FI. Probable duplicated entries (e.g., IRRI strains 2010/2022, 1005/1006/SWD, 1010/1016) and other enzymatically identical accessions (6007/6008, 6003/6005/6006, 2016/2018) were reported. Unique strains were recorded, both those positively identified to species such as FI and RU and others of the CA-ME-MI cluster (e.g., 2001, 2009, 2021, UCD 60 (probably IRRI 2005), 3001, 3008, 3016, 4033).

MI (from Paraguay and the Galapagos Islands) were all unique by their faster PGM-2 locus. However, a principal component analysis showed no discernible phenetic distance separating MI from the other species of the CA-ME-MI cluster (Fig. 1). The strains of these species commonly have two-celled leaf trichomes.

Incorrect labelling was found among many isolates of the 1000 series (FI) in the IRRI collection. Colombian strains within the 1014–1027 series, among others, have now been re-labelled as ME (2010–2022). An earlier series of growth experiments under high temperature had indicated that these strains were heat-tolerant

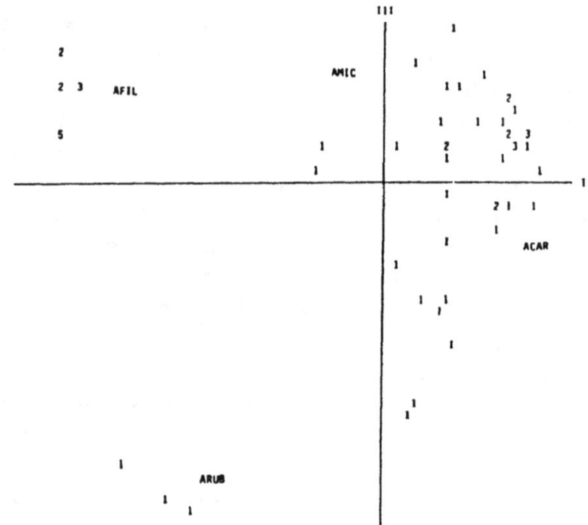

*Fig. 1.* Principal component analysis of 57 accessions of New World *Azolla* (Zimmerman et al., 1989b). AFIL = FI, ARUB = RV, AMIC = MI, ACAR = CA, ME.

(IRRI, 1986). Biomass production was similar to MI 4018 and unlike heat-sensitive FI. Possible natural hybrids in this series were noted for further genetic analysis.

## DNA polymorphisms

Results with DNA restriction fragment length polymorphisms (RFLPs) reinforced the conclusions derived from phenetic data and located additional polymorphisms among strains which were enzymatically related. Heterologous/ homologous gene probes and detailed hybridization results are described elsewhere (submitted manuscripts).

Hybridization patterns in several cases distinguished P″ (and sometimes PI from PP), the true RU, NI, and FI. The CA-ME-MI cluster remained highly coherent. None of the last three species were elucidated as a separate identity by genetic fingerprints, as was the case of MI with certain zymograms.

Distinctively polymorphic accessions, which had been previously observed to be enzymatically unique, included 2001, 3001, 6502, and 6503. New strains to be added to this list included 66, ENG, 3015, 4030, 4072, 5001, and 7001. Based on preliminary information, there is a high likelihood of additional unique accessions being identified as RFLP research is continued.

MI strains collected in Paraguay over ten years ago were not genetically identical to others collected in 1988 by Drs. Watanabe and Zimmerman. These recent strains were mostly taken from the same semi-arid region (Chaco Province) as the first survey. Certain fingerprints also helped distinguish the cold tolerant RU 6500 series. Further and more extensive trials are necessary in this regard.

The pAm7 MI homologous probe was useful by its discriminatory ($+/-$) effect with *Azolla*. The original claim (Plazinski et al., 1989) for MI species specificity was based on dot-blot hybridizations with a few *Azolla* strains of different species. Our hybridizations with dozens of *Azolla* accessions showed that the discrimination was not species-specific (Fig. 2). Successful hybridizations, however, have so far only been achieved with accessions from the CA-ME-MI cluster. On the other hand, accessions of this group also

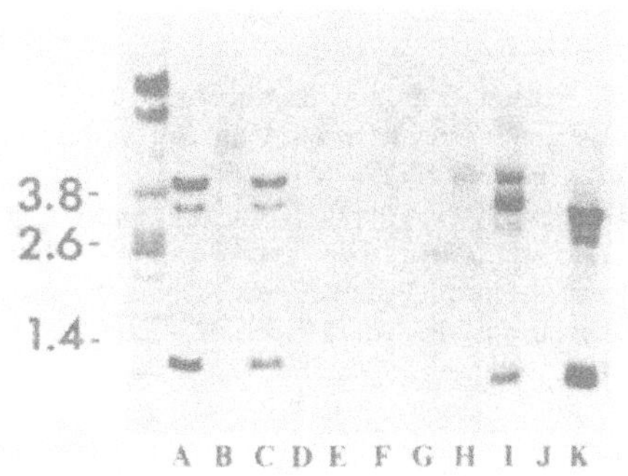

Fig. 2. Southern hybridization with the pAm7 probe and *Hin*dIII-digested DNA of CA and ME accessions.

formed the majority of strains which did not hybridize to pAm7. FI, NI, PI, and PP accessions did not hybridize, but their sample numbers were small. Furthermore, results were unclear for several strains (including RU 6502 and PI 72), so this probe may not be exclusively specific for a select CA-ME-MI subgroup.

## Breeding

Sexual hybrids have been realized from techniques developed for cross-breeding *Azolla* (Van Cat et al., 1989). FI × MI hybrids, successfully showing positive heterosis in field biofertilizer studies, are sterile. More recently, fertile hybrids bearing both mega- and microsporangia have been attained by ME × MI crosses (Watanabe, unpubl.) This achievement provided further evidence of the CA-ME-MI taxonomic relatedness.

Any taxonomic modification of New World species requires more empirical evidence. The answers found in our enzyme and RFLP results are exceeded by the number of new questions derived from our data. Major prerequisites to advancing *Azolla* genetics are successful utilization of molecular markers, the sporulation trigger, and sexual hybridizations.

## Acknowledgements

The work presented is condensed from a final report of research supported by the U.S. Agency for International Development, Grant no. DPE-5542-G-SS-6041-00.

## References


Maniatis T, Fritsch E F and Sambrook J 1982 Molecular Cloning: A Laboratory Manual. Cold Spring Harbor Press, Cold Spring Harbor, NY.

Plazinski J, Zheng Q, Taylor R, Rolfe B G and Gunning B E S 1989 Use of DNA/DNA hybridization techniques to authenticate the production of new *Azolla-Anabaena* symbiotic associations. FEMS Microbiol. Lett. 65, 199–204.

Van Cat D, Watanabe I, Zimmerman W J, Lumpkin T A and de Waha Baillonville T 1989 Sexual hybridization among *Azolla* species. Can. J. Bot. 67, 3482–3485.

Zimmerman W J, Lumpkin T A and Watanabe I 1989a Isozyme differentiation of *Azolla* Lam. Euphytica 42, 163–170.

Zimmerman W J, Lumpkin T A and Watanabe I 1989b Classification of *Azolla* spp., section *Azolla*. Euphytica 43, 223–232.

# Assessment and attempt to explain the high performance of Azolla in subdesertic tropics versus humid tropics

T. de WAHA BAILLONVILLE[1], H.F. DIARA[2], I. WATANABE[3], P. BERTHET[1] and C. VAN HOVE[1]
[1]*Université Catholique de Louvain, Laboratoire de Physiologie Végétale, Place Croix du Sud 4, B-1348 Louvain-la-Neuve, Belgium,* [2]*West Africa Rice Development Association, B.P. 96 Saint Louis, Senegal and* [3]*International Rice Research Institute, Soil Microbiology Division, P.O. Box 933, Manila, Philippines*

*Key words:* Azolla, cluster, covulation, natural conditions, productivity

## Abstract

Two similar experiments aimed at selecting performing *Azolla* strains at Richard Toll, Senegal (subdesertic tropics) and Los Baños, Philippines (humid tropics) show that annual mean productivity of the tested strains is higher in the first site than in the second, the best ecotypes in Senegal producing more than twice as much as in the Philippines.

Attempts to explain these differences by climatic or interaction of climatic variables (in the range explored) are unsuccessful. Further analyses are thus needed to understand the reasons, climatic or other, responsible for the different behaviour of *Azolla* under the two studied ecologies.

## Introduction

Attempts to introduce *Azolla* in various environments, as fertilizer and/or as food, require the selection of high yielding ecotypes and thus reference collections including *Azolla* strains from as many biotopes as possible.

The present work was primarily aimed at selecting high performance strains for very contrasted environments: tropical subdesertic represented by experimental plots of the West Africa Rice Development Association (WARDA) at Richard Toll, Senegal, and tropical humid represented by the experimental farm of the International Rice Research Institute (IRRI) at Los Baños, Philippines.

It is based on the *Azolla* reference collections of IRRI and of the Catholic University of Louvain, joined since 1987, and on a standardized methodology allowing easy, year round, biomass measurements of a large number of *Azolla* strains under natural conditions.

## Material and methods

### Azolla strains used

Both at IRRI and WARDA 37 strains, among the well adapted of the locally introduced strains, have been tested. At IRRI 12 *A. pinnata* var. *imbricata* (PI), 2 *A. pinnata* var. *pinnata* (PP), 3 *A. microphylla* (MI), 6 *A. mexicana* (ME) and 14 *A. caroliniana* (CA) strains and at WARDA 19 PI, 7 CA, 5 MI and 6 *A. filiculoides* (FI). Among them, only 7 strains are common to both sites.

### Experimental set-up

*Azolla* strains are grown, and their productivity measured weekly, as described by Van Hove et al. (1983) and de Waha et al. (1991). The experimental layout is a fully randomized design with 2 replicates. Productivity data are given as g FW m$^{-2}$ d$^{-1}$.

468

*Climatic data*

Climatic data are provided by the IRRI and by the Compagnie Sucrière Sénégalaise metero-lological stations. Only the less correlated variables ($r < 0.95$)) are considered: maximum air temperature (Max T.) minimum air temperature (Min T.), mean air temperature (Mean T.), daily range of temperature (Range), rainfall (RAIN) and minimum relative humidity (Min RH).

*Statistical analyses*

ANOVA, CORR, CLUSTER, GLM, and REG procedures of the SAS software are used (SAS, 1985).

**Results**

*Mean productivity*

At each site analysis of variance shows highly significant strain and time effects and highly significant interactions between the two, indicating different productivity patterns throughout the year among the different *Azolla* strains. Mean annual productivity has therefore been calculated for each strain; they are presented (Fig. 1) versus their standard deviation. Based on the centroid method which compares the euclidian distance between strains, 7 clusters are considered.

At IRRI (Fig. 1a) one cluster (A) with a mean productivity of 28.6 g FW $m^{-2}$ $d^{-1}$ includes all the PI and PP strains, whereas the other (B) with a mean productivity of 67.7 g FW $m^{-2}$ $d^{-1}$ groups all the CA, ME and MI strains.

In Senegal (Fig. 1b) all the strains have higher mean annual productivity than those tested in the Philippines, with standard deviations generally higher.

The most isolated clusters (C and G) comprise respectively 2 poorly adapted strains and one strain, very productive but with a very high standard deviation. Cluster F, which contains the majority of the PI strains and only PI strains, is characterized by high productivities and relative-

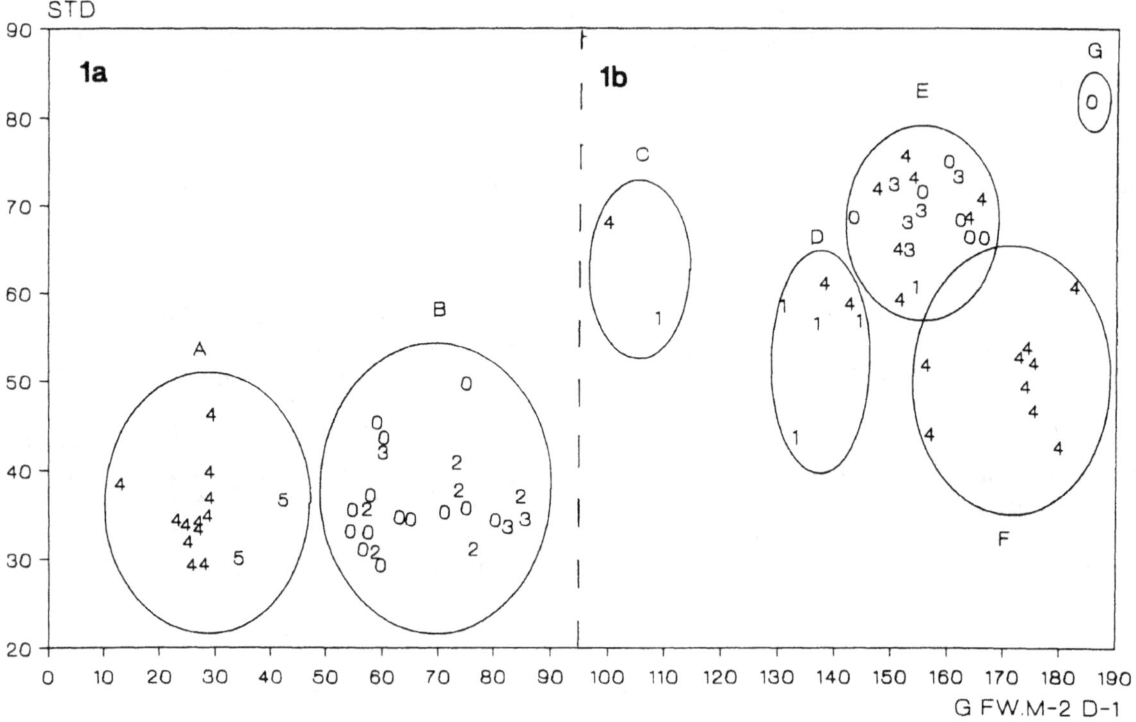


*Fig. 1.* Mean annual productivity: **a** at IRRI (4 Aug. 1987–26 July 1988) and **b** at WARDA (23 Dec. 1985–15 Dec. 1986). 0 = CA, 1 = FI, 2 = ME, 4 = PI and 5 = PP.

ly low standard deviations. Cluster D, like C, includes only FI and and PI strains, whereas strains from all the tested species are present in cluster E.

*Seasonal variation of productivity*

The productivity patterns on a monthly basis of all the tested strains, grouped according to the centroid method, are presented in Figure 2.

Statistical analysis shows a minimum in June and a maximum in August–September for the 2 clusters at IRRI. At WARDA the minimum for all groups is in January while June–July always belongs to the most productive period, which extends more or less from March to August according to the cluster.

Analysis of covariance shows that at IRRI all the climatic variables (except minimum temperature) significantly affect productivity which increases with decreasing maximum and mean temperature, daily range of temperature, solar radiation and increasing rainfall and relative humidity. Cluster B is more sensitive to those variables than A.

At WARDA productivity significantly increases with all climatic variables except daily range of temperature, rainfall and relative humidity which do not affect the productivity of the 5 groups significantly. Clusters E and G are the most responsive, followed by F and by C and D.

Simple correlations are computed to determine which climatic variables are affecting the most the productivity of the different clusters.

At IRRI none of the correlations are found significant at 0.05 level, whereas at WARDA they are significant and positive with maximum, minimum, mean temperature and solar radiation for all groups. The most correlated variable is solar radiation for clusters C, D, F and G ($0.72 < r < 0.82$) and maximum temperature for cluster E ($r = 0.85$). The second most correlated climatic variable is mean temperature ($0.70 < r < 0.82$) except for cluster G (maximum temperature, $r = 0.79$). The least correlated variable is minimum temperature for all groups ($0.54 < r < 0.68$) except group C (maximum temperature, $r = 0.68$). To consider interactions among the climatic variables, multiple regressions are

computed for all groups including the 3 most correlated climatic variables. At IRRI, determination coefficients do not exceed 31% (cluster B) and at WARDA, they vary from 59% (cluster F) to 77% (cluster E).

Addition of a fourth variable in the equations does not give any substantial increase of $r^2$.

## Conclusions

*Azolla* use in agriculture has long been limited to subtropical, humid, climates. Various *Azolla* strains were nevertheless introduced recently with success in the northern part of Senegal, characterized by a subdesertic climate (Van Hove et al., 1987). The present study confirms the high performances of *Azolla* in such ecosystems (Richard Toll), with a mean annual productivity of all tested strains of 157 g FW m$^{-2}$ d$^{-1}$ as compared to the 57 g FW m$^{-2}$ d$^{-1}$ at Los Baños. This raises the question of the influence of climatic variables on *Azolla* productivity.

With the aim of identifying climatic variables possibly responsible for such changes various statistical analyses have been used; when applied to individual strains they didn't lead to any convincing evidence. This confirms difficulties met by Lumpkin to relate the relative growth rate of 8 accessions, grown year round in Hangzhou (P.R. of China), to climate (Lumpkin et al., 1986). Further analyses were therefore realized on clusters of strains, grouped according to the level and stability throughout the year of their productivities, assuming that strains belonging to the same group react similarly to climatic changes. Two such clusters are monospecific: A: which contains only *A. pinnata* strains with very low productivities at IRRI and F, which contains only *A. pinnata* var. *imbricata* strains, with very high productivities at WARDA. Significant correlations between some individual climatic variables and productivity appear at WARDA. Multiple regressions with 3 variables could explain up to 77% of the variation of the productivity at WARDA whereas they are unsuccessful at IRRI. This is probably due to the shortest range of variation of the different climatic variables at this site.

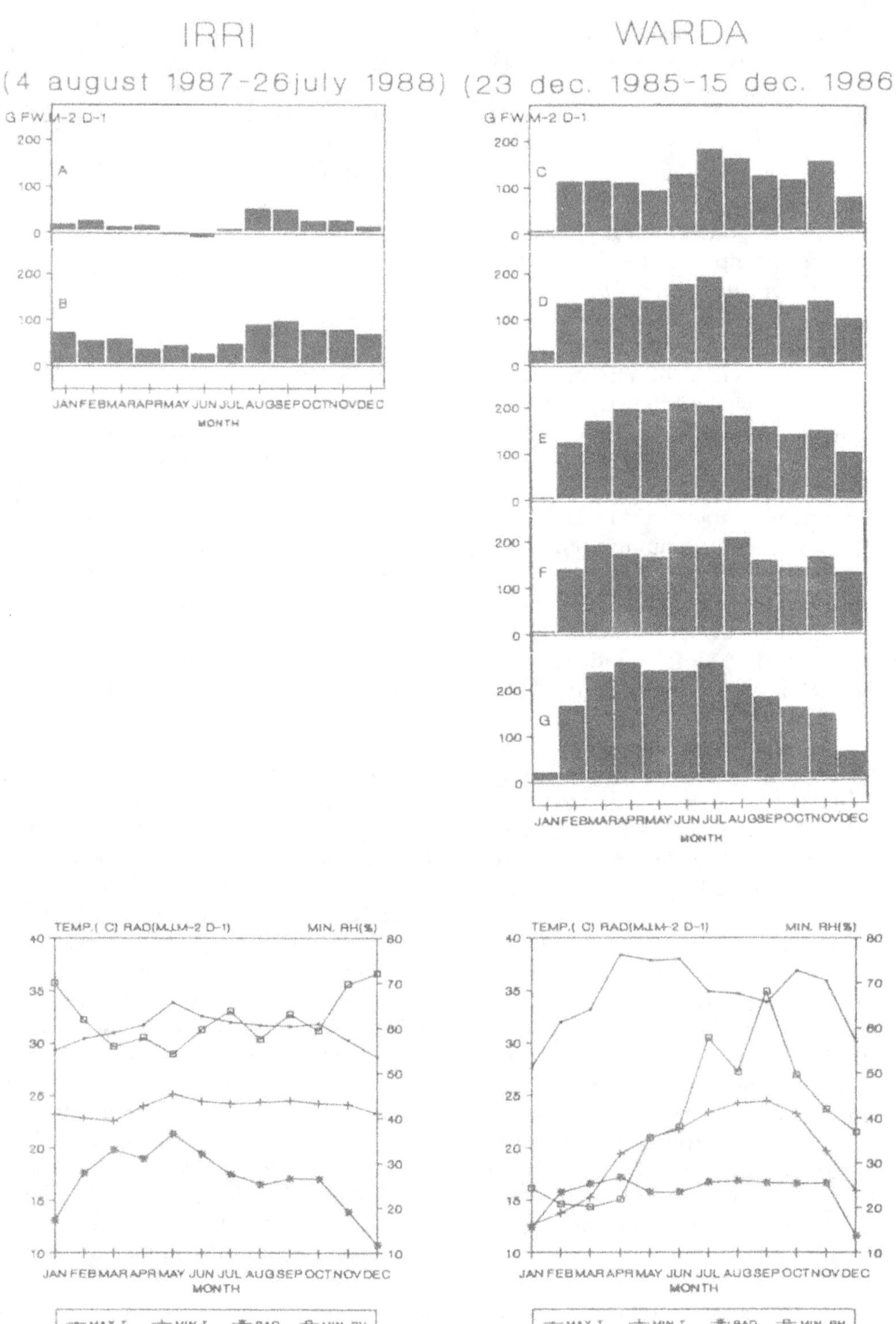



*Fig. 2.* Mean monthly climatic conditions and productivity of the different groups considered at IRRI and WARDA.

It remains that the present analyses do not give any satisfactory explanation to the very different performances of *Azolla* in the two biotopes and that new experiments will be necessary in order to identify the limiting factors involved at Los Baños.

## Acknowledgements

This research was supported by the Belgian Administration for Development Cooperation (AGCD) for special projects with WARDA (1980–1989) and with IRRI (1987–1992).

## References

de Waha Baillonville T, Diara H F, Godard P and Van Hove C 1991 A standardized methodology to measure *Azolla* biomass under natural conditions. International Rice Commission Newsletter June 1991.

Lumpkin T and Bartholomew D P 1986 Predictive models for the growth response of eight *Azolla* accessions to climatic variables. Crop Science 26, 107–111.

SAS 1985 SAS User's Guide: Statistics, Version 5 Ed. SAS Institute Inc. Cary, NC. 956 p.

Van Hove C, de Waha Baillonville T, Diara H F, Godard P, Mai Kodomi Y and Sanginga N 1987 *Azolla* collection and selection. *In Azolla* Utilization, pp 77–87. IRRI, Los Baños, Philippines.

Van Hove C, Diara H F and Godard P 1983 *Azolla* en Afrique de l'Ouest-in West Africa. Ed. C Van Hove. Ed Oleffe, Gourt St Etienne, Belgium. 56 p.

# CONSTRAINTS FOR THE WIDESPREAD USE OF AZOLLA IN RICE PRODUCTION


S.A. Kulasooriya
University of Peradeniya and Institute of Fundamental Studies
Kandy
Sri Lanka


ABSTRACT: The traditional use of Azolla in rice cultivation in China and Vietnam goes back to several centuries, but is has remained a mere botanical curiosity until the mid-seventies.
Intensive research carried out in several countries during the past 15 years, has demonstrated its potential as an attractive biofertilizer for rice. Inspite of this, Azolla has not yet been accepted widely by rice farmers for field use.
This is due to several constraints, which may be categorised broadly into:(a) fundamental & scientific, (b) economic & organizational and (c) social & political. The major fundamental constraints are limitation of water supply and phosphorus and susceptibility to temperature, pests and pathogens. Economic and organizational requirements revolve around the development of a simple technology that fits into existing rice production system, the production, storage and distribution of good quality material, the education and training of farmers and the derivation of maximum economic benefits by the multipurpose use of Azolla. Social and political constraints depends primarily upon the attitudes of farmers, administrators and politicians.
In this presentation, an attempt would be made to discuss ways and mean of overcoming these constraints in order to move this technology to the end user: the farmer.


## Historical

The use of Azolla for rice in China has been recorded in Chinese literature as early as 540 A.D. (Liu Chung Chu 1979), and its organised use in rice cultivation appears to have existed during the Myng Dynasty (1368-1644 AD). According to Dao and Tran (1979) its use in Vietnam dates back to the 11th century. Inspite of these ancient origins, this practice has been traditional and localised until the late 1960's when Government sponsored programs made it spread rapidly (Lumpkin & Plucknett 1982). Reports on the extent of Azolla use varies from 1.34 to 6.5 million ha for China and 500,000 ha for Vietnam (FAO 1988), but later reports indicates a declining trend in its use due to the availability of chemical fertilizer and the cultivation of more economical crops (Watanabe & Liu Chun Chu 1990). Except for the limited success in the Philippines in the 1980's, the use of Azolla has not really moved beyond experimental demonstrations in most other countries (FAO 1988). Its potential as a biofertilizer for rice has been demonstrated in Brazil, India, Philippines, Pakistan, Sri Lanka, Thailand, and West Africa (IRRI 1987). Inspite of all this research and field demonstrations, the use of Azolla has not yet moved onto the rice farmers. This is due to a number of constraints which have to be surmounted prior to any widespread acceptance of Azolla. These constraints may be broadly categorised as (a) fundamental & scientific, (b) economic & organizational and (c) social & political and for purposes of clarity its best that they are considered separately.

(a) Fundamental & Scientific

Being a living plant, the growth and rapid establishment of Azolla in any ecosystem would depend upon its interaction with the environment. Major environmental factors that influence its growth are availability of water and dissolved P, effect of temperature, solar insolation and humidity, and infections by pest and pathogens.

1. Water

Although Azolla grows well in contact with moist soil it requires standing water to disperse. Otherwise the roots stick to the mud and most plants get submerged and die when water is reimpounded. On the otherhand if the depth of the flood water is too high the plants could suffer nutrient deficiencies, particularly phosphorus. The ideal situation would be a level of water shallow enough to permit the roots to touch the soil occasionally. Another indirect, detrimental effect is the increase in temperature of stagnant water under high solar insolation. This could be minimised by having a slow flow through of water, across the field. Having flow through systems of irrigation water would be considered inefficient and wasteful water management. Drainage and reimpounding of water is also necessary to incorporate Azolla periodically in order to obtain its maximum fertilizer effect. All this means that to realise the full fertilizer potential of Azolla it is necessary to have a regular supply of controllable water. Such water supplies are not available in most developing countries, even under irrigated systems and is therefore a major constraint for the large scale use of Azolla.

In most situations water becomes available just before land preparation and this period is not long enough to produce biomass of Azolla adequate to provide a significant basal dressing of fertilizer. However, during the initial tillering period, standing floodwater is maintained to minimise weed growth, and dual culture of Azolla can be adopted during this period. Two incorporations of dual cultured Azolla along a rice cultivation cycle in Sri Lanka has been shown to provide grain yields equivalent to the addition of 55 to 85 KgN/ha of urea fertilizer (Kulasooriya et al 1987). Similar experiments done at other locations

have shown that on a  average the effect of two Azolla incorporations
is equivalent to 60 KgN/ha of urea fertilizer (Kulasooriya et al 1988).

## 2. Planting pattern of Rice

Soil incorporation of Azolla can be done most effectively only under
row transplanted rice.   However, such transplanting is rarely done
in developing countries.  For example 75% of the farmers in Sri Lanka
use direct seeding.  Even the few who adopt transplanting, do so ran-
domely.  Under such planting patterns, periodic Azolla incorporations
cannot be done without damaging rice plants.  It is unrealistic to
expect the farmers to change their rice planting patterns to accom-
modate Azolla, until they are fully convinced that such changes can
assure them substantial economic returns.  It would be more pragmatic
to adopt dual culture of Azolla under random transplanted and direct
seeded rice, without soil incorporations.  Such growth may have minimal
effects (if at all) on the standing crop, but could lead to a build
up of N-rich organic matter after several successive crop cycles.
An increase of CEC from 13.3 me to 27.2 me/100g has been reported in
Indonesia after successive growth of Azolla in rice fields (FAO 1988) .

## 3.  Phosphorus

Of the inorganic nutrients essential for the growth of Azolla, phosphorus appears to be the most limiting one under field conditions.
Azolla can grow without P-applications only in soils rich in available
P (Olsen P<25ppm) and those having a low P-sorption capacity of >1500
mg $P_2O_5$/100g (Watanabe and Ramirez 1984).  The floodwater P-levels
of rice soils are generally low because added P gets rapidly fixed
in the clay soils.  As long as the Azolla plants remain floating on
the floodwater, they encounter P-deficiency.  This may be overcome
by having a shallow level of standing water that permits the Azolla
roots to touch the soil periodically or by the moist soil culture
technique (Liu Chung Chu 1987), but such water control could be seldom
achieved in most rice fields.  Moist soil cultivation of A. caroliniana
has been shown to leave 11 to 15 t/ha of Azolla biomass after harvest
(Liu Chung Chu 1987) and this method has been suggested as a useful
new development suitable for the Asian countries which have problems
of high temperature.  There are however no recent reports of this technique been tried out in such countries.

An often asked question is whether Azolla requires P-fertilizers in
addition to what is provided to a rice crop.  The answer could be 'no'
because P added to Azolla eventually becomes available to the rice
crop together with a bonus of nitrogen.  However, for Azolla to assimilate P efficiently it has to be applied in several splits (Watnanbe
et al 1980) and this requires additional labour; a constraint to most
farmers.  This could be overcome by exploiting the ability of Azolla
for luxury consumption of P (Lumpkin and Plucknett 1982).  Such P-
enriched nursery material when used as inocula increase their biomass
3 to 8 times under P-deficient conditions, before they reach a critical

476

stage (Lumpkin 1987). Use  of P-loaded Azolla inoculum followed by
a P-fertilizer application schedule with one basal and two top dressings
similar to N-fertilizer applications, could integrate Azolla technology
to a fertilizer application practice already familiar to the farmers.

4.  Pests and pathogens

Another major environmental limitation for the widespread use of Azolla,
particularly in the humid tropics, is its susceptibility to pests and
pathogens (Mochida et al 1987).  Pest attacks were encountered in all
the localities in Sri Lanka where field growth of Azolla was attempted.
Pest damaged Azolla is easily infected by fungi leading to rapid death
and decomposition.  While it was necessary to use chemical pesticides
to control severe attacks, it was observed that vigorously growing
Azolla plants had some resistence to pest infestations.  If the environmental conditions, particularly temperature and P-availability were
favourable, pest attacks were less damaging.  However, the use of pesticides is ineveitable at present if large scale cultivation of Azolla
is to be successful in the humid tropics.

There have been some reports of non-chemical material such as neemcake,
tobacco leaf extract having pesticidal effects on Azolla pests (Kannaiyan
& Nandabalan 1987) and the exploitation of such sources could be quite
profitable.  During casual observations, it has been noticed that there
are certain predators including certain rice field frogs feeding on
Azolla pests.  Careful studies on such biological control agents could
pay rich dividends.

Inoculum production and preservation

As long as vegetative material is used as field inocula, year round
maintenance of Azolla cultures becomes an essential requirement for
its widespread use.  Depending on the locality and season, there could
be problems of either high  or low temperature, water availability,
time and labour, needed for such maintenance.  Techniques have been
developed in China for the harvesting and seeding of fields with Azolla
sporocarps (Lu Shuying 1987), but limited sporulation among most Azolla
species and the slow growth of young sporophytes are limitations for
the immediate application of these methods, Nevertheless, it is certainly worthwhile to pursue these studies as they may result in break-
throughs that could overcome the constraints of maintaining vegetative
cultures of Azolla and the transporation of bulky material to and within
the rice fields.

Another approach to overcome environmental limitations is to breed
new strains of Azolla that are resistent to such constraints.  An ideal
Azolla strain should not only grow rapidly and fix nitrogen efficiently,
but also be resistent to desiccation, high and low temperature, low
levels of phosphorus and pest and pathogens.  Success of conventional
breeding techniques to obtain such strains would remain rather remote

until the control mechanisms of its sexual cycle are well understood.
Recent successes in hybridization of Azolla (Do van Cat et al 1989)
have given renewed hope of deriving resistent Azolla strains.

b) Economic and Organizational

The current techniques available for Azolla use are all labour intensive,
and increases the drudgery of farmers who are already burdened with
a load of manual work.  Unless the work involved from inoculation,
P-additions to soil incorporation become less cumbersome, it is most
unlikely that the methods would be accepted voluntarily by farmers.
Simple innovations like setting up of Azolla nurseries at the highest
elevation of a field so that inoculation can be effected through the
flow of irrigation water, use of P-loaded nursery material to minimise
frequent P-additions and mere walking over an Azolla cover even with
less effective soil incorporation, could help in popularizing this
technology among farmers.

A promising development is the Azoll-rice-fish culture reported from
China (FAO 1988a).  This system not only accelerates the cycling of
nitrogen in a rice field, but improve the productivity per unit area
of land and enhances the income of farmers.  No technology would be
attractive to the farmers, unless it can bring them some monetary
rewards.

If any country is seriously considering the adoption of Azolla technology on a large scale, it is necessary to have a government sponsored
national plan.  A central organization should be entrusted with the
development and maintenance of a culture collection having diverse
germplasm.  This organization should ensure the provision of good quality inoculum material together with the technical know-how on their
use.  It should also continue research primarily directed towards solving field problems.  In addition there should be several location
specific "Azolla production and distribution units" which supply inoculum material as well as extension services to the farmers.  Field
multiplication of Azolla should be entrusted to motivated farmers
who should be financially rewarded for their efforts.  They should
be made to produce Azolla not only for rice but also as animal feed,
so as to elevate its position to a marketable product.

Incentives such as fertilizer subsidies should be curtailed to motivate
farmers to adopt alternative sources.  However, this should be done
gradually and cautiously so that grain production is not adversely
affected.  It is also necessary to emphasise that Azolla cannot replace
the entire chemical-N demand of modern high yielding varieties of rice.
Azolla is best used in combination with fertilizer nitrogen, as N-
recovery from  N-labelled fertilizer by rice plants was enhanced by
such combined application.

478

c) Social and Political

Since the introduction of new improved varieties of rice, together
with the agrochemical packages nearly 30 decades ago, an entire generation of rice farmers have got used to the "bagged synthetic inputs".
A radical change in their attitudes would be necessary to bring them
back to "organic farming". Policy decisions for such changes will
have to be taken at the highest levels. The escalating costs and the
uncertainty of an uninterrupted supply of petro-based fertilizers
may induce governments to decide on drastic changes. Prior to that
scientists will have to be ready with simple, easily adaptable techniques that could ensure the smooth transfer of this technology to
the rural farmers.


Conclusion

Azolla has a high potential as a nitrogen rich fertilizer for rice.
A number of environmental, organizational and social problems have
to be overcome to fully realise this potential. Further studies are
necessary to find solutions to problems connected with high temperature,
P-nutrition and pests and pathogens. Where water control is not possible
only a limited benifit can be derived. Dual cultured Azolla under
row transplanted rice when incorporated into the soil, can replace
½ the N-fertilizer requirement. Azolla grown under direct seeded or
random transplanted rice is best left unincorporated to provide a
residual effect. Use of Azolla with chemical fertilizer could improve
fertilizer uptake efficency. A large scale Azolla program should have
a centrally controlled organization to provide good quality material
and technical know-how. Radical changes in attitudes are necessary
to consider Azolla as a marketable product that could bring monetary
rewards to the farmer.

References


Dao, T.T. Tran Q.T. (1979). Use of Azolla in Rice Production in
    Vietnam. In Nitrogen and Rice. The International Rice Research
    Institute, Los Banos, Philippnes, 395-405.

Do Van, C., Watnanbe, I., Zimmerman, T., Lumpkin, T. and De Waha
    Baillonville, T. (1989). Sexual hybridization among Azolla species
    Can. J. Bot. (in press).

FAO (1988) Bio and organic fertilizer: Prospects and progress in Asia
    RAPA Publication: 1988/10, Food and Agricultural Organization of
    the United nations, Bangkok, 78p.

FAO (1988a) The "Rice-Azolla-Fish" system. RAPA Bulletin 4, 1988. Food
    and Agricultural Organization of the United Nations, Bangkok.

IRRI (1987) 'Azolla Utilization' International Rice Research Institute, Los Banos, Philippines 296p.

Kannaiyan, S. and Nandabalan, K. (1987). Influence of neem cake on blak root disease incidence in Azolla. In 'Azolla Utilization' International Rice Research Institute, Los Banos, Philippines 287.

Kulasooriya, S.A., Hirimburegama, W.K. and Abeysekera, S.W. (1987). Use of Azolla in Sri Lanka. In 'Azolla Utilization' International Rice Research Institute, Los Banos, Philippines, 131-140.

Kulasooriya, S.A., Seneviratne, P.R.G., de Silva, W.S.A.G., Abeysekera, S.W. and de Silva, P.A. (1988). Isotopic studies on N-Fixation and availability of N from Azolla under rice field conditions, Symbiosis 6, 151-166.

Liu Chung Chu (1979). Use of Azolla in rice production in China in Nitrogen and Rice. The International Rice Research Institute, Los Banos, Philippines.

Liu Chung Chu (1978). Reevaluation of Azolla Utilization in Agricultural Production. In Azolla Utilization, International Rice Research Institute, Los Banos, Philippines 67-76.

Lumpkin, T.A. and Plucknett, D.L. (1982). Azolla as a green manure use and management in crop production, Westview press, Boulder Colorado, 230p.

Lumpkin, T.A. (1987) Environmental requirements for successful Azolla growth. In 'Azolla Utilization' International Rice Research Institute, Los Banos, Philippines, 89-97.

Lu Shuying (1987). Methods for using Azolla filiculoides sporocarps to culture sporophytes in the field. In 'Azolla Utilization' International Rice Research Institute, Los Banos, Philippines 27-32.

Mochida, O., Yoshiyasu, Y. and Dimaano, D. (1987). Insect pests of Azolla in the Philippines. In 'Azolla Utilization' International Rice Research Institute, Los Banos, Philippines 207-221.

Watanabe, I., Berja, N.S. and Del Rosario, D.C. (1980). Growth of Azolla in paddy fields as affected by phosphorus fertilizer. Soil Science Plant Nutr. 26, 301-307.

Watanabe, I. and Ramirez, C. (1984). Relationship between soil Phosphrous availability and Azolla growth. Soil Science Plant Nutr. 30, 595-598.

Watanabe, I. and Liu Chung Chu (1990). Improvement of Nitrogen fixing systems and their integration into sustainable rice farming. Papers on Biological Nitrogen Fixation in Rice Fields. 14th International Congress of Soil Science. August 12-18 1990, Tokyo, Japan, 11-15.

# Heterotrophic metabolism and regulation of uptake hydrogenase activity in symbiotic cyanobacteria

M.C. MARGHERI, M.R. TREDICI, G. ALLOTTA and L. VAGNOLI
*Centro di Studio dei Microrganismi Autotrofi, CNR, P. le delle Cascine 27, I-50144 Firenze, Italy*

*Key words:* heterotrophy, nitrogenase, symbiotic cyanobacteria, uptake hydrogenase

## Abstract

The $H_2$ uptake activity of three cyanobionts isolated from *Cycas revoluta, C. circinalis* and *Azolla filiculoides* was shown to be related primarily to the growth rate and independent of the main mode of carbon nutrition. Significant $H_2$ uptake was found in the coralloid roots of *Cycas revoluta* and *Zamia furfuracea* (3 and 22 times higher than the respective $C_2H_2$ reduction activities). The results attained allow us to conclude that in cyanobacteria, in contrast to most nitrogen-fixing heterotrophs, uptake hydrogenase activity is not repressed by carbon substrates and that cyanobacteria in association seem to be endowed with sufficient $H_2$ uptake capacity to recover all of the $H_2$ released during the process of $N_2$-fixation.

## Introduction

$H_2$ evolution is an obligatory side reaction of $N_2$-fixation, and a minimum of 25% of the electron flux through nitrogenase is diverted to the reduction of protons during the reduction of $N_2$ to $NH_3$. This nitrogenase mediated $H_2$ production wastes, apparently without any benefit, substantial amounts of reductant and ATP (Houchins, 1984; Robson and Postgate, 1980; Yates, 1988).

However, many $N_2$-fixing microorganisms possess an effective $H_2$-oxidizing system, termed uptake hydrogenase, which can recycle the $H_2$ released by nitrogenase and which is thought to enhance significantly the efficiency of the $N_2$-fixation process by either recouping some ATP and reductant, protecting nitrogenase against inactivation by oxygen, or removing $H_2$ itself, which is considered to be a competitive inhibitor of $N_2$-reduction (Dixon, 1972, Houchins, 1984; Robson and Postgate, 1980).

In several heterotrophic diazotrophs, like *Rhizobium* and *Azotobacter*, organic carbon substrates act as repressors of uptake hydrogenase activity (Maier et al., 1979; Partridge et al., 1980) and a similar type of control of the enzyme seems to operate in heterocystous cyanobacteria as well. Kumar et al. (1986) observed complete inhibition of uptake hydrogenase activity in autotrophically grown *Anabaena cycadeae* following addition of 2% glucose to the growth medium. Almon and Böger (1988) reported that induction of uptake hydrogenase does not occur if a large pool of carbohydrate storage material is available. The repression of uptake hydrogenase by carbon substrates in heterocystous cyanobacteria assumes great significance since it implies that $H_2$ recycling is not operative in cyanobacteria living in symbiotic association, conditions under which the main nutrition mode of the cyanobiont is heterotrophy (i.e. the organism relics for carbon almost entirely on the supply of carbohydrates from the host plant). In confirmation of this, Kumar et al (1986) did not find any $H_2$ uptake activity in *Anabaena cycadeae* growing symbiotically in cycad roots. These authors concluded that the absence of activity in symbiosis was the consequence of an ample supply of carbohydrates from the surrounding cycad root tissue.

482

To assess the general validity of these findings we determined the $H_2$ uptake ability of three free-living cyanobionts grown under heterotrophic conditions and of two cyanobacterial symbioses.

## Materials and methods

### Organisms and culture conditions

*Nostoc* sp. Cc. (SAG 29.90) and *Nostoc* sp. Cr4 from coralloid roots of *Cycas circinalis* and *Cycas revoluta*, respectively, and *Anabaena azollae* AaS from *Azolla filiculoides* fronds, were isolated and rendered axenic in this laboratory. The cyanobacteria were grown photoautotrophically and heterotrophically under the conditions described previously (Tredici et al., 1988; 1990). The growth medium was supplemented with 1.3 $\mu M$ $NiCl_2$. Coralloid roots of *Cycas revoluta* and *Zamia furfuracea*, freshly collected from nursery plants, were used for the determinations on the symbiotic systems.

### Assay conditions

For the assays on the cyanobacterial isolates under heterotrophic conditions the carbon substrate (glucose for the two *Nostoc* strains and fructose for *A. azollae*) was added to the growth medium just before the start of the experiments to a concentration of 0.3, 0.8 and 2.0% (w/v) (chemo- and photo-heterotrophic cultures) and to a concentration of 0.3% (mixotrophic cultures). Experimental assays were done in borosilicate glass vials (26 mL) containing samples (10 mL) of cultures at an initial concentration ranging from 300–400 mg dry wt $L^{-1}$. The vials were sealed with butyl rubber septa (Pierce) and $H_2$ was injected (5% v/v in *Nostoc* Cc and *Nostoc* Cr4 cultures, and 15% in *A. azollae*

cultures) with a gas-tight syringe. Photoautotrophic cultures were also supplied with $CO_2$ (18% v/v of the gas phase) after having buffered the medium with 270 mg $L^{-1}$ $Na_2CO_3$. The vials were then incubated in an orbital incubator at $30 + 0.5°C$ under continuous illumination (80 $\mu E$ m$^{-2}$ s$^{-1}$) or in darkness.

For the assays on the symbioses, the coralloid roots of the two cycads, either washed repeatedly with sterile distilled water or surface sterilized with sodium hypochlorite (1.2%) and washed, were cut into sections of ca 1 mm thickness. The root sections (20–25 mg dry wt) were then introduced into glass vials (8 mL) which were sealed with butyl rubber septa. After sealing, the vials were supplied with $H_2$ (4% v/v) and incubated in the light under the conditions described above.

### Analytical methods

All the analytical procedures used in this work have been described elsewhere (Tredici et al., 1988; 1990).

## Results

### Effect of carbon substrates on the growth rate and nitrogenase activity of three free-living cyanobionts

In the three strains assayed the highest growth rates (i.e. lowest generation times) were observed during autotrophic and mixotrophic growth. The addition of the carbon substrate (0.3%) decreased the generation times of *A. azollae* and *Nostoc* sp. Cc grown photoautotrophically, while higher generation times were found in mixotrophic than in autotrophic cultures of *Nostoc* Cr4 (Table 1).

Under both chemo- and photo-heterotrophy *A. azollae* showed significantly shorter genera-

*Table 1.* Generation times of three free-living cyanobionts in photoautotrophic and mixotrophic cultures

| Growth conditions | Nostoc sp. Cc | Nostoc sp. Cr4 | A. azollae AaS |
|---|---|---|---|
| Autotrophy | $19.4 \pm 1.2$ | $14.4 \pm 1.8$ | $16.5 \pm 0.8$ |
| Mixotrophy | $16.5 \pm 0.9$ | $19.4 \pm 2.4$ | $12.0 \pm 0.8$ |

Values ($\pm SE$) reported are the means calculated for the first 9 hours after the start of the experiments.

tion times than the two *Nostoc* strains. Under these conditions, in neither of the strains did increasing the concentration of the carbon substrate from 0.3 to 2.0% have an appreciable effect on growth rate (Tables 2, 3 and 4).

Without carbon substrate addition (control cultures) no growth was observed in darkness, while some residual growth was present in cultures supplemented with DCMU in the light. This was most likely due to the fact that 20 $\mu M$-DCMU does not completely abolish oxygenic photosynthesis in the cyanobacterial strains assayed (Tredici et al., 1990).

The effect of carbon substrate addition on nitrogenase activity roughly paralleled the effect on growth rate (Tables 2–5). Maximum nitrogenase activities were measured under autotrophic and mixotrophic growth conditions (Table 5); very low or no activity was shown by cultures without carbon substrate addition in darkness and in the light with DCMU despite their significant $H_2$ consumption activity (Tables 2–4). Under each of the conditions experimented *A. azollae* cultures showed higher acetylene reduction rates than the two *Nostoc* strains. In neither of the strains did the addition of the carbon substrate enhance the nitrogenase activity of photoautotrophic cultures.

*Table 2.* Generation times, mean specific rates of $H_2$ consumption and mean $C_2H_2$ reduction activity of *Nostoc* sp. Cc grown under chemo- and photo-heterotrophic conditions

| Growth conditions | Carbon substrate addition | Generation time (h) | Mean specific rate of $H_2$ consumption (nmol mg$^{-1}$ h$^{-1}$) | | Mean $C_2H_2$ reduction activity (nmol mg$^{-1}$ h$^{-1}$) |
|---|---|---|---|---|---|
| | | Time interval (h) | Time interval (h) | | Time interval (h) |
| | | 0–44 | 0–20 | 20–44 | 0–44 |
| Chemoheterotrophy | none | no growth | 105.3 ± 11.0 | 0 | 0 |
| (darkness) | glucose 0.3% | 67.5 ± 5.5 | 113.5 ± 4.1 | 54.8 ± 4.7 | 57.0 ± 13.5 |
| | glucose 0.8% | 63.8 ± 3.9 | 94.5 ± 15.1 | 39.5 + 3.7 | 49.7 + 6.7 |
| | glucose 2.0% | 62.4 ± 5.2 | 92.5 ± 4.2 | 35.5 ± 3.1 | 48.2 ± 7.5 |
| Photoheterotrophy | none | 382.0 ± 38.6 | 69.3 ± 3.4 | 59.6 ± 6.5 | 14.0 ± 11.5 |
| (light + 20 $\mu M$ DCMU) | glucose 0.3% | 47.5 ± 3.1 | 60.2 ± 6.4 | 100.8 ± 6.2 | 110.0 ± 9.8 |
| | glucose 0.8% | 48.1 ± 2.4 | 55.8 ± 3.1 | 104.3 ± 3.0 | 100.5 ± 11.5 |
| | glucose 2.0% | 53.0 ± 2.5 | 53.0 ± 3.2 | 89.2 ± 6.4 | 82.4 ± 18.6 |

*Table 3.* Generation times, mean specific rates of $H_2$ consumption and mean $C_2H_2$ reduction activity of *Nostoc* sp. Cr4 grown under chemo- and photo-heterotrophic conditions

| Growth conditions | Carbon substrate addition | Generation time (h) | Mean specific rate of $H_2$ consumption (nmol mg$^{-1}$ h$^{-1}$) | | Mean $C_2H_2$ reduction activity (nmol mg$^{-1}$ h$^{-1}$) |
|---|---|---|---|---|---|
| | | Time interval (h) | Time interval (h) | | Time interval (h) |
| | | 0–44 | 0–20 | 20–44 | 0–44 |
| Chemoheterotrophy | none | no growth | 20.5 ± 7.0 | 0 | 0 |
| (darkness) | glucose 0.3% | 74.0 ± 7.6 | 71.2 ± 13.0 | 82.6 ± 6.8 | 39.4 ± 7.4 |
| | glucose 0.8% | 79.5 ± 6.0 | 92.5 ± 28.0 | 72.4 ± 3.7 | 49.5 ± 4.0 |
| | glucose 2.0% | 61.6 ± 5.5 | 58.5 ± 3.5 | 98.7 ± 2.5 | 49.2 ± 3.7 |
| Photoheterotrophy | none | 344.5 ± 41.0 | 61.8 ± 12.0 | 0 | 1.0 ± 0.5 |
| (light + 20 $\mu M$ DCMU) | glucose 0.3% | 59.2 ± 5.5 | 94.8 ± 4.6 | 192.6 ± 19.0 | 61.0 ± 17.0 |
| | glucose 0.8% | 52.6 ± 4.8 | 98.7 ± 8.0 | 148.6 ± 11.2 | 79.5 ± 22.5 |
| | glucose 2.0% | 56.3 ± 6.2 | 77.7 ± 24.0 | 142.3 + 9.2 | 70.5 + 17.5 |

*Table 4.* Generation times, mean specific rates of $H_2$ consumption and mean $C_2H_2$ reduction activity of *Anabaena azollae* AaS grown under chemo- and photo-heterotrophic conditions

| Growth conditions | Carbon substrate addition | Generation time (h) | Mean specific rate of $H_2$ consumption (nmol mg$^{-1}$ h$^{-1}$) | | Mean $C_2H_2$ reduction activity (nmol mg$^{-1}$ h$^{-1}$) |
|---|---|---|---|---|---|
| | | Time interval (h) | Time interval (h) | | Time interval (h) |
| | | 0–43 | 0–19 | 19–43 | 0–43 |
| Chemoheterotrophy (darkness) | none | no growth | 332.5 ± 43.0 | 0 | 2.1 ± 1.1 |
| | fructose 0.3% | 28.7 ± 0.6 | 364.7 ± 18.0 | 132.6 ± 7.9 | 71.7 ± 22.0 |
| | fructose 0.8% | 29.3 ± 1.0 | 371.5 ± 23.3 | 114.6 ± 5.0 | 66.5 ± 23.3 |
| | fructose 2.0% | 29.7 ± 0.3 | 299.1 ± 26.7 | 112.6 ± 8.4 | 58.5 ± 17.8 |
| Photoheterotrophy (light + 20 $\mu M$ DCMU) | none | 293.2 ± 21.5 | 285.0 ± 43.3 | 91.8 ± 41.5 | 3.1 ± 2.2 |
| | fructose 0.3% | 15.6 ± 0.1 | 316.5 ± 7.3 | 279.2 ± 8.0 | 170.5 ± 26.5 |
| | fructose 0.8% | 15.7 ± 0.1 | 325.2 ± 12.1 | 280.4 ± 11.1 | 183.5 ± 27.5 |
| | fructose 2.0% | 15.5 ± 0.1 | 306.1 ± 3.8 | 256.0 ± 5.3 | 185.0 ± 14.5 |

## *Effect of carbon substrates on the $H_2$ consumption rates of three free-living cyanobionts*

It appears evident from data shown in Tables 2–5 that the three cyanobionts do possess significant uptake hydrogenase activity during heterotrophic growth. Under chemo- and photo-heterotrophic growth conditions the mean rates of $H_2$ consumption of *Nostoc* Cr4 cultures were always higher than those measured in controls without glucose addition (Table 3). In the case of *Nostoc* Cc and *A. azollae* no great difference of activity was found between the cultures and the controls in the first 19–20 hours after glucose addition; in the subsequent 24 hour period, how-ever, a much higher activity was observed in growing cultures (Tables 2–4). Without carbon substrate addition in darkness (control cultures for chemoheterotrophy) the $H_2$ uptake activity dropped to zero after 20–24 hours of incubation. It is worth noting that a significant level of nitrogenase activity was recorded in growing cultures. Since a minimum of 25% of the electron flux to nitrogenase is necessarily diverted to the reduction of protons under $N_2$-fixing conditions, the real $H_2$ consumption rates of these cultures must be enhanced by 5–30%.

When the carbon substrate (0.3%) was added to photoautotrophically grown cultures (i.e. mixotrophy versus autotrophy was compared) an

*Table 5.* Mean specific rates of $H_2$ consumption and mean $C_2H_2$ reduction activity of three cyanobionts grown both auto- and mixo-trophically

| Organisms | Time interval (h) | Mean specific rate of $H_2$ consumption (nmol mg$^{-1}$ h$^{-1}$) | | Mean $C_2H_2$ reduction activity (nmol mg$^{-1}$ h$^{-1}$) | |
|---|---|---|---|---|---|
| | | Autotrophy | Mixotrophy | Autotrophy | Mixotrophy |
| *Nostoc* sp. Cc | 0.0 –3.0 | 317.6 ± 17.6 | 93.0 ± 41.7 | 125.2 ± 4.1 | 113.0 ± 4.2 |
| | 3.0– 6.0 | 117.4 ± 70.5 | 86.1 ± 16.0 | 124.0 ± 4.3 | 123.4 ± 6.8 |
| | 6.0– 9.0 | 127.3 ± 74.1 | 119.5 ± 51.4 | 115.6 ± 2.5 | 119.0 ± 5.3 |
| | 9.0–24.0 | N.D. | 53.3 ± 9.0 | N.D. | 125.5 ± 7.4 |
| *Nostoc* sp. Cr4 | 0.0– 3.0 | 435.7 ± 24.2 | 332.0 ± 21.9 | 136.6 ± 5.1 | 148.7 ± 6.3 |
| | 3.0– 9.0 | 319.2 ± 15.4 | 153.8 ± 59.2 | 158.7 ± 14.4 | 160.4 ± 9.5 |
| | 9.0–24.0 | N.D. | 88.0 ± 12.4 | N.D. | 86.3 ± 4.6 |
| *Anabaena azollae* AaS | 0.0– 3.0 | 1382.3 ± 290.5 | 180.6 ± 54.5 | 402.3 ± 3.5 | 388.7 ± 19.0 |
| | 3.0– 6.0 | 523.5 ± 40.6 | 689.8 ± 77.7 | 525.0 ± 22.9 | 450.9 ± 8.7 |
| | 6.0– 9.0 | 259.9 ± 52.8 | 449.4 ± 43.3 | 424.0 ± 5.6 | 405.2 ± 2.8 |
| | 9.0–24.0 | N.D. | 226.0 ± 21.5 | N.D. | 369.3 ± 12.4 |

initial, significant·reduction of $H_2$ consumption activity was observed in the three strains and particularly in *Nostoc* Cc and *A. azollae* (70% and 87% reduction of the activity, respectively) (Table 5). This inhibition was, however, partially mitigated with time.

The rapid decrease in activity with time exhibited by photoautotrophic cultures is not easy to explain. It is very likely related to the difficulty encountered in maintaining stable conditions of pH and $CO_2$ supply during the course of the experiment.

*$H_2$ consumption in coralloid roots of* Cycas revoluta *and* Zamia furfuracea

Coralloid roots of *Cycas revoluta* and of *Zamia furfuracea*, assayed in air, under $N_2$-fixing conditions, showed a significant $H_2$ uptake capability, the mean rates of $H_2$ consumption being 3 and 23 times higher than the respective rates of acetylene reduction (Table 6). Coralloid roots devoid of the cyanobiont exhibited neither $H_2$ consumption nor $C_2H_2$ reduction activity.

## Discussion

Three heterocystous cyanobacteria isolated from symbiotic associations showed significant uptake hydrogenase activity when grown heterotrophically (i.e. with an organic compound as the main carbon source for growth) in free-culture. This contrasts with the findings of Kumar et al. (1986) with *Anabaena cycadeae*, and we wonder if the omission of nickel (which has shown to be essential for the expression of uptake hydrogenase activity) from the growth medium in the experiments carried out by this group might explain the discrepancy. The activity of the enzyme was not affected substantially by the main mode of carbon nutrition of the organism but appeared rather to be correlated with the growth rate. The highest $H_2$ uptake activities were found at the highest growth rates (e.g. under autotrophy and mixotrophy), while very low, or no, activity was observed in non-growing cultures.

It can be concluded from our experiments that the assumption of Kumar et al. (1986) that 'the carbon limitation control of uptake hydrogenase which operates in *Rhizobium* and *Azotobacter* operates also in cyanobacteria' has no general validity. We cannot exclude, however, the existence of some kind of regulation of the enzyme by carbon substrates. It may well be that, as suggested by Houchins (1984), 'continuous consumption within the heterocyst of carbohydrates supplied by the vegetative cells may keep the heterocyst in a state of partial carbon starvation and thus allow expression of the enzyme in spite of catabolite repression'. This hypothesis seems to find confirmation in our experiments since the activity of the enzyme was shown to be higher at higher growth rates, i.e. when carbohydrate consumption within the heterocysts, which provides reductant and carbon skeletons for ammonia fixation, should be higher.

*Table 6.* Mean specific rates of $H_2$ consumption and mean $C_2H_2$ reduction activity of coralloid roots of *Cycas revoluta* and *Zamia furfuracea*

| | *Cycas revoluta* | | *Zamia furfuracea* | |
|---|---|---|---|---|
| | Mean specific rate of $H_2$ consumption (nmol mg$^{-1}$ h$^{-1}$) | Mean $C_2H_2$ reduction activity (nmol mg$^{-1}$ h$^{-1}$) | Mean specific rate of $H_2$ consumption (nmol mg$^{-1}$ h$^{-1}$) | Mean $C_2H_2$ reduction activity (nmol mg$^{-1}$ h$^{-1}$) |
| Washed coralloid roots | $8.3 \pm 2.1$ | $2.6 \pm 0.6$ | $6.9 \pm 1.8$ | $0.3 \pm 0.2$ |
| Surface sterilized coralloid roots | $6.2 \pm 1.7$ | $0.4 \pm 0.2$ | N.D. | N.D. |
| Washed coralloid roots devoid of the cyanobiont | 0.0 | 0.0 | 0.0 | 0.0 |

Values($\pm$ SE) reported are the means calculated for the first 23 hours after the start of the experiments.

In contrast to *A. cycadeae* grown in symbiotic state, the two symbioses examined in our study showed significant $H_2$ uptake activity which was amply capable of recycling all of the $H_2$ evolved by nitrogenase. We must therefore assume that, at least under certain physiological conditions, uptake hydrogenase is not repressed in cyanobacteria growing in association and that, unlike most legumes, cyanobacterial symbioses would not contribute to the massive release of $H_2$ into the atmosphere from $N_2$-fixing systems (Conrad, 1980).

## References


Almon H and Böger P 1988 Nitrogen and hydrogen metabolism: Induction and measurement. Meth. Enzymol. 167, 459–467.

Conrad R and Seiler W 1980 Contribution of hydrogen production by biological nitrogen fixation to the global hydrogen budget. J. Geophys. Res. 85, 5493–5498.

Dixon R O D 1972 Hydrogenase in legume root nodule bacteroids: Occurrence and properties. Arch. Microbiol. 85, 193–201.

Houchins J P 1984 The physiology and biochemistry of hydrogen metabolism in cyanobacteria. Biochim. Biophys. Acta 768, 227–255.

Kumar A P, Perraju B T V V and Singh H N 1986 Carbon nutrition and the regulation of uptake hydrogenase activity in free living and symbiotic *Anabaena cycadeae*. New Phytol. 104, 115–120.

Maier R J, Hanus F J and Evans H J 1979 Regulation of hydrogenase in *Rhizobium japonicum*. J. Bacteriol. 137, 824–829.

Partridge C D P, Walker C C, Yates M G and Postgate J R 1980 The relationship between hydrogenase and nitrogenase in *Azotobacter chroococcum*: Effect of nitrogen sources on hydrogenase activity. J. Gen. Microbiol. 119, 313–319.

Robson R L and Postgate J R 1980 Oxygen and hydrogen in biological nitrogen fixation. Annu. Rev. Microbiol. 34, 183–207.

Tredici M R, Margheri M C, Giovannetti L, De Philippis R and Vincenzini M 1988 Heterotrophic metabolism and diazotrophic growth of *Nostoc* sp. from *Cycas circinalis*. Plant and Soil 110, 199–206.

Tredici M R, Margheri M C, De Philippis R and Materassi R 1990 The role of hydrogen metabolism in photoheterotrophic cultures of the cyanobacterium *Nostoc* sp. strain Cc isolated from *Cycas circinalis* L. J. Gen. Microbiol. 136, 1009–1015.

Yates M G 1988 The role of oxygen and hydrogen in nitrogen fixation. *In* The Nitrogen and Sulphur Cycles. Eds. J A Cole and S J Ferguson. pp 383–416. Cambridge University Press, Cambridge, UK.

# Nitrogen fixing cyanobacteria potential uses

S. BOUSSIBA
*Micro-Algal Biotechnology Laboratory, The Jacob Blaustein Institute for Desert Research, Ben-Gurion University, Sede Boker 84990, Israel*

*Key words:* mass production, N fertilizers, $N_2$-fixing cyanobacteria, phycobiliproteins

**Abstract**

The large scale production of nitrogen fixing cyanobacteria is discussed and the use of ammonia excreting mutants of *Anabaena siamensis* is described. *Gloeotrichia natans* is considered for use as a biofertilizer and for the production of phycobiliproteins.

## Introduction

Large-scale production of nitrogen-fixing cyano-bacteria (blue-green algae) was aimed originally at expanding the traditional application of diazotrophic heterocystous cyanobacteria as a nitrogen source in rice paddies (Roger and Kulasooriya, 1980). Production was confined to non-lined open ponds, in which the resulting algal product contained a high percentage of soil, and the productivity was highly unpredictable in terms of the amount and kind of algal biomass that could be produced (Roger and Kulasooriya, 1980; Venkataraman, 1969).

Interest in the mass culture goes beyond their use as a nitrogen biofertilizer. The cyanobacteria also constitute a source of valuable products such as phycobiliproteins, polysaccharides and protein for feed and food (Cohen, 1986). One of the major problems limiting this biotechnology, however, is the fragmentary nature of the infor-mation available on the mass production of these microorganisms (Boussiba, 1988; Fontes et al., 1987). Some time ago, we began investigating the potential production of nitrogen-fixing cyanobacteria in outdoor cultures for application as N-fertilizer and as source of biochemicals.

*Nitrogen fixing cyanobacteria as nitrogen biofer-tilizers in rice fields*

The lack of available chemical fertilizers, espe-cially nitrogenous ones at economic prices, is one of the basic problems facing agriculture in de-veloping countries. N-chemicals account for as much as 30% of the total fertilizers needed for agriculture crops and are often regarded as the limiting factors in food production in developing countries. It is therefore not surprising that biological nitrogen fertilizers that efficiently transfer nitrogenous compounds from the media to the plants are of great interest in many coun-tries, particularly in tropical Asia where rice is one of the major agricultural crops. Unfortu-nately the increasing cost of N-fertilizer an the widening gap between supply and demand of nitrogen in the developing countries have placed heavy constraints on the farmers. Realizing the influence of energy cost on current and probable future prices of N-fertilizer, the need to stimu-late research on alternative sources of nitrogen for rice cultivation is thus imperative.

Rice fields continuously undergo environmen-tal changes during maturation of the rice plants (Roger and Kulasooriya, 1980). During the growth cycle of the rice plants, light becomes limiting due to tillering development, and there is an increase of pH from 6 to 7 in the inoculation stages, to 8–9.5 towards the end of growth. Also, due to evaporation, there is a constant increase in salt concentration, while temperature may also fluctuate over a wide range (Venkataraman, 1986). Clearly, these environmental factors may directly affect the growth and development of cyanobacteria (CBA) in rice fields. In particular, these factors may control nitrogenase activity and therefore affect the performance of these microorganisms as nitrogen biofertilizers.

The concept of using $N_2$ fixing CBA as nitrogen biofertilizers in rice fields is not yet fully explored, and some major problems are still limiting the wide utilization of this biofertilization technique:

The lack of an understanding of the environmental conditions prevailing in the rice ecosystem which in some cases affect the blooming of the algae (either indigenous species or inoculated ones)

The inability to produce good quality inocula at an economical price

The low efficiency of the transfer (utilization) of the fixed nitrogen by the CBA to the rice plants.

During the last five years, intensive work concerning the potential use of nitrogen-fixing cyanobacteria as biofertilizers in rice fields has been carried out in our laboratory.

Anabaena azollae *as a potential nitrogen biofertilizer*

### *Optimization of growth conditions*

*A. azollae* possesses several characteristic features which may be considered advantageous: it fixes nitrogen at almost optimal rates over a broad range of pH; tolerates a wide range of temperatures; and can withstand up to 1% NaCl in the growth medium without its growth or its nitrogenase activity being significantly affected. Indeed, these features of *A. azollae* have been described previously (Antarikanonda and Lorenzen, 1983) as the desirable ones, when considering natural isolates of $N_2$ fixing cyanobacteria to serve as nitrogen biofertilizer.

### *Outdoor production*

An important stage when considering application of $N_2$ fixing cyanobacteria as nitrogen biofertilizer is mass production of high quality inoculum of the desirable strains (Watanabe, 1984). Data concerning mass production of nitrogen fixing cyanobacteria are still limited, and the rate of reported production $6–8\ \mathrm{g\ m^{-2}\ day^{-1}}$ is relatively low. Recently Fontes et al. (1987) obtained higher rates of production 8 to 13 g (dry weight) $\mathrm{m^2\ day^{-1}}$ using *Anabaena variabilis*. It is imperative to consider these data with great caution since they were obtained in a small scale $(0.25\ \mathrm{m^2})$ and over a relatively very short period of time. The highest rate of production obtained in this work was $17.9\ \mathrm{g}$ (A.F.D.W.) $\mathrm{m^{-2}\ day^{-1}}$, obtained in a bigger reactor $(2.5\ \mathrm{m^2}$ in size), the rate of production calculated from a culture being at steady state of at least 25 days.

*Mutants which continuously release ammonia isolated from* Anabaena siamensis

The rice field ecosystem provides favorable conditions for the development of a large variety of free-living CBA and symbiotic strains which are capable of fixing atmospheric nitrogen (Antarikanonda, 1984; 1985; Antarikanonda and Lorenzen, 1982). It is generally assumed that the nutrients fixed by CBA are made available to the rice plants through exudation autolysis and microbial decomposition. Under field conditions, however, only a part of the fixed nitrogen is available to the rice plant, some being either reincorporated by the microflora or volatilized. Under these circumstances, it is difficult to control the flow of nitrogen compounds needed for the development of rice plants. A possible solution to this problem is to develop strains of cyanobacteria which release ammonium continuously.

In the present work we used a rice field isolate of *Anabaena siamensis* which has a fast growth rate $(0.123\ \mathrm{h^{-1}})$ and high nitrogen fixing capacity (Antarikanonda and Lorenzen, 1982). This strain also adapts well to temperature fluctuations (25 to 42°C) prevailing in rice fields to the salinity ranges (1 to 2%) existing in most of the tropical wetland soils (Antarikanonda, 1985). We hereby describe the characterization of an ammonium-excreting mutant of *A. siamensis* (Thomas et al., 1990).

An ammonium-excreting mutant (SS1) of the rice field nitrogen-fixing cyanobacterium *Anabaena siamensis* was isolated after ethyl methanesulfonate mutagenesis by selection on 500 $\mu M$ L-methionine-DL-sulfoximine (MSX) (Thomas et al., 1990). SS1 grew in the presence and absence of MSX at a rate comparable to that of the wild type strain, with a doubling time of 5.6 h. The rate of ammonium release by SS1 depended on cell density; it peaked at the 12th hour of growth with 8.7 $\mu$mol mg chl$^{-1}$ h$^{-1}$ (at a chlorophyll concentration of 5 $\mu$g mL$^{-1}$) and slowed down to almost nil at the fourth day of growth. Similar pattern of release by immobilized SS1 was observed between 12 to 20 h after loading the alginate beads in packed-bed reactors at the rate of 11.6 $\mu$mol mg chl$^{-1}$ h$^{-1}$. The rate was later reduced significantly due to the fast growth of SS1 on the substrate. Prolonged release of ammonium at the peak level was achieved only by maintaining SS1 under continuous cultivation at low chlorophyll levels (5 to 7 $\mu$g mL$^{-1}$). Under these conditions, nitrogen fixation in the mutant was 30% higher than that in its parent and glutamine synthetase (GS) activity was less than 50%. Immunoblot analysis revealed that SS1 and its parent have the same quantity of GS protein under ammonium excretion conditions. In addition, a protein with a molecular weight of 30,000-daltons seems to have been lost, as seen by electrophoretic separation of total proteins from SS1.

Fertilization of rice plants under laboratory conditions by application of another MSX-resistant mutant of *A. variabilis*, SA1, was found successful (Latorre et al., 1986). The shorter doubling time of SS1 and the lack of a lag period at the beginning of the growth cycle as compared with SA1 (Spiller et al., 1986), which exhibits a lag period, are obvious assets for mass cultivation of the former. These characteristics make possible the production of considerable amounts of inoculum material within a short period. The usefulness of SS1 as a biofertilizer to rice plants in the actual isolate location, Thailand, should be studied in order to scale up the technology transfer from laboratory to the rice field.

### Gloeotrichia natans – *a nitrogen fixing cyanobacterium – a possible source of phycobiliproteins*

*Gloeotrichia natans* a nitrogen fixing cyanobacterium is commonly found in rice fields in the Phillippines. Inspite of that no work has been carried out to establish its key feature of physiology nor its potential production in outdoor cultures. This study aimed therefore to provide some of this information (Palacpac et al., 1990).

Under optimal growth conditions (38°C, pH 8.0, no carbon enrichment) the specific growth rate of the rice-field isolate of *Gloeotrichia natans* was 0.076 h$^{-1}$. The pH of the medium (between 6.5 and 9.0) did not influence the growth rate, but it did affect phycobiliprotein content, as reflected by a change of the color of the cultures. At pH 7.0 the culture was green-brown with phycobiliproteins constituting up to 10% of the total protein, while at pH 9.0 the culture was brownish-black and the content of these pigments was as high as 28% of the total protein. In outdoor cultures the specific growth rate was directly related to cell density in the range of 0.7–1.5 g dry weight L$^{-1}$ at a rate of stirring of 30 rpm, and inversely related to cell density at half this rate. At a stirring rate of 30 rpm, daily production of outdoor cultures harvested to maintain cell densities of 0.7, 1.15 and 1.5 g L$^{-1}$ were 14.7, 17.1 and 18.1 g m$^{-2}$ d$^{-1}$, respectively. This rate of production was maintained for more than 45 days. Phycobiliprotein content in the culture kept at a density of 1.5 g L$^{-1}$ reached 14% of the total biomass.

The possibility of obtaining a monoalgal inoculum of *G. natans* at a reasonable rate of production will thus improve its efficiency as a nitrogen biofertilizer as well as facilitate the extraction of valuable products, such as phycobiliproteins. The red and blue phycobiliportein pigments, which together constitute 14% of the algal dry weight may be used as coloring agents for food and cosmetics, and as immunofluorescence probes in medical research (due to their high fluorescence).

## Conclusions

To date only a few works emphasis the potential of mass cultivation of nitrogen fixing cyanobacteria. This group of microalgae however, have some advantage when considering commercial enterprise in relation to mass production; members of this group can tolerate a wide range of temperatures and pH's; the growth medium is minimal and the main component is carbon which can be supplied as carbon dioxide or as a carbonate salt; the lack of combined nitrogen reduces drastically the chance for contamination with other algae. It is therefore imperative to further test the production of these strains as an alternative source for chemicals and food additive.

## References

Antarikanonda P. 1984 Production of extracellular free amino acids by cyanobacterium *Anabaena siamensis*. Curr. Microb. 11, 191–196.

Antarikanonda P 1985 A new species of the genus *Anabaena*: *Anabaena siamensis* sp nov. (Cyanophyceae) from Thailand. Nova Hedwigia 41, 343–352.

Antarikanonda P and Lorenzen H 1983 Highly efficient N$_2$ fixing blue green algae in Thailand: A possible remedy to nitrogen scarcity. Bangkok MIRCEN News Letters, 3, 1–6.

Antarikanonda P and Lorenzen H 1982 N$_2$ fixing blue green algae (Cyanobacteria) of high efficiency from paddy soils of Bangkok. Thailand: Characterization of species and nitrogen fixing capacity in the laboratory. Acta Hydrobiol. Suppl. 63, 53–70.

Boussiba S 1988 N$_2$-fixing cyanobacteria as nitrogen biofertilizer: A study with the isolate *Anabaena azollae*. Symbiosis 6, 129-138.

Cohen Z 1986 Products from Microalgae. *In* Handbook of Microalgal Mass Culture. Ed. A Richmond. pp 421–454. CRC Press Inc. Boca Raton, Florida.

Fontes A G, Vargas M A, Moreno J, Guerero M G and Losada M 1987 Factors affecting the production of biomass by a nitrogen-fixing blue-green alga in outdoor culture. Biomass 13, 33–43.

Lattore C, Lee J H, Spiller H and Shanmugam K T 1986 Ammonium ion-excreting cyanobacterial mutant as a source of nitrogen for growth of rice: A feasibility study. Biotech. Lett. 8, 507–512.

Palacpac N M, Martinez M and Boussiba S 1990. Mass cultivation of the nitrogen-fixing cyanobacterium *Gleotrichia natans*, indigenous to rice fields. J. Appl. Phycology 2, 319–325.

Roger P A and Kulasooriya S A 1980 Blue-green algae and rice. The International Rice Research Institute. Los Baños, Laguna, Philippines. 112 p.

Spiller H, Latorre C, Hassan M E and Shanmugam K T 1986 Isolation and characterization of nitrogenase derepressed mutant strains of cyanobacterium *Anabaena variabilis*. J. Bacteriol. 165, 412–419.

Thomas S P, Zaritsky and Boussiba S 1990 Amonium excretion by a sulfoximine-resistant mutant of the rice field cyanobacterium *Anabaena siamensis*. Appl. Environm. Microb. 56, 3499–3504.

Venkataraman G S 1969 The Cultivation of Algae. Indian Council of Agricultural Research. New Delhi. 317 p. Bennett A and Bogorad L (1973) Complementary chromatic adaptation in a filamentous blue-green alga. J. Cell Biol. 58, 419–435.

Venkataraman L V 1986 Blue-green algae as biofertilizers. *In* Handbook of Microalgal Mass Culture. Ed. A Richmond. pp 455–471. CRC Press Inc., Boca Raton, FL.

Watanabe I 1984 Use of symbiotic and free-living blue-green algae in rice culture Outl. Agric. 13, 166–172.

# METABOLIC REGULATION OF NITROGEN FIXATION IN ANOXYGENIC PHOTOTROPHIC BACTERIA


Stefan Nordlund
Department of Biochemistry
Arrhenius Laboratories
Stockholm University
S-106 91 Stockholm
Sweden


## Introduction

Nitrogen fixation in anoxygenic phototrophs was first demonstrated in Rhodospirillum rubrum by Kamen and Gest [1]. Shortly after this report in 1949, purple and green sulfur bacteria were also shown to fix nitrogen [2,3]. Over the last 40 years the list of nitrogen fixing phototropic bacteria has grown considerably (Table 1). Madigan et al. [4] made an extensive study on 18 species of the family Rhodospirillaceae, showing that all but one, Rhodocyclus purpureus, were able to grow with N$_2$ as the only nitrogen source. Furthermore they demonstrated that dark microaerobic nitrogen fixation occurred in all species except Rhodospirillum fulvum. In addition Rhodobacter capsulatus [5] and Rsp. rubrum [6] have been shown to grow under dark fermentative diazotrophic conditions.

The purple and green sulfur phototrophs have not been studied in great detail since the first reports [2,3]. However various aspects of nitrogen fixation have been investigated in Chlorobium [7], Chromatium [8,9,10], Thiocapsa roseopersicina [11] and Ectothiorhodospira [12].

## Nitrogen fixation

### Nitrogenase

After the milestone reports on reproducible nitrogenase activity in crude extracts from non-photosynthetic bacteria [13,14] there was a continuos progress in the characterization and purification of the enzyme from these organisms. Although some investigations on nitrogenase in cell free extracts from phototrophic bacteria were reported the activities were low and variable. As we know today the reason for the difficulties with these organisms was due to the inactivation of nitrogenase by the metabolic control system discussed below. Today nitrogenase has been purified from Rsp. rubrum [15,16] and

**Table 1. Nitrogen fixing anoxygenic phototrophic bacteria**

---

<u>Purple non-sulfur</u>

| | |
|---|---|
| Rhodospirillum rubrum | Rhodobacter capsulatas |
| Rsp. photometricum | R. sphaeroides |
| Rsp. fulvum | R. sulfidophila |
| Rsp. molischianum | |
| Rsp. salxeigens | |
| | Rhodomicrobium vannielli |
| Rhodopseudomonas palustris | |
| Rps. viridis | Rhodocyclus gelatinosus |
| Rps. acidophila | Rhodocyclus tenius |
| Rps. sulfoviridis | |
| Rps. blastica | Rhodopila globiformis |

<u>Purple and green sulfur</u>

| | |
|---|---|
| Chromatium vinosum | Clorobioum |
| Thiocapsa roseopersicina | |
| Ectothiorhodospira | |

---

<u>R. capsulatus</u> [17]. The properties of dinitrogenase (MoFe-protein) and dinitrogenase reductase (Fe-protein) from these organisms are essentially the same as those of nitrogenase from non-photosynthetic bacteria. In addition dinitrogenase from <u>Chromatium</u> has been characterized [9] and a purification procedure for both proteins from this organism has been published although no properties were reported [10].

**Electron transport to nitrogenase**

The carriers involved in electron transport to nitrogenase have been best established in <u>Klebsiella pneumoniae</u>. In this organism pyruvate is oxidized by the product of <u>nifJ</u>, a specific pyruvate: flavodoxin oxidoreductase, which transfers electrons to a flavodoxin. The enzyme is a dimer of two identical polypeptides with a $M_r$ of 120 000 [18], contains 8 [18] or 16 [19] Fe and about the same amount of acid-labile sulfide per dimer. The enzyme requires Coenzyme A and thiamine pyrophosphate for activity [18,19]. It is also highly specific for both the donor and the acceptor, i.e. pyruvate and flavodoxin. This nitrogen fixation specific flavodoxin , (the <u>nifF</u> product) has FMN as cofactor and a $M_r$ of 22 000 [20,21].

The electron transport to nitrogenase in phototrophic bacteria has not been studied in great detail. Ludden and Burris [22] showed that pyruvate or α-ketoglutarate supported nitrogenase activity in crude extracts of <u>Rsp. rubrum</u> and that this activity was dependent on added

Coenzyme A and greatly stimulated by addition of ferredoxin. We have recently purified and characterized a pyruvate: oxidoreductase from Rsp. rubrum, with properties similar to the enzyme from K. pneumoniae (E. Brostedt and S. Nordlund, this volume). The purified enzyme will transfer electrons from pyruvate to nitrogenase in the presence of Coenzyme A, thiamine pyrophosphate and ferredoxin.

The involvement of ferredoxin rather than flavodoxin in phototrophic bacteria is suggested by the demonstration that ferredoxin I and ferredoxin II, reduced by illuminated chloroplasts, will support nitrogenase activity in crude extracts from Rsp. Rubrum [23]. In agreement with these data we have not been able to get hybridization with a nifF probe from K. pneumoniae with Rsp. rubrum DNA (J. Knutsson Jansson and S. Nordlund, unpublished). Rsp. rubrum does however produce a flavodoxin, although only under Fe-limiting conditions [24]. R. capsulatus is capable of synthesizing 4 different ferredoxins and the genes for these ferredoxins have been cloned [25]. Ferredoxin I and II have been shown to transfer electrons to nitrogenase [26,27], but ferredoxin II is the only one not being repressed by high ammonia [25].

Although the electron transprot in Rsp. rubrum seems to be similar to that in K. pneumoniae with ferredoxin substituting for flavodoxin, this may not be the general system in phototrophic bacteria. Hybridization experiments indicate that there is no gene homologous to nifJ in R. capsulatus [28] and the involvement of the proton motive force has been suggested R. sphaeroides [29].

## nif-genetics

The model system for studying nif-gene organization and regulation has since long been K. pneumoniae. In this organism 20 genes are organized in one cluster of 24 kb [30]. However, when other organisms have been studied it has become clear that this organization is rather the exception than the general rule. The only phototroph in which the nif-genes have been studied in detail is R. capsulatus, in which the nif-genes are separated into three clusters on the chromosome [28,31,32,33]. Furthermore, the nif map of R. capsulatus includes 36 coding regions, of which 14 have their counterpart in K. pneumoniae [28]. It will be of great interest to learn about the function of the "extra" genes, especially with respect to possible couplings between nitrogen fixation and photosynthesis. The genetic regulation of nitrogen fixation in R. capsulatus has also been studied [34] and although differences were demonstrated, the general features are the same as in K. pneumoniae [35].

## Ammonia assimilation

In nitrogen fixing organisms ammonium ions are assimilated through the reactions catalyzed by glutamine synthetase (I) and glutamate synthase (II) [36]. The occurrence of both of these enzymes in a number of nitrogen fixing phototrophs has been reported [37,38].

$$NH_4^+ + glu + ATP \ \ ------> \ \ gln + ADP + P_i \hspace{2cm} (I)$$

$$gln + \alpha\text{-}KG + NAD(P)H \ \ ------> \ \ 2\ glu + NAD(P)^+ \hspace{1cm} (II)$$

Among phototrophic bacteria glutamine synthetase has been purified from Rsp. rubrum [39,40,41], Rps. palustris [42,43], R. capsulatus [44,45] and R. sphaeroides [46]. All these enzymes exhibit essentially the same properties as the enzyme from Escherichia coli [47]. With the possible exception of Rsp. rubrum, glutamine syunthetase is regulated through the adenylylation cascade mechanism well established in E. coli [47]. The regulation of this key-enzyme seems to be more complex in Rsp. rubrum although adenylylation certainly does occur as shown in a recent report [48].

Glutamate synthase has been less extensively studied in phototrophs. The only report concerns a partial purification of the enzyme from Rsp. rubrum [49]. We have however recently purified glutamate synthase from this organism to near homogeneity and made a partial characterization (Carlberg & Nordlund, unpublished). The properties of the enzyme are essentially the same as those of the E. coli enzyme [50]. As yet we have not been able to show if glutamate synthase from Rsp. rubrum is subjected to regulation.

**Metabolic regulation**

**The "switch-off" effect**

The physiological manifestation of the metabolic control of nitrogen fixation in phototrophic bacteria is the reversibel inhibition of nitrogenase activity by e.g. ammonium ions. This phenomena, which has been termed the "switch-off" effect [51], was first demonstrated in Rsp. rubrum already 1950 [52], but was "rediscovered" 20 years later [53,54]. Since then the "switch-off" effect has been studied in a number of organisms and in fact most of the phototrophs listed in Table 1 have been shown to exhibit the effect. Most of the more detailed studies have been done with Rsp. rubrum and the list of "switch-off" effectors has been extended to include asparagine, glutamine [54], darkness, oxygen, carbonyl cyanide m-chlorophenylhydrazone and phenazine methosulfate [55].

**The regulatory cycle of dinitrogenase reductase**

The discovery of a membrane associated activity that increased nitrogenase activity in extracts of Rsp. rubrum was the beginning of the elucidation of the molecular mechanism of the "switch-off" effect [56,57]. In these reports it was shown that the reason for loss of nitrogenase activity was the conversion of dinitrogenase reductase from an active to an inactive form. Through the work of Ludden and co-workers it was shown that the inactive form is modified in one of its subunits by an ADP-ribose moiety bound to arginine-101 [58]. Figure 1 shows the present model for the regulatory cycle of dinitrogenase reductase, with the two enzymes catalyzing the inactivation (DRAT) and the activation (DARG) of dinitrogenase reductase.

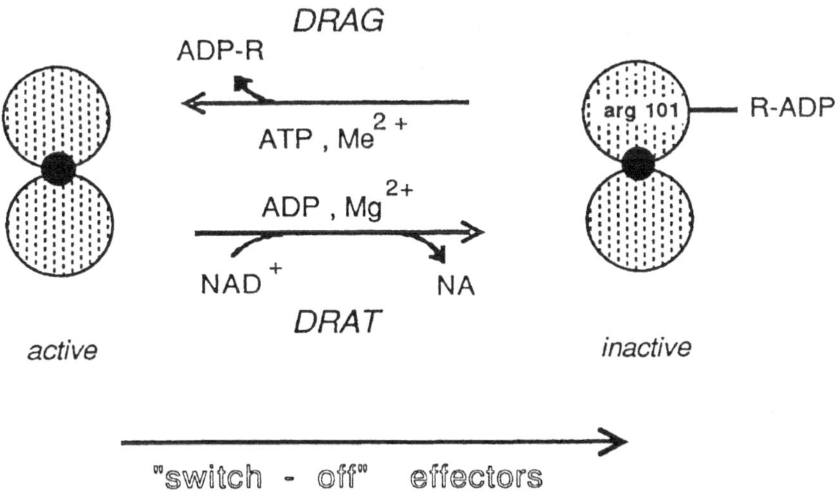


Figure 1. Regulatory cycle of dinitrogenase reductase

The ADP-ribosylation of dinitrogenase reductase is catalyzed by dinitrogenase reductase ADP-ribosyltransferase, DRAT, (inactivating enzyme)
which has been highly purified from Rsp. rubrum [59]. The enzyme is
a monomer with $M_r$ of 30 000, it is unstable during purification but
not oxygen labile and it is present in low amounts in the cell [59].
The $K_m$ for NAD$^+$ with Rsp. rubrum dinitrogenase reductase is remarkable
high (2 mM) and the enzyme requires MgADP for activity [59]. The reason for the inactivity of the modified dinitrogenase reductase has
been shown to be its inability to interact with dinitrogenase [60].
The crystal structure of dinitrogenase reductase from Azotobacter
vinelandi has recently been shown to 3Å resolution [61] and the ADP-
ribosylated residue, arginine 101, is located in the area which is
believed to be involved in the interaction with dinitrogenase. It can
thus be envisaged that the ADP-ribose groups will block this interaction.

Activation by removal of the ADP-ribose moiety is catalyzed by dinitrogenase reductase-activating glycohydrolase, DRAG (activating enzyme)
which has also been purified from Rsp. rubrum [62,63]. It has a $M_r$ of
32 000 [62], is extremely oxygen labile [56,57] and like DRAT it is
present in minute amounts in the cell [62]. It is dependent on the
presence of MeATP, which is not hydrolyzed, and a free divalent cation,
preferably Mn$^{2+}$ or Fe$^{2+}$ [56,57,64,65]. It is interesting that DRAG is
present in Rsp. rubrum grown under all conditions tested [66] indicating that it may have some other role in the metabolism of this phototroph. Since the genes for DRAG and DRAG, draT and draG, have been
cloned and sequence [67] it should be possible to study this intriguing aspect of these regulatory proteins.

In addition to the detailed studies of the metabolic regulation in
Rsp. rubrum also other phototrophs have been investigated. It has
been shown that dinitrogenase reductase in R. capsulatus also is ADP-
ribosylated on arginine-101 [68] and the activity of DRAG has been

demonstrated in this organisms [69,70]. Evidence for the covalent modification of dinitrogenase reductase in  Chromatium has also been presented but the modifying group was not identified [10]. Although the "switch-off" effect has been demonstrated in  R. sphaeroides it has not been possible to show that this is due to modification of dinitrogenase reductase [29,71]. This could be due to exceptionally high activity of DRAG or that some other system regulates nitrogenase activity in this organism, as suggested [29].

Control of metabolic regulation

One of the major questions concerning the metabolic regulation of nitrogenase in phototrophs has been the identiy of the "signal" between the "switch-off" effectors and DRAT/DRAG. A number of possible metabolic candidates have been suggested, like changes in the pools of amino acids [72,73], adenine nucleotides [73,74] and pyridine nucleotides [75,76]. However, in no case has a correlation between these pools and "switch-off" been demonstrated.

Since nitrogen fixation and ammonia assimilation are dependent on generation of ATP and NAD(P)H we have hypothesized that the $NAD^+$/NADH ratio may be at least part of the "signal". $NADP^+$ is produced in the glutamate synthase reaction and regeneration of NADPH is believed to occur in the transhydrogenase reaction [77] which leads to the generation of oxidized NAD: NADH is regenerated through reversed electron transport driven by the proton motive force [77]. Addition of ammonium ions, glutamine or asparagine as "switch-off" effector would cause an increased flow through the glutamine synthetase/glutamate synthase pathway, thus oxidizing NADPH and indirectly NADH. With respect to using change to darkness as a "switch-off" effector, Jackson and Croft has previously shown that placing Rsp. rubrum in the dark leads to a rapid oxidation of NAD(P)H [78]. Finally the effect of uncouplers would be that neither the transhydrogenase reaction nor the reversed electron flow, which both are dependent on the proton motive force, would be operating. Consequently, all these "switch-off" effectors would lead to a sudden increase in the concentration of $NAD^+$, thereby increasing the rate of the DRAT reaction. Since DRAT has a high $K_m$ for $NAD^+$ it can be assumed that this enzyme will be operating at low rate at "normal" $NAD^+$ concentration.

In order to test this hypothesis we have recently studied the effect of $NAD^+$ as "switch-off" effector in Rsp. rubrum [79]. It was shown that $NAD(P)^+$ but not NAD(P)H functioned as "switch-off" effectors and that the duration of "switch-off" was dependent on light intensity. Furthermore we could show that analoges of NAD had the same effect in vivo as has been demonstrated in vitro [59], i.e. analoges modified in the nicotinamide portion did not cause "switch-off" whereas those changed in the adenine moiety did. We also obtained "switch-off" by $NAD^+$ in N-starved cells. Our data indicate that NAD is taken up by the cells and by using $^3H$ as well as $^{32}P$ - NAD we have been able to demonstrate uptake, occurring on the same time scale as "switch-off" (Soliman & Nordlund, unpublished). Taken together, these results

support our model that changes in the NAD+/NADH ratio are involved in
the control of the regulatory cycle. However, for "switch-off" to occur also DRAG must be regulated and as yet no physiological effector
of this enzyme has been identified.

## ADP-ribosylation of glutamine synthetase in Rsp. rubrum

Glutamine synthetase in Rsp. rubrum has been shown to be inhibited
in vivo concomittantly with nitrogenase when ammonium ions [80] or
darkness [81] were used as "switch-off" effectors. The enzyme shows
low transferease activity in the absence of $Mg^{2+}$ when isolated from
cells subjected to "switch-off" prior to harvest [39,41,81] and can
not be reactivated by treatment with snake-venom phosphodiesterase.
Both of these observations are in contrast to the properties of adenylylated glutamine synthetase from E. coli [47]. These and other
characteristics have raised the question if glutamine synthetase in
Rsp. rubrum is regulated by ADP-ribosylation, possibly by DRAT. Recent
work by Ludden and coworkers has shown that indeed glutamine synthetase can by ADP-ribosylated in Rsp. rubrum, but that this reaction is
not catalyzed by DRAT [48]. This would indicate that yet another regulatory system operating by ADP-ribosylation is present in Rsp.
rubrum and previous reports [41,81] showing that the low-active form
of glutamine synthetase can not be activated by DRAG are in support
of this. Since it was also shown that adenylylation occurs in Rsp.
rubrum [48] the physiological role of ADP-ribosylation will have to be
further studied.

Nitrogen fixation has been studied for more than 100 years. The physiology, enzymology and genetics of this process have shown exciting
features that has contributed to our understanding of not only this
process but also bacterial metabolism and genetic regulation in general. The studies on nitrogen fixation in phototrophic bacteria have
added yet another fascinating feature metabolic regulation, which
again will help us to understand a more general phenomena, regulation
by covalent modification.

## Acknowledgement

Work from the author's lab was supported by grants from the Swedish
Natural Science Research Council.

## References

1. Kamen, M.D. and Gest, H. (1949) "Evidence for a nitrogenase system
   in the photosynthetic bacterium Rhodospirillum rubrum", Science
   109, 560.
2. Lindström, E.S., Burris, R.H. and Wilson, P.W. (1949) "Nitrogen
   fixation by photosynthetic bacteria", J. Bacteriol. 58, 313-315.
3. Lindström, E.S., Tove, S.R. and Wilson, P.W. (1950) "Nitrogen
   fixation by green and purple sulfur bacteria" Science 112,
   197-198.

4.  Madigan, M., Cox, S.S. and Stegman, R.A. (1984) "Nitrogen fixation and nitrogenase activities in members of the family Rhodospirillaceae" J. Bacteriol. 157, 73-78.
5.  Madigan, M.T., Wall, J.D. and Gest, H. (1979) "Dark anaerobic dinitrogen fixation by a photosynthetic microorganism" Science 204, 1429-1430.
6.  Schultz, J.E., Gotto, J.W., Weaver, P.F. and Yoch, D.C. (1985) "Regulation of nitrogen fixation in Rhodospirillum rubrum grown under dark fermentative conditions" J. Bacteriol. 162, 1322-1324.
7.  Heda, G.D. and Madigan, M.T. (1986) "Aspects of nitrogen fixation in Chlorobium" Arch. Microbiol. 143, 330-336.
8.  Winter, H.C. and Arnon, D.I. (1970) "The nitrogen fixation of photosynthetic bacteria" Biochim. Biophys. Acta 197, 170-179.
9.  Evans, M.C.W., Telfer, A. and Smith, R.V. (1973) "The purification and some properties of the molybdenum-iron protein of Chromatium nitrogenase" Biochim. Biophys. Acta 310, 344-352.
10. Gotto, J.W. and Yoch, D.C. (1985) "Regulation of nitrogenase activitiy by covalent modification in Chromatium vinosum" Arch Microbiol. 141, 40-43.
11. Rodinov, Y.V., Lebedeva, N.V. and Kondratieva, E.N. (1986) "Ammonia inhibition of nitrogenase activity in purple and green bacteria" Arch. Microbiol. 143, 345-347.
12. Bognar, A., Desrosiers, L., Libman, M. and Newman, E.B. (1982) "Control of nitrogenase in a photosynthetic autotrophic bacterium Ectothiorhodospira sp." J. Bacteriol. 152, 706-713.
13. Schneider, K.C., Bradbeer, C., Singh, R.N., Wang, L.C., Wilson, P.W. and Burris, R.H. (1960) "Nitrogen fixation by cell-free preparations from microorganisms" Proc. Natl. Acad. Sci. 46, 726-733.
14. Carnahan, J.E., Mortenson, L.E., Mower, H.F., Castle, J.E. (1960) "Nitrogen fixation in cell-free extracts of Clostridium pasteurianum" Biochim. Biophys. Acta 44, 520-535.
15. Nordlund, S., Eriksson, U. and Baltscheffsky, H. (1978) "Properties of the nitrognease system from the photosynthetic bacterium Rhodospirillum rubrum" Biochim. Biophys, Acta 504, 248-254.
16. Ludden, P.W. and Burris, R.H. (1978) "Purification and properties of nitrogenase from Rhodispirillum rubrum, and evidence for phosphate, ribose and an adenine-like unit covalently bound to the iron protein" Biochem. J. 175, 251-259.
17. Hallenbeck, P.C., Meyer, C.M. and Vignais, P.M. (1982) "Nitrogenase from the photosynthetic bacterium Rhodopseudomonas capsulata: Purification and molecular properties" J. Bacteriol. 149, 708-717.
18. Shah, V.K., Stacey, G. and Brill, W.J. (1983) "Electron transport to nitrogenase" J. Biol. Chem. 258, 12064-12068.
19. Wahl, R.C. and Orme-Johnson, W.H. (1987) "Clostridial pyruvate oxidoreductase and the pyruvate-oxidizing enzyme specific to nitrogen fixation in Klebsiella pneumoniae are similar enzymes" J. Biol. Chem. 262, 10489-10496.
20. Nieva-Gomez, D., Roberts, G.P., Klevickis, S. and Brill, W.J. (1980) "Electron transport to nitrogenase in Klebsiella pneumoniae" Proc. Natl. Acad. Sci. 77, 2555-2558.

21. Deistung, J., Cannon, F.C., Cannon, M.C., Hill, S. and Thorneley, R.N.F. (1985) "Electron transport to nitrogenase in Klebsiella pneumoniae" Biochem. J. 231, 743-753.

22. Ludden, P.W. and Burris, R.H. (1981) "in vivo and in vitro studies on ATP and electron donors to nitrogenase in Rhodospirillum rubrum" Arch. Microbiol. 130, 155-158.

23. Voch, D.C. and Arnon, D.I. (1975) "Comparison of two feredoxins from Rhodospirillum rubrum as electron carriers for the native nitrogenase" J. Bacteriol. 121, 743-745.

24. Ousanovich, M.A. and Edmondson, D.E. (1971) "The isolation and characterization of Rhodospirillum rubrum flavodoxin" Biochem. Biophys. Res. Commun 45, 327-336.

25. Jounneay, V., Grabau, C., Schatt, E., Duport, C., Meyer, C. and Vignais, P.M. (1990) "Properties of four ferredoxins from Rhodobacter capsulatus: biochemical and genetic studies of their role in $N_2$ fixation in Nitrogen fixation: achievements and objectives (Gresshoff, Roth, Stacey & Newton eds.) Chapman and Hall, New York, in press.

26. Hallenbeck, P.C., Jouanneau, Y. and Vignais, P.M. (1982) "Purification and molecular properties of a soluble ferredoxin from Rhodopseudomonas capsulata" Biochim. Biophys. Acta 681, 168-176.

27. Yakunin, A.F. and Gogotov, I.N. (1983) "Properties and regulation of synthesis of two ferredoxins from Rhodopseudomonas capsulata" Biochim. Biophys. Acta 725, 298-308.

28. Klipp, W. (1990) "Organization and regulation of nitrogen fixation genes in Rhodobacter capsulatus" see ref. 25.

29. Haaker, H., Laane, C., Hellingwerf, K., Houver, B., Konings, W.N. and Veeger, C. (1982) "Short-term regulation of the nitrogenase activity in Rhodopseudomonas sphaeroides" Eur. J. Biochem. 127, 639-645.

30. Arnold, W., Rump, A., Klipp, W., Priefer, U.B. and Puhler, A. (1988) "Nucleotide sequence of a 24 206-base-pair DNA fragment carrying the entire nitrogen fixation gene cluster of Klebsiella pneumoniae" J. Mol. Biol. 203, 715-738.

31. Haselkorn, R. (1986) "Organization of the genes for nitrogen fixation in photosynthetic bacteria and cyanobacteria" Ann. Rev. Microbiol. 40, 525-547.

32. Wall, J.D. and Braddock, K. (1984) "Genetic mapping of nif genes in Rhodopseudomonas capsulata" J. Bacteriol. 158, 404-410.

33. Willison, J.C., Ahombo, G., Chaberg, J., Magnin, J.-P. and Vignais, P.M. (1985) "Genetic mapping of the Rhodopseudomonas capsulata chromosomes show non-clustering of the genes involved in nitrogen fixation" J. Gen. Microbiol. 131, 3001-3015.

34. Kranz, R.G., Pace, V.M. and Caldicott, I.M. (1990) "Inactivation, sequence and lacZ fusion analysis of the regulatory locus required for repression of nitrogen fixation genes in Rhodobacter capsulatus" J. Bacteriol. 172, 53-62.

35. Dixon, R.A. (1984) "The genetic complexity of nitrogen fixation" J. Gen. Microbiol. 130, 2745-2755.

36. Nagatani, H., Shimizu, M. and Valentine, R.C. (1971) "The mechanism of ammonia assimilation in nitrogen fixing bacteria" Arch. Microbiol. 79, 164-175.

500

37. Brown, C.M. and Herbert, R.A. (1977) "Ammonia assimilation in the members of Rhodospirillaceae" FEMS Lett. 1, 43-46.
38. Brown, C.M. and Herbert, R.A. (1977) "Ammonia assimilation in purple and green sulphur bacteria" FEMS Lett. 1, 39-42.
39. Soliman, A., Nordlund, S., Johansson, B.C. and Baltscheffsky, H. (1981) "Purification of glutamine synthetase from Rhodospirillum rubrum by affinity chromatography" Acta Chem. Scand. B35, 63-64.
40. Yoch, D.C., Cantu, M. and Zang, Z.-M. (1983) "Evidence for a glutamine synthetase-chromatophore association in the phototroph Rhodospirillum rubrum: Purification, properties and regulation of the enzyme" J. Bacteriol. 154, 632-639.
41. Soliman, A. and Nordlund, S. (1989) "Purification and partial characterization of glutamine synthetase from the photosynthetic bacterium Rhodospirillum rubrum" Biochim. Biophys. Acta 994, 138-141.
42. Alef, K., Burkardt, H.-J., Horstmann, H.-J. and Zumft, W.G. (1981) "Molecular characterization of glutamine synthetase from the nitrogen-fixing phototrophic bacterium Rhodopseudomonas palustris" Z. Naturforsch. 36 c, 246-254.
43. Alef, K. and Zumft, W.G. (1981) "Regulatory properties of glutamine synthetase from the nitrogen-fixing phototrophic bacterium Rhodopseudomonas palustris" Z. Naturforsch. 36 c, 784-789.
44. Johansson, B.C. and Gest, H. (1977) "Adenylylation/deadenylylation control of the glutamine synthetase of Rhodopseudomonas capsulata" Eur. J. Biochem. 81, 365-371.
45. Caballero, F.J., Cejudo, F.J., Florencio, F.J., Cardenas, J. and Castillo, F. (1985) "Molecular and regulatory properties of glutamine synthetase from the phototrophic bacterium Rhodopseudomonas capsulata E1F1" J. Bacteriol. 162, 804-809.
46. Engelhardt, H. and Klemme, J.-H. (1982) "Purification and structural properties of the adenylylated and deadenylylated glutamine synthetase from Rhodopseudomonas sphaeroides" Arch. Microbiol. 133, 202-205.
47. Rhee, S.G., Chock, P.B. and Stadtman, E.R. (1989) "Regulation of Escherichia coli glutamine synthetase" in Advances in Enzymology (Meister, A. ed.) pp. 37-92, John Wiley & Sons, New York.
48. Woehle, D.L., Lueddecke, B.A. and Ludden, P.W. (1990) "ATP-dependent and NAD-dependent modification of glutamine synthetase from Rhodospirillum rubrum in in vitro" J. Biol. Chem. in press.
49. Yelton, M.M. and Yoch, D.C. (1981) "Nitrogen metabolism in Rhodospirillum rubrum: characterization of glutamate synthase" J. Gen. Microbiol. 123, 335-342.
50. Miller, R.e. (1974) "Glutamate synthase" in Microbiol. iron metabolism (Neilands, J.B. ed.) pp. 283-302, Academic Press, New York.
51. Zumft, W.G. and Castillo, F. (1978) "Regulatory properties of the nitrogenase from Rhodopseudomonas palustris" Arch. Microbiol. 117, 53-60.
52. Gest, H., Kamen, M.D. and Bregoff, H.M. (1950) "Studies on the metabolism of photosynthetic bacteria", J. Biol. Chem. 182, 153-170.
53. Shick, H.-J. (1971) "Regulation of photoproduction in Rhodospirillum rubrum by ammonia" Arch. Microbiol. 75, 110-120.

54. Neilson, A.H. and Nordlund, S (1975) "Regulation of nitrogenase synthesis in intact cells of Rhodospirillum rubrum. Inactivation of nitrogen fixation by ammonia, L-glutamine and L-asparagine" J. Gen. Microbiol. 91, 53-62.
55. Kanemoto, R.H. and Ludden, P.W. (1984) "Effect of ammonia, darkness and phenazine methosulfate on whole-cell nitrogenase activity and Fe-protein modification in Rhodospirillum rubrum" J. Bacteriol. 158, 713-720.
56. Ludden, P.W. and Burris, R.H. (1976) "Activation factor for the iron protein of nitrogenase from Rhodospirillum rubrum" Science 194, 424-426.
57. Nordlund, S., Eriksson, U. and Baltscheffsky, H. (1977) "Necessity of a membrane component for nitrogenase activity in Rhodospirillum rubrum" Biochim. Biophys. Acta 462, 187-195.
58. Pope, M.R., Murell, S.A. and Ludden, P.W. (1985) "Covalent modification of the iron protein of nitrogenase from Rhodospirillum rubrum by adenosine diphosphoribosylation of a specific arginine residue" Proc. Natl. Acad. Sci. 82, 3173-3177.
59. Lowery, R.G. and Ludden, P.W. (1988) "Purification and properties of dinitrogenase reductase ADP-ribosyltransferase from the photosynthetic bacterium Rhodospirillum rubrum" J. Biol. Chem. 263, 16714-16719.
60. Ludden, P.W., Hageman, R.V., Orme-Johnson, W.H. and Burris, R.H. (1982) "Properties and activities of "inactive" Fe protein from Rhodospirillum rubrum" Biochim. Biophys. Acta 700, 213-216.
61. Georgiadis, M.M. (1990) "Progress on the structure of the nitrogenase iron protein from Azotobacter vinelandii" see ref. 25.
62. Saari, L.L., Triplett, E.W. and Ludden, P.W. (1984) "Purification and properties of the activating enzyme for the iron protein of nitrogenase from the photosynthetic bacterium Rhodospirillum rubrum" J. Biol. Chem. 259, 15502-15508.
63. Ljungström, E., Yates, M.G. and Nordlund, S. (1989) "Purification of the activating enzyme for the Fe-protein of nitrogenase from Azospirillum brasilense" Biochim. Biophys. Acta 994, 210-214.
64. Nordlund, S. and Norén, A. (1984) "Dependence on divalent cations of the activation of inactive Fe-protein of nitrogenase from Rhodospirillum rubrum" Biochim. Biophys. Acta 791, 21-27.
65. Guth, J.G. and Burris, R.H. (1983) "The role of $Mg^{2+}$ and $Mn^{2+}$ in the enzyme-catalyzed activation of nitrogenase Fe protein from Rhodospirillum rubrum" Biochem. J. 213, 741-749.
66. Triplett, E.W., Wall, J.D. and Ludden, P.W. (1982) "Expression of the activation enzyme and Fe protein of nitrogenase from Rhodospirillum rubrum (J. Bacteriol. 152, 786-791.
67. Fitzmaurice, W.P., Saari, L.L., Lowery, R.G., Ludden, P.W. and Roberts, G.P. (1989) "Genes coding for the reversible ADP-ribosylation system of dinitrogenase reductase from Rhodospirillum rubrum" Mol. Gen. Genet. 218, 340-347.
68. Jouanneau, Y., Roby, C., Meyer, C. and Vignais, P.M. (1989) "ADP-ribosylation of dinitrogenase reductase in Rhodobacter capsulatus" Biochemistry 28, 6524-6530.

502

69. Yakunin, A.F. and Gogotov, I.N. (1984) "The activity of two forms
    of nitrogenase from Rhodopseudomonas capsulata in the presence
    of different electrons donors" FEMS Microbiol. Lett. 23, 217-229.
70. Gotto, J.W. and Yoch, D.C. (1985) "The regulation of ferredoxin-
    dependent nitrogenase activity in Rhodospirillum rubrum and Rhodo-
    pseudomonas capsulata" FEMS Microbiol. Lett. 28, 107-111.
71. Yoch, D.C., Li, J., Hu, C.-Z. and Scholin, C. (1988) "Ammonia
    switch-off of nitrogenase from Rhodobacter sphaerodies and Methylo-
    sinus trichosporium: no evidence for Fe protein modification"
    Arch. Microbiol. 150, 1-5.
72. Kanemoto, R.H. and Ludden, P.W. (1987) "Amino acid concentration
    in Rhodospirillum rubrum during expression and switch-off of
    nitrogenase activity" J. Bacteriol. 169, 3035-3043.
73. Li, J., Hu, C.-Z. and Yoch, D.C. (1987) "Changes in amino acid and
    nucleotide pools of Rhodospirillum rubrum during switch-off of
    nitrogenase activity initiated by $NH_4$ or darkness" J. Bacteriol.
    169, 231-237.
74. Paul, T.D. and Ludden, P.W. (1984) "Adenine nucleotide levels in
    Rhodospirillum rubrum during switch-off of whole-cell nitrogenase
    activity" Biochem. J. 224, 961-969.
75. Nordlund, S. and Höglund, L. (1986) "Studies of the adenylate and
    pyridine nucleotide pools during nitrogenase "switch-off" in
    Rhodospirillum rubrum" Plant & Soil, 90, 203-209.
76. Ludden, P.W. and Roberts, G.P. (1989) "Regulation of nitrogenase
    activity by reversible ADP-ribosylation" Curr. Top. Cell Regul.
    30, 23-56.
77. Amesz, J. and Knaff, D.B. (1988) "Molecular mechanisms of bacte-
    rial photosynthesis" in Biology of anaerobic microorganisms
    (Zehnder, A.J.B. ed.) pp. 113-178, J. Wiley & Sons, New York.
78. Jackson, J.B. and Crofts, A.R. (1968) "Energy-linked reduction
    of nicotinamide adenine dinucleotides in cells of Rhodospirillum
    rubrum" Biochem. Biophys. Res. Commun. 32, 908-915.
79. Soliman, A. and Nordlund, S. (1990) "Studies on the metabolic
    regulation of nitrogenase activity in Rhodospirillum rubrum" see
    ref. 25.
80. Falk, G., Johansson, B.C. and Nordlund, S. (1982) "The role of
    glutamine synthetase in the regulation of nitrogenase activity
    ("switch-off" effect) in Rhodospirillum rubrum" Arch. Microbiol.
    132, 251-253.
81. Nordlund, S., Kanemoto, R.H., Murell, S.A. and Ludden, P.W. (1985)
    "Properties and regulation of glutamine synthetase from
    Rhodospirillum rubrum" J. Bacteriol. 161, 13-17.

# ORGANIZATION OF THE GENES ENCODING UPTAKE HYDROGENASE IN THE PHOTOSYNTHETIC BACTERIUM *RHODOBACTER CAPSULATUS*


A. COLBEAU, P. RICHAUD, J.P. MAGNIN, J. CABALLERO, B. CAUVIN
AND P.M. VIGNAIS
*Biochimie Microbienne LBIO/DBMS (CNRS URA 1130)*
*CEN-G, 85X, 38041 Grenoble cedex, France*



ABSTRACT : Mutants of *Rhodobacter capsulatus* deficient in hydrogenase activity (Hup⁻ mutants) were used for genetic mapping. The *hup* markers were found to cluster on the chromosome in the proximity of the *his*-1 marker. A cluster of *hup* genes comprising the structural genes was isolated from a cosmid library of *R. capsulatus* B10 containing recombinant DNA with 40 kbp insert DNA and was sequenced. Downstream of the structural genes (*hupSL*) 15 open reading frames were identified, the function of many of them has not been identified yet. The product of the *hupR1* gene shares sequence and domain similarities with the transcriptional activators NifA and NtrC from *R. capsulatus* and *Klebsiella pneumoniae*, resp.. The *hupSL* promoter was fused to the *lacZ* gene and plasmid pAC142 carrying the fusion was introduced into the Hup⁻ mutants and into the wild type strain B10 to study comparatively the β-galactosidase activity and the chromosomally-encoded hydrogenase activity in response to environmental stimuli. In a mutant carrying a mutation in *hupR1*, both the β-galactosidase activity and the hydrogenase activity were reduced to 6 % of that found in the wild type strain, a result consistent with the assumption that HupR1 protein acts as a transcriptional activator.


## 1. Introduction

The nonsulfur purple photosynthetic bacterium *Rhodobacter capsulatus* produces and consumes molecular hydrogen. Hydrogen production is mediated by the nitrogenase complex while hydrogen consumption occurs via uptake hydrogenase located in the cytoplasmic membrane [for reviews see 15, 16]. Although the hydrogenase can catalyse the reversible oxidation of hydrogen to protons and electrons, the *R. capsulatus* enzyme functions physiologically as a hydrogen-uptake hydrogenase. It is membrane associated, and chromatophores were shown to oxidize hydrogen with coupled ATP synthesis [11] indicating that hydrogenase is part of the respiratory chain and, as such, plays an essential role in photoautotrophic and chemoautotrophic growth.

To evaluate the role of hydrogenase in hydrogen production (i.e. in hydrogen recycling), mutants deficient in hydrogenase activity (Hup⁻ mutants) were selected after chemical mutagenesis for lack of autotrophic growth (Aut⁻ phenotype) and hydrogenase activity (Hup⁻ phenotype)(see e.g. [19]). The isolated Hup⁻ mutants were subsequently used to isolate *hup* genes from *R. capsulatus* DNA libraries [5, 6]. One plasmid, pBC2, the insert of which contained several of the *R. capsulatus hup* genes identified so far was studied further. About 25 kbp of *R. capsulatus* DNA have now been sequenced and 17 open reading frames (ORFs) identified.

504

## 2. Results

### 2.1 MAPPING OF *aut/hup* GENES ON THE CHROMOSOME

Contrarily to what occurs in some species, e.g. *Alcaligenes eutrophus* [8] or *Rhizobium leguminosarum* [4] where hydrogenase genes reside on a megaplasmid, the hydrogenase (*hup*) genes of *R. capsulatus* are located on the chromosome [6]. This was demonstrated by 2 types of experiments : 1) by using derivatives of B10 cured of its endogenous plasmid; the plasmid-less strains still retained hydrogenase activity [21] ; 2) by showing that *hup* genes are linked to chromosomal markers.

To obtain the double mutants necessary for genetic mapping, the Hup⁻ strains were submitted to a metronidazole treatment which leads to a selection of spontaneous Nif⁻ mutants [17, 18], ninety five per cent of which are being in a *nif* regulatory gene [20], termed *nifA*-like by Ahombo et al. [1] and *nifR4* by Kranz and Haselkorn [9]. The product of *nifR4* shares amino acid sequence similarities with the product of *ntrA* (*rpoN*) genes of other bacteria [3, 7].

| Hup⁻ Strains | (metronidazole treatment) | Hup⁻ Nif⁻ Strains | Relevant genotype | | |
|---|---|---|---|---|---|
| JP57 | ⟶ | JP73 | *hup-8,* | *nif*R4 | (*nif*-122) |
| RCC12A | ⟶ | JP232 | *hup-9,* | *nif*R4 | (*nif*-129) |
| RS13 | ⟶ | JP241 | *hup-10,* | *nif*R4 | (*nif*-131) |
| IR4 | ⟶ | JP217 | *aut-4,* | *nif*R4 | (*nif*-126) |

Fig. 1 : Construction of Hup⁻ Nif⁻ double mutants and list of the strains used for genetic mapping (cf [6]).

The Hup⁻ strains listed on figure 1 gave rise to the Hup⁻ Nif⁻ double mutants mutated in *nifR4* since the cloned *nifR4* gene could restore nitrogenase synthesis in all of the Nif⁻ mutants [6]. The map location of the *aut/hup* genes was determined by conjugal transfer of chromosomal genes in *R. capsulatus*, mediated by the R-plasmid pTH10, a mutant of RP4, between donor strains carrying auxotrophic markers and containing pTH10 in a free state, and the Hup⁻ Nif⁻ double mutants as the recipient strains. Recombinants were first screened for autotrophic growth. After autotrophic growth, secondary markers, (e.g. *his*-1, Nif⁺) were analyzed after replication of the transconjugant clones on N free medium. The *hup-8, hup-9* and *hup-10* markers showed cotransfer frequencies of 25-30 % with *nifR4* and, on average, 55 % with *his-1* (Fig. 2).

This shows that all 3 markers are located in the same region of the chromosome in the proximity of *his-1*. On the other hand, the *aut-4* marker presented a cotransfer frequency of 98-99 % with the *nifR4* marker. The *aut-4* mutation carried by the hydrogenase-deficient strain IR4 is not located in the *nifR4* gene since mutant IR4 has a Nif⁺ phenotype while NifR4⁻ mutants have a Hup⁺ phenotype [6]. On the circular genetic map of *R. capsulatus* established by Willison and coworkers for strain B10 [20, 22], *hup* genes map in the region of the chromosome that contains genes for histidine biosynthesis, the photosynthesis gene cluster and the *nif* gene cluster containing the structural genes of nitrogenase.

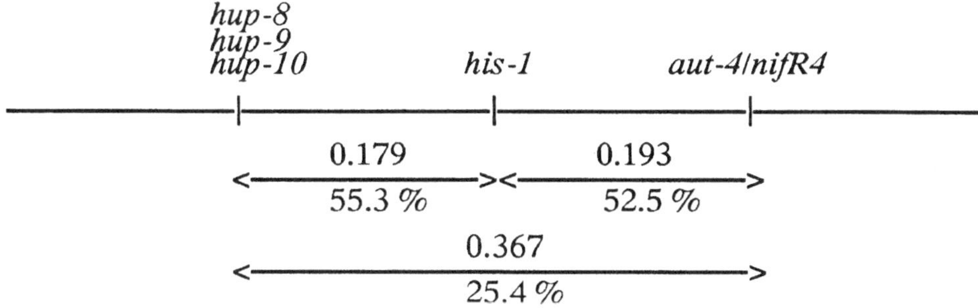


Fig. 2 : Gene order and cotransfer frequencies of *hup*, *nifR4*, *aut-4* and *his-1* markers in pTH10-mediated crosses between the donor strain JB90 (*ade-1*, *his-1*, pTH10) and the Hup⁻ Nif⁻ double mutants (from [6]).

## 2.2 PHYSICAL ORGANIZATION OF *hup* GENES IN *R. CAPSULATUS*

Several genes appear to be involved in the synthesis of hydrogenase since at least 3 different *Eco*RI DNA fragments from *R. capsulatus* of about 20 kbp were shown to complement 3 distinct series of our Hup⁻ mutants [5, 6, 10]. A fourth *Eco*RI DNA fragment from *R. capsulatus* with a restriction map totally different from the preceding ones and completementing Hup⁻ mutants was isolated in the laboratory of J. Wall [23].
The structural genes were first isolated and sequenced. They comprise *hupS* capable of encoding a 34 256 dalton polypeptide (small subunit) and *hupL* encoding the large subunit of 65 839 daltons [10]. *Rhodobacter capsulatus* hydrogenase shares a high degree of identity with other membrane-bound uptake-hydrogenases [cf. 12, 14]. A total of 17 open reading frames (ORFs) has now been identified. Once the nucleotide sequence was known, it was then possible to compare it with the physical map of the DNA fragments studied in the group of J. Wall [23] ; the three *Eco*RI fragments isolated by these authors could be aligned with the insert DNA of pBC2. Furthermore, from the complementation experiments after Tn5 mutagenesis [23], it can be inferred that some of the ORFs identified, whose function is still unknown, could be *hup* genes. Among the identified *hup* genes, one of them named *hupR1* is capable of encoding a protein presenting striking structure similarities with transcriptional activators belonging to a superfamily of regulators which become activated by phosphorylation in response to environmental stimuli (e. g. *ntrC*) [2, 13] (Fig. 3). In their N-terminal region the HupR1 protein and the product of *ntrC* share 26 % identical aminoacids. Both contain 3 highly conserved residues ( 2 aspartic and one lysine residues) typically found in the response regulator superfamily. The C-terminal regions exhibit an even higher degree of identity (37 %) and present a helix turn helix structure typical of DNA-binding proteins.

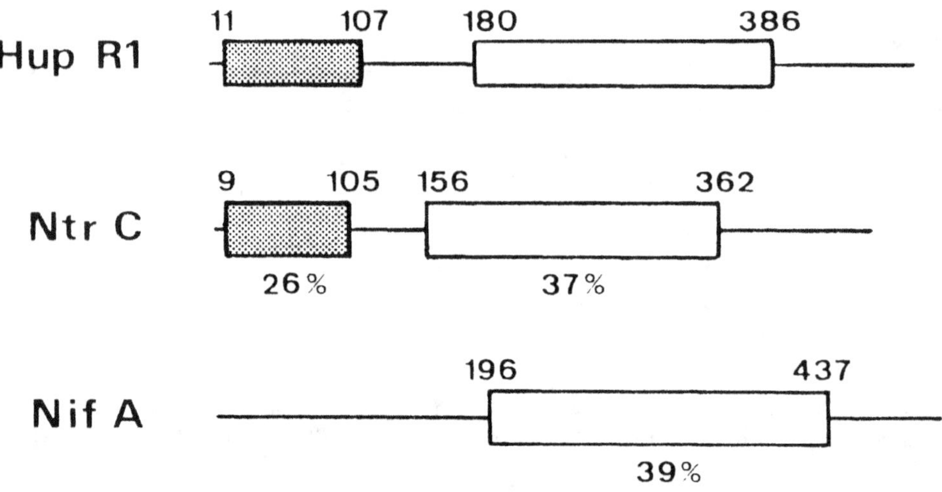

Fig. 3 Comparison of the domain organization of HupR1 and of two transcriptional regulators (NtrC and NifA from *K. pneumoniae* and *Rhodobacter capsulatus*, resp.) belonging to the response regulator superfamily [cf. 13].

To identify signals to which *R. capsulatus* cells respond by an increased or decreased hydrogenase synthesis, a transcriptional fusion of the *E. coli lacZ* gene to the promoter of the *hup* structural operon (*hupSL*) was constructed. The activity of the *hupSL* promoter is being determined in different genetic backgrounds, under different physiological conditions. Preliminary results indicate that in a mutant carrying a mutation in the *hupR1* gene, the activity of the *hupSL* promoter, either on the chromosome upstream of the *hupSL* genes and measured by hydrogenase activity, or on the plasmid carrying the *lacZ* fusion and monitored by β-galactosidase activity is reduced to 6 % of that in the wild type strain B10. These data are consistent with the assumption that HupR1 is a transcriptional activator.

Our current efforts aim at identifying the signals to which the *R. capsulatus* cells respond in modulating hydrogenase expression and the molecular components which contribute to signal transduction.

ACKNOWLEGMENTS : The help of Cécile Marillat in the preparation of this manuscript is gratefully acknowledged. The investigation was supported by research grants from the Centre National de la Recherche Scientifique (CNRS URA 1130 alliée à l'INSERM).

# REFERENCES


1. Ahombo, G., Willison J.C. and Vignais, P.M. (1986) 'The *nifHDK* genes are contiguous with *nifA*-like regulatory gene in *Rhodobacter capsulatus*', Mol. Gen. Genet. 205, 442-445.
2. Albright, L.M., Huala, E. and Ausubel, F.M. (1989) 'Prokaryotic signal transduction mediated by sensor and regulator protein pairs', Annu. Rev. Genet. 23, 311-36.
3. Alias, A., Cejudo, J., Chabert, J., Willison J.C. and Vignais, P.M. (1989) 'Nucleotide sequence of wild-type and mutant *nif*R4 (*ntr*A) genes of *Rhodobacter capsulatus* : identification of an essential glycine residue', Nucleic Acids Research 17, 5377.
4. Brewin, N.J., Dejong, T.M., Philipps, D.A. and Johnston, AWB (1980) 'Cotransfer of determinants for hydrogenase activity and nodulation ability in *Rhizobium leguminosarum*', Nature 288, 77-79.
5. Colbeau, A., Godfroy, A. and Vignais, P.M. (1986) 'Cloning of DNA fragments carrying hydrogenase genes of *Rhodopseudomonas capsulata*', Biochimie 68, 147-155.
6. Colbeau, A., Magnin, J.P., Cauvin, B., Champion, T. and Vignais, P.M. (1990) 'Genetic and physical mapping of an hydrogenase gene cluster from *Rhodobacter capsulatus*', Mol. Gen. Genet. 220, 393-399.
7. Jones, R. and Haselkorn, R. (1989) 'The DNA sequence of the *Rhodobacter capsulatus ntrA, ntrB* and *ntrC* gene analogues required for nitrogen fixation', Mol. Gen. Genet. 215, 507-516
8. Kortlücke, C., Hogrefe, C., Eberz, G., Pühler, A. and Friedrich, B. (1987) 'Genes of lithoautotrophic metabolism are clustered on the megaplasmid pHG1 in *Alcaligenes eutrophus*', Mol. Gen. Genet. 210, 122-128.
9. Kranz, R.G. and Haselkorn, R. (1985) 'Characterization of *nif* regulatory genes in *Rhodopseudomonas capsulata* using *lac* gene fusions', Gene 40, 205-215
10. Leclerc, M., Colbeau, A., Cauvin, B. and Vignais, P.M. (1988) 'Cloning and sequencing of the genes encoding the large and the small subunit of the $H_2$-uptake hydrogenase (*hup*) of *Rhodobacter capsulatus*', Mol. Gen. Genet. 214, 97-107. Correction 215, 368.
11. Paul, F., Colbeau, A. and Vignais, P.M. (1979) 'Phosphorylation coupled to hydrogen oxidation by chromatophores from *Rhodopseudomonas capsulata*', FEBS Lett. 106, 29-33.
12. Richaud, P., Vignais, P.M., Colbeau, A., Uffen, R. and Cauvin, B. (1990) 'Molecular biology studies of the uptake hydrogenase of *Rhodobacter capsulatus* and *Rhodocyclus gelatinosus*', FEMS Microbiol. Reviews 76 (in press).
13. Stock, J.B., Ninfa, A.J. and Stock, A.M. (1989) 'Protein phosphorylation and regulation of adaptive responses in bacteria', Microbiological Rewievs 53, 450-490.
14. Uffen, R.L., Colbeau, A., Richaud, P. and Vignais, P.M. (1990) 'Cloning and sequencing the genes encoding uptake hydrogenase subunits of *Rhodocyclus gelatinosus* hydrogenase', Mol. Gen. Genet. 221, 49-58.
15. Vignais, P.M., Colbeau, A., Willison, J.C. and Jouanneau, Y. (1985) 'Hydrogenase, nitrogenase and hydrogen metabolism in the photosynthetic bacteria'. In "Advances in Microbial Physiology" (A.H. Rose and D.W. Tempest, eds), 26, 155-234, Academic Press, Inc. (London) Ltd.
16. Vignais, P.M., Willison, J.C., Jouanneau, Y., Colbeau, A., Magnin, J.P., Leclerc, M. and Schatt, E. (1988) 'Nitrogen fixation, nitrogenase and hydrogenase in the photosynthetic bacteria'. In "Nitrogen fixation: hundred years after" (H. Bothe, F.J. de Bruijn and W.E. Newton, Eds) (Proceedings of the VII international Congress on Nitrogen Fixation, Köln (Cologne), FRG (1988) pp. 147-155. Gustav Fischer. Stuttgart. New York.
17. Wall, J.D., Love, J. and Quinn, S.P. (1984) 'Spontaneous Nif⁻ mutants of *Rhodopseudomonas capsulata*', J. Bacteriol. 159, 652-657.

18. Willison, J.C. and Vignais, P.M. (1982) 'The use of metronidazole to isolate Nif⁻ mutants of *Rhodopseudomonas capsulata* and the identification of a mutant with altered regulatory properties of nitrogenase', J. Gen. Microbiol. 128, 3001-3010.
19. Willison, J.C., Madern, D. and Vignais, P.M. (1984) 'Increased photoproduction of hydrogen by non-autotrophic mutants of *Rhodopseudomonas capsulata*', Biochem. J. 219, 593-600.
20. Willison, J.C., Ahombo, G., Chabert, J., Magnin J.P. and Vignais, P.M. (1985) 'Genetic mapping of the *Rhodopseudomonas capsulata* chromosome shows non-clustering of genes involved in nitrogen fixation', J. Gen. Microbiol. 131, 3001-3015.
21. Willison, J.C., Magnin, J.P., and Vignais, P.M. (1987) 'Isolation and characterization of *Rhodobacter capsulatus* strains lacking endogenous plasmids', Arch. Microbiol. 147, 134-142.
22. Willison, J.C., Ahombo, G. and Vignais, P.M. (1990) 'Genetic control of nitrogen metabolism in the photosynthetic bacterium *Rhodobacter capsulatus*', in W.R. Ulrich, C. Rigano, A. Fuggi and P.J. Aparicio (eds) "Inorganic Nitrogen in Plants and Microorganisms: Uptake and Metabolism" Springer-verlag, Berlin Heidelberg New York pp 312-319.
23. Xu, H.W., Love, J., Borghese, R. and Wall, J.D. (1989) 'Identification and isolation of genes essential for hydrogen oxidation in *Rhodobacter capsulatus*', J. Bacteriol. 171, 714-721.

# Diazotrophic mixed culture of *Azotobacter vinelandii* and *Rhodobacter capsulatus*


J. OELZE
*Institut für Biologie 2 (Mikrobiologie), Universität Freiburg, Schänzlestrasse 1, DW-7800 Freiburg Germany*



*Key words:* *Azotobacter vinelandii*, diazotrophic growth, mixed culture, *Rhodobacter capsulatus*


## Abstract


The question, whether *Azotobacter vinelandii* can provide fixed N for the growth of other organisms, was studied with mixed cultures of *A. vinelandii* and *Rhodobacter capsulatus*, grown with aeration in the light. $N_2$-fixation by *R. capsulatus* was prevented by growing the cultures on either mannitol, glycerol or ethanol, which cannot be used by this organism. In the course of growth with mannitol, cell numbers of both organisms increased largely in parallel and attained a maximal ratio of about one *A. vinelandii* per ten *R. capsulatus*. Prolonged growth of mixed cultures with mannitol did not lead to an adaptation of *R. capsulatus* to this compound. After growth on either one of the three alcohols, mixed cultures exhibited almost twice as high protein levels as pure cultures of *A. vinelandii*. Up to 80% of the protein of mixed cultures was incorporated into *R. capsulatus*. The results suggest that *A. vinelandii* provided an organic N-source for the growth of *R. capsulatus*.


## Introduction

It is generally believed that free-living diazotrophs like Azotobacters contribute only little to the input of nitrogen into the biosphere. Earlier observations of growth stimulation of plants by Azotobacter are now believed to result from growth factors rather than from a supply of fixed nitrogen (Postgate, 1982).

In order to answer the question whether Azotobacter can provide fixed nitrogen for the growth of other organisms, the behaviour of the phototrophic *Rhodobacter capsulatus* was studied in mixed culture with *Azotobacter vinelandii*. Since the former is a diazotroph as well, culture conditions were selected, which allow $N_2$-fixation in *A. vinelandii* but exclude this activity in *R. capsulatus*. According to the literature, these conditions should be fulfilled, if the cultures are supplied with certain alcohols like ethanol, glycerol or mannitol (Trüper and Pfennig, 1978). The results demonstrate that *R. capsulatus* performs significant growth in media containing such alcohols, provided *A. vinelandii* is present as well.

## Materials and methods

*A. vinelandii* was maintained on a modified Burks medium (Kleiner, 1975) containing either mannitol, glycerol or ethanol ($15\,\mathrm{g\,L^{-1}}$) as carbon sources and sodium cirate ($0{,}05\,\mathrm{g\,L^{-1}}$) complexing iron ions. The cultures were grown aerobically in 200 mL of medium at 150 rpm and 30°C in 1 L Erlenmeyer flasks. *R. capsulatus* was maintained phototrophically on medium MS (Lascelles, 1959). Mixed cultures of *A. vinelandii* and *R. capsulatus* were grown on Burks medium supplemented with nicotinic acid ($1.0\,\mathrm{mg\,L^{-1}}$), thiamine HCl ($1.0\,\mathrm{mg\,L^{-1}}$), and biotin ($10\,\mu\mathrm{g\,L^{-1}}$). The cultures (300 mL) were grown in the presence of air at 400 rpm, 3,000 lx and 30 °C in 1 L Erlenmeyer flasks.

Cell numbers were taken with samples after addition of formaldehyde (2%).

Nitrogenase activity was determined by acetylene reduction as well as by hydrogen evolution. In situ nitrogenase activity of mixed cultures was assayed after addition of acetylene (10%) to the head space of the culture vessel. Acetylene reduction by *R. capsulatus* was assayed anaerobically as described (Steinborn and Oelze, 1989). In situ $H_2$ production by *R. capsulatus* was estimated after collecting the gas with syringes injected into the hermetically sealed culture bottles.

Protein as well as bacteriochlorophyll (BChl) were determined as described.

## Results and discussion

Growth of mixed cultures of *A. vinelandii* and *R. capsulatus* on mannitol as carbon source are represented by the data depicted in Figure 1.

In the course of four days, cell protein increased by a factor of about one hundred to $1 \, g \, L^{-1}$. This was largely parallelled by the formation of the photopigment BChl by *R. capsulatus*. Determination of cell numbers revealed that growth of *R. capsulatus* proceeded largely in parallel to growth of *A. vinelandii*. Cell numbers of *R. capsulatus* were up to ten times higher than numbers of *A. vinelandii*.

Growth of *R. capsulatus* in the presence of *A. vinelandii* may be based on several possibilities:
1. Utilization of cell material derived from non-growing *A. vinelandii*;
2. Growth on substrates provided by growing *A. vinelandii*
   a. inorganically bound N,
   b. a suitable C source derived from mannitol,
   c. an organic N source;
3. Decreased ambient $O_2$ concentration, required for phototrophic $N_2$-fixation (Vignais et al., 1985), due to the high rate of $O_2$ consumption by *A. vinelandii*
4. Adaptation of *R. capsulatus* to diazotrophic growth on mannitol.

In order to decide, which of the possibilities might apply, a mixed culture was used to inoculate growth media with different combinations of mannitol, malate, and ammonium sulfate. The results obtained after 78 h of anaerobic growth in the light are compiled in Table 1.

Significant growth of *R. capsulatus* was possible only if the cultures were supplied with malate and ammonium sulfate. No growth occurred if the mannitol-medium was supplemented only with ammonium sulfate. $H_2$ production, however, representing the presence of an active nitrogenase (Vignais et al., 1985), could be observed after incubation in the presence of man-

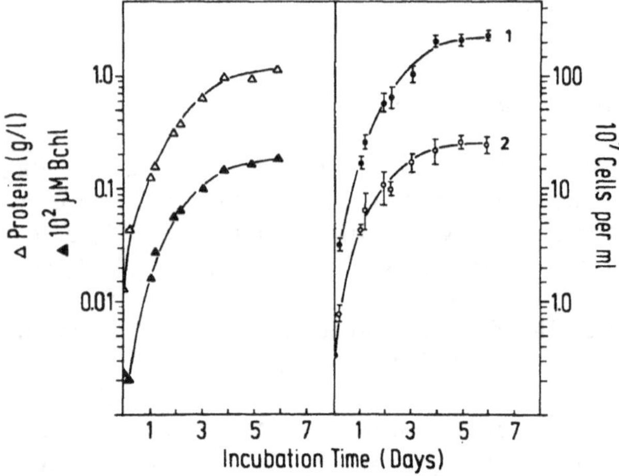

Fig. 1. Diazotrophic growth of *Azotobacter vinelandii* (2) and *Rhodobacter capsulatus* (1) in mixed culture on mannitol with aeration and illumination (3,000 lx).

Table 1. Growth behaviour of mixed cultures of *Azotobacter vinelandii* (b) and *Rhodobacter capsulatus* (a) under anaerobic conditions in the light on mannitol ($15 \, g \, L^{-1}$) and different combinations of $(NH_4)_2SO_4$ ($10 \, mM$) and malate ($3 \, g \, L^{-1}$)

| | $H_2$ Evolution (mL 100 mL of culture) | Bchl (nmol mL$^{-1}$) | Protein ($\mu g \, mL^{-1}$) | $10^7$ Cells per mL |
|---|---|---|---|---|
| $t = 0h$ | | 0.6 | 60 | $14.3^a/0.7^b$ |
| $t = 78h$ | | | | |
| Mannitol | – | 0.6 | 53 | 13.6/1.1 |
| Mannitol $(NH_4)_2SO_4$ | – | 0.8 | 80 | 18.2/1.4 |
| Mannitol Malate | 60 | 2.6 | 103 | 28.0/1.7 |
| Mannitol $(NH_4)_2SO_4$ Malate | – | 14.1 | 412 | 94.0/4.0 |

nitol plus malate.' These results do not support possibilities 1, 2a, 3, and 4, mentioned above.

Lack of adaptation of *R. capsulatus* to growth on mannitol was confirmed after reisolation of the phototrophic bacterium from a mixed culture, transferred regularly for a period of two months. As compared to the original pure culture, the re-isolated culture of *R. capsulatus* exhibited the same requirements, namely, growth only in the presence of malate plus ammonium sulfate and $H_2$ evolution in the presence of malate.

The data, presented in Table 1, were obtained with anaerobic cultures, suggesting that growth of *R. capsulatus* in mixed culture with mannitol was not the result of the generation of reduced $O_2$ concentrations by *A. vinelandii*. This conclusion is supported by results obtained after gassing mixed cultures with $N_2$. This treatment should inhibit the energy metabolism and, thus, nitrogenase activity of *A. vinelandii*, while it should simulate these activities in *R. capsulatus*. Nevertheless, total protein and BChl syntheses became completely inhibited. The inhibitory action of $N_2$, however, was reversible so that protein as well as BChl synthesis resumed as soon as the culture was aerated again.

Possibility 2b, i.e. metabolism of mannitol by *A. vinelandii* to a substrate more suitable for *R. capsulatus*, became unlikely after assaying acetylene reduction with mixed cultures in the course of a light-dark cycle. If *R. capsulatus* was able to exhibit nitrogenase activity, this should indicate the presence of a reduced carbon source more suitable than mannitol as an electron donor. Moreover, since nitrogenase activity of *R. capsulatus* largely depends on light, switching off the light should result in at least a partial switch-off of nitrogenase. Figure 2 demonstrates that application of a light-dark cycle was of no significant effect on the rate of in situ acetylene reduction by a mixed culture.

For control, acetylene reduction by the same culture was assayed anaerobically in the presence of malate. Under these conditions, preventing the activity of *A. vinelandii*, nitrogenase was switched off upon switching off the light (Fig. 2).

All of the results reported above lead to the suggestion that *A. vinelandii* provides an organic N source for the growth of *R. capsulatus*. At the

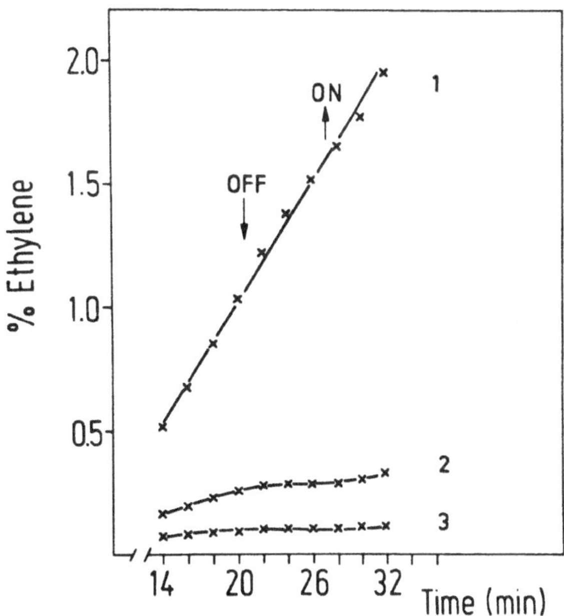


*Fig. 2.* Acetylene reduction by mixed cultures (protein, $0.6\,g\,L^{-1}$, BChl $6.5\,\mu M$) of *Azotobacter vinelandii* and *Rhodobacter capsulatus* in the light. At the times indicated, light was switched off and on, respectively. For growth conditions, see Fig. 1. In situ (1), under anaerobic conditions and in the presence of malate (2), under anaerobic conditions and no external C source (3).

present, however, we have no information on the chemical nature of such a substance. Nevertheless, the question arises whether growth in mixed culture influences total nitrogen fixation.

Determination of maximal yield values after growth either with mannitol, glycerol, or ethanol showed that pure cultures of *A. vinelandii* produced $607 \pm 40$, $620 \pm 50$, and $695 \pm 55\,\mu g$ of protein per mL, respectively, while the corresponding values obtained with mixed cultures were $1019 \pm 115$, $1009 \pm 87$, and $1332 \pm 244\,\mu g$ per mL. Experiments performed with the latter two substrates led to essentially the same results as those reported above for mannitol.

At first sight, the yield values suggest that differences in protein levels of pure and mixed cultures represent the amount of nitrogen incorporated into *R. capsulatus*. This would be about 40% of the total of mixed cultures. This percentage appears too low when it is remembered that mixed cultures contained up to ten times more cells of *R. capsulatus* than of *A. vinelandii*. In order to obtain a more precise value, the average protein content of cells of *R. capsulatus* was determined after counting pure culture samples of known protein concentrations. On this basis, it can be proposed that about 70–80% of the total protein of mixed cultures may be incorporated into *R. capsulatus*.

Taken together, the results of this report indicate that *A. vinelandii* is able to provide a considerable amount of its fixed nitrogen for the growth of other organisms. Low cell numbers of *A. vinelandii* do not necessarily represent the level of nitrogen input into biomass.

## Acknowledgements

This investigation was financially supported by the Deutsche Forschungsgemeinschaft. The expert assistance of B Söntgerath is gratefully acknowledged.

## References

Postgate J R 1982 The Fundamentals of Nitrogen Fixation. Cambridge University Press, Cambridge.

Trüper H G and Pfennig N 1978 Taxonomy of the Rhodospirillales. *In* The Photosynthetic bacteria. Eds R K Clayton and W R. Sistrom. pp 19–27 . Plenum Press, New York, London.

Kleiner D 1975 Ammonium uptake by nitrogen fixing bacteria. I. *Azotobacter vinelandii*. Arch. Microbiol. 104, 163–169.

Lascelles J 1959 Adaptation to form bacteriochlorophyll in *Rhodopseudomonas sphaeroides*: change in activity of enzymes in pyrrole synthesis. Biochem. J 72, 508–518.

Steinborn B and Oelze J 1989 Nitrogenase and photosynthetic activities in chemostat cultures of *Rhodobacter capsulatus* 37b4 grown under different illuminations. Arch. Microbiol. 152, 100–104.

Vignais P M, Colbeau A, Willison J C and Jouanneau Y 1985 Hydrogenase, nitrogenase, and hydrogen metabolism in the phototrophic bacteria. Adv. Microbial Phys. 26, 155–234.

# STRUCTURAL AND GENETIC STUDIES ON FOUR FERREDOXINS FROM THE PHOTOSYNTHETIC BACTERIUM *RHODOBACTER CAPSULATUS* ; PROBABLE INVOLVEMENT IN NITROGEN FIXATION


Y. Jouanneau, E. Schatt, C. Duport, C. Grabau, C. Meyer, and P. M. Vignais
Biochimie Microbienne/DBMS (CNRS URA 11300), C.E.N.-G., 85X,
38041 Grenoble Cedex, France



ABSTRACT : Four ferredoxins have been identified and chemically characterized in *Rhodobacter capsulatus*. The possible function of these ferredoxins has been investigated by means of *in vitro* biological assays and genetic studies. The primary structure of each of the four ferredoxins has been determined. The nature of their iron-sulfur clusters was established using UV-visible and EPR spectroscopy. FdI and FdIII contain two [4Fe-4S] clusters, FdII contains both one [3Fe-4S] and one [4Fe-4S] cluster, whereas FdIV has a [2Fe-2S] cluster similar to plant-type ferredoxins. The structural gene of FdI, *fdxN*, was cloned, sequenced and shown to be part of a group of *nif* genes. The gene encoding FdIV, *fdxC*, was localized just upstream of *fdxN*. The gene coding for FdIII, *fdxB*, has been found to be part of an operon containing three *nif* genes (*nifE, nifN, nifQ*) known to be involved in the FeMo-co synthesis. The transcription of *fdxB, fdxC* and *fdxN* is inhibited in cells grown with excess $NH_4^+$. The gene encoding FdII was cloned and sequenced. This gene, called *fdxA*, is regulated independently of *nif* genes.


## 1. Introduction

Ferredoxins are small electron carrier proteins that serve in a variety of metabolic processes, including nitrogen fixation. They contain one or two Fe-S clusters and have very low mid-point redox potentials, often below -400 mV. Ferredoxins, as well as flavodoxins, are the only known electron carriers having a redox potential low enough to serve as physiological electron donors to nitrogenase. The electron transport system to nitrogenase has been described in detail only in *Klebsiella pneumoniae*. A flavodoxin and an oxidoreductase, encoded by *nifF* and *nifJ* respectively, carry out the transfer of electrons from pyruvate to the nitrogenase complex [1, 2]. In photosynthetic bacteria, the presence of genes homologous to *nifF* and *nifJ* has not yet been proven. In fact, there probably exist in these bacteria a different pathway of electron transport to nitrogenase that is adapted to their particular metabolism.
In our laboratory, we have studied nitrogen fixation in *Rhodobacter capsulatus* by a variety of physiological, biochemical and genetic approaches [3]. Nitrogenase isolated from this bacterium was found to be similar in many respects to the enzyme of other diazotrophs [4], except that the component 2 (Fe protein) was found to be regulated by mono ADP-ribosylation [5, 6]. A ferredoxin (FdI) was isolated from nitrogenase-derepressed cells and found to transfer electrons to nitrogenase in an *in vitro* system [7]. A second ferredoxin (FdII), also capable of *in vitro* electron transfer to nitrogenase, has been identified in cells grown under both *nif* repressing and derepressing conditions [8, 9]. We have recently investigated further the structure and function of these two ferredoxins. We have also isolated and characterized two new ferredoxins, called FdIII and FdIV. In this report, the properties of the four ferredoxins from *R. capsulatus* are compared and

their possible function is evaluated based on biochemical and genetic studies.


## 2. Results

### 2.1. PROPERTIES OF THE FERREDOXINS FROM *R. CAPSULATUS*

The biochemical properties of the four ferredoxins from *R. capsulatus* are summarized in Table 1. We have determined the primary structure of each of them, as deduced from the nucleotide sequence of their respective genes as well as by partial amino-acid sequencing [10, 11, 12]. The sequence of the FdIII polypeptide was reported in [13] as the product of an open reading frame (ORF5). Sequence comparison indicated that the four ferredoxins had limited similarities with each other. By comparison to other known ferredoxin structures, the ferredoxins of *R. capsulatus* may be related to three distinct protein categories.

FdI is homologous to ferredoxins containing two [4Fe-4S] clusters exemplified by the *Peptococcus aerogenes* ferredoxin [14]. The FdI polypeptide is 10 residues longer than the *P. aerogenes* ferredoxin and resembles ferredoxins from a few other photosynthetic bacteria, featuring at the C terminal side an insertion of 6 residues between the second and the third cys in the conserved motif CXXCXXCXXXCP [10]. Our EPR analyses provide experimental evidence that FdI contains two [4Fe-4S] clusters.

FdII is composed of 111 amino-acids and is homologous to the 7Fe ferredoxin I of *Azotobacter vinelandii* [11]. That FdII contains both a [3Fe-4S] and a [4Fe-4S] has recently been shown by EPR spectroscopy [9].

Fd III consists of 100 residues, including 9 cysteines, 8 of which are adequately spaced to accomodate two [4Fe-4S] clusters. However, FdIII displays only limited sequence similarities with 2[4Fe-4S] bacterial ferredoxins. EPR spectroscopy revealed a prominant signal at $g = 1.94$ in the reduced form and showed a complex spectrum that could result from two interacting Fe-S clusters. Chromophore absorption at 390 nm and analysis of the iron and sulfur atom content were also consistent with FdIII being a 2[4Fe-4S] ferredoxin.

TABLE 1. Properties of four ferredoxins from *Rhodobacter capsulatus*.

| Properties | FdI | FdII | FdIII | FdIV |
|---|---|---|---|---|
| Molecular weight[a] | 6,728 | 12,450 | 10,690 | 10,156 |
| Type of Fe-S cluster | 2[4Fe-4S] | [3Fe-XS]<br>+<br>[4Fe-4S] | 2[4Fe-4S] | [2Fe-2S] |
| Structural gene | *fdxN* | *fdxA* | *fdxB* | *fdxC* |
| Synthesis repressed by $NH_4^+$ | yes | no | yes | yes |
| *In vitro* transfer of electrons to nitrogenase | yes | yes | no | no |

[a] Molecular weight of the apoprotein calculated from the aminoacid sequence.

The FdIV polypeptide was recently identified as the product of a gene that we called *fdxC* [12]. FdIV was isolated as the product of expression of *fdxC* in *Escherichia coli*. Its amino acid sequence resembles that of plant-type ferredoxins. EPR and UV-visible spectroscopy indicated that FdIV contained a [2Fe-2S] cluster. FdIV has not yet been isolated from *R. capsulatus* although transcription from *fdxC* was detected. The other three ferredoxins were isolated from bacteria grown in N-limited cultures, conditions that promote derepression of nitrogenase gene. FdI was found to be about 5-fold more abundant than FdII and FdIII, under these growth conditions. When bacteria were grown with excess $NH4^+$, which represses *nif* gene expression, only FdII appeared to be synthesized.
The interaction of each ferredoxin with nitrogenase was tested under identical experimental conditions. Only FdI and FdII were found capable of transferring electrons between illuminated chloroplasts and purified nitrogenase.

## 2.2. GENETIC STUDIES OF THE FERREDOXINS : PREDICTION OF THEIR ROLE IN $N_2$ FIXATION

The genes encoding the four ferredoxins described above have been cloned and sequenced. The FdI gene was called *fdxN* by analogy with other potential ferredoxin genes found next to *nif* genes in other diazotrophs [10, 15]. *FdxN* mapped in a region of the chromosome containing a major *nif* gene cluster (16) (Fig. 1). The *fdxC* gene encoding FdIV was identified 46 bp upstream of *fdxN* [12, 17]. Further upstream, an open reading frame was found only 18 bp before the start codon of *fdxC*. Considering the close proximity of these genes and the absence of promoter consensus sequences in front of either *fdx* gene, it is suggested that the ORF, *fdxC* and *fdxN* are part of the same operon. Transcriptional analysis using probes specific to *fdxC* and *fdxN* indicated that the expression of both genes is repressed in bacteria provided with an excess of fixed nitrogen ($NH4^+$) [10, 12]. These results indicate that FdI and FdIV are synthesized only under growth conditions allowing *nif* derepression, and therefore suggest that they are involved in nitrogen fixation.

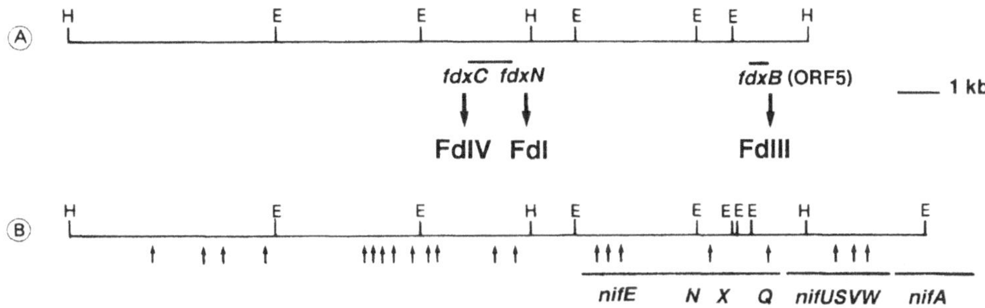


Figure 1. Physical map showing the relative position of *fdxN, fdxB* and *fdxC* encoding FdI, FdIII and FdIV respectively.

A Restriction map of a · 20 kb fragment carried on cosmid pES2. The relative position of *fdx* genes was determined by hybridization experiments and nucleotide sequence analysis [10, 12].

B Physical map of the *nif* region described by Klipp et al. [16]. Arrows indicate Tn5 insertions generating *nif* mutations. Position of the indicated *nif* genes was determined by the group of Klipp and Pühler. The gene coding for FdIII, previously called ORF5, was found immediately upstream of *nifQ* [13].

The gene encoding FdIII, initially called ORF5, was localized as part of a *nif* operon, including *nifE, nifN, nifX* and *nifQ* [13]. The FdIII gene, which we propose to call *fdxB*, is only about 6 kb away from *fdxCN*, which probably belongs to an adjacent operon (Fig. 1). The particular location of *fdxB* next to genes involved in FeMo-co synthesis makes it likely that FdIII somehow participates in the  processing of nitrogenase component 1.

The FdII gene was recently cloned from a DNA library of *R. capsulatus* and called *fdxA* [11, 18], like the structural gene of *A. vinelandii* FdI [19], because of the similarities between the two ferredoxins (see above).The restriction map of a 11 kb fragment carrying *fdxA* indicates that the FdII gene is located in a previously unstudied region of the chromosome. The expression of FdII in *R. capsulatus* was studied by using a *fdxA::lacZ* translational fusion. Results indicate that FdII is constituvely synthesized at a constant level [18].

## 3. Conclusions

Four ferredoxins have been biochemically and genetically characterized in *R. capsulatus*. These ferredoxins showed marked structural differences based on amino acid sequence comparison and on the nature of their Fe-S cluster(s). Four ferredoxins have also been isolated from *Rhodospirillum rubrum*, two of which were found to be membrane bound [20]. The two soluble ferredoxins of *Rs. rubrum* [21] have molecular properties similar to *R. capsulatus* FdI and FdII. Previously, FdII was described as a 18 kDa polypeptide containing a single [4Fe-4S] cluster [8]. Inaccuracy in the molecular weight or Fe and S atom content determinations presented in this earlier report might explain discrepancies with our data. Consistent with the finding that FdII was present in cells grown under various conditions [8], the transcription of *fdxA* appeared constant and unregulated. The other three ferredoxins FdI, FdIII and FdIV are expected to play a role in nitrogen fixation because of the location of their structural genes among *nif* genes. Also, the inhibition of transcription of *fdxB*, *fdxC* and *fdxN* by $NH4^+$ is analogous to *nif* gene regulation. Reduced FdI and FdII serve as reductants for nitrogenase *in vitro*. Since FdI is synthesized only under *nif* derepressing conditions and is about five fold more abundant than FdII, we suspect that FdI acts as the actual physiological electron donor to nitrogenase. In agreement, *fdxN* mutants were found to be severely impaired in the ability to fix nitrogen (W. Klipp, personal communication).

## AKNOWLEDGMENTS

This work was supported by grants from the Centre National de la Recherche Scientifique (CNRS UA1130 alliée à l'INSERM).

## REFERENCES

1. Nieva-Gomez D., Roberts G.P., Kleviebis S., and Brill W.J. (1980) 'Electron transport to nitrogenase in *Klebsiella pneumoniae*', Proc. Natl. Acad. Sci. USA. 77, 2555-2558.
2. Shah V.K., Stacey G., and Brill W. J. (1983) 'Electron transport to nitrogenase : purification and characterization of pyruvate : flavodoxin oxidoreductase, the *nifJ* gene product', J. Biol. Chem. 258, 12064-12068.
3. Vignais P.M., Colbeau A., Willison J.C., and Jouanneau Y. (1985) 'Hydrogenase, nitrogenase and hydrogen metabolism in the photosynthetic bacteria', Adv. Microb. Physiol. 26, 155-234.

4. Hallenbeck P.C., Meyer C.M., and Vignais P.M. (1982) 'Nitrogenase from the photosynthetic bacterium *Rhodopseudomonas capsulata* : purification and molecular properties', J. Bacteriol. 149, 708-717.

5. Jouanneau Y., Meyer C., and Vignais P.M. (1983) 'Regulation of nitrogenase activity through iron protein interconversion into an active and an inactive form in *Rhodopseudomonas capsulata*', Biochim. Biophys. Acta 749, 318-328.

6. Jouanneau Y., Meyer C., and Vignais P.M. (1989) 'ADP ribosylation of dinitrogenase reductase in *Rhodobacter capsulatus*', Biochemistry 28, 6524-6530.

7. Hallenbeck P.C., Jouanneau Y., and Vignais P.M. (1982) 'Purification and molecular properties of a soluble ferredoxin from *Rhodopseudomonas capsulata*', Biochim. Biophys. Acta 681, 168-176.

8. Yakunin A.F. and Gogotov I.N. (1983) 'Properties and regulation of synthesis of two ferredoxins from *Rhodopseudomonas capsulata*', Biochim. Biophys. Acta 725, 298-308.

9. Jouanneau Y., Meyer C., Gaillard J., and Vignais P.M. (1990) 'Purification and characterization of a 7Fe-ferredoxin from *Rhodobacter capsulatus*', Biochem. Biophys. Res. Commun 171, 273-279.

10. Schatt E., Jouanneau Y., and Vignais P.M. (1989) 'Molecular cloning and sequence analysis of the structural gene of ferredoxin I from the photosynthetic bacterium *Rhodobacter capsulatus*', J. Bacteriol. 171, 6218-6226.

11. Duport C., Jouanneau Y., and Vignais P.M. (1990) 'Nucleotide sequence of *fdxA* encoding a 7Fe ferredoxin of *Rhodobacter capsulatus*' Nucleic Acids Res. 18, 4618.

12. Grabau C., Schatt E., Jouanneau Y., and Vignais P.M. (1990) 'A new [2Fe-2S] ferredoxin from *Rhodobacter capsulatus* : Coexpression with a 2[4Fe-4S] ferredoxin in *Escherichia coli* J. Biol. Chem. (in press).

13. Moreno-Vivian C., Hennecke S., Pühler A., and Klipp W. (1989) 'Open reading frame 5 (ORF5) encoding a ferrredoxin-like protein and *nif*Q are cotranscribed with *nifE, nifN, nifX* and ORF4 in *Rhodobacter capsulatus*', J. Bacteriol. 171, 2591-2598.

14. Adman E.T., Sieker L.C., and Jensen L.H. (1973) 'The structure of a bacterial ferredoxin' J. Biol. Chem. 248, 3987-3996.

15. Jouanneau, Y. (1990) 'Properties of three ferredoxins from the photosynthetic bacterium *Rhodobacter capsulatus* : studies on their possible role in nitrogen fixation, in Ullrich W.R., Rigano C., Fuggi A., and Aparicio P.J. (eds.), Inorganic nitrogen in plants and microorganisms : uptake and metabolism. Springer, Berlin, Heidelberg, New York, pp. 248-253.

16. Klipp W., Masepohl B., and Pühler A. (1988) 'Identification and mapping of nitrogen fixation genes of *Rhodobacter capsulatus* : duplication of the *nifA-nifB* region', J. Bacteriol. 170, 693-699.

17. Saeki K., Miyatake Y., Young D.A., Marrs B.L., and Matsubara H. (1990) 'A plant ferredoxin like gene is located upstream of ferredoxin I gene (*fdxN*) of *Rhodobacter capsulatus*', Nucleic Acids Res. 18, 1060.

18. Duport C., Jouanneau Y., and Vignais P.M. (1990) 'Cloning and transcriptional analysis of the *fdxA* gene coding for the 7Fe ferredoxin II from *Rhodobacter capsulatus*' (submitted).

19. Morgan T.V., Lundell D.J., and Burgess B.K. (1988) '*Azotobacter vinelandii* ferredoxin I: cloning sequencing, and mutant analysis', J. Biol. Chem. 263, 1370-1375.

20. Yoch D.C., Carithers R.P., and Arnon D.I. (1977) 'Isolation and characterization of bound iron-sulfur proteins from bacterial photosynthetic membranes. Ferredoxin III and IV from *Rhodospirillum rubrum* chromatophores', J. Biol. Chem. 252, 7453-7460.

21. Yoch D.C., Arnon D.I. and Sweeney W.V. (1975) 'Characterization of two soluble ferredoxins as distinct from bound iron-sulfur proteins in the photosynthetic bacterium *Rhodospirillum rubrum*' J. Biol. Chem. 250, 8330-8336.

GROWTH AND AMINO ACID LIBERATION BY A MUTANT STRAIN OF *Anabaena variabilis* RESISTANT TO THE AMINO ACID ANALOGUE AZETIDINE 2-CARBOXYLIC ACID.


J. S. YUNES, P. ROWELL and N. W. KERBY
*Department of Biological Sciences,*
*University of Dundee,*
*Dundee DD1 4HN,*
*Scotland, U.K.*



ABSTRACT. Amino acid liberation by a mutant strain of *Anabaena variabilis*, ACA 15, was stimulated by increased irradiance, $CO_2$ limitation and by supplying nitrate to $N_2$-fixing cultures of the mutant strain. ACA-resistance in the mutant strain may be attributable to the overproduction of alanine which can inhibit ACA uptake. Nitrite liberation and effects of nitrate on $CO_2$ fixation and of $CO_2$ on nitrogenase activity are altered in the mutant strain. Compared to the parent strain, ACA 15 may be impaired in the ability to adjust its metabolism in response to changes in nitrogen, carbon and energy supply.


## 1. Introduction

A number of mutant strains of cyanobacteria which liberate amino acids have been obtained [see 1], although the metabolic changes leading to amino acid liberation are often not identified. In this paper, we discuss the properties of an analogue-resistant mutant of *A. variabilis*, in which the balanced assimilation of carbon and nitrogen appears to be disrupted. Characterization of such strains may contribute to a better understanding of factors which affect amino acid liberation and help to identify the mechanisms by which pathways of carbon and nitrogen metabolism are interrelated.

## 2. Results and Discussion

Mutant strains of the cyanobacterium *A. variabilis* ATCC 29413 were produced by NTG mutagenesis, selected for resistance to the amino acid analogue azetidine 2-carboxylic acid (ACA) and screened for their ability to liberate amino acids. Under $N_2$-fixing conditions, one mutant strain (ACA 15) showed a similar growth pattern to the parent strain, although in nitrate-containing medium (bubbled with air and at an irradiance of 80 $\mu E.m^{-2}.s^{-1}$) its growth was markedly inhibited and the phycobiliprotein/chlorophyll ratio decreased. A wide spectrum of amino acids was liberated by ACA 15 in all phases of growth; alanine was the most abundant, followed by phenylalanine, leucine, isoleucine, arginine and several others in smaller quantities. In contrast, glutamate, the most abundant free pool amino acid, was liberated in only small quantities. When ACA 15 was grown under $N_2$-fixing conditions the spectrum of amino acids liberated was unchanged, although the quantities liberated were smaller than in the presence of nitrate.

Factors affecting the liberation of amino acids were then further investigated and, since amino acid liberation was higher in the presence of nitrate, the production of nitrite and ammonium was also tested. Only low levels of ammonium were detected, and this is not discussed further. Increased light (from 15 to 80 $\mu E.m^{-2}.s^{-1}$) together with air bubbling of cultures

520

resulted in an increased rate of amino acid liberation and decreased rate of growth of ACA 15. In the short term, increased light resulted in a rapid liberation of amino acids by both the parent and, to a greater extent, the mutant strain, whereas an increased supply of $CO_2$ prevented this amino acid liberation. Nitrite was initially present in the growth medium at approximately 60 $\mu$M. Increased light resulted in further nitrite liberation which was more rapid in the case of ACA 15, whereas an increased $CO_2$ supply resulted in rapid nitrite uptake in both parent and mutant strains. The amino acids released into the medium could have resulted either from the free-pool of amino acids in response to shock, protein breakdown or newly assimilated nitrogen. The mutant strain also showed an increased glutamine synthetase activity under the conditions where greatest amino acid liberation occurred. When light (energy supply) is increased together with the provision of fixed nitrogen (as nitrate) an adequate supply of carbon ($CO_2$) is needed in order to sustain an increased rate of nitrogen assimilation. However, in the mutant strain ACA 15 the responses to light and nitrate are clearly altered.

Nitrogenase activity was stimulated up to two-fold when $CO_2$ (5% $CO_2$ in air) was supplied to the parent strain, whereas this effect was not seen in the case of ACA 15. This effect of $CO_2$ in the parent strain may be related indirectly to the provision of fixed carbon to supply reductant for nitrogenase and to the supply of carbon skeletons for assimilation of fixed nitrogen [2]. The reduction of nitrite, derived from nitrate, is dependent on the availability of $CO_2$ in *Synechococcus* 6301 [3] and this was also the case in *A. variabilis* and the mutant ACA 15. The data on nitrite liberation by air-grown cultures, together with the effects of $CO_2$ on nitrogenase activity in ACA 15, indicate that the regulatory effects of $CO_2$ on nitrogen assimilation are altered. In addition, the rates of photosynthetic $CO_2$ fixation are affected by the provision of nitrate; preliminary data showed that $^{14}$C-incorporation by ACA 15 at an irradiance of 80 $\mu$E.m$^{-2}$.s$^{-1}$ is inhibited by nitrate, in contrast to the parent strain where nitrate stimulated $^{14}$C-incorporation. Rates of $^{14}CO_2$ fixation were comparable in both strains when cells were grown under $N_2$-fixing conditions. This is consistent with inhibition of the growth of ACA 15 by nitrate.

Alanine is the amino acid liberated in the largest quantity by ACA 15 and may be transported into the cells by a system which is also responsible for proline, glycine and serine uptake as well as their liberation [see 1]. This system may also transport ACA, a proline analogue. $^{14}$C-alanine was taken up more rapidly by ACA 15 than by the parent strain and uptake was inhibited by the uncoupler CCCP (50 $\mu$M). The uptake of alanine by ACA 15 showed two $K_m$ values (5 $\mu$M and 70 $\mu$M) with corresponding $V_{max}$ values of 33.3 and 116.7 pmol.$\mu$g chl.min$^{-1}$. The parent strain showed corresponding $K_m$ values of 25 $\mu$M and 88 $\mu$M and $V_{max}$ values of 13.9 and 20.5 pmol.$\mu$g chl$^{-1}$.min$^{-1}$. The differences in kinetics of alanine uptake between the parent strain and ACA 15 may be related to the higher rates of alanine liberation by ACA 15.

## 3. REFERENCES


(1) Kerby, N.W., Rowell, P. and Stewart W.D.P. (1989). The Transport, Assimilation and Production of Nitrogenous Compounds by Cyanobacteria and Microalgae, in R.C. Cresswell, T.A.V. Rees and N. Shah (eds.), *Algal and Cyanobacterial Biotechnology*, Longman, U.K., pp. 51 90.

(2) Ernst, A., Kirschenlohr, H., Diez, J. and Boger, P. (1984). Glycogen Content and Nitrogenase Activity in *Anabaena variabilis*, *Arch Microbiol.* 140: 120-125.

(3) Kramer, E. and Schmidt, A. (1989). Nitrite Accumulation by *Synechococcus* 6301 as a Consequence of Carbon or Sulfur Deficiency. *FEMS Microbiol. Lett.* 59: 191-196.

# INTERACTION B-Ca ON THE NITROGENASE ACTIVITY IN NITROGEN-FIXING CYANOBACTERIA (Anabaena PCC 7119)


BOLAÑOS, L.; MATEO, P.; BONILLA, I.
Dept. Biología
Universidad Autónoma de Madrid
28049 - MADRID
Spain


## INTRODUCTION

The requirement for boron is not a general feature in cyanobacteria. Anabaena PCC 7119, a dinitrogen-fixing cyanobacterium, cultured in presence of combined nitrogen was not affected by boron deficiency. However, when this microorganism was grown under dinitrogen-fixing conditions lacking in boron, a distinct inhibition of growth and a decrease in photosynthetic pigments and protein contents were observed. Nitrogenase activity of boron deficient cultures was reduced about 40% that of the controls within the first 2h of culture (1).

A similar calcium requirement for aerobic nitrogen fixation by heterocystous blue-green algae has also been shown likewise the role of calcium seems to be related to protection of nitrogenase against oxygen-inhibition (2).

In the present study we show the interaction B-Ca on the growth, nitrogenase activity and photosynthesis in nitrogen-fixing cyanobacteria.

## MATERIALS AND METHODS

Culture conditions, media and cells free of boron were prepared and process as described in previous works (1).

Nitrogenase activity and photosynthesis were determined as previously described (1).

## RESULTS AND DISCUSSION

In order to determinate the interaction B-Ca on the nitrogen fixation and photosynthesis in Anabaena PCC 7119 0.21 mM, 0.42 mM or 0.84 mM Ca with boron deficiency cultures were tested.

Inhibition of growth resulting from boron deficiency was reversible when the cultures were supplemented with twice Ca concentration than control cultures (data not shown). A similar recovery of photosynthesis was observed when boron-deficient cultures were supplemented with Ca (Fig. 1 B).

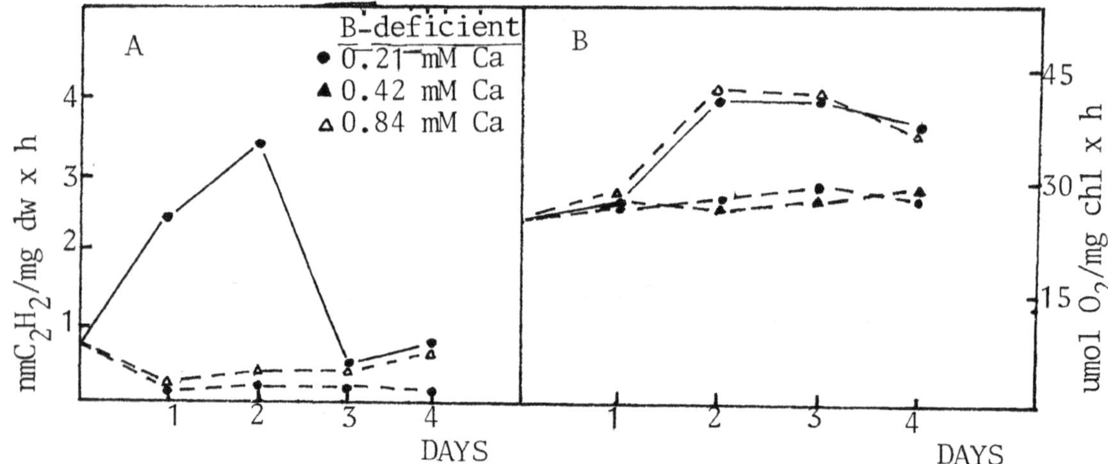


Figure 1. A. Nitrogenase activity. B. Photosynthesis (measured as oxygen evolution). Control (●——●), B-deficient (----).

Nitrogenase activity of all cases was reduced to about 95 % than the control within the first 24h of culture. However, after 24h of deficiency the cultures with twice calcium concentration showed a gradual recovery of nitrogenase activity (Fig. 1A). In contrast, photosynthesis resulted altered after 24h of growth in cultures not supplemented with calcium (Fig. 1 B). The results being similar in control and calcium cupplemented cultures.

These results suggest that calcium are involved in nitrogen fixation. In previous papers, we presented evidence indicating that boron was esential for dinitrogen fixation in _Anabaena_ PCC 7119 (3,4). A possible role of calcium could be the stabilization of heterocyst envelope together with boron. This hypothesis could explain the gradual recovery of nitrogenase activity in boron deficient cultures supplemented with calcium.

REFERENCES

1. MATEO, P.; BONILLA, I.; FERNANDEZ-VALIENTE, E.; SANCHEZ-MAESO, E. 1986. Essentiality of boron for dinitrogen fixation in _Anabaena_ sp. PCC 7119. Plant Physiol. 81:430-433.

2. RODRIGUEZ, H.; RIVAS, J.; GUERRERO, M.G.; LOSADA, M. 1990. Calcium requirement for aerobic nitrogen fixation by heterocystous blue-green algae. Plant Physiol. 92(4):886-890.

3. GARCIA-GONZALEZ, M.; MATEO, P.; BONILLA, I. 1988 Boron protection for oxygen diffusion in heterocysts of _Anabaena_ sp. PCC 7119. Plant Physiol. 87:785-789.

4. GARCIA-GONZALEZ, M.; MATEO, P.; BONILLA, I. 1990. Effect of boron deficiency on photosynthesis and reductant sources and their relationship with nitrogenase activity in _Anabaena_ sp. PCC 7119. Plant Physiol. 93:560-565.

# THE EFFECT OF AMMONIA-EXCRETING CYANOBACTERIA ON THE GROWTH OF RICE


J. R. MILAM[1], S. L. ALBRECHT[2,3], F. KAMURU[2] AND K. T. SHANMUGAM[1].
*Department of Microbiology and Cell Science[1], Department of Agronomy[2] & United States Department of Agriculture-Agricultural Research Service[3], University of Florida, Gainesville, FL 32611 (USA)*


Most crops require nitrogen fertilizer for peak productivity. However, the increasing production costs for nitrogen fertilizer mandates the exploitation of biological nitrogen fixation. Cyanobacteria make a valuable contribution to the nitrogen content of many soils and nitrogen fixation by cyanobacteria may be responsible for continued rice production in areas of the Orient where little or no fertilizer has ever been added to the soil.

Free-living, nitrogen-fixing organisms do not excrete nitrogen. The fixed nitrogen is released only after the microorganisms die. However, mutant strains of cyanobacteria that cannot assimilate, but excrete the ammonia reduced by nitrogenase, have been constructed (Spiller, H., C. Latorre, M. E. Hassan and K. T. Shanmugam. 1986. J. Bacteriol. 165:412-419). Preliminary studies suggest that these strains excrete ammonia into the environment and can supply a significant amount of nitrogen to growing rice plants.

<u>Anabaena variabilis</u> Kutz (ATCC 29413), strains SA-0 and SA-1, were grown in half strength Allen and Arnon medium (Allen, M. B., and D. I. Arnon. 1955. Plant Physiol. 30:366-372) containing fructose and ammonia. Rice seedlings were transplanted after five days and the plants were grown to maturity in large plastic containers at two different locations at the University of Florida. Containers were flooded at the 3 leaf stage and inoculated with the cyanobacteria immediately after flooding. Other nutrients were applied as recommended by the University of Florida Soil Testing Laboratory. Nitrogen, as $NH_4Cl$, was applied to the appropriate treatments at three intervals during the growing season. The containers were arranged in a randomized complete block design with border rows. The ammonia concentration in the flood water was monitored weekly using Nessler's method. Nitrogenase activity in the flood water and upper levels of soil were estimated by the acetylene reduction technique. At the end of the growing season the plants were separated into grain, shoots and roots then dried to a constant weight.

Both the parent strain (SA-0) and the ammonia-excreting mutant strain (SA-1) could be found in the flood water after inoculation. The populations of both strains declined rapidly during the first two weeks of the experiment, however, the rate of decline was reduced over the next 6 weeks. Overall, SA-0 declined more rapidly than SA-1. Nitrogenase activity and ammonia levels were the greatest in soil/water cores taken from plants that had been inoculated with SA-1. The nitrogenase activity of containers inoculated with SA-0 were about five times greater than the control while the activity of SA-1 was commonly 10 to 15 times greater. The addition of combined nitrogen decreased nitrogenase activity of both strains. The mean ammonia concentration in the flood water in containers inoculated with SA-1 were generally similar to nitrogen fertilized treatments while the concentration in the containers inoculated with SA-0 were similar to the non-fertilized controls. Dry weights were determined for three years. There was a linear response to nitrogen fertilization every year. In 1987 inoculation with SA-1 at any of three different inoculation levels was roughly equal to fertilization with 50 Kg N/ha. Inoculation with SA-0 was slightly better than, but not statistically different from the non-inoculated, non-fertilized controls. In 1988 plants inoculated with SA-1 had a greater dry weight than plants inoculated with SA-0 or the non-inoculated controls and were nearly equivalent to plants grown with 75 Kg N/ha. In 1989, plants inoculated with SA-1 again outyielded the non-inoculated controls and plants inoculated with SA-0 and had a dry weight almost equal to plants fertilized with 50 Kg N/ha.

The population dynamics of the inoculated strains SA-0 and SA-1 suggests that, even though these organisms are lost from the environment, sufficient numbers remain in the flood water to contribute to the nitrogen economy of any associated rice plants. Ammonia excretion data and nitrogenase activities of these strains are consistent with laboratory findings and support the hypothesis that ammonia is excreted from SA-1. Dry weight data from the field over a three year period show that there is little difference between inoculation with SA-0 and the non-inoculated controls, however, inoculation with SA-1 generally provides a dry weigh yield similar to fertilization with approximately 50 Kg N/ha. These studies provide evidence that ammonia-excreting strains of cyanobacteria grown in coculture with rice can be beneficial for grain production.

PHOTOAUTOTROPHIC AND PHOTOHETEROTROPHIC N$_2$ FIXATION IN *OSCILLATORIA* SPP EXPOSED TO ALTERNATING LIGHT AND DARKNESS


J.R. GALLON, M.A. HASHEM, O.A. AL-AMOUDI & K.J. FLYNN
School of Biological Sciences
University College of Swansea
Swansea
SA2 8PP
U.K.



ABSTRACT. When grown photoautotrophically or photoheterotrophically under alternating 12h light and 12h darkness, *Oscillatoria* sp UCSB8 fixed N$_2$ maximally during the period of darkness. However, intracellular nitrogen may be more significant than intracellular carbon in regulating N$_2$ fixation in these cultures, in contrast to the situation in the unicellular cyanobacterium *Gloeothece*.


## 1. INTRODUCTION

The filamentous non-heterocystous cyanobacteria, *Oscillatoria* spp UCSB8 and UCSB25 can fix N$_2$ aerobically. When grown under alternating 12h light and 12h darkness, both strains confine N$_2$ fixation exclusively to the dark phase. In this they resemble most other non-heterocystous cyanobacteria [3,4]. However, unlike many other cyanobacteria, *Oscillatoria* spp UCSB8 and UCSB25 can grow diazotrophically in the presence of DCMU and a suitable carbon source (glucose or fructose). Here we describe the behaviour of photoheterotrophically grown cultures of *Oscillatoria* sp UCSB8 under alternating 12h light and 12h darkness.

## 2. METHODS

*Oscillatoria* sp UCSB8 and *Gloeothece* sp CCAP 1430/3 were grown as described in [4] and [2] respectively. For photoheterotrophic growth, the medium was supplemented with 20 µM DCMU and 20 mM glucose. Intracellular amino acids were determined by HPLC [1].

## 3. RESULTS AND DISCUSSION

When grown photoheterotrophically under 12h light and 12h darkness, cultures of *Oscillatoria* sp UCSB8 continued to show peaks of N$_2$ fixation about 7h into each dark phase. The cyclic pattern of N$_2$ fixation imposed by this illumination regime therefore appears to be independent of Photosystem II activity, which is blocked by DCMU. Furthermore, since exogenous glucose was available at all times during the light: dark cycle, it is unlikely that these cells ever became carbon-limited.

Figure 1.  GLN:GLU in photoheterotrophic *Oscillatoria* sp UCSB8 (●) and photoautotrophic *Gloeothece* sp CCAP 1430/3 (■) grown under alternating 12h light and 12h darkness.

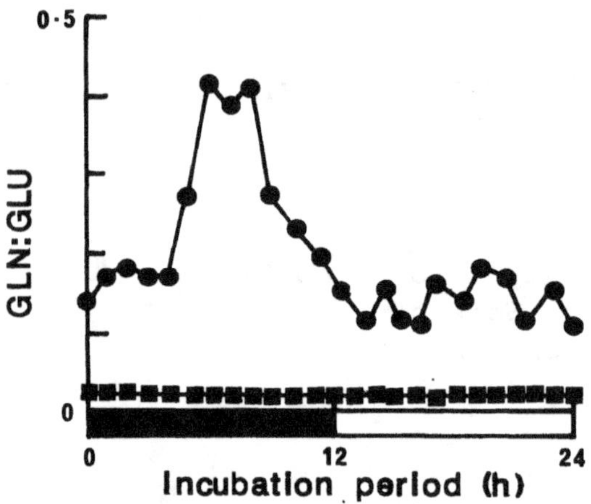


During photoheterotrophic growth of *Oscillatoria* sp UCSB8 under alternating light and darkness, fluctuations were seen in various intracellular amino acids and the ratio of glutamine to glutamate (GLN:GLU) had its highest values 6-8 h into the dark period (Fig. 1), coincident with maximum rates of $N_2$ fixation.  Values of GLN:GLU of 0.5 are typical of nitrogen-replete eukaryotes.

A similar pattern was seen in photoautotrophically grown cultures of *Oscillatoria* sp UCSB8.  In contrast, in photoautotrophically grown *Gloeothece*, there was little fluctuation in the intracellular concentrations of amino acids, and GLN:GLU remained constant at less than 0.05 (Fig. 1), suggestive of permanent nitrogen-stress.

In *Oscillatoria* sp UCSB8, the pattern of $N_2$ fixation seen under alternating light and darkness is therefore not apparently imposed by variations in the availability of intracellular carbon.  Rather, it may reflect a transient period of nitrogen repleteness, which could serve to inhibit nitrogenase synthesis.  In this respect, *Oscillatoria* sp UCSB8 differs from *Gloeothece*, in which the pattern of $N_2$ fixation is imposed by an inhibition of nitrogenase synthesis that is probably caused by carbon limitation during the latter half of the dark period [2].

## 4.  REFERENCES


[1] Flynn, K.J. (1988) 'Some practical aspects of measurements of dissolved free amino acids in natural waters and within microalgae', Chem. Ecol. 3, 269-293.

[2] Gallon, J.R., Perry, S.M., Rajab, T.M.A., Flayeh, K.A.M., Yunes, J.S. and Chaplin, A.E. (1988) 'Metabolic changes associated with the diurnal pattern of $N_2$ fixation in *Gloeothece*', J. General Microbiol. 134, 3079-3087.

[3] Gallon, J.R. (1989) 'The physiology and biochemistry of $N_2$ fixation by non-heterocystous cyanobacteria', Phykos 28, 18-46.

[4] Khamees, H.S., Gallon, J.R. and Chaplin, A.E. (1987) 'The pattern of acetylene reduction by cyanobacteria grown under alternating light and darkness', Br. Phycol. J. 22, 55-60.

# OCCURRENCE AND LOCALIZATION OF AN UPTAKE HYDROGENASE IN THE FILAMENTOUS HETEROCYSTOUS CYANOBACTERIUM NOSTOC PCC 73102


Peter Lindblad [1] & Anita Sellstedt [2]

[1] *Department of Physiological Botany*
*University of Uppsala*
*Box 540*
*S-751 21 Uppsala, Sweden*

[2] *Department of Plant Physiology*
*University of Umeå*
*S-901 87 Umeå, Sweden*



ABSTRACT. Free-living $N_2$-fixing *Nostoc* PCC 73102 cells, a filamentous heterocystous cyanobacterium originally isolated from coralloid roots of the cycad *Macrozamia* sp., were examined for the presence of an uptake hydrogenase ($H_2$ase) enzyme. *In vivo* and *in vitro* hydrogen uptake measurements were used to study activities and SDS-PAGE and Western immunoblots to reveal occurrence of the $H_2$ase protein. Also, transmission electron microscopy and immunocytological labelling were used to study the cellular and subcellular distribution of $H_2$ase in the *Nostoc* cells.


*In vivo* measurements (Sellstedt & Lindblad 1990) demonstrated an active uptake of $H_2$ in both light and darkness. Light stimulated *in vivo* $H_2$ uptake with approximately 100 %, and this was further doubled by increasing the $pH_2$ from 56 to 208 $\mu M$ $H_2$. An *in vitro* $H_2$ uptake of 1.1 $\mu mol$ $H_2 \cdot$ mg (protein)$^{-1} \cdot$ h$^{-1}$ was observed when using phenazinemethosulphate as e$^-$-acceptor.

Western immunoblots (Braun-Howland *et al.* 1988, Lindblad & Sellstedt 1989) revealed that a polypeptide with a MW of about 55 kDa was immunologically related to uptake $H_2$ase holoenzyme purified from *Alcaligenes latus*.

Immunolocalization (Bergman *et al.* 1985, Lindblad & Sellstedt 1989) demonstrated that the $H_2$ase protein was located both in heterocysts (Fig 1) and vegetative cells. A higher specific labelling was associated with the cytoplasmic membranes where the vegetative cells are in contact with each other and where they actually are dividing into two vegetative cells. Using the particle analysis of an image processor, approximately equal $H_2$ase-gold labelling per cell area was observed in the $N_2$-fixing heterocysts compared to the vegetative cells.

Figure 1. Immuno-gold localization of uptake $H_2$ase in a heterocyst of *Nostoc* PCC 73102 using (i) polyclonal rabbit-anti--uptake $H_2$ase antiserum, and (ii) goat-anti--rabbit IgG conjugated to 5 nm collodial gold particles. Bar = 1.0 $\mu$m.

ACKNOWLEDGEMENTS. We are grateful to Dr DJ Arp (Univ California, USA) for providing us with antiserum of *Alcaligenes latus* hydrogenase. We also would like to thank A Axén for excellent technical assistance. This work was financially supported by grants from the Swedish Natural Science Research Council, Carl Tryggers Foundation, and Marianne & Marcus Wallenberg Foundation.

REFERENCES

Bergman B, Lindblad P, Pettersson A, Renström E & Tiberg E (1985) Planta 166: 329-334

Braun-Howland EB, Lindblad P, Nierzwicki-Bauer SA & Bergman B (1988) Planta 176: 319-332

Lindblad P & Sellstedt A (1989) FEMS Microbiology Letters 60: 311-316

Sellstedt A & Lindblad P (1990) Plant Physiology 92: 809-815

# THE EFFECT OF PHYTOHORMONES ON NITROGENASE ACTIVITY AND GROWTH OF *NOSTOC MUSCORUM* AGARDH


B. MARSALEK, M. SIMEK, A.LUKESOVA
*Institute of Soil Biology*
*Czechoslovak Academy of Sciences*
*Na sadkach 7*
*370 05 Ceske Budejovice*
*Czechoslovakia*


## 1.INTRODUCTION

Growth and nitrogen fixation of cyanobacteria can be influenced by a variety of environmental factors. However, little is known about the possible effects of gibberellins ⟨GAs⟩ or abscisic acid ⟨ABA⟩ on the cyanobacteria. In this paper we describe the effects of GA₃ and ABA in the concentration range of $0.001 - 10$ mg.l$^{-1}$ on the growth and nitrogenase activity of cyanobacterium *Nostoc muscorum* Agardh.

## 2.MATERIAL AND METHODS

Cyanobacterium was cultivated on agar with Bold's Basal Medium without nitrogen,complemented with phytohormones of a certain concentration. In some cases ABA was applied only when cyanobacterium developed on the surface of agar. Biomass was calculated on the assumption that content of chlorophyll <u>a</u> is 0.3 % of fresh biomass weight. Nitrogenase activity was determined by a modified acetylene reduction assay and expressed as amount of ethylene produced, either as a total nitrogenase activity ⟨TNA⟩ of cyanobacteria in one tube or a specific nitrogenase activity ⟨SNA⟩ of 1 mg of fresh biomass of cyanobacteria.

## 3.RESULTS AND DISCUSSION

Our experiments confirmed the hypothesis that GA₃ acts as a stimulator of some enzymes. In the first experiment, long-term influence of 5 different concentrations of GA₃ was studied after 3 weeks of cultivation. GA₃ stimulated growth at all the concentrations tested, the lowest concentration being most effective. Higher GA₃ concentrations decreased TNA, while the lowest one increased it significantly. SNA was reduced by GA₃, although not significantly. In the second experiment the effect of the lowest and the highest concentrations of GA₃ was studied after 1, 2, and 3 weeks of cultivation. The biomass of

cyanobacterium was not found to be significantly affected by GA$_3$.
However both TNA and SNA were enhanced.
Four experiments were taken with ABA. After 3 weeks of cultivation all
ABA concentrations used surprisingly stimulated growth of
cyanobacterium, the increases of 0.001, 0.01 and 1.0 mg ABA.l$^{-1}$ being
significant. These concentrations also increased TNA, while
concentrations of 0.1 and 10.0 mg ABA.l$^{-1}$ lowered it slightly. SNA was
enhanced by 0.01 mg ABA.l$^{-1}$, the other concentrations lowered it. In
the second experiment, a slight stimulatory effect of ABA in the lowest
and the highest concentrations on growth of cyanobacterium was found at
all periods of measurement after 1, 2, and 3 weeks of cultivation.
However, TNA was increased by both ABA concentrations only after 1 week
of cultivation. After 2 and 3 weeks, the lower ABA concentration
retained its stimulatory effect, while the higher one did not
practically affect TNA. SNA was not significantly affected by any ABA
concentrations.
Stimulatory effect of ABA was confirmed in a short-term experiment,
when ABA was applied on cyanobacterium in the exponencial phase of
growth, after 2 weeks of cultivation. Both concentrations ⟨0.001 and
10.0 mg ABA.l$^{-1}$⟩ stimulated TNA after 1 and 2 days of further
cultivation, the higher one significantly. In another short-term
experiment, ABA was applied on cyanobacterium in the stationary phase
after 4 weeks of cultivation, and TNA was measured 2, 24, 48, and 120 h
after addition of phytohormone. At all periods of measurement ABA highly
stimulated TNA, and at the end of cultivation enhanced both biomass and
SNA ⟨Table 1⟩.
Our findings were examined with a view to the stress theory. In this
connection the stimulatory effect of ABA on growth and nitrogenase
activity of *Nostoc muscorum* may be explained as a stress response of
cyanobacterium activating metabolism and resulting in enhanced growth
and nitrogenase activity. In the light of these findings, ABA does not
seem to act as a typical phytohormone in cyanobacteria. Besides, in most
experiments biomass and TNA showed no significant correlations.

TABLE 1. The influence of ABA on growth, total nitrogenase
activity ⟨TNA⟩ and specific nitrogenase activity ⟨SNA⟩ of
N.muscorum ⟨means of 10 replicates, standars deviations in
parentheses⟩.

| ABA | TNA after incubation ⟨h⟩ [1] | | | | biomass [2] | SNA [3] |
|---|---|---|---|---|---|---|
| | 2 | 24 | 48 | 120 | ⟨after 120 h⟩ | |
| 0 | 0.22 | 0.84 | 0.93 | 1.32 | 0.54 | 0.02 |
| | ⟨0.09⟩ | ⟨0.65⟩ | ⟨0.92 ⟩ | ⟨1.38⟩ | ⟨0.12⟩ | ⟨0.02⟩ |
| 10 mg.l$^{-1}$ | 0.72 | 46.9 | 124.1 | 906.9 | 0.92 | 8.60 |
| | ⟨0.32⟩ | ⟨22.9⟩ | ⟨115.0⟩ | ⟨389.3⟩ | ⟨0.09⟩ | ⟨3.70⟩ |

1⟩ nmol C$_2$H$_4$.tube$^{-1}$.time$^{-1}$ ; 2⟩ mg fresh weight.tube$^{-1}$
3⟩ nmol C$_2$H$_4$.mg$^{-1}$.h$^{-1}$

EFFECT OF AMMONIUM ON THE NITROGENASE ACTIVITY OF <u>ANABAENA VARIABILIS</u>


M. Jha*, A.F. Yakunin, Chan Van Ni and I.N. Gogotov
Institute of Soil Science and Photosynthesis
USSR Academy of Sciences, Pushino
142292 Moscow Region, USSR
* R.B. College, Dalsingsarai, Bihar, India



ABSTRACT. Concentration and pH dependent inhibition of nitrogenase in
<u>A. variabilis</u> by ammonium in light and dark has been observed. The
degree of nitrogenase inhibition was decreased by preincubation in
methionine sulfoximine. Obtained data indicate that inhibition of
nitrogenase activity of <u>A. variabilis</u> cells by ammonium had other
nature than that of nitrogenase switch off effect in purple bacteria.


INTRODUCTION

As compared to purple bacteria (1) the regulation of nitrogenase by
ammonium has been less studied in cyanobacteria. It is known that
ammonium inhibits acetylene reduction. The inhibitory effect of
ammonium on acetylene reduction by cyanobacteria has been observed at
alkaline (pH 10) condition (2) or under low light (3). The aim of our
work was to study the effect of ammonium in more detail on nitrogenase
activity of <u>A. variabilis</u> in different conditions.

MATERIALS AND METHODS

<u>A. variabilis</u> (ATCC 29413) was grown for 2 days in Allen and Arnon
medium (4) in the light ($25$ W/m$^2$) with bubbling by 1% $CO_2$ in air. The
cells were resuspended in fresh medium containing HEPES-K buffer
(final conc. 10mM) at different pH (6-10). Acetylene reduction was
measured in the light (10 W/m$^2$, Argon) or in the dark (Argon, 10% $O_2$).

RESULTS AND DISCUSSION

At the neutral pH 7 ammonium did not inhibit acetylene reduction by <u>A.
variabilis</u> cells under high light intensity (10 W/m$^2$) but this process
was completely suppressed under low light (1.3 W/m$^2$). At pH 8 the rate

of acetylene reduction was decreased two fold after addition of 1 mM ammonium. More strong inhibitory effect of ammonium on nitrogenase was observed at pH 9 and 10. The rate of ammonium uptake by A. variabilis cells was observed to be highest in the light at pH 10. The light grown A. variabilis cells reduced acetylene in dark aerobic conditions with lower rate as compared to light. This dark nitrogenase activity was inhibited by ammonium at wide pH range (6-10). The inhibitory effect of ammonium on dark nitrogenase activity was decreased immediately after light switch on. In contrast to purple bacteria (1) the degree of nitrogenase inhibition by ammonium was dependent on ammonium concentration and was highest in the presence of 0.2 to 0.5 mM and 5 mM ammonium respectively. Other contrasting peculiarities of A. variabilis were the low rate of ammonium inhibition (after 15-20 min of ammonium addition) and the absence of recovery of initial rate of acetylene reduction after 4-5 h of incubation. The degree of ammonium inhibition of nitrogenase activity of A. variabilis remarkably decreased after 2-3 h preincubation of cells with methionine sulfoximine (0.1 mM), but when ammonia and MSX were added simultaneously, MSX did not protect the inhibition of nitrogenase significantly. In contrast to ammonium, urea and nitrate did not inhibit nitrogenase activity of A. variabilis in the light or in the dark. It is probably connected with the low rate of uptake and reduction of these nitrogen sources to ammonium by nitrate grown cells of A. variabilis. Glutamine (5 mM), product of ammonia assimilation, inhibited acetylene reduction by cells of this cyanobacterium in the dark but not in the light.

Obtained data indicate that ammonium inhibits activity of nitrogenase in the cells of A. variabilis in both light and dark at physiological pH. This regulation has nature other than nitrogenase switch off by ammonium in purple bacteria and probably is a result of decrease of ATP or reductant pool during ammonium assimilation.

REFERENCES

(1) Hallenbeck, P.C. (1987), CRC Critical Rev. Microbiol., 1-48.
(2) Reich, S., Almon, H. and Boger, P. (1986), FEMS Microbiol. Lett., 34, 53-56.
(3) Yoch, D.C. and Gotto, J.W. (1983), J. Bacteriol., 151, 800-806.
(4) Allen, M.B. and Arnon, D.J. (1955), Plant Physiol., 30, 366-372.

# MODIFICATION AND IN-VITRO ACTIVATION OF DINITROGENASE REDUCTASE FROM ANABAENA VARIABILIS


*Ines Böhm, Andrea Halbherr and Peter Böger*
Lehrstuhl für Physiologie und Biochemie der Pflanzen
Universität Konstanz, D-7750 Konstanz, Germany


## Abstract


Cell-free preparations of inactivated nitrogenase from *Anabaena variabilis* were reactivated by a membrane fraction from *A. variabilis and Rhodospirillum rubrum*, known to contain the dinitrogenase reductase activating glycohydrolase (DRAG). These results imply that the modifying group at the cyanobacterial Fe-protein (dinitrogenase reductase) is ADP-ribose as originally described for *R. rubrum*. The reactivating activity could only be found in membranes of heterocysts and when an active nitrogenase was present. A nitrogenase inactivating activity was observed in the supernatant from *A. variabilis* extracts, which was strongest in cells with highly regulated nitrogenase activity.


## Introduction

The Fe-protein from the heterocystous cyanobacterium *A. variabilis* was shown to be modified due to different external factors (1,2). Modification was visualized by Western blotting using a specific antibody prepared against the Fe-protein of *A. variabilis* (3). In *R. rubrum* a similar modification occurs by an ADP-ribosylation at the Arg-100 of the dinitrogenase reductase. This reaction is catalyzed by the dinitrogenase reductase ADP-ribosyltransferase (DRAT), the reverse step by the dinitrogenase reductase activating glycohydrolase (DRAG) (4).

## Materials and Methods

In-vivo inactivated nitrogenase of *A. variabilis* and *R. rubrum* was partially purified by DEAE cellulose chromatography according to (3). The reaction mixture for reactivation contained (0.5 ml assay volume): 5 mM ATP, 25 mM creatine phosphate, 0.05 mg creatine kinase, 10 mM $MgCl_2$, 0.5 mM $MnCl_2$, 0.5 mg BSA, 40 mM TEA, pH 7.5, 10 mM $Na_2S_2O_4$, and the indicated amounts of partially purified nitrogenase. For inactivation 2 mM ADP, 1 mM $NAD^+$ and 5 mM $MgCl_2$ had to be added to the soluble fraction of *A. variabilis* ( 0.1 ml assay volume), which was anaerobically prepared according to (3); also nitrogenase was assayed after (3) in a 0.5 ml assay volume.

## Results and Discussion

REACTIVATION.   Physiologically inactivated nitrogenase was partially purified from *A. variabilis* and *R. rubrum*. For the reactivation tests dinitrogenase reductase was titrated against resuspended membranes from *A. variabilis* or *R.rubrum*, to achieve optimum reactivation.

After a short lag phase nitrogenase from *A. variabilis* was reactivated (62.2 nmol $C_2H_4$/mg prot. x min) by a membrane fraction from *A. variabilis* (3.5 mg protein) and even better by that from *R. rubrum* (0.74 mg protein) exhibiting a specific activity of 101.3 nmol $C_2H_4$/mg prot. x min. *R. rubrum* nitrogenase is about 4-fold more activated by membranes of *R. rubrum* (0.74 mg protein, yielding 106.6 nmol $C_2H_4$/mg prot. x min) than by a preparation from *A. variabilis* (2.35 mg protein). For these experiments, however, the reactivation conditions, adapted from *R. rubrum*, possibly were not yet at optimum for *A. variabilis*.

After 5 minutes of boiling or treatment with pure oxygen the membranes of both organisms could no longer reactivate the modified Fe-protein, proving that the reactivation is an enzymatic reaction.

*A. variabilis* is a heterocystous cyanobacterium. Under aerobic conditions nitrogen fixation takes place exclusively in the heterocysts. The question arises, whether the activating enzyme is found in heterocysts only or is present in all cells of the filament. Therefore, the reactivating activity was tested with membranes from filaments, vegetative cells as well as from heterocysts. With the membrane fraction from vegetative cells no reactivating activity was found. Based on protein the membranes from isolated heterocysts produced a 50% higher reactivation than those from filaments.


INACTIVATION.   The soluble fraction of 1 day-grown cells of *A. variabilis* were incubated anaerobically with ADP, $NAD^+$ and $MgCl_2$ to inactivate nitrogenase. After different periods of incubation aliquots were assayed for nitrogenase activity. After a short lag phase the nitrogenase activity decreased to 50% of the initial activity within 30 minutes.

Inactivation was also assayed in the soluble fraction of filaments from A. variabilis, grown for 1, 2 or 3 days. During 70 minutes of anaerobic incubation with ATP and $NAD^+$ present, nitrogenase activity of 1 day-grown cells exhibited 10% of the activity versus the sample without ATP and $NAD^+$. The nitrogenase activity of 2 or 3 days grown cells even decreased to 7% or 1%, respectively. This agrees with the decrease of nitrogenase activity found in the intact filaments (see  1  for details).


## Conclusions

Both, cross reaction of nitrogenase with the membrane fractions of the two organisms, and a similar time course of the reactivation demonstrate, that the mechanism of reactivation is similar in the two organisms. In *A. variabilis* the reactivating enzyme is only attached to the membranes of heterocysts, the cells which are capable of fixing nitrogen.

In *A. variabilis* an enzyme is present corresponding to the DRAT-enzyme from *R. rubrum*. Seemingly, this enzyme activity relates to nitrogenase activity found in the cells. It is more effective in cultures grown for 2 or 3 days than in a 1-day culture.


## References

(1) Ernst A., Reich S., Böger P. (1990) "Modification of Dinitrogenase Reductase in the Cyanobacterium *Anabaena variabilis* Due to C-Starvation and Ammonia", J. Bacteriol. 172, 748-755.
(2) Ernst A., Liu Y.-D., Reich S., Böger P. (1990) "Diurnal Nitrogenase Modification in the Cyanobacterium *Anabaena variabilis*", Botan. Acta 103, 183-189.
(3) Reich S., Böger P. (1989) "Regulation of Nitrogenase Activity in *Anabaena variabilis* by Modification of the Fe Protein", FEMS Microb. Lett. 58, 81-86.
(4) Ludden P.W., Roberts G.P. (1989) "Regulation of Nitrogenase Activity by Reversible ADP-Ribosylation", in Curr. Topics Cell. Regulation 30, 23-56.

# NITROGEN FIXATION IN SPANISH RICEFIELDS.


A.QUESADA, E.SANCHEZ MAESO and E.FERNANDEZ
VALIENTE
Dpt.Biology. Autonomous University of Madrid
Cantoblanco. 28049-Madrid.
Spain.



ABSTRACT.A method for the "in situ" measurement of nitrogenase activity in shallow-water rice fields was developed (Quesada et al. 1989). The method was applied to measure nitrogen fixation in the ricefields of Valencia (Spain) in June 1988 and in January, May, June and July 1989. Field assays were conducted in 25 locations selected at random. Estimates of acetylene-reducing activity in the light at different locations and months showed a wide variability. Most locations showed, however, nitrogenase activity in the light in most of the sampling periods. Values ranged from 4 to 400 umol ethylene.m$^{-2}$.h$^{-1}$. Eventually, when assays were performed on cyanobacterial blooms values of 1200 umol ethylene.m$^{-2}$.h$^{-1}$ were recorded. No nitrogenase activity was found in the dark. Daily evolution of nitrogenase activity showed a maximal value in the afternoon.


## INTRODUCTION.

$N_2$-fixing cyanobacteria constitute a major group of micro-organisms in paddy field ecosystems. Their potential use as biofertilizers in rice cultivation have been recognized for many years (Roger and Kulasooriya 1980). Most reports concerning cyanobacteria and rice arise from studies of asiatic rice fields. Little work has been done, however, in european ricefields, which markedly differs from the asiatic ones, particularly in that concenning to the input of fertilizers and pesticides.
In this report we present data of a study conducted to investigate biological nitrogen fixation and its relation-ship with $N_2$-fixing cyanobacteria in ricefields of Valencia (Spain).

## MATERIALS AND METHODS.

Field assays of biological nitrogen fixation were made following the method reported previously (Quesada et al. 1989). Water analysis were made using the methodology given in the Hach Water Analysis Handbook (1985), with a portable laboratory Drel-5. Total $N_2$-fixing cyanobacteria were evaluated by dilution and plating using agarized N-free medium.

## RESULTS AND DISCUSSION.

Soil and water samples were collected, to determine the abundance of cyanobacteria. In spite of the large variability of the fields with respect to physical and chemical parameters, we have recorded the consistent presence of $N_2$-fixing cyanobacteria in all samples collected. Major genera observed were <u>Nostoc</u>, <u>Anabaena</u>, <u>Calothrix</u>, <u>Scytonema</u> and <u>Nodularia</u>. Mean values of soil samples in 1989 ranged from $12.27 \times 10^4$ CFU/cm$^2$ to $4.7 \times 10^4$ CFU/cm$^2$. The lowest values were recorded in June during the drying of fields that take place in the middle of the crop season, and when most pesticides were added to the fields. Data from water samples were more variable, mean values in 1989 ranged from $1.17 \times 10^4$ CFU/l to $6.38 \times 10^5$ CFU/l. Due to their wide variability, it was impossible to establish a general pattern of evolution of the number of $N_2$-fixing cyanobacteria in water during the crop season.
  Estimates of nitrogen fixation in the light at different location and months showed a wide variability. This is a reflection of the heterogeneous distribution of cyanobacteria and of the differences in water conditions. No nitrogenase activity was observed in any time in a 25% of the locations. Most locations showed, however, nitrogenase activity in most of the sampling periods. Values ranged from 4 to 400 umol ethylene.m$^{-2}$.h$^{-1}$. Eventually when assays were performed on cyanobacterial blooms values of 1200 umol ethylene.m$^{-2}$.h$^{-1}$ were recorded. No nitrogenase activity was found in the dark, suggestting that nitrogen fixation in this area was mainly due to cyanobacteria. Daily evolution of nitrogenase activity followed a curve of the third type of that described by Roger and Kulasooriya (1980), with the maximal activity in the afternoon. Statistical study of the data by factorial analysis and Spearman correlation, indicated a positive correlation between nitrogenase activity and the mineralization factor, which fundamentally include Calcium, Hardness and Conductivity.

## REFERENCES.

Quesada A, Sanchez Maeso E and Fernandez Valiente E (1989) New incubation device for in situ measurement of acetylene reducing activity. J.Appl.Phycol.1,195-200.
Roger PA and Kulasooriya SA (1980) Blue-green algae and rice. The International Rice Research Institute. Los Baños. Laguna. Philippines.

PRODUCTION OF DINITROGEN-FIXING CYANOBACTERIA FOR SOIL INOCULATION


L. TOMASELLI, L. GIOVANNETTI, L. FALCHINI and R. MATERASSI
Centro di Studio dei Microrganismi Autotrofi, CNR e
Dipartimento di Scienze e Tecnologie Alimentari e
Microbiologiche, Università. P.le delle Cascine, 27 - 50144
Firenze, Italy.

ABSTRACT. In order to obtain a biomass production of valuable
dinitrogen-fixing cyanobacteria for soil inoculation, controlled
cultures of selected <u>Nostoc</u> strains were realized outdoors. Dried,
viable, clay based inocula were prepared and assayed.


## 1. INTRODUCTION

Dinitrogen-fixing cyanobacteria are particularly attractive for
production of soil inoculants. Nevertheless, little has been done with
regard to the achievement of controlled mass cultures and procedures
for obtaining dried, selected, viable and high-quality inoculants
(Metting, 1988; Roger and Kulasooriya, 1980). The aim of this work
concerns both these aspects.


## 2. MATERIALS AND METHODS

### 2.1 Culture system

Outdoor cultures of selected <u>Nostoc</u> strains (105V, 1C0, 1/19, OP25 and
OS8) were carried out in Bg-11o medium supplemented with $NaHCO_3$ (200
mg $l^{-1}$), in round miniponds (1-$m^2$ surface, 9 cm depth), equipped with
a pH-stat system and provided with a rotatory system which ensures
mixing.

### 2.2. Dried inocula and evaluation

Fresh harvested biomasses of the selected <u>Nostoc</u> strains and of
<u>Anabaena</u> R84 and <u>Calothrix</u> 20A were mixed with montmorillonite at 1:1

(w/w) ratio and air-dried (40°C). The efficiency of these dried inocula in soil colonization was estimated after one month of development under favourable conditions and their growth on soil was compared with that of fresh viable inocula. Quantity of inocula applied in the soil corresponded to about $3-5 \times 10^3$ $N_2$-fixing CFU $cm^{-2}$ soil.


## 3. RESULTS AND DISCUSSION

_Nostoc_ 1/19 gave the highest output rate in semi-continuous culture. The mean productivity value calculated over one month in temperate climatic conditions ($20.5 - 32.4°C$, mean solar irradiance 24 MJ $m^{-2}$ $day^{-1}$) was about $12.6$ g $m^{-2}$ $day^{-1}$ at dilution rate of 0.3 $day^{-1}$ (areal density of 46 g $m^{-2}$). The strains 1/19, 105V and OP25 showed a biochemical composition expecially significant for the good protein content ( 50% of dry weigth).

Dry inoculants mixed with montmorillonite were still viable after a three-year storage. The efficiency of growth on soil, compared with that of fresh viable inocula was significantly high for _Nostoc_ strains 105V, 1CO, 1/19 and OS8. Whereas, _Anabaena_ R84 and _Calothrix_ 20A which in a previous experiment were found to be not much suitable for soil colonization did not show a satisfactory growth on soil (Tomaselli et al., 1990). _Nostoc_ OP25 showed a relative low development as well (about 40% of that obtained with fresh inoculated biomass).

The existence of a significant variability among strains in the recovery of an efficient growth, following soil inoculation, indicates that we must also consider this quality in the selection of dinitrogen-fixing cyanobacteria for soil inoculation, besides the features of fast growth on soil, high productivity rates in outdoor cultures, resistence to adverse conditions and high nitrogenase activity.


## 4. REFERENCES

Metting, B. (1988) Microalgal Biotechnology, Cambridge University Press, Cambridge.

Roger, P.A., Kulasooriya, S.A. (1980) Blue-green algae and Rice, IRRI, Los Banos, Philippines.

Tomaselli, L. et al. (1990) Ann. Microbiol. 40, in press.

PRODUCTIVITY OF THE MARINE NITROGEN FIXING CYANOBACTERIUM NODULARIA
SP. IN OUTDOOR CULTURE


B. PUSHPARAJ, E. PELOSI, G. TORZILLO and A. SACCHI
Centro di Studio dei Microrganismi Autotrofi, CNR.
P.le delle Cascine, 27 - 50144 Firenze, Italy


1. INTRODUCTION. Nitrogen fixing cyanobacteria are potentially useful both industrially and agriculturally as sources of reduced nitrogen, as well as for biomass production for chemicals and pharmaceuticals. Nevertheless the information on marine nitrogen fixing cyanobacteria, with regard to mass cultivation are relatively scarce. In the present study the productivity of the marine cyanobacterium Nodularia sp., cultivated both in open ponds and closed photobioreactors are reported.

2. MATERIALS AND METHODS. Raceway ponds were oval shaped with two channels forming a single loop (4 m$^2$) and stirred by a paddle wheel. Tubular photobioreactor was described in detail elsewhere (1). Nitrogenase activity was measured by acetylene reduction assay.

3. RESULTS. The biomass yield obtained in outdoor experiments performed with the marine cyanobacterium Nodularia sp. reached 11.6 g/m$^2$ in tubular photobioreactors and 8.8 g/m$^2$ in open ponds as an average from June to September (Fig.1).

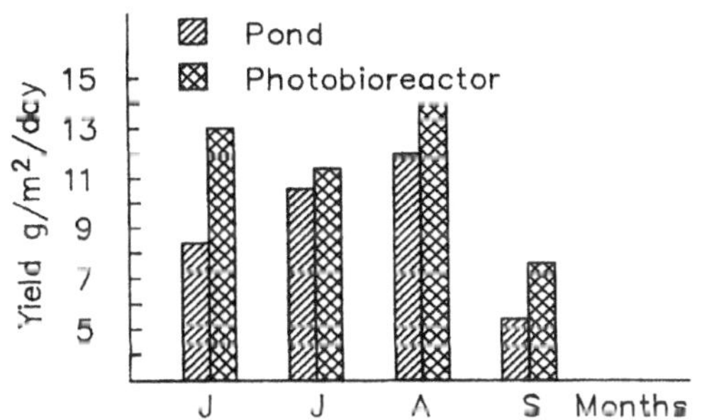


Fig. 1 — Comparison of the mean yield of *Nodularia* sp. grown in ponds and photo—bioreactors.

Table 1 — Biochemical and elemental composition of *Nodularia* sp. cells grown in a photobioreactor

|  | % of dry weight | | |
| --- | --- | --- | --- |
| Crude protein | 63.98 | C | 49.3 |
| True protein | 50.9 | H | 6.8 |
| Carbohydrates | 20.69 | N | 10.2 |
| Lipids | 9.98 | O | 28.39 |
| Ash | 5.34 | | |

The nitrogenase activity of _Nodularia_ sp. cells grown in a photobioreactor, sampled at different hours of the day and incubated at 28°C under solar light and in the dark for one hour, reached values of about 150 and 45 nm $C_2H_4$/mg d.w./h respectively (Fig.2). Table 1 shows biochemical and elemental composition of _Nodularia_ sp. cells grown outdoors. The high nitrogen content (about 10% of dry weight) and true protein (about 51%) are most interesting. These levels are in the range of those reported for _Spirulina_.

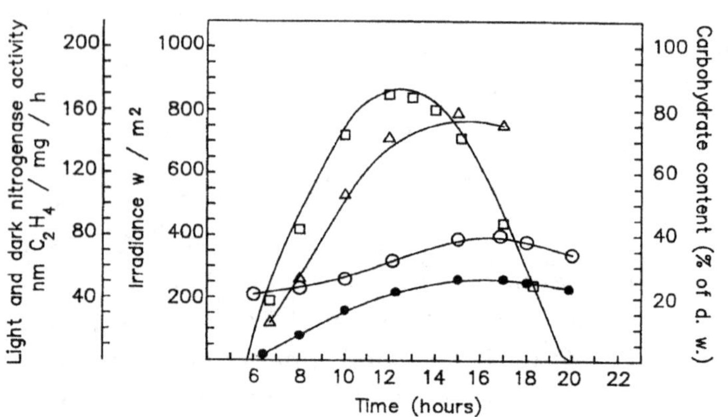

Fig. 2— Nitrogenase activity of _Nodularia_ sp. cells grown in a outdoor photobioreactor and incubated in the dark and under sunlight conditions. □ Light irradiance △Nitrogenase activity under sunlight ●and in the dark.O Carbohydrate content.

4. DISCUSSION. Our investigation on the productivity of _Nodularia_ sp. grown outdoors have shown that both open and closed systems are suitable for its mass cultivation. The results shown here proved that with the closed system the yield achieved was 34% higher than that obtained with open pond. This difference can be explained with a better temperature condition realized in the photobioreactor. The maximum nitrogenase activity observed was postponed after the maximum light intensity and continued to maintain high values even in the afternoon. A same trend was found in the carbohydrate content of the cells which increased until late afternoon. It could be supposed that nitrogenase activity in the afternoon was supported by a high carbohydrate content of the cells. When the cells were incubated in the dark, the nitrogenase activity was not switched off immediately to zero level but continued at a decreasing rate. After overnight incubation, a nitrogenase activity of 40-50 nm $C_2H_4$/mg/h was observed.

## 5. REFERENCES

1. Bocci, F., Torzillo, G., Vincenzini, M. and Materassi, R. In Algal Biotechnology. Stadler, T. (Ed.). Elsevier Applied Science Pub., London (1988) pp. 219-228.

# STUDIES ON RICE PRODUCTION USING AZOLLA AS BIOFERTILIZER IN GUINEA-BISSAU


JACINTO RODRIGUES DIAS and FRANCISCO CARRAPIÇO
*Dep. de Pesquisa Agrícola, Ministério do Desenvolvimento Rural e Agricultura, C.P. 71
Bissau, Guiné-Bissau and Dep. de Biologia Vegetal, Fac. de Ciências de Lisboa, Bloco C2,
Campo Grande, 1700 Lisboa, Portugal.*



ABSTRACT. The preliminary results obtained with the rice culture in Guinea-Bissau, using different fertilizer mixtures, indicates that *Azolla* can be used for increasing rice yield at low cost, without significant loss of productivity .


## Introduction

The *Azolla-Anabaena* symbiosis is the only fern-cyanobacteria association that presents agricultural interest (Lumpkin and Plucnett, 1980; Kannaiyan, 1986; Van Hove and Diara, 1987; Shi and Hall, 1988). In the case of Guinea-Bissau, the peasants subsistence type of agriculture prevents the existence of the necessary funds to buy the main productive factors. These farmers prefer not to buy chemical fertilizers, but to increase the area of rice cultivation (Dias, 1989, personal communication). In some regions of the country, *Azolla* presents a well developed pattern and can be found during several months of the year. Previous work done on some of the Western African countries by ADRAO, namely by the Azolla Project (Van Hove and Diara, 1987; Diara *et al.*, 1987; Van Hove, 1989) and by local researchers of DEPA indicates the possibility of using this fern on rice culture. Based on this data, several assays were made to study the effect of *Azolla* as green manure on rice yield and compare with the effect of urea fertilizer in Guinea-Bissau.

## Materials and methods

The *Azolla* and rice experiments were developed in Contuboel (25 km from Bafata) where the agronomical field station from the Agronomic Research Department (DEPA) is located. *Azolla pinnata* was collected in the Geba river and maintained in the field station in small basins using superphosphate and furadan granules. The preparation of the field for *Azolla* nursery was done by the method suggested by Kannaiyan (1986) and also developed by ADRAO. The rice variety used during the years of 1987 and 1988 (drought campaign) was the BG 90-2. Different amounts of *Azolla* compost (7 t/ha or 14 t/ha), used 1 week before rice transplanting, were compared with urea fertilizer (87 kg N/ha). Some other experiments were conducted using a mixture of *Azolla* compost and urea - *Azolla* 7 t/ha + 43.5 kg N/ha (urea) and *Azolla* 14 t/ha + 43.5 kg N/ha (urea) -. All assays were compared with a blank treatment (no chemical nitrogen and no *Azolla* inoculation).

## Results and discussion

Guinea-Bissau is a Western African country with an area of 36,125 $km^2$ and a population of 1 million inhabitants, with 850,000 people working on agriculture. The rice is one of the most important crops in the country and one of the main food sources of the population (Dias, 1988,

542

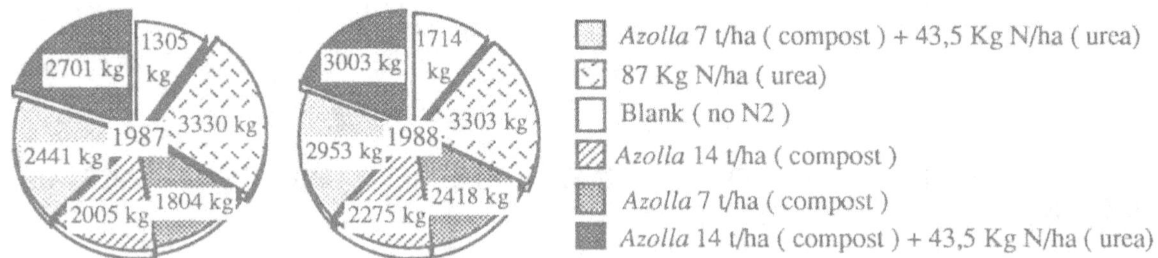



Fig. 1 - Production (Kg ha$^{-1}$) of the BG 90-2 rice variety, using different fertilizers during 1987 and 1988.

personal data). *Azolla pinnata* is localized in the Eastern region, throughout the Geba river, namely on the sectors of Contuboel, Sonaco and Bafata. This fern presents a vegetative cycle during several months of the year (August to March) and a sporulation period between November and January (Carrapiço and Dias, 1989, unpublished data). The experiments carried out on the rice fertilization (Fig.1), show the best results, using only chemical fertilizer, with a production of 3,300 kg/ha of this crop in the year of 1987 and a production of 3,303 kg/ha in 1988. The second best production was obtained using a mixture of *Azolla* and urea, in the proportions of *Azolla* 14 t/ha + 43.5 kg N/ha (urea). The results of this assay shows, in 1987, a rice production of 2,701 kg/ha and in the year of 1988 a production of 3,033 kg/ha. The results obtained with the blank treatment are in the year of 1987, 1,305 kg/ha and in 1988, 1,714 kg/ha. In spite of the best results being obtained with urea, as was expected, the prices of this chemical fertilizer and other productive factors strongly disagree with the exclusive use of it on rice fertilization and suggest the use of *Azolla* compost combined with chemical fertilizer for increasing rice yield at low cost in this country. If these results can be confirmed by others experiments, it is possible to save about 40-50% of chemical fertilizer without significant loss of productivity in the rice yield. These preliminary data agree with the conclusions of other authors working on *Azolla* as green manure (Watanabe, 1982; Kannaiyan, 1986; Van Hove, 1989). Several experiments are now in course to compare the potential use of fresh *Azolla* with compost, including a program to study and improve the use of this biological system by local farmers.

## Acknowledgements

The authors are grateful to Engº Scharwz da Silva for his cooperation in the development of the Azolla program, to Mr. Amadú Embaló for the field support and to FCG and JNICT for the financial support for the work.

## References


Diara, H.F., Van Brandt, H., Diop A.M. e Van Hove, C. (1987) '*Azolla* and its use in rice culture in West Africa', in IRRI (ed.), Azolla Utilization - Proceedings of Workshop on Azolla Use, Fujian, China, 1985., Manila, Philippines, pp. 147-152.

Lumpkin, T.A. and Plucnett, D.L. (1980) '*Azolla* : botany, physiology and use as green manure'. Econ. Bot. 34, 111- 153.

Kannaiyan, S. (1986) 'Studies on *Azolla pinnata* for rice crop', Res. J. Pl. Envir. 3, 1-16.

Shi, D.-J. and Hall, D.O. (1988) 'The *Azolla-Anabaena* association: historical perspective, symbiosis and energy metabolism', The Botanical Review 54, 353-386.

Van Hove, C. and Diara H. F. (1987) '*Azolla* introduction in African agriculture progress and problems', Int. Rice Comm. Newsl. 36, 1-4.

Van Hove, C. (1989) 'Azolla. Ses emplois multiples. Son intérêt en Afrique', FAO, pp. 1-53.

Watanabe, I. (1982) '*Azolla-Anabaena* symbiosis - its physiology and use in tropical agriculture', in Y.R. Dommergues and H.G. Diem (eds.), Microbiology of tropical soils and plant productivity, Martinus Nijhoff/Dr W. Junk Publishers, The Hague, pp. 169-185.

THE ROLES OF ANABAENA AZOLLAE IN ANABAENA-AZOLLA ASSOCIATION

Cang Lin, Y.D. Yang, X.L. Zhang and C.C. Liu
Azolla Research Centre,
Fujian Academy of Agricultural Sciences,
Fuzhou, 350013, Fujian,
China.

ABSTRACT. The roles of Anabaena in the symbiosis were studied by using
Azolla strains with heterologous _Anabaena azollae_ or without Anabaena.
As a symbiont, Anabaena has a strong effect on the performances of Azolla either tolerance to high temperature or low temperature. Anabaena
changed the absorption pattern of mineral nutrients of Azolla from the
medium. The effect of Anabaena on the ammino acid metabolism of Azolla
plant is also discussed.

## 1. INTRODUCTION

_Anabaena azollae_, a heterocystous, nitrogen-fixing blue-green alga occurs
as a symbiont in the leaf cavity  of the water fern, _Azolla_.Up to  now,
a little is known about the role of _Anabaena azollae_ in Anabaena-Azolla
association, except its fixation of $N_2$ and $CO_2$ from atmosphere.Recently
we succeed in making normal growth of Anabaena-free Azolla from decapit-
aed megasporcarp of Azolla, (1) and reconstituting heterologous Anabaena with Anabaena-free Azolla (2,3), opened a new way to study the roles
of Anabaena in symbiotic association. This paper reports the results of
experiments related to the roles of symbiont in symbiotic association
other  than $N_2$-fixation.

## 2. MATERIALS AND METHODS

The Azolla species and strains used in the experiments were a)Anabaena-
free Azolla which was from decapitated megasporocarp of _Azolla filiculo-
ides_ and _Aozlla microphylla_;b) _Azolla filiculoides_ (1034), _Azolla mic_
rophylla (4031); c) Newly reconstituted _Azolla filiculoides_ (NRAF,1035)
and newly reconstituted _Azolla microphylla_ (NRAM 4033).The _Anabaena_ in
NRAF was from _Azolla microphylla_ while the _Anabaena_ in NRAM was from _A._
_filiculoides_.
     All Azolla strains were grown in IRRI's water culture solution (4).
Anabaena-free Azolla was grown in the same medium with 80 mg $NH_4NO_3$/L.
All Azolla strains tested were pre-cultured for 1-2 weeks.The experiments
for the responses to high temperature of the symbiont were conducted in
a KG cabinet where the temperature regime, light intensity and relative

humidity was 37°C/29°C (day/night), 50 WM$^{-2}$ and 75% respectively. The other experiments were conducted under green house with air temperature of 18°-28°C. At the end of experiment the fresh weight of Azolla was determined following gentle drying of the Azolla between layers of tissue paper for 30 min. K, Zn, Fe, Cu, Ca was digested by the mixture of Nitric acid and Perchloric acid (5:1 v/v), and measured by shimeadju AA-64013 atomic absorption spectrometer. Total ammino acid of oven-dried Azolla was extracted by hydrochloric acid and measured by Hitachi 8350-50 Automatic Ammino acid Analyzer.

## 3. RESULTS

At 37°C/29°C (day/night), the temperature-sensitive species Azolla filiculoides had almost stopped growth at 2nd week and had no-acetylene reduction acitivity after four weeks treated. Symbiotic A. filiculoides produced less biomass comparison with Anabanena-free Azolla filiculoides grown in N-containing medium. However, at the same temperature regime, the performances of newly reconstituting Azolla filiculoides where the Anabaena from high temperature-tolerance species, Azolla microphylla showed great improvement either in biomass production or in N$_2$-fixation. On the other hand, the biomass production of Azolla microphylla with the Anabaena from A. filiculoides (NRAM) was reduced greatly comparied with original A. microphylla. However, under low temperature regime, NRAM showed better performance than that of original one.These indicated that Anabaena was one of the components involved in the performance of Azolla under various temperature regimes.

Results obtained from atomic absorption spectrometer showed that the concentration of Fe in symbiotic A. filiculoides (1034, 1035) was significant higher than that of Anabaena-free A. filiculoides.While the concentration of K,Zn and Ca in Anabaena-free A. filiculoides plant was much higher than that of symbiotic Azolla plants. Newly reconstituted A. filiculoides (1035) had greater demand in P and higher concentration of Ca in sporophyte comparied with A. filiculoides. Under the same condition the accumulation of Fe, Mn, Cu, and Zn in sporophyte was about 14-20%, 18-32%, 10-20%, 10-24% more than that of original A. filiculoides respectively. The effect of Anabaena on ammino acid metabolism was observed. For example, total ammino acids in symbiotic Azolla plants were much higher comparied with Anabaena-free Azolla, particularly aspartic acid, glutamic acid, leucine and isoleucine. These information adds much in our present knowledge of the relationship between host and symbiont.

## 4. REFERENCES.
(1) Lin C.and I. Watanabe (1988) New Phytologist 108:341-344.
(2) Lin C.and I. Watanabe, C.C. Liu, D.Y. Zheng, L.F. Tan (1988) In.Bothe et al.(eds) Nitrogen Fixation:Hundred Year After P:223-227.
(3) Lin C. C.C. Liu, D.Y. Zheng, L.F. Tan and I. Watanabe (1989) Science in China (Series B) 32(5):551-559.
(4) Watanabe I. Et al.(1977) IRRI Res. Pap. Ser. 11.

FIELD STUDIES USING [15]N DILUTION TECHNIQUES ON AZOLLA N CONTRIBUTION
TO RICE IN NIGERIA.


N. ESIOBU[1], K. MULONGOY[2], & C. VAN HOVE[3]
1  Department of Biological Sciences, University of Lagos,
   Akoka, Nigeria.
2  Soil Microbiologist, IITA, PMB 5320, Ibadan, Nigeria.
3  Laboratory of Plant Physiology, UCL, 1348
   Louvain-la-Neuve, Belgium.


ABSTRACT.  The nitrogen contribution and effects of one Azolla caroliniana strain to rice (Oryza sativae ITA 212) was evaluated using the
nitrogen – 15 dilution techniques.  About 80 kg N/ha Azolla increased
rough grain yield over control by 1.3 tons/ha (34.6%) and was equivalent to yields from about 50 kg N/ha ammonium sulphate.  Fertilizer
use efficiency was enhanced by 129% and 144% for ammonium sulphate and
Azolla respectively, when application was made at 2 weeks before panicle initiation as compared to basal dressing.  There was no significant
residual effect on the second rice crop in any of the treatments.  This
demonstrates the relatively rapid mineralisation of Azolla in tropical
soils.  Application time of Azolla biofertilizer seems to be crucial in
optimising its effects.  It is important to determine the best application time in any locality prior to Azolla use as this factor is greatly
influenced by soil type, weather conditions, rice maturing age, Azolla
strain used and its growth conditions.


INTRODUCTION

Despite the fact that nitrogen is the major limiting nutrient in wetland rice production, N fertilizers continue to be applied at rates not
higher than 40% of that needed for optimum rice yield (Stangel 1979).
The problem of low rate of fertilization would be minimized if all the
supplied N was efficiently utilized by the crop.  Presently, the poor
recovery of N applied by conventional techiques exacerbates the situation (Cao et al. 1983).  There is a general agreement that pointdeep
placement of N fertilizer enhances its recovery in rice (Joint FAO/
IAEA, 1978) but there is a dearth of information on the several factors
influencing the [15]N recovery from the age-long studied inorganic
fertilizers.

The case of biofertilizers is further complicated by the complex web

of factors affecting mineralisation.  <u>Azolla-Anabaena</u> is now well acknowledged as a biological nitrogen fixing system with huge potentials
in rice fertilization.  To enhance the impact of <u>Azolla</u> technology,
mechanistic information on the fate of the N2 fixed in the <u>Azolla</u> is
indispensable.  This paper summarizes the first field $^{15}$N studies on
Azolla use in rice fertilization in Nigeria.  Objectives included (1)
to assess effects of <u>Azolla</u> on yield as compared to inorganic fertilizers (2) to delineate the influence of application time on the biofertilizer use efficiency.

MATERIALS AND METHODS

The experiments were carried out at the irrigated paddy E fields of the
International Institute of Tropical Agriculture (IITA) situated in
Ibadan, on the subhumind zone of Nigeria with average precipitation
rate of 1200 mm/yr.  The study lasted from December to April representing the late dry season to early rainy season.

Experimental Soil

The soil (85% sand, 7% silt and 7.6%clay) had an average PH of 6.5,
0.74% C, 0.069% N.  The cation exchange capacity (CEC) was 6.37 meq/
100g and 0.0224 meq/100g total acidity.

Rice Variety and Fertilizers Employed

A medium age rice variety - <u>Oryza sativa</u> L. var.  ITA 212 was obtained
from Rice Program unit IITA, and employed in the studies.  <u>Azolla</u> <u>car-
oliniana,</u> strain ADUL - 44 CA, of Brazilian origin was collected from
the ADUL Azolla stock maintained at UCL, Belgium.  The inorganic N
fertilizer used was ammonium sulphate.  Others were mariate of potash
(80% K2O) and single superphosphate.

Preparation of Fertilizers

<u>Azolla</u> was propagated on monoculture on soil water culture in ponds.
Single superphosphate (30kg/ha) was added to enhance <u>Azolla</u> growth. A
separate <u>Azolla</u> pond to which nitrogen - 15 was added as ammonium sulphate 77.57% $^{15}$N a.e (50g) served for labelling the fern.  The isotopic
content of inorganic fertilizer was diluted to 7.49% and 15.75% $^{15}$N a.e.
before application in the treatments.

Experimental plan and Treatments

The set-up constituted a randomized completed block design with 3
replications.  The isotope plots measured 1m$^2$ and were separated by
thick bunds lined with plastic sheets to avoid nitrogen drift.  A total
of six treatments were set-up viz; (1) No nitrogen control (2) Labelled
<u>Azolla</u> applied at transplanting and a non-labelled split added at 49

days after transplanting (DAT) (3) Non-labelled <u>Azolla</u> added at transplant and a $^{15}$N labelled <u>Azolla</u> at 49 DAT : (; 75 kg N/ha) (4) 21.216g labelled ammonium sulphate at transplant and 21.216g non-lablled ammonium sulphate at 49 DAT: (90 kg N/ha). (5) As in (4) but labelled ammonium sulphate applied at 49 DAT (90 kg N/ha). (6) Labelled N (ammonium sulphate) was added at transplant and also at 40 DAT: (120 kg N/ha).

All fertilizers were incorporated at 5cm depth in furrows between rice rows – at the appropriate times. Incorporation times were selected to represent as closely as possible, a practical application of <u>Azolla</u> in dual cropping (monocrop and inter crop).

Analysis

Total nitrogen and $^{15}$N enrichment were determined in the biofertilizer fern used at each incorporation. At harvest (113 DAT), rice plants were manually uprooted and separated into roots, straw, and grains. Then routine analysis and measurements of yeild, N content, $^{15}$N enrichment were carried out. Total N was done by the micro k jeld ahl procedure (IITA 1982) while $^{15}$N enrichment was determined following the Dumas combustion (Fielder and Proksch, 1975). All the results were tested statistically by General Linear Model ANOVA statistics and Duncan's Multiple Range Test Procedures.

Residual Effect Experiments

Two weeks after harvesting the first crop, rice seedlings were transplanted into the microplots in exactly the same manner as the former. No fertilizers were added. At 60 DAT, the isotope plots were sampled. Three  rice plants were randomly uprooted and treated as described above.

RESULTS AND DISCUSSIONS

Yield Response

Table 1 shows the yield response of ITA 212 rice to the various N sources. Grain yield was increased by 29.21 to 49.97% over the control due to N fertilization.

TABLE 1 Yield of ITA 212 under the various fertilizer treatments in the microplot.

| *Treatment (Basal + 49 DAT) | N applied g/m$^2$ | Grain g/m$^2$ | Straw g/m$^2$ | Root g/m$^2$ | Whole plant g/m$^2$ |
|---|---|---|---|---|---|
| Control | – | 328.67c | 277.30a | 69.52b | 675.49b |
| 15Azo' Azo | 7.201 | 431.40b | 396.17a | 83.86a | 911.42a |
| Azo + 15Azo | 7.538 | 424.67b | 416.73b | 84.46a | 925.86a |
| 15AM + AM | 9.036 | 456.13ab | 370.43a | 108.89a | 935.45a |
| AM + 15AM | 9.037 | 452.20ab | 365.30a | 103.67a | 921.17a |
| 15AM + 15AM | 12.032 | 492.90a | 437.93a | 113.19a | 1041.02a |

548

* AM = Ammonium sulphate, Azo – Azolla
Column means with the same letter are not significantly different at
p = 0.05 by GLM AMOVA and DMRT.

At a glance, yield in all the inorganic N fertilized  plots  was higher
than that in Azolla plots (though not statistically significant).
However calculations of yield per unit nitrogen added; is higher for
Azolla than for ammonium sulphate.

Watanabe et al.(1989) observed a similar phenomemon and attributed it to
a better uptake of Azolla N than urea N.  In general improved plant
nutrient availability due to priming effect is one of the advantages of
green manures.  The role of Azolla as a K and P source to rice has been
recently emphasized (Liu, 1987).  Hence the yield improvement (34%) by
Azolla is this study is a response to N contribution and enhanced
supply of other plant nutrients.

Nitrogen Uptake and Influence of Application Time

Nitrogen recovery rates in different parts of the rice plant is shown
in table 2.  Inspite of high and positive yield responses from fertilized plots, $^{15}$N recovery rates were quite low.  This strongly suggests
priming effects and added nitrogen interactions (ANI) with the high soil
$^{14}$N pool.  Nitrogen recovered from ammonium sulphate and Azolla applied
just before transplant were 6.81% & 4.13% respectively and 15.65% and
10.1% when the fertilizers were top dressed at 49 DAT.

TABLE 2. Nitrogen Uptake and Fertilizer Efficiency for Azolla and
ammonium Sulphate applied at two growth stages.

| Treatment | Plant Port | %$^{15}$N.a.e. | $^{15}$N recovery rate (fertilizer efficiency) |
|---|---|---|---|
| $^{15}$Azo + Azo | Grain | 0.440 | 2.15 |
| | Straw | 0.309 | 1.38 |
| | Whole Plant | 0.410 | 4.13 |
| Azo + $^{15}$Azo | Grain | 0.878 | 4.40 |
| | Straw | 0.989 | 5.00 |
| | Whole Plant | 0.931 | 10.10 |
| $^{15}$Am  + Am | Grain | 0.466 | 3.04 |
| | Straw | 0.376 | 2.58 |
| | Whole Plant | 0.467 | 6.81 |
| Am + $^{15}$Am | Grain | 1.400 | 8.51 |
| | Straw | 1.022 | 6.15 |
| | Whole Plant | 1.148 | 15.65 |
| $^{15}$AM + $^{15}$Am | Grain | 0.776 | 2.46 |
| | Straw | 1.057 | 2.73 |
| | Whole Plant | 0.910 | 5.69 |

Uptake and efficiency of the fertilizers were enhanced by 129.81% and
144.55% for ammonium sulphate and Azolla respectively.  The proportion

of assimilated $^{15}N$ used in grain production was essentially same for (Azo + $^{15}$Azo) and ($^{15}$Am + $^{15}$Am) at 43.44 to 44.59%. This emphasizes the high efficeincy fo the Azolla N supplied just before panicle formation. Improved $^{15}N$ recovery from Azolla fertilizers applied just before panicle formation is consistent with reports from Watanabe et al. (1989) among others. The relatively rapid mineralization of Azolla in tropical soils (over 65% in 10 days) coupled with other well estabblished phenomena such as reduced immobilization, well developed root system, exhaustion of soil N pool at the peak N demand by plant could explain the augmented efficiency of the nitrogen fertilizers applied at 49 DAT.

Residual Effects

At 60 DAT, the dry matter yield of the control and treatment plots (135.75 to $217.00g/m^2$) did not differ significantly. Similarly, no differences were observed in root yield and N content.

In general yield was very poor due certainly to lack of plant nutrients. Percent fertilizer recovered ranged from 0.29 - 0.63% for all the fertilizers. This clearly indicates that there were no residual effects.

REFERENCES


Cao Z.H., De Datta S.K. and Fillery I.R.D (1983). Effect of placement methods on floodwater properties and recovery of applied nitrogen ($^{15}$N-labelled urea) in wet land rice. Soil Sci. Soc. Am. J. 48,196 - 203.

Fielder R. and Proksch G. (1975). The determination of nitrogen - 15 by emission and mass spectroscopy in biochemical analysis. Anal. Chim. Acta. 78, 1 - 62.

IITA (1982). Automated and Semiautomated methods for soil and plant analysis. Manual Series No. 7, IITA, Ibadan, Nigeria.

Joint FAO/IAEA (1978) lsotopic studies on rice fertilization. IAEA Tech. Report Series No. 181. 133 pp.

Liu C.C. (1987) Re-evaluationof Azolla utilization in agricultural production. In Azolla utilization. Proc. Workshop on Azolla use, 31 March - 5 April 1985; China. IRRI Publ. pp 68 - 76.

Stangel P.J. (1979) Nitrogen requirement and adequacy of supply for rice production. In: Nitrogen and Rice, IRRI publishers, Phillippines. pp. 45 - 72.

Watanabe I., Ventura W. Mascarina G. and ESkew D.L. (1989) Fate of Azolla spp and urea nitrogen applied to wet land rice (Oryza sativa L.) Boil Fertil Soils 8, 102 - 110.

# LOCALIZATION OF NITROGEN IN THE *AZOLLA - ANABAENA* SYMBIOSIS BY EFTEM


P. ALBERTANO and M. GRILLI CAIOLA
*Department of Biology*
*II University of Rome 'Tor Vergata'*
*Via E. Carnevale, 00173 Rome*
*Italy*



ABSTRACT. The localization of nitrogen within the leaf cavities of *Azolla filiculoides* Lam., where *Anabaena azollae* Strasb. and bacteria are living, had been obtained by means of electron spectroscopic imaging (ESI) performed on ultrathin unstained sections of different leaves. A net distribution of nitrogen localized inside the epidermal hairs of *Azolla*, the vegetative cells and heterocysts of *A. azollae* and bacteria has been visualized by image analysis. By electron energy loss spectroscopy (EELS) nitrogen peaks have been evidenced in spectra recorded in the observed areas.


## 1. Introduction

The basic biology of the *Azolla-Anabaena* symbiosis has been extensively investigated in the last two decades by different methodological approaches with special attention to the nitrogen fixation processes (6). The nitrogen fixed as ammonium by *A. azollae*, and released in the leaf cavities, is assimilated as glutamate by the fern (5), and then translocated apo- or symplastically to the other parts of *Azolla* and probably from the apexes to the vegetative cells of *Anabaena* (2). While the transfer cell ultrastructure in the terminal cells of the primary branched hairs of *Azolla* could be functional in supplying nitrogen to the apical colonies, it has been hypothesized that the ammonium assimilation could be due to the body cells of the secondary branched hairs (2). Nevertheless, at present the role of the epidermal hairs of the fern in the nitrogen exchanges among the symbionts has to be confirmed, and the different nitrogen compounds arising from the nitrogenase activity have to be localized in the symbiotic system.

To this purpose, conventional transmission electron microscopy and cytochemistry gave some indications on the presence of peptidic materials, positive to tannic acid fixation, located in the mucilaginous matrix and inside the epidermal hair cells (1, 4), but the evidence for nitrogen compounds distributed in the leaf cavities could not be achieved by these techniques. Therefore, high sensitive methods based on energy-filtering transmission electron microscopy (EFTEM), which offers the possibility to localize at subcellular level the elements with low atomic number (7) have been employed to test the possibility to visualize and identify nitrogen containing structures.

552

## 2. Materials and Methods

Apexes (meristemes and leaf primordia) of *Azolla filiculoides* Lam. cultivated as previously, reported (3) and mature leaf cavities of III (9-12) and IV (13-16) groups have been fixed in 2.5% glutaraldehyde in phosphate buffer pH 7.2, post-fixed in 1% osmium tetroxide, dehydrated in ethanol series, and embedded in Epon. Ultrathin (<40 nm) unstained sections, collected on uncoated 6OO mesh copper grids, have been observed by a CEM 902 Zeiss electron microscope at 80 kV. Electron spectroscopic images (ESI) of specimens have been taken just above the edge of the electron absorption specific for the nitrogen ($N_K$ edge 401 eV) at E= 410 eV and, as reference carrying out information on the background, below the edge at E= 370 eV. The net nitrogen distribution images were obtained by computer-assisted image processing. Electron energy loss spectra (EELS) of $N_K$ were acquired on 200 or 300 nm diameter zone in the areas were nitrogen was localized.

## 3. Results and Discussion

The image analyses obtained after ESI images of the fern apexes evidenced some amounts of nitrogen localized under and inside the cell wall of the primary branched hairs (PBH) of *Azolla* as well as in the cytoplasm of the vegetative cells of *Anabaena* (Fig. 1).

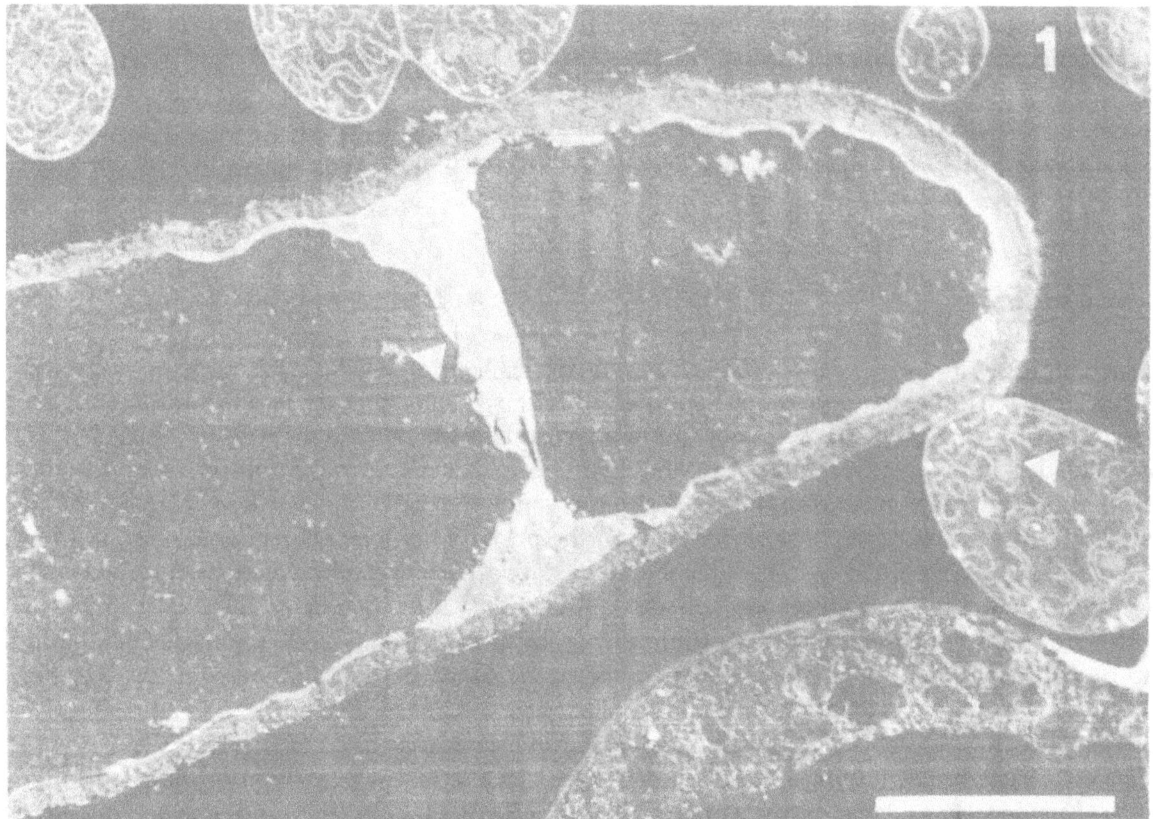

Fig. 1. ESI image of a longitudinal section of the terminal cell of a PBH and of vegetative cells of *A. azollae* in the apex of *A. filiculoides*. The arrows indicate the structures where nitrogen has been localized. Bar = 5 μm.

In the mature cavities of group III the presence of nitrogen in the epidermal hairs of the fern appeared reduced, whereas large cyanophycin granules and N-amounts were present inside the cytoplasm of vegetative cells and heterocysts of the phycobiont. Nitrogen could be also found inside the cytoplasm of *Arthrobacter sp.* cells. lying near the *Anabaena* heterocysts. In the leaf cavities of group IV no N-compounds have been localized inside the simple hairs of *Azolla*, while very small quantities were present inside the vegetative cells and heterocysts of *Anabaena* often vacuolized. Therefore in the apexes the presence of nitrogen compounds bound to the cell structure of the PBH and inside the vegetative cells of *Anabaena* would support the hypothesis of a N-transport through the hairs (2). On the other hand, the reduced bound nitrogen inside the epidermal hairs of the leaf cavities, where there is the highest $N_2$-fixation activity, could be due to a very efficient ammonium assimilation, glutamate synthesis and translocation processes by the fern. No or very few aggregates containing nitrogen have been found inside the mucilaginous matrix, even inside the basal leaf cavities of the same strain of *A. filiculoides*, where it was shown a ten fold higher concentration of soluble ammonium if compared to the apexes and to the group III (3). The impossibility, in our case, to localize nitrogen in soluble compounds arised from the washing of this fraction during conventional fixation, as could be confirmed by the results of tannic acid fixation (4). In conclusion, the use of ESI method provides helpful results in the localization of nitrogen at subcellular level. Moreover, the EELS spectra taken at the $N_K$ edge of cyanophycin granules inside vegetative cells and heterocysts of *A. azollae* appeared characterized by a double peak, that differed from the N-peaks obtained in the 'honey-comb' cytoplasmic area of the heterocysts of *Anabaena* as well as from those under the cell wall of the epidermal hairs of *Azolla*.

## 4. Acknowledgments

The authors are grateful to Prof. Dr. W. Probst for his expert technical assistance and to Carl Zeiss Co. (Oberkochen , FRG) that kindly made available the instrumentation.

## 5. References


1. Albertano, P. and Grilli Caiola , M. (1988) 'Ultrastructural study of *Azolla microphylla* Kaulf.', Giorn. Bot. Ital. 122, 84-85.
2. Calvert, H. E., Pence, M. K. and Peters, G. A. (1985) 'Ultrastructural ontogeny of leaf cavity trichomes in *Azolla* implies a functional role in metabolite exchange', Protoplasma 129, 10-27.
3. Canini, A., Grilli Caiola, M. and Mascini, M. (1990) 'Ammonium content, nitrogenase activity and heterocyst frequency within the leaf cavities of *Azolla filiculoides* Lam.', FEMS Microbiol. Lett. (in press).
4. Grilli Caiola, M. and Albertano, P. (1988) 'Recognition mechanisms in the *Azolla-Anabaena* symbiosis', in S. Scannerini et al. (eds.), Cell to cell signals in plant, animal and microbial symbiosis, NATO ASI Series H17, Springer-Verlag, Berlin Heidelberg, pp. 27-38.
5. Meeks , J. C., Steinberg, N. A., Enderlin, C. S., Joseph, C. M. and Peters, G. A.(1987) '*Azolla- Anabaena* relationships. XIII. Fixation of $(^{13}N)N_2$', Plant Physiol. 84, 883-6.
6. Peters, G. A. and Meeks, J. C. (1989) 'The *Azolla-Anabaena* symbiosis: basic biology', Ann. Rev. Plant Physiol. Plant Mol. Biol. 40, 193-210.
7. Reimer, L., Fromm, I. and Rennekamp, R. (1988) 'Operation modes of electron spectroscopic imaging and electron energy-loss spectroscopy in a transmission electron microscope', Ultramicroscopy 24, 339-354.

# THE PATTERN OF RELEASE OF AZOLLA-NITROGEN IN FLOODED AND UNFLOODED SOILS AT THREE TEMPERATURES.


SHERIFF O. SANNI
Federal Department of Agricultural Land Resources,
141, Ahmadu Bello Way, Victoria Island, P.M.B. 12613,
Lagos, Nigeria.



ABSTRACT. The patterns of release of azolla-nitrogen from two different soil types, Apomu and Ngala soils were compared. Both soil types contained portions of wet *Azolla pinnata* var *pinnata* and dried azolla in separate pots. The pots were incubated at three temperatures; viz 25°C, 30°C and 35°C. At weekly intervals, the soils were extracted with 2NKC1 and $NH_4$-N were determined. In both types of soil, temperature, time and treatment had significant effects on the amounts of $NH_4$-N determined. Flooding had significant effects on amounts of nitrogen over unflooded soils at 1% level in both soils. Sporadic bursts of nitrogen were more marked in Apomu soil than Ngala at 30°C and 35°C; and the effects of temperature on N-release were more marked in Apomu than Ngala soil.


## Introduction

In order to obtain full benefits from Azolla amendment of soils, it is essential to know the behavious of dead azolla in such soils. For Azolla to be useful soil ameliorant, its nitrogen release must, in a partical soil, coincide with critical stages of growth of rice. Watanabe, Espinas, Beja and Alimago (1977) incubating both fresh and dried Azolla with soil found that amonia release was more rapid from fresh Azolla than dried Azolla. After six weeks, 62–75% of total nitrogen in Azolla becomes available slowly.

Since rice is grown under wet conditions in different types of soil in Nigeria, this paper aims at comparing the patterns of Azolla-nitrogen release in a southern loamy sand and a northern loamy clay soils.

## Materials and Methods

A northern loamy clay soil, Ngala soil series used for growing rice in Gongola State, and a southern loamy sand used for the experiment. 3.5g of fresh *Azolla pinnata* var pinnata were weighed into 200 ml polysterene cups — 40.0gm of the soils that have been air-dried and passed through 2mm sieve. The pots were flooded with distilled water. In other sets of experiment, 10g. of the two types of soil were divided into two, one portion was placed into 25ml McCathney bottles and 0.25g. dried *Azolla* was placed on the soil. The other portion of the soil was placed on the *Azolla*. The soils were wetted with distilled water, but without flooding. Both the pots and McCathney bottles were placed in three incubators at 25°C, 30°C and 35°C respectively. The following treatments were used: (1) Flooded, (2) Unflooded, (3) Control Flooded, and (4) Control Unflooded. At one week intervals, the samples were shaken for 30 minutes with 2NKC1 and filtered through No 42 Whatman filter paper. The filtrates were analyzed for $NH_4-N$ in an automalyzer.

## Results

Quantities of $NH_4-N$ determined in both soils are shown in Tables 1 and 2. The following observations could be made from the Tables:

(1) Effect of flooding was significant in both soils at 1% level.

(2) Time (week) was significant at 1% level for Apomu soil but at 15% level in Ngala soil.

(3) Temperature had a significant effect in both soils: 1% for Apomu and 5% for Ngala.

(4) Interaction between temperature and treatment (flooding) was significant in Apomu soil but was not significant for Ngala soil.

(5) There was significant interaction between flooding and time in both soil types.

In Apomu soil, there were sporadic bursts of $NH_4-N$ in 30°C and 35°C incubated soil. This was not so marked in Ngala soils.

## Discussion

From the results of these experiments it could be seen that greater amounts of Azolla-nitrogen are released in flood soils than unflooded ones irrespective of the soil type. However, Azolla-nitrogen determined in Ngala soil in the three temperature range appeared generally lower than the amounts determined for Apomu soil. Explanations for this depressed figure are not easily available besides the fact that since Ngala soil appeared inherently to possess a higher fertility regime, Azolla amendment might not significantly affect rice yield, see Tables 3 and 4. Further research is required using other soil types of equally high fertility.

Table 1: Amounts of $NH_4$–$N$(ppm) in Apomu soil incubated
with Azolla at three temperatures

| Treatments | Week 1 | Week 2 | Week 3 | Week 4 | Week 5 | Week 6 | Week 7 |
|---|---|---|---|---|---|---|---|
| $25^\circ$C | | | | | | | |
| Flooded | 38.0 | 34.6 | 42.1 | 51.2 | 75.3 | 58.0 | 75.7 |
| Unflooded | 25.0 | 25.0 | 26.4 | 30.4 | 36.6 | 35.8 | 35.7 |
| Cont. Fld. | 22.6 | 25.7 | 28.6 | 30.7 | 30.8 | 33.2 | 31.1 |
| Cont. Unfld. | 20.2 | 20.2 | 20.7 | 21.3 | 23.3 | 22.7 | 22.3 |
| $30^\circ$C | | | | | | | |
| Flooded | 38.1 | 41.1 | 47.2 | 59.7 | 89.6 | 93.8 | 82.5 |
| Unflooded | 21.8 | 28.2 | 39.0 | 43.6 | 51.5 | 45.3 | 45.9 |
| Cont. Fld. | 23.7 | 30.0 | 32.4 | 32.1 | 32.4 | 35.2 | 32.1 |
| Cont. Unfld. | 20.1 | 21.2 | 25.5 | 29.6 | 24.8 | 30.2 | 29.3 |
| $35^\circ$C | | | | | | | |
| Flooded | 25.0 | 48.7 | 53.0 | 60.5 | 59.3 | 67.4 | 66.8 |
| Unflooded | 27.1 | 29.3 | 31.2 | 31.0 | 39.7 | 41.1 | 40.1 |
| Cont. Fld. | 20.0 | 19.8 | 25.5 | 24.0 | 26.6 | 24.9 | 23.7 |
| Cont. Unfld. | 17.4 | 20.0 | 17.8 | 19.7 | 19.2 | 19.4 | 20.5 |

Analysis of Variance

| | S.E. | LSD (1%) |
|---|---|---|
| Temperature | 0.65 | 2.454 |
| Week | 1.004 | 3.748 |
| Treatment | 0.759 | 2.834 |
| Temperature x Treatment | 1.315 | 4.908 |

558

Table 2: Amounts of $NH_4$–N(ppm) in soil incubated
with azolla at three temperatures in Ngala soil.

| Treatments | Week 1 | Week 2 | Week 3 | Week 4 | Week 5 | Week 6 |
|---|---|---|---|---|---|---|
| **25°C** | | | | | | |
| Flooded | 24.2 | 30.9 | 38.5 | 41.43 | 41.3 | 52.2 |
| Unflooded | 21.8 | 23.8 | 25.2 | 30.3 | 21.0 | 34.2 |
| Cont. Fld. | 15.5 | 20.1 | 23.3 | 24.2 | 21.0 | 20.9 |
| Cont. Unfld. | 19.1 | 20.6 | 21.1 | 19.4 | 19.2 | 20.4 |
| **30°C** | | | | | | |
| Flooded | 29.0 | 31.9 | 38.5 | 45.7 | 45.2 | 48.0 |
| Unflooded | 19.9 | 21.7 | 31.2 | 31.5 | 30.8 | 31.6 |
| Cont. Fld. | 20.3 | 20.9 | 20.0 | 20.8 | 21.4 | 21.4 |
| Cont. Unfld. | 19.5 | 16.8 | 19.9 | 20.4 | 20.8 | 21.0 |
| **35°C** | | | | | | |
| Flooded | 22.0 | 28.2 | 39.4 | 41.9 | 42.5 | 44.5 |
| Unflooded | 19.3 | 20.3 | 28.5 | 31.2 | 34.6 | 33.6 |
| Cont. Fld. | 17.4 | 19.4 | 19.4 | 20.4 | 20.8 | 21.0 |
| Cont. Unfld. | 16.5 | 18.9 | 19.2 | 20.0 | 19.5 | 20.1 |

Analysis of Variance

| | DF | SS | MS | F | LSD |
|---|---|---|---|---|---|
| Temperature | 2 | 279.21 | 339.61 | 79.90 | 1.81* |
| Treatment | 3 | 15150.71 | 5050.26 | 37.84 | 6.16** |
| Treatment x Week | 15 | 4077.57 | 271.84 | 5.33 | |

Table 3: Chemical analysis of Apomu Soil

| Organic Matter (%) | Total Nitrogen (%) | Available Phosphorus (ppm) | Na | K | Ca | Mg | pH |
|---|---|---|---|---|---|---|---|
| | | | m.e. % | | | | |
| 0.86 | 0.038 | 15.5 | 0.02 | 0.21 | 0.77 | 0.96 | 6.5 |

Table 4: Chemical analysis of Ngala Soil

| | | | Exchangeable Cations | | | | |
|---|---|---|---|---|---|---|---|
| Organic Matter (%) | Total Nitrogen (%) | Available Phosphorus (ppm) | Na | K | Ca | Mg | pH |
| | | | (m.e. %) | | | | |
| 1.72 | 0.092 | 35.4 | 0.5 | 1.4 | 4.45 | 2.46 | 7.8 |

## References

Watanabe, I., Espinas, C.R., Beja, N.S., and Alimago, B.U. (1977). Utilization of the Azolla — anabaena complex as a nitrogen fertilizer for rice. International Rice Research Institute, IRRI Res. Pap. Ser. No 11, 1 – 15.

FACTORS CONTROLLING MORHPHOLOGICAL CHANGES AND THE LIFE CYCLE OF
NITROGEN-FIXING CYANOBIONTS IN DIFFERENT _AZOLLA_ SPECIES.



R. Caudales[1], A.D. Antoine[1] and A.C. Vasconcelos[2]
[1]Department of Biochemistry and Microbiology
and [2]Biological Sciences
Rutgers - The State University
New Brunswick, New Jersey, 08903-0231, USA



ABSTRACT.  _Azolla_ plants grown under controlled conditions were dissected
from the main axes to the 20th leaves and cyanobacteria were observed
under phase contrast microscopy.  Acetylene reduction was assayed weekly
in order to correlate color of plants with cyanobacterial morphology.
Morphology, life cycle and nitrogen fixation of the cyanobacteria were
regulated by age, light intensity and chromatic changes of the host.
Hormogonia were produced as a result of akinete germination with cells
smaller than parental trichomes and by multiple differentiation of
vegetative cells adjacent to single or double heterocysts.  Based on the
developmental life cycle (hormogonia formation), cyanobionts in _Azolla_
should be classified in the genus _Nostoc_ rather than _Anabaena_.


Introduction.  _Azolla_ is the host of the nitrogen-fixing cyanobacterium
know as _Anabaena azollae_ Strasberger.  Since the first description of the
cyanobacteria in _Azolla_ cavities by Strasberger (1873) its taxonomic
status has long been questioned.  Due to the ambiguous concept of sheath
surrounding the trichomes, lack of akinetes, position of akinetes with
respect to heterocysts and life cycle.  The goal of this research was to
establish the true identity of the cyanobacteria found in the cavities of
different _Azolla_ species and factors that regulate the cyanobiont
morphology, life cycle and nitrogen fixation.

Materials and Methods.  Plants were grown on nitrogen-free medium at
approximately 100 uE m$^{-2}$.s$^{-1}$, (14:10) using incandescent and fluorescent
light at temperature of 24:16 $^{o}C$.  Acetylene reduction was assayed at
different weekly intervals in order to correlate color of plants with
cyanobacterial morphology.  To study the morphology and development cycle
of the cyanobacteria, the apex, 5th, 10th, 15th, 20th leaves were
dissected from the main axes of 2, 4, 6, and 8 week-old _Azolla_ plants.
The cyanobacteria were observed under phase-contract microscopy.

Results.  Plant age, light intensity and pigment composition regulated
the morphology and life cycle of the cyanobacteria.  Cyanobionts inside
the cavities of _Azolla_ received a light spectrum influenced by
physiological and chromatic changes in the plant.  Green light increased

heterocyst frequency and acetylene reduction.  Yellow light stimulated akinete germination and hormogonial differentiation.  In red light, trichomes were short and vegetative cells were reduced in width, lacked normal pigmentation, and had negligible acetylene reduction.  Morphological differentiation of vegetative cells and heterocysts were synchronized with growth in young *Azolla* plants.  In the apex of *Azolla*, trichomes were composed entirely of small, vegetative cells (hormogonia).  In older leaves the cyanobacteria were enclosed in the cavity and embedded in mucilagenous materials, and trichome sheaths were reduced in size.  Vegetative cells enlarged in width and length as a function of leaf age.  Heterocysts were intercalary, terminal and multiple.  Akinetes, which occurred in all species of *Azolla*, when present, were regulated by age, light intensity and chromatic changes of the plant, and were found between vegetative cells or adjoining one or two heterocyst.  Akinetes tended to occur in older leaves of younger pants, but reverted to the apex of older plants, and were single or in chains with variable shapes and sizes depending on *Azolla* species.  Akinete germination gave rise to a hormogonium composed of cells smaller than the parental trichome.  Hormogonia were also formed adjacent to single or double heterocysts or between heterocysts.  The occurrence position of single and multiple heterocysts related to akinetes depend on *Azolla* species.

Discussion.  Morphology, life cycle and nitrogen fixation of the cyanobacteria are regulated by age, light intensity and chromatic changes of the host.  Based on microphotographic analysis and the above observation, cyanobionts in *Azolla* should be classified in the genus *Nostoc* rather than *Anabaena*.  The reasons for this, in recapitulation, are (a) akinete germination given rise to hormogonia, with cells smaller that parental trichomes; (b) hormogonia formed adjacent to single or double heterocysts by multiple differentiation of vegetative cells; (c) multiple heterocysts occurring as reported from Cycads; (d) akinetes, when present, between vegetative cells of adjacent to one or both sides of heterocysts, and single or in chains.

Strasburger, E. (1873). Weber *Azolla*. Verlag von Ambr. Abel. Jena, Leipzig.

*FATTY ACID COMPOSITION OF FREE-LIVING ANABAENA AND NOSTOC CYANOBACTERIA AND OF CYANOBIONTS FROM DIFFERENT AZOLLA SPECIES.*

*R. Caudales and J. Wells*
*Dept. of Biochem. and Microbiol. Rutgers University*
*New Brunswick, NJ 08903-0231 USA and U.S. Dept. of*
*Agriculture, 600 E. Mermaid Lane, Philadelphia, PA 19118 USA ·*

*Abstract. Total cellular fatty acids of 10 strains each of free-living, nitrogen-fixing Nostoc and Anabaena cyanobacteria grown for 12 days on $BG_{11}$-N-free medium, and of 10 cyanobionts isolated from seven species of 14 to 18-day old Azolla plants grown under controlled conditions were determined by GLC-mass spectroscopy. Significant differences existed between free-living Anabaena and Nostoc isolates. Ratios of total unsaturated 16-carbon fatty acids/total saturate acids (Classes A+B) was greater than 0.5 for Anabaena and less than 0.5 for Nostoc. Free-living strains of Nostoc could also be distinguished from cyanobionts of Azolla by fatty acid composition. Ratios of Class C/A+B (unsaturated/saturated) were greater than 2 for free-living isolates and less than 2 for cyanobionts. Many of these fatty acids may be reflective of evolutionary trends in the development of the cyanobacteria, and can be used for taxonomic differentiation.*

*Introduction. Cyanobacteria have been subdivided taxonomically in five morphological sections. Section IV is comprised of seven genera: Anabaena, Nodularia, Cylindrospermum, Nostoc, Scytonema, Calothrix, and Tolypothrix. Due to morpholgical similarities, however, the division between Anabaena and Nostoc is problematic. Rippka et al. (1979) recently suggested the use of the developmental cycle (hormogonia formation) as the major criterion to differentiate the genera Nostoc/Anabaena. However, as many Nostoc isolates do not show a development cycle, new and reliable toxonomic techniques should be explored. Fatty acids in Eubacteria and microalgae have been shown to be specific of specie level and a good tool for bacteria identification. We have found that fatty acid composition determined by GLC-MS is and easy and reliable technique to separate the genera Nostoc from Anabaena. In addition we have found unknown fatty acids of taxonomic value.*

*Materials and methods. Fatty acids were identified by co-chromatography with reference standards, when available. Major fatty acid peaks were confirmed chemically, by methods previously described, in terms of being saturated straight chains, unsaturated, hydroxy-substituted, or cyclic methyl esters of fatty acids. Equivalent carbon chain length (ECL) was calculated for each peak, and provided further confirmation of identity by reference to published reports (Casano et al. 1988)*

*Results. Cellular fatty acid composition of free-living Nostoc and Anabaena isolates included 9 major components, or groups of components, each over 1% of the total; 22 minor components, each 0.01 to 1% of total; and 8 trace-level components. Major fatty acids (and average percent of total for all free-living isolates) included the saturated straight chains (Class A) 16:0 (27.4%) and 18:0 (1.1%); unsaturated fatty acids (Class C) 14:1-(7) (4.1%), 16:4-(4) (1.2%), 16:1-all forms (20.7%), 18:3 (11.7%), 18:2 (17.7%), and 18:1-all forms (5.7%); and the branched-chain (Class E) iso-16:0 (2.1%). The unsaturated 16-carbon fatty acids were a diverse group, with 8 different isomeric forms detected based on ECL calculations and on co-chromatography with standards, the most important being 16:1 cis-9 and 16:1 trans-9. Significant differences existed between the fatty acid profiles of free-living isolates of Anabaena and Nostoc. The average percentages of Class C was significantly lower for Anabaena (59.4%) than for Nostoc (67.5%), with no overlapping of the ranges. Free-living isolates (of both genera) could generally be distinguished from Nostoc cyanobionts by the presence of a partly-characterized unsaturated 15-carbon fatty acid with an ECL of 14.3. The component was present beyond the trace level in 80% of all free-living isolates examined, but absent in cyanobionts.*

*Discussions. Significant taxonomic distinctions between groups of bacteria are generally reflected in the composition of cellular fatty acids. Such is true of Anabaena and Nostoc, where basic compositional differences can be demonstrated in fatty acids composition, particularly in the physiologically-important ratios of saturated to unsaturated acids. Similarly, the cyanobionts from Azolla spp. could be distinguished from all free-living strains of Nostoc and Anabaena. A practical ramification of this analysis is the prospect of identifying Nostoc/Anabaena isolates without the need for detailed morphological studies. Fatty acids profiles of representatives of all 5 groups of cyanobacteria can be developed for rapid identification. Precise fatty acid profiles, moreover, can confirm identifications made on the basis of morphological criteria. Regarding the taxonomic position of the cyanobionts from Azolla, further analysis of morphological and physiological observations are required (in progress) to confirm what is suggested from the fatty acid composition-namely, that they are sufficiently different from free-living forms to constitute a distinct taxon.*

*Casano, F., J. Wells and T. Van Der Zwet. (1988). Fatty acid profiles of Erwinia amylovora as influenced by growth medium, physiological age and experimental conditions. J. Phytopathology 121:267-274.*

*Rippka, R., J. Dervelles, J.B. Waterbury, M. Herdman and R.Y. Stanier (1979). Genetic assignments, strain histories and properties of pure cultures of cyanobacteria. J. Gen. Micorbiol. 111:1-61.*

# DINITROGEN FIXATION (ARA) IN <u>LOBARIA PULMONARIA</u>

M.A. MONSALVE AND F. BERMUDEZ DE CASTRO
*Univ. Complutense*
*Fac. Biología*
*Dpt. Ecología*
*28040 Madrid. Spain.*

## Introduction

Dinitrogen fixation in lichens has been largely proved (Kershaw and Millbank, 1970; Pike et al., 1972) and the fixing rates in different ecosystems like tundra (Kallio, 1976) and <u>Pseudotsuga menziesii</u> forests (Denison, 1979) are known, although its role in nitrogen retraining is not completely demonstrated. A part of a more extensive study about the nitrogen cycle in deciduous forests it is shown the ARA of <u>Lobaria pulmonaria</u> ,a diazotrophic lichen growing on the trunks of <u>Quercus pyrenaica</u> and <u>Fagus silvatica</u> in a mixed forets on Pyrenean oak and a common beech.

## Materials and Methods

The studied area is a mixed forest of Pyrenean oak and common beech situated in the north of Madrid (Spain), called "Hayedo de Montejo", which is one of the southernmost beech forests in Europe. Several lichens grow on the trunks. The total lichen cover on the trees of oak and beech is 40.15% and 44.95%, respectively, and the <u>L. pulmonaria</u> cover is 22.88% and 4.48%.
Samplings were carried out monthly, for 18 months in the trunks of <u>Q. pyrenaica</u> and for 6 months in the trunks of <u>F. silvatica</u>, taking thallus from different trees at random. Dics of 2 cm in diameter were cut and 10 dics were incubated in 10 ml vials for 1 hour in a 10% acetylene atmosphere. The incubation was carried out in the same forets, at foot of the trees, in order to maintain better the environmental conditions. The formed ethylene was measured in a gas chromatograph KONIC 2000 C-series, with column of Porapak R.

## Results

The obtained results are expressed in nM $C_2H_2$ per incubation hour and dry thallus gram and in nM $C_2H_2$ per incubation hour and thallus square centimetre. The average of ARA in <u>L. pulmonaria</u> on <u>Q. pyrenaica</u> was $97.45 \pm 53.42$ nM $C_2H_2$/h.g and $1.99 \pm 1.07$ nM $C_2H_2$/h.cm$^2$ and of the lichens on $\underline{F. \ silvatica}$ 218.68 ± 211.79 and 5.43 ± 5.25 nM $C_2H_2$/h and g or cm$^2$, respectively. The annual fluctuation is obverved by the maxima in July and October and the minima in August and September.

## Discussion

The ARA of $\underline{L. \ pulmonaria}$ is smaller than that of $\underline{L. \ pulmonaria}$, $\underline{L. \ oregana}$, $\underline{Peltigera \ aphthosa}$, $\underline{Pseudocyphellaria \ anthraspis}$ y $\underline{Sticta \ weigelii}$ of the epiphyte community found in the Douglas fir canopy (Denison, 1979). Although the absolute values of ARA are not very high, they represent an important input due to abundance of lichens in the forest, taking into consideration that a very advanced stage of the succession approaches a stable balance and the external nitrogen requisitions are minima and the nutrients are retrained mainly through the litter fall.

The very high standard errors approaching the average are surprising, in spite of the fact that each one has been obtained with 10 replicas. It must be taken into account that to the variability imposed by the nature of biological material it added the fact that the cyanobacterium responsible for dinitrogen fixation is not in the thallus but confined in nodule-like cephalodia, which increases data dispersion. It must be added to this that dinitrogen fixation in lichens is very sensitive to humidity (Henriksson and Simu, 1971), so the orientation and position of $\underline{L. \ pulmonaria}$ on the trees trunks and the microclimates in the forest created by the relative position of the trees to streams or slopes distort the measures.

## References

Denison, W.C. (1979) '$\underline{Lobaria \ oregana}$, a nitrogen-fixing lichen in old-growth Douglas fir forest', in J.C. Gordon, C.T.Wheeler and D.A. Perry (eds.) Symbiotic Nitrogen Fixation in the Management of Temperate Forests, Oregon St. Univ. Corvallis, pp. 266-275.

Henriksson, E. and Simu, B. (1971) 'Nitrogen fixation by lichens', Oikos 22, 119-121.

Kallio, S. (1976) 'Studies on Elemental Nitrogen Fixation in Lichens in North Finland', Univ. of Turku.

Kershaw, K.A. and Millbank, J.W. (1970) 'Nitrogen metabolism in lichens', New Phytol. 69, 75-79.

Pike, L.H., Tracy, D.M., Sherwood, M.A. and Nielsen, D. (1972) 'Estimates of biomass and fixed nitrogen of epiphytes from old-growth Douglas fir', in J.F. Franklin, L.J. Dempster and R.H. Waring (eds.) Research on Coniferous Forests Ecosystems, Pac. NW Forests and Range Exp. Sta. Portland, pp. 177-187.

# BIOLOGICAL NITROGEN FIXATION IN ARCTIC SOIL AND VEGETATION DETERMINED BY THE ACETYLENE REDUCTION METHODE


Arild Endal and Bjørn Solheim,
Dept. of Plant Physiology and Microbiology, IBG,
University of Tromsø, 9000 Tromsø, Norway



ABSTRACT. To evaluate biological nitrogen fixation a total of 45 species of plants and a few soil samples from Spitzbergen, ($79^{o}N$) have been tested for nitrogenase activity by the acetylene reduction method. Activity was low in soil, roots and underground parts of mosses, higher in upper parts of moss. The highest level of activity was found in *Nostoc commune* collonies, $1.012\mu molh^{-1}$ **g dry weight**$^{-1}$. It is concluded that cyanobacteria are the main source of biologically fixed nitrogen in arctic soils.


## Introduction
In Arctic and Antartic tundra biological nitrogen fixation is the only or main nitrogen source for plant growth. In Spitzbergen, Svalbard there are no known legumes. Three types of special microbial communities can visually be distinguished. The dryest, a black crust, is probably lichens living on dead moss, the intermediate a slimy layer of microorganisms and decaying organic material, and the wetter have collonies of *Nostoc commune*. *N. commune* has been found to fix up to 1400umol $N_2$/g dry weight and hour (1).

## Material and methods
A total of 7 dicotyledones, 12 monocotyledones, 17 mosses, 9 lichens, and a number of soil and undefined soil cover were tested in acetylene reduction assay (ARA). Incubation was done under different conditiones with samples of 1 to 5 g fresh weight. Vegetation samples were incubated with 0.5 ml destilled water in 10 ml sealed glass vials for two weeks and streaked on nitrogen free mineral agar. Several cyanobacteria were isolated that showed ARA in pure culture.

## Results
The ARA ranged from 0 to 1umol etylene per g dry weigt and hour (Table 1). The highest ARA levels were in *Drepanocladus uncinatus*, *Polytricum alpinum*, *Calliergon richardsonii*, some lichens and collonies of *Nostoc commune*. The maximum ethylene production in *N. commune* was 1.01umol $h^{-1}$g dry weight$^{-1}$ or 0.25umol $h^{-1}cm^{-2}$ at $18^{o}C$. Others have found ARA of 0.12 – 0.14 nmol g dry weight$^{-1}h^{-1}$ and up to 150nmol cm $^{1}$24h$^{-1}$ in antarctic moss (2,3). In *Nostoc muscorum* levels of 119–200 nmol g dry weight$^{-1}h^{-1}$ was reported (2). Reports from Arctic Canada give values up to 8.87 umol $m^{-2}$ $h^{-1}$ (4).

There was a clear difference between different mosses. *Paludella squarosa* and *Ptilidium ciliare* showed no ARA while *Polytricum alpinum* and *Drepanocladus unsinatus* showed high ARA. Roots of mono- and dictyledones, soil and peat showed little or no activity. This includes *Dryas octopetala* roots which has been reported to show ARA in Colorado U.S.A.(5). The highest ARA in mosses was in the upper 1 cm.
In *Drepanocladus* clusters of cyanobacteria were found in the U formed by the leaf, and in *P.alpinum*, between lamella of cell rows. Cyanobacteria were not found in the youngest or oldest parts of the mosses. The unvegetated areas with slimy layer and the dead moss with lichens have high numbers of *Oscillatoria*.

Table 1: Range of acetylene reduction measured as etylene production ($nmol\ h^{-1}g$ dry weight$^{-1}$).

| | | |
|---|---|---|
| Nostoc collonies | 0.04 | – 1012 |
| Mosses | 0 | – 500 |
| Lichens | 0.04 | – 44.2 |
| Soil crust | 0.2 | – 9.4 |
| Roots ( monocot., dicot.) | 0 | – 0.25 |
| Soil, peat, silt, loam | 0 | – 0.13 |

## Discussion

Cyanobacteria capable of growing on nitrogen free medium was isolated from all locations. The difference in ARA showed that cyanobacteria had a preference for certain mosses. Moss probably gives some protection against drying. The restriction of assosiation to the green and living parts of the moss may be due to exchange of nutrients between the organisms or a result of light conditions lower down in the moss cover.
The high values found in the surface crust on the soil and upper parts of mosses compared to peat, soil and plant roots show that cyanobacteria are the main source of fixed nitrogen.
Activity in wetted moss and *Nostoc* were up to 100 times higher than in dry samples. This indicates that the cyanobacteria are in a resting state when dry.
High numbers of *Oscillatoria sp.* in the wet unvegetated areas may be responsible for the activity in these locations. In the dryer type of unvegetated soil lichens of the genus *Collemma* may also be responsible.
In a goose nesting area the differences in ARA between mosses were high. The highest activities were measured in the very wet moss close to a small lake. Available nitrogen in ground and water was not measured.

## Refferenses

1: Lennihan, R. and Dickson, L. G. (1989) "Distribution, abundance and physiological aspects of *N. commune* in a high arctic ecosystem", Journal of Phycology, 25 (2) 16.
2: Christie, P. (1987) "Nitrogen in two contrasting Antarctic bryophyte communities", Journal of Ecology, 75, 73–93.
3: Nakatsubo, T. and Ino, Y. (1986) "Nitrogen cycling in an Antarctic ecosystem. 1. Biological nitrogen fixation in the vicinity of Syowa station", Mem. Natl. Inst. Polar Res. Ser. E, 37, 1–10.
4: Karagatsides, J. D., Lewis, M. C. and Schulman, H.M. (1985) "Nitrogen fixation in the high arctic tundra at Sarcpa Lake, Northwest Territoties", Can. J. Bot.,63, 974–979.
5: Wojciechowski, M. F. and Heimbrook, M. E. (1984) "Dinitrogen fixation in alpine tundra, Niwot Ridge, Front Range, Colorado, U.S.A." Arctic and Alpine Research, 16 (1) 1–10.

THE <u>NOSTOC</u>-<u>GUNNERA MAGELLANICA</u> SYMBIOSIS: DEVELOPMENTAL PATTERNS
RELATED TO NITROGEN FIXATION.

Erik Soderback, Peter Lindblad, Birgitta Bergman
Depatment of Botany, Stockholm University
S-106 91 Stockholm
Sweden

## ABSTRACT

Several parameters in the development of the nitrogen-fixing <u>Nostoc-
Gunnera magellanica</u> symbiosis were studied. All parameters studied
showed a developmental profile from young towards older symbiotic
tissues, with the highest activities and most functional symbiosis located in relatively young parts of the plant. After leaf shedding of the
plant the symbiotic cells showed signs of degeneration, and nitrogenase
activity declined gradually to zero.

## INTRODUCTION

All species of <u>Gunnera</u> studied has been shown to develop nitrogen-
fixing symbioses with filamentous heterocystous cyanobacteria of the
genus <u>Nostoc</u>. The cyanobacteria are located intracellularly in distinct
symbiotic tissues along the stem of the plant. Infection takes place at
the growing apex. The symbiotic <u>Nostoc</u> differentiate an extremely high
frequency of heterocysts, up to 80% has been reported, and a high nitrogenase activity. We have studied the development of cyanobacterial biomass, heterocyst frequency and nitrogenase protein level in relation to
the nitrogenase activity along the stem of the plant. Nitrogen assimilation pathway are also discussed in connection to preliminary studies
of the glutamine synthetase protein level and activity.

## MATERIAL AND METHODS

Plantlets of <u>Gunnera magellanica</u>, grown outdoors on free-land, and a
free-living axenic <u>Gunnera</u> isolate of <u>Nostoc</u> were used in these studies.
Nitrogenase activity was assayed with the acetylene reduction technique, glutamine synthetase activity with the transferase assay. Protein
levels of both enzymes were determined by immuno-gold labelling with
antisera raised against purified proteins, and analyzed with a computer-
ized image analysis system connected to an electron microscope. Tissue
preparations for immunolabelling and electron microscopy were as in
Braun-Howland et al. 1988.

## RESULTS AND DISCUSSION

Cyanobacterial biomass development showed an approx. linear increase from the apex to a plateau, reached in older symbiotic tissues. The maximal heterocyst frequency of <u>Nostoc</u>, ca 60%, were located just proximal to the maximal biomass. The nitrogenase activity, however, showed no correlation to these parameters; instead were the activity peak located in young symbiotic tissues with intermediate biomass and heterocyst frequency of <u>Nostoc</u>. The nitrogenase protein level of <u>Nostoc</u> heterocysts were approx. constant throughout the plant. Hence must an inactivation of heterocysts and the nitrogenase enzyme occur.

The structural development of the symbiotic tissue explained some of the results obtained. The decline in nitrogenase total activity coincided with the appearence of degenerated <u>Nostoc</u> cells. It also coincided quite well with leaf shedding of the <u>Gunnera</u> plant. It is most probable that the relocation of plant photosynthates has a major impact on the development of the symbiotic <u>Nostoc</u>. When the leaf connected to the symbiotic tissue is shed, no nutrient transportation or translocation occurs, why <u>Nostoc</u> development and growth stops and cells start degenerate due to starvation.

In the oldest symbiotic tissues, located at the base of the plant, no intact cells were found, and nitrogenase activity was zero. However, immunolabelling of the tissue showed an substantial antigenic answer in the tissue, why a major part of the nitrogenase protein must have been inactivated but otherwise stayed intact in the old <u>Gunnera</u> cells.

The glutamine synthetase activity of plant tissue showed a gradual increase in young parts. The maximal activity was reached in relatively young tissue, and remained constant throughout the plant. Infected and uninfected stem tissues were easily separated, and showed the same specific activity. Though the infected symbiotic tissues contributes relatively little to the total amount of tissue, is probably the bulk GS activity of plant origin. Nevertheless showed immunolabelling that GS level is as high or higher in symbiotic <u>Nostoc</u> compared to a free-living isolate. Hence is probably the primary assimilation pathway for fixed nitrogen as in free-living isolates, that is glutamine via glutamine synthetase, and ammonia not the product translocated to the host cyto-plant.

## REFERENCES

Braun-Howland EB, Lindblad P, Nierzwicki-Bauer SA, Bergman B (1988) Dinitrogenase reductase (Fe-protein) of nitrogenase in the cyanobacterial symbionts of three Azolla species: Localization and appearence during heterocyst differentiation. Planta 176: 319-332

MORPHOLOGICAL AND PHYSIOLOGICAL CHARACTERIZATION OF $N_2$-FIXING
SYMBIOTIC CYANOBACTERIA


L. VAGNOLI, M.C. MARGHERI, G. ALLOTTA and R. MATERASSI
C.S. dei Microrganismi Autotrofi, CNR, e Dip. di Scienze e
Tecnologie Alimentari e Microbiologiche, Università.
P.le delle Cascine, 27, 50144 Firenze - Italia


## 1. INTRODUCTION

Nitrogen-fixing cyanobacteria which establish symbiotic associations
with different plant groups have been identified mainly as Nostoc
species, except for the cyanobionts of Azolla, which generally have
been referred to the genus Anabaena. Nevertheless, on the basis of the
proposal of Rippka (5), the cyanobionts of Azolla as well should be
assigned to the genus Nostoc (6, 7). Except for Azolla, the degree of
specificity between the host plant and the cyanobiont in the
plant-cyanobacteria associations is rather low (1, 2, 4). However the
capability to enter into symbiotic association is not a common
character of the strains of the genus Nostoc. Some properties, such as
facultative photoheterotrophic growth and production of actively
gliding hormogonia, are considered to be important for the
reconstitution of the symbiosis (3, 6). In order to obtain further
information on this matter, we have investigated the main
morphological and physiological characters of several strains isolated
from different plant-cyanobacteria associations.

## 2. MATERIALS AND METHODS

The cyanobacterial strains were isolated and made axenic by standard
microbiological techniques. The morphological properties of the
strains were determined by microscopic observation of cultures growing
in Bo.-11o medium (5). The ability of photoheterotrophic growth was
verified as described by Rippka (5). Phycobiliproteins were extracted
and identified according to Bryant (2).

## 3. RESULTS AND DISCUSSION

All the cyanobacterial strains showed heterocystous immotile filaments and motile hormogonia, typical of the genus <u>Nostoc</u>. Besides, all the strains were capable of heterotrophic growth. The isolates from <u>Azolla</u>, in accordance with the close specificity of the symbiosis, appeared to be very similar in their micromorphological properties, developmental life cycle, assimilation pattern (Table 1) and pigment composition (absence of phycoerythrin). The cyanobionts of <u>Anthoceros</u>, <u>Gunnera</u> and cycads showed several characters in common, as presence of <u>punctiforme</u> stage, marked tendency to produce motile hormogonia, and synthesis of phycoerythrin. Such analogies among these <u>Nostoc</u> isolates should indicate that only some species or strains, endowed with particular properties, are involved in the establishment of these symbioses. However, some differences observed among isolates from the same plant-host species or genus, concerning colony morphology, shape of akinetes, assimilation patterns and PE/PC ratio (Table 1), are a further confirmation of the rather low degree of specificity of these plant-cyanobacteria associations.

Table 1 – Facultative photoheterotrophy and pigment composition in the cyanobacterial symbiotic strains.

| Strain | Host plant | Photoheterotrophic growth using*: | | | | | C-PE/PC ratio** |
| | | Glc | Fru | Suc | Mal | Rib | |
|---|---|---|---|---|---|---|---|
| AaL | *Azolla caroliniana* | − | + | − | − | − | |
| AaN | *Azolla caroliniana* | − | + | − | − | − | |
| AaR | *Azolla filiculoides* | − | + | − | − | − | |
| AaS | *Azolla filiculoides* | − | + | − | − | − | |
| Al1 | *Anthoceros laevis* | + | + | + | + | − | 0,37 |
| Al2 | *Anthoceros laevis* | + | + | + | + | − | 0,31 |
| Al3 | *Anthoceros laevis* | + | + | + | + | (+)*** | 0,40 |
| Cc | *Cycas circinalis* | + | − | (+) | (+) | − | 0,50 |
| Cc2 | *Cycas circinalis* | − | − | (+) | − | (+) | 0,49 |
| Cc3 | *Cycas circinalis* | + | + | − | (+) | (+) | 0,51 |
| Cr1 | *Cycas revoluta* | (+) | + | + | − | + | 0,34 |
| Cr2 | *Cycas revoluta* | + | + | + | + | + | 0,54 |
| Cr3 | *Cycas revoluta* | + | − | (+) | − | (+) | 0,36 |
| Cr4 | *Cycas revoluta* | + | − | + | + | + | 0,54 |
| Cr5 | *Cycas revoluta* | − | + | (+) | − | + | 0,44 |
| Cr6 | *Cycas revoluta* | (+) | + | + | − | + | 0,38 |
| Ef1 | *Encephalartos ferox* | (+) | + | + | − | + | 0,85 |
| Ev1 | *Encephalartos villosus* | + | + | + | + | + | 0,89 |
| M1 | *Macrozamia* sp. | + | + | (+) | (+) | (+) | 0,43 |
| Mm1 | *Macrozamia moorei* | + | + | (+) | (+) | − | 0,46 |
| De1 | *Dioon edule* | + | (+) | − | − | − | 0,73 |
| Ds1 | *Dioon spinolosum* | + | (+) | − | − | (+) | 0,42 |
| Lp1 | *Lepidozamia peroffskyana* | + | + | − | − | + | 0,32 |
| Z1 | *Zamia* sp. | + | + | + | + | + | 0,45 |
| Gm | *Gunnera manicata* | (+) | + | − | − | − | 0,33 |

* Glc, glucose; Fru, fructose; Suc, sucrose; Mal, maltose; Rib, ribose.
** C-PE, C-phycoerythrin; PC, phycocyanin. *** (+), weak growth.

## 4. REFERENCES


1) Bonnett, H.T. and Silvester, W.B. (1981), New Phytol. 89, 121–128.
2) Bryant, D.A. (1982), J.Gen.Microbiol., 128, 835–844.
3) Enderlin, C.S. and Meeks, J.C. (1983), Planta 158, 157–165.
4) Lindblad, P., Haselkorn, R., Bergman, B. and Nierzwicki-Bauer, S.A. (1989), Arch. Microbiol. 152, 20–24.
5) Rippka, R., Deruelles, J., Waterbury, J.B., Herdman, M. and Stanier, R.Y. (1979), J. Gen. Microbiol. 111,1–61.
6) Rippka, R. (1988), Methods Enzym. 167, 28–67.
7) Tomaselli, L., Margheri, M.C., Giovannetti, L., Sili, C. and Carlozzi, P. (1988), Ann. Microbiol. 38,157–161.

# GENETIC ORGANIZATION OF RHODOBACTER SPHAEROIDES CHROMOSOME REGION CONTAINING GENE GLNA


S.Shestakov, Y.Churin, V.Zinchenko and M.Bibikova.
Department of Genetics, Moscow State University, Moscow 119899, USSR.



**ABSTRACT.** The Rhodobacter sphaeroides DNA fragments containing gene glnA have been cloned and investigated. The gene glnB linked with glnA in a single operon have been identified on the basis of hybridization analysis and detection of protein product. There are some indications that gene glnD is located upstream of glnBA-operon and translated in the opposite direction to glnBA.


**INTRODUCTION.** Regulation of nitrogen fixation requires GLNB and GLND proteins which sense and respond to the concentration of fixed nitrogen (Gussin et al., 1986). It was suggested that genes glnB and glnD are involved into the nitrogen-sensing circuit in photosynthetic bacteria (Kranz et al. 1990). In Rhodobacter capsulatus gene glnB is upstream of glnA in one operon. Mutations in glnBA operon lead to glutamine auxotrophy and derepression of nitrogenase (Kranz, Haselkorn, 1988; Kranz et al., 1990). We have obtained the same type of $Gln^-Nif^{con}$ mutants in Rhodobacter sphaeroides (Shestakov et al., 1988). Study of the R.sphaeroides DNA fragments which complement these mutants have revealed two regulatory genes located very close to the gene glnA.

**RESULTS AND DISCUSSION.** DNA fragments containing gene glnA have been cloned from the R.sphaeroides library using a probe with the R.capsulatus gene glnA. Introduction of vector plasmid containing the complete gene glnA or of multicopy plasmid with a part of glnA (390 bp) into R.sphaeroides $Gln^-$ mutant cells restores glutamine synthetase (GS) activity and regulation of nitrogenase by ammonia.

Mini Mu phage has been inserted into DNA fragments containing gene glnA (fig.1). When introduced into $Gln^-$ mutants derivatives, with insertions 600 and 520, did not complement these mutants. Pvu II subfragment (where insertion 600 and 520 are localized) has cloned in a vector of expression. The introduction of this hybrid into the E.coli strain producing mini-cells has provided the synthesis of two polypeptides. The protein of 55 KD corresponds to GS. The existence of the glnBA-operon is suggestive of the polypeptide of 13,5 KD being protein II encoded by the gene glnB. Availability of the gene _glnB_ in the Pvu II subfragment has also been confirmed by hybridization with a DNA probe containing only the _R.capsulatus_ gene _glnB_.

Insertional mutations have been transfered from plasmids into the _R.sphaeroides_ chromosome by interposon technique. Insertions 520 and 600 have resulted in Gln⁻ phenotype. The appearance of "ntr"-like type of mutants caused by insertions 25 and 49 indicated the existence of a new gene located upstream the _glnBA_-operon (fig.1). These mutations lead to a phenotype similar to the _glnD_ mutations: the low level of GS and inability to use some aminoacids for growth. The introduction of SmaI-subfragment of 2,5 kb cloned in pVC19 in particular orientation results in the Gln⁻ phenotype of _E.coli_. This indicate that element revealed could be _glnD_ since the GlnD mutants of _E.coli_ are glutamine auxotrophs. Transcription of the element occurs in the direction opposite to the _glnBA_ transcription. Insertional mutation 25 has been introduced into _E.coli_ chromosome by interposon method. The mutant obtained showed the Gln⁻ phenotype. This mutation has been mapped on the 4 min of _E.coli_ chromosome that corresponds to the location of the _glnD_. The data presented allow a supposition that mutations caused by insertions 25 and 49 lie in the gene _glnD_ This gene is presumably identical to the ORF which is located close to the _glnBA_ operon in the _R.capsulatus_ chromosome and translated in the opposite direction (Kranz _et al._, 1990).

**Figure 1.** Maps of the _Rhodobacter sphaeroides_ chromosome region containing gene _glnA_

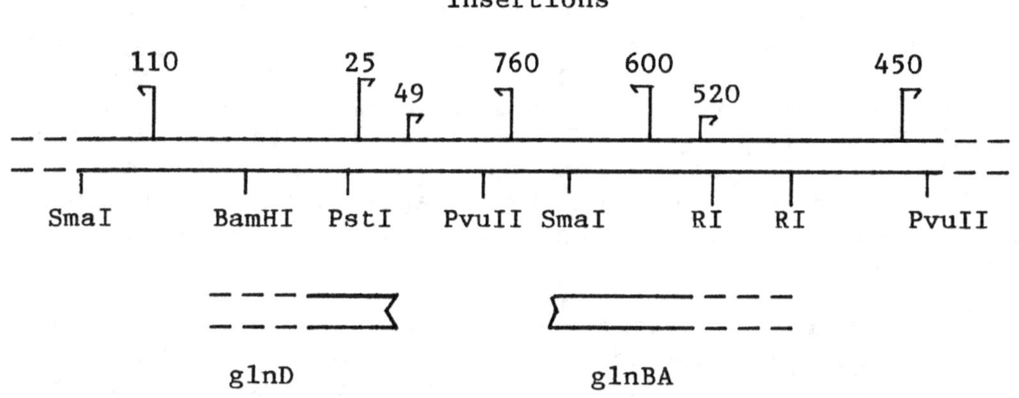


## REFERENCES

1. Gussin G.N., Ronson C.W., Ausbel F.M. (1986). Annu. Rev. Genet. **20**, 567-591
2. Kranz R.G., Maselkorn H. (1988). Gene **71**, 65-74
3. Kranz R.G., Pace V.M., Caldicott I.M. (1990). J.Bacteriol. **172**, 53-62
4. Shestakov S., Zinchenko V. _et al._, (1988) in: Nitrogen Fixation Hundred Years After. Ed. M.Bothe, F.J. de Bruijn, W.E. Newton, Gustav Fisher, Stuttgartt-New York, 163-169.

# FACTORS INVOLVED IN SELECTION OF Nif⁻ MUTANTS IN N-LIMITED CULTURES OF THE PHOTOSYNTHETIC BACTERIUM *Rhodobacter capsulatus*.


J. PIERRARD, J.C. WILLISON and P.M. VIGNAIS
Biochimie Microbienne / DBMS/LBIO
Centre d'Etudes Nucléaires de Grenoble 85 X
38041 Grenoble cedex
France


When *Rhodobacter capsulatus* strain B10 is grown photosynthetically in N-limited continuous culture, spontaneous Nif⁻ mutants accumulate, and eventually ouytgrow the wild-type strain (1). These spontaneous Nif⁻ mutants contain lesions in the *nifR4* regulatory gene, which is homologous to the *ntrA* gene of enteric bacteria. Experiments in mixed chemostat cultures, using defined *nifR4* and *nifH* mutants, have shown that selection is related to the loss of nitrogenase synthesis and activity, and is not due to an additional regulatory effect of the *nifR4* mutations (2). In order to identify the factors responsible for this selection, we have compared in detail the behaviour of the wild-type and a Nif⁻ strain during growth in N-limited chemostat cultures.

At dilution rate (D) of 0.06 h⁻¹, cultures of the mutant strain, AP29, contained about 30% more biomass and twice as many colony forming units as did cultures of the wild-type strain, B10. However, microscopic observation indicated that cultures of both strains contained similar number of cells, with few aggregates or chains. In B10, the cellular content of nitrogenase was estimated to represent 13% of the total cell protein. Loss of nitrogenase synthesis should therefore result in a significant "$NH_4^+$-sparing" effect, with a concomitant increase in cell density. However, a more accurate quantification of cell numbers will be necessary to establish this.

At dilution rate of 0.21 h⁻¹, the rate of selection of Nif⁻ mutants in cultures of B10 was observed to be more than six times lower than at D = 0.06 h⁻¹, as shown in figure 1. In contrast, the cellular nitrogenase content and the rate of nitrogenase-mediated $H_2$ production by the culture, were similar at the two dilution rates (see table 1).

Table 1. Nitrogenase content and $H_2$ production in chemostat cultures of strain B10 grown at different dilution rates

| Dilution rate (D) (h⁻¹) | Nitrogenase content | | | | $H_2$ production |
| | Rc1 | Rc2 | N₂ase | Rc2/Rc1 | |
| | (% of total cell protein) | | | (mol/mol) | (ml/mg dry wt/h) |
|---|---|---|---|---|---|
| 0.057 | 10.1 | 3.6 | 13.7 | 1.24 | 94 |
| 0.205 | 9.5 | 3.1 | 12.6 | 1.19 | 103 |

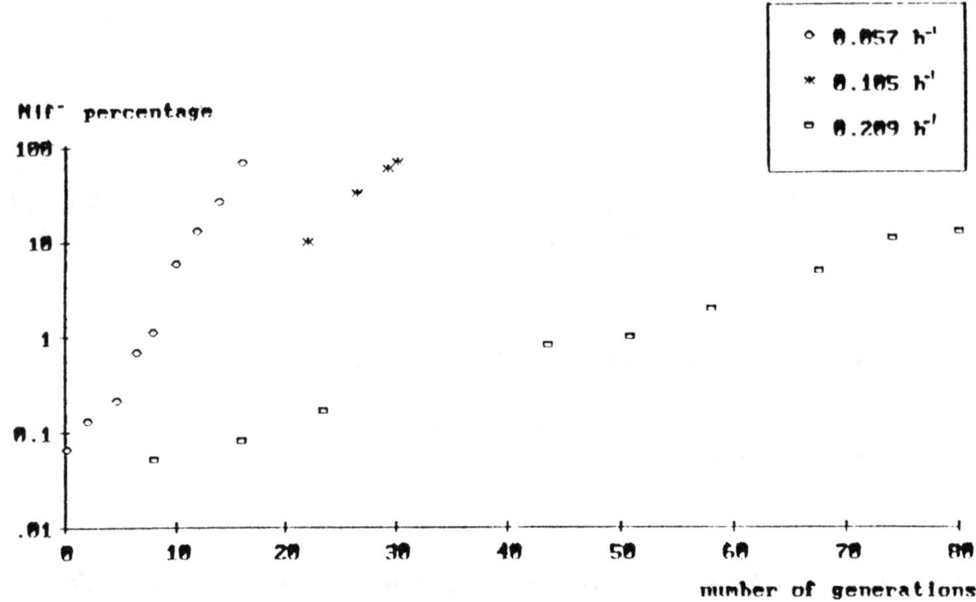



figure 1. Effect of dilution rate on the selection of Nif⁻ mutants.

This suggests that the selection of Nif⁻ mutants may be related in some way to ATP synthesis, since the specific rate of ATP production will be higher at $D = 0.21$ h$^{-1}$ than at $D = 0.06$ h$^{-1}$ (to support the increased growth rate), whereas the rate of ATP hydrolysis linked to the production of $H_2$ is apparently the same (assuming a constant stoichiometry of $ATP/H_2$). At both dilution rates, the concentration of free $NH_4^+$ was below the limits of detection ($< 1$ µM), showing that growth was limited by the supply of N source.
We propose, therefore, that under these conditions, the rate of ATP synthesis is limited by growth rate, so that loss of nitrogenase synthesis and activity results in an "ATP-sparing" effect. This, in turn, might lead to an increase in cell density and/or cell viability, and possibly to an increase in the rate of $NH_4^+$ transport at sub-micromolar concentrations.


(1) Allibert, P., Odom, J.M., Wall, J.D. & Vignais, P.M. (1984) "Phenotypic changes in hydrogen evolving chemostat cultures of Rhodopseudomonas capsulata", FEMS Microbiol. Lett. 23, 221-226.
(2) Willison, J.C. & Lac, B. (1988) "Behaviour of Nif⁻ mutants of Rhodobacter capsulatus in photosynthetic ammonia-limited chemostat cultures" FEMS Microbiol. Lett. 56, 61-66.

GENETIC CHARACTERIZATION OF THE GENE ENCODING FERREDOXIN II FROM THE NITROGEN FIXING PHOTOTROPHIC BACTERIUM *RHODOBACTER CAPSULATUS*

C. DUPORT, Y. JOUANNEAU and P.M. VIGNAIS
Laboratoire de Biochimie Microbienne/LBIO/DBMS
C.E.N.G. 85X
38041 Grenoble cedex, France

ABSTRACT : The structural gene encoding ferredoxin II (*fdxA*) from the photosynthetic bacterium *Rhodobacter capsulatus* was isolated from a cosmid library using two oligonucleotide probes corresponding to the N-terminal sequence of FdII. The nucleotide sequence of *fdxA* was determined. The deduced polypeptide composed of 111 amino acids, has a calculated molecular weight of 12, 540. Sequence comparison indicates that FdII shares common structural features with ferredoxins containing one [3Fe-4S] and one [4Fe-4S] including 9 conserved cysteines. Northern hybridization using a 329 bp DNA fragment internal to the *fdxA* coding region revealed the existence of a mRNA transcript of approximately 400 bases under both *nif*-repressing and *nif*-derepressing growth conditions. The expression of FdII in *R. capsulatus* was studied as a function of changes in the growth conditions by using a *fdxA::lacZ* translational gene fusion. FdII appeared to be constitutively expressed.

## Introduction

Three ferredoxins have been so far isolated from the photosynthetic bacterium *Rhodobacter capsulatus*. Ferredoxin I (FdI) and FdIII are synthetized only under nitrogen-fixing growth conditions whereas FdII appeared to be constitutively expressed [1]. The nucleotide sequence of the genes coding for FdI (*fdxN*) and for FdIII (*fdxB*) has been previously reported [2, 3]. We now report the molecular cloning, nucleotide sequence determination and transcriptional analysis of the gene *fdxA* coding for FdII.

## Results and Discussion

The structural gene for ferredoxin II (FdII), *fdxA*, has been isolated from a genomic library of *R. capsulatus* DNA consisting of 40 kb *BamH*I fragments. Two sets of 20-mer oligonucleotides corresponding to two regions of the N-terminal sequence of FdII were synthetized and used as hybridization probes to identify a clone containing the coding sequence of *fdxA*. From this clone, a 3.6 kb *Hind*III fragment containing sequence complementary to both probes was subcloned into pUC19 to give pCD5.

The nucleotide sequence of *fdxA* was determined using pCD5 and oligonucleotide probes as primers. 649 bp were sequenced, 336 bp of which composed the *fdxA* gene [4]. Sequence analysis showed a motif similar to the *E. coli* -10/-35 promoter 33 bp upstream of the initial ATG, and a putative stem and loop structure 33 bp downstream of the last *fdxA* codon. The deduced amino acid sequence was found to be in accordance with the N-terminal sequence  of FdII, except for the initial methionine. The FdII polypeptide composed of 111 residues had a theoritical molecular weight of 12, 540, in good agreement with the value experimentally determined for the purified protein [5]. The amino acid sequence of FdII was found to exhibit significant similarity with all published 7Fe ferredoxin sequences from other bacteria. FdII appeared to be 65 % homologous to *Azotobacter vinelandii* FdI, the gene of which was called fdxA [6] ; we therefore proposed to give to the *R. capsulatus* FdII gene the same name.

Northern blot analysis revealed a single mRNA species of approximately 400 bases homologous to a 329 bp *fdxA* specific probe under both *nif*-repressing and *nif*-derepressing conditions. The size of the transcript suggested that *fdxA* was transcribed as a monocistronic message.

The expression of *fdxA* in *R. capsulatus* was measured by using an extrachromosomal *fdxA-lacZ* translational gene fusion [7]. In *R. capsulatus* wild type, *fdxA* appeared to be expressed at a constant level throughout growth in rich or minimal media under light aerobic or dark anaerobic conditions suggesting that FdII was constitutively synthetized. Nevertheless, in anaerobic nitrogen-limited cultures, expression of *fdxA* appeared to be affected upon nitrogenase derepression. Since this effect was not seen in a *nifHDK* mutant strain, we assumed that the decrease of the *fdxA* transcription was somehow linked to the synthesis of nitrogenase.


## References


1. Jouanneau, Y., Grabau, C., Schatt, E., Duport, C., Meyer, C., and Vignais, P.M. (1990) In : Gnesshoff Roth, Stacey and Newton (eds). Nitrogen fixation : Achievements and objectives. Chapman and Hall, New York, London (in Press).
2. Schatt, E., Jouanneau, Y., and Vignais, P.M. (1989) J. Bacteriol. 171, 6218-6226.
3. Moreno-Vivian, C., Hennecke, S., Pühler, A., and Klipp, W. (1989) J. Bacteriol. 171, 2591-2598.
4. Duport, C., Jouanneau, Y., and Vignais, P.M. (1990) Nucleic Acids Res. 18, 4618.
5. Jouanneau, Y., Meyer, C., Gaillard, J., and Vignais, P.M. (1990) Biochem. Biophys. Res Commun (in press).
6. Morgan, T.V., Lundell, D.J., and Burges, B.K. (1988) J. Biol. Chem. 263, 1370-1375
7. Duport, C., Jouanneau, Y., and Vignais, P.M. (1990) Mol. Gen. Genet. (submitted).

# DNA SEQUENCE OF A NEW FERREDOXIN GENE MAPPING IN A *nif* REGION OF *RHODOBACTER CAPSULATUS*, AND EXPRESSION IN *ESCHERICHIA COLI*


C. GRABAU, E. SCHATT, Y. JOUANNEAU, AND P. M. VIGNAIS
Laboratoire de Biochimie Microbienne/LBIO/DBMS
CENG 85X
38041 Grenoble cedex France



ABSTRACT   We have sequenced and identified a new ORF (*fdxC*) upstream of one of the ferredoxin genes (*fdxN*) of *Rhodobacter capsulatus*. The deduced amino acid sequence is similar to plant-type [2Fe-2S] ferredoxins and exhibits conserved spacing of the ligand-binding cysteine residues. Northern blot analysis showed that the expression of *fdxC* and *fdxN*, like *nif* genes, is repressed by the presence of ammonium in the growth medium. To study the *fdxC* gene product, we cloned a DNA fragment carrying both the *fdxC* and *fdxN* genes into the vector pUC18, downstream of the *lacZ* promoter. *E. coli* cells harbouring this plasmid contained a brown protein corresponding to the *fdxN* gene product and a red protein corresponding to the *fdxC* gene product. This indicates that the *R. capsulatus* genes are expressed in *E. coli* and that *E. coli* is able to synthesize and insert both the [2Fe-2S] and [4Fe-4S] clusters in the proteins correctly.


## Introduction

*R. capsulatus* is a purple non-sulfur photosynthetic bacterium capable of fixing atmospheric nitrogen. Two ferredoxins from this organism have been isolated which may play a role in nitrogen fixation (a third ferredoxin has been isolated but its role, if any, in nitrogen fixation is unknown)(1). Both ferredoxin genes map near each other in *nif* gene clusters on the chromosome (2,3). One gene, *fdxB* (encoding FdIII)(1), has been identified as an open reading frame (ORF5) in the *nifENXQ* operon, and may play a role in the synthesis of the iron-molybdenum cofactor (3). The other Fd gene, *fdxN* (encoding FdI), is located upstream of the *nifENXQ* operon (2). Although the role of FdI is not known, it has been shown to catalyze the *in vitro* electron transfer to nitrogenase (4) and therefore has been implicated in nitrogen fixation. Further suggestion that FdI may play a role in nitrogen fixation comes from transcriptional analysis; *fdxN* mRNA transcripts were found only in cells grown under *nif*-derepressing conditions (2). We have recently identified a new ferredoxin gene that is located immediately upstream of the *fdxN* gene (5). We report here the sequence and characterization of this gene, which we call *fdxC*, and show that it encodes a fourth ferredoxin of *R. capsulatus*.

## Results and Discussion

We have identified a new ORF (*fdxC*) upstream of the *fdxN* gene (which encodes FdI). The gene is 285 bp in length and encodes a 95 amino acid polypeptide with a predicted Mr of 10,156. A potential ribosome binding site was found upstream of the gene. No upstream sequence was found which matched either the *nif* (sigma$^{54}$-dependent) promoter or the consensus *E. coli* promoter. The deduced amino acid sequence contains 5 cysteine residues, four of which exhibit spacing similar to plant-type [2Fe-2S] ferredoxins. Comparison to plant-type ferredoxins shows approximately 25% amino acid sequence similarity.

Expression of the 2 ferredoxin genes in *E. coli* was accomplished by subcloning them in vector pUC18 under control of the *lac* promoter. Extracts of cells carrying the *fdxCN* plasmids contained a brown protein, which was shown to correspond to the *fdxN* gene product (FdI), and a red protein corresponding to the *fdxC* gene product (FdIV). EPR and UV-visible spectroscopy confirmed that the FdxC protein contained a [2Fe-2S] cluster and the FdxN protein contained two [4Fe-4S] clusters, and that the centers were correctly assembled and inserted in the ferredoxins expressed in *E. coli*.

Northern (mRNA) blot analysis showed that *fdxC* and *fdxN* expression was derepressed in cells grown under conditions of nitrogen limitation and repressed when excess ammonium ions were present in the growth medium. Thus, the *fdxC* and *fdxN* genes appear to be regulated like *nif* genes. The absence of identifiable *fdxC* and *fdxN* promoters, and the location of the genes in a region of the chromosome where *nif* genes map, suggest that both genes may be part of a larger operon, perhaps including *nif* genes. Indeed, we have found another ORF immediately upstream of the *fdxC* gene, and others (6) have generated *nif* mutations both upstream and downstream of the region where we have found the *fdxC* and *fdxN* genes to map.

## References


1. Jouanneau, Y., Grabau, C., Schatt, E., Duport, C., Meyer, C., and Vignais, P. M. (1990) In Gresshoff, Roth, Stacey, and Newton (eds.), Nitrogen Fixation: Achievements and Objectives, Chapman and Hall, New York, London.
2. Schatt, E., Jouanneau, Y., and Vignais, P. M. (1989) J. Bacteriol. 171, 6218-6226.
3. Moreno-Vivian, C., Hennecke, S., Pühler, A., and Klipp, W. (1989) J. Bacteriol. 171, 2591-2598.
4. Hallenbeck, P. C., Jouanneau, Y., and Vignais, P. M. (1982) Biochim. Biophys. Acta 681, 168-176.
5. Grabau, C., Schatt, E., Jouanneau, Y., and Vignais, P. M. (1990) submitted.
6. Klipp, W., Masepohl, B., and Pühler, A. (1988) J. Bacteriol. 170, 693-699.

# ANALYSIS OF DNA RESTRICTION PATTERN IN THE GENUS *ECTOTHIORHODOSPIRA*


A. GORI, L. GIOVANNETTI, S. VENTURA, C. VITI and R. MATERASSI
*Dip. di Scienze e Tecnologie Alimentari e Microbiologiche Univ. Firenze and Centro di Studio dei Microrganismi Autotrofi CNR*
*Piazzale delle Cascine, 27*
*I-50144 FIRENZE*


## Introduction

The restriction pattern of total DNA is a widely diffused method for grouping strictly related microorganisms mainly for medical purpose. An improvement of this method, which consists in the use of SDS-PAGE and silver staining for pattern revelation and in a numerical analysis of data clearly showed that the restriction pattern constitutes a unique fingerprint of each strain that could be used to identify clusters at about the species level (Degli-Innocenti *et al.*, 1990; Giovannetti *et al.*, 1990). In the present work this method was applied to two strains of nitrogen-fixing phototrophic bacteria, EST4 and EST8, isolated from a saltern basin which were attributed to the genus *Ectothiorhodospira* but whose species assignement was impossible with a detailed phenotypical study (Ventura *et al.*, 1988).

## Materials and Methods

Total DNA was extracted from the twelve red *Ectothiorhodospira* strains reported in Figure 1. After purification with CTAB (trimethylcetylammonium bromide), all DNAs were digested with restriction endonuclease *Bgl*II and the fragments were separated with SDS-PAGE and stained with silver nitrate. Boehringer DNA molecular weight marker VI was added at each sample to be used in the rescaling procedure. Traces were scanned and peaks were detected with a LKB Ultroscan laser densitometer connected to a personal computer. A dedicated program was used to rescale band patterns and to calculate Dice similarity coefficients ($S_D$) between pairs of patterns. Strains were clustered by the UPGMA method using the SAS package.

## Results and Discussion

The analysis of the restriction profiles obtained with SDS-PAGE showed that each strain has a unique band pattern which constitutes its genomic fingerprint. The matrix of $S_D$ coefficients, analysed with the UPGMA, gave the dendrogram shown in Figure 1. Four aggregate clusters plus two single member clusters were found. Each of clusters 1 and 3 matches the species level for *E.*

582

*halophila* and *E. vacuolata*, while cluster 2 contains two strains of *E. mobilis* plus the type strain of *E. shaposhnikovii*, thus confirming the relatedness of these two species. The position of *E. mobilis* DSM 237 cannot be easily explained with the present data. It is worthwhile to note that species with a low salt requirement housed in clusters 2 and 3 merge at a $S_D$ value slightly minor than the cut-off one. Strain EST8 forms cluster 4 with *E. marismortui*: the good degree of genomic similarity between these two strains is in agreement with lipid data obtained by B. Tindall (personal communication). The restriction profile of strain EST4, which has a low correlation with the others, placed this strain far from all the red *Ectothiorhodospira* species described at present, in accord with previous phenotypic data.

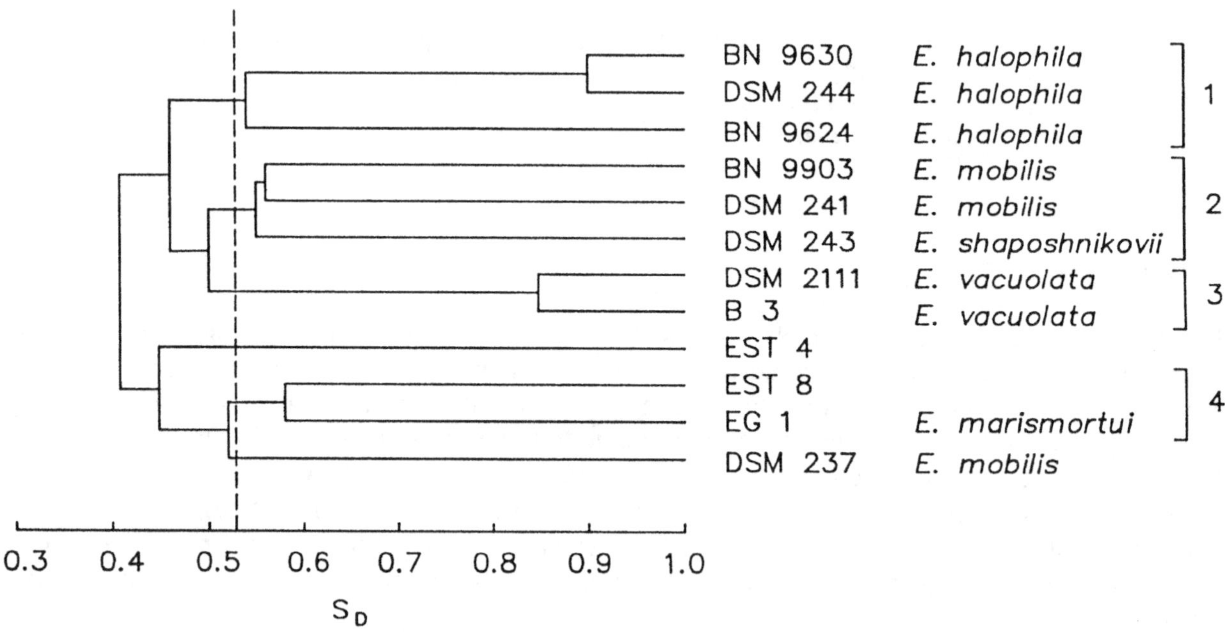


Figure 1. UPGMA clustering of the $S_D$ matrix obtained from *Bgl*II pattern analysis.

# References


Degli-Innocenti F., Ferdani E., Pesenti-Barili B., Dani M., Giovannetti L. and Ventura S. (1990). J. Biotechnol. 13, 335-346.

Giovannetti L., Ventura S., Bazzicalupo M., Fani R. and Materassi R. (1990). J. Gen. Microbiol. 136, 1161-1166.

Ventura S., De Philippis R., Materassi R., and Balloni W. (1988). Arch. Microbiol. 149, 273-279.

EFFECT OF Mo, V AND W ON THE GROWTH AND NITROGENASE SYNTHESIS IN PHOTOTROPHIC BACTERIA



A.F. Yakunin, A.A. Tsygankov, I.N. Gogotov and M. Jha
Institute of Soil Science and Photosynthesis
USSR Academy of Sciences, Pushino
142292 Moscow Region, USSR




ABSTRACT. The influence of Mo, V and W on the growth and nitrogenase synthesis by purple bacteria Rhodobacter capsulatus, R. sphaeroides and cyanobacteria Anabaena variabilis and A. azollae have been studied. Obtained data indicate that these microorganisms excluding R. sphaeroides are capable of synthesizing alternative nitrogenase(s).



INTRODUCTION
Chemotrophic bacteria from genus Azotobacter are capable to synthesize two alternative nitrogenases (containing and not containing V) besides conventional Mo containing enzyme (1). But there is only one communication about occurence of alternative nitrogenase in phototrophic microorganism (2). It is known that in purple bacteria Rhodobacter capsulatus sequences homologous to the nitrogenase genes are present in multiple copies (3). In this work the influence of Mo, V and W on the growth and nitrogenase synthesis by several phototrophic bacteria has been studied.


MATERIAL AND METHODS
Turbidostat cultures of purple bacteria Rhodobacter capsulatus B10 and R. sphaeroides 2R and batch cultures of cyanobacteria Anabaena variabilis ATCC 29413 and A. azollae (from Azolla caroliniana) were grown as described previously (4, 5) with dinitrogen as nitrogen source. Nitrogenase activity (production of ethylene, ethane and hydrogen) was measured by gas chromatography (5).


RESULTS AND DISCUSSION
The rate of acetylene reduction by Mo-grown cultures of R. capsulatus, R. sphaeroides, A. variabilis and A. azollae was at least two fold higher than that of hydrogen evolution. The reduction of acetylene to ethylene was not accompanied with production of ethane. It indicates

the existance of Mo-containing nitrogenase in these cells. The growth rate and nitrogenase activity of dinitrogen fixing cultures of these microorganisms decreased significantly after exclusion of Mo from the medium. The ratio of rates of ethylene production and hydrogen evolution by Mo deficient cultures of R. sphaeroides remained unchanged. Vanadium inhibited growth and nitrogenase activity of Mo deficient cultures of this bacterium, which did not produce ethane during acetylene reduction. Mo deficiency did not change the ratio of the ratio of ethylene production and hydrogen evolution by cells of A. variabilis and A. azollae. The ratio of hydrogen evolution by R. capsulatus cells remained at the previous level, but the rate of ethylene production was reduced significantly by Mo limitation. Mo deficient cultures of A. variabilis, A. azollae and R. capsulatus produced ethane (less than 3% of the amount of ethylene) during acetylene reduction. Vanadium stimulated the growth and nitrogenase synthesis by Mo deficient cells of these cyanobacteria but had no effect in the case of R. capsulatus. The process of hydrogen evolution by Mo deficient and V grown cultures of these microorganisms had been more resistant to inhibitory effect of acetylene than in Mo grown cultures. Ethane producing capacity and the resistance of hydrogen evolution to inhibition by acetylene disappeared after Mo addition in the medium. Tungstate inhibited growth and nitrogenase synthesis by cyanobacterial cells under Mo suficient conditions stronger than in Mo deficient conditions.

Obtained data indicate that R. capsulatus, A. variabilis and A. azollae but not R. sphaeroides may synthesize alternative nitrogenases in Mo deficient conditions. We suppose that R. capsulatus is capable to synthesize only second alternative nitrogenase (not containing Mo and V) because vanadium did not stimulate growth and nitrogenase synthesis by Mo deficient cultures. In contrast A. variabilis and A. azollae may synthesize both alternative nitrogenases.

REFERENCES


1) Goerger, R.D. and Bishop, P.E. (1988), CRC Critical Rev. Microbiol. 16, 1-14.
2) Kentemich, T., Danneberg, G., Hundeshagen, B. and Bothe, H. (1988), FEMS Microbiol. Lett. 51, 19-24.
3) Scolnik, P.A. and Haselkorn, R. (1984), Nature, 307, 289-292.
4) Yakunin, A.F., Tsygankov, A.A. and Gogotov, I.N. (1989), Proc. USSR Academy Sci. 281, 482-485.
5) Yakunin, A.F., Chan Van Ni and Gogotov, I.N. (1989), Proc. Acad. Sci. 307, 1269-1271.

# NITROGEN METABOLISM IN *RHODOBACTER CAPSULATUS*


R. BORGHESE AND J.D. WALL
Biochemistry Department
117 Schweitzer Hall
University of Missouri-Columbia
Columbia, MO 65211  -  USA


## ABSTRACT


A new class of *R. capsulatus* mutants was isolated that lacks GOGAT but grows well on $N_2$. Because this species has no detectable GDH activity, this finding suggests that an unidentified ammonium assimilatory pathway may exist.


## INTRODUCTION

Assimilation of inorganic nitrogen in prokaryotes has been shown to proceed via the action of glutamate dehydrogenase (GDH) and/or  the coupled action of glutamine synthetase (GS) and glutamate synthase (GOGAT).  The phototrophic diazotroph *Rhodobacter capsulatus* possesses only the GS/GOGAT pathway. Extrapolating from the observations with enteric bacteria, it would be inferred that GS or GOGAT mutants should be auxotrophs for glutamine or glutamate, respectively.  While glutamine auxotrophs have been isolated, have not been described bona fide glutamate auxotrophs in this species.  Here preliminary evidence is presented for the operation of an alternative to GOGAT for the assimilation of ammonium in *R. capsulatus*.

## MATERIALS AND METHODS

Enzymatic assays, ammonium and methylammonium (MA) uptake, and B-galactosidase assays followed published procedures (Johansson and Gest, 1976; Rapp et al., 1986; Miller, 1972; respectively).

## RESULTS

A series of mutants (e.g. ADI-1 and ADI-5) were isolated, after EMS mutagenesis of the wild-type strain B100, that grew more slowly with ammonium as the nitrogen source under aerobic conditions.  In contrast, a number of amino acids, including glutamine and glutamate, supported rapid aerobic growth.  Under photosynthetic/anaerobic conditions these mutants grew close to or at wild-type rates with all the nitrogen sources tested including ammonium and dinitrogen.

Ammonium assimilatory enzymes were assayed in crude extracts prepared from the mutants grown under aerobic or anaerobic/photosynthetic conditions.  GS was found at wild-type levels while GOGAT was always undetectable, regardless of the growth conditions.  Alanine dehydrogenase (ADH), a candidate for ammonium assimilation under certain nutritional conditions (Caballero et al.,1989), showed activities that were  similar or lower as compared to the wild-type, suggesting that ADH does not compensate for the loss of GOGAT activity in this strain.

Ammonium uptake rates, an indication of the total ammonium assimilatory capacity of the cells, were consistent with the growth rate data for the wild-type and mutants grown on ammonium.  Under aerobic conditions, the mutants showed  uptake rates that were about 20% of the wild-type; whereas under anaerobic/photosynthetic and nitrogen-fixing conditions, rates that were 55% and 75% compared to the wild-type strain B100 were obtained.

To determine whether ADI-1 and ADI-5 were altered in their global nitrogen regulatory response, they were also assayed for MA (an ammonium analog) uptake and for their ability to derepress a *nifH::lacZ* fusion.  MA uptake is regulated by a two component signalling system analogous to NtrBC in enteric bacteria; i.e., *R.capsulatus* NRII (NtrC) mutants show no MA uptake under *nif*-derepressing conditions.  ADI-1 and ADI-5 grown photosynthetically either on $N_2$ or glutamate, showed MA uptake levels that were between 70% and 90% of the wild-type.  Wild-type cells grown aerobically on glutamate, also nitrogen-derepressing conditions, showed uptake values similar to photosynthetically  *nif*-derepressed cells.  In contrast, the mutants grown under aerobic conditions had no measurable MA uptake.  Finally a *nifH::lacZ* fusion was introduced into the strains.  The B-galactosidase activities  were the same for all the strains under both nitrogen-repressing and nitrogen-derepressing conditions, indicating no impairment in the pathway that regulates the expression of the *nif* genes.

## DISCUSSION

The *R. capsulatus* phenotype described here suggests that an ammonium assimilatory pathway other than GS and GOGAT may exist in this species.  This pathway apparently operates more efficiently under anaerobic nitrogen-fixing conditions and would account for the almost normal growth rate of these mutants on $N_2$ in the absence of GOGAT. Under aerobic conditions either this putative pathway is operating at low level or it is not functioning and other secondary routes (e.g. ADH) may slowly assimilate ammonium.  The fact that these mutants show a pattern of MA uptake different from the wild-type suggests that either some regulatory component of nitrogen metabolism has also been mutated or that the free ammonium pools may be increased by slower assimilation.  Because *nifH::lacZ* expression appeared unaffected, any regulatory feature altered must be specific to uptake functions. Alternatively MA uptake is probably a more sensitive indicator of ammonium pools than is *nif* expression since ammonium is a competitive inhibitor and the preferred substrate for uptake.

In conclusion, mutants of *R. capsulatus* have been isolated that lack GOGAT activity yet grow quite well with $N_2$ as sole nitrogen source.  An alternative route for glutamate synthesis must be present.

## REFERENCES

Caballero F.J., Cardenas J. and Castillo F. (1989)  Purification and properties of L-alanine dehydrogenase of the phototrophic bacterium *Rhodobacter capsulatus* E1F1. J. Bacteriol. **171**:3205-3210

Johansson B.C. and Gest H. (1976)  Inorganic nitrogen assimilation by the photosynthetic bacterium *Rhodopseudomonas capsulata*. J.Bacteriol. **128**:683-688

Miller J. (1972)  Experiments in molecular genetics. Cold Spring Harbor Laboratory. Cold Spring Harbor, NY

Rapp B.J., Landrum D.C. and Wall J.D. (1986)  Methylammonium uptake by *Rhodobacter capsulatus*. Arch. Microbiol. **146**:134-141

PURIFICATION OF A PYRUVATE: OXIDOREDUCTASE FROM NITROGEN FIXING
RHODOSPIRILLUM RUBRUM

ERICA BROSTEDT AND STEFAN NORDLUND
Department of Biochemistry
Arrhenius Laboratories
Stockholm University
S-106 91 Stockholm
Sweden

## ABSTRACT

We have purified and partially characterized a pyruvate:oxidoreductase
from Rsp. rubrum. We have also been able to reconstitute the electron
transport from pyruvate to nitrogenase in presence of a crude preparation of ferredoxin.

## Introduction

In Klebsiella pneumoniae the proteins transferring electrons from an
organic electron donor (pyruvate) to nitrogenase have been identified
as pyruvate: flavodoxin: oxidoreductase (nifJ gene product) and flavodoxin (nifF gene product) [1]. It is, so far, the only diazotroph
where the physiological electron carriers are known in detail.
Pyruvate has earlier been shown to support low nitrogenase activity in
Rsp. rubrum extracts [2].

## Materials and Methods

Pyruvate: oxidoreductase activity was measured spectrophotometrically
as methyl viologen reduction at room temperature [3]. 1U = 1micromole
methyl viologen reduced/min. Nitrogenase activity was measured as ace-
thylene reduction. Iron was determined colorimetrically [4].

Rsp. rubrum was grown photoheterotrophically in Ormerod's medium with
molecular nitrogen. The cells were suspended in 0.1 M Tris-HCl pH 7.8
with 1 mM pyruvate, 1 mM $MgCl_2$, 1 mM TPP, 0.5 mM dithionite, 2 mM
dithiothreitol and 0.43 mM PMSF and broken in a Ribi cell fractionator.
The broken cells were centrifuged at 20 000 g for 25 min. and the
supernatant was centrifuged at 170 000 g for 90 min. to give the cell
extract used for purification. The purification was performed anaerobically and all solutions were evacuated and Ar flushed several cycles
25 mM Tris-HCl pH 7.5 with 1 mM $MgCl_2$, 1 mM TPP, 2mM dithiothreitol
and 0.25 mM dithionite was the buffer used throughout the purification.

1. Q-Sepharose Fast Flow, equilibrated in 0.15 M NaCl in buffer. The enzyme was eluted with 0.25 M NaCl in buffer. The active fractions were pooled and concentrated by ultrafiltration.
2. Sephacryl S-300 Fast Flow, equilibrated and eluted with 0.17 M NaCl in buffer.
3. Red Sepharose CL-6B, equilibrated with 0.2 M NaCl in buffer, through which the active, pooled fractions were passed. This "eluate" was pumped directly on the step 4 column.
4. Q-Sepharose Fast Flow, equilibrated with 0.2 M NaCl in buffer and eluted with a gradient of 0.2 to 0.3 M NaCl in buffer.

## Results and Discussion

The purification method described here resulted in a 50 fold purification and 4% recovery in comparison with the cell extract. In spite of the high specific activity (15 U/mg) of the enzyme, SDS-PAGE shows some impurities. The molecular weight of the native enzyme was estimated to 250 000  Da using gel filtration. SDS-PAGE showed a molecular weight of 120 000  Da/subunit. The iron content was estimated to 9-11 nmoles iron/nmole enzyme. The ability of the purified enzyme to transfer electrons from pyruvate to nitrogenase was investigated by substituting dithionite in the standard nitrogenase assay for pyruvate, pyruvate:oxidoreductase and a crude preparation of ferredoxin. The activity obtained was about 45% of that with dithionite. In absence of CoA or pyruvate the activity was almost completely lost (1% remaining). $K_m$ for CoA and pyruvate were estimated to 8.3 and 156 µM, respectively. Further studies are clearly needed to determine which ferredoxin is most active and to optimize the ratios between the enzymes.

## Acknowledgements

This work was supported by grants from the Swedish Natural Sciences Research Council to Stefan Nordlund.

## References

1. Shah, V.K., Stacey, G. and Brill, W.J. (1983) J. Biol. Chem. 258, 12064-12068.
2. Ludden, P.W. and Burris, R.H. (1981) Arch. Microbiol. 130, 155-158.
3. Wahl, R.C. and Orme-Johnson, W.H. (1987) J. Biol. Chem. 262, 10489-10496.
4. Fish, W.W. (1988) Methods in Enzymol. 27, 357-364.

# DEREPRESSION OF NITROGENASE BY 2-KETOGLUTARATE IN A GLUTAMATE AUXOTROPH OF RHODOBACTER SPHAEROIDES


H.Y.SONG, Y.Q.WU, X.WANG AND J.H.SUN
Shanghai Institute of Plant Physiology
Academia Sinica
300 Fenglin Road, Shanghai 200032
People's Republic of China


## 1. A Glutamate Synthase Mutant Lacking Nitrogenase Acativity in R. sphaeroides

A glutamate auxotroph was isolated from R. sphaeoides. Biochemical analysis revealed that the mutant was deficient in glutamate synthase (GOGAT) activity. In comparison with parent strain the mutant had a low activity of glutamine synthetase (GS), but changed to a much higher level of adenylylation state. The mutant was unable to grow with respect to the utilization of a wild range of nitrogen sources, such as proline, histidine, arginine, serine, adenine, etc. Measurements of intracellular amino acid pool showed that the glutamine content of mutant was much higher than in the parent strain. It seems that a highly responsive feedback system was operated for modulation of nitrogenase in response to the changes in pool of metabolic intermediates, and the intracellular level of glutamine was a key of the feedback modulation of nitrogenase synthesis. Nitrogenase was found in GOGAT mutant of Rhodospirillum rubrum (1), however, we were unable to detect the nitrogenase in GOGAT mutant of R.sphaeroides, which failed to fully derepress GS. In view of these facts, there seem to be two possiblilities: The loss of GOGAT activity increases the glutamin pool which inhibits the activity and the deadenylylation of GS. The low level of unadenylylated GS, inturn, represses nitrogenase. Alternatively, GOGAT itself affects the regulation of either GS or a nitrogen control factor, or the asm mutations affect a commom regulator needed for both GS and GOGAT function.

## 2. Derepression of Nitrogenase in a Glutamate Auxotroph of R.sphaeroides

Nitrogenase was derepressed by the treatment with methionine-DL-sulfoximine (MSX) in the GOGAT mutant of R. sphaeroides, which, as mentioned above, exhibits a Nif⁻ phenotype. However, the pleiotropic nature of the mutant remained unchanged. The derepression of nitrogenase by MSX could be repressed by glutamine, showing that glutamine was a factor triggering nitrogenase repression. On the other hand, the nitrogenase of the GOGAT mutant can also be derepressed by 2-ketoglutarate with the level of enzyme proportional to the concentration of 2-ketoglutarate. Analysis of glutamine pool indicates that the mutant accumulates glutamine. Incubation of the mutant in N-limited medium lowered the glutamine pool. But at the moment when the nitrogenase was derepressed, the glutamine in the mutant was still double that of the wild strain. It was proposed that not the absolute level of glutamine, but the ratio of glutamine to 2-ketoglutarate modulates the derepression of nitrogenase in the GOGAT mutant of R. sphaeroides.

## 3. Reciprocal Effects of Glutamine and 2-Ketoglutarate on Adenylylation State of GS in R. sphaeroides

It is noteworthy that in R.sphaeroides the state of adenylylation of GS (n value) varied with the changes in the concentration of metabolites that are either the substrate (ATP, UTP, Pi) or the allosteric effector (2-ketoglutarate, glutamine) of cascade enzymes (2). UTP, the substrate of UTase, affected the n value of GS, thus proving the interconversion of $P_{II}$-UMP and $P_{II}$ proteins and the existence of UTase (2). Variation of the n value of GS with increasing concentration of glutamine at different fixed concentrations of 2-ketoglutarate and vice versa were observed. The reciprocal effects of glutamine and 2-ketoglutarate were identified. It was a ratio of glutamine, not the absolute concentration of those effectors determining the state of GS in vivo. It was proposed that the synthesis of nitrogenase in Klebsiella pneumoniae be transmitted by the bicyclic cascade control system of GS and, more precisely, by its $P_{II}$ regulatory protein (3). The intracellular signal, the ratio of glutamine to 2-ketoglutarate, which is generated by GS in response to the enviromental stimulus (4). Here a problem arise as to which component receives the signal of glutamine/2-ketoglutarate ratio in the GOGAT mutant of R. sphaeroides. Is it the UTase as proposed for E. Coli? To answer these questions, a detailed study of the cascade control system of GS in photosynthetic bacteria is needed.

## 4. References


1. Weare NM. (1978) Biochim Biophys Acta 502:486-494
2. Wang X, Song HY. (1989) Science in China (Series B) 32:960-969
3. Gussin GN, Ronson CW, Ausubel FM. (1986) Ann Rev Genet 20:567-591
4. Magasanik B. (1989) Biochimie 71:1005-1012

OUTDOOR HYDROGEN PHOTOEVOLUTION BY IMMOBILIZED
CELLS OF RHODOPSEUDOMONAS PALUSTRIS


G. TASSINATO, E. CAPOLINO, C. SILI and M. VINCENZINI
Dipartimento di Scienze e Tecnologie Alimentari e
Microbiologiche - Università - e Centro di Studio dei
Microrganismi Autotrofi - CNR.
P.le delle Cascine, 27, I - 50144 Firenze, Italy



ABSTRACT. $H_2$ photoevolution by immobilized cells of Rhodopseudomonas palustris was assayed outdoors with a bioreactor covering a surface of 1.8 $m^2$. The $H_2$-evolving activity of the system was found to be unstable, about 90% of the initial activity being lost after 30 days.


1. INTRODUCTION

Yield and duration of $H_2$ photoevolution by purple non-sulphur bacteria, frequently investigated in recent years (Vincenzini et al., 1982 a, 1982 b, 1986), were highly improved by whole cell immobilization techniques so that the use of these microorganisms for a large scale production of $H_2$ has been widely suggested. However, the feasibility of the process under outdoor conditions has been not yet extensively verified. The performances of a photobioreactor of significant dimensions for outdoor hydrogen production by immobilized bacterial cells are here described.

2. MATERIALS AND METHODS

Rhodopseudomonas palustris 42 OL was entrappped in agar beads. The beads, manufactured at a low cell density, were maintained in growth medium in order to cause cell proliferation inside the matrix. This preincubation lasted until a homogenous red colour of the beads was achieved. The photobioreactor was made from plexiglass sheets and consists of eight modules (55 x 27 x 5 cm) in a container (covering a surface of 1.8 $m^2$) provided with a system for circulation of cooled air. Diffusers for $N_2$ bubbling, a temperature sensor, a gas outlet and an inlet/outlet for substrate solution are present in each module. Under operative conditions, each module contained 2.2 l of agar beads and 4.8 l of substrate solution (30 mM malate). The photobioreactor was

592

operated avoiding, by the use of a shading net, incident light intensities too high and reactor overheating. Nitrogen gas was bubbled into the modules before each dark period; substrate solution was substituted every other day.

## 3. RESULTS AND DISCUSSION

Hydrogen production reached the highest value ($7.3 \, l \, H_2 \, m^{-2}$) during the second day, with a maximum evolution rate of about $1.6 \, l \, H_2 \, m^{-2} \, h^{-1}$, and then decreased markedly reaching a steady state where the volume of the gas daily produced maintained in the range $2.5$-$3.5 \, l \, H_2 \, m^{-2}$ for about 15 days. After this period the activity of the system decreased again and only $0.7 \, l \, H_2 \, m^{-2}$ were obtained during the 30th day. Typical daily kinetics of hydrogen photoevolution, shown in Fig.1, was characterized by an initial lag period which was the more long the lower the total amount of $H_2$ daily produced was. The highest $H_2$ production rates occurred at high light intensities with temperatures

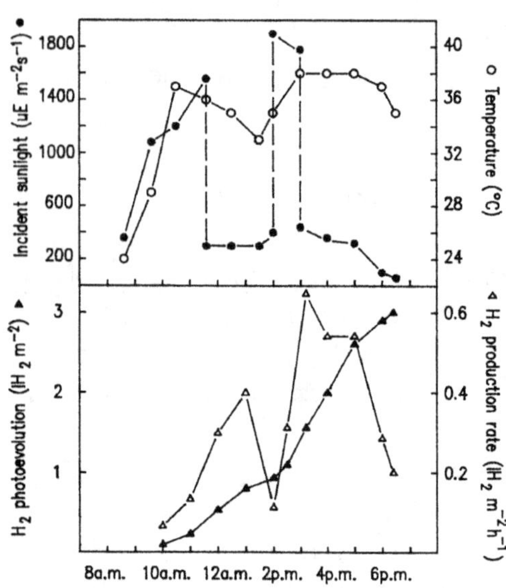


Fig. 1— Time course of H$_2$ photoevolution

of 32-38°C. The results pointed out that, under outdoor conditions, the process of hydrogen production by immobilized cells of purple non-sulphur bacteria is characterized by unstable performances which seem to have in the temperature control their determining factor. Further studies are therefore required for a large scale exploitation of the process.

## 4. REFERENCES


Vincenzini, M., Materassi, R., Tredici, M.R. and Florenzano, G. Int. J. Hydrogen Energy 7, 231-236, 1982.
Vincenzini, M., Materassi, R., Tredici, M.R. and Florenzano, G. Int. J. Hydrogen Energy 7, 725-728, 1982.
Vincenzini, M., Materassi, R., Sili, C. and Florenzano, G. Int. J. Hydrogen Energy 11, 623-626, 1986.

AZOLLA BIOMASS AS GREEN MANURE FOR RICE IN CALABRIA


F. MILICIA and F. FAVILLI*
Ente Regionale di Sviluppo Agricolo della Calabria, Cosenza
* Dipartimento di Scienze e Tecnologie Alimentari e
Microbiologiche, Università. P.le delle Cascine, 27 - 50144
Firenze, Italy.

ABSTRACT. A mean yield of 10.5 g (d.w.) m$^{-2}$ day$^{-1}$ of biomass of the local strain of <u>Azolla filiculoides</u>, recently discovered in Calabria, was easily obtained in outdoor mass culture. The use of <u>Azolla</u> biomass as green manure for rice allowed to obtain rice yield significantly higher then those obtained with the same amounts of nitrogen as ammonium sulphate.

## 1. INTRODUCTION

The discovery of a local strain of <u>Azolla filiculoides</u> (Florenzano et al., 1986) in channels surrounding the paddy fields in the "Piana di Sibari" (Calabria, Southern Italy) opened the way to a research programme aimed to evaluate the potential for <u>Azolla</u> biomass production and the perspectives of its utilization as green manure for rice culture in southern Italy.

## 2. METHODS

<u>Azolla filiculoides</u> biomass was produced in outdoor cultures in concrete and earthern ponds of 100 and 600 sq. metres. The outdoor mass cultures were operated in a semicontinuous regimen collecting the biomass every 2-3 days (Carlozzi et al., 1988). Rice field experiments were conducted for two consecutive years in plots of 50 sq. metres. The effect of <u>Azolla</u> as green manure was compared with the recommended chemical nitrogen fertilization (Favilli et al., 1988).

## 3. RESULTS

Table 1 shows the yield of <u>Azolla filiculoides</u> biomass in outdoor mass culture.

Table 1 – Productivity and nitrogen content of the local strain of <u>Azolla filiculoides</u> mass culture in 100 sq. m. earthern pond.

| Months | Biomass $g(d.w.)\ m^{-2}\ day^{-1}$ | N content % of d.w. |
|---|---|---|
| May | 14.0 | 3.6 |
| June | 14.5 | 4.0 |
| July | 15.3 | 4.3 |
| August | 12.5 | 3.9 |
| September | 14.0 | 3.8 |
| October | 11.0 | 3.8 |
| November | 7.5 | 3.0 |
| December | 6.3 | 3.1 |
| January | 4.5 | 3.1 |
| February | 4.0 | 3.8 |
| March | 9.6 | 3.7 |
| April | 13.9 | 3.1 |
| Average | 10.5 | 3.6 |

The semicontinuous regimen used for plants harvesting allowed to obtain high biomass yield. The average nitrogen content of the biomass was about 3.6% (d.w.) for all the experimental period.

Table 2 shows the effects of <u>Azolla</u> biomass as green manure on rice yield.

The high agronomic efficiency obtained with the <u>Azolla</u> biomass indicates a good efficient utilization of the biologically fixed nitrogen in the rice-<u>Azolla</u> system.

Table 2 – Comparative effects of <u>Azolla</u> biomass and mineral nitrogen fertilization on rice yield.

| Fertilization Kg N ha$^{-1}$ | | Grain yield Kg ha$^{-1}$ | | Increase over control: Kg ha$^{-1}$ | | Agronomic efficiency[*] | |
|---|---|---|---|---|---|---|---|
| as $(NH_4)_2 SO_4$ | as <u>Azolla</u> | 1987 | 1988 | 1987 | 1988 | 1987 | 1988 |
| 0 | 0 | 2084 | 2930 | 0 | 0 | 0 | 0 |
| 25 | 0 | 3057 | 3810 | 973 | 880 | 38.9 | 25.2 |
| 50 | 0 | 3819 | 5500 | 1735 | 2570 | 34.7 | 51.4 |
| 100 | | | 7810 | | 4880 | | 48.8 |
| 0 | 25 | 6018 | 4870 | 3934 | 1948 | 157.3 | 77.6 |
| 0 | 50 | 7798 | 7430 | 5714 | 4500 | 114.2 | 90.0 |
| | 100 | | 9190 | | 6260 | | 62.5 |
| 50 | 50 | 8089 | 8620 | 6005 | 5690 | 60.0 | 56.9 |

[*] Agronomic efficiency is calculated dividing the yield increase by the N applied.

The results demonstrate the great potential of Azolla filiculoides as green manure for rice production in areas, as southern Italy, characterized by climatic conditions favourable for the fern growth and for its maintenance during the winter season. Field experiments are in progress in larger plots in order to confirm the results obtained.

## 4. REFERENCES

Carlozzi, P. et al. (1988) Proceed. of the 1986 Intern.Congr.on Renewable Energy Sources 1, 175–182, Madrid, May 1986.

Favilli et al. (1988) Proceed. of the First Intern. Symp. on Paddy Soil Fertility 2, 841–846, Chiang May, Thailand, December 1988.

Florenzano et al. (1986) Agric. Ital. 122, 57–77.

# Molecular ecology of *Frankia*: Advantages and disadvantages of the use of DNA probes

ANTOON D.L. AKKERMANS, DITTMAR HAHN and M. SAJJAD MIRZA
*Department of Microbiology, Agricultural University, Hesselink van Suchtelenweg 4, 6703 CT Wageningen, The Netherlands*

*Key words:*   actinorhizal plants, nitrogen fixation, oligonucleotide probes, ribosomal RNA

## Abstract

Ecological studies on the actinomycete *Frankia* are often influenced by the difficulty to isolate and identify this microorganism. The application of molecular biological techniques offers possibilities to detect microbes without isolation and cultivation. *Nif* genes or whole plasmids can serve as targets for the design of specific probes. Alternatively, ribosomal RNA (rRNA), commonly used in modern phylogenetic studies, can be used as a target molecule in ecological studies. This paper gives an overview of new developments on the use of 16S rRNA as a target molecule for oligonucleotide probes. Group-specific sequences in the 16S rRNA of *Frankia* can be used as targets for oligonucleotide probes that a) recognize ineffective *Frankia* strains on *Alnus*, b) recognize effective strains on *Alnus*, c) recognize all *Frankia* strains tested so far. The present paper summarizes the essential steps needed for the use of these probes for the detection of *Frankia* strains in soil without isolation and cultivation.

## Introduction

Microbial ecology has been dependent on the use of conventional microbiological techniques for a long time. Isolation and subsequent characterization of pure cultures on selective media or in situ measurements of microbial activity were the main parameters used in ecological studies. Most soil microorganisms, including our model organism *Frankia*, are still difficult to isolate and to grow on plates, because they occur in low numbers in soil and do not have selective isolation media. Previous attempts to quantify populations of *Frankia* in soil were almost exclusively based on nodulation tests (van Dijk, 1984; Smolander, 1990). Although this method is highly sensitive (i.e. lower limit of detection 0.07 nodulation units/cm$^3$ of soil (Smolander 1990) which approximates 1–100 *Frankia* cells/cm$^3$), it is time-consuming (1.5–2 months), does not dis-

criminate between different *Frankia* strains, excludes non-infective strains (Hahn et al., 1988) and neglects the effects of competition between strains (Hahn et al., 1990b). Quantification of *Frankia* using colony-forming units (CFU) on plates can only be done under axenic conditions. The use of CFU as a quantitative parameter also underestimates the *Frankia* population, because clusters of mycelium of this hyphae forming organism are regenerated rather than single cells.

Today, the introduction of new techniques like immunofluorescence microscopy enables detection of specific types of microbes with labeled antisera without isolation of the organisms. Although this method looks generally promising, the application is limited by the specificity of the applied antisera and by the microscopial techniques, that need about $10^1$ cells per mL for reliable detection.

To obtain more information on the development of an optimal actinorhizal plant-*Frankia* symbiosis, reliable isolation and/or sensitive detection and identification methods have to be developed. Marker molecules that can be used to identify *Frankia* strains introduced to natural environments are necessary to follow the establishment of these strains in nodules and in soil. Identification problems can be solved by applying specific probes that can be used to identify *Frankia* strains without isolation and cultivation. *Nif*-genes have been shown to be possible candidates to act as specific markers in nitrogen-fixing strains (Simonet et al., 1988). The development of quantitative DNA extraction methods from natural environments, i.e. soil or aquatic environments, makes DNA markers an attractive tool for ecological research (Fuhrman et al., 1988; Holben et al., 1988; Ogram et al., 1988; Sommerville et al., 1989; Steffan et al., 1988). Methods that can be used to amplify minute quantities of target sequences (Polymerase Chain Reaction) are currently developed (Saiki et al., 1988; Steffan and Atlas, 1988) and could be used in combination with specific probes to detect bacteria that are present in small amounts.

Stable, universal markers can also be found within sequences of 16S rRNA. Differences in the rRNA molecule are not distributed randomly across the entire molecule, but are rather clustered in specific regions. It is therefore not always necessary to determine the entire nucleic acid sequence in order to obtain a suitable probe sequence. Analyses of variable regions of 16S rRNA of closely related organisms indicate sufficient variation to design the probe of interest, despite the fact that DNA/DNA homology studies suggested these two species might actually be one and the same. This makes rRNA sequences an attractive target for diagnostic research (Kohne et al., 1986; Viscidi and Yolken, 1987). Ribosomal RNA is preferable to DNA as a probe target because of its relative abundance and stability in the cell.

Synthetic complementary oligonucleotides serve as specific probes in the DNA/rRNA hybridization reaction. Group- or species-specific sequences that can be used as specific targets for probes in hybridization experiments are found within various microorganisms (Giovannoni et al., 1988; Goebel et al., 1987; Stahl et al., 1988).

Because of the high specificity of the probe-target system, the detection limit becomes very low. Under certain circumstances, e.g. pure cultures, as little as a single bacterium can be detected (Giovannoni et al., 1988; DeLong et al., 1989). Even in mixed cultures where target organisms may occur in low numbers these hybridization techniques form a powerful tool for detection of microorganisms, especially for pathogens, symbionts and parasites, where isolation and cultivation are difficult (Tenover, 1988). Although the potential for molecular probes in ecology looks highly promising, current extraction methods of nucleic acids are far from optimal and do not allow us to measure microbial biomass quantitatively. This is particularly the case with filamentous organisms, e.g. *Frankia*.


## General features of Frankiaceae and *Frankia*

Although the knowledge about the interaction of woody plants and the actinomycete *Frankia* has a long history, pure cultures of this actinomycete have been obtained only recently. First attempts to taxonomically characterize the microsymbiont *Frankia* have been made by using its ability to enter a symbiotic association with higher plants and its morphological and structural differences from other actinomycetes (Becking, 1970). The combination of these criteria with cytochemical characteristics has led to the emendation of the family Frankiaceae with the genus *Frankia* (Becking, 1970; 1974). First successful isolations were reported about twelve years ago (Callaham et al., 1978). Since that time hundreds of isolates have been obtained using different isolation techniques (Baker et al., 1979a,b; Baker and O'Keefe, 1984; Berry and Torrey, 1979). However, detailed information on the phylogenetic position of pure cultures of *Frankia* and the inter- and intrageneric relationships of strains is still fragmentary.

The first classification within the genus *Frankia* was based on host plant relationships using crushed nodules as inocula sources (Becking, 1970). However, further investigations with pure cultures demonstrated that the host specificity groups were quite different (Baker, 1987; Normand and Lalonde, 1986). All isolates obtained from nodules were assigned to the genus *Frankia* on the basis of (i) morphological features, such as sporangia and vesicle formation in submerged liquid culture, (ii) chemical composition of certain cell constituents such as cell wall type III, phospholipid type PI and the presence of the diagnostic sugar 2-O-methyl-mannose and (iii) the ability to fix nitrogen and to nodulate plants (Lechevalier, 1984; Lechevalier and Lechevalier, 1984).

Molecular methods like restriction enzyme analysis of total DNA (Dobritsa, 1985) or determination of DNA base composition and DNA/DNA homology (An et al., 1985; 1987; Fernandez et al., 1989) also demonstrate the genetic diversity among *Frankia* strains because only low levels of homology are obtained. Ratios of homology in genomic DNA/DNA hybridizations between 67 and 94% within one compatibility group or levels lower than 50% between different compatibility groups are obtained. However, DNA/DNA homology studies can characterize the compatibility groups and have led to the emendation of *Frankia alnii* (Fernandez et al., 1989). It has been demonstrated that *Nif* genes have a conserved character (Ruvkun and Ausubel, 1980) and can therefore be used to investigate large phylogenetic relationships (Henneke et al., 1985). Different restriction sites within these genes can supply specific hybridization patterns of digested total DNA with *Nif* gene probes indicating a possible usefulness of these patterns in taxonomical investigations and as specific markers in ecology oriented research (Meesters, 1988; Normand and Lalonde, 1986; Normand et al., 1988; Simonet et al., 1988).

### *Frankia* and *Frankia*-like strains

In spite of the successful identification of many *Frankia* strains on the basis of morphological features (vesicles, sporangia), the occurrence of nitrogenase and the ability to nodulate plants, there is still little or no attention focused on isolates that lack some of the above-mentioned features. During the last few years we have isolated actinomycetes from different host plants (*Alnus, Coriaria*) that do not form vesicles and do not fix nitrogen. Some of these strains form ineffective nodules on certain clones of *Alnus glutinosa*, others fail to form nodules (Hahn et al., 1988) due to the variable degree of resistance of the host (van Dijk and Sluimer, 1990).

As the ineffective *Frankia* strains lack some morphological and physiological characteristics of typical *Frankia* strains and as nodule formation on actinorhizal plants is reduced or even absent, detection of the ineffective strains and studies on their competitive abilities are quite difficult. Reliable markers are required to detect both types of *Frankia* in nodules and in soil without reisolation. An attempt was made to find specific markers in a molecule which was commonly used to unravel evolutionary relationships: the 16S ribosomal RNA (Hahn, 1990; Hahn et al., 1989a). New sequencing techniques allowed the rapid determination of total or almost total 16S rRNA sequences (Embley et al., 1988; Lane et al., 1988). Total 16S rRNA sequences indicated the presence of conserved and variable regions. Conserved regions had been used to investigate quantitative evolutionary relationships among bacteria. The conserved regions of the total 16S rRNA sequence of the effective *Frankia* strain Ag45/Mut15 were compared with aligned sequences of other actinomycetes and used to determine the position of the family Frankiaceae in the phylogenetic tree of the actinomycetes (Hahn et al., 1989b). Analyses of variable regions of 16S rRNA of closely related organisms indicated sufficient variation, despite the fact that DNA/DNA homology studies suggested these two species might actually be one and the same. Large differences in DNA/DNA homology studies of *Frankia* which were also obtained between strains of one compatibility group suggested chances of large variation within the variable regions of different strains. Sequence analyses of variable regions of 16S rRNA of two ineffective *Frankia* strains (i.e. AgB1.9 and AgW1.1) and the effective strain Ag45/Mut15, all belonging to the *Alnus*-compatibility group, showed large differences in base composition. These sequences were used to design complementary synthetic oligonucleotides that could act as specific probes in hybridization experiments. The specificity of these probes was

598

shown in hybridization experiments against immobilized rRNA from 23 *Frankia* strains belonging to different compatibility groups and of several related soil actinomycetes (Hahn et al., 1989a). The probes were able to distinguish between Nif⁺ and Nif⁻ strains, between several Nif strains and between several *Alnus* compatible Nif⁻ strains and strain AgKG'84/4 also belonging to the *Alnus*-compatibility group. Strong strain specific sequences, however, were not obtained. The design of oligonucleotide probes opens up the possibility to investigate competitive abilities of selected strains under defined conditions, e.g. in model systems with perlite and defined *Frankia* strains. The question whether competition studies under these controlled conditions are ecologically relevant needs further investigation because little basic knowledge on *Frankia* population dynamics is yet available. The application of probes to identify introduced strains in soil remains restricted, due to the low specificity for strains. Up to now we are not able to design reliable strain specific probes that can be used to follow the establishment of introduced *Frankia* strains in natural environments. A much more promising application of probes towards rRNA is concerned with the development of a genus-specific oligonucleotide probe against *Frankia* that theoretically enables quantitative detection of total *Frankia* populations. This *Frankia* probe shows stringent hybridization to rRNA of all ineffective and non-infective isolates, indicating a relationship with *Frankia*. The restricted specificity of the probe, however, can only give an indication of the identity of the isolates. Reliable identification will only be achieved after the analysis of the sequences of 16S rRNA.

## Extraction of RNA from nodules and soil

The application of oligonucleotide probes in the detection of specific *Frankia* strains not only depends on specificity of the probes but also on the development of a reliable isolation method for target sequences. Ribosomal RNA is preferable to DNA as target because of its relative abundance in metabolically active cells. Actinorhizal nodules represent enrichments of *Frankia*, which are metabolically highly active and consequently contain large amounts of *Frankia* RNA. Our investigations resulted in the development of a rapid RNA extraction method that was sensitive enough to investigate strain composition also from very small nodules or lobes. The detection of target sequences, however, remained limited by the design of specific probes and the ratios of different target sequences in one sample. For reliable signal expression in hybridization experiments quite similar amounts of target sequences per sample were needed.

So far, the usefulness of rRNA sequences as targets for oligonucleotide probes was only shown in combination with pure cultures of *Frankia* or in metabolically highly active enrichments, e.g. nodules (Hahn et al., 1990b). Terrestrial environments like soil contain populations of many different microorganisms. These populations normally grow under suboptimal nutrition conditions. Bacteria adapt to these conditions by forming special starvation cells, which are metabolically inactive and contain only low amounts of rRNA. The starvation response often results in viable, but non-culturable populations. The recalcitrant character of *Frankia*, which are difficult to isolate, makes it a useful model microorganism of soil bacteria. The application of oligonucleotide probes for detection of *Frankia* in soil depends on the development of an extraction method for RNA (Hahn et al., 1990a). RNA directly isolated from soil as target for *Frankia* specific oligonucleotide probes was useful in detection of *Frankia*. Quantification of the obtained signals, however, is still unreliable because *Frankia* is a hyphae forming organism. It is also quite difficult to correlate cell numbers because *Frankia* is a filamentous organism. It is also quite difficult to correlate cell numbers (theoretical estimation) to the amount of RNA. The concentration of these molecules in an organism is a function of the activity of the individual cell. Quantification of hybridization signals therefore depends on the availability of basic information of the metabolic activity of *Frankia* cells in soil. This information, however, is very difficult to obtain for recalcitrant microorganisms like *Frankia*. It is much easier for other microorganisms, e.g. for *Streptomyces*. *Streptomyces* spores are quite easy to isolate from soil and the establishment of *Streptomyces* cells, i.e. as spores or as mycelium in soil, is well studied. Quantification based on hybridization signals must be possible when this basic knowledge is available. In the case of *Frankia* methods that enable quantification must still be developed. Similar to *Streptomyces* these quantification methods for *Frankia* can be of direct character, e.g. quantitative extraction of spores, or of indirect character, e.g. determination of mycelium by phage counts.

## Concluding remarks

So-called 'non-culturable' or difficult to cultivate cells, including *Frankia*, have frustrated soil microbiologists for a long time. Our current knowledge of the ecology of *Frankia* is far from complete. The development of rapid and sensitive methods to detect *Frankia* on the basis of rRNA sequences opens up new ways to study other recalcitrant microorganisms in the environment. This probably will become most relevant to study: a. pathogens, which create safety problems when cultivated, or which grow too slowly for conventional plating techniques, b. genetically modified microorganisms, after release in the environment and their impact on natural populations of bacteria, c. 'non-culturable' cells or organisms which do not grow on selective media and which occur in low numbers in soil and water. This molecular approach in microbial ecology can definitely be explored further when the advantages of rRNA as a stable target and the rapid extraction of RNA from soil can be combined with in vitro amplification methods commonly used with DNA or mRNA. Promising approaches can also be expected in in situ studies using hybridization signal intensity of fluorescent dye labelled oligonucleotides (Amann et al., 1990; DeLong et al., 1989) and the amount of rRNA as criterium for bacterial activity.

## Acknowledgements

These investigations were financially supported by the Foundation for Fundamental Biological Research (BION), which is subsidized by the Netherlands Organization for the Advancement of Pure Research (NWO), by the Commission of the European Communities (EEC) (No. EN3B-0043-NL (GDF)), by the Deutsche Forschungsgemeinschaften, and by the Netherlands Integrated Soil Research Programme. The authors are thankful to Marjo Starrenburg and Wilma van Vliet for technical assistance.

## References

Amann R I, Krumholz I, and Stahl D A 1990 Fluorescent oligonucleotide probing of whole cells for determinative, phylogenetic, and environmental studies in microbiology. J. Bacteriol. 172, 762–770.

An C S, Riggsby W S, and Mullin B C 1985 Relationship of *Frankia* isolates based on deoxyribonucleic acid homology studies. Int. J. System. Bacteriol. 35, 140–146.

An C S, Riggsby W S and Mullin B C 1987 DNA relatedness of *Frankia* isolates ArI4 and EuI1 to other actinomycetes of cell wall type III. The Actinomycetes 20, 50–54.

Baker D, Kidd G H and Torrey J G 1979a Separation of actinomycete nodule endophytes from crushed nodule suspensions by Sephadex fractionation. Bot. Gaz. 140 (Suppl.), S49-S51.

Baker D, Torrey J G and Kidd G H 1979b Isolation by sucrose-density fractionation and cultivation in vitro of actinomycetes from nitrogen-fixing root nodules. Nature 281, 76–78.

Baker D 1987 Relationships among pure cultured strains of *Frankia* based on host specificity. Physiol. Plant. 70, 245–248.

Baker D and O'Keefe D 1984 A modified sucrose fractionation procedure for the isolation of frankiae from actinorhizal root nodules and soil samples. Plant and Soil 78, 23–28.

Becking J H 1970 Frankiaceae Fan. nov. (Actinomycetales) with one new combination and six new species of the genus *Frankia* Brunchorst 1886 174. Int. J. Syst. Bacteriol. 20, 201–220.

Becking J H 1974 The Family Frankiaceae. *In* Bergey's Manual of Determinative Bacteriology. Eds. R E Buchanan and N E Gibbons. pp 701–706. Williams and Wilkins Co., Baltimore, MD.

Berry A and Torrey J G 1979 Isolation and characterization in vivo and in vitro of an actinomycetous endophyte from *Alnus rubra* Bong. *In* Symbiotic Nitrogen Fixation in the Management of Temperate Forests. Eds. J C Gordon, C T Wheeler and D A Perry. pp 69–83. Oregon University Press.

Callaham D, Del Tredici P and Torrey J G 1978 Isolation and cultivation in vitro of the actinomycete causing root nodulation in Comptonia. Science 199, 899–902.

DeLong E F, Wickham G S and Pace N R 1989 Phylogenetic strains: ribosomal RNA-based probes for the identification of single cells. Science 243, 1360–1363.

Dorbitsa S V 1985 Restriction analysis of the *Frankia* spp. genome. FEMS Microbiol. Lett. 29, 123–128.

Embley T M, Smida J and Stackerbrandt E 1988 Reverse transcriptase sequencing of 16S ribosomal RNA from *Faenia rectivirgula*, *Pseudonocardia thermophila* and *Saccharopolyspora hirsuta*, three Wall type IV actinomycetes which lack mycolic acids. J. Gen. Microbiol. 134, 961–966.

Fernandez M P, Meugnier H, Grimont P A D and Bardin R 1989 Deoxyribonucleic acid relatedness in the genus *Frankia*. Int. J. System. Bacteriol. 39, 424–429.

Fuhrman J A, Comeau D E, Hagstrom A and Chan A M 1988 Extraction from natural planctonic microorganisms of DNA suitable for molecular biological studies. Appl. Environ. Microbiol. 54, 1426–1429.

Giovannoni S J, DeLong E F, Olsen G J and Pace N R 1988 Phylogenetic Group-specific oligonucleotide probes for identification of single microbial cells. J. Bacteriol. 170, 720–726.

Goebel U B, Geiser A and Stanbridge E J 1987 Oligonucleotide probes complementary to variable regions of ribosomal RNA discriminate between *Mycoplasma* species. J. Gen. Microbiol. 133, 1969–1974.

Hahn D 1990 16S rRNA as molecular marker in ecology of *Frankia*. Ph.D. Thesis, Agricultural University, Wageningen. The Netherlands.

Hahn D, Starrenburg M J C and Akkermans A D L 1988 Variable compatibility of cloned *Alnus glutinosa* ecotypes against ineffective *Frankia* strains. Plant and Soil 107, 233–243.

Hahn D, Dorsch M, Starrenburg M J C and Akkermans A D L 1989a Synthetic oligonucleotide probes for identification of *Frankia* strains. Plant and Soil 118, 210–219.

Hahn D, Kester R, Starrenburg M J C and Akkermans A D L 1990a Extraction of ribosomal RNA from soil for detection of *Frankia* with oligonucleotide probes. Arch. Microbiol. 154, 329–335.

Hahn D, Lechevalier M P, Fischer A and Stackebrandt E 1989b Evidence for a close phylogenetic relationship between members of the genera *Frankia*, *Geodermatophilus*, and *Blastococcus* and emendation of the family Frankiaceae. Syst. Appl. Microbiol. 11, 236–242.

Hahn D, Starrenburg M J C and Akkermans A D L 1990b Oligonucleotide probes that hybridize with rRNA as a tool to study *Frankia* strains in root nodules. Appl. Environ. Microbiol. 56, 1342–1346.

Hennecke H, Kaluza K, Thony B, Fuhrmann M, Ludwig W and Stackebrandt E 1985 Concurrent evolution of nitrogenase genes and 16S rRNA in *Rhizobium* species and other nitrogen fixing bacteria. Arch. Microbiol. 142, 342–348.

Holben W E, Jansson J K, Chelm B K and Tiedje J M 1988 DNA probe method for the detection of specific microorganisms in the soil bacterial community. Appl. Environ. Microbiol. 54, 703–711.

Kohne D, Hogan J, Jonas V, Dean E and Adams T H 1986 Novel approach for rapid and sensitive detection of microorganisms: DNA probes to rRNA. *In* Microbiology-1986. Ed. L Leive. pp 110–12, American Society for Microbiology, Washington DC.

Lane D J, Pace B, Olsen G J, Stahl D A, Sogin M L and Pace N R 1985 Rapid determination of 16S rRNA sequences for phylogenetic analyses. Proc. Natl. Acad. Sci. USA 82, 6955–6959.

Lechevalier M P 1984 The taxonomy of the genus *Frankia*. *In* Frankia Symbiosis and Actinorhizal Plants. Eds. A D L Akkermans, D Baker, K Huss-Danell and J D Tjepkema. pp. 1–6. Nijhoff/Junk Publishers, The Hague.

Lechevalier M P and Lechevalier H A 1984 Taxonomy of *Frankia In* Biological, Biochemical and Biomedical Aspects of Actinomycetes. Eds L Orttiz–Ortiz, L F Bojalil and V Yakoleff. Academic Press, New York. pp 575–582.

Meesters T M 1988 The function of vesicles in the actinomycete *Frankia*. Ph.D. Thesis, Agricultural University, Wageningen. The Netherlands.

Normand P and Lalonde M 1986 The genetics of actinorhizal *Frankia*: A review. Plant and Soil 90, 429–453.

Normand P, Simonet P and Bardin R 1988 Conservation of Nif sequences in *Frankia*. Mol. Gen. Genet. 213, 238–246.

Ogram A, Sayler G S and Barkay T 1988 DNA extraction and purification from sediments. J. Microbiol. Methods 7, 57–66.

Ruvkun G B and Ausubel F M 1980 Interspecies homology of nitrogenase genes. Proc. Natl. Acad. Sci. USA 77, 191–195.

Saiki R K, Gelfand D H, Stoffel S, Scharf S J, Higuchi R, Horn G T, Mullis K B and Erlich H A 1988 Primer-directed enzymatic amplification of DNA with a thermostable DNA polymerase. Science 239, 487–491.

Simonet P, Le N T, Teissier du Cros E and Bardin R 1988 Identification of *Frankia* strains by direct DNA hybridization of crushed nodules. Appl. Environ. Microbiol. 54, 2500–2503.

Sommerville C C, Knight I T, Straube W L and Colwell R R 1989 Simple, rapid method for direct isolation of nucleic acids from aquatic environments. Appl. Environ. Microbiol. 55, 548–554.

Smolander A 1990 *Frankia* populations in soils under different tree species – with special emphasis on soils under *Betula pendula*. Plant and Soil 121, 1–10.

Stahl D, Flesher B, Mansfield R H and Montgomery L 1988 Use of phylogenetically based hybridization probes for studies of ruminal microbial ecology. Appl. Environ. Microbiol. 54, 1079–1084.

Steffan R J and Atlas R M 1988 DNA amplification to enhance detection of genetically engineered bacteria in environmental samples. Appl. Environ. Microbiol. 54, 2185–2191.

Steffan R J, Goksoyr J, Bej A K and Atlas R M 1988 Recovery of DNA from soils and sediments. Appl. Environ. Microbiol. 54, 2908–2915.

Tenover F C 1988 Diagnostic deoxyribonucleic acid probes for infectious diseases. Clin. Microbiol. Rev. 1, 82–101.

Van Dijk C 1984 Ecological aspects of spore formation in the *Frankia*-Alnus symbiosis. Ph.D. Thesis, Univ. Leiden. The Netherlands.

Van Dijk C and Sluimer-Stolk A 1990 An ineffective strain type of *Frankia* in the soil of natural stands of Alnus glutinosa (L.) Gaertn. Plant and Soil 127, 107–121.

Viscidi R P and Yolken R G 1987 Molecular diagnosis of infectious diseases by nucleic acid hybridization. Molecular Cellular Probes 1, 3–14.

# EXPRESSION OF ACTINORHIZINS IN THE DEVELOPMENT OF THE *FRANKIA-ALNUS* SYMBIOSIS.


A. SÉGUIN AND M. LALONDE
*Centre de recherche en biologie forestière*
*Faculté de foresterie et de géomatique*
*Université Laval, Sainte-Foy*
*Québec, Canada, G1K 7P4*



ABSTRACT. Nitrogen-fixing root nodules of actinorhizal species develop as the result of the symbiotic invasion of host root tissue by a compatible microorganism from the actinomycetal genus *Frankia*. A comparative molecular study of the cellular mechanisms involved in root tissue infection and nodule (actinorhiza) formation may permit the identification of the genes required for the establishment of this symbiosis, when expressed by each symbiont. The products of gene expression have been termed actinorhizins. Analysis of soluble proteins by SDS-PAGE and 2D-PAGE showed many differences between non-infected roots and actinorhizae. In all, these findings demonstrated that protein synthesis is altered in response to morphological and physiological changes resulting from the establishment of actinorhizae. An overview of the recent data obtained on the molecular characterization of actinorhiza development and possible strategies for future research will be given.


## 1. Introduction

Actinorhizal plants are comprised mainly of woody perennial angiosperms which develop a root symbiosis with nitrogen-fixing actinomycetes of the genus *Frankia*, a group of filamentous prokaryotes which are classified in the order Actinomycetales (for review, see Normand and Lalonde, 1986; Mullin and An, 1990; Simonet *et al.*, 1990). These tree and shrub species play an important ecological role as pioneer plants on nitrogen-poor soils. Many studies have demonstrated the value of the *Frankia*-actinorhizal symbiosis to reforestation and land reclamation projects (Dawson, 1990; Diem and Dommergues, 1990; Hibbs and Cromack, 1990; Wheeler and Miller, 1990). Besides the applied uses of actinorhizal plants, the study of this symbiosis is also of much interest to those investigating the biological basis of nitrogen-fixing symbioses.

In the *Rhizobium*-legume association, microsymbionts will fix nitrogen symbiotically only with a host species of the Leguminosae family (with the exception of *Parasponia*) and are thus highly host specific. Actinorhizal microsymbionts have a much wider host range; actinomycete-induced nitrogen-fixing nodules (actinorhizae) may be found among over 25 genera of dicotyledonous plants belonging to 8 different families. Although obvious differences exist between the legume symbiosis and actinorhizae, the significant advances made in the understanding of different molecular aspects of the *Rhizobium*-legume symbiosis may be helpful in elucidating *Frankia*-actinorhizal interactions.

In this report, we describe different approaches used to study the expression of actinorhizins in the development of *Alnus* symbioses. Distinctive features of this actinorhizal system, by analogy to legume nodules, will also be briefly discussed.

## 2. Analysis of the protein content of uninfected *Alnus* roots and actinorhizae

Formation of actinorhizae occurs following a series of specific developmental steps involving *Frankia* hyphae and host root cells. Morphogenesis and fine structure of actinorhizae, as well as the infection process and actinorhiza development, have been considered in recent reviews (Newcomb and Wood, 1987; Berry and Sunell, 1990). It is clear that among actinorhizal plants, *Alnus* represents the genus for which the largest amount of information on different aspects of this symbiosis is available. This includes several different topics, such as the genetics of the host (Bousquet and Lalonde, 1990), the development of cell, tissue culture systems and genetic transformation (Séguin and Lalonde, 1990a), studies of host-*Frankia* interactions, nodule physiology (Huss-Danell, 1990) and the initiation of many studies at the DNA level. For these reasons, we consider *Alnus* as a model system for the molecular study of actinorhiza development.

The molecular analysis of the *Frankia*-actinorhizal symbiosis has so far been mainly concerned the microsymbiont. Both the lack of knowledge about host genetics and the techniques required to study its molecular biology has hampered most the progress of the role of host plant genes in this symbiosis. Many efforts have been made over the past 20 years to describe the molecular interactions between *Rhizobium* and legumes and these have resulted in a better understanding of this symbiosis. Legume nodules are often considered to be unique organs and this characteristic is also supported by the expression of a set of specific genes, the nodulin genes (van Kammen, 1984). These plant proteins, or nodulins, are specifically induced during symbiotic nitrogen fixation and play both metabolic and structural roles within infected and uninfected nodule cells. Similarly, specific polypeptides are present in actinorhizae, but are undetectable in uninfected roots and free-living *Frankia;* these are named actinorhizins (Tremblay *et al.,* 1986). In the latter case, it is important to mention that no distinction was made between host plant and microsymbiont origin of these products, as was done in studies of the Legume-*Rhizobium* symbiosis.

As a first step in elucidating the expression of specific genes underlying the molecular process of actinorhiza development we have compared the polypeptide changes observed at different stages of the *Frankia-Alnus* symbiosis by gel electrophoresis. To use this approach, a reliable method for the extraction of total protein from actinorhiza and root tissue is required before proceeding with the analysis of polypeptide changes during the development of the actinorhiza. Extensive studies on the physiology of actinorhizae have contributed to the development of methods for the extraction and purification of the enzymes involved in different metabolic processes, including nitrogen fixation and assimilation (reviewed in Huss-Danell, 1990). However, the approach used to study actinorhizins is quite different. For such studies, the analytical techniques must be further refined to permit the detection of subtle differences between actinorhiza and uninfected root proteins. Such techniques have been described recently for the extraction of total protein from root tissue and actinorhiza of *Alnus* (Séguin and Lalonde, 1988; Belefant and Mullin, 1990). Protein analysis of root exudates by isoelectric focusing has also been described (Vergnaud *et al.,* 1987).

Phenolic compounds present in the root and actinorhiza of *Alnus* may interact with proteins and have hampered their complete separation by gel electrophoresis. We found that direct extraction in cold acetone, containing 2-mercaptoethanol as an antioxidant, minimized the degradation of protein in the preparation. The dry powder produced following acetone precipitation was then redissolved in a buffer solution containing polyvinylpyrolidones and polystyrene ion-exchangers to adsorb the phenols, which were then removed by high speed centrifugation. This method we have developed for protein separation may be used in concert with sodium dodecyl sulfate polyacrylamide gel electrophoresis (SDS-PAGE) or two-dimensional polyacrylamide gel electrophoresis (2D-PAGE), after complete resolubilisation in the appropriate buffer (see Fig. 1.).

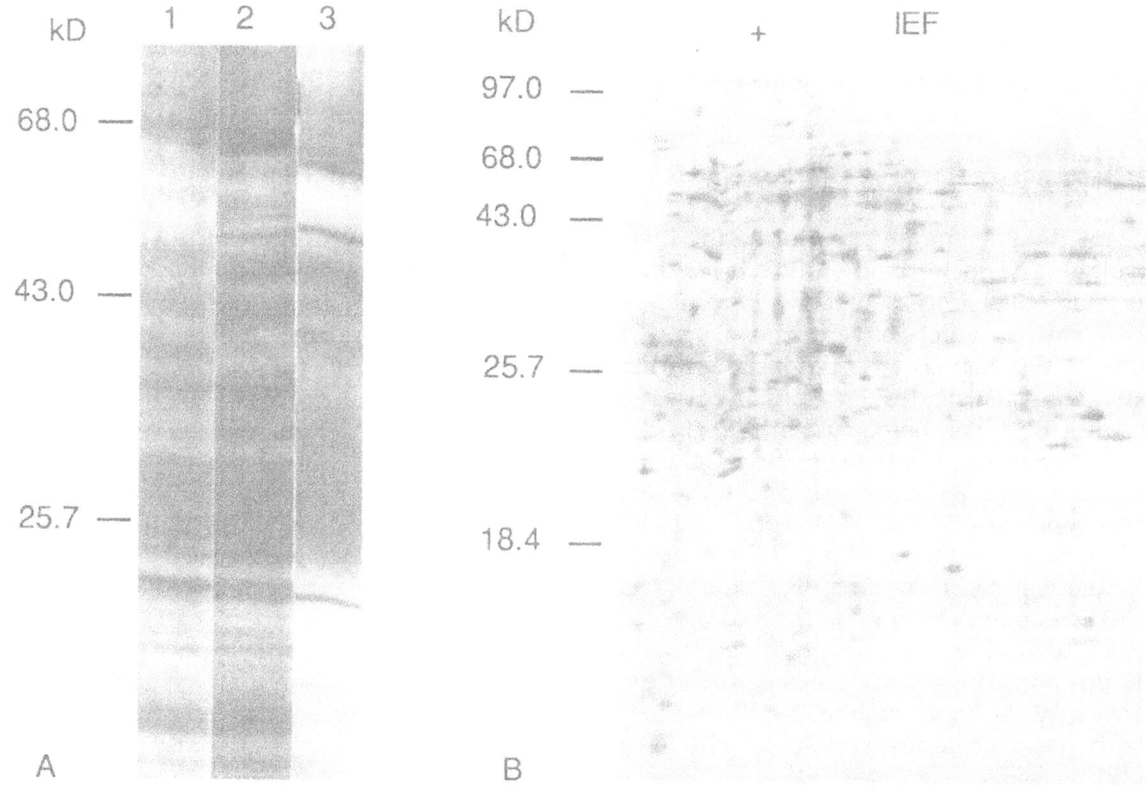

Figure 1. Analysis of the protein content of *Alnus* roots, actinorhiza and *Frankia* culture. (A) SDS-PAGE gradient (8-23%) after silver staining. Lane 1, non-infected roots; 2, actinorhizae; 3, *Frankia* culture. (B) 2D-PAGE of proteins extracted from actinorhizae after silver staining. The sizes of the markers are given on the left of the gel.

Analysis of soluble proteins from actinorhizae and non-infected roots of *A. glutinosa* by SDS-PAGE showed an increased accumulation in actinorhizae of three polypeptides of molecular weight 36, 50 and 61 kD (Séguin and Lalonde, 1990b). The further characterisation of some of these polypeptides, using an immunological technique, will be presented in the following section. The comparison of 2D-PAGE patterns from uninfected *A. glutinosa* roots and actinorhizae have revealed many differences in soluble protein content (Séguin and Lalonde, 1990b). This indicates that extensive changes in gene expression and protein turnover occur during actinorhiza development. However, 2D-PAGE patterns were too complex to be easily interpreted.

The use of radiolabelled proteins synthesized by the *in vitro* translation of RNA is a technique which has been used in several fields of biology. This technique has also been shown to be very useful in the study of plant gene expression in legume root nodules (Govers *et al.*, 1987; Verma *et al.*, 1986). A major advantage of this technique is that the microsymbiont RNA is not translated. The observed modifications of the *in vitro* synthesized proteins patterns, resulting from the development of actinorhizae, will be limited to the products of plant gene expression. Unfortunately, the isolation of RNA from woody plant tissue may been hampered by the presence of pigments, mainly polyphenols which bind to nucleic acids. Different methods for the extraction of RNA from roots and actinorhizae have been tested in our laboratory and we will briefly discuss in section 4 the variability in the quantity and quality of the RNA obtained from actinorhiza and root tissue. In our own study, highly purified RNA was obtained from unifected roots and actinorhizae, but was not translated *in vitro*, using a commercial preparation of rabbit reticulocyte lysate. These results have been observed in another laboratory (Dr. T. Bisseling, personal communication) and may be due to the presence of inhibitors of polypeptide synthesis, which were copurified with the RNA (Merino *et al.*, 1990) .

## 3. Characterization of actinorhiza specific proteins by Western blotting.

A considerable amount of information about the biochemistry of the actinorhizal nodule is now available. As mentioned earlier, it has been shown that some nodulins play essential roles in the metabolic processes of the *Rhizobium*-legume symbioses. Some of the enzymes found in actinorhizae which have already been described in the literature (for review, see Huss-Danell, 1990), might be potential candidates as plant or microsymbiont actinorhizins. An exhaustive list of the enzymes involved in different metabolic processes in the actinorhiza will not be given here, but rather a brief overview of some enzymes will be discussed with respect to their possible specificity to actinorhiza.

As a microsymbiont, *Frankia* may use various sources of carbon derived from plant photosynthates. Metabolic adaptation of the microsymbiont may occur to sustain the supply of carbon required for growth and the energy needed for nitrogen fixation. As well as the enzymes involved in carbon metabolism, the malate-aspartate shuttle suggested by Akkermans *et al.* (1981) between the host plant and the microsymbiont may also be specifically found in the actinorhiza; the proteins involved in this transport mechanism may represent potential actinorhizins. In addition, nitrogen fixation and metabolism depend on many specific enzymes. Obviously the first enzyme that comes to mind is the nitrogenase enzyme complex of *Frankia*, which has a structure similar to that of other nitrogen-fixing microorganisms (Benson *et al.*, 1979).

In order to characterize the actinorhiza specific proteins, we tested different antisera specific for purified or partially purified protein preparations. Immunoblot analysis with an anti-nitrogenase (Fe protein) antibody, raised against purified *Rhodospirillum rubrum* nitrogenase, revealed a protein of 36 kD. A second anti-nitrogenase (MoFe protein) antibody, raised against the purified MoFe protein of *Azotobacter vinelandii* nitrogenase, reacted with a protein of 61 kD. These results suggest that the 36 kD and 61 kD proteins, specific to *A. glutinosa* actinorhiza extracts, are constituents of the dinitrogenase reductase enzyme of *Frankia* (Séguin and Lalonde, 1990b).

The use of immunological assays has been useful for the identification of nodulins and the analysis of nodulin gene expression in the legume nodule. Nodule-specific antiserum preparations may be obtained by titration of an antiserum directed against total plant protein of the nodule, using proteins from uninfected roots. Using Western blot analysis of pea nodule proteins, treated with a nodule-specific antiserum preparation, Bisseling *et al.* (1983) were able to detect 30 different nodule-specific polypeptides. Antisera directed against legume nodulins have been tested for cross-reactivity with actinorhiza proteins. Attempts to detect a leghemoglobin-like protein in *A.glutinosa* using antibodies reacting with pea and soybean leghemoglobin were unsuccessful (Simonet *et al.*, 1990). The absence or the low concentration of hemoglobin (Tjepkema, 1983) in the nodule of many actinorhizal species (e.g. *Alnus* and *Elaeagnus*) may explain those results. However, antibodies directed against the glutamine synthetase enzyme of pea reacted on Western blots with proteins isolated from nodules of actinorhizal plants (Simonet *et al.*, 1990). Glutamine synthetase (GS) activity has been found in root nodules of *A. glutinosa* and is associated with the plant cytoplasm (Blom *et al.*, 1981; Hirel *et al.*, 1982). GS activity has also been recently detected in pure cultures of *Frankia* (Edmands *et al.*, 1987). However, it remains to be determined whether a plant-specific GS is found strictly in the actinorhiza and should thus be considered as an actinorhizin. A glutamate dehydrogenase (GDH), also involved in the assimilation of fixed ammonia, is a potential candidate as an actinorhiza-specific protein. The transport of nitrogen compounds from the nodules to the host plant tissue occurs mainly in the form of citrulline for most alder species (reviewed in Huss-Danell, 1990). Much evidence suggests that citrulline production is catalysed by the plant enzyme, ornithine carbamyl transferase (OCT) which may be specifically induced or overproduced in the actinorhizae (Blom *et al.*, 1981; Martin *et al.*, 1983). The process of nitrogen fixation is accompanied by the formation of $H_2$. The elimination of this $H_2$ is accomplished by the action of an uptake hydrogenase that oxidizes $H_2$. An antiserum directed against the hydrogenase of *Azotobacter chroococcum* (kindly provided by Dr. Daniel Arp, Corvallis, OR, USA) was also tested in Western blotting experiments, but failed to detect a similar enzyme in actinorhizae.

Pectolytic activity was detected in actinorhizae and in pure cultures of *Frankia* (Séguin and Lalonde, 1989). Cellulolytic activity was also detected in pure cultures of *Frankia* (Safo-Sampah and Torrey, 1988). These enzymes may be involved at the level of the actinorhiza for nutritional interactions by permitting the encapsulation material (Lalonde and Knowles, 1975) to be used as a metabolic substrate. In order to identify polypeptides responsible for the pectolytic activity we used different antisera directed against various preparations of pectate lyase enzymes from *Erwinia* spp. in Western blotting experiments. Rabbit polyclonal antibody obtained after immunization with a preparation of pectate lyase from *Erwinia caratovora* (kindly provided by Drs. L. Ward and S. De Boer, Vancouver, Canada) and a mouse monoclonal antibody (Ward and De Boer, 1989) were tested. Two monoclonal antibodies directed against a preparation of pectate lyase from *Erwinia chrysanthemi* (kindly provided by Dr. Y. Bertheau, Paris, France) were also tested in Western blotting experiments. In spite of the sequence similarity found between the *Erwinia* pectate lyase gene and *Frankia* DNA, we were not able to detect any homologous protein in actinorhiza. However, an antiserum prepared against a culture filtrate of *Frankia*, exhibiting pectolytic activity, detected the expression of a 53 kD polypeptide in the actinorhiza (Séguin and Lalonde, 1990b).


## 4. Strategy for identification of specific actinorhiza genes in the development of the actinorhizal symbiosis.

Several developmental similarities are observed between the *Rhizobium*-legume and the *Frankia*-actinorhizal plant symbiosis. Strategies and molecular techniques used to study *Rhizobium*-legume symbiosis represent attractive approaches to the study of actinorhiza development. For example, molecular genetic and immunological techniques have been used to identify specific mRNAs and proteins in legume root nodules that are not detectable in uninfected roots (Bisseling *et al.*, 1984; Legocki and Verma, 1980). General knowledge about the involvement of host genes in *Rhizobium*-legume symbiosis has increased markedly in recent years. Several recent reviews describe in detail various aspects of the *Rhizobium*-legumes symbiosis (Verma *et al.*, 1986; Govers *et al.* 1987; Long, 1989; Appelbaum, 1990). The similarities between *Frankia*-actinorhizal and *Rhizobium*-legume symbiosis are numerous and it is possible that some host plant genes are conserved as for microsymbiont genes.

### 4.1. UTILISATION OF NODULIN GENES AS HETEROLOCOUS PROBES FOR THE IDENTIFICATION OF NODULIN-LIKE SEQUENCES

Different techniques for the isolation of nucleic acids from actinorhizal plants have been developed. These techniques are an essential prerequisite for the study of the molecular genetics of the host. Isolation of high molecular weight genomic DNA from different tissues of actinorhizal plants has been reported. In many cases leaves were used for DNA extraction (Roberts *et al.*, 1985; Bousquet *et al.*, 1989), but seeds (Hattori and Johnson, 1985) and material from cell and tissue cultures (Giasson and Lalonde, 1987; Mackay *et al.*, 1988) were also found suitable. The isolation of restrictable DNA from some actinorhizal plants has allowed the use of techniques such as restriction fragment length polymorphism (RFLP) analysis for the characterization of their genomes at the DNA level (Bousquet *et al.* 1989). Molecular cloning of DNA sequences from a genomic library of alder has been reported by different laboratories (Lévesque *et al.*, 1989; B.C. Mullin, personnal communication). As a first step in the molecular characterization of *Alnus* as a symbiont, different nodulin genes from legumes may be used as heterologous probes for the screening of genes with related function. Leghemoglobin cDNA clones from soybean have been used with success to identify sequences in the genomic DNA of actinorhizal plants (Roberts *et al.*, 1985). A similar procedure was also used successfully to detect leghemoglobin-like DNA sequences in actinorhizal as well as non-symbiotic plants (Hattori and Johnson, 1985). However, the sequence homologies identified at the DNA level are not necessarily maintained at the protein level. Moreover, the presence of a leghemoglobin-like protein may be linked to a normal plant process which does not relate to the nitrogen-fixation capability of the actinorhiza (Appleby *et al.*, 1988). It remains to be determined whether these sequences are transcribed and code for the hemoglobin that has been found in actinorhizal nodules.

An alternative to this approach would be to find sequence similarity at the RNA level, on Northern blots. In this way, discrimination between sequences that are expressed specifically in the actinorhiza might be obtained. As previously mentioned, we have encountered many difficulties in obtaining pure preparations of RNA from roots and actinorhizae of *A. glutinosa*. We have found that methods using phenolic solvents, as described by Goldberg *et al.* (1981), Logemann *et al.* (1987) and Mohapatra *et al.*, (1987), are not well suited for *Alnus*. However, the method described by Hughes and Galau (1988) for the preparation of RNA from cotton leaves provided good results (see Fig. 2.).

This method is based on the use of a low temperature and detergents to lyse the cell, without the use of phenol or other organic denaturants during the first steps of extraction. Using this method, RNA isolated from uninfected roots and actinorhizae of *A. glutinosa* was used for Northern hybridizations. Preliminary experiments, in which Northern blots of RNA extracted from actinorhizae were probed with different early nodulin genes (work done in collaboration with Dr. T. Bisseling, Wageningen, The Netherlands), revealed sequence similarities with the ENOD2 gene from soybean (Franssen *et al.*, 1987). ENOD2 is a hydroxyproline-rich (HRGP) protein that is most likely a cell wall component and is formed in legume nodule parenchyma "inner cortex" cells (van de Wiel *et al.*, 1990). Using the ENOD2 gene, we are now screening a cDNA library prepared from poly A(+) RNA isolated from actinorhizae of *A. glutinosa*.

A special HRGP gene that is expressed exclusively at the site of lateral root formation has been recently characterized (Keller and Lamb, 1989). Since the products of many early nodulin are hydroxyproline-rich cell wall proteins, it would be interesting to compare the HRGPnt3 gene sequences with other nodulins genes. In a fashion similar to the development (morphogenesis) of lateral roots, the primary nodule lobe primordia arise endogenously from the pericycle (Callaham and Torrey, 1977). Inoculation of *Alnus* roots with *Frankia* resulted in an increasing number of lateral root primordia (Angulo Carmona, 1974). In fact, meristematic activity at two specific sites is the result of *Frankia* inoculation. In the first instance, meristematic activity has been observed in the cortical tissue of the prenodule as an early consequence of microsymbiont infection, and secondly with the initiation of lateral root primordia. Analysis of phytohormone concentrations in actinorhizae has revealed elevated levels of auxins, cytokinims and gibberellins (Wheeler *et al.*, 1979). Recent findings that *Frankia* is able to secrete auxins and a cytokinin (Stevens and Berry, 1988; Berry *et al.*, 1989) may indicate that the microsymbiont is directly involved in the development of the actinorhiza, by the production of plant growth regulators.

Finally, Hahn *et al.*, (1990) have shown that pigment-free RNA can be isolated from *Alnus* nodules by a guanidine-HCl method (Chirgwin *et al.* 1979). Also, Twigg and Mullin

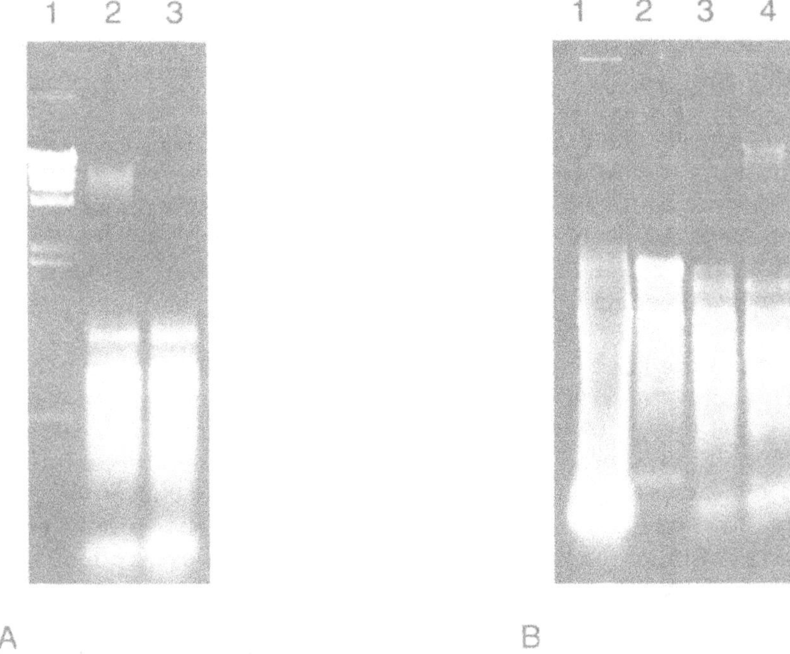

Figure 2. Comparison of different methods for the extraction of total RNA from *Alnus* roots and actinorhizae. (A) RNA separated on a 1.4% agarose gel and stained with ethidium bromide. Lane 1, lambda DNA digested with *Hind*III; 2, RNA extracted from actinorhizae using the method of Hughes and Galau (1988); 3, same RNA preparation after treatment with DNAse (BRL). (B) RNA separated on a 1.4% agarose gel and stained with ethidium bromide. Lane 1, RNA isolated from actinorhizae by a method using phenolic solvents (Mohapatra *et al.*, 1987); 2, RNA isolated from *Alnus* roots by the method of Hughes and Galau (1988); 3 and 4, RNA isolated from actinorhizae (Hughes and Galau 1988) obtained after inoculation with cultures of *Frankia* ACN1$^{AG}$ and CPN1 respectively.

(1990) reported a modification of the method described by Dellaporta *et al.* (1983) for isolation of RNA from *Alnus* roots and nodules and the construction of a cDNA library from poly A (+) RNA. Recently, actinorhiza-specific clones have been identified from a cDNA library, prepared from a poly A (+) RNA preparation isolated from actinorhizae of *Alnus rugosa*, by differential hybridization (Lévesque *et al.*, 1989). These nodule-specific genes showed different patterns of expression similar to early and late nodulin gene expression in legumes. Two of these "nodulin" cDNA sequences seem to be present on a single genomic clone. These clones require further characterization to confirm the gene structure and to determine gene function.

## 4.2. ALTERNATIVE STRATEGIES FOR THE MOLECULAR CLONING OF ACTINORHIZIN GENES

It is well known for the *Rhizobium*-legume symbiosis that host-specific nodulation (*nod*) genes determine both the infection and nodulation of specific hosts (Long, 1989). These *nod* genes are regulated by plant signals, known to include various flavonoid compounds (Zaat *et al.*, 1988). It is likely in actinorhizal symbiosis that one or many factors exuded by the host may be involved in the regulation of early nodulation development processes. Root hair deformation is an early characteristic of the *Frankia-Alnus* interactions, a phenomenon that is also observed with legumes. Some evidence suggests that molecular interactions between the symbionts occur at a very early stage of the infection process. Filtrates from the co-culture of axenic roots of alder and *Frankia* were found to induce extensive root hair deformation in *A. glutinosa* (Prin and Rougier, 1987). The authors suggested that plant exudates, released in the culture medium, are involved in the induction of root hair deformation. These plant compounds may stimulate the synthesis of an active deforming factor by *Frankia*, or may modify a deforming factor already released by *Frankia* in the culture medium.

Mutants of *Frankia* with impaired function in the sucessful development of nitrogen-fixing actinorhizae would be useful in the identification and analysis of the successive steps leading to the formation of the actinorhiza. Unfortunately, the practice of developing mutant strains unable to induce nodulation, as done previously with *Rhizobium*, is not yet feasible with *Frankia*. Single cell plating, which is required for the isolation of mutants, is difficult due to the filamentous nature of *Frankia*. The lack of an endogenous cloning system in *Frankia* also restricts the number of approaches available to study the molecular genetics of the microsymbiont. Re-introduction of the DNA fragment of interest into the bacteria may be used to confirm the proposed gene function and to test specific gene promotors using a reporter gene.

An effective way to overcome the problem of plating single cells for mutant analysis and transformation system development is to produce protoplasts from *Frankia* cultures. Protoplast isolation was first described by Faure-Raynaud *et al.* (1984), but the regeneration of cultures was not demonstrated. Normand *et al.* (1987) and Tisa and Ensign (1987) have since described both protoplast formation and regeneration for *Frankia*. Finally, recent developments in the use of electroporation for bacterial transformation represent an attractive avenue. Conditions for the uptake of marker DNA in undigested *Frankia* hyphae have recently been described using this method and these results appear promising (Cournoyer and Normand, 1990).

At the molecular level, root hair deformation and its successful penetration by the microsymbiont (cf. *Rhizobium* or *Frankia*), is a process that may involve the expression of specific genes from both symbionts. Freeze-fracture methods have been developed for the selective removal of root hairs from white clover, alfalfa (Gerhold *et al.*, 1985) and pea (Röhm and Werner, 1987). These procedures yield enough material for detailed biochemical studies of the root hair system. Changes that occur in the gene expression of root hairs, due to interaction with *Rhizobium*, have been recently investigated with success (Gloudemans *et al.*, 1989). Large numbers of seedlings of actinorhizal plants with a relatively high germination rate are required to undertake the biochemical characterization of root hairs. The use of the polymerase chain reaction (PCR) technique (Mullis and Faloona, 1987) should reduce the quantity of material required for such analyses, since the construction of cDNA libraries from a few cells has been described (Belyavsky *et al.*, 1989; Welsh *et al.*, 1990). It has been recently shown that this method is applicable to a wide variety of vegetative tissues of trees including actinorhizal plants (Bousquet *et al.*,1990). The amplification procedure may represent an attractive avenue for the study of gene expression at the root hair level, in relation to the infection process of *Frankia*. Assuming that the first level of specificity for the recognition of the *Frankia* by the host plant is at the root hair level, comparison of root hair proteins and the RNA populations from actinorhizal and non-actinorhizal plants might provide some information about host-*Frankia* specificity. The genera of *Alnus* spp. (actinorhizal) and *Betula* spp. (non-actinorhizal), which are closely related phylogenetically (Bousquet and Lalonde, 1990) may be a model of choice for this kind of approach.

Besides *Rhizobium*-legume symbiotic interactions, plant-pathogen interactions have aroused much interest and many laboratories have been working to describe the mechanisms by which plants defend themselves against pathogenic infection. It is conceivable that the recognition process in plant-microbe symbiotic interactions may share common messengers with pathogenic interactions. In the latter case, a cascade of events leading to the production of phytoalexins occurs in response to microbial invasion (reviewed in Albersheim *et al.*, 1986). On the other hand, the microsymbiont may have special keys to switch off the plant defense mechanism that will ensure the proper invasion of specific tissue for the establishment of the symbiosis. Oligosaccharides, produced after the degradation of the cell wall, may act as chemical messengers (oligosaccharins) in the regulation of compatibility for the symbiotic association.

Davis *et al.* (1986) have shown that a pectin-degrading enzyme (alpha-1,4-D-endopolygalacturonic acid lyase) from the phytopathogenic bacterium *Erwinia carotovora* induced phytoalexin accumulation in cotyledons of soybean. Recent findings that *Frankia* has sequence homology with pectate lyase genes from *Erwinia chrysanthemi* (Séguin and Lalonde, 1989; Simonet *et al.*, 1990) and the detection of pectolytic activity in pure culture (Séguin and Lalonde, 1989) raise the question, as suggested by Berry and Sunell (1990), that one or many chemical messengers (like oligosaccharins) may be involved in regulating the establishment of the symbiosis.

Finally, it must be kept in mind that the development of actinorhizae follows a sequence similar to that of lateral root formation. It is probable that the formation of actinorhizae does not require the expression of actinorhiza specific genes, but is regulated by genes expressed in the normal process of root development. The knowledge of the cellular processes and mechanisms governing actinorhiza development may provide some information which will improve our understanding of the processes regulating root development.

## Acknowledgments

We thank Paul de la Bastide and Louis Bernier for critical reading of the manuscript and Madeleine Pelchat for editing work. This work was supported by a Natural Sciences and Engineering Research Council of Canada operating grant to M. L. (No. A-2920).

## References

Akkermans, A.D.L., Huss-Danell, K. and Roelofsen, W. (1981). Enzymes of the tricarboxylic acid cycle and the malate-aspartete shuttle in the $N_2$-fixing endophyte of *Alnus glutinosa*. Physiol. Plant. **53**, 289-294.

Albersheim, P., Darvill, A.G., Sharp, J.K., Davis, K.R. and Doares, S.H. (1986). Studies on the role of carbohydrates in host-microbe interactions. *In* Recognition in Microbe-Plant Symbiotic and Pathogenic Interactions (B.J.J. Lugtenberg, ed.) NATO ASI Series, Vol. H4. Springer-Verlag. Berlin pp. 297-308.

Angulo Carmona, A.F. (1974). La formation des nodules fixateurs d'azote chez *Alnus glutinosa* (L.) Vill. Acta Bot. Neerl. **23**, 257-303.

Appelbaum, E. (1990). The *Rhizobium/Bradyrhizobium*-legume symbiosis. *In* Molecular Biology of Symbiotic Nitrogen Fixation. Gresshoff, P.M. (ed.). CRC Press, Boca Raton. pp. 132-158.

Appleby, C.A., Bogusz, D., Dennis, E.S. and Peacock, W.J. (1988). A role for heamoglobin in all plant roots? Plant Cell Environment **11**, 359-367.

Belefant, H. and Mullin, B.C. (1990). The identification of proteins specific to symbiosis in the *Alnus-Frankia* actinorhiza. 8th International Congress on Nitrogen Fixation, Knoxville, Tennessee, May 20-26 (Abstract G-22).

Belyavsky, T., Vinogradova, T. and Rajewsky, K. (1989). PCR-based cDNA library construction: general cDNA libraries at the level of a few cells. Nucleic Acids Res. **17**, 2919-2932.

Benson, D.R., Arp, D.J. and Burris, R.H. (1979). Cell-free nitrogenase and hydrogenase from actinorhizal root nodules. Science **205**, 688-689.

Berry, A. and Sunell, L.A. (1990). The infection process and nodule development. *In* The Biology of *Frankia* and Actinorhizal Plants. Schwintzer, C.R. and Tjepkema, J.D. (eds). Academic Press. San Diego. pp. 61-81.

Berry, A.M., Kahn, R.K.S. and Booth, M.C. (1989). Identification of indole compounds secreted by *Frankia* HFPArI3 in defined culture medium. Plant Soil. **118**, 205-209.

Bisseling, T., Been, C. Klugkist, J. van Kammen, A. and Nadler, K. (1983). Nodule-specific host proteins in effective and ineffective root nodules of *Pisum sativum*. EMBO J. **2**, 961-966.

Bisseling, T., Govers, F. and Stiekema, W. (1984). The identification of proteins and their mRNAs involved in the establishment of an effective symbiosis. Oxford Surveys of Plant Mol. Cell. Biol. **1**, 53-83.

Blom, J., Roelofsen, W. and Akkermans A.D.L. (1981). Assimilation of nitrogen in root nodules of alder (*Alnus glutinosa*). New Phytol. **89**, 321-326.

Bousquet, J., Simon, L. and Lalonde, M. (1990). DNA amplification from vegetative and sexual tissues of trees using polymerase chain reaction. Can. J. For. Res. **20**, 254-257.

Bousquet, J. and Lalonde, M. (1990). The genetics of actinorhizal Betulaceae. *In* The Biology of *Frankia* and Actinorhizal Plants. Schwintzer, C.R. and Tjepkema, J.D. (eds). Academic Press. San Diego. pp. 239-261.

Bousquet, J., Girouard, E., Strobeck, C., Dancik, B.P. and Lalonde, M. (1989). Restriction fragment polymorphisms in the rDNA region among seven species of *Alnus* and *Betula papyrifera*. Plant Soil **118**, 231-240.

Callaham, D. and Torrey, J.G. (1977). Prenodule formation and primary nodule development in roots of *Comptonia* (Myricaceae). Can. J. Bot. **51**, 2306-2318.

Chirgwin, J.M., Przybyla, A.E., MacDonald, R.J. and Rutter, W.J. (1979). Isolation of biologically active ribonucleic acid from sources enriched in ribonuclease. Biochemistry, **18**, 5296-5299.

Cournoyer, B. and Normand, P. (1990). Electroporation of *Frankia* spp. 8th International Congress on Nitrogen Fixation, Knoxville, Tennessee, May 20-26 (Abstract E-81).

Davis, K.R., Darvill, A.G., Albersheim, P. and Dell, A. (1986). Host-pathogen interactions. XXIX Oligagalacturonides released from sodium polypectate by endopolygalacturonic acid lyase are elicitors of phytoalexins in soybean. Plant Physiol. **80**, 568-577.

Dawson, J.A. (1990). Interactions among actinorhizal and Associated plant species. *In* The Biology of *Frankia* and Actinorhizal Plants. Schwintzer, C.R. and Tjepkema, J.D. (eds). Academic Press. San Diego. pp. 299-316.

Dellaporta, S.L., Wood, J. and Hicks, J.B. (1983). A plant DNA minipreparation: version II. Plant Mol. Biol. Reporter **1**, 19-21.

Diem, H.G. and Dommergues, Y.R. (1990). Current and potential uses and management of Casuarinaceae in the tropics and subtropics. *In* The Biology of *Frankia* and Actinorhizal Plants. Schwintzer, C.R. and Tjepkema, J.D. (eds). Academic Press. San Diego. pp. 317-342.

Edmands, J., Noridge, N.A. and Benson, D.R. (1987). The actinorhizal root nodule symbiont *Frankia* sp. strain CpI1 has two glutamine synthetase. Proc. Natl. Acad. Sci. USA **84**, 6126-6130.

Faure-Raynaud, M., Bonnefoy, M. A., Perradin, Y., Simonet, P. and Moiroud, A. (1984). Protoplast formation from *Frankia* strains. Microbios **41**, 159-166.

Franssen, H.J., Nap, J-P, Gloudemans, T., Stiekema, W., van Dam, H., Govers, F., Louwerse,J., van Kammen, A. and Bisseling, T. (1987). Characterization of cDNA for nodulin-75 of soybean: A gene product involved in early stages of root nodule development. Proc. Natl. Acad. Sci. USA **84**, 4495-4499.

Gerhold, D.L, Dazzo, F.B. and Gresshoff, P.M. (1985). Selective removal of seedling root hairs for studies of the *Rhizobium*-legume symbiosis. J. Microbiol. Methods **4**, 95-102.

Giasson, L. and Lalonde, M. (1987). Restriction pattern analysis of desoxyribonucleic acid isolated from callus and suspension of actinorhizal and non-actinorhizal Betulaceæ. Physiol. Plant. **70**, 304-310.

Gloudemans, T., Bhuvaneswari, T.V., Moerman, M., van Brussel, T., van Kammen, A. and Bisseling, T. (1989). Involvement of *Rhizobium leguminosarum* nodulation genes in gene expression in pea root hairs. Plant Mol. Biol. **12**, 157-167.

Goldberg, R.B., Hoschek, G., Tam, S.H., Ditta, G.S. and Breidenbach, R.W. (1981). Abundance, diversity, and regulation of mRNA sequence sets in soybean embryogenesis. Dev. Biol. **83**, 201-217.

Govers, F. Nap, J-P, van Kammen, A. and Bisseling, T. (1987). Nodulins in the developing root nodule. Plant Physiol. Biochem. **25**, 309-322.

Hahn, D., Starrenburg M.J.C. and Akkermans, A.D.L. (1990). Oligonucleotide probes that hybridize with rRNA as a tool to study *Frankia* strains in root nodules. Appl. Environ. Microbiol. **56**, 1342-1346.

Hattori, J. and Johnson, D.A. (1985). The detection of leghemoglobin-like sequences in legumes and non-legumes. Plant Mol. Biol. **4**, 285-292.

Hibbs, D.E. and Cromack, K. (1990). Actinorhizal plants in Pacific Northwest forests. *In* The Biology of *Frankia* and Actinorhizal Plants. Schwintzer, C.R. and Tjepkema, J.D. (eds). Academic Press. San Diego. pp. 343-363.

Hirel, B., Perrot-Rechenmann, C., Maudinas, B. and Gadal, P. (1982). Glutamine synthetase in alder (*Alnus glutinosa*) root nodules. Purification, properties and cytoimmunochemical localization. Physiol. Plant. **55**, 197-203.

Hughes, D. and Galau, G. (1988). Preparation of RNA from cotton leaves and pollen. Plant Mol. Biol. Reporter **6**, 253-257.

Huss-Danell, K. (1990). The physiology of actinorhizal nodules. *In* The Biology of *Frankia* and Actinorhizal Plants. Schwintzer, C.R. and Tjepkema, J.D. (eds). Academic Press. San Diego. pp. 129-156.

Keller, B. and Lamb, C.J. (1989). Specific expression of a novel cell wall hydroxyproline-rich glycoprotein gene in lateral root initiation. Genes Devel. **3**, 1639-1646.

Lalonde, M. and Knowles, R. (1975). Ultrastructure, composition and biogenesis of the encapsulation material surrounding the endophyte in *Alnus crispa* var. *mollis* root nodules. Can. J. Bot. **53**, 1951-1971.

Legocki, R. P. and Verma, D.P.S. (1980). Identification of "nodule-specific" host proteins (nodulins) in soybean involved in the development of *Rhizobium*-legume symbiosis. Cell **20**, 153-163.

Lévesque, M., Gleeson, T.D. and Johnson, D.A. (1989). Structure et expression de gènes (nodulins) impliqués dans la fixation de l'azote chez l'aulne, *Alnus incana*. 57e Congrès de l'ACFAS, Montréal, Canada, du 15 au 19 mai (Abstract).

Logemann, J., Schell, J. and Willmitzer, L. (1987). Improved method for the isolation of RNA from plant tissues. Anal. Biochem. **163**, 16-20.

Long, S.R. (1989). Rhizobium-legume nodulation: Life together in the underground. Cell, **56**, 203-214.

Mackay, J., Séguin, A. and Lalonde, M. (1988). Genetic transformation of 9 *in vitro* clones of *Alnus* and *Betula* by *Agrobacterium tumefaciens*. Plant Cell Rep. **7**, 229-232.

Martin, F., Hirel, B. and Gadal, P. (1983) Purification and properties of ornithine carbamyl transferase 1 from *Alnus glutinosa* root nodules. Z. Pflanzenphysiol. **111**, 413-422.

Merino, M.J., Ferreras, J.M., Munoz, R., Iglesias, R. and Girbés, T. (1990) Plant species containing inhibitors of eukaryotic polypeptide synthesis. J. Exp. Botany. **41**, 67-70.

Mohapatra, S.S., Poole, R.J. and Dhindsa, R.S. (1987). Changes in protein patterns and translatable messenger RNA populations during cold acclimation of alfalfa. Plant Physiol. **84**, 1172-1176.

Mullin, B.C. and An, C.S. (1990). The molecular genetics of *Frankia*. *In* The Biology of *Frankia* and Actinorhizal Plants. Schwintzer, C.R. and Tjepkema, J.D. (eds). Academic Press. San Diego. pp. 195-214.

Mullis, K.B. and Faloona, F.A. (1987). Specific synthesis of DNA *in vitro* via a polymerase catalyzed chain reaction. *In* Methods in Enzymology. Wu, R. (ed) Academic Press, Orlando. pp. 335-350

Newcomb,W. and Wood, S.M. (1987). Morphogenesis and fine structure of *Frankia* (Actinomycetales) : The microsymbiont of nitrogen-fixing actinorhizal root nodules. Int. rev. Cytol. **109**, 1-88.

Normand, P. and Lalonde, M. (1986). The genetics of actinorhizal *Frankia*, a review. Plant Soil. **90**, 429.

Normand, P., Simonet, P., Prin, Y. and Moiroud, A. (1987). Formation and regeneration of *Frankia* protoplasts. Physiol. Plant. **70**, 259-266.

Prin, Y. and Rougier, M. (1987). Preinfection events in the establishment of *Alnus-Frankia* symbiosis: Stydy of the root hair deformation step. Plant Physiol. (Life Sci. Adv.) **6**, 99-106.

Roberts, M.P., Jafar, S. and Mullin, B.C. (1985). Leghemoglobin-like sequences in the DNA of four actinorhizal plants. Plant Mol. Biol. **5**, 333-337.

Röhm, M., and Werner, D. (1987). Isolation of root hairs from seedlings of *Pisum sativum*. Identification of root hair specific proteins by in situ labeling. Physiol. Plantarum **69**, 129-136.

Safo-Sampah, S. and Torrey, J.G. (1988). Polysaccharide-hydrolyzing enzymes of *Frankia* (Actinomycetales). Plant Soil. **112**, 89-97.

Séguin, A. and Lalonde, M. (1988). Identification of host proteins involved in the development of *Frankia-Alnus* symbiosis. 2th International Congress on Plant Molecular Biology, Jerusalem, November 13-18 (Abstract 450).

Séguin, A. and Lalonde, M. (1989). Detection of pectolytic activity and *pel* homologous sequence in *Frankia*. Plant Soil. **118**, 221-229.

Séguin, A. and Lalonde, M. (1990a). Micropropagation, tissue culture, and genetic transformation of actinorhizal plants and *Betula*. *In* The Biology of *Frankia* and Actinorhizal Plants. Schwintzer, C.R. and Tjepkema, J.D. (eds). Academic Press. San Diego. pp. 215-238.

Séguin, A. and Lalonde, M. (1990b). Characterization of actinorhizins involved in the development of *Frankia-Alnus* symbiosis. 8th International Congress on Nitrogen Fixation, Knoxville, Tennessee, May 20-26 (Abstract E-77).

Simonet, P., Normand, P., Hirsch, A.M. and Akkermans, A.D.L. (1990). The genetics of the *Frankia*-actinorhizal symbiosis. *In* Molecular Biology of Symbiotic Nitrogen Fixation. Gresshoff, P.M. (ed.). CRC Press, Boca Raton, pp. 77-109.

Stevens, G.A. and Berry, A.M. (1988). Cytokinin secretion by *Frankia* sp. HFPArI3 in defined medium. Plant Physiol. **87**, 15-16.

Tisa, L.S. and Ensign, J.C. (1987). Formation and regeneration of protoplasts of the actinorhizal nitrogen-fixing actinomycete *Frankia*. Appl. Environ. Microbiol. **53**, 53-56.

Tjepkema, J.D. (1983). Hemoglobins, iun the nitrogen-fixing root nodules of actinorhizal plants. Can. J. Bot. **61**, 2924-2929.

Tremblay, F.M., Périnet, P. and Lalonde, M. (1986). Tissue culture of *Alnus* spp. with regard to symbioses. *In* Biotechnology in Agriculture and Forestry. Bajaj, Y.P.S. (ed). Springer-verlag, Berlin and New York. pp. 87-100.

Twigg, P. and Mullin, B.C. (1990). Extraction and cDNA cloning of mRNA from roots and nodules of European Black Alder (*Alnus glutinosa* L. Gaetrn.). 8th International Congress on Nitrogen Fixation, Knoxville, Tennessee, May 20-26, 1990 (Abstract G-22).

van de Wiel, C., Scheres, B., Franssen, H., van Lierop, M-J., van Lammeren, A., van Kammen, A. and Bisseling, T. (1990). The early nodulin transcript ENOD2 is located in the nodule parenchyma (inner cortex) of pea and soybean root nodules. EMBO J. **9**, 1-7.

van Kammen, A. (1984). Suggested nomenclature for plant genes involved in nodulation and symbiosis. Plant Mol. Biol. Reporter **2**, 43.

Vergnaud, L., Chaboud, A. and Rougier, M. (1987). Preliminary analysis of root exudates of *in vitro*-micropropagated *Alnus glutinosa* clones. Physiol. Plantarum **70**, 319-326.

Verma, D.P.S., Fortin, M.G., Stanley, J., Mauro, V.P. Purohit, S. and Morrison, N. (1986). Nodulins and nodilin genes of *Glycine max*. Plant Mol. Biol. **7**, 51-61.

Ward, L.J. and De Boer, S.H. (1989). Characterization of a monoclonal antobody against active pectate lyase from *Erwinia carotovora*. Can. J. Microbiol. **35**, 651-655.

Welsh, J., Liu, J.P. and Efstratiadis, A. (1990). Cloning of PCR-amplified total cDNA: construction of a mouse oocyte cDNA library. Genet. Anal. Techn. Appl. **7**, 5-17.

Wheeler, C.T. and Miller, I.M. (1990). Current and potential uses of actinorhizal plants in Europe. *In* The Biology of *Frankia* and Actinorhizal Plants. Schwintzer, C.R. and Tjepkema, J.D. (eds). Academic Press. San Diego. pp. 365-389.

Wheeler, C.T., Henson, I.E. and McLaughlin, M.E. (1979). Hormones in plant bearing actinomycete nodules. Bot. Gaz. (Chicago), Suppl. **140**, S52-S57.

Zaat, S.A.J., Wuffelman, C.A., Mulders, I.H.M., van Brussel, A.A.N. and Lugtenberg, B.J.J. (1988). Root exudates of various host plants of *Rhizobium leguminosarum* contains different sets of inducers of *Rhizobium* nodulation genes. Plant Physiol. **86**, 1298-1303.

# Rapid, exponential growth and increased biomass yield of some *Frankia* strains in buffered and stirred mineral medium (BAP) with phosphatidyl choline


J. SCHWENCKE
*Laboratoire d'Enzymologie du CNRS, F-91198 Gif-sur-Yvette, France. Present address: Laboratoire de Biotechnologie les Symbioses Forestières Tropicales. ORSTOM/CTFT–CIRAD. 45 bis Ave. de la Belle Gabrielle, F-94736 Nogent-sur-Marne Cedex, France*


*Key words:*  actinomycete, Casuarina, Frankia

**Abstract**


Rapid, exponential growth of *Frankia* sp.BR (ORS 020608) isolated from *Casuarina equisetifolia* was found to require stirring (magnetic bar) and adequate buffering (MES or MOPS buffers). We used a mineral medium (BAP) at 28°C with propionate and $NH_4^+$ as the sole carbon and nitrogen source respectively. Sporulation was enhanced by stirring, but the addition of phosphatidyl choline was found to impair sporulation and slightly improve growth. In these conditions a generation time of 14–16 hours was observed with a maximal biomass yield of 108 $\mu$g of protein $\times$ mL$^{-1}$ (Bicinchoninic acid) at day 4 when growth ceased. In stirred conditions *Frankia* growth as microcolonies (30–200 $\mu$m diameter) showing an 'open mesh' type of mycelium and also a random distribution of growing hyphae inside and in the periphery of the microcolonies (acridine orange vital staining). In static conditions macrocolonies and only peripheric growth can be observed. Optimal growth in buffered and stirred conditions in mineral medium has been also observed with Cj and CeF *Frankia* isolates. The reasons for rapid and balanced hyphal growth in stirred conditions are discussed.


## Introduction

The very slow rates and low biomass yield of *Frankia* isolates grown in static conditions are well documented. Growth periods of weeks are usually needed to obtain adequate amount of biomass and doubling periods of 2 to 3 days are common. Static growth produces macrocolonies (1000 $\mu$m or more) with hyphae showing marked heterogeneous physiological conditions making physiological experiments unreliable (Benson and Schultz, 1990; Lechevalier and Lechevalier, 1990; Normand and Lalonde, 1986). Reasons for the unbalanced growth of filamentous organisms are discussed in (Benson and Schultz, 1990). When starting a study on the proteolytic system of *Frankia* spBR we were confronted with these drawbacks. Few papers have been published on the growth of *Frankia* under stirring conditions. Thus work by Diem and Dommergues indicate that the obtention of inoculum of *Frankia* spBR was enhanced by low speed shaking for 10 days periods (Diem and Dommergues, 1990). Also cells of *Frankia* sp ArI3 (Murray et al., 1984) and CpI$_1$ (Noridge and Benson, 1986) have been reported to grow under shaking but, to the best of our knowledge, no systematic study has been published on this matter. Here we report the conditions under which three *Frankia* isolates from *Casuarina* trees can be grown under stirring to overcome most of the afore mentioned problems of static growth.

## Materials and methods

### Organism and culture media

The origin and designation of the *Frankia* strains are indicated in Table 1. All strains were maintained in solid BAP medium (Fontaine et al., 1986). They were grown at 28°C using 500 mL flasks for 250 mL of medium and stirred at 200 rpm with a cross-type magnetic bar (29 mm diam). Cultures were started with 4 $\mu$g of mycelial protein per mL from an exponentially growing culture, syringe homogenized 6 times using a $0.7 \times 30$ mm sterile needle. For protein determination aliquots of the culture were centrifuged at 2000 rpm in a swinging-bucket rotor using plastic (conical bottom) tubes, the pellet washed twice with BAP medium then, resuspended in water, sonicated on ice $5 \times 10$ seconds prior to determining protein by the bicinchoninic acid method (Pierce enhanced protocol) according to manufacturer's instructions.

### Vital staining and microscopic techniques

Aliquots of the cultures were added to the same volume of a $0.05$ mg mL$^{-1}$ solution of acridine orange in $50$ m$M$ Na$_2$HPO$_4$ pH 6.8, (stored at 4°C in the dark) and incubated at room temperature for at least 25 min. For epifluorescence observation and photography we used a combination of a 450–495 nm excitation, 510 nm emission filters in a Reichert-Jung Polyvar microscope. Fluorescent bead markers were from Fullbright.

### Reagents

Crude soybean phosphatidyl choline is a product P-3644 Type IV-S from Sigma. A stock solution was prepared dissolving 500 mg in 50 ml absolute ethanol at 60°C, filter sterilized and added to 15 mL of sterile water. Mes (M-8250) and ac-

ridine orange (A-6014 lot 26C-01941) were from Sigma. Absolute ethanol is product 983 from Merck. All other chemicals were of analytical grade.

## Results

### Conditions for fast growth of Frankia *spBR*

Optimal growth shown in Figure 1 was obtained by adding $10$ m$M$ Mes/Tris (pH 6.8) and $2$ $\mu$g mL$^{-1}$ of crude soybean phosphatidyl choline (sPC) and stirring at 200 rpm. Growth obtained in static conditions is also shown for comparison. In these conditions rapid, exponential growth started 12 hours after inoculation and was maintained for about 60 hours. *Frankia* was found to grow as microcolonies (30–200 $\mu$m) showing an 'open-mesh' type of mycelia. Maximal mycelial growth ($108$ $\mu$g protein mL$^{-1}$) was obtained at day 4 and thereafter a lytic phase was observed.

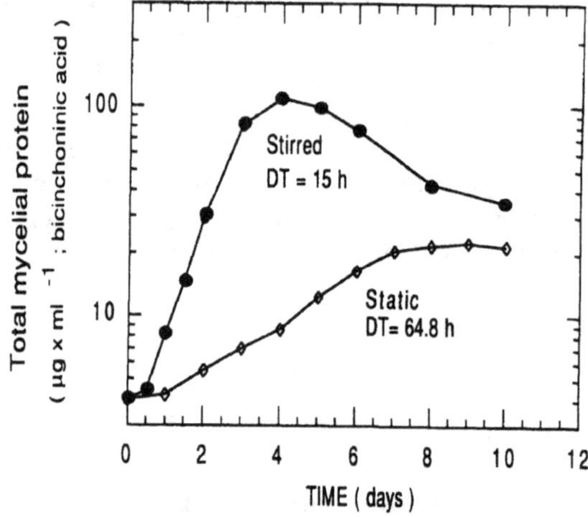


Fig. 1. Growth curve of *Frankia* sp. BR in stirred (●) and static (◇) conditions. DT = doubling time. For details see Material and methods.

Table 1. Origin of *Frankia* strains used in this study

| Trivial designation | *Casuarina* Original host | Strain N° | Geographycal Origin | Reference |
|---|---|---|---|---|
| BR | *C. equisetifolia* | ORS 020608 | Brasil | Diem, unpubl. |
| CeF | *C. equisetifolia* | ORS 020607 | Florida, USA | Diem and Dommergues, 1983 |
| Cj | *C. junghuhniana* | ORS 021001 | Bangkok, Thailand | Diem et al., 1982 |

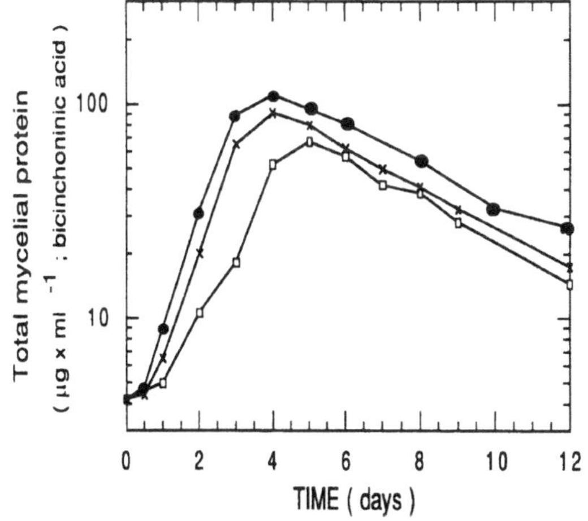

*Fig. 2.* Growth of *Frankia* sp. BR under stirring conditions in BAP alone (□), BAP + 10 m$M$ Mes/Tris pH 6.8 (×), BAP + 10 m$M$ Mes/Tris pH 6.8, 2 $\mu$g mL$^{-1}$ crude soybean phosphatidyl choline (●).

Stirring is the most important parameter in promoting growth (Fig. 2). Buffering with Mes/Tris (pH 6.8) as well as, addition of 2$\mu$g mL$^{-1}$ of crude soybean phosphatidyl choline (sPC) slightly increases biomass yield. Stirring enhance the appearance of an 'open mesh' type of mycelia but also sporangia. The addition of sPC clearly impairs sporangia formation up to day 6 and enhances the 'open-mesh' structure (Fig. 3a, b, c). Only small sporangia were observed starting from day 6 up to day 14. In static growth conditions large colonies and sporangia were observed already at day 6–7 (Fig. 3d).

*Growth under stirring of* Frankia *sp. Cj and CeF*

As for *Frankia* sp. BR, strains Cj and CeF were found to grow rapidly and with a clear exponential phase in BAP + sPC-Mes under shaking. A summary of these results is shown in Table 2.

Strain CeF grows rapidly showing a quite particular filiform morphology (Fig. 3e) and a longer lag period (Table 2). Strain Cj grows also as microcolonies with an 'open mesh' structure but with a clearly different aster-like morphology (Fig. 3f).

**Discussion**

We describe here that the mycelia of three different *Frankia* isolates from *Casuarina* can be grown under stirring in defined conditions showing a rapid exponential growth, no sporangia, no RTH forms (Diem and Dommergues, 1985) and no lysed hyphae. Results using a 99% pure phosphatidyl cholines mixture from poultry origin (Schwencke, unpublished results) suggest that the delayed appearance of sporangia may be due to phosphatidyl cholines itself and not to another specific component of vegetal origin. In this context it is interesting that lecithin is a component of Qmod, a *Frankia* growth medium (Lalonde and Calvert, 1979) and that Quispel et al. (1983) have reported that diglycerides and phospholipids from *Alnus* root extracts are stimulating substances for the isolation of some *Frankia* strains. However, no rapid growth was detected under stirring conditions using the rich Qmod medium indicting that a defined mineral medium as BAP (Fontaine et al., 1986) is essential for good results with the strains studied here.

Besides the lack of sporangia, RTH and lysed hyphae, the physiological homogeneity of strains grown in our conditions is suggested by the observation of "active growing" cells inside microcolonies after acridine orange vital staining. Phosphatidyl cholines may improve balanced growth and, as a consequence, delay sporangia

*Table 2.* Comparative data of three *Frankia* strains growing in stirred, buffered BAP medium, with sPC[a]

| Strain | Initial lag period (hours) | Exponential growth | | Maximal biomass obtained[b] $\mu$g protein mL$^{-1}$ |
| --- | --- | --- | --- | --- |
| | | Duration (days) | Doubling time (hours) | |
| BR | 12 | 2.5 | 15–16 | 109 |
| Cj | 18 | 2.0 | 14–15 | 121 |
| CcF | 36 | 2.5 | 14–15 | 131 |

[a] 5 m$M$ Na propionate, 10 m$M$ NH$_4$Cl, 10 m$M$ Mes/Tris (pH 6.8), 2 $\mu$g crude soybean phosphatidyl choline (sPC).
[b] Measured at day of maximal growth using the micro bicinchoninic acid assay at 60°C.

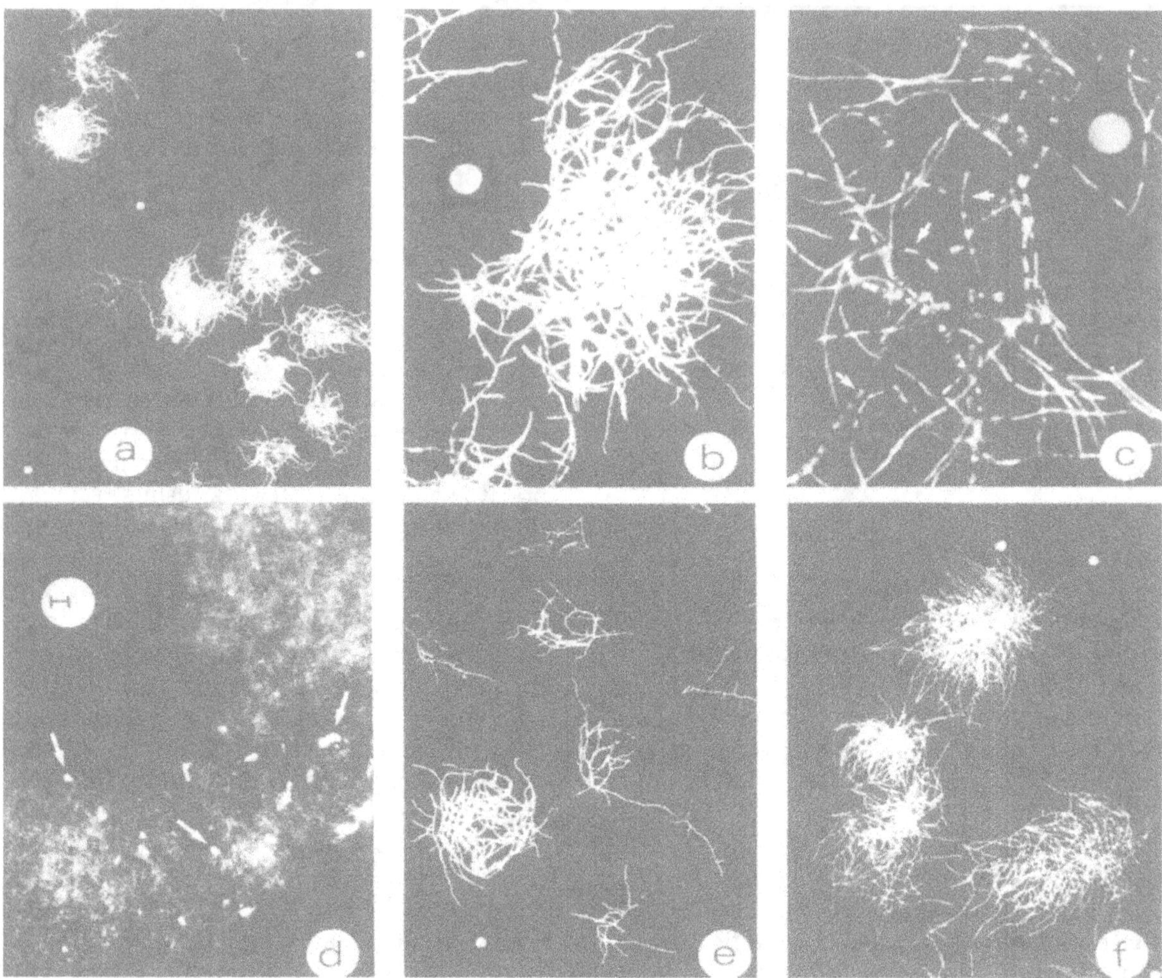

*Fig. 3.* Morphology of various *Frankia* microcolonies grown under stirring conditions as visualized by epifluorescence after acridine orange vital coloration. **a**: sp. BR, 68 hrs, fluorescent beads of 10 $\mu$m; **b**: sp. BR, 68 hrs fluobead marker of 10 $\mu M$; **c**: higher magnification of a single microcolonie of sp. BR, 68 hrs, fluobead marker 5.3 $\mu$m. The clear hyphal zones correspond to yellow-orange fluorescence indicating actively growing cells (arrow heads); **d**: sp. BR, 130 hrs static growth, note sporangia inside the macrocolonie (arrows), bar = 10 $\mu$m; **e**: sp. CeF 48 hrs, fluobead marker 10 $\mu$m, **f**: sp. Cj, 48 hrs, fluobead marker 10 $\mu$m.

appearance if they are utilized for membrane biosynthesis (see also Quispel et al., 1983) and may co-jointly with stirring, facilitate the formation of the "open-mesh" type of mycelia. Most probably this type of structure is a reflection of facilitated nutrient and $O_2$ accessibility to the hyphae as well as efficient elimination of waste products. All these factors may collaborate to facilitate rapid exponential growth and homogeneous physiological conditions in the whole mycelium.

The question of whether other strains may similarly grow in optimized stirring conditions is under study. Preliminary results indicate that isolates from *Allocasuarina*, *Alnus* and *Colletia* can be efficiently grown under shaking in buffered mineral media with phosphatidyl cholines although, according to the strain, changes in the carbon source are necessary. However, balanced growth under stirring will certainly be limited to isolates having the adequate degree of microaerophilicity. In spite of this limitation this opens the way for the generalization of the method to a variety of *Frankia* strains and also, quite importantly, to the development of reliable biochemical studies of *Frankia*.

## Acknowledgements

I am grateful to Dr Y Dommergues and Dr H G Diem for their constant encouragement and helpful discussions. I am indebted to Dr S Brown (ISV, CNRS) for facilities to use the Reichert-Jung microscope. I thank also P Benoist for Fig. 3d, A. Müller for help with some experiments and Mme Nappé for her patient collaboration in the preparation of this camera-ready typescript. This research was supported in part by Grant MRT 88T0807 from the Ministère de la Recherche et de la Technologie.

## References

Benson D R and Schultz N 1990 Physiology and biochemistry of *Frankia* in culture. *In* The Biology of *Frankia* and Actinorhizal Plants. Eds. C R Schwinizer and J D Tjepkemo. pp 107–127. Academic Press, San Diego, CA.

Diem H G, Gauthier D and Dommergues Y R 1982 Isolement et culture in vitro d'une souche infective et effective de *Frankia* isolée de nodules de *Casuarina* sp. C. R. Acad. Sc. (Paris) 295, 759–763.

Diem H G and Dommergues Y R 1983 The isolation of *Frankia* from nodules of *Casuarina*. Can. J. Bot. 61, 2822–2825.

Diem H G and Dommergues Y R 1985 In vitro production of specialized reproductive torulose hyphae by *Frankia* strain ORS 021001 isolated from *Casuarina junghuhniania* root nodules. Plant and Soil 87, 17–29.

Noridge N A and Benson D R 1986 Isolation and nitrogen-fixing activity of *Frankia* sp CpI$_1$ vesicles. J. Bacteriol. 166, 301–305.

Diem H G and Dommergues Y R 1990 Current and potential uses and management of *Casuarinaceae* in the tropics and subtropics. *In* The Biology of *Frankia* and Actinorhizal Plants. Eds. C R Schwintzer and J D Tjepkema. pp 317–342. Academic Press, San Diego, DA.

Fontaine M S, Torrey J G and Young P H 1986 Effects of long term preservation of *Frankia* strains on infectivity, effectivity, and in vitro nitrogenase activity. Appl. Environ. Microbiol. 51, 694–698.

Lalonde M and Calvert H E 1979 Production of *Frankia* hyphae and spores as an infective inoculant for *Alnus* species. *In* Symbiotic Nitrogen Fixation in the Management of Temperate Forest. Eds. J C Gordon, C T Wheeler and D A Perry. pp 95–110. Oregon State University, Corvallis, OR.

Lechevalier M and Lechevalier H A 1990 Systematics, isolation and culture of *Frankia*. *In* The Biology of *Frankia* and Actinorhizal Plants. Eds. C R Schwintzer and J D Tjepkema. pp 35–60. Academic Press, San Diego, CA.

Murry M A, Fontaine M S and Torrey J G 1984 Growth kinetics and nitrogenase induction in *Frankia* sp. HFPA.I3 grown in batch culture. Plant and Soil 78, 61–78.

Normand F and Lalonde M 1986 The genetics of actinorhizal *Frankia*: A review. Plant and Soil 90, 429–453.

Quispel A, Burggraaf A J P, Borsie H and Tak T 1983 The role of lipids in the growth of *Frankia* isolates. Can. J. Bot. 61, 2801–2806.

# ENHANCING NODULATION AND GROWTH OF *CASUARINA GLAUCA* PLANTED IN SUBSTRATE ALREADY  INFECTED WITH *FRANKIA*


M.G. Girgis and Y. Ishac, Unit of Bioferilizers, Faculty of Agriculture, Ain Shams University, Cairo.
H. El-Lakany, Desert Development Center, American University in Cairo. Y.R. Dommergues, BSSFT (ORSTOM/CTEF-CIRAD) 45 bis Av. de la Belle Gabrielle, 94736, Nogent-sur-Marne.



**ABSTRACT:**

*Casuarina glauca* seedlings were grown in a compost infected by crushed nodules of the same species of *Casuarina*.

**First experiment:** When the plants  were 3 months old, three treatments were used: no polymeric inoculant applied (treat. 1); inoculation with *Frankia* (strain ORS 021001) type A (alginate) polymeric inoculant (treat. 2); inoculation with *Frankia* (strain ORS 021001) type AK (alginate ++ kaolin) polymeric inoculant (treat. 3). When the plants were 6 months old, the effect of the different treatments could be classified as follows: nodule number 1 = 2 <3; nodule dry weight, 1 = 2 = 3; basal stem diameter (tree volume) 1 = 2 <3; plant height and dry weight, 1 = 2 = 3. In other words, the addition of kaolin to the polymeric inoculant appeared to significantly improve nodulation and tree volume seedlings grown in an already *Frankia* infected substrate.

**Second Experiment:** When the plants were 3 months old, five treatments were used: no polymeric inoculant applied (treat. 1); inoculation with *Frankia* (S19) polymeric inoculant alone (treat 2); inoculation with a mixture of *Frankia* (S19) and *Bradyrhizobium* (treat. 3), or Frankia (S19) and *Azorhizobium* (treat. 4) polymeric inoculants. When plants were 6 months old, nodule dry weight and basal stem diameter in treat. 4 were significantly lower than in other treat., suggesting an inhibitory effect of *Azorhizobium*. In treatment 3 nodule number was significantly increased as compared to other treatments, suggesting a favourable effect of *Bradyrhizobium*  on root infection of plants grown in an already infected substrate.

## INTRODUCTION:

Polymeric inoculants (Dommergues et al., 1979; Bashan, 1986; Diem and Domemrgues, 1990) based on the entrapment of *Frankia* in alginate beads have been successfully used in the field (Sougoufara et al., 1989). However, it was thought that their performance could still be improved through two approaches : addition of clay to the polymer (alginate); association with helper bacteria, whose role was suggested by Knowlton et al. (1980).

Certain clay amendments are known to dramatically protect soil microorganisms exposed to adverse conditions especially drought (Dommergues, 1964; Marshall, 1970; Dommergues and Mangenot, 1970). It is with this view that a first experiment was carried out to examine the possible effect of kaolinite addition to alginate in the preparation of Frankia polymeric inoculants to be applied to *Casuarina glauca*.

There have been a number of reports that various bacterial and fungal species can promote nodulation by rhizobia (e.g. Ascon - Aguilar and Barea, 1981; Yahalom et al., 1987; Nishijima et al., 1988; Alagawadi and Gaur, 1988). However, little work has been done in the case of *Frankia*, except the pioneer work of Knowlton et al. (1980) and that of Diem and Gauthier (1982) on the dual inoculation of *Casuarina equisetifolia* by *Glomus mosseae* and *Frankia* . The second experiment reported here was undertaken to examine the associative effect of two rhizobium strains used as rhizospheric bacteria.

## MATERIAL AND METHODS:

The seedlings of *Casuarina glauca* were provided by the Desert Development Center (DDC), Sadat City, Egypt. The seedlings were cultivated for 3 months according to the procedure adopted in the DDC nursery. Seeds were placed in polythene containers filled with a substrate made of a mixture of compost and sandy soil (1 : 1) inoculated with crushed nodules of *Casuarina glauca* collected in the vicinity. Superphosphate was applied at the surface of the containers once a week. N fertilization was applied with irrigation water. The compost composition was as follows: alfalfa hay 50%; peanut shells 20%; farmyard manure 10%, crop residues 15%; urea 2.5% ammonium sulphate 2.5%. The 3 month-old plants thus obtained were used in the following experiments:-

**First Experiment:**

Three treatments were applied:-

- Treatment 1: Inoculation with crushed nodules only; no polymeic inoculant applied; addition of amount of buffer solution equivalent to that applied in other treatments;

- Treatment 2: Inoculation with crushed nodules and by polymeric inoculant A (alginate without kaolinite) with *Frankia* strain ORS 021001 (Diem et al., 1983).

- Treatment 3: Inoculation with crushed nodules and by polymeric inoculant AK (alginate with Kaolinite) with same *Frankia* strain as in treatment 2. Inoculant A was prepared according to usual procedure by entrapping a 15 day - old culture of *Frankia* in a 4% alginate pseudo-solution (Diem et al., 1989). Inoculant AK was obtained by entrapping the same culture in a mixture of alginate (4%) and Kaolinite (10%). Before being applied to the seedlings, air-dried beads of inoculants A and AK were rehydrated in a phosphate buffer pH 7.4 (15 g $K_2HPO_4$ and 2 g $KH_2PO_4$ g per liter of distilled water) using 10 g of dried inoculant A and 35 g of inoculant AK per liter of buffer. After 10 - 14 hours the resulting slurry was hormogenized and applied to the plants at the rate of 5ml per container, the penetration of the slurry along the root system being facilitiated by digging holes in the substrate using a stick 0.5cm diameter. Number of plants per treatment : 36.

**Second Experiment:**

Four treatments were applied:

- Treatment 1: Inoculation with crushed nodules only; no polymeric inoculant applied; addition of amount of buffer solution equivalent to that applied in other treatments;

- Treatment 2 : Inoculation with crushed nodules and by polymeric inoculant A with *Frankia* strain S 19 (Girgis et al., 1990);

- Treatment 3: Inoculation with crushed nodules and by two polymeric inoculants A: Inoculant with *Frankia* strain S 19 and inoculant with *Bradyrhizobium japonicum* strain G 49 (USDA 122) (Lagacherie et al., 1977).

- Treatment 4: Inoculation with crushed nodules and by two polymeric inoculants A : inoculant with *Frankia* strain S 19 and inoculant with *Azorhizobium caulinodans* strain ORS 571 (Dreyfus et al., 1988).

The preparation of polymeric *Bradyrhizobium* and *Azorhizobium* inoculants was similar to that of *Frankia* inoculant except that the cultures were 8 and 4 days old respectively.

Number of plants per treatment : 36.

## Observations:

When the plants were 9 months old, the following measurements made: nodule number and dry weight, plant height, basal stem diameter, plant dry weight.

## RESULTS AND DISCUSSIONS

In all cases, the substrate used to grow the *Casuarina glauca* seedlings was already infected with *Frankia* since, according to the usual procedure adopted in the Desert Development Center, crushed nodules were always incorporated to the substratum. This fact must be taken into account to interprete the experimental results reported in Tables 1 and 2.

## First Experiment:

The ranking of the three treatments was as follows (Table 1): nodule number 1 < 2 < 3; nodule dry weight 1 = 2 < 3; basal stem diameter (a characteristic related to tree volume) 1 = 2 < 3; plant height and dry weight 1 = 2 = 3. In other words adding kaolinite to alginate for the preparation of the inoculant significantly improved nodulation, which can be explained by the fact that kaolinite would directly protect the *Frankia* cells against unfavourable edaphic factors (Dommergues, 1964; Dommergues and Mangenot, 1970), or indirectly affect the infection process. The effect of kaolinite addition to alginate on plant growth was less spectacular and concerned only the basal diameter. Plant dry weight was increased but not significantly. Height was not affected by the treatment, a result that is not surprising since it is known that the use of height as the sole criterion for assessing the effect of inoculation may not be sufficient (Sougoufara et al., 1989). The reason for the weak response of the plant, as far as dry weight is concerned, is probably because the plants were tightly packed on racks, which hindered the growth.

## Second Experiment

Adding *Bradyrhizobium Azorhizobium* to *Frankia* significantly affected nodulation, i.e. increased nodule number in the case of *Bradyrhizobium*, and decreased nodule weight in the case of *Azorhizobium*. A significant effect of either rhizobia strains could be expected since rhizobia are known to be excellent root colonizers (Schmidt, 1978), a characteristic enabling them to interact with Frankia. However, the resulting type of interaction (favourable or unfaovurable)

Table (1): First experiment. Response of 6-month old *Casuarina glauca* to inoculation with crushed nodules alone or crushed nodules plus two types of *Frankia* (ORS 021001) polymeric inoculants; inoculant A (alginate without Kaolin) and inoculant AK (Alginate + Kaolin).

| INOCULATION WITH | NODULE NO. PER PLANT | NODULE DRY WEIGHT (g/Plant) | PLANT HEIGHT (cm) | BASAL STEM DI-AMETER (cm) | PLANT DRY WEIGHT (g/Plant) |
|---|---|---|---|---|---|
| 1. Crushed nodules | 10.2 a | 2.06 a | 103 a | 2.79 a | 20.1 a |
| 2. Crushed nodule + inoculant A. | 26.1 a | 2.25 a | 101 a | 2.85 a | 22.5 a |
| 3. Crushed nodules + inoculant AK | 63.9 b | 3.59 a | 99 a | 3.37 b | 27.2 a |

Polymeric inoculants were applied when plants were 3 months old.

Date in each column, followed by same letter are not significantly different according to the Newman-Keuls test (P < 0.01).

Remark: nodule dry weight is significantly higher in treatment 3 than in treatment 2 for  P < 0.05 .

Table (2) : Second experiment. Response of 6-month old *Casuarina glauca* to four inoculation treatments: treatment 1, inoculation with crushed nodules alone; other treatments inoculation with crushed nodules plus different polymeric inoculants containing *Frankia* (S19); treatment 2, *Frankia* alone; treatment 3, *Frankia* plus *Bradyrhizobium* japonicum G49; treatment 4, *Frankia* plus *Azorhizobium caulinodans* ORS 571.

| INOCULATION WITH | NODULE NO. PER PLANT | NODULE DRY WEIGHT (g/Plant) | PLANT HEIGHT (cm) | BASAL STEM DI-AMETER (cm) | PLANT DRY WEIGHT (g/Plant) |
|---|---|---|---|---|---|
| 1. Crushed nodules alone. | 10.6 a | 2.22 a | 103 a | 2.9 a | 18.6 a |
| 2. Crushed nodule + *Frankia*. | 7.7 a | 3.93 b | 122 b | 3.2 a | 29.1 bc |
| 3. Crushed nodule + *Frankia*. + Bradyrhizobium. | 18.5 b | 2.64 a | 105 a | 3.1 a | 22.9 ab |
| 4. Crushed nodule + *Frankia*. + Azorhizobium. | 11.4a | 0.90 c | 103 a | 2.3 b | 16.8 a |

Polymeric inoculants were applied when plants were 3 months old.

Date in each column, followed by same letter are not significantly different according to the Newman-Keuls test (P < 0.01).

could not be predicted nor was it explained.

Adding *Azorhizobium* to *Frankia* significantly reduced plant growth expressed either as height, dry weight or basal stem diameter. This unexpected deleterious effect of *Azorhizobium* on the *Casuarina - Frankia* symbiosis certainly deserves further investigations.

**ACKNOWLEDGEMENTS:**

The authors wish to thank Medhat H. Abaas from NRC, Cairo and Eiman Yasin from ARC, Cairo, for their kind assistance.

**REFERENCES**

Alagawadi, A.R. & Gaur A.C. (1988). Associative effect of *Rhizobium* and phosphate-solubilizing bacteria on the yield and nutrient uptake of chickpea. Plant and Soil 105, 241 - 245.

Azcon-Aguilar C. & Barea J.M. (1981). Field inoculation of *Medicago* with VA mycorrhiza and rhizobium in phosphate - fixing agricultural soil. Soil BIol. & Biochem. 13, 19 - 22.

Bashan Y. (1986). Alginate beads as synthetic inoculant carriers for slow-release of bacteria that affect plant growth. Appl. Environ. Microbiol. 51, 1089 - 1098.

Diem, H.G. & Gauthier D. (1982). Effect de l'infection mycorhizienne (*Glomus mosseae*) sur la nodulation et la croissance de *Casuarina equisetifolia*. C.R. Acad. Sc. Paris 294, Ser. 3, 215 - 218.

Diem H.G., Gauthier D. & Dommergues Y.R. (1983): An effective strain of *Frankia* from *Casuarina* sp. Can. J. Bot. 61, 2815 - 2821.

Diem H.G. & Dommergues Y. (1990). Current potential uses and management of Casuarinaceae in the tropics and subtropics in The biology of *Frankia* and actinorhizal plants (C.R. Schwintzer and J.D. Tjepkema eds.) pp. 317 - 342. Academic Press, New York.

Diem H.G., Ben Khalifa K., Neyra M. and Dommergues Y.R. (1989): Recent advances in the inoculant technology with special empahsis on plant symbiotic microorganisms. In Proc. Workshop on Advanced Technologies for increased Agriculture Production : Actual Situation, Future Prospects and Concrete Possibilities of Application in Developing Countries. Santa Margherita Ligure, Italy, 25 - 29 Sept. 1988 (U. Leone, G. Rialdi and R. Vanore eds.) pp. 196 - 209, Consiglio Nazionale delle Ricerche, Roma & Universita degli Studi di Genova, Genova, Italy.

Dommergues Y. (1964): Etude de quelques facteurs influant sur le comportement de la microflore du sol au cours de la dessiccation. Science du Sol., 141 - 145.

Dommergues Y. & Mangenot F. (1970): Ecologie microbienne du sol. pp. 473 - 474. Masson, Paris.

Domemrgues Y. Diem H.G. & Divies C. (1979): Polyacrylamide-entrapped *Rhizobium* as an inoculant for legumes. Appl. Environ. Microbiol. 37, 779 - 781.

Dreyfus B., Garcia J.L. & Gillis M. (1988): Characterisation of *Azorhizobium caulinodans* gen. nov., sp. nov. a stem - nodulating nitrogen - fixing bacterium isolated from *Sesbania rostrata Inter. J. System. Bacteriol. 38, 89 - 98.*

Girgis, M.G.Z., Ishac, Y.Z., El Haddad M., Saleh, E.A., Diem H.G. and Dommergues Y.R. (1990: First report on isolation and culture of effective *Casuarina* - compatible strains of *Frankia* from Egypt. Proc. Second International Workshop, Cairo 15 - 25 Jan. 1990 (in press).

Knowlton S.K., Berry, A. & Torrey J.C. (1980): Evidence that associated soil bacteria may influence root hair infection of actnorhizal plants by *Frankia*. Can. J. MIcrobiol. 26, 971 - 977.

Lagacherie B., Hugot R. & Amarger N. (1977): Selection de souches de *Rhizobium japonicum* d'aprés leur compétitivité pour l'infection. Ann. agron. 28, 379 - 389.

Marshall K.C. (1970): Methods of study and ecological significance of rhizobium - clay interaction In Proceedings of the UNESCO/IBP Symposium on" Methods of Study in Soil Ecology" (J. Phillipson ed.) pp. 107 - 110. UNESCO, Paris.

Nishijima, F., Evans W.R. & Vesper S.J. (1988): Enhanced nodulation of soybean by *Bradyrhizobium* in the presence of *Pseudomonas fluorescens*. Plant and Soil 111, 149 - 150.

Schmidt E.L. (1978). Ecology of the legume root nodule bacteria. In *Interactions* between non-pathogenic microorganisms and plants (Y.R. Dommergues and S.V. Krupa eds.) pp. 269 - 303 Elsevier, Amsterdam.

Sougoufara B., Diem H.G. & Dommergues Y.R. (1989): Response of field grown *Casuarina equisetifolia* to inoculation with *Frankia* strain ORS 021001 entrapped in alginate beads. Plant and Soil 118, 133 - 1437.

Yahalom E., Okon, Y. & Dovrat A. (1987). Azospirillum effects on susceptibility to *Rhizobium* nodulation and on nitrogen fixation of several forage legumes. Can. J. Microbiol. 33, 510 - 514.

STRUCTURE OF THE GENUS *FRANKIA*

M.P. FERNANDEZ, S. NAZARET, P. SIMONET, B. COURNOYER,
P. NORMAND
*Laboratoire de Biologie des sols . URA CNRS 697,Bat 741,
Université Lyon 1, 43 bd du 11 Novembre 1918,
F-69622 Villeurbanne cedex, France*

ABSTRACT. The lack of a reliable taxonomy limits studies on the nitrogen-fixing symbiotic actinomycetal genus *Frankia*. DNA-DNA hybridizations which has become the method for defining species were made on more than 40 strains isolated from 7 actinorhizal plant genera belonging to the three major cross-inoculation groups. Nine genomic species were distinguished. RFLPs of the rDNA confirmed this grouping. 16S rDNA sequencing on PCR amplified total DNA from 35 strains mainly chosen among the nine genomic species permitted to establish the relationship of these groups by clustering them phylogenetically. This method of characterization was shown to be efficient for positioning non isolated strains with DNA extracted from a nodule.

The question of definition of species in the nitrogen-fixing symbiotic actinomycetal genus *Frankia* has been addressed by several authors. It has important implications mainly for the choice of representative strains in studies requiring a large spectrum of strains, and also whether or not host infectivity should be considered in strains definition.
The original taxonomy proposed by Becking (1974) based on cross inoculation groups obtained with crushed nodules (i.e. not with pure culture which did not exist at the time) was found to be deficient after pure cultures were found to have a different host range from that of crushed nodules. Using pure cultures, Baker (1987).redefined host infectivity groups. Four overlapping groups were described. Recent studies have focussed on lipids (Simon et al. 1989) and sugars (St. Laurent et al. 1987) composition, *nif*-RFLPs (Normand et al. 1988; Nazaret et al. 1989) and isoenzymes polymorphism (Gardes et al. 1987; Faure-Reynaud et al. 1990) and nodulation speed (Nesme et al. 1985), trying to find patterns among strains. The three major nodulation groups were generally well separated but in all cases tight clustering was noted, meaning that similar or closely related strains were tested repeatedly. This waste of effort was due to the lack of a reliable taxonomy.
A proposal by Lalonde et al. (1988) based on several criteria (lipids, nif-RFLPs and mainly host range) did not covered *Casuarina*-infectives strains and proposed to group together into subspecies those strains which only had in common the characteristic of being divergent from other strains and from one another.
The method which has become the standard, recommended by most bacteriologists (Wayne et al. 1987) is DNA-DNA hybridizations with a 70% threshold for defining species which remains the only taxonomic level precisely defined. There being no reason not to treat *Frankia* in a standard manner, we decided to apply this technique to more than 40 strains. Most of the results of this study (Fernandez et al. 1989) are shown in Table 1.
The *Alnus*-infectivity group was found to consist of at least three genomic species, one of which (genomic species 1) was designated *Frankia alni*. The Elaeagnaceae-infectivity group consist of at least five genomic species, and the Casuarinaceae-infectivity group of one genomic species.Nine strains did not fall into any of these genomic groups and thus may represent one or more additional groups.

TABLE 1. DNA-DNA hybridizations between *Frankia* strains. Numbers on top of the columns refer to the reference DNA from strains 1: ACoN24d, 2: AVc22c, 3: ARgP5$^{AG}$, 4: Ea1$_{12}$, 5: TXHR31e$^{HR}$, 6: EUN1f, 7: HRN18a, 8: Ea50$_1$ and 9: ORS020606.

| Source of unlabeled DNA | % of Reassociation at 75°C with labeled DNA from: | | | | | | | | |
|---|---|---|---|---|---|---|---|---|---|
| | 1 | 2 | 3 | 4 | 5 | 6 | 7 | 8 | 9 |
| **Genomic group 1** | | | | | | | | | |
| ACoN24d | 100 | 13 | 22 | 7 | - | 0 | 6 | 0 | - |
| Ar24H3 | 81 | - | - | - | - | 0 | - | - | - |
| Ar24O2 | 97 | - | - | - | - | - | - | - | 3 |
| Ag24$_{251}$ | 60 | - | - | - | - | - | - | - | - |
| ArI3 | 82 | - | - | - | - | - | - | - | - |
| ARgN22d | 77 | - | 32 | - | - | - | - | - | - |
| ACN1$^{AG}$ | 87 | - | - | - | - | - | - | - | - |
| CpI1 | 81 | - | - | - | 8 | 0 | - | - | - |
| A2J | 64 | - | 18 | - | - | - | - | - | - |
| **Genomic group 2** | | | | | | | | | |
| AV22c | 1 | 100 | 13 | - | 0 | 0 | - | 0 | 8 |
| AVN17 | 2 | 89 | - | - | - | - | - | - | - |
| Ac23$_{40}$ | - | 80 | - | - | - | - | - | - | - |
| AI43$_1$ | 7 | 110 | 21 | - | - | 0 | - | - | 0 |
| **Genomic group 3** | | | | | | | | | |
| ARgP5$^{AG}$ | 2 | 8 | 100 | 2 | - | - | 0 | - | - |
| **Genomic group 4** | | | | | | | | | |
| Ea1$_{12}$ | - | - | - | 100 | - | 12 | 16 | - | - |
| Ea1$_2$ | 0 | 0 | - | 71 | - | 0 | - | - | 0 |
| Ea2$_6$ | - | - | - | 81 | - | - | - | - | - |
| Ea3$_3$ | - | - | - | 69 | - | 2 | 20 | - | 0 |
| Eacm5$_1$ | - | - | - | 85 | - | - | - | - | - |
| HR27$_{14}$ | 4 | 6 | - | 103 | - | 2 | 22 | - | 0 |
| **Genomic group 5** | | | | | | | | | |
| TX31e$_{HR}$ | - | - | - | - | 100 | - | 0 | - | - |
| EAN1pec | - | - | - | 38 | 81 | - | - | 0 | - |
| HRX40$_1$ | - | - | - | 49 | 79 | 0 | 4 | - | - |
| **Genomic group 6** | | | | | | | | | |
| EUN1f | 0 | - | 0 | - | 4 | 100 | 3 | - | - |
| **Genomic group 7** | | | | | | | | | |
| HRN18a | - | - | - | - | - | - | 100 | - | - |
| **Genomic group 8** | | | | | | | | | |
| Ea50$_1$ | - | - | - | 19 | - | 12 | 16 | 100 | - |
| **Genomic group 9** | | | | | | | | | |
| ORS020606 | 1 | - | - | - | - | 0 | - | - | 100 |
| ORS022602 | - | 2 | - | 3 | - | - | 1 | 0 | 83 |
| ORS020607 | - | - | - | - | - | - | - | - | 87 |
| CcI3 | - | - | - | - | - | - | - | - | 107 |
| ORS020608 | - | - | 0 | - | 0 | - | - | - | 88 |
| HFP022801 | - | - | - | - | - | - | - | - | 80 |
| ORS021001 | - | - | - | - | - | - | 0 | - | 79 |
| ORS020609 | - | - | - | - | - | - | - | - | 69 |

Besides, in a work aimed at finding strains- and species-specific oligonucleotide probes, we isolated and sequenced the whole ribosomal operon of one *Frankia* strain, the *Casuarina*-infective *Frankia* strain ORS020606.(Normand et al., submitted), This sequence was compared with the corresponding one in *Streptomyces ambofaciens* and the results expressed in a histogram (Fig. 1).

Fig. 1. Histogramme of sequence similarity between ribosomal RNA of *Frankia* ORS020606 and *Streptomyces ambofaciens*. P, E2, L1, L2 and L3 correspond to hypervariable zones in 16S and 23S rRNA. ITS1 and ITS2 correspond to internal transcribed spacers.

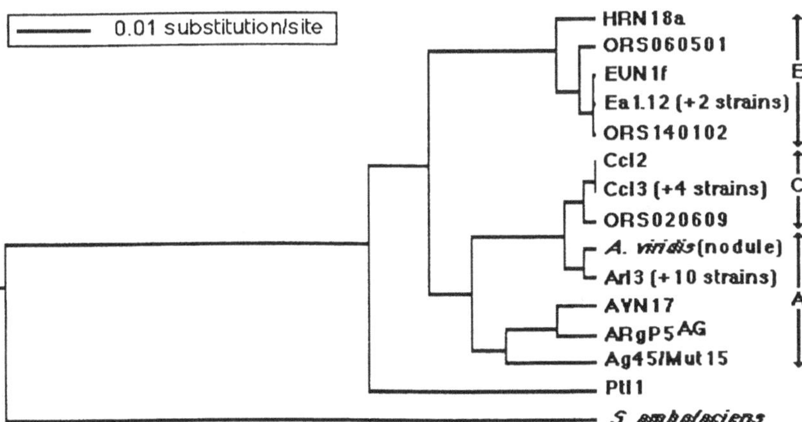


One of the two hypervariable zones in the 16S gene (corresponding to zone E2) was selected using Embley et al. (1988). Two oligonucleotides corresponding to regions bordering that region were synthesized and the 274 nucleotides of 35 *Frankia* strains mainly chosen among the nine genomic species were amplified by PCR (Nazaret et al. submitted).

Fig. 2. Evolutionary tree with contemporary tips

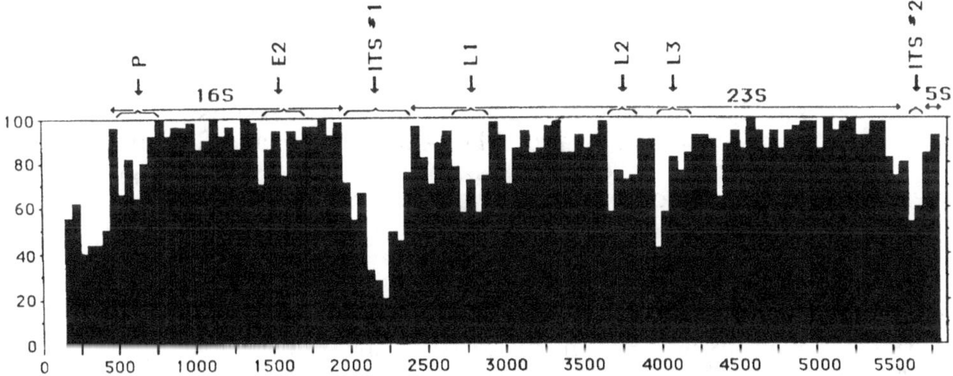

Amplification on DNA directly extracted from nodules of *Alnus viridis* allowed us to obtain the corresponding sequence of a non-isolated strain.The sequences were then aligned and the proportion of pair-wise matches recorded. The resulting distance matrix was analysed and a contemporary tips tree (Fig. 2) was built using the Fitch-Margoliash algorithm (Felsenstein, 1985).This tecnique not only permits grouping together strains but also orders these groups relative to one another.

Six strains from the genomic species 1, ACoN24d, ArI3, Ar24H3, Ar24O2, ARgN22d and ACN1$^{AG}$, were sequenced and found to have an identical sequence. Strains AV22c and AVN17 from genomic species 2 were found to have the same 16S sequence, different from that of *F. alni*. Strain ARgP5$^{AG}$ was also quite different from the first two groups. Thus, the three genomic *Alnus*-infective species were confirmed by 16S sequencing.

In the Elaeagnaceae-infective strains, genomic species 4 represented by Ea1$_{12}$ and genomic species 5 represented by HRX40$_1$, which hybridized at 49% were found to have an identical 16S sequence. EUN1f represents a similar situation hybridizing at low level with the other Elaeagnaceae-infective strains tested, thus forming a genomic species by itself. Nevertheless it is almost similar (except for an undefined position) to the 16S sequence of the two other groups. Strains HRN18a and ORS060501, which were found to hybridize at a significant level with no other strain would constitute further genomic species, were found to differ marquedly in their 16S sequences (5 and 4 nucleotides respectively different from Ea1$_{12}$, 4 nucleotides different from one another).Thus three genomic species in this group were similar in their 16S sequences while two strains remain markedly different.

The *Casuarina*-infective strains were found to form one coherent genomic species and their 16S sequences were identical (or nearly so in the case of ORS020609) supporting the status of species given to this group. The status of strain CcI2 infective on Elaeagnaceae and *Gymnostoma* (Casuarinaceae) (Torrey and Racette 1989) but hybridizing only at 38% with ORS020606 Fernandez et al. 1989) remains to be clarified. The use of single cell isolates may help elucidate it.

RFLPs using the *Steptomyces ambofaciens* ribosomal cluster as probe confirmed all these groupings (Fernandez et al. in preparation). No instance of inconsistency was noted.

All these techniques show a high level of convergence, the 16S sequences permitting to establish the relationship of these species relative to one another and validating the results of the DNA-DNA hybridization which is the established method of defining species. Other techniques may help but should not take precedence over it. These results support the contention that there are more than one species in the *Alnus*-infective strains and more than one species in the Elaeagnaceae-infectives strains as suggested by Lalonde et al. 1988 and thus that host infectvity is not a valid criterion. The discrepancies between the two sets of results all derive from the fact that 16S sequences are more conserved than total DNA and thus that 16S sequencing may not be an appropriate method for comparing closely related strains as in the Elaeagnaceae-infectivity group. There was no instance of inconsistency between the two techniques.

## References


Baker, D.D. (1987) 'Relationships among pure cultured strains of *Frankia* based on host specificity', Physiol. Plant. 70, 245-248.

Becking, J.H. (1974) 'Family III Frankiaceae', in R.E. Buchanan and N.E. GIbbons (eds.), Bergey's Manual of determinative Bacteriology, 8th ed., The Williams and Wilkins Co., Baltimore, pp. 701-706.

Embley, T.M. and Stackebrandt, E. (1988) 'Reverse transcriptase sequencing of 16S ribosomal RNA from *Faenia rectivirgula, Pseudonocardia thermophila* and *Saccharopolyspora hirsuta*, three wall type IV actinomycetes which lack mycolic acids', J. Gen. Microbiol. 134, 961-966.

Faure-Raynaud, M., Bonnefoy-Poirier, M.A. and Moiroud, A. (1990) 'Diversity of *Frankia* strains isolated from actinorhizae of a single *Alnus rubra* cultivated in nursery', Symbiosis 8, 147-160.

Felsenstein, J. (1985) 'Confidence limits on phylogenies: an approach using the bootstrap', Evolution 39, 783-791.

Fernandez, M.P., Meugnier, H., Grimont, P.A.D. and Bardin, R. (1989) 'Deoxyribonucleic acid relatedness among members of the genus *Frankia'* Int. J. Syst. Bacteriol. 39, 424-429.

Gardes, M., Bousquet, J. and Lalonde, M. (1987) 'Isosyme variation among 40 *Frankia* strains' Appl. Environ. Microbiol., 53, 1596-1603.

Lalonde, M., Simon, L., Bousquet, J. and Séguin, A. (1988) 'Advances in the taxonomy of *Frankia*: recognition of species *alni* and *elaeagni* and novel subspecies *pommerii* and *vandijkii* in H. Bothe, F.J. de Brujin and W.E. Newton (eds.), Nitrogen fixation: hundred years after, Gustav Fisher, Stuttgart, pp. 671-680.

Nazaret S., Simonet, P., Normand, P. and Bardin, R. (1989) 'Genetic diversity among *Frankia* isolated from *Casuarina* nodules' Plant and Soil 118, 241-247.

Nesme, X., Normand, P. and Tremblay, F.M. (1985) 'Nodulation speed of *Frankia* sp. on Alnus glutinosa, Alnus crispa and Myrica gale' Can. J. Bot. 63, 1292-1295.

Normand, P., Simonet, P. and Bardin, R. (1988) 'Conservation of *nif* sequences in *Frankia'* , Mol Gen. Genet. 213, 238-246.

St-Laurent, L., Bousquet, L., Simon, L. and Lalonde, M. (1987) 'Separation of various *Frankia* strains in the *Alnus* and *Elaeagnus* host specificity groups using sugar analysis' Can. J. Microbiol. 33, 764-772.

Simon, L., Jabaji-Hare, S., Bousquet, J. and Lalonde, M. (1989) 'Confirmation of *Frankia* species using cellular fatty acids analysis' System. Appl. Microbiol. 11, 229-235.

Torrey, J.G. and Racette, S. (1989) 'Specificity among the Casuarinaceae in root nodulation by *Frankia*' Plant and Soil 118, 157-165.

Wayne, L.G., Brenner, D.J., Colwell, R.R., Grimont, P.A.D., Kandler, O., Krichevsky, M.I., Moore, L.H., Moore, W.E.C.; Murray, R.G.E., Stackebrandt, E., Starr, M.P. and Trüper, H.G. (1987) 'Report of the Ad Hoc Committee on Reconciliation of Approaches to Bacterial Systematics' Int. J. Syst. Bacteriol. 37, 463-464.

# GROUPING OF FRANKIA STRAINS BY DNA-DNA HOMOLOGY: HOW MANY GENOSPECIES ARE IN THE GENUS FRANKIA?


V.N. AKIMOV, S.V. DOBRITSA, and O.S. STUPAR
*Institute of Biochemistry and Physiology of Microorganisms*
*USSR Academy of Sciences*
*Pushchino, Moscow Region, 142292*
*U.S.S.R.*


The genus *Frankia* comprises actinomycetes forming nitrogen-fixing symbioses with diverse non-leguminous plants. Various attempts have been made to delineate groups (or species) within the genus; however, no consistently useful taxonomic criteria for proper classification of *Frankia* isolates have been found.

We used DNA hybridization (membrane filter method) to estimate diversity among 28 *Frankia* strains (Table). Nine genospecies were differentiated, including five genospecies among 19 strains compatible with members of the plant genus *Alnus* (isolates from *Alnus* and *Comptonia*) and four genospecies among 9 strains compatible with the plant family Elaeagnaceae (isolates from *Elaeagnus, Hippophae, Shepherdia,* and *Colletia*). These data do not support the proposal of Lalonde *et al.* [2] that strains of *Frankia* compatible with *Alnus* be considered as a single species, *Frankia alni*, with two subspecies, *pommerii* and *vandijkii* and strains of *Frankia* from the *Elaeagnus* host-specificity group be considered as a single species, *Frankia elaeagni*. The levels of DNA relatedness among the strains representing a genospecies ranged from 64 to 100%. Three genospecies compatible with *Alnus* clustered at the level of about 30% homology. Three genospecies compatible with the Elaeagnaceae clustered at the level of 25 to 35% homology. Members of these two clusters as well as the other three genospecies containing single strains were only 5 to 15% related to each other. The distinct genospecies delineated in this study cannot be differentiated from each other on the basis of phenotypic properties and, thus, cannot be named at present.

Our experimental work had been finished when the paper by Fernandez *et al.* [1] appeared, which also demonstrated a high degree of heterogeneity among *Frankia* strains. Since our set of strains was almost totally different, it is impossible to say at present which of our genospecies (if any), except for genospecies 1, corresponds to the genospecies delineated by Fernandez *et al.* Thus, an agreement on reference strains is required between research groups.

**TABLE.** Levels of DNA relatedness among *Frankia* strains

| Genospecies | Strain | G+C content (mol%) | % Homology with labeled DNA from | | | | | | |
|---|---|---|---|---|---|---|---|---|---|
| | | | CpI1 | Ai1c | AGI5 | AirI2 | EaI10 | Ccl.17 | S15 |
| *Alnus* compatibility group | | | | | | | | | |
| 1 | CpI1 | 71.5 | 100 | - | 14 | 13 | - | 21 | - |
| | ArI3 | 71 | 100 | - | - | - | - | - | - |
| | ArI4 | 73 | 74 | - | - | - | - | - | - |
| | AGPM2.8 | - | 94 | - | - | - | - | - | - |
| | AvcI1 | 72 | 64 | - | - | - | - | - | - |
| | AvcI1.R1 | 73 | 66 | - | - | - | - | - | - |
| | AirI1 | 73 | 77 | - | - | - | - | - | - |
| | An2.1 | 72 | 85 | - | - | - | - | - | - |
| | An2.24 | - | 81 | - | - | - | - | - | - |
| | A7 | - | 100 | - | - | - | - | - | - |
| | A43 | - | 77 | - | - | - | - | - | - |
| | A51 | - | 93 | - | - | - | - | - | - |
| | A153 | - | 100 | - | - | - | - | - | - |
| 2 | Ai1c | 69.5 | 29 | 100 | 17 | - | - | - | - |
| | Ai15a | 70 | 29 | 91 | - | - | - | - | - |
| | Ag9b | 69 | - | 90 | - | - | - | - | - |
| 3 | Ai6b | 70 | 28 | 30 | 13 | 10 | - | 16 | - |
| 4 | AGI5 | - | 14 | 7 | 100 | 7 | - | 10 | 7 |
| 5 | AirI2 | 72 | 13 | 7 | 10 | 100 | - | 11 | - |
| Elaeagnaceae compatibility group | | | | | | | | | |
| 6 | EaI10 | 72 | 20 | - | 16 | 14 | 100 | 35 | 18 |
| | EaI11 | 72 | - | - | - | - | 90 | - | 22 |
| 7 | Ccl.17 | 71 | 15 | - | 11 | 9 | - | 100 | - |
| 8 | S15 | 70 | 22 | - | 12 | 14 | - | 36 | 100 |
| | S14 | 69.5 | - | - | - | - | - | - | 100 |
| | S13 | 70 | - | - | - | - | - | - | 66 |
| | Hr5-o | - | - | - | - | - | 15 | - | 87 |
| | H109 | - | - | - | - | - | 21 | - | 82 |
| 9 | EuI1b | - | 3 | - | 7 | 4 | - | 5 | 4 |

## References


1. Fernandez, M.P., Meugnier, H., Grimont, P.A.D., and Bardin, R. (1989) 'Deoxyribonucleic acid relatedness among members of the genus *Frankia*', Int.J.Syst.Bacteriol., 39, 424-429.
2. Lalonde, M., Simon, L., Bousquet, J., and Seguin, A. (1988) 'Advances in the taxonomy of *Frankia*: Recognition of species *alni* and *elaeagni* and novel subspecies *pommerii* and *vandijkii*', in H.Bothe, F.J.de Bruijn and W.E.Newton (eds.), Nitrogen Fixation: Hundred Years After, Gustav Fisher, Stuttgart, pp. 671-680.

# GENETIC DIVERSITY OF *ELAEAGNUS*-ISOLATED *FRANKIA* STRAINS INFECTIVE ON BOTH *ALNUS* AND *ELAEAGNUS*


M. BOSCO[1,2], M. P. FERNANDEZ[1], R. MATERASSI[2] and R. BARDIN[1]
[1]*Laboratoire d'ecologie Microbienne, C.N.R.S. U.R.A. 697, Université Lyon I Bat. 741, 43 bd du 11 Novembre 1918, 69622 Villeurbanne, France, and*
[2]*Dipartimento di Scienze e Tecnologie Alimentari e Microbiologiche, Università degli Studi di Firenze, Piazzale delle Cascine 27, 50144 Firenze, Italy.*


## Introduction

The genus *Frankia* (Actinomycetales) is to date well described by morphology, physiology, infectivity, cell biochemistry, serology and molecular biology, but it is not yet possible to define species within it.. Although Elaeagnaceae-isolated *Frankia* strains have been strictly separated from *Alnus*-compatible strains by various studies, Margheri *et al.* [5] and Dobritsa *et al.* [3] reported on *Frankia* strains isolated from *Elaeagnus* spp. that were capable to nodulate plants belonging to genera *Alnus*, *Elaeagnus* and *Hippophaë*. Flexible [7] *Frankia* strain EanII57 induced effective nodules on *Elaeagnus* sp. and ineffective nodules on *Alnus glutinosa* [5] and *A. incana*.[6]. Its further characterization was needed as a fondamental step to promote research on the "molecular dialogue" that allows plant-microsymbiont recognition.

The aim of this work was to examine the genetic diversity between *Frankia* strain EanII57 and two of its co-isolates, previously described as morphologically and physiologically uniform. We also evaluated the genomic relationships between them and *Frankia* reference strains belonging to type species 1 (*Frankia alni*) [1] and genomic species 4 (Elaeagnaceae) [4].

## Materials and Methods

The following pure *Frankia* strains were grown at 28°C in FP [2] medium: EanII10 (UFI 13270210), EanII50 (UFI 13270250), EanII57 [5] (UFI 13270257), Eal$_{12}$ [4] (ULF130100112), Ea2$_6$ [4] (ULF130100206), HR27$_{14}$ [4] (ULF140102714) and Ar24O$_2$ [12] (ULF0131024152).

DNA extractions and total DNA Restriction Fragment Lenght Polymorphism (R.F.L.P.) analysis were performed as described by Simonet *et al.* [12], using *Bgl* II, *Bam* H I, *Sst* I, *Sma* I and *Sal* I endonucleases, *Frankia* ArI3 $^{32}$P labelled *nif*HDK and *nif*AB [8] genes probes and *Streptomyces ambofaciens* rRNA [10] genes probe. Total DNA homologies (DNA-DNA hybridizations) were carried out as previously described [4], using $^3$H labelled DNAs from *Frankia* strains EanII10 and Eal$_{12}$ as references. The following isozyme activities were tested according to Selander *et al.* [11] and Pasteur *et al.* [9]: DIA, IBDH, MDH, PGI, and IA-EST.

## Results

Using RFLPs, strains EanII10, EanII50 and EanII57 displayed almost similar DNA patterns irrespective of the restriction endonucleases and gene probes employed. However, *Sma* I restriction patterns of total DNAs clearly showed that EanII57 lacks a 8.6 Kpb band, comparing with EanII10 and EanII50. Five hybridization fragments (6.68, 5.70, 3.96, 3.7 and 2.8 Kpb) were detected in EanII50 and EanII57 *Sst* I patterns hybridized with *nif*AB probe. EanII10 showed only the last two bands (3.7 and 2.8 Kpb). The five isozyme systems tested showed no differences of banding patterns among the three strains. DNA-DNA hybridization trials revealed, on the one hand, that *Frankia* strain EanII10 was 100% related to EanII50 and EanII57, 22-25% to Ea2$_6$ and HR27$_{14}$ (genomic species 4) and 8.5% to Ar24O$_2$ (*Frankia alni*, type species 1). On the other hand, *Frankia* Ea1$_{12}$ (genomic species 4) labelled DNA showed an homology of 13.7% with EanII10, 15.5% with EanII50, 14.4% with EanII57, 81% with Ea2$_6$ and 100% with HR27$_{14}$.

## Discussion

According to our results on DNA-DNA hybridization, *Frankia* strains EanII10, EanII50 and EanII57 constitute a specific genogroup, clearly distinct from both *Frankia alni* (type species 1) and genomic species 4. However, the fact that EanII10 showed a relative higher DNA relatedness with Ea2$_6$ and HR27$_{14}$ than with Ar24O$_2$, if correlated with its host range, effectivity and cultural characteristics, might indicate that these three flexible *Frankia* strains belong to the Elaeagnaceae-compatibility group rather than to the *Alnus*-specificity one.

The identical isoenzyme results and the homogeneity of most of the *nif* and rRNA RFLPs indicate that *Frankia* strains EanII10, EanII50 and EanII57 might have a common origin. However, differences in *Sma* I restriction fingerprints and *Sst* I patterns hybridized with *nif*AB probe allowed to separate the three tested strains from each other.

Sequence analysis of conserved zones of the genome, as rRNA genes, should improve our knowledge on the relationships among these strains. Any specific variation in their genes sequences will then allow the elaboration of specific strain probes, to be used as markers for autoecological studies (survival, infectivity and competitivity), both in natural and controlled environments.

## References


1. Becking, J.H. (1974) in R. E. Buchanan and N. E. Gibbons (eds.), Bergey's Manual of Determinative Bacteriology, 8th ed. The Williams and Wilkins Co., Baltimore, pp.701-706.
2. Bosco, M. (1990) Doctoral Thesis, University of Florence, Italy.
3. Dobritsa, S. V., Novik, S. N. and Stupar, O. S. (1989) Proceedings of the Fifth International Symposium on Microbial Ecology (ISME 5) O-15-8, p.84. Kyoto.
4. Fernandez, M. P., Meugnier, H., Grimont, P. A. D. and Bardin, R. (1989) Int. J. Sistem. Bacteriol. 39, 424-429.
5. Margheri, M. C., Vagnoli, L., Favilli, F. and Sili, C. (1985) Ann. Microbiol., 35, 143-153.
6. Margheri, M. C., Bosco, M., Vagnoli, L. and Favilli, F. (1989) Ann. Microbiol. 39,247-255.
7. Miller, I. M. and Baker, D. D. (1986) Protoplasma 131, 82-91.
8. Nazaret, S., Simonet, P., Normand, P. and Bardin, R. (1989) Plant and Soil 118, 241-247.
9. Pasteur, N., Pasteur, G., Bonhomme, F., Catalan, J. and Britton-Davidian, J. (1987) Lavoisier, Paris.
10. Pernodet, J. L., Boccard, F., Alegre, M. T., Blondelet-Rouault, M. H. and Guérineau, M. (1988) EMBO J. 7,277-282.
11. Selander, R. K., Caugant, D. A., Ochman, H., Musser, J., Gilmour, M. N. and Whittam, T. S. (1986) Appl. Env. Microbiol. 51, 873-884.
12. Simonet, P., Thi Le, N., Moiroud, A. and Bardin, R. (1989) Plant and Soil 118, 13-22.

THE INFECTIVITY AND SYMBIOTIC PERFORMANCE OF A Frankia STRAIN AS
AFFECTED BY ITS GROWTH MEDIUM TOWARDS Hippophae rhamnoides L.


E. CERVANTES and C. RODRIGUEZ-BARRUECO
IRNA-CSIC
Apartado 257
37071 Salamanca
Spain

ABSTRACT. We have investigated the effect of two different culture
media on the infectivity and symbiotic performance of a Frankia
strain UGL140101 towards the actinorhizal plant Hippophae rhamnoides.
Frankia grown in BuCT medium resulted in a more efective symbiotic
association and increased growth of the plants. Also, we report here
the effect of a co-inoculation with Azospirillum brasilense strain
INTA AZ39 on H. rhamnoides plants from the same experiment. Azospirillum inoculation resulted in an increased growth of the plants and in
the number of nodules formed.


## INTRODUCTION

The practical improvement of the symbiotic associations requires the
search of the more efficient host plant-endophyte combination and
also the accurate determination of the best growing conditions for
both plant and microbe symbionts. In this contribution we report the
infectivity and symbiotic performance of a previously described Frankia strain (UGL140101) kindly provided by CT Wheeler, Glasgow University, on the actinorhizal plant H. rhamnoides L. as affected by the
composition of the Frankia growth medium. H. rhamnoides is a shrub
whose ability to thrive in poor soils makes it a good candidate as
pioneer plant. Besides, we have investigated the effect of a joint
inoculation with Azospirillum brasilense strain INTA Az39 kindly provided by E. Rodríguez-Cáceres from the University of Bs.As. Argentina
on the growth of part of the plants from above experiment.


## MATERIALS AND METHODS

Frankia strain UGL140101 was grown in BuCT and BS media respectively
of the following composition:
BuCT (g.l$^{-1}$): $K_2HPO_4$, 1.0; $NaH_2PO_4.2H_2O$, 0.67; $MgSO_4.7H_2O$, 0.2;
$CaCl_2.2H_2O$, 0.1; FeNaEDTA, 0.01; Propionate, 0.5; casein hydrolisate,
1.0; Biotin, 0.002; Tween 80, 0.5 ml, and 1 ml of trace elements soltn.

BS medium (g.l$^{-1}$): $K_2HPO_4$, 0.3; $NaH_2PO_4$, 0.2; $MgSO_4.7H_2O$, 0.2; $CaCl_2.2H_2O$, 0.1; FeNaEDTA, 0.01; glucose, 7.5; propionate, 0.5; casein hydrolisate, 5.0; pyridoxin, 0.0005; Thiamin, 0.0001; Nicotinic acid, 0.0005 Vit. B12, 0.0016; and 1 ml trace elements solution (1).
_Azospirillum_ strain INTA Az39 was cultured according to Okon (2) to a density of 7 x $10^7$ cells/ml. One ml of the suspension was applied to the roots of each plant.

## RESULTS AND DISCUSSION

Table 1. Growth data and ARA of _H.r._ plants inoculated with _Frankia_ cultured in BuCT or BS medium. Age at harvest: 7 months

| _Frankia_ medium | Number of nodules | Root d.w. g. | Shoot d.w. g. | Plant d.w. g. | ARA nmol.h$^{-1}$.g$^{-1}$ |
|---|---|---|---|---|---|
| BuCT | 22 | 0.15 | 0.69 | 3.16 | 369.5 |
| BS | 17 | 0.09 | 0.12* | 0.65* | 107.5* |

*) Significant difference at p=0.05

Table 2.shows the effect of co-inoculating the plants with _Azospirillum_ and _Frankia_ grown in BS and BuCT media indistinctly.

Table 2. Growth and ARA data of _H. rhamnoides_ inoculated with _Frankia_ with or without _Azospirillum_

| _Azospirillum_ | Number of nodules | Root d.w. g. | Shoot d.w. g. | Plant d.w. g. | ARA nmol.h$^{-1}$.g$^{-1}$ |
|---|---|---|---|---|---|
| + | 27 | 0.15 | 0.54 | 2.57 | 204.7 |
| − | 19* | 0.12 | 0.41 | 1.91 | 238.5 |

*) Significant difference at p=0.05

The general effect of _Azospirillum_ was an increase in plant d.w. ARA of detached nodules was not different between treatments. A significant increase in nodulation due to the presence of this bacterium, may explain its positive effect on overall symbiotic effectiveness and pl. growth. Similar effects have been obtained in _Casuarina_ (C.R.B. submitted), and should be considered with other actinorhizal plants in the field.

## REFERENCES


1) Burggraaf AJP and Shipton WA. 1982. Plant and Soil 69, 135-147.
2) Okon Y, Albrecht SL and Burris RH. 1976. J. Bacteriol. 127, 1248-1254

# UTILISATION AND METABOLISM OF C1 COMPOUNDS BY *FRANKIA*


ABDELHAMID BRIOUA and C.T. WHEELER
Department of Botany
University of Glasgow
Glasgow G12 8QQ
Scotland
United Kingdom



ABSTRACT. *Frankia* growth was improved up to 30% by increasing $CO_2$ to 10%. Carboxylase enzymes detected were phosphoenolpyruvate carboxylase, phosphoenolpyruvate carboxykinase with lesser activity of pyruvate and propionyl CoA carboxylases. Ribulosebisphosphate carboxylase was not detected and the inability of several strains to grow with $H_2+CO_2$ as sole C source suggests *Frankia* does not grow chemoautotrophically. Methane supported slight growth of a few strains but most strains grew slowly on methanol and formate. The pattern of incorporation of radio-activity from $H^{14}CHO$ by *Frankia* cell-free extracts suggests metabolism via the hexulose monophosphate pathway.


## INTRODUCTION

Investigation of utilisation of C1 compounds by *Frankia* is confined to demonstration of growth stimulation by additional $CO_2$ (Akkermans et al., 1983) and suggestions that *Frankia* can utilise $H_2$ and $CO_2$ for chemo-autotrophic growth (Shipton & Burggraaf, 1983). Metabolism of other C1 compounds such as methane and methanol, common habitats of some actino-rhizal species, has not been reported previously.

## MATERIALS AND METHODS

Growth was determined by protein assay 6 weeks after transfer of washed mycelium to medium containing appropriate C sources (Hooker & Wheeler, 1987). Cultures of carboxylase enzyme assay were grown in propionate medium in air + 5% $CO_2$ and for $H^{14}CHO$ metabolism, in medium with $CH_3OH$.
Carboxylase enzymes in cell-free supernatants of washed, sonicated *Frankia* were assayed (pH 8.0) by incorporation of label from $H^{14}CO_3$ (Simpson et al., 1979; Utter & Keech, 1963). Hexose phosphate synthetase, catalysing the condensation of ribose-5-phosphate and HCHO was assayed (pH 7.0) by incorporation by cell-free extracts of label from $H^{14}CHO$ into barium acetate precipitable fractions (Lawrence et al., 1970).

## RESULTS AND DISCUSSION

642

Growth of *Frankia* strains DDB 01310210 and UGL 013103 was enhanced by 9, 20 and 30% following enrichment of the incubation atmosphere with 1, 3 and 10% $CO_2$ respectively. The main carboxylase activities detected were PEP carboxylase and PEP carboxykinase, with lesser activity of pyruvate and propionyl CoA carboxylases. Ribulosbisphosphate carboxylase (RuBPc) was not detected.

None of the *Frankia* strains tested (Table 1) grew in air with 5% ($H_2+CO_2$) and 1.5, 5, 10 or 20% $O_2$. These data and the failure to detect RuBPc, necessary for the utilisation of $CO_2$ via the pentose pathway, suggest that *Frankia* cannot grow chemoautotrophically.

Most strains used methanol and 3 tested utilised formate as sole C source for growth (Table 1). Two strains showed slight growth with methane. The growth rate of all strains with C1 compounds was low (<15% of that on propionate medium) and was not stimulated by addition of $H_2$.

Table 1.  *Frankia* growth (μg protein / 10-ml) with C1 carbon sources.

| Strains | Control (C-free) | Methane 5% in air | Methanol 0.15M | Na-formate 75 mM |
|---------|------------------|-------------------|----------------|------------------|
| ArI3 | 15.8±1.78 | 14.3±1.15 | 14.6±0.94 | |
| ArI4 | 11.1±0.92 | 13.0±0.51* | 16.0±1.36** | 16.2±0.46** |
| 1.2.5Q | 10.5±1.19 | 10.9±2.83 | 11.6±1.36 | |
| 1.1.5F | 7.3±0.89 | 8.0±1.60 | 11.2±1.00** | |
| 1.1.7F | 9.6±1.40 | 11.3±1.09 | 13.7±0.90** | |
| 1.1.8Bu | 8.6±0.94 | 10.4±1.41 | 11.5±1.26** | |
| 3.1.5P | 9.1±0.71 | 11.5±0.66* | 14.1±0.28*** | 14.7±0.67*** |
| 3.1.10P | 8.0±0.66 | 8.0±1.14 | 13.8±0.96*** | 14.9±1.17*** |

Means ± SE (n=3) are significantly different from controls at P<0.1*, P<0.05**, P<0.01***.

Little radioactivity from $H^{14}CHO$ was incorporated by cell-free extracts into the amino acids glycine and serine, suggesting that metabolism does not proceed via the serine pathway (Large & Bamforth, 1988). Stimulation by ribose-5-phosphate of incorporation of radioactivity from $H^{14}CHO$ into a sugar phosphate, barium acetate precipitable fraction, the mobility of which changed on TLC following digestion with phosphatase, suggests the presence of hexose phosphate synthetase. This indicates that methane and methanol are assimilated via the hexulose phosphate pathway.

REFERENCES


Akkermans, A.D.L. et al., (1983) Can. J. Bot. **61**, 2793-2800.
Hooker, J.E. & Wheeler, C.T. (1987)  Physiol. Plant., **70**, 333-341.
Large, P.J. & Bamforth, C.W. (1988)  Methylotrophy & Biotechnology Longman, England.
Lawrence, A.J. et al. (1970) Biochem. J., **116**, 631-639.
Shipton, W.A. & Burggraaf, A.J.P. (1983) Can. J. Bot. **61**, 2774-2782.
Simpson, F.B., Maier, R.J. & Evans, H.J. (1979) Arch. Microb. **123**,1-8.
Utter, M.F. & Keech, D.B. (1963) J.Biol. Chem., **238**, 2603-2608.

# NITROGENASE ACTIVITY AND NITROGENASE PROTEIN LEVELS IN A *Frankia-Alnus incana* SYMBIOSIS SUBJECTED TO DARKNESS


P.-O. Lundquist and K. Huss-Danell
Dept of Plant Physiology
University of Umeå
S-901 87 UMEÅ
SWEDEN



ABSTRACT. Nitrogenase activity of symbiotic *Frankia* decreased when host plants (*Alnus incana* (L.) Moench) were exposed to darkness for up to 2 days. The amount of nitrogenase protein (MoFe-protein) as determined from Western immunoblotting decreased in parallel to the decrease of nitrogenase activity. During recovery form dark treatment nitrogenase activity increased in parallel to the amount of MoFe-protein. However, after 4 days in normal light/dark conditions nitrogenase activity *in vivo* per nodule FW was not fully recovered to control level, suggesting that additional effects on nodule functioning had occured during dark treatment.


## Introduction

In the symbiosis between the nitrogen-fixing *Frankia* and *Alnus incana*, *Frankia* depends on the plant for the carbon that it needs for growth and nitrogen fixation. In earlier studies of regulation of the nitrogen-fixing symbiosis plants were exposed to various forms of environmental changes (water stress, darkness and addition of ammonium (1)). In all cases nitrogenase activity of intact plants (*in vivo*) decreased as well as nitrogenase activity *in vitro*. In this study plants were exposed to prolonged darkness as a way to study the regulation of the nitrogen-fixing symbiosis and the aim of the study was to investigate if a loss of nitrogenase activity could be explained by loss of nitrogenase protein.

## Material and methods

Cloned plants of *Alnus incana* (L.) Moench incoulated with a local source of *Frankia* (spore (+), hup (-)) were grown in a climate chamber with an 18/6 cycle of light/dark and 25/15 ᵒC. The dark treatment was carried out in the climate chamber by placing the plants under black cloth. Nitrogenase activity of intact root systems and of anaerobically prepared *Frankia* vesicle clusters supplemented with dithionite and Mg-ATP, was measured as acetylene reduction activity. Amounts of nitrogenase protein (MoFe-protein) and nitrogenase reductase protein (Fe-protein) were determined using Western immunoblotting with antisera originally made against the Fe-protein of *Rhodospirillum rubrum* and against the MoFe-protein of *Azotobacter vinelandii* (2). Precast 10-15 % gradient SDS-polyacrylamide gels were run using the Pharmacia PhastSystem and stained with Coomassie R 350 and silver staining.

## Results and discussion

Nitrogenase activity *in vivo* and nitrogenase activity *in vitro* of anaerobically prepared *Frankia* vesicle clusters, decreased in parallel during a dark treatment for up to 2 days. This was interpreted as loss of active nitrogenase and has earlier been shown for other types of plant treatments (water stress, darkness and addition of ammonium) (1). By using Western immunoblotting with antisera against both the MoFe-protein and the Fe-protein of nitrogenase it was possible to show good correlation between loss of activity and loss of MoFe-protein (2). In addition, protein stained SDS-polyacrylamide gels revealed a loss of two polypeptides corresponding in size with the subunits of the MoFe-protein and the Fe-protein, but no other obvious changes in protein patterns were observed. The reason for loss of nitrogenase activity could be inactivation and degradation of nitrogenase protein, and inhibited synthesis of nitrogenase, possibly as a consequence of increased oxygen concentration in the root nodule due to reduced respiratory consumption following the carbon starvation.

During recovery from dark treatment, nitrogenase activity *in vitro* and the amount of MoFe-protein, both expressed on a protein basis, increased in parallell and had recovered to the control level after four days in normal light/dark conditions (3). Nitrogenase activity *in vivo* per nodule fresh weight was however not fully recovered to control level after four days in normal light/dark conditions. This suggests that nitrogenase activity was metabolically limited or that the nodules of dark treated plants contained less amount of *Frankia* than the control plants.

References


1. Huss-Danell, K., Vikman, P.-Å., Lundquist, P.-O., (1990) Physiology of symbiotic *Frankia* with emphasis on *Alnus* nodules. in Endocyto-biology IV, Nardon,P., Gianinazzi-Pearson, V., Grenier, A.M., Margulis, L., Smith, D.C. (eds.), INRA, Paris, pp. 39-44
2. Lundquist, P.-O. and Huss-Danell, K., Nitrogenase activity and amounts of nitrogenase proteins in a *Frankia-Alnus incana* symbiosis subjected to darkness. Plant Physiology (in press).
3. Lundquist, P.-O. and Huss-Danell, K., (manuscript).

SECRETION OF AMINOPEPTIDASES AND PROTEINASES BY *FRANKIA* sp BR.

A. Müller[1,2], P. Benoist[2], H.G. Diem[2] and J. Schwencke[1]
[1]Laboratoire d'Enzymologie du CNRS, 91198 Gif-sur-Yvette, France.
[2]Laboratoire des Biotechnologies des Systèmes Symbiotiques Forestiers Tropicaux 94736 Nogent-sur-Marne, France.

**ABSTRACT**. Using cultural conditions allowing rapid, exponential growth of *Frankia* sp BR a wide variety of proteins were found to be secreted into the medium. Among them six aminopeptidases (FAP) and a number of endoproteinases (PF) were revealed. Analysis by native agarose-polyacrylamide gel electrophoresis (1) indicated that all the exocellular FAP and the majority of secreted PF showed electrophoretic mobilities corresponding to apparently similar intracellular enzymes. Only five of the secreted PF were found to be specifically exocellular. These findings opens the question of the significance of these enzymes in the infection step or during the symbiotic life of *Frankia*.

**INTRODUCTION**. Exocellular *Frankia* enzymes may play an important role in the penetration of the host-tree hair-roots and/or in the normal ongoing of the symbiosis. Few studies have been reported on this matter. However, the exocellular pectinolytic activity found in some *Frankia* isolates has been proposed to participate in the infection process (2). Also, cellulolytic activities have been reported for some *Frankia* strains (3). Interestingly, extracellular proteases have been identified as bacterial pathogenicity factors and proteinase inhibitors as plant defense response to microbial attack (review, 4). Also besides polysaccharides, glycoproteins are well known structural components of plant cell walls (5) and root-hair specific proteins have been found to be located in their outer surface (6). These reasons and our current interest in the proteolytic system of *Frankia* prompted us to study its exocellular proteinases. We report here that *Frankia* spBR is able to secrete an imposing number of exocellular proteins and among them six aminopeptidases and a variety of endoproteinases.

**MATERIALS AND METHODS. Organism and culture conditions**. We have used *Frankia* spBR (ORS 020608) isolated from *Casuarina equisetifolia* by Dr. H.G. Diem. Cells were grown in stirred, Mes buffered BAP medium with 2 µg/ml phosphatidyl choline as described (7).

**Obtention of exocellular proteins**. Filtered culture medium was treated with DEAE-cellulose. The resin washed once with 20 mM Hepes/Tris pH 8, and exocellular proteins extracted with 2 M NaCl. The extract was desalinized with Biogel P6DG and then immediately used for the experiments or stored at -80°C.

**Electrophoretic development of proteases**. Aminopeptidases were revealed as in (1) and endoproteinases by gelatine-PAGE as in (8).

RESULTS. Analysis of *Frankia* secreted proteins by denaturing PAGE showed up to 76 different peptides from 10 kDa to about 150 kDa. Among them, six aminopeptidase (FAP) were detected after native-Agarose-PAGE (1) and found to show the same mobility as their intracellular counterparts. Among the FAPs, FAP-3 specific for proline terminal residues (1) appears to be secreted only during exponential growth. All the others are secreted during exponential growth and even 10 days after growth arrest. The activity of various secreted PF appears to increase after growth arrest. Only five secreted PF were exclusively exocellular, all the others appears to share similar mobilities with intracellular PF as revealed by gelatine-PAGE.

DISCUSSION. We report here for the first time that *Frankia* spBR presents a quite remarkable protein secretion system. Protein secretion is active during exponential growth. Analysis of the secreted proteins by SDS-PAGE shows up to 76 different peptides which amounts up to 10 % of the total mycelial protein. Among these proteins at least 15 PF and six FAP were detected. However, the variety of exocellular *Frankia* proteinases found is not surprising *per se* since the ability of *Actinomycetes* to produce and secrete a number of proteolytic enzymes is well known (review, 9). The physiological role that exocellular proteinases may play in the life cycle of *Frankia in vitro* or under symbiotic conditions is an open question. They may facilitate saprophytic survival in the soil. They may also be essential either to counteract the action of induced host defensive proteins or to provide amino acids as additional carbon sources. Whether *Frankia* produce proteinases in the rizosphere of the host cell plant is not known. However, since a number of glycoproteins are well established structural components of plant cell walls (review, 5) and some specific proteins appears to be present in the outer surface of root-hairs (6), the possibility that exocellular *Frankia* exocellular proteinases may participate in some way in the infection process is also a challenging question.

REFERENCES

1. Benoist, P. and Schwencke, J. (1990) Anal. Biochem. **187**, 337-344.
2. Seguin, A. and Lalonde, M. (1989) Plant and Soil **118**, 221-229.
3. Safo-Sampah, S. and Torrey, J.G. (1988) Plant and Soil 112, 89-97.
4. Lamb, C.J., Lawton, M.A., Dron, M. and Dixon, R. (1989) Cell 55, 215-224.
5. Varner, J.E. and Lin, L.-S. (1989) Cell **56**, 231-239.
6. Röhm, M. and Werner, D. (1987) Physiol. Plantarum **69**, 129-136.
7. Schwencke, J. (1990) This Symposium.
8. Heussen, C. and Dowdle, E.B. (1987) Anal. Biochem. 102, 196-202.
9. Peczynska-Czoch, W. and Mordarski, M. (1988) "Actinomycete enzyme" in M. Goodfellow, S.T. Williams and M. Mordarsky (eds.). Actinomycetes in Biotechnology. Academic Press, San Diego, pp. 220-283.

Interaction between VAM and actimorhizal symbionts in <u>Alnus nepalensis</u>.


D.K. Jha, G.D. Sharma and R.R. Mishra
Department of Botany, North-Eastern Hill University, Shillong-793014 (India)


The effect of tripartite association on P-uptake was studied in <u>Alnus nepalensis</u> seedlings grown in pots. The nodulation and nitrogenase activity in plants with dual inoculation of <u>Glomus</u> and <u>Frankia</u> was higher than in seedlings with either <u>Glomus</u> or <u>Frankia</u> Symbiosis. Dual inoculum improved the nitrogen economy. The non-mycorrhizal plants showed more dependence on external supply of phosphorus than those with dual inoculation. Plants with tripartite association showed better growth response at low P-level. The improvement in nodulation and nitrogenase activity, however, occurred at higher P-level i.e. at a level twice of normal P-level in soil. Nevertheless, the highest P-dose was inhibitory to growth as well as nitrogenase activity.

Introduction

The Hill soils are deficient in N and P, which are most common but essential elements for plant growth. Use of chemical fertilizers is the only alternative to improve soil fertility but it has other implications like short supply and their cost etc. Therefore, the biological nitrogen fixation offers a better alternative to the use of commercial fertilizers in order to maintain adequate plant productivity. The actinorhizal plants typically colonize nutrient deficient habitat. There are various studies on interaction between <u>Rhizobium</u> and VAM fungi Barea <u>et al</u>, 1989) but very few exist for actinorhizal plants (Russo, 1989). The present experiment was designed to investigate the efficiency of tripartite association in P-mobilization and seedling growth.

Materials and Methods

Potassium dihydrogen phosphate was used as P source. Five levels i.e. 0 kg ha (Po), 30 kg ha (p$\frac{1}{2}$), 60 kg ha (p1), 120 kg ha (P2) and 240 kg ha (P3) were used. <u>Glomus etunicatus</u> was used as mycorrhizal inoculum. The seedlings were inoculated with mycorrhizal spores, whereas the non-mycorrhizal set received the same quantity of autoclaved inoculum. Fifty ml of nodule homogenate was used for inoculation of <u>Frankia</u>. All together three sets were maintained i.e. Frankia + VAM, only <u>Frankia</u> and only VAM.

The seeds of <u>A.nepalensis</u> were soaked in water for 24 hours, placed on sterilized moist filter paper in petridishes and incubated in BOD incubator at 25±1$^{\circ}$C. After two weeks the germinated seeds (approximately 3 cm) were transferred into pots with different treatments viz. VAM + <u>Frankia</u>, VAM without <u>Frankia</u> and <u>Frankia</u> without VAM. Six seedlings were maintained in each pot.

The experiment was carried out in green house under natural conditions. The maximum and minimum temperature ranged between 20$^{\circ}$ to 11$^{\circ}$C. The average photoperiod was 28.5 x 10 lux with an average humidity of 79%.

The percentage root infection was assessed by phillips and Hayman's technique (1970). The nitrogenase enzyme was measured by a gas liquid chromatograph (Chemito 300) following the method of McNabb and Geist (1979).The shoot dry weight and the nodule dry weight were measured after drying them at 60$^{\circ}$C in a hot air oven for 48 hours.

Results

The non-mycorrhizal seedlings responded more to the external P-fertilization than mycorrhizal ones (Fig.1). The growth of nodulated mycorrhizal plants was more at low P-level as compared to highest P level, where the growth of mycorrhizal and non-mycorrhizal seedling did not vary significantly (Fig.1). The mycorrhizal infection markedly affected the nodulation and nitrogenase activity. The mycorrhizal seedlings have an edge over their non-mycorrhizal counterparts interms of nodule dry weight and nitrogenase activity (Fig.1). The intermediate level of P fertilization (2P) markedly affected the nodule dry weight as well as the nitrogenase activity (Fig.1). Nevertheless, these traits were significantly retarded at highest P doze.

Discussion

The results indicated that the VA mycorrhizal inoculation had a marked influence on plant growth and nitrogen fixation. The non-mycorrhizal plants required more P than modulated mycorrhizal plant to achieve the same growth rate (Medina <u>et al</u>, 1988). The improved growth of mycorrhizal nodulated plants at low P-level confirmed the mycorrhizal efficiency in P stress condition, which could be an ecological aid to the plants to adapt to low fertile soils, thus reducing the cost of fertilization (Plenchette <u>et al</u>, 1983). The improved growth of seedlings was the result of increased supply of N and P in tripartite symbiotic association than bipartite or in absence of any symbionts. An adequate supply of phosphate is an essential requirement for biological nitrogen fixation (Gardner <u>et al</u> 1984) which was achieved by VAM symbionts through improved P-uptake. The high soil-P, which was found stimulatory to nitrogen fixation may be owed to high P-requirement in nitrogen fixation.

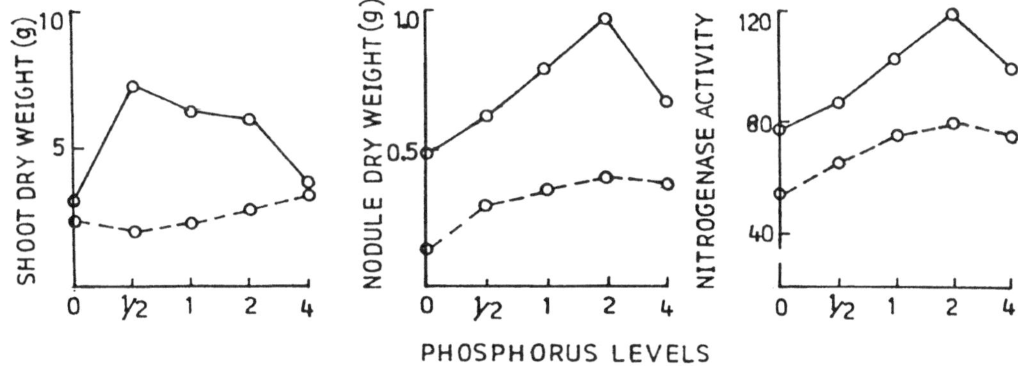

PHOSPHORUS LEVELS

Fig.1. Shoot dry weight(g), nodule dry weight(g)and nitrogenase activity($\mu$M $C_2H_2$ reduced $h^{-1}g^{-1}$ fresh weight of nodule) of <u>A.nepalensis</u> seedlings growing at different levels of phosphorus.(o—o, mycorrhizal+Frankia and o--o,non-mycorrhizal)

References


Barea, J.M. Azc'on, R. and Azc'on-Ayuibar, C (1989) Time course of $N_2$-fixation (15N) in the field dover growing alone or in mixture with ryegrass to improve pasture productivity and inoculated with vesicular - arlanscular mycorrhizal fungi; New phytologist 112, 399-404.

Gardner, I.C., Oelland, D.M. and Scotts, A. (1984) ' Mycorrhizal improvement in non-leguminous nitrogen fixing associations with particular reference to <u>Hippophae rhamnoides</u> L.' Plant soil 78, 189-200.

McNabb, D.H. and Geist, J.M. (1979) 'Acetylene reduction assay of symbiotic $N_2$ fixation under field conditions, Ecology 60, 1070-1072.

Medina, O.A. Sylvia, D.M. and Kretschmer Jr. A.E. (1988) 'Response of siratro to vesicular - arbuscular mycorrhizal fungi II. Efficacy of selected vesicular - arbuscular fungi at different phosphorus levels', Soil Science Society American Journal 52, 416-419.

Phillips, J.M. and Hayman, D.S. (1970) 'Improved procedure for claning roots and staining parasitic and vesicular arbuscular mycorrhizal fungi for rapid assessment of infection', Transcation of British mycohegical society 55, 158-161.

Plenchelle, C., Furlan, V. and Fortin, J.A. (1983) 'Responses of endomycorrhizal plants grown in a calcined mont-morillonite day to different levels of soluble phosphorus I. effect on growth and mycorrhizal development, Canadian Journal of Botany 61, 1377-1383.

Russo, O.R. (1989) 'Evaluating alder-endophyte (<u>Alnus acuminata</u> - <u>Frankia</u> - Mycorrhizae) interactions. I. Acetylene reduction in seedlings inoculated with <u>Frankia</u> Strain Ar13 and <u>Glomus intraradices</u>, under three phosphorus levels', Plant soil 118, 151-155.

# COMPARISON BETWEEN *QUERCUS ROBUR / ALNUS CORDATA* MIXED PLANTATION AND *QUERCUS ROBUR* IN MONOCULTURE

E. BURESTI[1], A. M. DOMENACH[2], M. BOSCO[3] and A. MOIROUD[2]
[1]Istituto Sperimentale per la Selvicoltura, V.le S. Margherita 80, 52100 Arezzo, Italy,
[2]Laboratoire d'ecologie Microbienne, C.N.R.S. U.R.A. 697, Université Lyon I Bat. 741, 43 bd du 11 Novembre 1918, 69622 Villeurbanne, France and
[3]Dipartimento di Scienze e Tecnologie Alimentari e Microbiologiche, Università degli Studi di Firenze, Piazzale delle Cascine 27,  50144 Firenze, Italy.

## Introduction

Actinorhizal and leguminous plants, due to their capacity to fix nitrogen in symbiosis with bacteria, often act as pioneer in the colonization of eroded and spoilt soils. In temperate regions, actinorhizal shrubs and trees also have a great potential as intercrops for other tree species.
Since 1982, series of field experiments on the reclamation of lignite spoil banks by mixed plantations have been undertaken along the valley of Arno river, (AR, Italy), by the "Istituto Sperimentale per la Selvicoltura".
In this work, we report preliminary results obtained in a mixed plantation of *Quercus robur* and *Alnus cordata* on lignite spoil bank. The effect of interplanted italian alders (*Alnus cordata*) on growth and shape of peduncolate oaks (*Quecus robur*) has been examined, and its correlation with nitrogen input due to nitrogen-fixing trees has been evaluated.

## Materials and Methods

*Study site.* The plantation is located at Santa Barbara (AR) on lignite spoil banks; topsoil is very heterogeneous, but it is often a silty clay, sub-acid.soil (clay 30%; sand 29%; silt 40%; pH= 6.4).

*Plantation pattern.* Monoculture: square planting m 3x3, pure oaks planted in 1983. Association: square planting m 3x3, alternated oaks and alders, planted in 1984. Total surface:  3 ha.

*Estimation of annual $N_2$ fixation in alder leaves* (*N input*). The method based on the $^{15}N$ natural abundance seems to be a more suitable one to estimate the $N_2$ fixation for the trees (Shearer and Kohl 1986, Domenach et al 1989).
This method is based on the difference in the $^{15}N$ natural abundance between fixed nitrogen and assimilated nitrogen. Twenty leaves were collected at different heights in mid October from each of 6 randomly chosen alders and 6 oaks.

$$\% \text{ Ndfa } = 100 \; \frac{\delta^{15}N \text{ non fixing plant} - \delta^{15}N \text{ fixing plant}}{\delta^{15}N \text{ non fixing plant} - \delta^{15}N \text{ fixing plant on N free medium}}$$

with $\delta^{15}N = 1000$ (atom $\%^{15}N$ sample $-$ atom $\%^{15}N$ air) / atom $\% \; ^{15}N$ air. Values were obtained by Mass Spectrometry (VG Sira 12, Manchester, England).

## Results

*Tree growth (cm):*
Monoculture:   Oak       $\bar{H}$=485;    $\bar{D}_{base}$= 9.6;    $\bar{D}_{130}$= 6.1;    $\bar{D}$ of four bigger branches = 2.2.
Association:    Oak       $\bar{H}$= 565;    $\bar{D}_{base}$= 8.3;    $\bar{D}_{130}$= 6.4;    $\bar{D}$ of four bigger branches = 1.7.
                  Alder     $\bar{H}$= 771;    $\bar{D}_{base}$= 15.3;  $\bar{D}_{130}$= 9.6.

From data collection carried out in automn 1989, oaks in mixed stands showed a remarkable increase of size, better stem shape and smaller lateral branches than those grown in pure stands.

*Annual fixation by alder growing associated with oak:*               <u>N fixed</u>

$\delta^{15}N$ fixing plant growing on N-free medium                  -2±0.4
gave the isotopic identity of N fixed.

$\delta^{15}N$ fixing plant                                -0.2±0.6      76%
was the isotopic value of alder growing in association with oak

$\delta^{15}N$ non fixing plant                        2.6±0.9
was the isotopic identity of N assimilated by oaks

*N input in the system by the alder litter:*
Tree density :                      560 alder/ ha
Alder leaf biomass :             4kg/tree
N concentration in the leaves :     3%
% N fixed :                       76%
from these values , N input in the system can be estimated: 560 x 4 x 3/100 x 76/100
<u>Input                           50kg N/ha</u>

Remark: the alder litter ($\delta^{15}N$ = -0.2) has an effect upon the isotopic value of N assimilated:
$\delta^{15}N$ *Quercus* in association (= 2.6±0.9 )is lower than $\delta^{15}N$ *Quercus* in monoculture (=3.7±0.9) showing an important mineralization of alder leaves.


## Discussion

Utilization of alders as interplant crop for broad-leaved species had already been employed in Europe, but the benefits for growth of the timber species in mixed plantations were not always so obvious, depending on nodulation efficiency.

$\delta^{15}N$ method is a good way to determine the ability of trees to fix atmospheric nitrogen under field conditions. Although in this experiment the presence of nitrogen-fixing symbiotic nudules was not controlled, $\delta^{15}N$ methodprovided tevidence that 80% of leaf nitrogen content derived from biological fixation. Our results showed also that *Alnus cordata* produce an abundant nitrogen-rich litter, which can be easily mineralized. So, italian alder, in association with oaks, increases available nitrogen for associated timber plants both by its high capacity to fix atmospheric nitrogen (80%) and its low nitrogen uptake from soil (20%).

The fact that this association promotes better growth and shape of the oak trunk may increase the economic stature and commercial value of wood.


References


Buresti, E. (1990) 'Gil impianti forestali delle discariche dell' ENEL miniera di Santa Barbara nel Valderno' ENEL.

Dawson, J.O. (1986) 'Actinorhizal plants: their use in forestry and agriculture', Outlook on Agriculture, 15, 202-208.

Domenach, A.M., Kurdali, F. and Bardin, R. (1989) 'Estimation of symbiotic dinitrogen in alder forest by the method based on natural [15]N abundance'. Plant Soil ,118, 51-59.

Kurdali, F., Domenach, A.M. and Bardin, R. (1990) 'Alder-poplar associations: determination of plant nitrogen by isotopes techniques'. Biol. Fertil. Soils ,9 (in press).

Shearer, G. and Kohl, D.H. (1986) '$N_2$ fixation in field settings: estimations based on natural [15]N abundance'. Aust. J. Physiol., 13, 699-756.

# EEC REGULATORY REQUIREMENTS FOR THE USE AND RELEASE OF GENETICALLY MODIFIED ORGANISMS


G. DEL BINO,
Commission of the European Communities,
Environmental control of products,
industrial installations and biotechnology,
Rue de la Loi 200,
B-1049 Brussels



ABSTRACT. The European Community, in its delicate role to promote research and industrial activities in the promising sector of biotechnology while ensuring the protection of public health and the environment, has recently adopted two Directives concerning the Regulation of Genetically Modified Organisms.
 Directive 90/219/EEC on the Contained Use of Genetically Modified Micro-organisms addresses all activities, from research and development to production, where genetically modified micro-organisms are used. The purpose of this Directive is:
- to establish a case-by-case risk assessment for all operations;
- to identify the cases where the contained use could entail a risk;
- to establish working practices and containment measures corresponding to the hazard the micro-organism presents;
- to prevent, in case of hazardous micro-organisms, their accidental release and limit the consequences of such accidents where they occur.
 A flexible notification procedure is proposed, based both on the kind of micro-organism used and on the scale of the operation envisaged.
 Directive 90/220/EEC on the Deliberate Release to the Environment of Genetically Modified organisms covers all genetically modified organisms (micro-organisms, plants and animals) through all the stages of release to the environment from small scale to large scale experimental introductions, as well as release through product marketing - although some products may be exempted from the provisions of this Directive if they are covered by other Community legislation which includes a similar environmental risk assessment. The main elements of this Directive are:
-An environmental risk assessment must always be carried out before any release to the environment whether in an experiment or product.
-No releases may be carried out without the consent of the competent authorities.
-A national approval procedure is foreseen for experimental releases.
-A Community approval procedure is foreseen for releases made by marketing a product.
 Once approved, a product can circulate freely throughout the Community.

## Introduction

The new discoveries that nitrogen fixation research has led to, in genetics, biochemistry, physiology and chemistry have opened up the possibility of applying genetic engineering techniques for the creation of more efficient nitrogen fixing bacteria or of constructing plants which can fix their own nitrogen thus doing away with their dependence on nitrogen fixing microbes or chemical nitrogen fertilizers.

The technique of genetic engineering has allowed the identification of many "useful" genes and their transfer to other organisms which didn't possess them before, by passing biological barriers and thus creating "new" organisms, with novel properties. Genetically modified nitrogen fixing bacteria and nitrogen fixing plants are only an example of genetically modified organisms which may be released to the environment in the future.

The expressed concerns are not only related to the possible impact on human health, but on the impact on the total environment. It is feared that the widespread use and release of genetically modified organisms might upset the delicate balance existing in nature or even have evolutionary impacts.

Given on the one hand the complex scientific issues involved and on the other hand the gaps in knowledge on fields such as the potential for gene transfer among organisms and the fate of engineered organisms in natural environments, it is evident that a case-by-case examination and an environmental risk assessment is appropriate for the use of genetically modified organisms. Concern over the fate of the engineered organisms in natural ecosystems and the possible environmental impacts is clearly more focussed on cases of deliberate release of GMOs to the environment, since the organisms are designed to survive and replicate in the environment at least for a certain period.

In response to these concerns, important Community legislation has been recently adopted by the Council of Ministers introducing measures to protect the environment and assure a single unified market for biotechnology in Europe.

The two Directives — Directive 90/219/EEC on the Contained Use of Genetically Modified Microorganisms, and Directive 90/220/EEC on the Deliberate Release to the Environment of Genetically Modified Organisms — introduce regulatory oversight of both experimental and commercial activities using organisms obtained by genetic modification, covering respectively, either activites where modified bacteria are intended to be kept under physical containment or activites where modified bacteria, plants or animals are intended to be used in the open environment.

The Directives introduce the obligation for an environmental risk assessment to be carried out, and for different approval procedures to be followed, as appropriate, in order to carry out experimental work, industrial production or market products throughout the Community under harmonised rules.

DIRECTIVE 90/220/EEC ON THE DELIBERATE RELEASE TO THE ENVIRONMENT OF GMOS

This Directive covers all genetically modified organisms (micro-organisms, plants and animals) through all the stages of release to the environment from small scale to large scale experimental introductions, as well as release through product marketing – although some products may be exempted from the provisions of this Directive if they are covered by other Community legislation which includes a similar environmental risk assessment.

The procedures foreseen in this Directive are based on the consensus reached within the OECD in 1986, that all deliberate releases should be reviewed on a step–by–step and case–by–case basis before being approved, because of very limited experience with releases. At present, the possible risks are difficult to be fully evaluated and any generic approach in a regulatory scheme would not, in the Commission's view, ensure adequate protection for the environment and the public. This approach was endorsed by the European Parliament in the opinion it adopted in May 89, and was clearly approved by the Council of Ministers in its agreement of the Directive.

The main elements of this Directive can be summarised as follows:

–An environmental risk assessment must always be carried out before any release to the environment whether in an experiment or product.

–No releases may be carried out without the consent of the competent authorities.

–A national approval procedure is foreseen for experimental relases.

–A Community approval procedure is foreseen for releases made by marketing a product. Once approved, a product can circulate freely throughout the Community.

For releases for "experimental purposes", a notification needs to be submitted to the competent authority and an environmental risk assessment must be carried out based on the parameters listed in the annex to the Directive. Only the competent authority on whose territory the release will take place is in charge of the review. No other Member State can intervene although other Member States

are informed, and may make comments if they so wish. A regular exchange of information between competent authorities is in fact extremely important to ensure uniform application of the procedures foreseen in the Directive.

As well as the information in the annexes and a risk assessment, the notifier must also provide information and data from similar releases carried out by him and may also refer to data or results from other notifiers, provided they have given their agreement. He also has the obligation to provide any new information which has consequences on the risks and which may become available either during the period the notification is being examined or after the release has been approved, and must adopt measures necessary to protect human health and the environment.

On receipt of the notification the competent authority must examine its conformity to the Directive, and evaluate the risk posed by the release. It may also, if it considers it appropriate, consult groups or the public. Having considered, where appropriate, any comments made by other Member States, it must respond to the notifier within 90 days of receipt of the notification indicating whether the release may proceed.

If further information or opinions have been requested, the period during which the competent authority is waiting is added to the 90 days.

On the completition of a release the notifier must send to the competent authority the result of the release in respect of any risk to man or the environment.

In the procedure for "placing on the market of products" containing GMOs, a notification must be submitted by the manufacturer or the importer to the competent authority in the Member State where the product will be placed on the market first. The competent authority concerned will examine the notification, carry out a risk assessment, and review the recommended precautions for the safe use of the product, and must decide within 90 days whether to put forward the application in the Community procedure with a favourable opinion or reject it.

The Community procedure envisages consultation with the other Member States during a 60-day period. During this time Member States can ask for further information or propose modifications to the approval. On completition of this period, and if no insoluble difficulties have arisen, the product will be approved by the competent authority to which the application was made and will be able to circulate freely throughout the European Community, under the conditions endorsed, without any further need of approval and without the possibility of limitations being placed on marketing by other Member States.

It is foreseen that the Commission, with the assistance of a Committee composed of Representatives of the Member States, will have the role of an arbitrator in case any disagreements arise between competent authorities as regards the community approval for a product. It is envisaged that this Committee will take its decision by qualified majority.

As with any proposal under Article 100a, it is also envisaged that Member States will have the possibility of adopting temporary measures in the case of emergencies where new evidence on the risks of a product are found. However, a final decision on this product must be made within three months, under the Committee procedure with a qualified majority.

Finally, the Directive places some general obligations on Member States, such as ensuring that all appropriate measures are taken to avoid adverse effects on human health and the environment which may arise from the release, designating the competent authority, and ensuring that inspections and other control measures are carried out.

It is also clearly stated that confidential information must be kept secret by the competent authorities and the Commission, but at the same time, that minimum set of non-confidential information relevant to any potential risk of the release, will always be made available to the public. Given the public concern about the potential risks from the deliberate releases of genetically modified organisms, it has been considered essential that the administrative procedures should be sufficiently transparent to ensure public confidence in the decision taken. Finally, exchange of information will play an important role in gaining experience. Provision is therefore made for such exchange between Member States via the Commission.

DIRECTIVE 90/219/EEC ON THE CONTAINED USE OD GENETICALLY MODIFIED MICRO-ORGANISMS

Directive 90/219/EEC on contained use addresses the use of genetically modified micro-organisms in all systems where barriers are used to restrict dissemination of live material to the external environment. The Directive is very largely based on the 1986 OECD report on safety in biotechnology and addresses all activites, from research and development to production, where genetically modified micro-organisms are used.

The purpose of this proposal is:

-to establish a case-by-case risk assessment for all operations;

-to identify the cases where the contained use could entail a risk;

-to establish working practices and containment measures corresponding to the hazard the micro-organism presents;

-to prevent, in case of hazardous micro-organisms, their accidental release and limit the consequences of such accidents where they occur.

A flexible notification procedure is proposed, based both on the kind of micro-organism used and on the scale of the operation envisaged.

In order to distinguish between low risk micro-organisms and high-risk or unknown-risk micro-organisms, two risk groups are established, group I and group II respectively, on the basis of a set of criteria defined by the OECD. All persons wishing to undertake work with a genetically modified micro-organism should classify it in one group according to the criteria provided and carry out a risk assessment of the proposed operation taking into account the factors outlined in the annex to the Directive.

If carrying out work for the first time in a particular installation, the user must notify the competent authority so as to register the building and allow any appropriate controls to be carried out. For group I (low risk) operations, the installation is considered, registered and appropriate for GMO operations to be carried out if no indication to the contrary is received within 90 days of notification. However, for installations designed for group II (high risk) operations, the competent authority must give its explicit consent within 90 days of notification.

Once the installation has been registered or approved, a simplified procedure is envisaged in the case of research activity to group I (low risk) micro-organisms. Here, only rules of good hygiene and safety practices are envisaged. Records of work are only to be kept and made available in case the competent authorities wish to see them. For industrial scale activities with group I organisms, a notification must be forwarded to the competent authorities containing a rather minimal set of information. The operation can take place if no objection is raised by the competent authority within 60 days of notification.

In the case of micro-organisms of group II, however, the user will have to demonstrate rather more explicitly the safe of the operation envisaged: firstly, more stringent principles of containment will have to be applied; secondly, both research and industrial activities will have to be notified, with proposed waiting periods of 60 and 90 days respectively, to allow the competent authority to check the content of the notification, and make comments if considered appropriate. For group II research activities, the notified operations can go ahead if no indication to the contrary is received from the competent authority within the 60 day period.

In the case of industrial scale operations for group II, the user will have to provide more details on micro-organisms, the operation and the installation in question, and an explicit consent must be obtained from the competent authorities within the 90-day waiting period.

The Directive also provides for the competent authority to be notified should an accidental release occur in its territory. Competent authorities must ensure that the necessary emergency, medium and long term measures are taken and alert any other Member State which could be affected by an accident; in parallel a system of exchange of information to ensure continuous monitoring of the situation throughout the European Community.

The scope of the Directive does not include micro-organisms which have not been obtained through genetic modification, and the Directive in addition explicitly excludes from its scope micro-organisms which have been obtained by techniques with which there is more experience, such as mutagenesis, self-cloning (for non-pathogenic naturally occurring micro-organisms) or certain kinds of cell fusion. It also excludes the construction and use of somatic animal hybridoma cells.

The Directive places some general obligations on Member States, such as ensuring that all appropriate measures are taken to avoid adverse effect on human health and the environment which may arise from the contained use of GMMs, designating the competent authority, and ensuring that inspections and other control measures are carried out.

It also clearly states that confidential information must be kept secret by the competent authorities and the Commission but, at the same time, that a minimum set of non-confidential information relevant to any potential risks of an operation should be made available; given the public concern about potential risks, it has been considered essential that specific groups or the public may be consulted, if considered appropriate by the competent authority.


**Conclusions**

Acknowledging the great potential of the new techniques of genetic modification for new products processes and the protection of the environment, the European Communities adopted the above mentioned Directives in order to provide a common regulatory framework which is a prerequisite for the development of the biotech industry.

It is hoped that the two Directives, with their provisions for uniform measures throughout the Member States, will help the development of the Biotechnology industry in a secure environment, without distortion of competition and in a way that provides the necessary safety for human health and the environment.

# STUDIES OF THE POTENTIAL FOR EXPRESSION OF NITROGENASE Fe-PROTEIN IN CELLS OF HIGHER PLANTS

M.J. DOWSON-DAY, J.L. ASHURST, J.WATTS*, R.A. DIXON
and M.J. MERRICK
AFRC Nitrogen Fixation Laboratory
University of Sussex
Brighton BN1 9RQ, UK

ABSTRACT. The potential for expression of the nitrogenase Fe-protein in <u>Nicotiana tabacum</u> has been examined by construction of hybrid <u>nifH</u> and <u>nifM</u> genes which should target their polypeptide products to the plant chloroplast. <u>In vitro</u> import of these proteins into isolated chloroplasts has been demonstrated both aerobically and anaerobically. Initial attempts to express the engineered genes in transformed plants were unsuccessful, possibly because of the low translational efficiency of the constructs used. Modified plant vectors have now been constructed using viral leader sequences. After electroporation into tobacco mesophyll protoplasts these vectors show significantly improved expression as measured by transient expression assays with ß-glucuronidase as the reporter. The improved vectors are now being assessed in transformed plants.

## 1. Introduction

Nearly two decades have now elapsed since it was first demonstrated that the nitrogen fixation (<u>nif</u>) genes of the facultative anaerobe <u>Klebsiella pneumoniae</u> could be transferred to a non-nitrogen fixing organism, namely <u>Escherichia coli</u>, so as to confer on <u>E.coli</u> the ability to fix nitrogen (Dixon and Postgate, 1972). Ever since then the proposition that it might one day be possible to transfer a <u>nif</u> gene package to an agronomically important higher plant such as wheat or rice has repeatedly appeared in articles concerned with practical applications of genetic engineering. However until relatively recently much of the fundamental knowledge which is arguably required before setting out on such a project was not available. The complete nucleotide sequence of the <u>K.pneumoniae</u> <u>nif</u> gene cluster was only determined in 1988 (Arnold et al., 1988) and at present the precise roles of a number of the <u>nif</u> gene products are unknown (Merrick, 1988). Likewise techniques for genetic transformation of cereal crops such as wheat and rice are still in their infancy. Nevertheless we believe that it is now possible to begin investigating the potential for <u>nif</u> expression in plants using a relatively simple model system.

* John Innes Institute, Norwich NR4 7UH, U.K.

## 2. A Model System

In choosing a model system as the basis of our work we have taken a number of factors into account. Firstly the _nif_ gene cluster of _K.pneumoniae_ comprises a total of 20 genes. It is therefore clearly neither possible nor desirable to consider the transfer of a complete _nif_ package to the plant genome as a first step.

The nitrogenase enzyme comprises two proteins; the tetrameric MoFe protein (Kp1) which contains a large number of metal clusters including the iron-molybdenum cofactor (FeMoco), and the dimeric Fe protein (Kp2) which contains a single $Fe_4S_4$ cluster. The Fe protein transfers electrons from an electron donor (in _K.pneumoniae_ the flavodoxin encoded by _nifF_) to the MoFe protein and can be purified independently of the MoFe protein.

Genetic evidence has implicated only two _nif_ genes in Kp2 biosynthesis. _NifH_ is the structural gene for Kp2 and _nifM_ is required for the synthesis of active Kp2 (Howard et al., 1985; Paul and Merrick, 1989). The precise role of _nifM_ is unknown but it may be required for the efficient insertion of the iron-sulphur cluster into apo-Kp2. The synthesis of active Kp2 in _E.coli_ carrying only _nifH_ and _nifM_ (Paul and Merrick, 1989) suggests that these two genes should probably be sufficient for synthesis of Kp2 in other foreign backgrounds including a plant cell. We therefore chose the synthesis of Kp2 from _nifH_ and _nifM_ as the simplest model system to test _nif_ gene expression in plants.

We have discussed previously the possible sub-cellular locations both for _nif_ genes and for nitrogenase in the plant cell (Merrick and Dixon, 1984, Merrick, 1986). In the context of the long term goal of transferring a _nif_ gene package into the plant genome the plastid has the attraction of an organelle with a transcriptional apparatus quite similar to that of prokaryotes (Hanley-Bowdoin and Chua, 1987). Consequently the amount of engineering which might be required to allow _nif_ gene expression in this environment could be considerably less than that required for expression in the nucleus. With regard to the synthesis of an active enzyme the oxygen sensitivity of both nitrogenase proteins has often been cited as precluding expression in the plastid. However it may ultimately be possible to ensure that the genes were only expressed in non-photosynthetic plastids e.g in the root. Furthermore the actual oxygen tensions produced in an actively photosynthesising chloroplast are not well defined, and recent evidence suggests that Kp2 is actually capable of reducing molecular oxygen to water if the protein is in greater than 4-fold excess over the concentration of oxygen (Thorneley and Ashby, 1989).

Bearing these factors in mind we have attempted to manipulate the _nifH_ and _nifM_ genes so as to allow targeting of their products to the plant chloroplast. In all cases the recipient plant was _Nicotiana tabacum_.


## 3. Engineering Nif Genes

The majority of chloroplast proteins are encoded in the nuclear genome. These proteins are synthesised as precursor molecules that are post-translationally imported into chloroplasts (Keegstra, 1989; Smeekens et al., 1990). The precursor proteins contain an N-terminal extension of some

50 amino acids which is proteolytically cleaved upon import by a protease activity located in the chloroplast stroma. The transit peptide contains the information for chloroplast specific import and such peptides have been used previously to direct bacterial proteins into the chloroplast (Van den Broeck et al., 1985). The transit peptide is also a major determinant in the subsequent localisation of the protein within the chloroplast i.e. in the stroma, thylakoid membrane or thylakoid lumen (Smeekens et al., 1986, 1990).

```
                              M     T     M     R
KpnifH          GTCACC      ATG   ACC   ATG   CGT


                              M     P     M     R
KpnifH(SphI)    GTCAGC      ATG   CCC   ATG   CGT
                            SphI


                Q     C     M     P     M     R
rbcS'-nifH     CAA   TGC   ATG   CCC   ATG   CGT
                            |


                              M     N     P     W
KpnifM          AGAAGA      ATG   AAC   CCA   TGG


                              M     H     P     W
KpnifM(SphI)    AGAAGC      ATG   CAC   CCA   TGG
                            SphI


                Q     C     M     H     P     W
rbcS'-nifM     CAA   TGC   ATG   CAC   CCA   TGG
                            |
```

Figure 1. Nucleotide sequences and translation products of K.pneumoniae nifH and nifM, and of the derived mutant and hybrid genes. | indicates the transit peptide cleavage site.


The transit peptide chosen for this work was that of the Nicotiana small subunit of ribulose bisphosphate carboxylate oxygenase (RUBISCO), encoded by rbcS. In order to link the coding sequence for the transit peptide to the 5' end of the nifH or nifM genes it was necessary to introduce an SphI restriction site (which normally flanks the first methionine codon of the mature RbcS polypeptide; Mazur and Chui, 1985) around the initiating ATG of these coding sequences. This was done using oligonucleotide mutagenesis and resulted in a change of a single amino acid immediately after the initiating methionine of each protein (Fig.1).

TABLE 1. Complementation of <u>K.pneumoniae</u> <u>nifH</u> and <u>nifM</u> mutants by modified genes.

| Strain | Acetylene reduction (as % of wild-type) |
|---|---|
| Wild-type | 100.0 |
| <u>nifH</u>⁻2213 | <1.0 |
| <u>nifH</u>⁻2213(<u>nifH</u>SphI⁺) | 135.0 |
| <u>nifM</u>⁻2104 | 15.0 |
| <u>nifM</u>⁻2104(<u>nifM</u>SphI⁺) | 96.0 |

The mutant genes were cloned under control of their normal <u>nif</u> promoters and then tested for their ability to complement appropriate chromosomal mutations in <u>K.pneumoniae</u>. As judged by the acetylene-reducing activities of the partial diploids, neither of the mutations significantly affected the activity of the gene products (Table 1). Surprisingly when comparable experiments were carried out using the hybrid genes encoding the complete RbcS transit peptide these were also found to complement appropriate chromosomal mutants. However Western blot analysis of extracts from strains carrying the hybrid genes suggested that there may be significant degradation of the hybrid proteins and it is possible that the proteins active in the complemetation are degradation products rather than the full length polypeptides.

## 4. <u>In vitro</u> import of nif proteins into chloroplasts

Import of foreign proteins into chloroplasts can be studied both <u>in vivo</u> and <u>in vitro</u> (Smeekens et al., 1986; de Boer et al., 1988) and we therefore chose to examine the <u>in vitro</u> import of the engineered <u>nifH</u> and <u>nifM</u> gene products to gain some insight into the likelihood that these proteins might be processed correctly <u>in vivo</u>.

The <u>rbcS'-nifH</u> and <u>rbcS'-nifM</u> coding sequences were placed under the control of the T7 promoter and transcribed <u>in vitro</u> using T7 RNA polymerase. Transcripts were translated in a wheat germ translation system, labelling the products with $^{35}$S-methionine. The hybrid proteins thus formed were incubated with illuminated, isolated pea chloroplasts and the fate of the hybrid proteins was analysed by SDS polyacrylamide gel electrophoresis. In both cases the hybrid proteins appeared to be both imported and cleaved to their normal size (Fig.2). Furthermore when the chloroplast membrane fraction was separated from the stroma by centrifugation the imported NifH and NifM were both present in the non-membrane fraction as was the RbcS control.

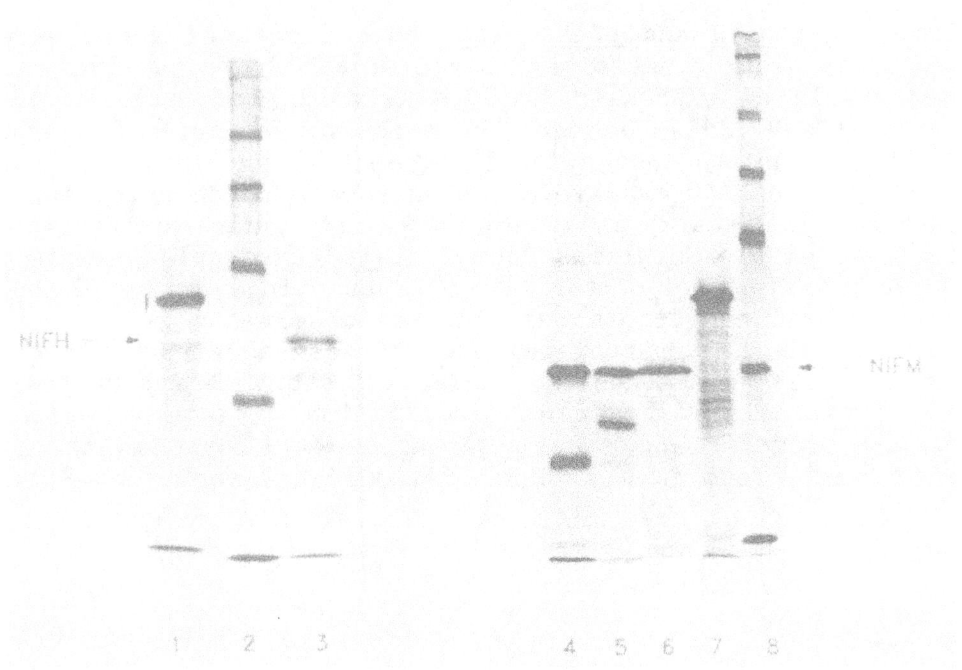


Figure 2.  *In vitro* import of <u>rbcS'-nifH</u> and <u>rbcS'-nifM</u> products into isolated pea chloroplasts.  Lane 1, RbcS-NifH pre-import; Lane 3, RbcS-NifH post-import; Lane 5, NifM immunoprecipitated from bacterial extract; Lane 6, RbcS-NifM post-import; Lane 7, RbcS-NifM pre-import; Lanes 2,4,8 molecular weight markers.

The time course of import of the <u>rbcS'-nifH</u> product was compared with that of an <u>rbcS</u> hybrid product encoded by the pea <u>rbcS</u> protein coding sequence fused to the same <u>Nicotiana</u> <u>rbcS</u> transit peptide used in the hybrid <u>nif</u> genes. This experiment revealed that the steady state level of both proteins in the chloroplast was attained after about 20 mins. and was maintained for at least an hour after that.

As discussed earlier it might ultimately be possible to regulate <u>nif</u> gene expression so that it occurs only in the dark or in non-photosynthetic plastids. Chloroplast import experiments of the type described above have usually been carried out using illuminated chloroplasts, although light is not essential for <u>in vitro</u> import (Cline, 1986). We were therefore interested to examine the potential for import in the dark, in the presence of added ATP, and in the presence of reduced glutathione and D-isoascorbic acid both of which contribute <u>in vivo</u> to the protection of the chloroplast from oxygen. The hybrid tobacco/pea RbcS protein was used for these studies and import could be demonstrated in chloroplasts incubated under nitrogen or in the presence of glutathione or ascorbate hence a reduction in oxygen levels does not inhibit import.

In summary these <u>in vitro</u> experiments suggest that the hybrid proteins encoded by <u>rbcS'-nifH</u> and <u>rbcS'-nifM</u> can be expected to be imported into chloroplasts when they are expressed in a plant cell, and that import should occur in either the light or the dark, in either aerobic or anaerobic conditions.

## 5. Expression of nif products in transformed plants

The hybrid <u>rbcS′-nifH</u> and <u>rbcS′-nifM</u> genes described above were cloned into the binary plant transformation vector Bin19 (Bevan, 1984) such that the genes should be expressed from the strong constitutive cauliflower mosaic virus (CaMV) 35S promoter. These constructs were then transformed into <u>N.tabacum</u> using conventional <u>Agrobacterium tumefaciens</u>-mediated transformation. Some 40 transformed plants were regenerated and shown by Southern blotting to carry an intact <u>rbcS′-nifH</u> gene. However when either whole leaf proteins or proteins from isolated chloroplasts were screened by Western blotting for material cross-reacting to a Kp2 polyclonal antibody no significant cross-reaction was observed.

In the light of the apparent failure of these constructs to express at significant levels we checked the nucleotide sequence around the cap site and the 5′ untranslated leader. A combination of the sequences derived from the 35S promoter and the <u>rbcS</u> leader meant that a strong hairpin could potentially form in the untranslated leader sequence (Fig.3).

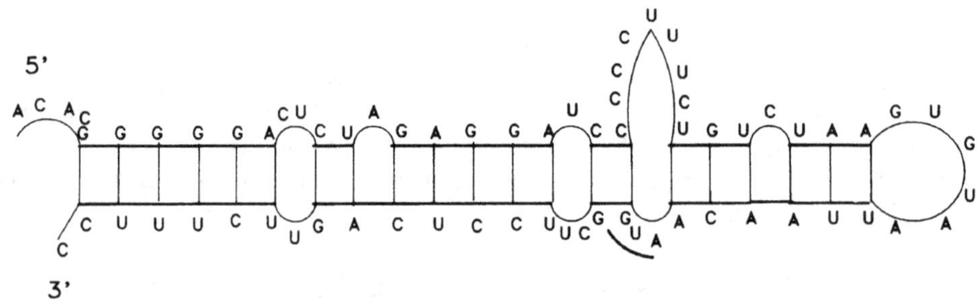

Figure 3. Untranslated leader sequence in the p35S <u>rbcS′-nifH</u> construct and potential hairpin stucture which is predicted to form in this sequence. The initiating methionine codon is underlined.

Most cytoplasmic messenger RNAs have untranslated leaders of 20 to 100 bases which tend to form only loose secondary structures (Kozak,1987). The 40s ribosome complex is believed to scan this leader sequence until reaching the most 5′ ATG where 80s ribosome assembly and translation initiation occur, although this event may be dependent on the base sequence around the ATG and on secondary structure. We therefore considered that the failure of our initial constructs to express at high levels could be due to poor translational efficiency caused by hairpin formation in the leader sequence. With these considerations in mind we decided to attempt to construct a modified vector which might have enhanced levels of expression.

## 6. Modified vectors for expression of foreign genes in plants

In vivo and in vitro studies have shown that a number of viral RNAs have a particularly high translational efficiency. Comparison of the translational efficiency of various heterologous mRNAs with substituted leaders suggests that this is due to their 5′ leaders (Jobling and Gehrke, 1987; Gallie et al., 1987; Browning et al., 1988), one of the most efficient of which is the 68 base sequence from the $U_1$ strain of tobacco mosaic virus RNA (termed $\Omega$) (Gallie et al., 1987; Wilson et al., 1990). We have therefore compared the effects of TMV $\Omega$ with other 5′ leaders on the expression of a reporter gene in plant cells.

The first 66 and 31 bases respectively of the RNA sequences of TMV$\Omega$ ($\Omega*$) and the 35S RNA of CaMV (35S*) were abutted into the 35S cap site at +1 (Fig.4). The leaders were designed to have an identical sequence around the initiating ATG to eliminate differences at the point of ribosome assembly. A transcriptional fusion vector, pBI221.2 (Jefferson,1987), was used as the standard in each of three experiments. pBI221.2 is a pUC-based plasmid containing the 35S promoter (−831 to +5) and an 18 base untranslated leader followed by the coding sequence for ß-glucuronidase (GUS) and the nopaline synthase (nos) terminator sequences.

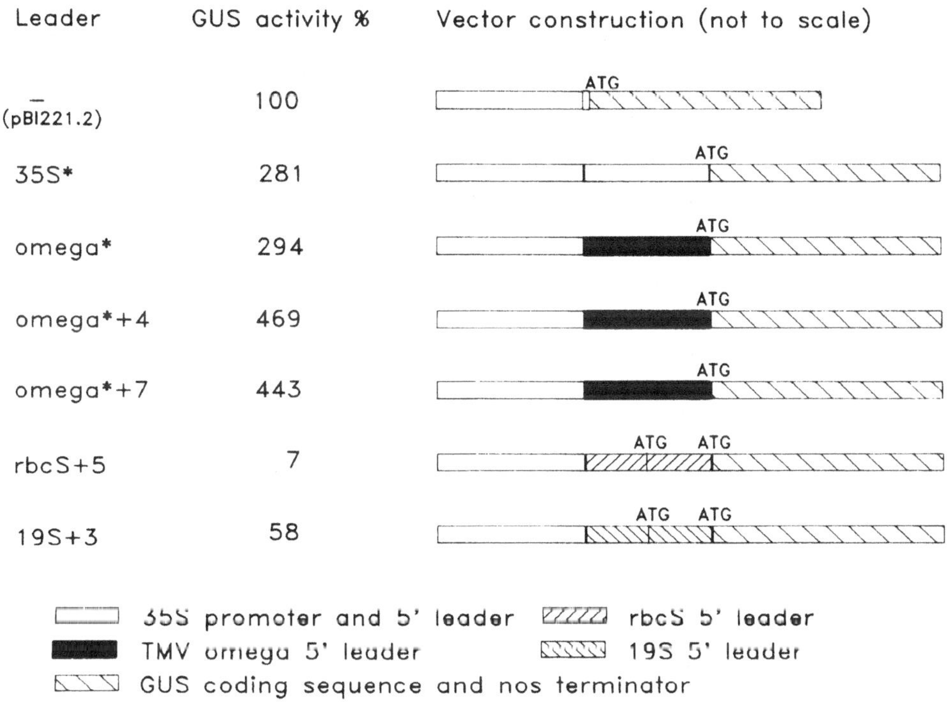

| Leader | GUS activity % |
|---|---|
| —<br>(pBI221.2) | 100 |
| 35S* | 281 |
| omega* | 294 |
| omega*+4 | 469 |
| omega*+7 | 443 |
| rbcS+5 | 7 |
| 19S+3 | 58 |



Figure 4. Structures of vectors constructed with alternative untranslated leaders. All vectors contain the 35S promoter, GUS coding sequence and nos terminator.

To assess the relative activities of the different constructs we electroporated tobacco mesophyll protoplasts with linear plasmid DNA carrying the various 5' leader sequences, and assayed transient expression of GUS activity 40 hours later. The Ω* and 35S* leaders each resulted in a 2.5 to 3.0 fold greater level of expression than was obtained with the control plasmid and there was no significant difference between the two leaders.

Since TMV RNA is not normally transcribed from DNA but is replicated by viral replicase it is not apparent whether the Ω 5' sequence will permit efficient transcription initiation when inserted directly into the 35S cap site. Therefore we constructed plasmids containing the Ω* sequence 3' to either four or seven bases of the original 35S leader, the fifth and eighth G of the 35S leader being substituted by the 5'G of Ω* (Fig.4). Both these additions resulted in significant increases in GUS activity relative to Ω* of approx. 1.5 fold.

For comparative purposes we also analysed expression of two translational fusions in comparable constructs to those using the viral leaders. These fusions used either the N.tabacum rbcS leader or the CaMV 19S ORFVI leader with the respective amino terminal coding sequences linked to polylinkers so as to maintain the ATG initiation codon of GUS in frame (Fig.4). The resultant number of additional 5'amino acids for RbcS-GUS and ORFVI-GUS were 20 and 16 respectively. The RbcS-GUS fusion constructed in this way is derived from the original RbcS-NifH fusion described earlier (Fig.1).

The GUS levels from the rbcS fusion were not above background and those from the 19S fusion were similar to pBI221.2. Hence the two viral leaders gave significantly enhanced GUS expression levels compared to other leaders and the most highly expressed vectors were those comprising Ω* with either +4 or +7 bases of the original 35S leader. Computer predictions of the secondary structures of the expected mRNAs were generated and the predictions for the Ω*+4 and Ω*+7 leaders showed the same patterns and the same ˆG values as those for Ω* indicating that the enhanced expression levels are not attributable to changes in RNA secondary structure. Hence it is possible that the advantage conferred by the extra 5' bases is the result of more efficient transcription initiation.

Having demonstrated the effects of these improved vectors in transient expression assays we are now examining their effects in transgenic plants. In these experiments we will correlate GUS expression in whole leaf tissue with that observed in protoplasts and we will also determine whether the modified vectors allow improved expression of our original rbcS'-nifH hybrid gene.


## 7.Summary

It will be apparent from the experiments described above that we are still some way from determining whether or not it is possible to express the nitrogenase Fe protein in a plant cell. However we have made some considerable advances. We have constructed hybrid rbcS'-nif genes and we have shown that the products of such genes can be correctly processed by the plant chloroplast and cleaved to produce a mature protein product of the predicted size. This import occurs both in the light and in the dark and we are hopeful that comparable processing could occur in intact plant cells.

The major obstacle to further analysis of this system is currently the inability to obtain sufficiently high levels of expression in transgenic plants. However in the course of these studies we have constructed modified vectors which, in transient expression assays, show considerably improved levels of expression. The simple model we are using, based on the use of just two _nif_ gene products, is very suitable for this initial exploratory work and until this system has been demonstrated to function there appears to be no good reason to study more complex models involving Kp1 biosynthesis. In the long term, methods for chloroplast transformation of higher plants may be developed in which case the prospect of introducing _nif_ genes directly into the plastid genome could be considered. In the meantime much fundamental work on the biosynthesis of nitrogenase remains to be done and the information coming from those studies may well aid attempts to express _nif_ genes in foreign hosts.


## 8. References


Arnold, W., Rump, A., Klipp, W., Priefer, U.B., and Puhler, A. (1988) Nucleotide sequence of a 24,206 base-pair DNA fragment carrying the entire nitrogen fixation gene cluster of _Klebsiella pneumoniae_. _J Mol Biol_ **203**: 715-738.

Bevan, M. (1984) Binary _Agrobacterium_ vectors for plant transformation. _Nucl Acids Res_ **12**: 8711-8721.

Browning, K.S., Lax, S.R., Humphreys, J., Ravel, J.M., Jobling, S.A., and Gehrke, L. (1988) Evidence that the 5'-untranslated leader of mRNA affects the requirement for wheat germ initiation factors 4A,4F, and 4G. _J Biol Chem_ **263**: 9630-9634.

Cline, K. (1986) Import of proteins into chloroplasts. _J Biol Chem_ **261**: 14804-14810.

de Boer, D., Cremers, F., Teertstra, R., Smits, L., Hille, J., Smeekens, S., and Weisbeek, P. (1988) _In vivo_ import of plastocyanin and a fusion protein into developmentally different plastids of transgenic plants. _EMBO J_ **7**: 2631 2635.

Dixon, R., and Postgate, J.R. (1972) Genetic transfer of nitrogen fixation from _Klebsiella pneumoniae_ to _Escherichia coli_. _Nature_ **237**: 102-103.


Gallie, D.R., Sleat, D.E., Watts, J.W., Turner, P.C., and Wilson, T.M.A. (1987) A comparison of eukaryotic viral 5'-leader sequences as enhancers of mRNA expression _in vivo_. _Nucl Acids Res_ **15**: 8693 8711.

Hanley-Bowdoin, L., and Chua, N-H. (1987) Chloroplast promoters. <u>Trends Biochem Sci</u> **12**: 67-70.

Howard, K.S., McLean, P.A., Hansen, F.B., Lemley, P.V., Koblan, K.S., and Orme-Johnson, W.J. (1985) <u>Klebsiella pneumoniae</u> <u>nifM</u> gene product is required for stabilization and activation of nitrogenase iron protein in <u>Escherichia coli</u>. J Biol Chem **261**: 772-778.

Jefferson, R.A. (1987) Assaying chimeric genes in plants: the GUS gene fusion system. <u>Plant Mol Biol Rep</u> **5**: 387-405.

Jobling, S.A., and Gehrke, L. (1987) Enhanced translation of chimaeric messenger RNAs containing a plant viral untranslated leader sequence. <u>Nature</u> **325**: 622-625.

Keegstra, K. (1989) Transport and routing of proteins into chloroplasts. <u>Cell</u> **56**: 247-253.

Kozak, M. (1987) An analysis of 5'-noncoding sequences from 699 vertebrate messenger RNAs. <u>Nucl Acids Res</u> **15**: 8125-8132.

Mazur, B.J., and Chui, C-F. (1985) Sequence of a genomic DNA clone for the small subunit of ribulose bis-phosphate carboxylase-oxygenase from tobacco. <u>Nucl Acids Res</u> **13**: 2373-2386.

Merrick, M., and Dixon, R. (1984) Why don't plants fix nitrogen. <u>Trends Biotech</u> **2**: 162-166.

Merrick, M.J. (1986) Prospects for the transfer of nitrogen fixation genes to crop plants. <u>J Roy Agricult Soc Engl</u> **147**: 202-207.

Merrick, M.J. (1988) Organisation and regulation of nitrogen fixation genes in <u>Klebsiella pneumoniae</u>. In <u>Nitrogen fixation: Hundred years after</u>. Bothe, H., de Bruijn, F.J., and Newton, W.E. (eds). Gustav Fischer, Stuttgart and New York, pp. 293-302.

Paul, W., and Merrick, M. (1989) The roles of the <u>nifW</u>, <u>nifZ</u> and <u>nifM</u> genes of <u>Klebsiella pneumoniae</u> in nitrogenase biosynthesis. <u>Eur J Biochem</u> **170**: 259-265.

Smeekens, S., Bauerle, C., Hageman, J., Keegstra, K., and Weisbeek, P. (1986) The role of the transit peptide in the routing of precursors toward different chloroplast compartments. <u>Cell</u> **46**: 365-375.

Smeekens, S., Weisbeek, P., and Robinson, C. (1990) Protein transport into and within chloroplasts. <u>Trends Biochem Sci</u> **15**: 73-76.

Thorneley, R.N.F., and Ashby, G.A. (1989) Oxidation of nitrogenase iron protein by dioxygen without inactivation could contribute to high respiration rates of <u>Azotobacter</u> species and facilitate nitrogen fixation in other aerobic environments. <u>Biochem J</u> **261**: 181-187.

Van den Broeck, G., Timko, M.P., Kausch, A.P., Cashmore, A.R., van Montagu, M., and Herrera-Estrella, L. (1985) Targeting of foreign proteins to chloroplasts by fusion to the transit peptide from the small subunit of ribulose-1,5-bisphosphate carboxylase. Nature **313**: 358-363.

Wilson, T.M.A., Saunders, K., Dowson-Day, M., Sleat, D.E., Trachsel, H., and Mundry, K.W. (1990) Effects of the 5'-leader sequence of tobacco mosaic virus RNA, or derivatives thereof, on foreign mRNA and native viral gene expression. In Post-transcriptional control of gene expression. McCarthy, J.E.G., and Tuite, M.F. (eds). NATO-ASI, (in press).

Index

*A. amazonense*  89, 155
*A. azollae*  583
*A. brasilense*  89, 139, 155, 269, 285, 349, 351
*A. halopraeferens*  155
*A. irakense*  205
*A. lipoferum*  89
*A. vinelandii*  37, 509
*A.lipoferum*  139
abscisic acid  529
ACA-resistance  519
Acaricide  73
*Acetobacter diazotrophicus*  89, 105, 155, 299
acetylene reduction assay  270
*Achaeta* species  50
acid  161
actimorhizal symbionts  647
actinomycetal  601
actinomycete  615
actinorhiza  601, 602
actinorhizae  3, 4, 602
adenine nucleotides  496
Adhesion of bacteria  7
ADP-ribose  494
ADP-ribosylation  495, 497
Agricultural crops  337
*Agrobacterium*  2
*Agrobacterium rhizogenes*  7
*Alcaligenes faecalis*  195
*Allolobophora chlorotica*  50
*Alnus*  4, 601
*Alnus*-compatible  637
*Alnus cordata*  651
*Alnus glutinosa*  4
*Alnus incana*  643
*Alnus nepalensis*  647
alternative nitrogenase(s)  583
Amino acid liberation  519
Aminopeptidases  645
*aminotransferase*  302
ammino acid metabolism  543
Ammonia-excreting cyanobacteria  523
ammonium  4, 5
*Anabaena*  2, 4, 359, 431, 521, 563
*Anabaena Azolla* association  463
*Anabaena azollae*  389, 447, 453, 543, 551
*Anabaena azollae* AaS  482

*Anabaena variabilis*  425, 519, 523, 531, 533
angiosperms (*Gunnera*)  4
*Anthoceros*  572
*Aporrectodea caliginosa*  50
*Aporrectorea rosea*  50
ARA  63, 565
Arctic soil  567
aromatic aminoacid aminotransferases  321
*Arthrobacter*  457
asparagine  494
associated BNF  315
associative nitrogen fixation  97, 212
Associative diazothrops  337
Associative dinitrogen fixers  317
Autoregulation  303
auxanographic  293
axenic rice plants  339
Azetidine 2-carboxylic acid  519
*Azoarcus*  235
*Azobacter paspalii*  5
azolla-nitrogen  555
*Azolla*  4, 5, 389, 457, 467, 473, 541, 543, 561, 563, 571
*Azolla-Anabaena Association*  389
*Azolla* biomass  593
*Azolla caroliniana*  545
*Azolla filiculoides*  447, 453, 481, 551
*Azolla pinnata*  555
Azorhizobium  3
Azospirilla  343
*Azospirillum*  2, 5, 6, 7, 8, 67, 79, 89, 139, 169, 181, 211, 219, 291, 303, 307, 317, 347, 355, 357
*Azospirillum brasilense*  113, 127, 133, 147, 161, 171, 175, 187, 231, 243, 249, 255, 283, 287, 289, 295, 321, 323, 325, 327, 345
Azospirillum-Gramineae interaction: effect of indole-3-acetic  161
*Azospirillum* in cereals  9
*Azospirillum* inoculation  182, 71, 73, 147, 171, 201, 295, 297, 301, 311, 313
*Azospirillum*-root interaction  113
Azotobacter  329, 333
*Azotobacter chroococcum*  57
*Azotobacter* spp  317
*Azotobacter vinelandii*  13, 59, 495

672

bacterial release 7, 8
Bacterial attachment 187
bacterization 357
bacteroid 8
bacteroids 2, 8
Baggio C. 161
Barbieri P. 161
Bazzicalupo M. 161
bidirectional transport 8
biochemicals 487
Biofertilization 69
biofertilizer 473, 545
Bioinoculants 327
bioluminiscence 275
*Bradyrhizobium* 2, 3, 6

*C. circinalis* 481
Caballero J. 503
calcium content of *Nostoc* 6720 405
Calcium 405
Calcofluor 287
calmodulin 405
canopy 63
capsular-polysaccharides 147
captan 355
carbohydrate content 540
carbon source affinity 231
carbonyl cyanide m-chlorophenylhydrazone 494
carboxylase enzyme 641
Cassette mutagenesis 55
*Casuarina* 6, 615
*Casuarina glauca* 621
*Casuarinaceae* 4
Cauvin B. 503
cell dedifferentiation 6
cell-surface polysaccharides 8
cell-wall degrading enzymes 7, 8
cephalodia 566
Cervantes E. 639
chemo-autotrophic growth 641
chickpea 181
*Chlorobium* 491
chloroplast 3
choline 155
*Chromatium* 491, 496
*Chromatium vinosum* 492
*Cicer arietinum* 181
citrulline 605

*Clorobioum* 492
$CO_2$ 641
$CO_2$ production 255
Colbeau A. 503
common *nod* genes 6
compartment 8
competition 3
Compound 48/80 406
computer analysis 170
contained use 653
conventional molybdenum nitrogenase 13
coralloid roots 527
corn 1, 6
culture medium 303
*Cyanobacteria* 4, 367, 369, 377, 411
*cyanobacterial mats* 443
cyanobionts 561, 571
cycads 572
*Cycas* 4
*Cycas revoluta* 481
cytochrome c oxidase 411

2,4-D 7, 269
deciduous forests 565
deliberate release 653
developmental profile 569
dice coefficient 171
dice similarity coefficients 581
Dicofol 71, 73
diffusion barriers 8
dinitrogen fixation 405, 566
Dinitrogen-fixing cyanobacteria 537
dinitrogenase (MoFe-protein) 492
dinitrogenase reductase (Fe-protein) 492
dinitrogenase reductase 457, 494
dinitrogenase reductase activating glycohydrolase (DRAG) 89, 495, 533
dinitrogenase reductase ADP-ribosyltransferase (DRAT) 89, 494, 495, 496, 497
dipterocarpol 4, 8, 9
DNA 169, 581
DNA-coated microprojectiles 2
DNA-DNA hybridizations 629
DNA hybridization 635
DNA relatedness 635
DNA:rRNA and DNA:DNA hybridizations 293
165 rDNA sequencing 629
dried 537

Duport C. 513
durum wheat 161, 327

*E. coli* 494
ecological 1
ecological aspects 9
economic 1, 9
economical problems 9
ectoine 155
*Ectothiorhodospira* 491, 492, 581
EF group 1 293
effectivity 3
*Eisenia fetida* 50
*Elaeagnus angustifolia* 63
Elaegnaceae-isolated 637
electron spectroscopic imaging 551
electron transport 513
*Enchytraeus buchholzi* 50
*Enchytronia* species 50
endo-rhizosphere 5
endo-symbiotic 5
endosymbiont 7, 8, 9
endosymbioses 1, 8
endosymbiosis 6, 8
energy coupling 61
engineered genes 659
*Enterobacter* 305
*Enterobacter agglomerans* 25
*Enterobacter cloacae* 25, 305
Enterobacteriaceae 211
epiphytic life 77
*Erwinia* 605, 609
erythromycin 457
*Escherichia coli* 2, 494
ethanol 303
*exo-Genes* 8
exo-polysaccharides 7, 147
exopolysaccharides 287, 295
external signal oxygen 2
extra-cellular signal 6
extracellular polysaccharides 219, 325

Fatty acid composition 563
*fdxA* 513
*fdxB* 513
*fdxC* 513
*fdxN* 513
FE$^{++}$ accumulation 327

ferredoxin 431, 493
ferredoxin-NADP* reductase 431
Ferredoxin 579
Ferredoxins 513
field trial 357
finger millet 243, 581
*fix genes* 6, 8
Fix L/Fix J 2
flavodoxin 431, 492, 493
flavonoids 6, 8
Flexible 637
formate 641
*Frankia* 2, 3, 4, 6, 8, 9, 595, 615, 621, 629, 635, 637, 639, 641, 643, 645, 647
*Frankia* growth 641
Frankia polymeric inoculants 622
free fixing microorganisms 63
Free-living bacteria 65
*Fridericia* species 50
fungicides 57, 355
*Fusarium Solani* 75
*Fusarium sp* 305

Galli E. 161
gel electrophoresis 602
gene expression 219, 289
gene *glnA* 573
Gene encoding ferredoxin 577
genetic diversity 637
genetic map 59
genetic mapping 504
genetic transformation 602
Genetically Modified organisms 653
genetics 3
genomic relationships 637
gibberellins 297, 529
Giovannetti L. 169, 581
GLN:GLU 526
*glnBA*-operon 573
*Gloeothece* 525
*Gloeothece* sp. 437
*Glomus clarum* 299
*Glomus etunicatus* 647
glutamate dehydrogenase 585
glutamate synthase 493, 496, 585
Glutamate auxotroph 589
Glutamate synthase 494
glutamine 494

674

glutamine synthetase 493, 494, 585, 497
glycine betaine 155
Gori A. 169, 581
Grabau C. 513, 579
Grain production 69
green manure 593
GroEL 48
growth chamber 339
growth rate 231
growth stimulating substances 5
Growth kinetics 313
*Gunnera* 572
Gymnosperms 4

hairy roots 7
halotolerant bacteria 155
heat production 255
hemoglobin 606
*Henlea* species 50
*Herbaspirillum seropedicae* 133, 225, 293
heterocyst frequency 405
heterocysts 4, 359, 447
heterotrophic metabolism 481
hexulose monophosphate pathway 641
Hippophae rhamnoides L. 639
hormogonia formation 561
hormone-like substance 3, 297
hormone production 347
host 6
host-microbe interactions 4
host-range 6, 9
host specific nodulation (*hsn*) genes 6
host specificity 7
HPLC 301
$H_2$ photoevolution 591
*hupR1* gene 503
$H_2$ uptake capacity 481
hydrogen peroxide 419
hydrogenase (*hup*) genes 504
hydrogenase 503, 605
hydrogenase-nitrogenase relationship 367
Hydrogenases 369
hydroxyproline-rich 606

IAA 301, 345
IAA biosynthesis 321
identification of strains 169
identification tests 169

indole-3-acetic acid 161
indole-3-acetic acid metabolism 113
Indole-3-Acetic Acid 301
indoleacetic acid 323
ineffective nodules 637
infection 3, 8
infection of grass leaves 225
infection thread 6
infection thread formation 7
Infectivity 639
inoculants 537
integration host factor 34
interaction B-Ca 521
Intracellular amino acids 525
iodonitrotetrazolium 270
iron-sulfur clusters 513
isotope dilution method 5
Jouanneau Y. 513, 579

*K. pneumoniae* 493
kallar grass 5, 235
Kelthane 71, 73
*Klebsiella* 2
*Klebsiella nif* L 2
*Klebsiella pneumoniae* 2, 55, 492
*Klebsiella pneumoniae nif* gene expression 43
*Klebsiella* R15 97

lateral roots 161, 606
Lea P.J. 405
leaf cavities 551
lecithin 4
lectins 8
leghemoglobin 3, 7, 8, 604
legumes 1
legume-*Rhizobium* symbioses 3, 7
lichen cover 565
lichens 565
light: dark 525
lignite spoil bank 651
Lipo Polysaccharides 7
*Lobaria pulmonaria* 565
*Lolium multiflorum* 341
luciferase activity 275
luciferase expression 275
*Lumbricus castaneus* 50
*Lumbricus rubellus* 50
*Lumbricus terrestris* 50

*Macrozamia* 4, 527
Magnin J.P. 503
maize 69, 357
maize roots 305
Maize rhizosphere 343
Marine 539
Maruca S. 169
Materassi R. 581
Melanin 311
membrane activity 175
membrane systems 8
metabolic regulation 496
Methane 641
methanol 641
methionine sulfoximine 531
Meyer C. 513
micro-projectiles 3
Microbial density 63
Microbial ecology 595
microcalorimetry 255
mineral uptake 161
mixed cultures 509
mixed plantation 651
Mixotrophy 425
molecular chaperones 43
monoclonal antibodies 37
Montaini P. 169
mottled stripe disease 225
MPN microorganisms 64
mucigel-like substance 187
Mutants 291

$^{15}N_2$ 5
N-assimilation 5
N-release 555
$N_2$-dependent growth 225
$N_2$ fixation 525
$N_2$-fixation activity 457
$N_2$-fixing ability 347
$N_2$-fixing cyanobacteria 535
N2 fixing bacteria 341
$NAD^+/NADH$ 496
$NADP^+$ 496
Napier grass 225
natural $^{15}N$ abundance 651
necrotic root cells 8
*Nicotiana tabacum* 659
*nif* 6, 8

*nif* A 2, 8
Ntr B/Ntr C 2
*nif* expression 659
*nif*-gene 493
*nif* gene region 359
*nif* genes 2, 3, 9. 513
*nif* L 2
*nif*-plasmids 30
*nif* regulatory gene 139
Nif-mutants 575
NifA 33, 43, 45, 503
*nifF* 25, 492, 493
*nifHDK* 127
*nifJ* 25, 492, 493
*nifJF*-cotranscript 29
*nifL* 25
NifT 25
NifW 25
NifZ 25
nitrate reductase 249
nitrate uptake 5
Nitrite excretion 249
nitrogen assimilation 399
nitrogen biofertilizer 487
nitrogen fixation 77
nitrogen fixing bacteria 353
nitrogen-fixing cyanobacteria 487, 521, 539
nitrogen-fixing phototrophic bacteria 581
nitrogen-fixing symbiotic actinomycetal 629
nitrogen-regulated promoter regions 79
nitrogen uptake 411
nitrogen fixation 25
nitrogenase 3, 37, 61, 513, 604
nitrogenase activity 269, 437, 569
nitrogenase assembly 43
nitrogenase genes 13
nitrogenase inhibition 531
nitrogenase 368, 491
nitrogenase activity 353
nitrogenase protein levels 643
*nod* box 6
*nod* D 6
*nod* D genes 6
*nod* D hybrids 6
*nod* E 7
*nod* genes 6, 8
nod genes 8
NodRM-1 7, 8

*Nodularia* Sp. 539
nodulation 3, 8, 621, 647
nodule organogenesis 6
nodulin genes 602
nodulins 3, 6, 7, 8, 604
non-heterocystous cyanobacteria 525
non-molybdenum nitrogenases 13
non-motile forms 311
non-symbiotic nitrogen fixation 65
*Nostoc* 377, 405, 527, 563, 571
*Nostoc commune* 567
*Nostoc-Gunnera magellanica* symbiosis 569
*Nostoc muscorum* 529
*Nostoc* sp. Cc. 482
*Nostoc* sp. Cr4 482
novobiocin 457
NtrC 503
Nucleotide sequence 283, 285
Nuti M.P. 161

*Octolasion lacteum* 50
oligonucleotide probes 595
oligosaccharins 609
Onek L.A. 405
open ponds 539
organic-C 64
*Oscillatoria* 525
*Oscillatoria chalybea* 419
*Oscillatoria limosa* 443
osmotic stress 155
Osmotolerance 155
outdoor cultures 593
oxidoreductase 492
oxygen 3
oxygen concentration 8
oxygen evolving system 419
oxygen uptake 411
oxygenic photosynthesis 437

*Parasponia* 3, 6, 7, 8
*Paspalum notatum* 5
PCR 629
pectate lyase 205, 605
pectin methylesterase 205
Pectolytic 605
performance of Azolla 467
peribacteroid membrane 8
pH 64, 303

phenazine methosulfate 494
phosphoenolpyruvate carboxykinase 641
phosphoenolpyruvate carboxylase 641
photoautotrophic 525
photobioreactors 539
photoheterotrophic 525
photosynthesis 4, 425
photosynthetic bacterium 503
phototrophic bacteria 583
phyllosphere 77
phytoalexins 8
Phytohormones 529
plant and plant-growth promoting rhizobacteria (PGPR) 181
plant-cyanobacteria symbioses 377
plant growth promoting rhizobacterium 148
plant lectins 7
plant pathogen 225
plasmid 312
pO$_2$ 8
poly-ß-hdyroxybutyrate metabolism 113
polymerase chain reaction 608
polyphenol oxidase activity 312
polysaccharides 147
24,-12-promoter 25
propionyl CoA carboxylases 641
proteases 8
Proteinases 645
proton efflux 175
proton motive force 496
Protoplast 608
*Pseudomonas* 293
*Pseudomonas rubrisubalbicans* 225
purple non-sulphur bacteria 591
pyridine nucleotides 496
pyruvate 641
pyruvate: flavodoxin 492
pyruvate: oxidoreductase 587

*Quercus robur* 651

R. *capsulatus* 492, 493, 494, 495
R. *leguminosarum* bv *trifolii* 6
R. *Leguminosarum* 7
R. *Leguminosarum* bv *viciae* 7
R. *meliloti* 6
R. *sphaeroides* 493, 494, 496, 583
R. sulfidophila 492

recombinant plasmid 133, 139
reforestation 601
regulation 127
Regulation of Genetically Modified Organisms 653
regulatory regions 283
reinoculated plants 235
respiration 425
respiratory protection 411
restriction pattern 169, 581
rhizobacteria 97
*Rhizobium*, 2, 3, 4, 6, 7, 8, 9
*Rhizobium* bacteroid 2, 8
*Rhizobium* host specificity 6
*Rhizobium*-legume 601
*Rhizobium*-legumes symbioses 181
rhizosphere 5, 8, 25, 63, 211, 341
rhizosphere competence 347
*Rhodobacter capsulatus* 491, 492, 503, 509, 513, 575, 577, 579, 583, 585
*Rhodobacter sphaeroides* 573, 589
*Rhodocyclus gelatinosus* 492
*Rhodocyclus purpureus* 491
Rhodocyclus tenius 492
Rhodomicrobium vannielli 492
Rhodopila globiformis 492
*Rhodopseudomonas palustris* 492, 591
*Rhodospirillaceae* 491
*Rhodospirillum fulvum* 491
*Rhodospirillum rubrum* 491, 492, 587
Ri vectors 2
Ribulosebisphosphate carboxylase 641
rice 1, 6, 7
rice cultivation 473
rice fields 5
rice production 523
rice roots 195
rice yield 541
rice field 5, 535, 593
rice genotypes 315
Richaud P. 503
risk assessment 653
rizospheric organism 172
RNA-polymerase-6$^{54}$-complex 33
Rodriguez Barrueco C. 639
root colonization 169
root development 5
root growth 297

root hair curling 7
root hair deformation 6
root hairs 161, 608
root lipid 4
root nodule symbioses 3
root nodules 3, 6, 605
root symbiosis 601
root-associated diazotrophs 345
root exudates 65
*Rps. acidophila* 492
*Rps. blastica* 492
*Rps. pallustris* 494
*Rps. sulfoviridis* 492
*Rps. viridis* 492
*Rsp. fulvum* 492
*Rsp. molischianum* 492
*Rsp. photometricum* 492
*Rsp. rubrum* 491, 493, 494, 495, 496, 497
*Rsp. salxeigens* 492
*Rubrisubalbicans* 293
Russianolive 63
ryegrass 341

S. sphaeroides 492
saline calcareous soil 243
Saline-sodic soils 211
salt stress 243
*Sambucus nigra* 50
Schatt E. 513, 579
selftransmissible plasmids 29
senescence 8
short roots 7
siderophores 5
Siderophores 291
sites for N$_2$-fixation 235
Smith R.J. 405
sodium chloride 155
soil inoculation 75, 537
soil organic matter 181
Soil redox potential 67
sorghum 225
sorghum rhizosphere 349, 351
species attribution 170
*Spirillum lipoferum* 5
sporangia 4
Spore-negative 4
Spore-positive 4
spores 4

succession 566
*sucrose* 155
sugar beet 329, 333
sugar cane 105, 225
sunflower rhizosphere 201
superoxide dismutase (SOD) 447
*sym* plasmids 7
"switch-off" effect 494
symbiotic cyanobacteria 481
*Synechococcus* 399

Taxonomic 293
terrestrial oligochaetes 49
thick 7
*Thiocapsa roseopersicina* 491, 492
thiram 355
Ti 2
*Tilia platyphyllos* 50
Tillandsia 77
tissue culture 602
tobacco roots 353
tomato plants 317
total-N 64
transcription 127
transcriptional activator 503
transformed plants 659
transhydrogenase 496
trehaloseglutamate 155
*Trema* 6, 7
Trifluoroperazine 406

triphenyltetrazolium 270
tryptophan 302
Tween 80 4

UPGMA 169, 581
upstream activator sequence 33
uptake 327
uptake hydrogenase 481, 527

VA fungi 8
VA-Mycorrhizae 299
vegetative cells 447
Ventura S. 169, 581
Verapamil 406
*Vicia sativa* 7
Vignais P.M. 503, 513, 579
Viti C. 581

$W_7$ 406
wheat
wheat 1, 6
wheat germ agglutinin 8
wheat root nodules (*p*-nodules) 269

*Xanthomonas maltophilia* 201

yield 231

*Zamia furfuracea* 481
Zanetti G. 161

# Developments in Plant and Soil Sciences


1.  J. Monteith and C. Webb (eds.): *Soil Water and Nitrogen in Mediterranean-type Environments.* 1981
    ISBN 90-247-2406-6
2.  J. C. Brogan (ed.): *Nitrogen Losses and Surface Run-off from Landspreading of Manures.* 1981
    ISBN 90-247-2471-6
3.  J. D. Bewley (ed.): *Nitrogen and Carbon Metabolism.* 1981        ISBN 90-247-2472-4
4.  R. Brouwer, I. Gašparíková, J. Kolek and B. C. Loughman (eds.): *Structure and Function of Plant Roots.*
    1981        ISBN 90-247-2510-0
5.  Y. R. Dommergues and H. G. Diem (eds.): *Microbiology of Tropical Soils and Plant Productivity.* 1982
    ISBN 90-247-2624-7
6.  G. P. Robertson, R. Herrara and T. Rosswall (eds.): *Nitrogen Cycling in Ecosystems of Latin America and
    the Caribbean.* 1982        ISBN 90-247-2719-7
7.  D. Atkinson, K. K. S. Bhat, M. P. Coutts, P. A. Mason and D. J. Read (eds.): *Tree Root Systems and Their
    Mycorrhizas.* 1983        ISBN 90-247-2821-5
8.  M. R. Sarić and B. C. Loughman (eds.): *Genetic Aspects of Plant Nutrition.* 1983        ISBN 90-247-2822-3
9.  J. R. Freney and J. R. Simpson (eds.): *Gaseous Loss of Nitrogen from Plant-Soil Systems.* 1983
    ISBN 90-247-2820-7
10. United Nations Economic Commission for Europe (ed.): *Efficient Use of Fertilizers in Agriculture.* 1983
    ISBN 90-247-2866-5
11. J. Tinsley and J. F. Darbyshire (eds.): *Biological Processes and Soil Fertility.* 1984        ISBN 90-247-2902-5
12. A. D. L. Akkermans, D. Baker, K. Huss-Danell and J. D. Tjepkema (eds.): Frankia *Symbioses.* 1984
    ISBN 90-247-2967-X
13. W. S. Silver and E. C. Schröder (eds.): *Practical Application of* Azolla *for Rice Production.* 1984
    ISBN 90-247-3068-6
14. P. G. L. Vlek (ed.): *Micronutrients in Tropical Food Crop Production.* 1985        ISBN 90-247-3085-6
15. T. P. Hignett (ed.): *Fertilizer Manual.* 1985        ISBN 90-247-3122-4
16. D. Vaughan and R. E. Malcolm (eds.): *Soil Organic Matter and Biological Activity.* 1985
    ISBN 90-247-3154-2
17. D. Pasternak and A. San Pietro (eds.): *Biosalinity in Action.* Bioproduction with Saline Water. 1985
    ISBN 90-247-3159-3
18. M. Lalonde, C. Camiré and J. O. Dawson (eds.): Frankia *and Actinorhizal Plants.* 1985
    ISBN 90-247-3214-X
19. H. Lambers, J. J. Neeteson and I. Stulen (eds.): *Fundamental, Ecological and Agricultural Aspects of
    Nitrogen Metabolism in Higher Plants.* 1986        ISBN 90-247-3258-1
20. M. B. Jackson (ed.): *New Root Formation in Plants and Cuttings.* 1986        ISBN 90-247-3260-3
21. F. A. Skinner and P. Uomala (eds.): *Nitrogen Fixation with Non-Legumes* (Proceedings of the 3rd Sym-
    posium, Helsinki, 1984). 1986        ISBN 90-247-3283-2
22. A. Alexander (ed.): *Foliar Fertilization.* 1986        ISBN 90-247-3288-3
23. H. G. v.d. Meer, J. C. Ryden and G. C. Ennik (eds.): *Nitrogen Fluxes in Intensive Grassland Systems.* 1986
    ISBN 90-247-3309-X
24. A. U. Mokwunye and P. L. G. Vlek (eds.): *Management of Nitrogen and Phosphorus Fertilizers in Sub-
    Saharan Africa.* 1986        ISBN 90-247-3312-X
25. Y. Chen and Y. Avnimelech (eds.): *The Role of Organic Matter in Modern Agriculture.* 1986
    ISBN 90-247-3360-X
26. S. K. De Datta and W. H. Patrick Jr. (eds.): *Nitrogen Economy of Flooded Rice Soils.* 1986
    ISBN 90-247-3361-8
27. W. H. Gabelman and B. C. Loughman (eds.): *Genetic Aspects of Plant Mineral Nutrition.* 1987
    ISBN 90-247-3494-0
28. A. van Diest (ed.): *Plant and Soil: Interfaces and Interactions.* 1987        ISBN 90-247-3535-1

# Developments in Plant and Soil Sciences


29. United Nations Economic Commission for Europe and FAO (eds.): *The Utilization of Secondary and Trace Elements in Agriculture*. 1987 ISBN 90-247-3546-7
30. H. G. v.d. Meer, R. J. Unwin, T. A. van Dijk and G. C. Ennik (eds.): *Animal Manure on Grassland and Fodder Crops. Fertilizer or Waste?* 1987 ISBN 90-247-3568-8
31. N. J. Barrow: *Reactions with Variable-Charge Soils*. 1987 ISBN 90-247-3589-0
32. D. P. Beck and L. A. Materon (eds.): *Nitrogen Fixation by Legumes in Mediterranean Agriculture*. 1988 ISBN 90-247-3624-2
33. R. D. Graham, R. J. Hannam and N. C. Uren (eds.): *Manganese in Soils and Plants*. 1988 ISBN 90-247-3758-3
34. J. G. Torrey and J. L. Winship (eds.): *Applications of Continuous and Steady-State Methods to Root Biology*. 1989 ISBN 0-7923-0024-6
35. F. A. Skinner, R. M. Boddey and I. Fendrik (eds.): *Nitrogen Fixation with Non-Legumes* (Proceedings of the 4th Symposium, Rio de Janeiro, 1987). 1989 ISBN 0-7923-0059-9
36. B. C. Loughman, O. Gašparíková and J. Kolek (eds.): *Structural and Functional Aspects of Transport in Roots*. 1989 ISBN 0-7923-0060-2; Pb 0-7923-0061-0
37. P. Planquaert and R. Haggar (eds.): *Legumes in Farming Systems*. 1990 ISBN 0-7923-0134-X
38. A. E. Osman, M. M. Ibrahim and M. A. Jones (eds.): *The Role of Legumes in the Farming Systems of the Mediterranean Areas*. 1990 ISBN 0-7923-0419-5
39. M. Clarholm and L. Bergström (eds.): *Ecology of Arable Land – Perspectives and Challenges*. 1989 ISBN 0-7923-0424-1
40. J. Vos, C. D. van Loon and G. J. Bollen (eds.): *Effects of Crop Rotation on Potato Production in the Temperate Zones*. 1989 ISBN 0-7923-0495-0
41. M. L. van Beusichem (ed.): *Plant Nutrition – Physiology and Applications*. 1990 ISBN 0-7923-0740-2
42. N. El Bassam, M. Dambroth and B.C. Loughman (eds.): *Genetic Aspects of Plant Mineral Nutrition*. 1990 ISBN 0-7923-0785-2
43. Y. Chen and Y. Hadar (eds.): *Iron Nutrition and Interactions in Plants*. 1991 ISBN 0-7923-1095-0
44. J. J. R. Groot, P. de Willigen and E. L. J. Verberne (eds.): *Nitrogen Turnover in the Soil-Crop System*. 1991 ISBN 0-7923-1107-8
45. R. J. Wright, V.C. Baligar and R. P. Murrmann (eds.): *Utilization of Acidic Soils for Crop Production*. 1991 ISBN 0-7923-1105-1
46. J. Kolek and V. Kozinka (eds.): *Physiology of the Plant Root System*. 1992 ISBN 0-7923-1205-8
47. A. U. Mokwunye (ed.): *Alleviating Soil Fertility Constraints to Increased Crop Production in West Africa*. 1991 ISBN 0-7923-1221-X; Pb 0-7923-1222-8
48. M. Polsinelli, R. Materassi and M. Vincenzini (eds.): *Nitrogen Fixation*. 1991 ISBN 0-7923-1410-7



Kluwer Academic Publishers – Dordrecht / Boston / London

GPSR Compliance
The European Union's (EU) General Product Safety Regulation (GPSR) is a set
of rules that requires consumer products to be safe and our obligations to
ensure this.

If you have any concerns about our products, you can contact us on

ProductSafety@springernature.com

In case Publisher is established outside the EU, the EU authorized
representative is:

Springer Nature Customer Service Center GmbH
Europaplatz 3
69115 Heidelberg, Germany